MW00844327

ELECTRONIC DEVICES:

SYSTEMS AND APPLICATIONS

Robert Diffenderfer
DeVry University
Kansas City, MO

THOMSON
DELMAR LEARNING™

Australia Canada Mexico Singapore Spain United Kingdom United States

Electronic Devices: Systems and Applications
Robert Diffenderfer

Vice President, Technology and Trades SBU:
Alar Elken

Editorial Director:
Sandy Clark

Senior Acquisitions Editor:
Steve Helba

Senior Developmental Editor:
Michelle Ruelos Cannistraci

Marketing Director:
Dave Garza

Channel Manager:
Fair Huntoon

Marketing Coordinator:
Casey Bruno

Production Director:
Mary Ellen Black

Production Manager:
Larry Main

Senior Project Editor:
Christopher Chien

Art/Design Coordinator:
Francis Hogan

Technology Project Manager:
Kevin Smith

Technology Project Specialist:
Linda Verde

Senior Editorial Assistant:
Dawn Daugherty

Printed in the United States of America
1 2 3 4 5 XX 06 05 04

For more information, contact
Thomson Delmar Learning
Executive Woods
5 Maxwell Drive, PO Box 8007,
Clifton Park, NY 12065–8007
Or find us on the World Wide Web at
http://www.delmarlearning.com

Library of Congress Cataloging-in-Publication Data

ISBN: 1401835147

Contents

Lab Manual Contents

Lab Manual to Accompany *Electronic Devices: Systems and Applications* is available on the accompanying Lab.*Source* CD

Preface

The purpose of *Electronic Devices: Systems and Applications* is to provide foundational knowledge of electronic devices by combining an easy-to-read presentation of the theory with practical system applications. The existence of electronic devices has made it possible to develop a vast array of modern conveniences, which most of us take for granted. Therefore, a good understanding of the different types of electronic devices, the theory regarding each device, and the application of each device is essential to a successful electronics career. The goal of *Electronic Devices: Systems and Applications* is to give students a broad understanding of electronic devices, their characteristics, and their applications. This knowledge will enable students to explore the ever-expanding world of electronic systems.

At times it can be difficult visualizing how or why a circuit is used, so to help readers understand and apply semiconductor theory, numerous examples of how devices are used in common, everyday items are illustrated throughout the text. The text uses a concise approach that is strong on fundamentals, while integrating a wide variety of real-world applications. The text's strong troubleshooting emphasis allows students to apply their newly acquired skills to real-life system applications. Ideally, students who study this book will already know the basics of algebra, trigonometry, and circuit theory.

Textbook Organization

The book begins with an overview of basic circuit theory. To maximize comprehension of the text, students should have a good understanding of the material presented in Chapter 1. Building on fundamental electronics concepts, the text then covers solid-state devices, including diodes, bipolar transistors, field-effect transistors, thyristors, operational amplifiers, and optoelectronic components. The text provides the DC and AC characteristics of the devices, along with opportunities to troubleshoot circuits containing the devices. The text also presents information on the usage of these devices within amplification, filtering, adder, subtractor, comparator, oscillator, and many other circuits. In addition to the device characteristics, practical application of the devices is emphasized to enhance the analysis, troubleshooting, and usage of the devices. Practical system applications utilizing the devices are presented throughout the text.

Chapter 1 is a review of basic circuit theory. Since students who are studying electronic devices should have a working knowledge of basic circuit theory, the material within this chapter can be assigned for student reading or can be highlighted as a starting point for teaching your course.

Chapter 2 provides an introduction to semiconductor theory.

Chapter 3 covers series AC and DC circuits using switching diodes, rectifier diodes, and zener diodes. Signal waveforms, component voltage, current, and power characteristics are evaluated. This chapter also presents troubleshooting techniques that can be used to find failed components.

Chapter 4 presents power supply circuit concepts, placing primary emphasis on AC-to-DC and DC-to-AC converter systems. Half-wave, full-wave, and full-wave bridge rectifier circuits are evaluated, along with their signal waveforms. Voltage regulators and filtering systems are shown.

Chapter 5 covers NPN and PNP BJTs, concentrating on the device characteristics, biasing circuits, amplifier circuits, and switching circuits. This chapter provides equations to calculate the DC operating voltages and currents as well as the AC gain and frequency response.

Chapter 6 provides the characteristics and usage of the n-channel and p-channel JFET, and the n-channel and p-channel MOSFET (enhancement and depletion mode). Biasing circuits, amplifier circuits and switching circuits are given.

Chapter 7 presents the operational amplifier. The inverting, noninverting, adder, subtractor, integrator, differentiator, and comparator circuits are evaluated. Digital interface circuits (the ADC and DAC) are discussed.

Chapter 8 considers the practical limitations of the operational amplifier due to unity gain bandwidth product and slew rate. Common AC circuits included in this section are the audio amplifier, precision voltage regulators, and peak detectors.

Chapter 9 evaluates first- and second-order low-pass, high-pass, bandpass, and bandstop filters.

Chapter 10 covers oscillators and multivibrators.

Chapter 11 presents the thyristor devices including the diac, SCR, and Triac, and shows practical applications using these components.

Chapter 12 covers optoelectronic devices. This chapter presents the photoemitter, photodetector, and optoisolator, and provides practical system applications of these devices.

Features

Key instructional design elements used within the book's structure include items such as:

Key Terms, Chapter Outline, Preview, and Objectives. These elements occur at the beginning of each chapter. A list of key terms presents important terminology that will be defined in the chapter. A chapter outline lists main topics covered in the chapter. A chapter preview sets the stage and presents the importance of the material in electronics. Finally, critical learning objectives describe the competencies students should achieve upon understanding the chapter material.

In-Process Learning Checks. These provide the reader with special hints, practical techniques, and information.

General Electronics

Computers

Industrial

Communications

Application Problems. These are located in each chapter and provide practical examples from a myriad of technical areas, such as the industrial, computer, and communications fields. These applications are highlighted in the text with special headings.

Component Datasheets. These show the operational characteristics of electronic devices.

Datasheet Questions. These encourage individuals to become familiar with the content of component datasheets and how to use them.

Safety Hints. These highlight requirements for safety, both while learning and when applying electronic principles in "hands-on" situations.

Troubleshooting Hints. These offer experience-based information about circuit problems, such as what to look for and how to fix them.

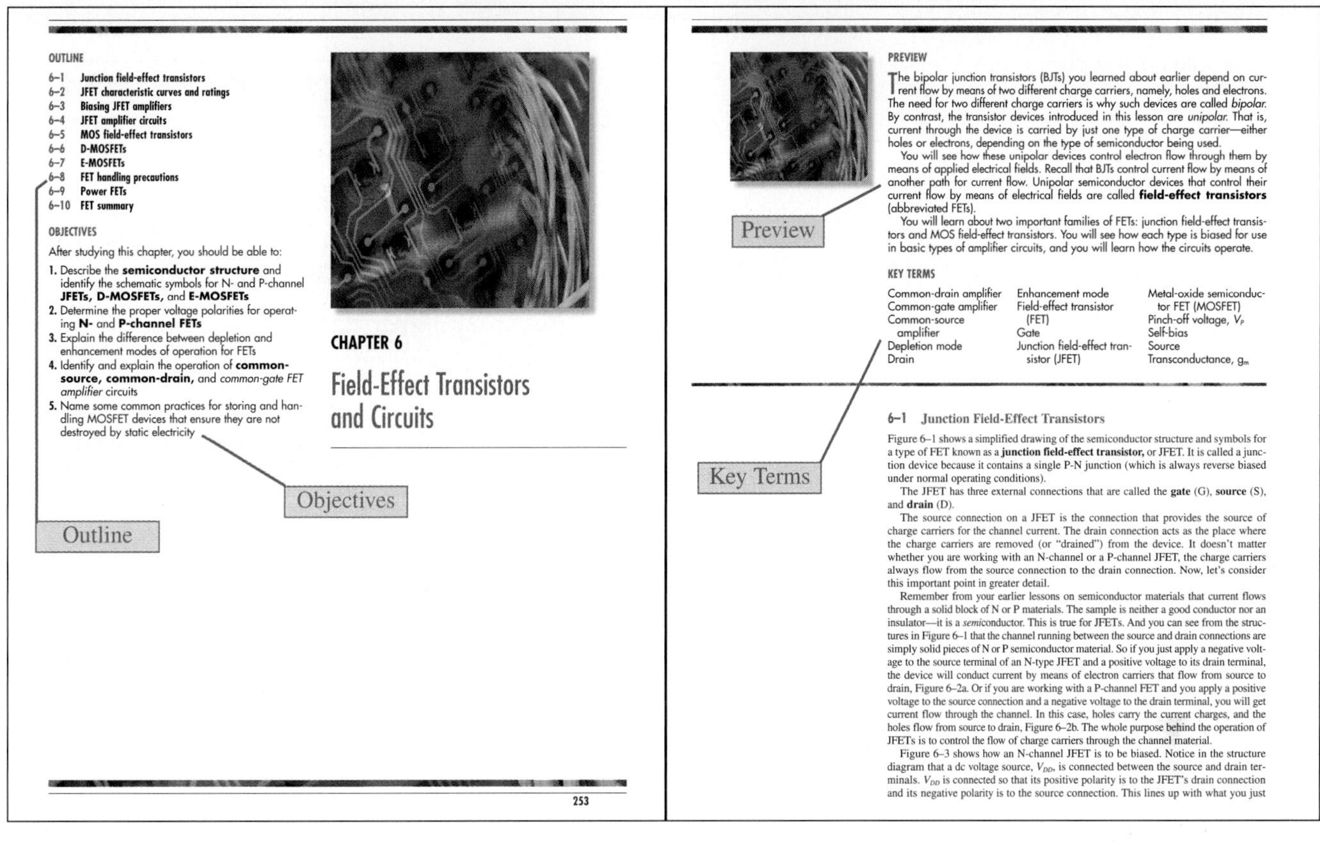

OUTLINE

6–1 Junction field-effect transistors
6–2 JFET characteristic curves and ratings
6–3 Biasing JFET amplifiers
6–4 JFET amplifier circuits
6–5 MOS field-effect transistors
6–6 D-MOSFETs
6–7 E-MOSFETs
6–8 FET handling precautions
6–9 Power FETs
6–10 FET summary

OBJECTIVES

After studying this chapter, you should be able to:

1. Describe the **semiconductor structure** and identify the schematic symbols for N- and P-channel **JFETs, D-MOSFETs,** and **E-MOSFETs**
2. Determine the proper voltage polarities for operating **N-** and **P-channel FETs**
3. Explain the difference between depletion and enhancement modes of operation for FETs
4. Identify and explain the operation of **common-source, common-drain,** and *common-gate FET amplifier* circuits
5. Name some common practices for storing and handling MOSFET devices that ensure they are not destroyed by static electricity

CHAPTER 6

Field-Effect Transistors and Circuits

253

PREVIEW

The bipolar junction transistors (BJTs) you learned about earlier depend on current flow by means of two different charge carriers, namely, holes and electrons. The need for two different charge carriers is why such devices are called *bipolar.* By contrast, the transistor devices introduced in this lesson are *unipolar.* That is, current through the device is carried by just one type of charge carrier—either holes or electrons, depending on the type of semiconductor being used.

You will see how these unipolar devices control electron flow through them by means of applied electrical fields. Recall that BJTs control current flow by means of another path for current flow. Unipolar semiconductor devices that control their current flow by means of electrical fields are called **field-effect transistors** (abbreviated FETs).

You will learn about two important families of FETs: junction field-effect transistors and MOS field-effect transistors. You will see how each type is biased for use in basic types of amplifier circuits, and you will learn how the circuits operate.

KEY TERMS

Common-drain amplifier
Common-gate amplifier
Common-source amplifier
Depletion mode
Drain
Enhancement mode
Field-effect transistor (FET)
Gate
Junction field-effect transistor (JFET)
Metal-oxide semiconductor FET (MOSFET)
Pinch-off voltage, V_P
Self-bias
Source
Transconductance, g_m

6–1 Junction Field-Effect Transistors

Figure 6–1 shows a simplified drawing of the semiconductor structure and symbols for a type of FET known as a **junction field-effect transistor,** or JFET. It is called a junction device because it contains a single P-N junction (which is always reverse biased under normal operating conditions).

The JFET has three external connections that are called the **gate** (G), **source** (S), and **drain** (D).

The source connection on a JFET is the connection that provides the source of charge carriers for the channel current. The drain connection acts as the place where the charge carriers are removed (or "drained") from the device. It doesn't matter whether you are working with an N-channel or a P-channel JFET, the charge carriers always flow from the source connection to the drain connection. Now, let's consider this important point in greater detail.

Remember from your earlier lessons on semiconductor materials that current flows through a solid block of N or P materials. The sample is neither a good conductor nor an insulator—it is a *semi*conductor. This is true for JFETs. And you can see from the structures in Figure 6–1 that the channel running between the source and drain connections are simply solid pieces of N or P semiconductor material. So if you just apply a negative voltage to the source terminal of an N-type JFET and a positive voltage to its drain terminal, the device will conduct current by means of electron carriers that flow from source to drain, Figure 6–2a. Or if you are working with a P-channel FET and you apply a positive voltage to the source connection and a negative voltage to the drain terminal, you will get current flow through the channel. In this case, holes carry the current charges, and the holes flow from source to drain, Figure 6–2b. The whole purpose behind the operation of JFETs is to control the flow of charge carriers through the channel material.

Figure 6–3 shows how an N-channel JFET is to be biased. Notice in the structure diagram that a dc voltage source, V_{DD}, is connected between the source and drain terminals. V_{DD} is connected so that its positive polarity is to the JFET's drain connection and its negative polarity is to the source connection. This lines up with what you just

Good Idea. These boxes reinforce key practical ideas.

Examples and Practice Problems. These show how to apply theory step by step, and allow immediate practice after learning concepts.

Used in a System. These features provide insight of how devices are utilized in the field of electronics. As you observe practical applications of the electronic devices, greater value is placed upon learning about the components and how they operate.

Summary, Review Questions, Problems, and Analysis Questions. These elements occur at the end of each chapter; the Summary presents a list of topic highlights, and is followed by a variety of questions that progress in difficulty from basic review questions to more advanced critical thinking problems. In some chapters, a **Circuit Summary** offers an illustrative review of key circuits and their corresponding waveforms and formulas. This can be a useful tool when comparing differences between the circuits covered within the chapter.

Formulas and Sample Calculator Sequences. These provide a quick reference of the chapter's most important formulas, followed by an example of how to solve them using a calculator.

Using Excel. This is a feature that encourages problem solving using spreadsheet exercises and Excel templates, which are available on the accompanying CD. Filenames are placed beside Excel screens in the text for easy reference to the CD.

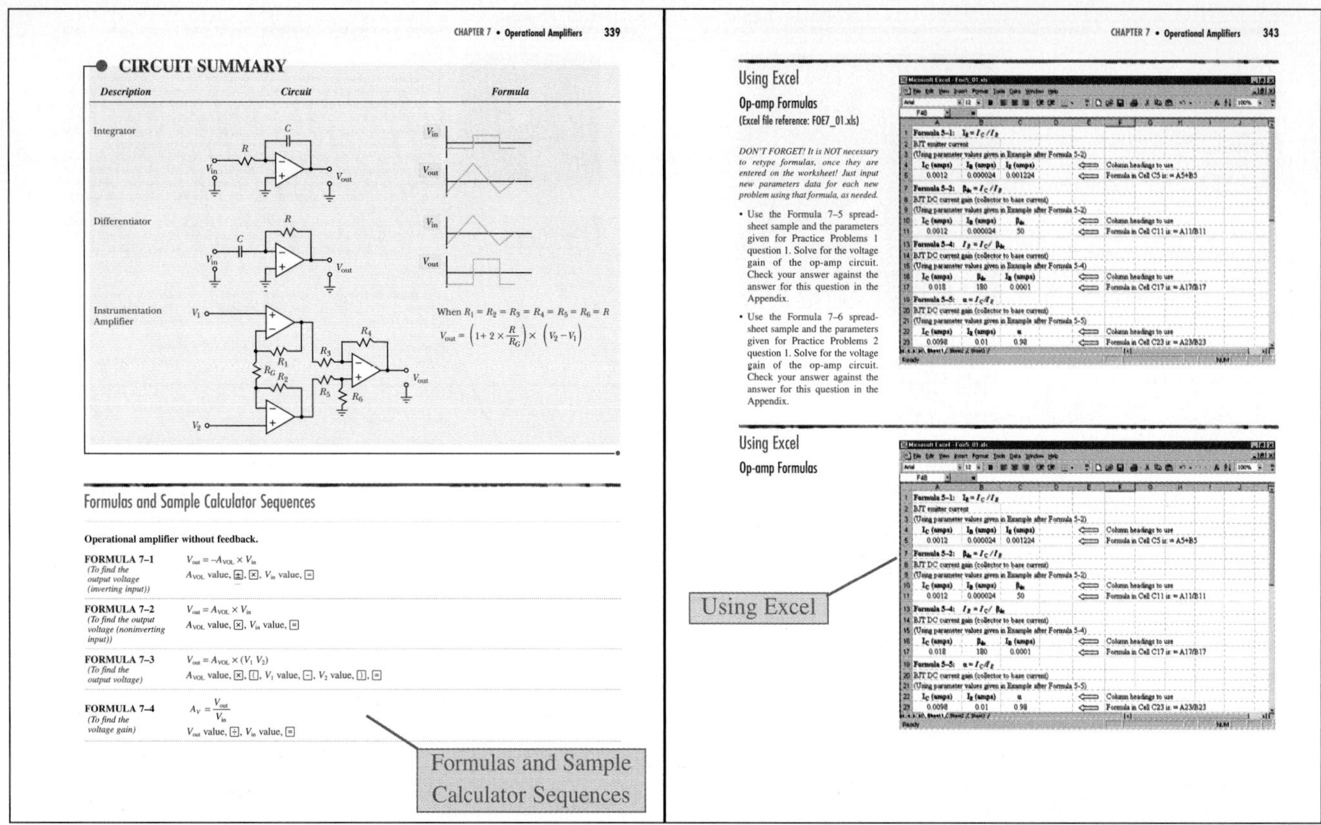
CHAPTER 7 • Operational Amplifiers 339

CIRCUIT SUMMARY

Description	Circuit	Formula
Integrator		
Differentiator		
Instrumentation Amplifier		When $R_1 = R_2 = R_3 = R_4 = R_5 = R_6 = R$ $V_{out} = \left(1 + 2 \times \frac{R}{R_G}\right) \times (V_2 - V_1)$

Formulas and Sample Calculator Sequences

Operational amplifier without feedback.

FORMULA 7–1 *(To find the output voltage (inverting input))*
$V_{out} = -A_{VOL} \times V_{in}$
A_{VOL} value, [±], [×], V_{in} value, [=]

FORMULA 7–2 *(To find the output voltage (noninverting input))*
$V_{out} = A_{VOL} \times V_{in}$
A_{VOL} value, [×], V_{in} value, [=]

FORMULA 7–3 *(To find the output voltage)*
$V_{out} = A_{VOL} \times (V_1\ V_2)$
A_{VOL} value, [×], [(], V_1 value, [−], V_2 value, [)], [=]

FORMULA 7–4 *(To find the voltage gain)*
$A_V = \frac{V_{out}}{V_{in}}$
V_{out} value, [÷], V_{in} value, [=]

CHAPTER 7 • Operational Amplifiers 343

Using Excel

Op-amp Formulas

(Excel file reference: FOE7_01.xls)

DON'T FORGET! It is NOT necessary to retype formulas, once they are entered on the worksheet! Just input new parameters data for each new problem using that formula, as needed.

- Use the Formula 7–5 spreadsheet sample and the parameters given for Practice Problems 1 question 1. Solve for the voltage gain of the op-amp circuit. Check your answer against the answer for this question in the Appendix.
- Use the Formula 7–6 spreadsheet sample and the parameters given for Practice Problems 2 question 1. Solve for the voltage gain of the op-amp circuit. Check your answer against the answer for this question in the Appendix.

Using Excel

Op-amp Formulas

multiSIM

MultiSIM Exercises. This feature explores computer simulation techniques and confirms operational characteristics through the use of pre-created MultiSIM circuit files.

Performance Projects Correlation Chart. These charts cross-reference specific projects from the Laboratory Manual that relate to chapter material. All of these labs are available on the accompanying *Lab.Source* CD.

SIMPLER Sequences and Troubleshooting Challenges. These features use a unique troubleshooting approach that integrates critical thinking and encourages you to solve both systems-level and component-level troubleshooting problems logically and efficiently.

Find the Fault. This feature provides students with the opportunity to evaluate a circuit and to identify the failed component by following different troubleshooting methods.

Practical Use of Color. Another important feature occurring throughout the book is the systematic use of color to clarify the illustrations. For example, each resistor pictorial displays an actual color code, adding a practical, real-world dimension to these illustrations. Voltage sources are always highlighted with yellow, resistance values are highlighted with brown, capacitive elements with blue, and so forth. The practical use of color is designed to tangibly improve the value of the illustrations for students.

Appendixes. These provide reference tools including color codes, common resistor values, answers, schematic symbols, how to use a scientific calculator, a comprehensive glossary, and more.

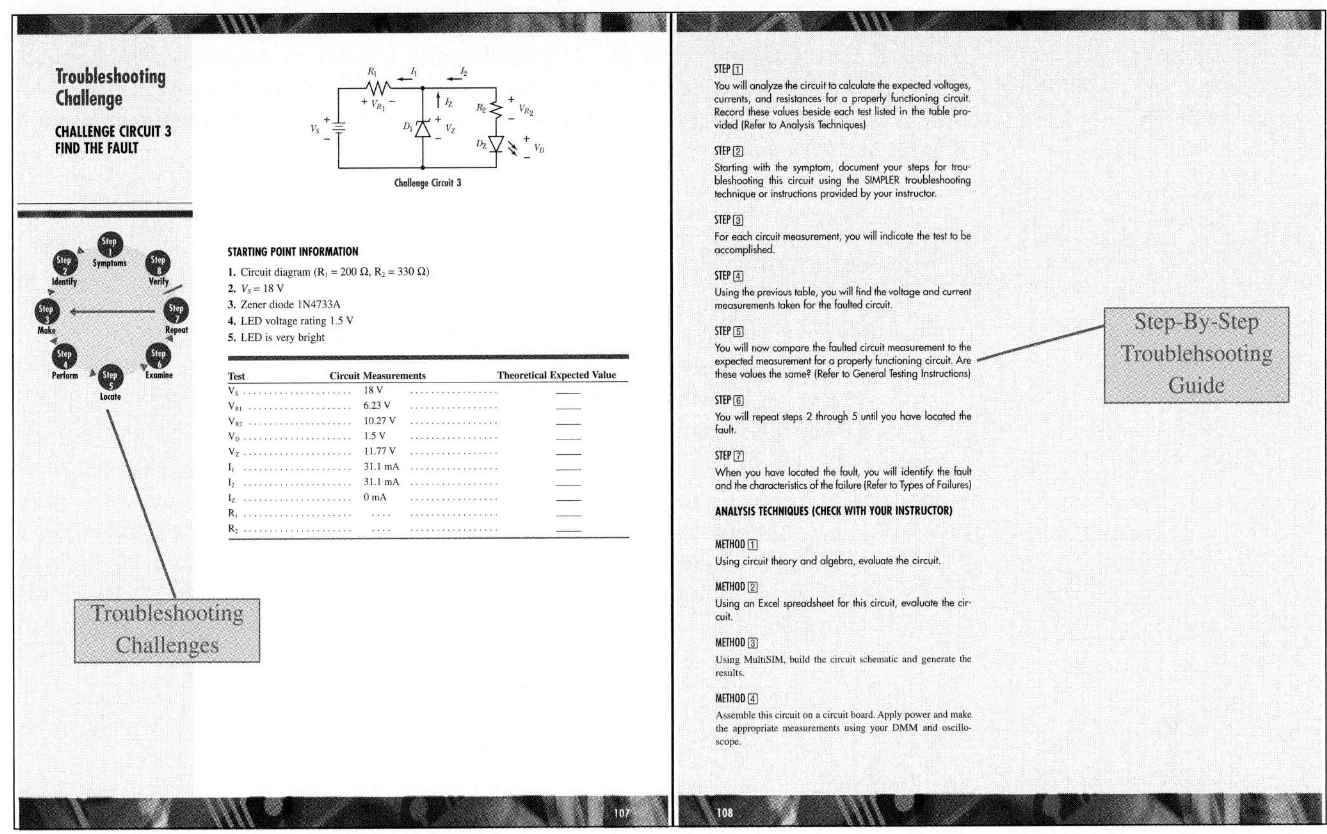

Troubleshooting with the SIMPLER Sequence

To become logical troubleshooters, technicians must develop good critical thinking skills. The SIMPLER troubleshooting technique (developed by Russell Meade) is a useful troubleshooting approach.

Troubleshooting Challenge

As students move through this book, they will encounter a series of troubleshooting problems called "Troubleshooting Challenges". Each Troubleshooting Challenge includes a challenge circuit schematic and starting point symptoms information. Using the SIMPLER sequence, students will step through and solve the simulated troubleshooting problems.

Using the SIMPLER troubleshooting method, students will be able to guide themselves through each Troubleshooting Challenge circuit and think critically about the problems they need to solve. One possible approach to solving each Troubleshooting Challenge is provided in step-by-step illustrations following each challenge.

The illustrations on pages 107–108 show a Troubleshooting Challenge and give instructions on how to complete the challenge. We hope that students and their instructors will have great success with the SIMPLER sequence technique and that these Troubleshooting Challenges will make both learning and teaching troubleshooting skills easier and more enjoyable.

Selecting Tests and Finding the Results

To simulate making a parameter test on a particular portion of the challenge circuit, students may select a test from the test listing provided with the Troubleshooting Challenge. Students may then look up the results of the test they selected by looking in Appendix C, which contains the identifier numbers assigned to each test on the Troubleshooting Challenge pages. Next to each identifier number, students will see the parameter value or condition that would be present if they made that same test on the actual circuit simulated by the Troubleshooting Challenge. *Please note:* The tolerances of components and of test instruments will create differences between the theoretical values and those actually shown in the answers. The theoretical values would be achieved only if all components were precisely rated and the test instruments had zero percent error.

Find the Fault

This is a troubleshooting feature included within the text that allows students to analyze the circuit, evaluate test data, and identify the failed component. Using one of many methods, such as a traditional mathematical approach, computer software (such as Excel spreadsheet or MultiSIM), laboratory experimentation, or the SIMPLER method, the student will analyze the circuit, determining the expected voltages and currents for the properly functioning circuit. Data is provided for the measurements acquired from the failed circuit. The student must document the steps used in the process of determining the failed component. Several troubleshooting challenge circuits are available, giving students many chances to evaluate the circuit and to identify the failed component. Students will have the opportunity to improve their troubleshooting skills by practicing a number of different troubleshooting techniques.

What Is on the CD?

Lab.Source

- Introducing new free lab exercises with Lab.*Source*™. Save your students from the added expense of purchasing a separate printed lab manual. Lab.*Source* is a CD that is bundled with the text and offers free laboratory exercises, which provide learners with the valuable hands-on experience they need to reinforce all key devices topics covered in the text. In today's economy, where every dollar counts, Lab.*Source* is a sensible alternative to expensive printed lab manuals.
- Lab.*Source* provides the entire *Electronic Devices: Systems and Applications Laboratory Projects Manual* on CD and includes 32 easy-to-use projects that provide hands-on experience with the concepts and principles covered in the text. While performing the projects, students will be asked to collect data, note observations, and draw conclusions. The "Story Behind the Numbers" projects are opportunities for students to apply their analytical, technical writing, and communication skills. The projects are grouped together by topic and can be performed by either individuals or groups of students.

Textbook Edition of MultiSIM

A free copy of the Textbook Edition of MultiSIM 7 allows students to create circuits, analyze pre-built circuits, simulate circuit behavior, and troubleshoot for faults.

Pre-built MultiSIM Circuit Files

These files allow students to immediately simulate circuits and transform schematics into live, interactive circuits. A special icon is placed beside selected schematics to direct students to the CD.

Excel Tables

Excel spreadsheet templates are another tool which allows students to manipulate data and calculate formulas more easily to solve electronics problems.

www.electronictech.com

This link directs you to the text's website. Please check the Online Companion for any text updates.

Supplements

e.resource

Available on one CD, the *e.resource* contains many tools and resources that can be used by instructors to enrich their classrooms and to make preparation time shorter. The *e.resource* features the following ancillaries:

(ISBN: 1401835163)

- **Instructor's Guide.** This comprehensive Instructor's Guide provides all the answers for the text and lab manual, along with teaching hints and sample course schedules.
- **PowerPoint Presentation Slides.** Key points and concepts are illustrated and can be used as a basis for lectures.
- **ExamView Computerized Testbank.** This computerized testbank holds over 1000 true/false, multiple-choice, and short-answer questions to assess student comprehension.
- **Image Library.** This tool allows instructors to customize PowerPoint slides, exams, transparencies, or class notes by importing selected full-color images from the text.
- **Lab.Builder.** This gives the instructor the advantage of being able to create their own labs electronically.

Acknowledgments

The author and Thomson Delmar Learning would like to thank the following individuals for their valuable insight and feedback during the development of this book:

Reviewers

Don Abernathy,
DeVry University, Irving, TX

Shaikh Ali,
City College, Fort Lauderdale, FL

Dave Barth,
Edison Community College, Piqua, OH

Eugene Evancoe,
Saddleback College, Mission Viejo, CA

William Hirst,
DeVry University, Tinley Park, IL

Mark Hughes,
Cleveland Community College, Shelby, NC

Chenggang Mei,
Southeast Missouri State University,
Cape Girardeau, MO

J.D. Neglia,
Mesa Community College, Mesa, AZ

Robert Pruitt,
DeVry University, Kansas City, MO

Max Rabiee,
University of Cincinnati, Cincinnati, OH

Robert Reaves,
Durham Technical Community College,
Durham, NC

Lee Rosenthal,
Fairleigh Dickinson Univ, Teaneck, NJ

William Routt,
Wake Tech Community College,
Raleigh, NC

Tim Staley,
DeVry University, Irving, TX

Jack Williams,
Remington College, Tampa, FL

Matt Porubsky,
Milwaukee Area Technical College,
Milwaukee, WI

Tony Smigin,
DeVry University, Kansas City, MO

Also, a special thanks to those who participated in a survey during the initial development of this text:

Sohail Anwar, Penn State University, Altoona College; Mark Fitzgerald, Piedmont VA Community College; John Fitzen, Idaho State University, College of Technology; Yolanda Guran-Postelthwaite, Oregon Institute of Technology; Mark Hughes, Cleveland Community College; Tulin Mangir, California State University; Elifuraha Mmari, DeVry Institute of Technology; Mark Oliver, Monroe Community College; Esteban Rodriquez-Marek, Eastern Washington University; Jeff Salehi, Southern Utah University; Robert Scoff, The University of Memphis; Tom Selis, Rhodes State; Hesham Shaalan, Texas A & M University; Lloyd Stallkamp, Montana State University; John R. Wall, East Carolina University; Randy Winzer, Pittsburgh State University; Jamie Zipay, Oregon Institute of Technology

Thank you to Russell Meade who developed the SIMPLER sequence and part of Chapter 4 on power supply fundamentals.

Contributors:

Carl Hill,
DeVry University, Kansas City, MO

Roman Stemprok,
University of North Texas, Denton, TX

Bill Kist,
New England Institute of Technology,
West Palm Beach, FL

I would like to express appreciation to the publishing team at Thomson Delmar Learning for working with me in the development of a quality textbook covering Electronic Devices. For the publishing endeavors, this team includes, Dave Garza, Steve Helba, Michelle Ruelos Cannistraci, Larry Main, Christopher Chien, Francis Hogan, and Dawn Daugherty. Special thanks to COBY and Speco Technologies for granting permission for the Used in a System photos.

About the Author

Robert Diffenderfer has over 28 years of professional experience in the aerospace industry (The Boeing Company and Eldec Corporation), the consumer electronics industry, and in teaching (University of Illinois, Cogswell College, John Brown University, and DeVry University). He received a Bachelor of Science in Electrical Engineering (BSEE) from John Brown University and a Master of Science in Electrical Engineering (MSEE) from the University of Illinois. He is currently a senior professor with DeVry University, teaching in the Bachelor of Science in Electronic Engineering Technology program at the Kansas City, Missouri campus. In 2002, he was nominated to Who's Who Among America's Teachers.

CHAPTER 1

Circuit Theory Fundamentals

OUTLINE

OBJECTIVES

After studying this chapter, you should be able to:

1. Identify each electrical quantity by its letter symbol
2. Indicate the unit of measure and its abbreviation for each electrical quantity
3. Determine the value for the common prefixes
4. Define branch and node
5. Identify series, parallel, and series-parallel circuits
6. Define and list some active and passive components
7. Calculate total resistance for series or parallel resistors
8. Calculate total capacitance for series or parallel capacitors
9. Calculate total inductance for series or parallel inductors
10. Write circuit laws and theorems in circuit analysis
11. State the measurement equipment to be used in obtaining circuit readings

PREVIEW

Prior to beginning our study of electronic devices, a review of the basic circuit theory will be provided. It is essential that you have an understanding of the basic circuit laws, theorems, and analysis techniques. The emphasis in this chapter is an overview of the fundamentals of electronics. A thorough coverage of this material is provided in *Foundations of Electronics: Circuits and Devices.* The information of this chapter should be familiar to you. The principles provided in this chapter will be used throughout the text.

KEY TERMS

Active device
Branch
Circuit
Conductor
Current
Insulator
Node
Parallel circuit
Passive device
Power
Series circuit
Series-parallel circuit
Voltage

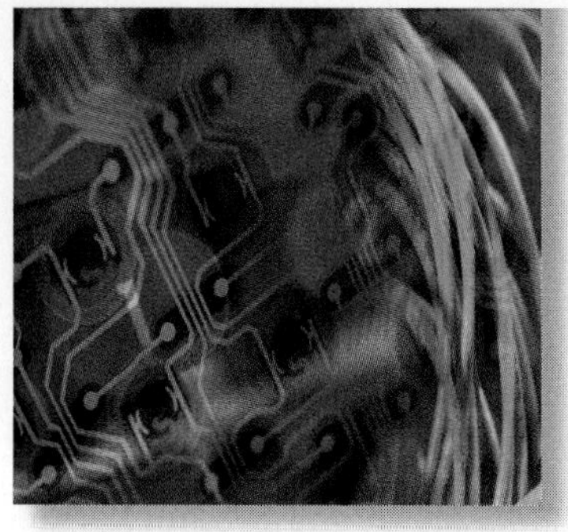

1–1 Electrical Quantities

A list of common electronic quantities is provided in Table 1–1. The first column is a listing of the electronic quantities. Not all quantities included in the list may be used in the text. The second column provides the symbol associated with each item. Uppercase symbols generally correlate with either a direct current (dc) quantity or a constant value associated with a time varying quantity. Lowercase symbols are typically associated with a time varying or instantaneous function such as alternating current (ac) signals. Subscripts are often used with these quantities to distinguish different elements and circuit location voltages, currents, power, etc. To indicate or emphasize that a voltage is a time varying value, the notation v(t) and i(t) may be used. This notation indicates that the voltage is a function of time and the current is a function of time t. The third and fourth columns provide the unit of measure with its abbreviation. The units and their abbreviations are given in accordance with the International System of Units (SI—System International). Table 1–2 is a listing of common prefixes used in electronics.

The following definitions will be beneficial in understanding the electronic quantities and concepts presented in this chapter.

- An **active device** is a component or device that either supplies energy to the circuit (such as a battery or power supply) or converts energy from one form to another form (such as a transformer or a transistor).
- A **branch** consists of a single component, or two or more components connected in series. (Refer to Figure 1–1.)
- A **circuit** is the combination of components or elements that are connected to provide paths for current flow to perform some useful function.

TABLE 1–1 Common electronic quantities

Electrical Quantity	Letter Symbol	Unit of Measure	Unit Abbreviation
Voltage (dc)	V	Volt	V
Voltage (ac, instantaneous)	v	Volt	V
Current (dc)	I	Ampere	A
Current (ac, instantaneous)	i	Ampere	A
Power (apparent)	S	Volt-amperes	VA
Power (reactive)	Q	Voltampere-reactive	VAR
Power (real, average, true)	P	Watt	W
Resistance	R	Ohm	Ω
Conductance	G	Siemens	S
Capacitance	C	Farad	F
Inductance	L	Henry	H
Impedance	Z	Ohm	Ω
Reactance	X	Ohm	Ω
Admittance	Y	Siemens	S
Susceptance	B	Siemens	S
Frequency	f	Hertz	Hz
Period	T	Seconds	s
Angular velocity	ω	radians/second	rad/s
Time	t	second	s

- A **conductor** is a material that permits the movement of electrons (such as gold, copper, silver, and aluminum).
- **Current** is a measure of the rate of flow of electrical charge through a circuit. $\left(I = \frac{Q}{t}\right)$
- An **insulator** is a material that prevents the movement of electrons (such as Teflon, and polyethylene).
- A **node** is the junction point between two or more components, or is the junction point of two or more branches. (Refer to Figure 1–1.)
- A **parallel circuit** is a circuit where two components or elements have both nodes in common. (Refer to Figure 1–3.)
- A **passive device** is a component or device that only absorbs energy (such as a resistor) or absorbs energy and later returns that energy to the circuit (such as a capacitor or an inductor).

TABLE 1–2 Common prefixes

Value	Prefix	Abbreviation
1×10^{12}	tera	T
1×10^{9}	giga	G
1×10^{6}	mega	M
1×10^{3}	kilo	k
1×10^{-3}	milli	m
1×10^{-6}	micro	µ
1×10^{-9}	nano	n
1×10^{-12}	pico	p
1×10^{-15}	femto	f

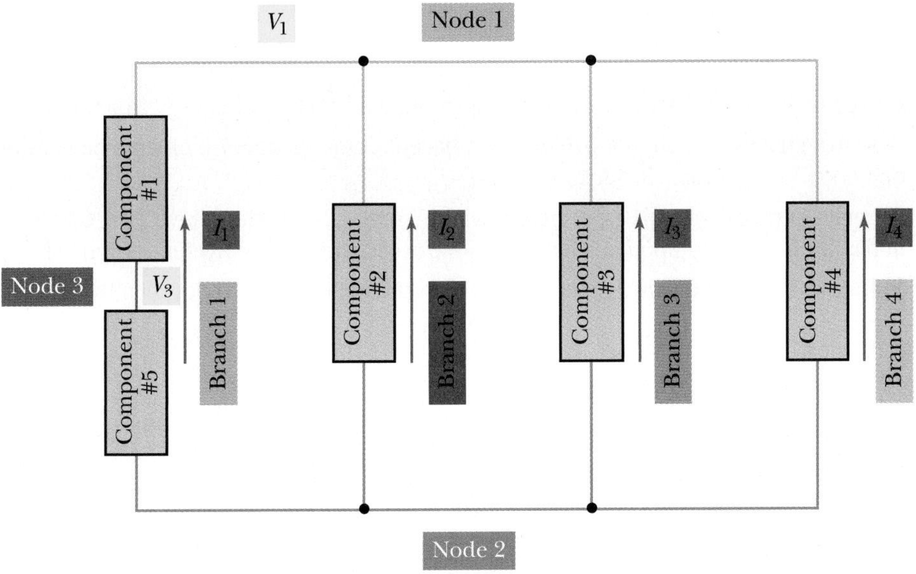

FIGURE 1–1 Branches and nodes

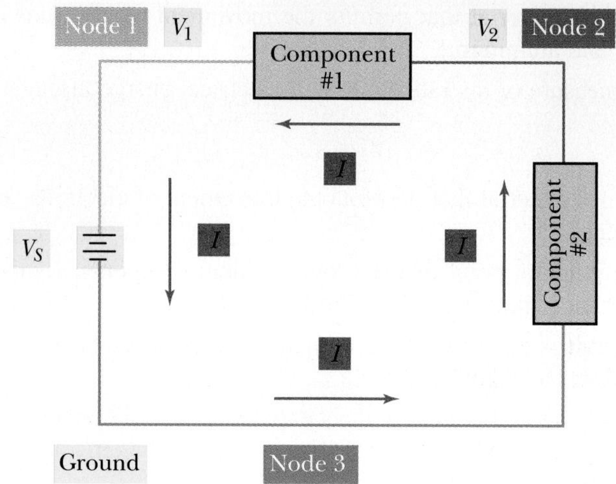

FIGURE 1–2 Series circuit

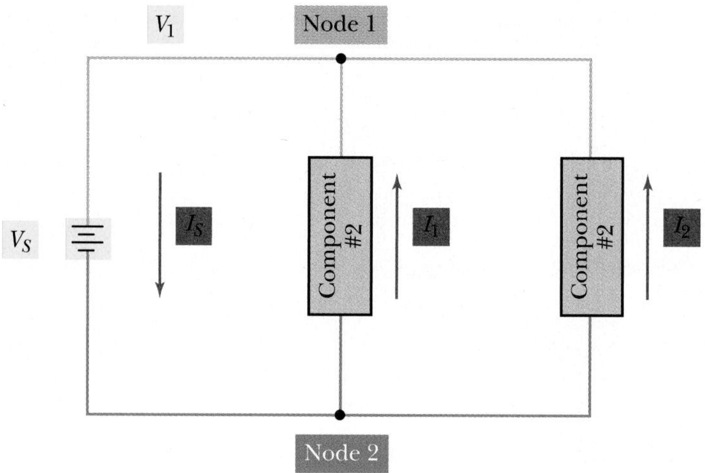

FIGURE 1–3 Parallel circuit

- **Power** is the rate of performing work or the rate of change of energy in a circuit.
- A **series circuit** is a circuit where two adjacent components or elements have only one node in common. (Refer to Figure 1–2.)
- A **series-parallel circuit** is a circuit where some components have two common nodes while other components have only one node in common. (Refer to Figure 1–4.)
- **Voltage** is the potential difference or the electromotive force (emf) between two points.

1–2 Electrical Components

Power Sources

Power sources are used to provide voltage and current to a circuit. The power source may provide a dc voltage to the circuit represented by the schematic symbol in Figure 1–5. An internal chemical reaction causes the battery to provide a dc output voltage. A dc power supply can also provide a dc output voltage by converting an ac voltage into a dc voltage. A power source can be an ac voltage shown by the schematic symbol in Figure 1–6. An alternator or generator will produce a sinusoidal waveform as shown in Figure 1–7. The ideal voltage source provides a fixed, constant voltage to all loads (0 Ω to infinite resistance—infinite current to 0 A). The ideal voltage source has a source resistance of 0 Ω.

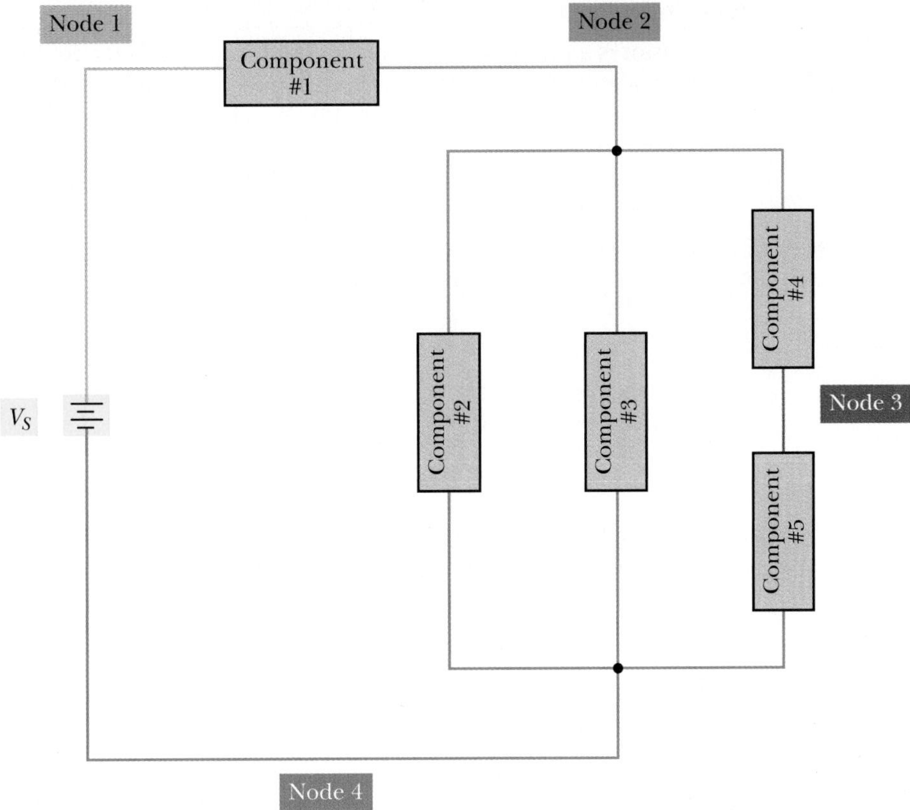

FIGURE 1–4 Series-parallel circuit

FIGURE 1–5 DC power supply or battery schematic symbol

FIGURE 1–6 AC voltage source schematic symbol

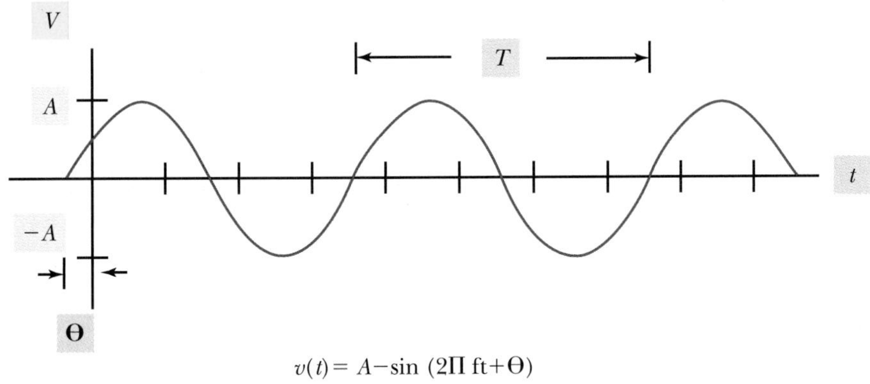

FIGURE 1–7 The sine wave

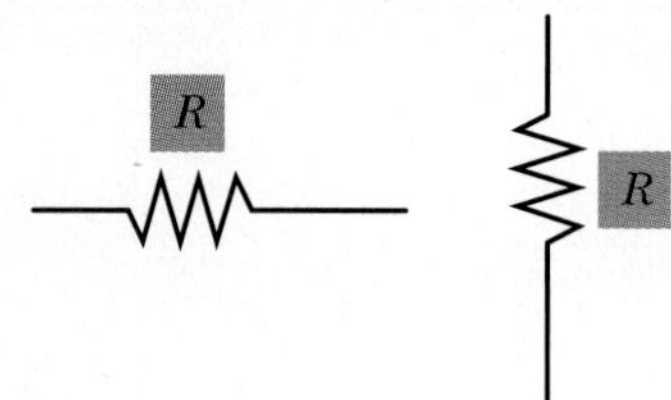

FIGURE 1–8 Fixed value resistor

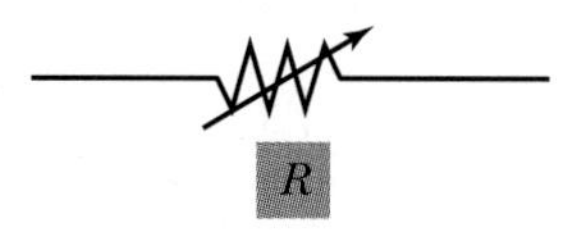

FIGURE 1–9 Variable or adjustable resistor

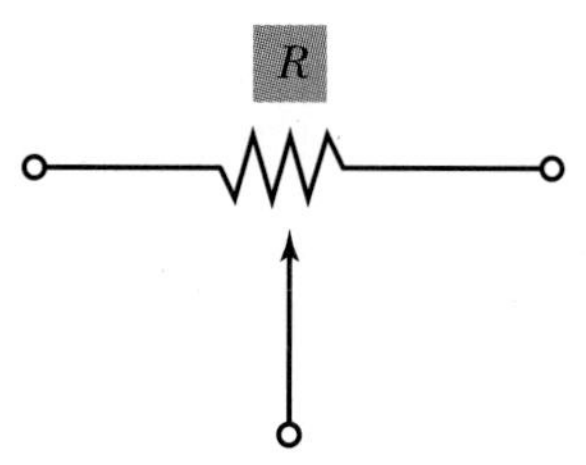

FIGURE 1–10 Potentiometer

Resistor

The resistor is a passive component that limits or controls the current flow in a circuit. Resistors are available as fixed value (Figure 1–8), variable (Figure 1–9) or three-terminal potentiometers (Figure 1–10). Traditional construction includes carbon, metal film (precision), and wire wound. By now you should have memorized the resistor color code (Table 1–3), which is utilized to identify the resistor value.

A general formula to calculate resistance is Formula 1–1.

FORMULA 1–1 $R = \frac{\rho l}{A}$

R—resistance in ohms (Ω)
ρ—resistivity in CM-Ω/ft
l—length in feet
A—cross-sectional area in CM (circular mils)

Resistors dissipate energy in the form of heat. Resistors are available with different power ratings (1/10 W, 1/8 W, 1/4 W, 1/2 W, 1 W, etc.) designed to dissipate the required heat. The power that a resistor must dissipate in a circuit can be calculated using Formula 1–2.

FORMULA 1–2 $P = V \times I$

Resistors can be connected in series as shown in Figure 1–11. The total resistance for series resistors is calculated using Formula 1–3.

FORMULA 1–3 $R_T = R_1 + R_2 + R_3 + \ldots + R_n$

R_n—represents the last series resistor

TABLE 1–3 Resistor color code (5, 10, 20% tolerance)

Color	Band 1 Value	Band 2 Value	Band 3 Multiplier	Band 4 Tolerance	Band 5 Reliability Factor
■ Black	0	0	10^0 (1)	—	—
■ Brown	1	1	10^1 (10)	20%	1%
■ Red	2	2	10^2 (100)	—	0.1%
■ Orange	3	3	10^3 (1 k)	—	0.01%
□ Yellow	4	4	10^4	—	0.001%
■ Green	5	5	10^5	—	—
■ Blue	6	6	10^6 (1 M)	—	—
■ Violet	7	7	10^7	—	—
■ Gray	8	8	10^8	—	—
□ White	9	9	10^9 (1 G)	—	—
□ Gold	—	—	10^{-1} (0.1)	5%	—
□ Silver	—	—	10^{-2} (0.01)	10%	—
No color	—	—	—	20%	—

Resistors can also be connected in parallel as shown in Figure 1–12. The total resistance for the parallel resistor circuit can be determined with Formula 1–4.

FORMULA 1–4 $$R_T = \frac{1}{\frac{1}{R_1} + \frac{1}{R_2} + \frac{1}{R_3} + \ldots + \frac{1}{R_n}}$$

If all of the parallel resistors were of the same value *R*, the total resistance can be calculated using Formula 1–5 where n is the number of resistors.

FORMULA 1–5 $$R_T = \frac{R}{n}$$

When only two resistors are in parallel, the total resistance for the parallel resistor circuit can be determined with Formula 1–6.

FORMULA 1–6 $$R_T = \frac{R_1 \times R_2}{R_1 + R_2}$$

Practical Notes

Resistors are classified as linear devices since the voltage across the resistor is equal to the current flowing through the resistor multiplied by the resistance of the resistor ($V = I \times R$).

Capacitor

The capacitor is a passive component consisting of two metal plates separated by a dielectric (nonconductive) material. Capacitors have the capacity to store energy. Ideally, electrons do not move from one metal plate through the dielectric material to the other metal plate. The condition where electrons move between the metal plates through the dielectric material is called leakage. Capacitors are available as fixed

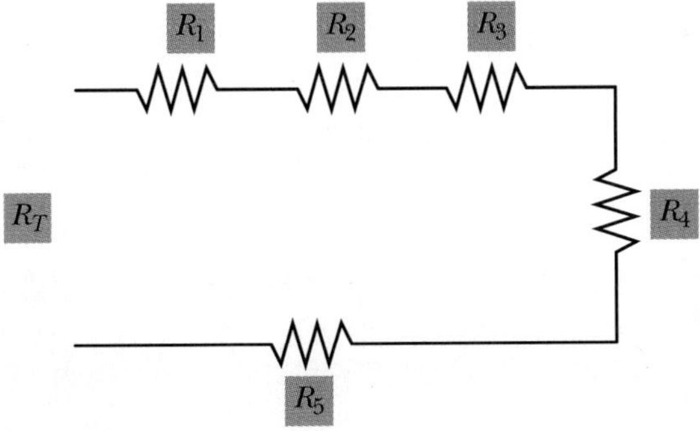

FIGURE 1–11 Series resistors

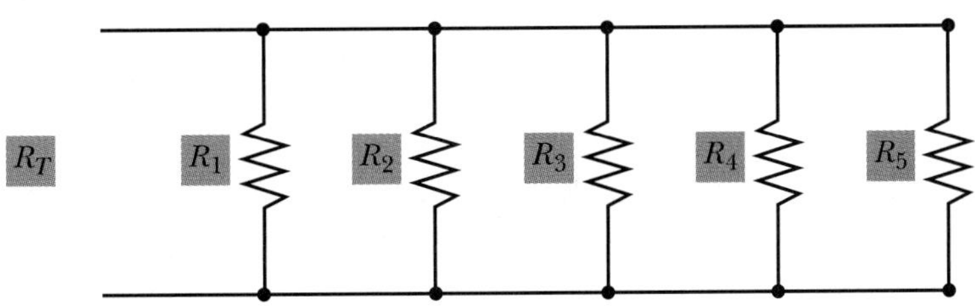

FIGURE 1–12 Parallel resistors

value, or variable. Figure 1–13 and Figure 1–14 provide schematic symbols for the capacitor. Capacitors are also categorized based on their construction. Ceramic and mica capacitors are precision nonpolarized capacitors having tolerances of 1 to 20% with values from 10 pF to 2.2 μF being available. Electrolytic capacitors (such as Aluminum and Tantalum) are polarized capacitors having tolerances of +20 to –85% with values from 1 μF to over 3.3 F. Formula 1–7 will find the capacitance of two plates separated by a dielectric material.

FORMULA 1–7 $C = \varepsilon_R \times \varepsilon_0 \times \frac{A}{d}$

C—capacitance in Farads (F)
$\varepsilon_R \times \varepsilon_0$—dielectric constant of nonconductive material between the plates
A—cross-sectional area in CM (circular mils)
d—distance between the plates

Capacitors store energy in the form of an electric charge. Formula 1–8 provides the relationship between the charge of the capacitor with its capacitance and the voltage across the capacitor. Using Formula 1–9 the energy stored in the capacitor can be calculated knowing the capacitor capacitance and the voltage across the capacitor.

FORMULA 1–8 $Q = C \times V$

FORMULA 1–9 $W = 0.5 \times C \times V^2$

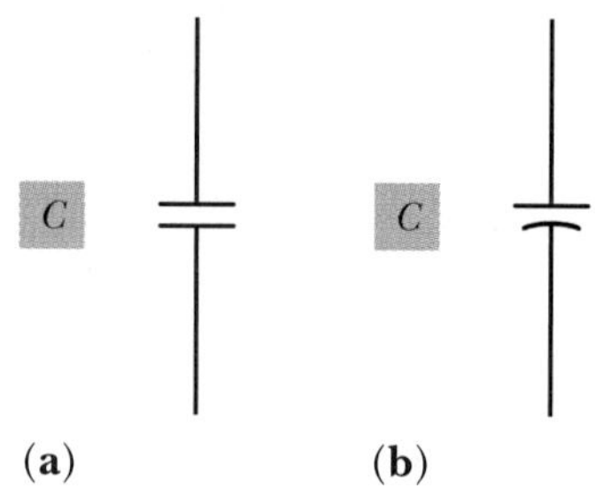

FIGURE 1–13 (a) Fixed value capacitor (nonpolarized). (b) Fixed value capacitor (polarized)

Q—charge in Coulombs
C—capacitance in Farads
V—voltage in Volts
W—Energy in Joules

Capacitors can be connected in series as shown in Figure 1–15. When capacitors are connected in series, the total capacitance of the series circuit is determined by using Formula 1–10.

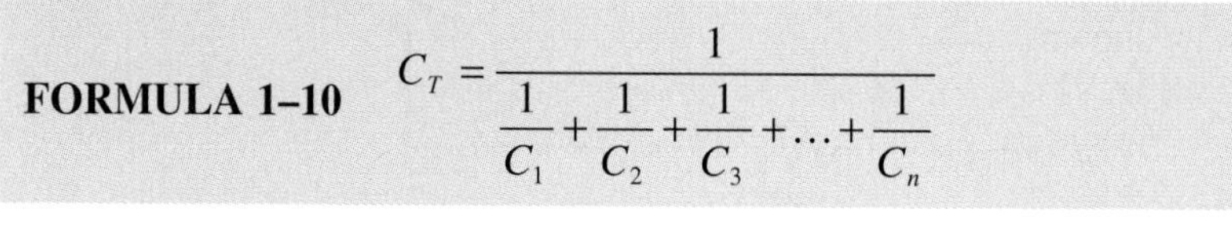

FORMULA 1–10 $$C_T = \frac{1}{\frac{1}{C_1} + \frac{1}{C_2} + \frac{1}{C_3} + \ldots + \frac{1}{C_n}}$$

C_n—represents the last series capacitor

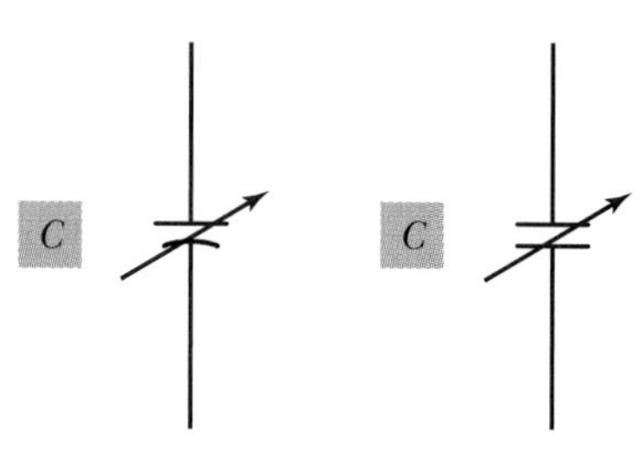

FIGURE 1–14 Variable capacitor

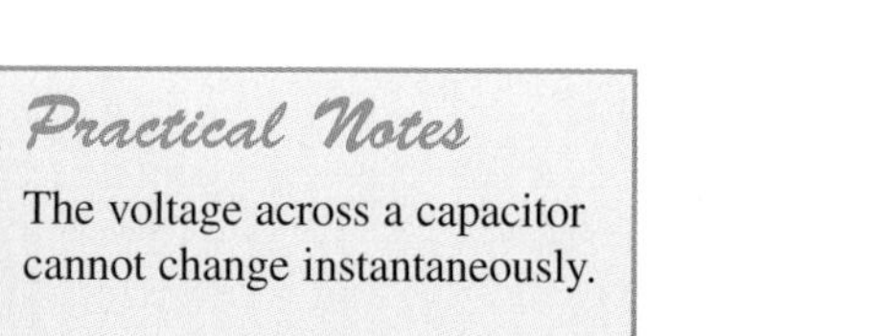

Practical Notes

The voltage across a capacitor cannot change instantaneously.

$$I = C\frac{\Delta V_C}{\Delta t}$$

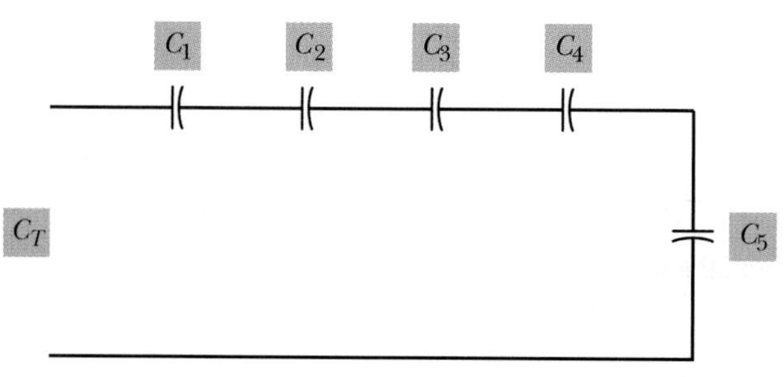

FIGURE 1–15 Capacitors in series

If all of the series capacitors are of the same value C, the total capacitance can be calculated using Formula 1–11 where n is the number of capacitors.

FORMULA 1–11 $C_T = \dfrac{C}{n}$

When only two capacitors are connected in series, the total capacitance for the series capacitor circuit can be found using Formula 1–12.

FORMULA 1–12 $C_T = \dfrac{C_1 \times C_2}{C_1 + C_2}$

Capacitors can also be connected in parallel as shown in Figure 1–16. The total capacitance for parallel capacitors can be determined with Formula 1–13.

FORMULA 1–13 $C_T = C_1 + C_2 + C_3 + \ldots + C_n$

Practical Notes The voltage applied across the capacitor must not exceed the manufacturer's specified capacitor working voltage parameter.

When a sinusoidal waveform is applied to a circuit containing a capacitor, the reactance of the capacitor is calculated using Formula 1–14.

FORMULA 1–14 $X_C = \dfrac{1}{2\pi f C}$

At low frequencies the capacitor's reactance is very large. For high frequencies the capacitor's reactance is small.

Inductor

The inductor is a passive component consisting of a wire coil. The schematic symbols for the inductor are shown in Figure 1–17. The inductor is available as fixed value, and adjustable or variable. There are air core inductors and iron core inductors.

Inductors store energy in the magnetic field surrounding the inductor. The energy stored is calculated with Formula 1–15.

FORMULA 1–15 $W = 0.5 \times L \times I^2$

L—inductance in Henrys
I—current in Amps
W—Energy in Joules

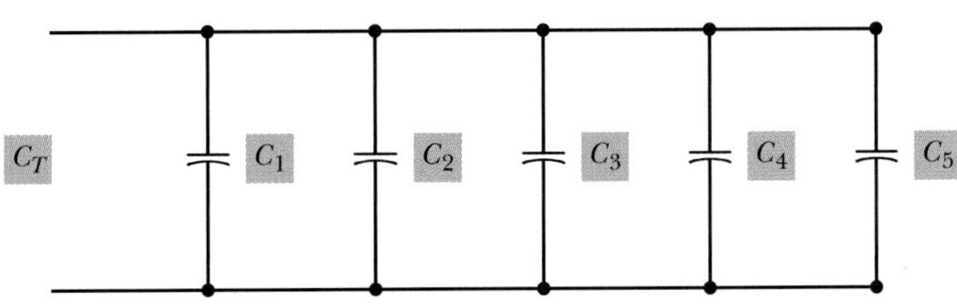

FIGURE 1–16 Capacitors in parallel

FIGURE 1–17 A fixed iron-core inductor

Practical Notes

The current flowing through an inductor cannot change instantaneously.

$$V_L = L\frac{\Delta I}{\Delta t}$$

The total inductance for inductors connected in series assuming no mutual inductance is calculated using Formula 1–16.

FORMULA 1–16 $L_T = L_1 + L_2 + L_3 + . . . + L_n$

L_n—represents the last series inductor

The total inductance of inductors connected in parallel assuming no mutual inductance is found with Formula 1–17.

FORMULA 1–17 $$L_T = \frac{1}{\frac{1}{L_1}+\frac{1}{L_2}+\frac{1}{L_3}+\ldots+\frac{1}{L_n}}$$

All inductors contain a dc coil resistance in addition to the coil inductance.

When a sinusoidal waveform is applied to a circuit containing an inductor, the reactance of the inductor is calculated using Formula 1–18.

FORMULA 1–18 $X_L = 2\,\pi\,f\,L$

At low frequencies the inductor's reactance is small. For high frequencies the inductor's reactance is large.

Practical Notes

Proper circuit protection and care must be taken to minimize the back emf when using switches in circuits containing inductors.

Transformers

The transformer (Figure 1–18) is an active component that transfers energy from the transformer's primary to the transformer's secondary by electromagnetic induction. Ideally, the power of the secondary is equal to the power of the primary as stated in Formula 1–19 and Formula 1–20.

FORMULA 1–19 $P_S = P_P$

FORMULA 1–20 $V_S\,I_S = V_P\,I_P$

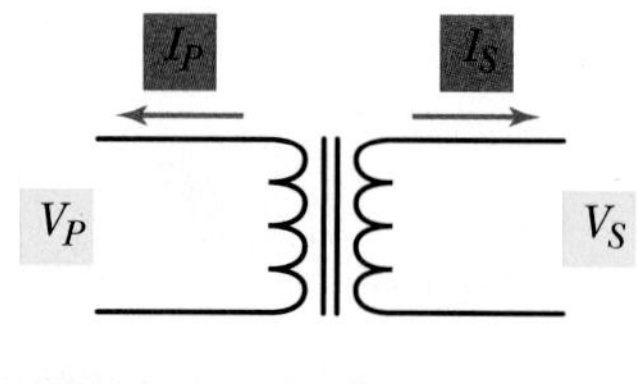

FIGURE 1–18 Transformer (iron-core)

The ratio of the primary voltage to the secondary voltage is directly proportional to the ratio of the primary coil turns to the secondary coil turns as shown in Formula 1–21.

FORMULA 1–21 $\frac{V_P}{V_S} = \frac{N_P}{N_S}$

1–3 Circuit Laws and Theorems

Several fundamental relationships exist that identify the relationship of voltages and currents in circuits.

Ohm's Law

Ohm's law is a mathematical equation describing the relationship of voltage, current, and resistance in a circuit. Ohm's law states that the voltage across a resistor is equal to the current flowing through that resistor multiplied by the resistance of that resistor (Formula 1–22).

FORMULA 1–22 $V = I \times R$

If the voltage across a resistor and the resistance of that resistor are known, Ohm's law can be rewritten to calculate the current flowing through that resistor (Formula 1–23).

FORMULA 1–23 $I = \dfrac{V}{R}$

When the voltage across a resistor and the current flowing through that resistor are known, the resistance of that resistor can be determined using Formula 1–24.

FORMULA 1–24 $R = \dfrac{V}{I}$

The Ohm's law triangle in Figure 1–19 can be a helpful tool to remember the three equations. For ac circuits, Ohm's law can be modified by replacing the resistance R with the impedance Z as shown in Formula 1–25 and Formula 1–26 where Z is a combination of resistance and reactance.

FORMULA 1–25 $V = I \times Z$

FORMULA 1–26 $V = I \times (R + jX)$

> *Practical Notes*
>
> Remember *ELI the ICE* man or *VLI the IC Vendor*. Restated: the voltage across an inductor leads the current flowing through that inductor by a phase angle of 90°, whereas, the current flowing through a capacitor leads the voltage across that capacitor by a phase angle of 90°.

Power

The power delivered from a power source and/or the power dissipated by a circuit component can be calculated by multiplying the voltage across a circuit element by the current flowing through that circuit element (Formula 1–27).

FORMULA 1–27 $P = I \times V$

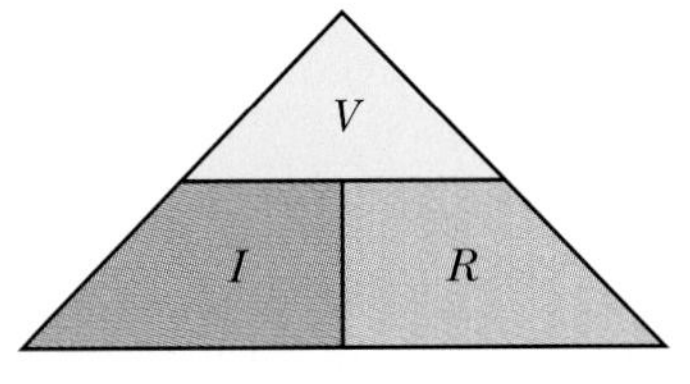

FIGURE 1–19 Ohm's law triangle

> **Practical Notes**
>
> For ac applications the apparent power S is equal to the voltage V multiplied by the current I.

If the power and voltage are known, the current can be determined using Formula 1–28.

FORMULA 1–28 $I = \dfrac{P}{V}$

When the power and current are known, the voltage can be found using Formula 1–29.

FORMULA 1–29 $V = \dfrac{P}{I}$

The power triangle of Figure 1–20 can be useful in remembering the three power equations.

FORMULA 1–30 $S = P + jQ$

The power equations can be combined with Ohm's law to calculate the power when the voltage across a resistor and the resistance of that resistor are known (Formula 1–31), or when the current flowing through a resistor and the resistance of that resistor are known (Formula 1–32).

FORMULA 1–31 $P = \dfrac{V^2}{R}$

FORMULA 1–32 $P = I^2 \times R$

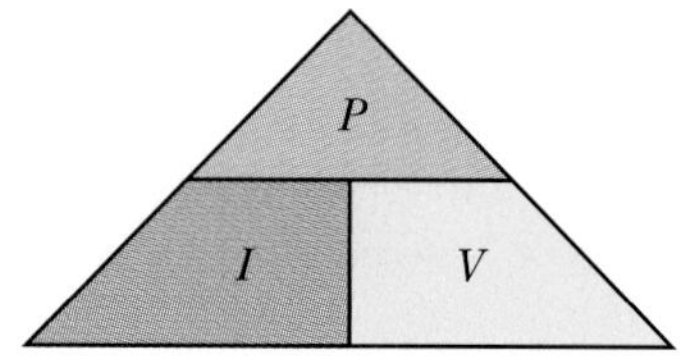

FIGURE 1–20 Power law triangle

The PIVR wheel (Figure 1–21) can be useful in recalling all of the power, current, voltage, and resistance equations.

FIGURE 1–21 PIVR wheel

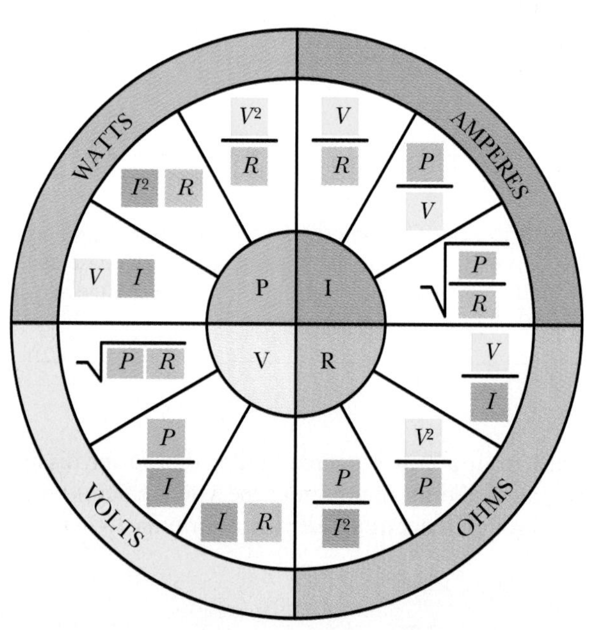

Kirchhoff's Voltage Law

Kirchhoff's voltage law (KVL) states that the sum of the voltages around a closed loop is equal to zero (Formula 1–33). Kirchhoff's voltage law can also be expressed as the sum of all voltage rises around a closed loop is equal to the sum of all voltage drops around the same loop. (Refer to Figure 1–22.)

FORMULA 1–33 $\sum_n V_n = 0$

Kirchhoff's Current Law

Kirchhoff's current law (KCL) states that the sum of the currents at a node is equal to zero (Formula 1–34). Kirchhoff's current law can also be expressed as the sum of all currents entering a node is equal to the sum of all currents leaving that node. (Refer to Figure 1–23.)

FORMULA 1–34 $\sum_n I_n = 0$

Voltage Divider Rule

The voltage divider rule states that the voltage across a resistor in a series resistor circuit can be determined by dividing the multiplication of the source voltage and the resistor's resistance with the total circuit resistance. For the Figure 1–24 circuit, the voltage across R_2 is found using Formula 1–35.

FORMULA 1–35 $V_2 = V_S \times \frac{R_2}{R_1 + R_2}$

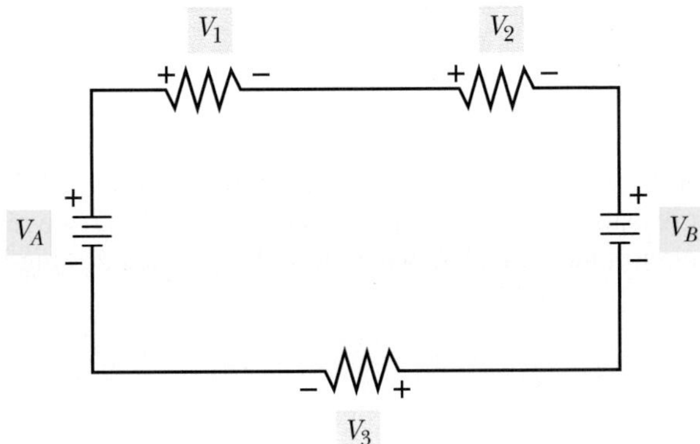

FIGURE 1–22 Kirchhoff's voltage law example

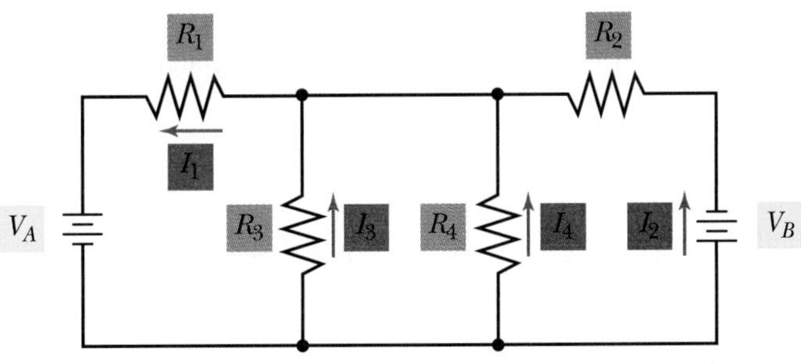

FIGURE 1–23 Kirchhoff's current law example

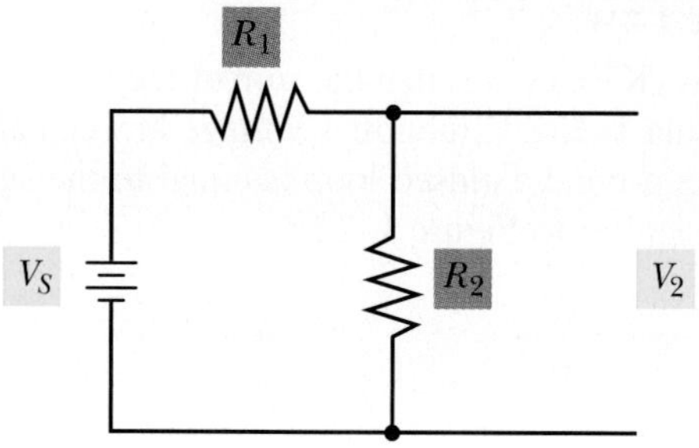

FIGURE 1–24 Circuit for voltage-divider rule

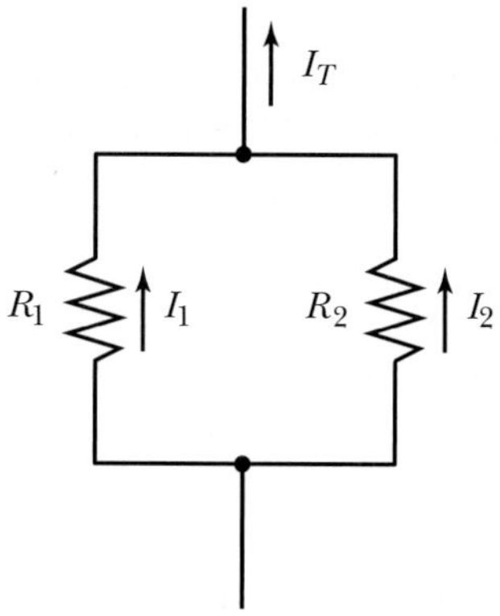

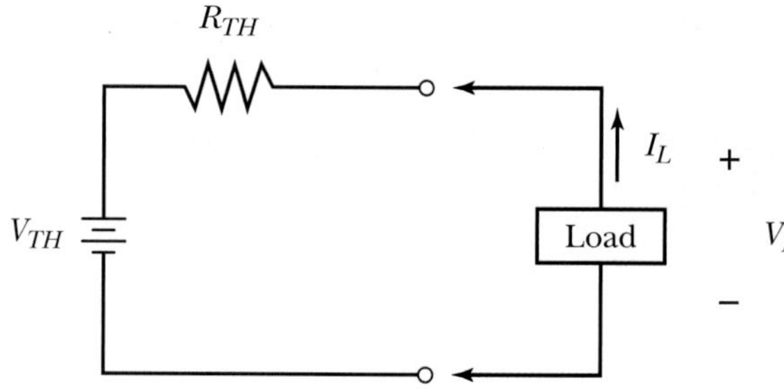

FIGURE 1–25 Current-divider rule circuit

FIGURE 1–26 Thevenin equivalent circuit

Current Divider Rule

The current divider rule states that the current flowing through a resistor in a parallel resistor circuit can be calculated by dividing the multiplication of the total circuit current and the other resistor's resistance with the summation of the two resistor values. For the Figure 1–25 circuit, the current flowing through R_2 is found using Formula 1–36.

FORMULA 1–36 $I_2 = I_T \times \dfrac{R_1}{R_1 + R_2}$

Thevenin's Theorem

Thevenin's theorem states that any circuit can be represented as a single voltage source known as the Thevenin voltage V_{TH} and a single series resistor known as the Thevenin resistance R_{TH} as shown by Figure 1–26.

The following procedure can be used to determine the values of the Thevenin voltage source and the Thevenin resistance:

1. The Thevenin voltage is equal to the open—circuit voltage across the two reference terminals after removing the circuit load.
2. The Thevenin resistance is equal to the equivalent resistance as viewed from the two reference terminals after removing the circuit load with all internal sources de-energized. (Voltage sources replaced with their series source resistance—which

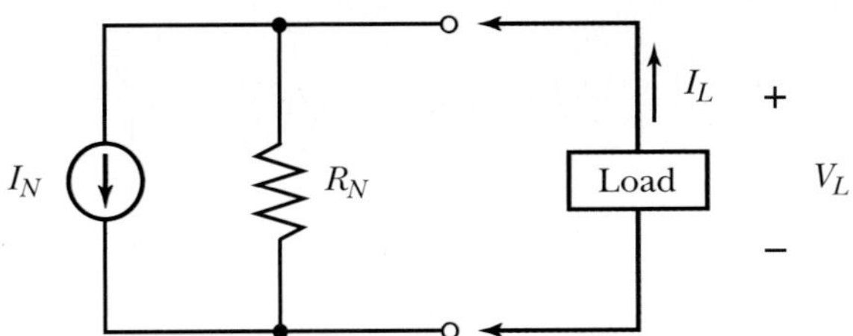

FIGURE 1–27 Norton equivalent circuit

is 0 Ω for ideal supplies, current sources replaced with their parallel source resistance—which is infinite for ideal supplies.)

Norton's Theorem

Norton's theorem states that any circuit can be replaced with a single current source known as the Norton current I_N and a parallel resistor known as the Norton resistance R_N as shown by Figure 1–27. The following procedure can be used to determine the values of the Norton current source and the Norton resistance:

1. The Norton current is equal to the current flowing through a wire connected between the two reference terminals (the short—circuit current) after removing the circuit load.
2. The Norton resistance is equal to the equivalent resistance as viewed from the two reference terminals after removing the circuit load with all internal sources de-energized. (Voltage sources replaced with their series source resistance—which is 0 Ω for ideal supplies, current sources replaced with their parallel source resistance—which is infinite for ideal supplies.)

Note that the Thevenin voltage is equal to the Norton current multiplied by the Thevenin resistance.

> *Practical Notes*
>
> Using a Thevenin or Norton equivalent circuit in place of the original circuit simplifies the analysis allowing an individual to evaluate the voltage and current of a component or circuit connected to the two reference terminals.

1–4 Circuit Analysis

Using the circuit laws and theorems, several circuit analysis techniques can be used to determine the circuit voltages and currents.

Loop Analysis

Loop analysis is a method to calculate the voltage and current for each component by writing the KVL equations for each circuit loop and the KCL equations for each circuit node after the assignment of branch currents.

The KVL equations for Figure 1–28 are:

FORMULA 1–37 $V_{S_1} - I_1 R_1 - I_3 R_3 = 0$

FORMULA 1–38 $I_3 R_3 - I_2 R_2 + V_{S_2} = 0$

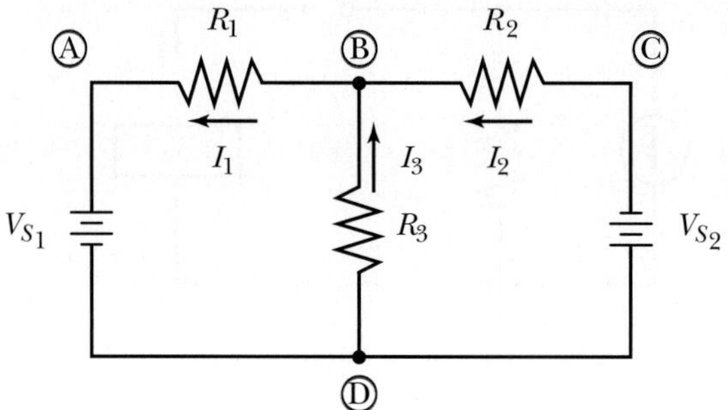

FIGURE 1–28 Example circuit for loop analysis

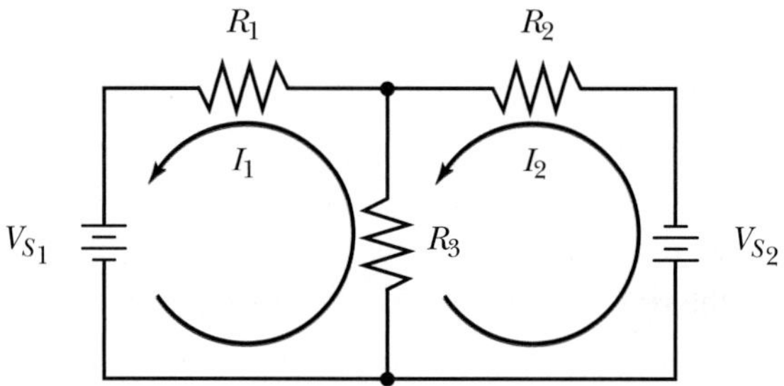

FIGURE 1–29 Example circuit for mesh analysis

The KCL equation for Node B in Figure 1–28 is:

FORMULA 1–39 $I_1 - I_2 - I_3 = 0$

Mesh Analysis

Mesh analysis is a method to calculate the voltage and current for each component by writing the KVL equations for each circuit loop after the assignment of loop currents.

The KVL equations for Figure 1–29 are:

FORMULA 1–40 $V_{S_1} - I_1 R_1 - I_1 R_3 + I_2 R_3 = 0$

FORMULA 1–41 $I_1 R_3 - I_2 R_3 - I_2 R_2 + V_{S_2} = 0$

Nodal Analysis

Nodal analysis is a method to calculate the voltage and current for each component by writing the KCL equations for each circuit node after the assignment of nodal voltages and branch currents.

The KCL equation for Node B in Figure 1–30 is:

FORMULA 1–42 $I_1 - I_2 - I_3 = 0$

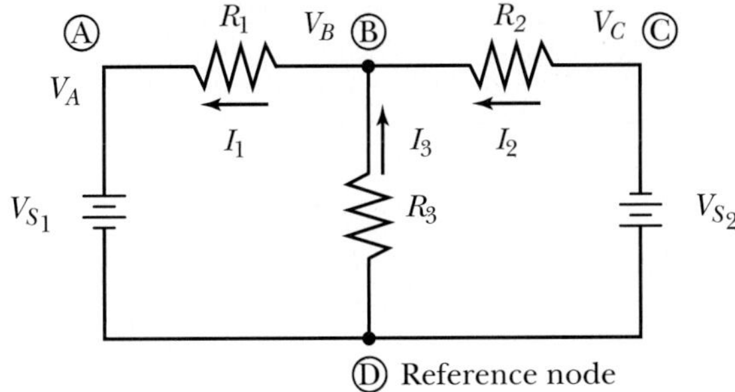

FIGURE 1–30 Example circuit for nodal analysis

The nodal voltages for super nodes of the Figure 1–30 circuit are:

Node A: $V_A = V_{S_1}$

Node C: $V_C = -V_{S_2}$

With Ohm's law, the branch currents can be written using the nodal voltages:

FORMULA 1–43 $I_1 = (V_A - V_B)/R_1$

FORMULA 1–44 $I_3 = V_B/R_3$

FORMULA 1–45 $I_2 = (V_B - V_C)/R_2$

Superposition

Superposition states that any voltage or current in a linear circuit with several voltage and/or current sources may be determined by first considering the voltage and currents produced by each source individually, and then combining all the individual voltage and current responses. Refer to Figure 1–31.

Solution Methods

Once the circuit equations are written the unknown variables must be found. Mathematical methods are available to accomplish the task of determining the solution for the unknown variables. When two or more equations exist having two or more unknown variables, the mathematical solution methods include substitution, simultaneous equations, and matrix algebra. Software programs exist for calculators and computers that will provide the variable solutions after the equations are entered into the calculator or computer. Electronics Workbench (EWB) and PSpice are programs that will provide the circuit voltages and currents at circuit locations selected by the user after the user enters the circuit schematic into the computer. The experimental method requires an individual to assemble the circuit after which power may be applied to circuit and measurements taken of the voltages and currents at specified locations.

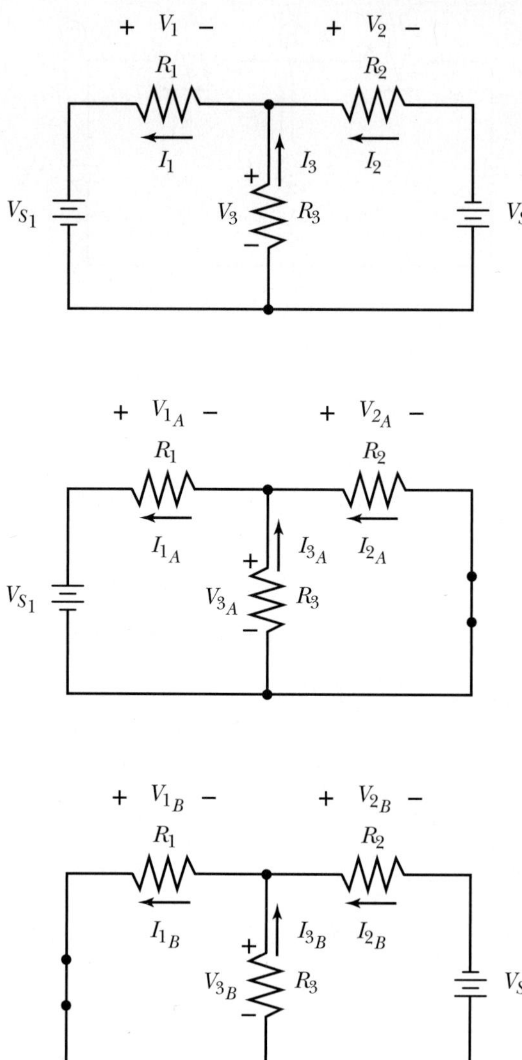

FIGURE 1–31 Example circuit for superposition

1–5 Measurement Equipment

Test equipment has been developed to evaluate electrical components and circuitry. Two instruments with which you should be extremely knowledgeable are the digital multimeter and the oscilloscope. You must know the intended usage of the instrument and how to operate that instrument. Some test equipment is designed as handheld units for field applications, while other equipment is designed for laboratory conditions.

Digital Multimeter (DMM)

The digital multimeter (Figure 1–32) is a versatile tool. The DMM contains an ohmmeter, a voltmeter, and an ammeter. Some DMMs include other meter functions such as continuity test, diode test, transistor test, and temperature measurement.

The DMM is able to measure the resistance of components indicating the value in ohms. The component must be removed from the circuit, the DMM placed into the ohmmeter mode, and the DMM leads connected to the component. The component's

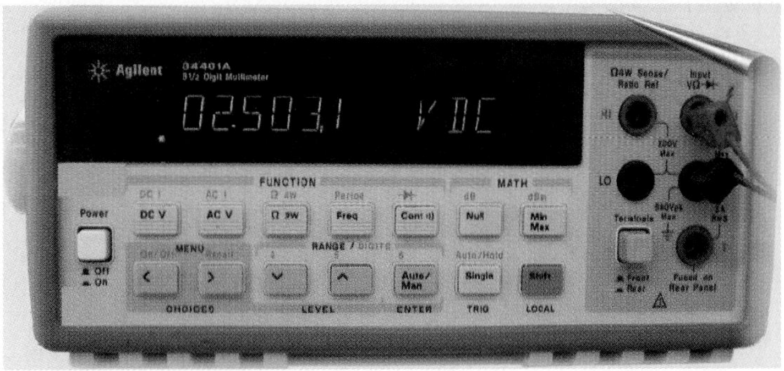

FIGURE 1–32 Digital multimeter

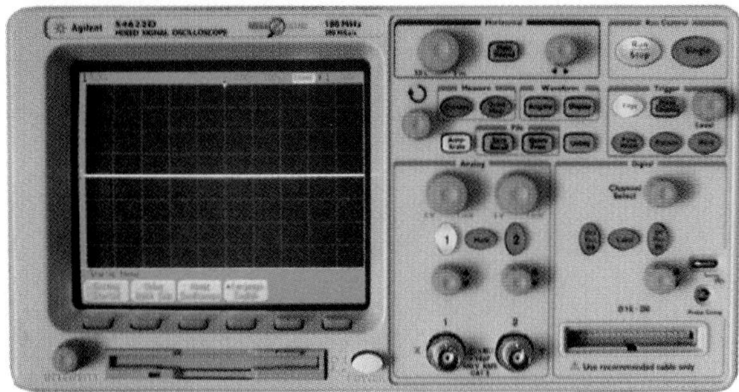

FIGURE 1–33 Agilent mixed mode oscilloscope (54600 series)—logic analyzer setup

resistance will be shown on the display. If the component is open or the component's value exceeds the DMM range, the display will indicate the out of range value (OL). Check the DMM user's manual to determine the open or out of range value for the DMM that you are using.

The DMM is able to measure the ac or dc voltages within a circuit. After placing the DMM into the ac or dc voltmeter mode, connect the DMM leads to the circuit locations of interest. With the power applied to the circuit, the DMM display will indicate the voltage between those two points.

The DMM is able to measure the ac or dc currents within a circuit. Extreme care must be taken in using the DMM's ammeter to prevent damage to the circuit and/or the ammeter. After placing the DMM into the ac or dc ammeter mode, the ammeter must be connected in series with the components for which the current is desired to be measured. With the power applied to the circuit, the DMM display will indicate the current flowing through that section of the circuit.

The Oscilloscope (Scope)

The oscilloscope (Figure 1–33) is a powerful instrument used to display the voltage in a circuit as time passes. Scopes are available as an analog scope and as a digital storage oscilloscope (DSO). Some scopes have two input channels with an external trigger input, while other scopes have four input channels with an external trigger input. The multimode scope is designed with two analog input channels and sixteen digital input channels (logic analyzer function). The oscilloscopes are also categorized as to the bandwidth limitations (such as 100 MHz). The bandwidth limitation indicates the maximum frequency that can be shown accurately on the display.

> **Practical Notes**
>
> All of the scope grounds are connected together internal to the scope chassis ground. The scope chassis ground is connected to the third prong of the power cord. Extreme care must be taken when connecting any of the scope ground leads to any circuitry. Connect the scope grounds to the circuit grounds. If this principle is not followed, severe damage can occur to the scope, the scope lead, and/or the circuitry being evaluated.

Review Questions

Select Ohms Law/Power Formulas:

1. Power = **a.** I/V
 b. IR
 c. IV
 d. V/I
2. Power = **a.** I^2V
 b. I^2R
 c. V^2/I
 d. I/V^2
3. Power = **a.** I/V^2
 b. V^2R
 c. V^2/R
 d. I^2/R
4. Current = **a.** V/P
 b. P/R
 c. R/P
 d. V/R
5. Current = **a.** $\sqrt{P/R}$
 b. $\sqrt{PR}$
 c. $\sqrt{V/R}$
 d. $\sqrt{V/R}$
6. Current = **a.** V/P
 b. P/V
 c. PR
 d. VR
7. Voltage = **a.** P/I
 b. P/I^2
 c. IP
 d. I/R
8. Voltage = **a.** $\sqrt{P/R}$
 b. $\sqrt{R/P}$
 c. $\sqrt{IR}$
 d. $\sqrt{PR}$
9. Voltage = **a.** IR
 b. RP
 c. IP
 d. I/P
10. R = **a.** IV
 b. V/P
 c. V/I
 d. IP
11. R = **a.** P/I^2
 b. P/I
 c. I/P
 d. I^2/P
12. R = **a.** V/P
 b. P/V
 c. I^2/P
 d. V^2/P

Answer True or False

13. Resistance is the opposition to current flow in a circuit.
14. One product of resistance in a circuit is heat.
15. The sum of the voltage drops around any closed path is equal to the voltage applied.
16. Each resistor in a series circuit gets its own share of the current.
17. Total power is the sum of the resistors.
18. The color code for a 1k ohm resistor is black, brown, orange.
19. A rheostat is used to control voltage in a circuit.
20. If resistance in a series circuit increases, current increases.
21. Voltage is measured across a resistor.
22. When one resistor in a three resistor series circuit shorts, power total will increase.
23. Ohm's law shows the relationships of voltage, power, and resistance.
24. If current is to be held at a constant value, an increase in resistance will require an increase in applied voltage.
25. Micro stands for 1×10^6.
26. Kilo stands for 1×10^3.
27. Tera stands for Earth.
28. Pico stands for 1×10^{-9}.
29. Mega stands for 1×10^6.
30. Milli stands for 1×10^{-3}.
31. Giga stands for 1×10^{12}.
32. Nano is the third moon of Saturn.
33. Electric current is able to flow through all conductive paths, including your body.

Series

34. Write Kirchhoff's two voltage laws:
 a. ______________________________
 b. ______________________________
35. Write the voltage divider formula:
 ____
36. Write the formula for RT: ______________________________
37. Write the formula for PT: ______________________________

Parallel

38. Write Kirchhoff's two current laws:

a. ______________________

b. ______________________

39. Write the current divider formula: ______________________

40. Write two formulas (reciprocal and product-sum) for RT in parallel:

a. ______________________

b. ______________________

41. Current in a parallel circuit does what? ______________________

42. Voltage in a parallel circuit ______________________

Problems

1. Convert the color code to the size of the resistor.

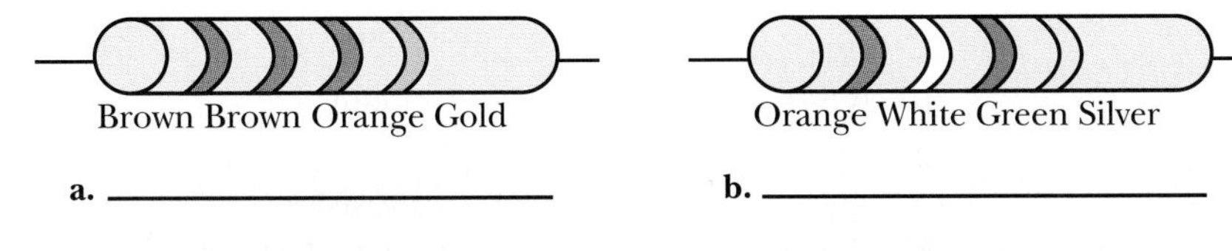

a. __________ **b.** __________ **c.** __________

Red Orange Green Silver

Green Blue Yellow Silver

Brown Red Yellow Silver

d. __________ **e.** __________ **f.** __________

2. Convert the size of the resistor to the color code.

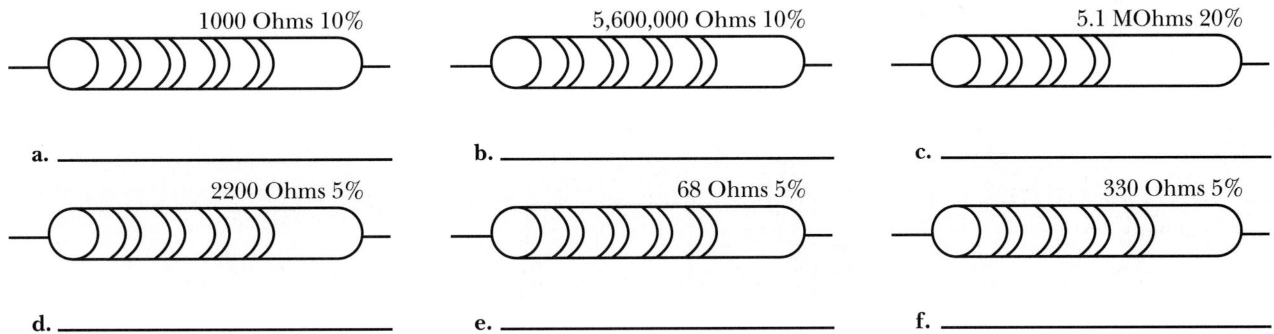

PIVR

3. Solve for PIVR Totals. Solve for PIVR for each resistor.

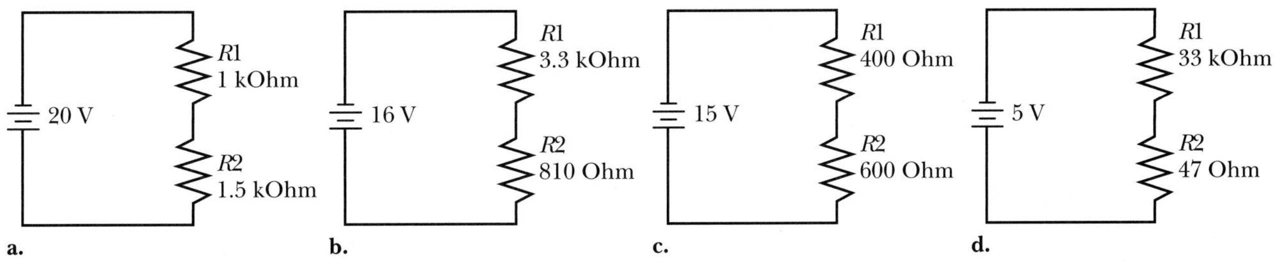

PIVR

4. Solve for PIVR Totals. Solve for PIVR for each resistor.

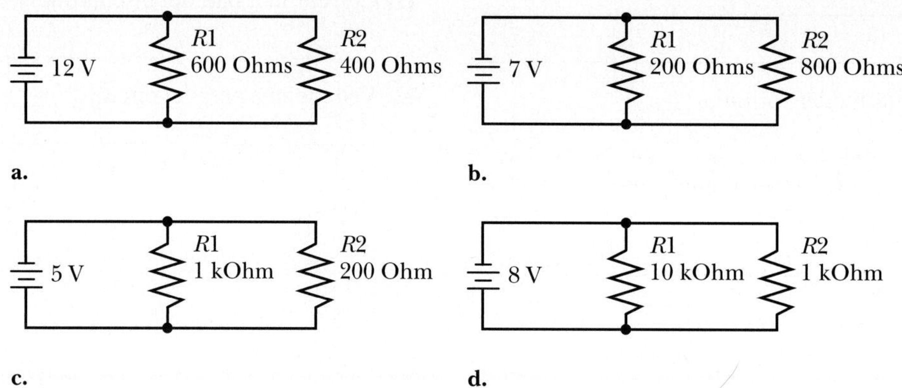

PIVR

5. Solve for PIVR Totals. Solve for PIVR for each resistor.

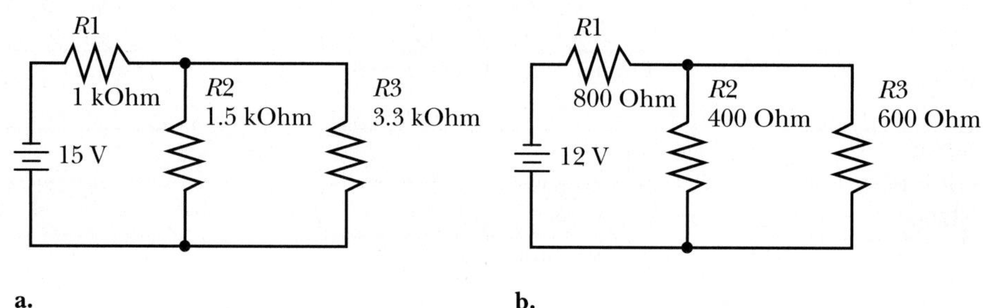

6. For each circuit:

If R1 opens, VR2 = ________

If R2 shorts, IR3 = ________

If R2 opens, VR3 = ________

If R1 shorts, IR2 = ________

If R2 shorts, VR1 = ________

If R3 opens, VR1 = ________

IR2 = ________

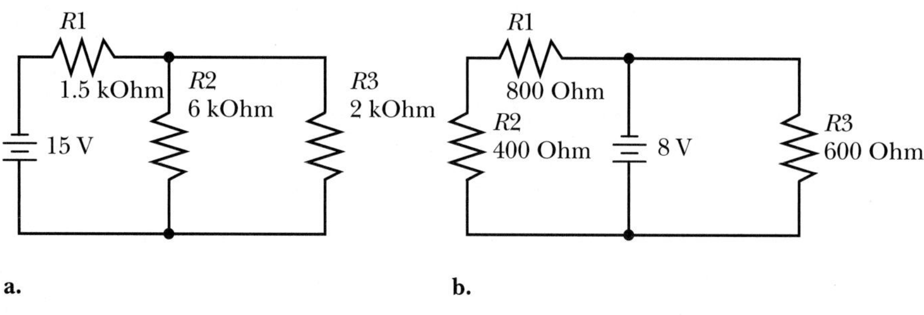

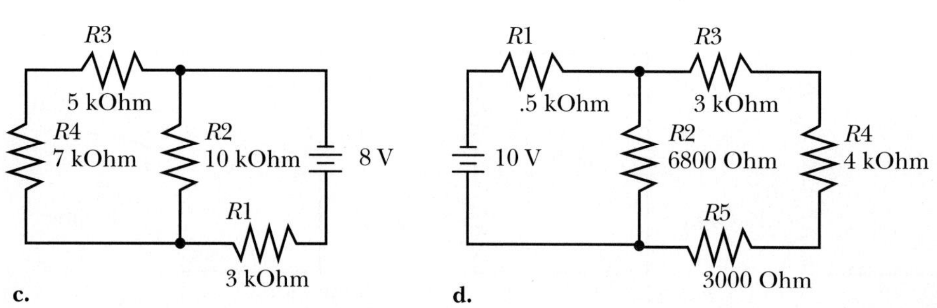

7. Solve for CT.

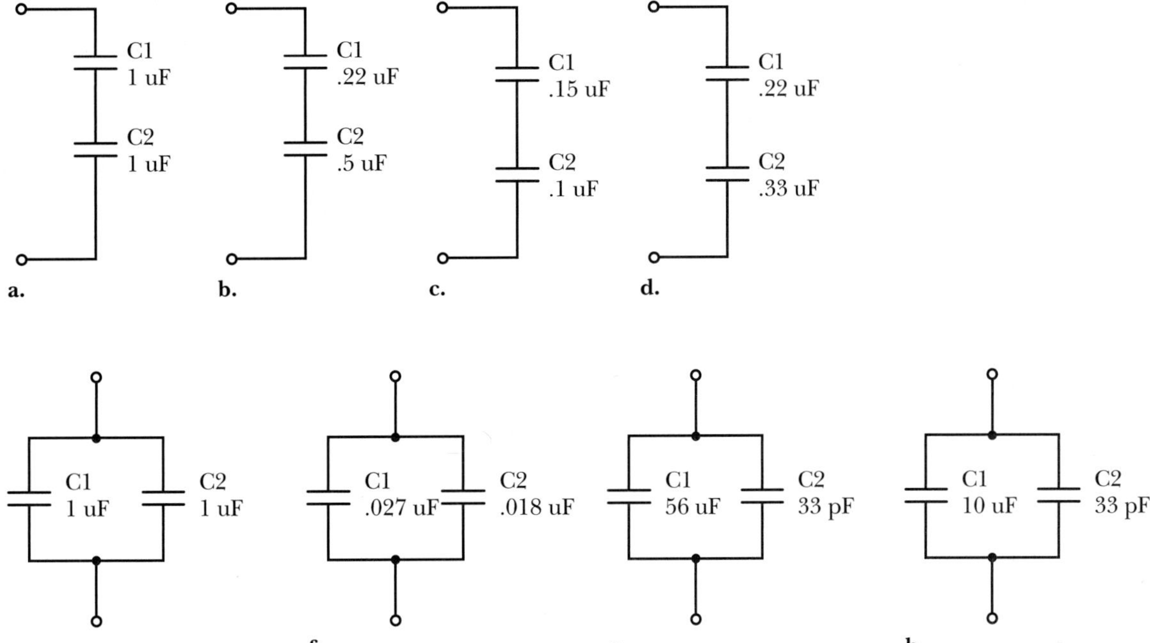

8. Solve for true currents in all resistors. Solve for true voltages across all resistors. Solve for Thevenin's voltage and resistance. Solve for Norton's current.

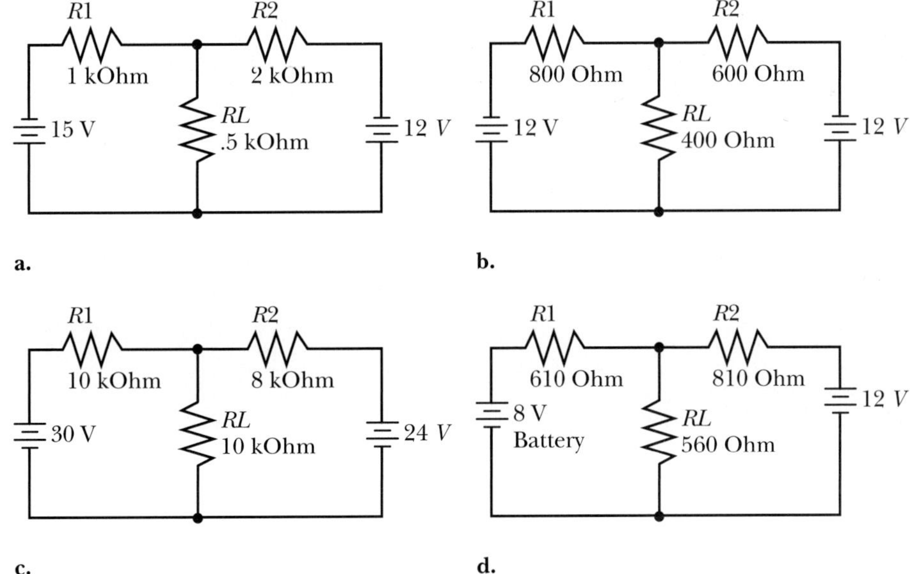

OUTLINE

2–1 **Semiconductor materials**
2–2 **The P-N junction**

OBJECTIVES

After studying this chapter, you should be able to:

1. Describe the difference between **valence electrons** and **conduction-band electrons**
2. Describe the main difference between **N-type** semiconductor materials and **P-type** semiconductor materials
3. Draw a diagram of a **P-N junction,** including the depletion region
4. Draw a P-N junction that shows the polarity of applied voltage for **forward biasing** the junction
5. Draw a P-N junction that shows the polarity of applied voltage for **reverse biasing** the junction
6. Explain the difference between the **barrier potential** and **reverse breakdown voltage** for a P-N junction
7. Sketch the **I-V curve** for a typical P-N junction, showing both the forward and reverse bias parts of the curve

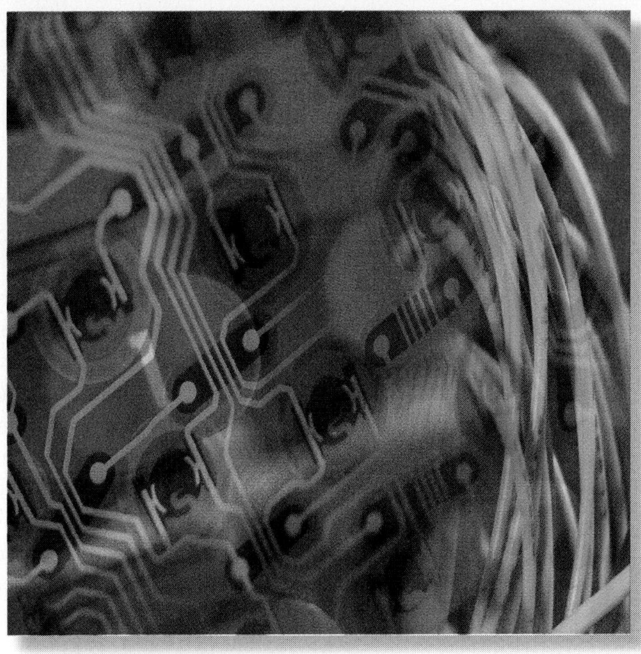

CHAPTER 2

Semiconductor Materials and P-N Junctions

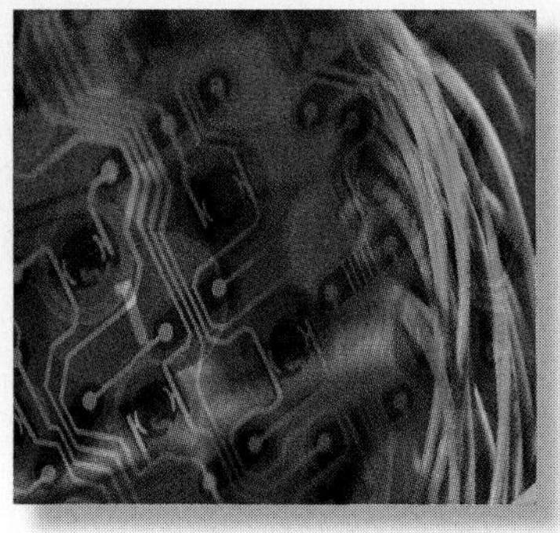

PREVIEW

Semiconductor devices have changed our world. Progress in electronics was once limited by having to use fragile, bulky, and power-gobbling vacuum tubes instead of tiny semiconductor devices. Transistors, diodes, integrated circuits, Complex Programmable Logic Devices (CPLD), and Application Specific Integrated Circuits (ASIC) are the most common examples of modern-day semiconductors. Pocket calculators, desktop computers, digital watches, video games, VCRs, portable electronic keyboards, DVDs, and cellphones have been developed due to semiconductor technology. Trying to imagine a world without the gadgets and appliances made possible with semiconductors is like trying to imagine modern society functioning without the automobile.

In this chapter, you get a look at the very basic inner workings that are common to all kinds of semiconductor devices. You will see why atoms of silicon are so important for making semiconductor devices, and you will learn about a few of the other important elements such as germanium, arsenic, gallium, and phosphorus that are applied in semiconductor technology. You will be introduced to the unique way that current flows through semiconductor materials, and you will get a first look at some graphs that are commonly used for describing the operation of simple semiconductor devices.

KEY TERMS

Barrier potential
Conduction band
Covalent bond
Depletion region
Diffusion current
Doping material
Forward bias
Majority carriers
Minority carriers
N-type material
Pentavalent atoms
P-N junction
P-type material
Reverse bias
Reverse breakdown voltage
Tetravalent atoms
Trivalent atoms
Valence electrons

2–1 Semiconductor Materials

Recall from Chapter 1 that conductors, nonconductors, and semiconductors offer different levels of resistance to current flow. Also remember that conductors such as copper, gold, and silver offer little opposition to current flow, whereas good insulators such as glass, mica, and most plastics offer a great deal of opposition to current flow. Semiconductor materials, such as silicon combined with very small amounts of aluminum or arsenic, oppose current flow at levels somewhere between that of the best conductors and best insulators.

In this chapter, you will discover how it is possible to adjust a semiconductor's ability to pass electrical current. You will see how the same semiconductor device can be a fairly good conductor under certain conditions and yet function as a rather good insulator under a different set of conditions. This ability to control the conductance of semiconductor material makes it suitable for use in devices that can control the flow of current through a circuit and amplify electrical signals.

Review of Atomic Structure

Referring to Figure 2–1, remember that the three primary particles of atoms are protons, neutrons, and electrons. The protons (positive particles) and neutrons (neutral particles) are located in the center, or nucleus, of the atom. The electrons (negative charges) are arranged in orbits that surround the nucleus. In a normal, stable atom, the number of electrons in the orbits exactly equals the number protons in the nucleus. This means that the total electrical charge of the atom is zero.

There is a definite limit to the number of electrons that can be included in each electron orbit, or shell. If you number the shells such that 1 is the innermost, 2 is the second from the nucleus, and so on, you can use the following formula to determine the maximum number of electrons that can occupy each shell, where n is the shell number.

FORMULA 2–1 $2n^2$

Formula 2–1 shows that the innermost shell (n = 1) can contain no more than 2 electrons, the second shell (n = 2) can hold up to 8, the third (n = 3) up to 18, and so on.

Each shell represents a different energy level for the electrons that occupy it. As shown in Figure 2–2, the energy levels increase with increasing distance from the nucleus. So you can see that electrons located in the innermost shell possess the least amount of energy, while those in the outermost shell have the greatest amount of energy. The electronic behavior of an atom mainly depends on the relative number of electrons in its outermost shell, or valence shell. Electrons included in the valence shell are called **valence electrons.**

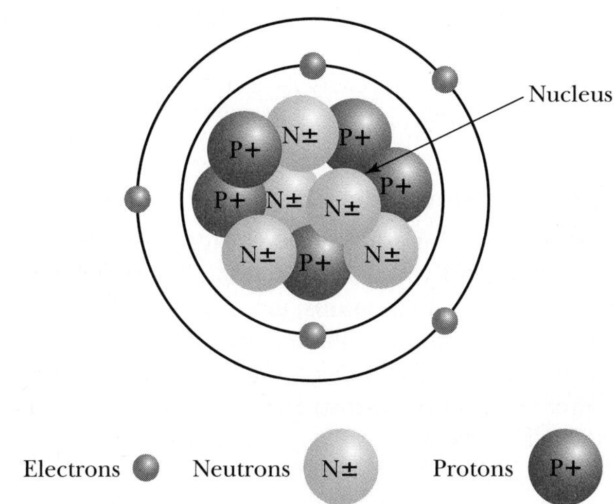

FIGURE 2–1 Atoms are composed of protons, neutrons, and electrons

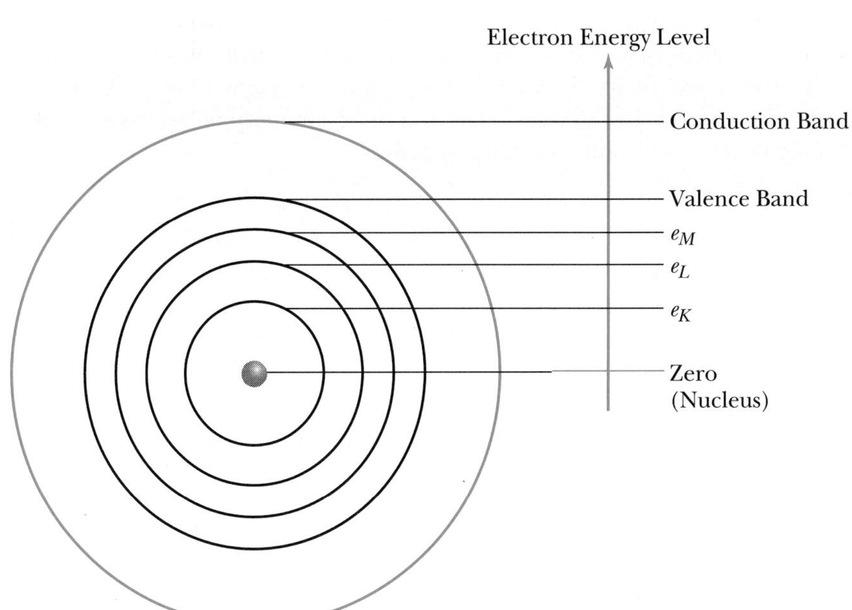

FIGURE 2–2 Energy levels for atoms

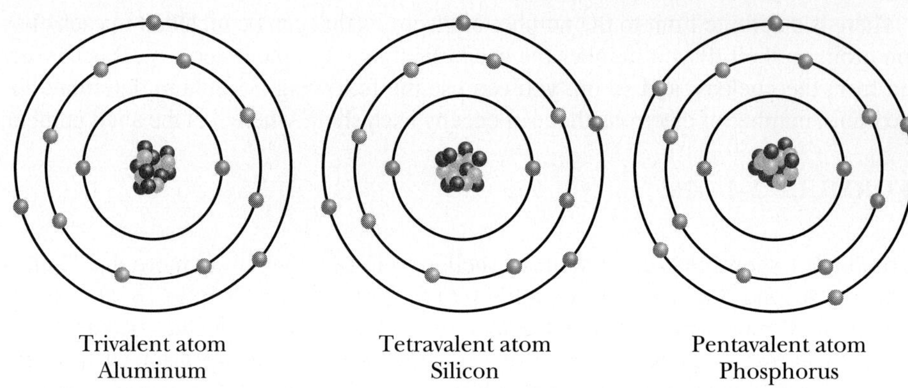

FIGURE 2–3 Trivalent, tetravalent, and pentavalent atoms

Exposing an atom to certain outside sources of energy (notably electrical energy) gives the valence electrons additional energy, which allows them to break away from their valence band and enter a **conduction band.** This is an important effect in electronics because electrons at the conduction-band energy levels are the only ones that are free to take part in the process of electron current flow. When a valence-band electron absorbs sufficient energy, it jumps up into a conduction band. Sooner or later that same electron has to fall back down to the valence-band level, and when that happens, the electron gives up its extra load of energy—usually in the form of heat, but sometimes as a particle of light energy.

Atomic Structures for Semiconductors

You will soon see that the most useful semiconductor materials are those made from atoms that have three, four, or five valence electrons. Atoms that normally have three valence electrons are called **trivalent atoms,** those normally having four valence electrons are called **tetravalent atoms,** and those normally having five outer electrons are called **pentavalent atoms.** Like any other atom, these are electrically neutral when they have their normal number of valence electrons. Figure 2–3 shows examples of each of these three kinds of atoms.

The valence electrons in these semiconductor materials are apt to absorb energy from heat and light sources as well as electrical sources. You will find through your studies of semiconductor devices that they respond to changes in temperature. In many instances, a relatively small increase in temperature can drive millions of valence electrons into conduction bands, thereby causing a dramatic decrease in electrical resistance.

These descriptions of atoms—shells, energy levels, valence electrons, and conduction-band electrons—are vital to your understanding of semiconductor materials and the behavior of semiconductor devices. Here is a brief summary of some key facts about the atomic structures of semiconductor materials.

1. "Tri-" means *three,* so a trivalent atom is one that has three electrons in the valence shell. Typical trivalent semiconductor materials are boron, aluminum, gallium, and indium.
2. "Tetra-" means *four,* so a tetravalent atom is one that has four electrons in its valence shell. Typically tetravalent semiconductor materials are germanium and silicon. Silicon is by far the more common (which explains why the center of semiconductor technology in northern California is nicknamed "Silicon Valley").
3. "Penta-" means *five,* so a pentavalent atom is one that has five electrons in its valence shell. Typical pentavalent semiconductor materials are phosphorus, arsenic, and antimony.
4. Valence electrons in semiconductor materials can absorb energy and jump to conduction bands by energy sources that include electricity, heat, and light.
5. Conduction-band electrons in semiconductor materials usually emit heat energy when they fall back to their valence bands. In certain semiconductors, particularly gallium arsenide, electrons dropping down to valence-band levels emit light energy.

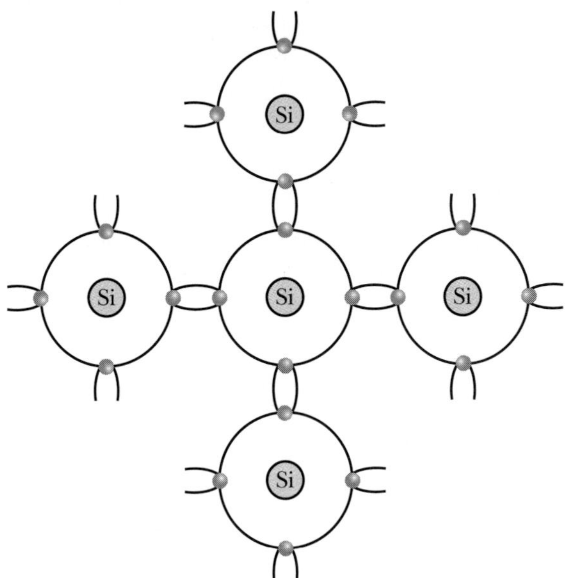

FIGURE 2–4 Covalent bonds for tetravalent atoms (silicon: Si)

Covalent Bonds

You have already learned from an earlier lesson that matter is composed of molecules, which are, in turn, composed of atoms that represent the basic chemical elements. Although every kind of matter in the universe is composed of atoms, there are a number of different ways atoms can be assembled to create molecules. The materials used for making semiconductors form what is known as a **covalent bond** between the individual atoms. In a covalent bond, two or more atoms share valence electrons.

Figure 2–4 shows how a single tetravalent atom forms covalent bonds with four other atoms of the same type. The bonds are actually three-dimensional, thus forming a crystalline molecule, or lattice, that is shaped like a cube. A single grain of common beach sand is made up of millions upon millions of tetravalent silicon atoms that are bonded in this fashion.

Each atom of a trivalent material forms covalent bonds with three other atoms; and by the same token, pentavalent atoms form covalent bonds with five other identical atoms. Semiconductors are manufactured from these trivalent, tetravalent, and pentavalent elements that are first highly purified, then carefully combined to produce the necessary electronic effects. Because of their solid, crystalline makeup, you often hear semiconductor devices (diodes, transistors, and integrated circuits, for instance) called *solid-state devices.*

Practical semiconductor materials are not made from just one kind of atom, however. Most are made from a highly purified tetravalent atom (silicon or germanium), which is combined with extremely small amounts of trivalent and pentavalent atoms that are called the **doping material.** The atoms that are chosen for this task have the ability to combine with one another as though they were atoms of the same element. In fact, a tetravalent atom can be "fooled" into making a covalent bond with a trivalent or pentavalent atom.

Figure 2–5 shows covalent bonds between four tetravalent atoms and one trivalent atom. The bond between the leftmost tetravalent atom and the trivalent atom is missing one electron, but the covalent bonding takes place in spite of the shortage of one electron. Likewise, a pentavalent atom can bond with a tetravalent atom. As shown in Figure 2–6, this situation leaves an extra electron in the bonding arrangement.

The theory of operation of semiconductor devices rests heavily upon the notions of gaps, or *holes,* that are left when creating covalent bonds between trivalent and tetravalent atoms, and the excess electrons that result from covalent bonds between tetravalent and pentavalent atoms.

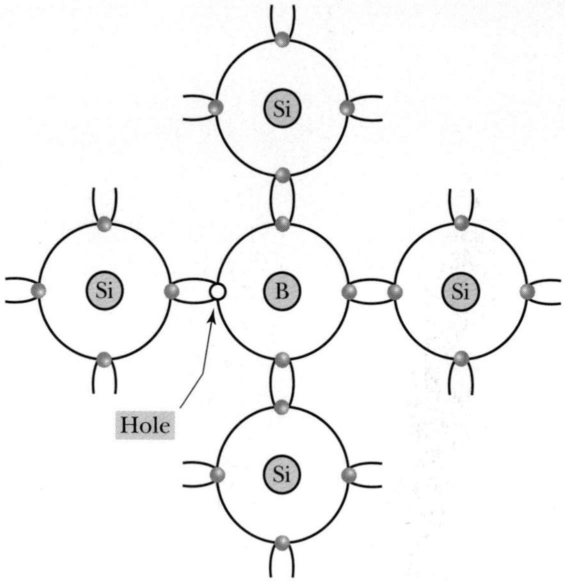

FIGURE 2–5 Covalent bonds between a trivalent atom (boron: B) and four tetravalent atoms (silicon: Si)

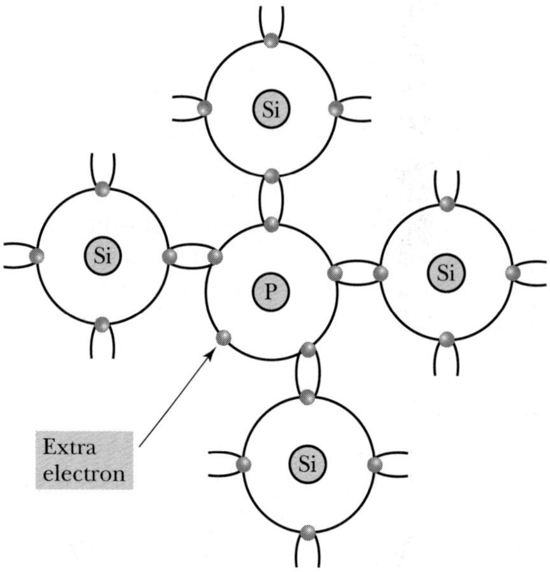

FIGURE 2–6 Covalent bonds between a pentavalent atom (phosphorus: P) and four tetravalent atoms (silicon: Si)

N-Type Semiconductors

An **N-type semiconductor material** is one that has an excess number of electrons. This means that one out of every million or so atoms has five electrons in covalent bonds instead of four. A block of highly purified silicon, for example, has four electrons available for covalent bonding. Arsenic is a similar material, but has five electrons available for covalent bonding. So when a minute amount of arsenic is mixed with a sample of silicon, the arsenic atoms move into places normally occupied by silicon atoms. The "fit" is a good one except for the fact that there is no place in the covalent bonds for the fifth electron. So the "fifth" electrons contributed by arsenic impurity are free to wander through the semiconductor material under external influences such as heat, light, and electrical energy.

Applying a source of electrical energy, or voltage, to a semiconductor material that has an excess number of electrons causes those electrons to drift through the material. These free electrons are repelled by the negative terminal of the applied potential and are attracted to the positive terminal. See Figure 2–7. Electron flow (current) is thus

Application Problem

Industrial

PN JUNCTION DIODES—LED CURRENT LIMIT

An indication LED (light emitting-diode) is widely used throughout the industry. For example, kitchen appliances have ON/OFF indication. Computers, videos, TVs, and satellite receivers also have several LEDs indicating different states. The following figure shows an LED circuit. The current limiting resistor placed in series with the LED is extremely important, otherwise the LED will burn out. If the desired current flow through an LED is 20 mA and a bias voltage is 5 V, calculate the resistance of such a resistor.

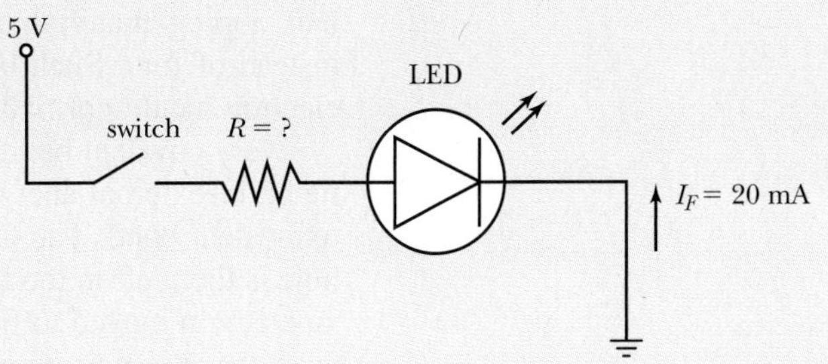

LED circuit

Solution An LED has a typical ON voltage of 2 V when conducting a forward current of 20 mA.

$$R = \frac{V_{cc} - 2.0\text{ V}}{I_F} = \frac{5\text{V} - 2.0\text{ V}}{0.020\text{A}} = 150\ \Omega$$

Choose the next resistor commercially available to increase a life expectancy of the LED.

$$R = 220\ \Omega$$

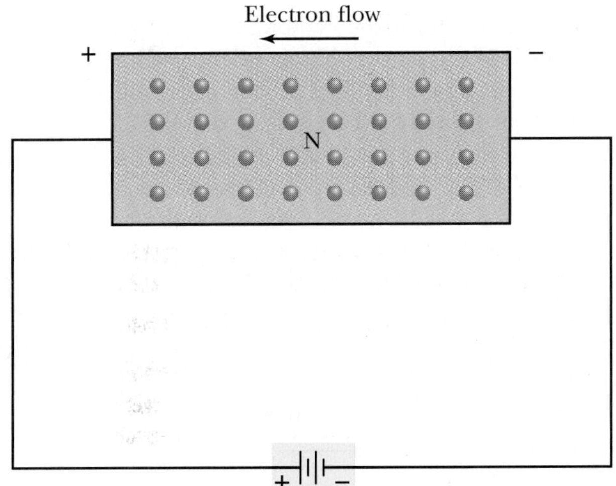

FIGURE 2–7 Current flow through an N-type semiconductor material

established through the semiconductor and any external circuitry connected to it. The current is carried through the semiconductor by electrons. Because electrons have a negative electrical charge, this type of semiconductor is called an N-type (negative-type) semiconductor.

The physical principle of conduction of electrons through an N-type semiconductor is slightly different from electron flow through a good conductor. For practical purposes, though, you can think of electron flow through an N-type semiconductor as ordinary electron flow through a common conductor.

P-Type Semiconductors

A **P-type semiconductor material** is one that has a shortage, or deficiency, of electrons. As with N-type semiconductors, the basic material is a highly purified tetravalent semiconductor such as silicon or germanium. The impurity, or doping material, for a P-type semiconductor is a trivalent element such as gallium or antimony. A trivalent doping material contributes covalent bonds that are made of three electrons instead of four. Such bonds are missing one electron, leaving a hole where a fourth electron would normally reside.

Every covalent bond that contains a hole is an unstable bond. This means that a little bit of external energy can cause the hole to be filled by an electron from a nearby tetravalent bond. The original hole is thus filled and its bond made more stable, but a hole is then left in the bond that gave up its electron. See Figure 2–8. We can say that an electron moved to fill a hole, but left a hole behind. It is more proper to say, however, that the hole moved. Applying an external voltage to a P-type semiconductor causes holes to drift from the point of positive charge to the point of negative charge. Holes, in other words, behave as positive charges. This is why such semiconductors are called P-type (positive-type) semiconductors.

More About N- and P-Type Semiconductor Materials

A highly purified semiconductor material that has not yet been doped is known as an *intrinsic semiconductor.* Once the doping material is added, the material becomes an *extrinsic semiconductor.*

N-type semiconductor materials are formed by adding minute amounts of a pentavalent element to the intrinsic semiconductor. This forces a few of the material's pentavalent atoms to provide a spare electron. The doping atoms in this instance are called *donor atoms* because they "donate" extra electrons to the material. The charge carriers in an N-type semiconductor are electrons, which are said to be the **majority carriers,** while holes are the **minority carriers.**

P-type semiconductors are formed by adding a tiny amount of one of the trivalent doping materials. This leaves a portion of the valence shells with a shortage of electrons (or excess holes). Because a trivalent doping material leaves holes that can subsequently accept electrons from other bonds, it is called an *acceptor atom.* The charge carriers in a P-type material are positively charged holes. Holes are the majority charge carriers in this instance, and electrons are said to be the minority carriers.

It is important to realize that N- and P-type materials are not electrically charged. One might suppose that an N-type material would possess a negative charge because it contains an excess number of electrons; and by the same token, a P-type material

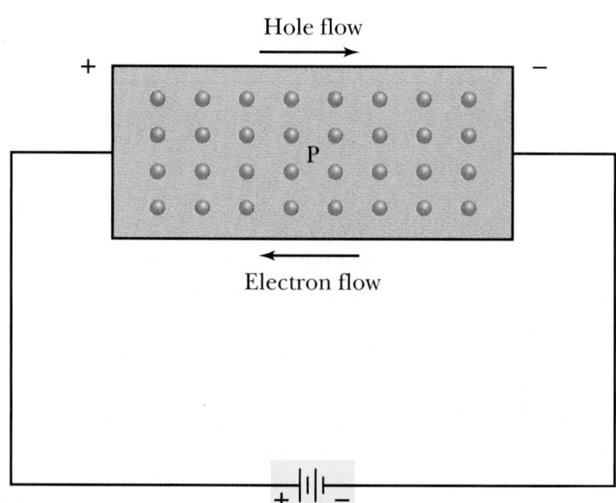

FIGURE 2–8 Current flow through a P-type semiconductor material

would have an inherent positive charge because of its shortage of electrons (or oversupply of positively charged holes). However, semiconductor materials cannot supply negative and positive charges to an external circuit as batteries do.

■ **IN-PROCESS LEARNING CHECK 1** Fill in the blanks as appropriate.

1. Before doping (in their pure form) semiconductor materials are sometimes called _______ semiconductors. Once they are doped with tiny amounts of an impurity atom, they are called _______ semiconductors.
2. N-type materials are formed by doping a _______ valent semiconductor material with a _______ valent material. P-type materials are formed by doping a _______ valent semiconductor material with a _______ valent material.
3. The doping atoms for N-type materials are called _______ atoms because they donate extra electrons to the covalent bonds. The doping atoms for P-type materials are called _______ atoms because they accept electrons that will fill the holes in the covalent bonds.
4. The majority carriers in an N-type material are _______, and the minority carriers are _______. The majority carriers in a P-type material are _______, and the minority carriers are _______. ■

2–2 The P-N Junction

N- and P-type semiconductor materials are rarely used alone. Practical semiconductor devices use both materials. Semiconductor *diodes,* for example, use a section of N-type material that is chemically and electrically fused to a section of P-type material (see Figure 2–9a). As you will discover in Chapter 5, common transistors are composed of one type of semiconductor sandwiched between sections of the other type: an N-type material between two sections of P material (called a *PNP transistor*), or a P-type material fit between two sections of an N material (called an *NPN transistor*). See the examples in Figures 2–9b and 2–9c.

The region where a P-type material is chemically and electrically fused with an N-type material is called a **P-N junction.** At the P-N junction within a semiconductor device, an excess number of electrons from the N-type material come into contact with an excess number of holes from the P-type material (see Figure 2–10a). Since there are free electrons in the N-type material near the junction of the P- and N-type materials, they are attracted to the positive holes near the junction in the P-type material. Some electrons leave the N-type material to fill the holes in the P-type material. And while electrons are leaving the N-type material in this fashion, they leave behind positive hole charges. In other words, equal numbers of electrons and holes flow across th P-N junction. Electrons move from the N-type material into the P-type material, and holes from the P-type material to the N-type material.

The result of this activity is that the N-type material now has an electron deficiency and is positively charged because of positive ions near the P-N junction, and the P-type material has a surplus of electrons and is negatively charged because of negative ions near the junction. This causes a potential difference near the junction that is known as the **barrier potential** or *contact potential* (V_B). See Figure 2–10b. The barrier potential for silicon materials is 0.7 V, while that of germanium-based semiconductors is about 0.3 V.

Current flowing through the junction without an externally applied source of energy is called the **diffusion current.** Diffusion current flows only long enough to establish the barrier potential and maintain it when an external source of energy—such as heat, light, or an electrical potential—is applied.

The junction region where the current carriers of both materials have balanced out is called the **depletion region.** This indicates that the region has been depleted of majority carriers because of the diffusion current. And because there are virtually no majority carriers (excess holes or excess electrons) in this region of the semiconductor, the region acts as a good insulator. These details are shown in Figure 2–10b.

Application Problem

Communications

SEMICONDUCTOR DEVICES

The semiconductor devices are a development of the last several decades; however, the diodes and amplifiers were commercially made for almost one hundred years. List some examples of previous utilization of the barrier potential and where they were used.

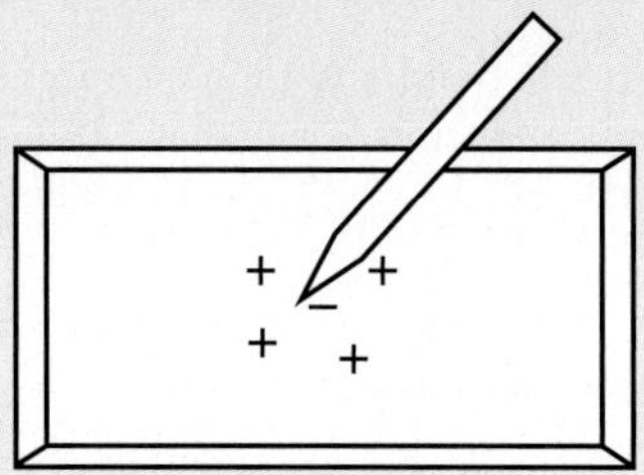

Solution

Radio—Utilized vacuum tubes.

"Fox hole radio"—soldiers in World War II built a radio receiver that did not need an external power supply. They wound several hundred turns of a telephone wire around their canteen or a gas mask. That created an LC oscillator circuit. They removed a speaker from a telephone receiver and used a safety pin to create the "N-P" junction forming the barrier potential. They simply pointed a sharp point into a flat conducting object—that caused a charge separation forming a diode. Then they only had to make an antenna. They used a several hundred foot long barbwire running along their trenches.

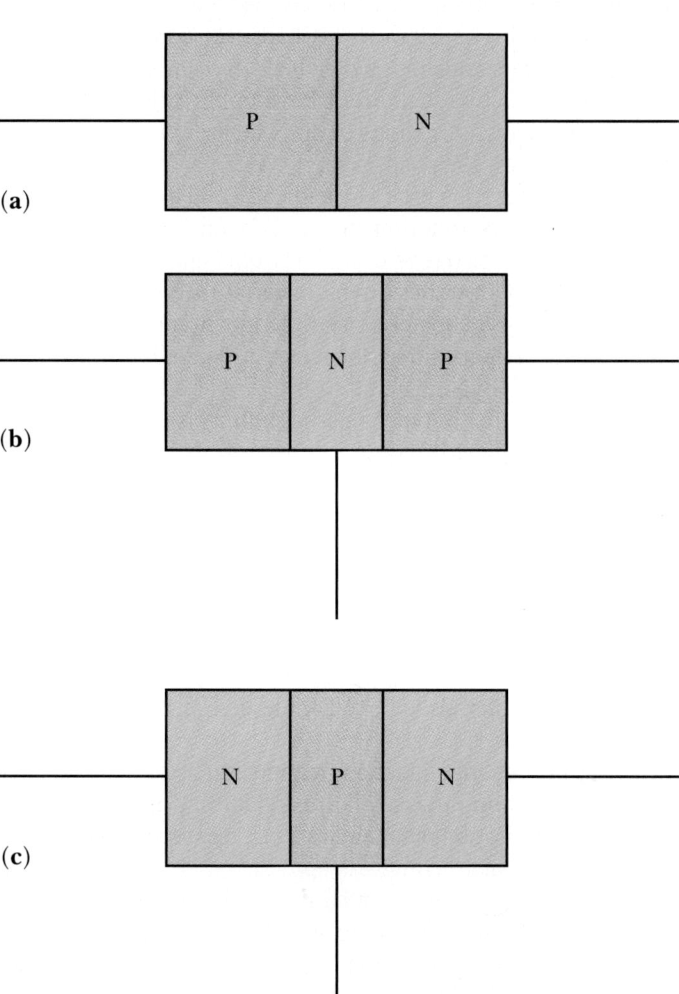

FIGURE 2–9 Practical devices made from combinations of N- and P-type semiconductor materials: (a) diode; (b) PNP junction transistor; (c) NPN junction transistor

The barrier potential between the N and P layers in a semiconductor device is usually between 0.3 V and 0.7 V, depending on the exact nature of the materials as well as outside influences such as temperature. Although it is not possible to measure the barrier potential directly, you will learn how it influences the operation of the semiconductor devices in later chapters. Also, it is important to realize that the depletion region is quite thin, measuring only micrometers (millionths of a meter) across. The actual thickness of this insulating region depends on a number of factors, but the most significant factor for our purpose is the polarity and the amount of voltage that is being applied to the semiconductor.

Biasing P-N Junctions

Figure 2–11 shows a voltage source applied to the ends of a two-layer (P-N) semiconductor device. In this instance, the positive terminal of the source is applied to the P-type region and the negative terminal to the device's N-type region. In the first figure, Figure 2–11a, the switch is open and no potential is applied to the material. There is no current flow in the depletion region. Any motion of majority carriers (electrons in the N-type material and holes in the P-type material) is low level and random, largely because of heating effects of the surrounding air.

The instant the switch is closed, Figure 2–11b, a negative potential is applied to the N-type material and a positive potential to the P-type material. Both the electron majority carriers in the N-type material and the hole majority carriers in the P-type material are forced toward the P-N junctions.

If the applied voltage is greater than the contact (barrier) potential of that particular P-N junction, the oncoming flood of majority carriers eliminates the charges in the depletion region. See Figure 2–11c. Positive hole charges in the depletion region are

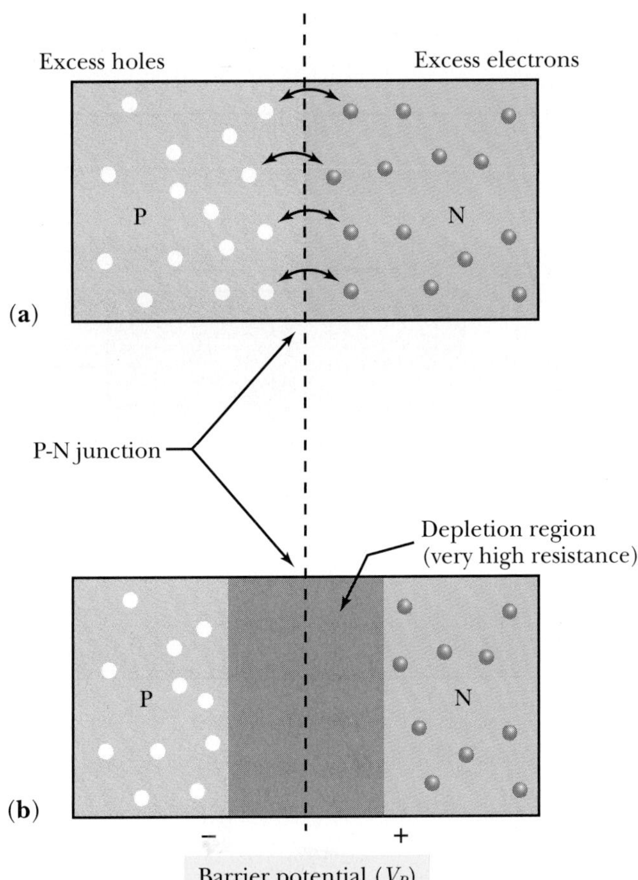

FIGURE 2–10 The depletion region of a P-N junction: (a) Holes and electrons recombine at the junction; (b) the recombination of holes and electrons quickly forms the depletion region

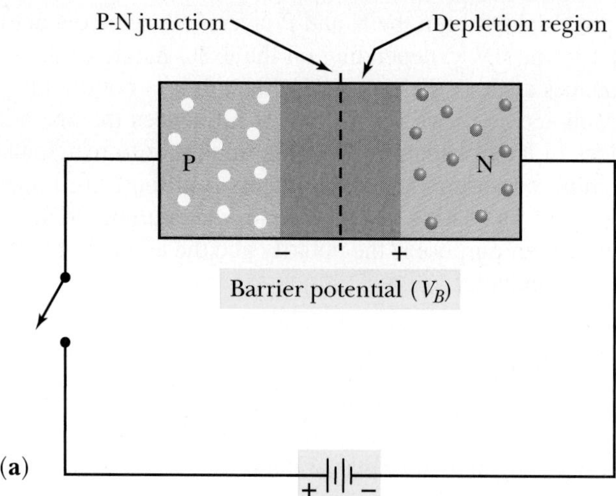

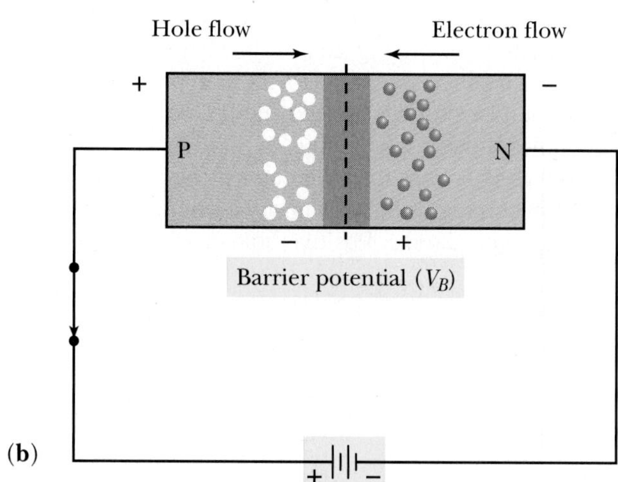

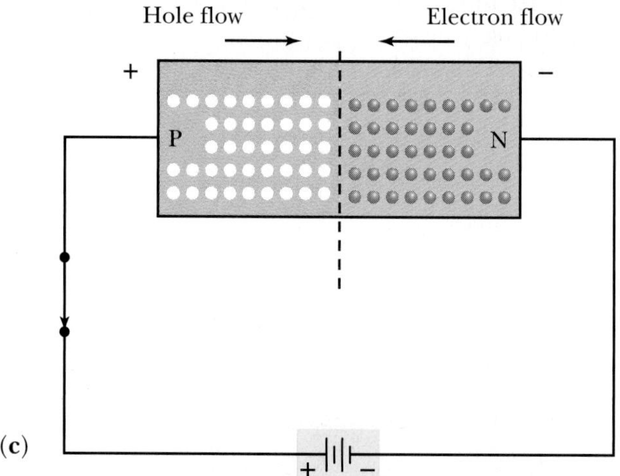

FIGURE 2–11 A forward-biased P-N junction: (a) no voltage applied to the semiconductor; (b) immediate reaction to closing the switch; (c) current flow established through the circuit

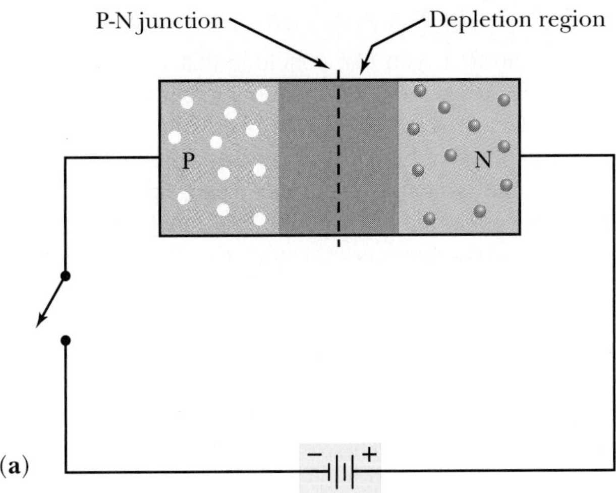

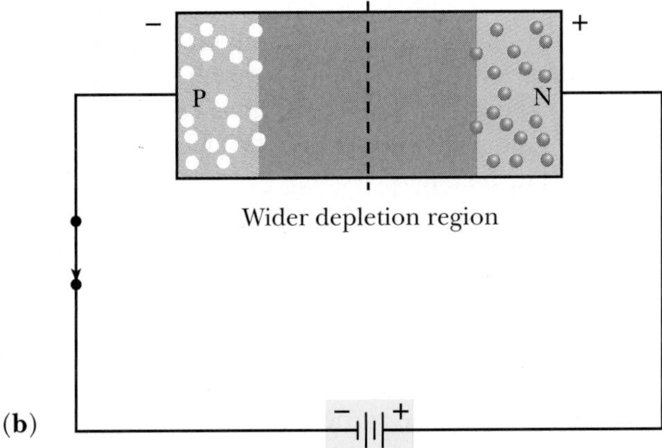

FIGURE 2–12 A reverse-biased P-N junction: (a) no voltage applied to the semiconductor; (b) current cannot flow through the device

fully occupied with electrons from the N-type material, and electrons in the depletion region are all fit into holes coming from the P-type material. The result is the total elimination of the non-conductive depletion region and the establishment of a steady-state electron current flow through the device.

A P-N junction that is conducting in this fashion—with electron current flow from the N-type material to the P-type material—is said to be **forward biased.** This is possible only when an external power supply is connected so that its negative supply is connected to the semiconductor's N-type material and the positive side of the power supply is connected to the P-type material.

If an external power source is connected with the polarities reversed from the direction just described, current cannot flow through the junction. Figure 2–12 shows a P-N junction that is **reversed biased.** When a positive potential is applied to the N-type material, electron majority carriers are drawn away from the junction's depletion region; and by the same token, a negative potential applied to the P-type material draws holes away from the junction. The overall result is that the depletion region is actually widened, Figure 2–12b, thus becoming a very good barrier to current flow through the device.

When you take a moment to compare the effects of forward and reverse bias on a P-N junction, you may begin to see one of the most important properties of P-N junctions: *A P-N junction conducts current in only one direction.*

Characteristic Curves for P-N Junctions

From the previous discussion, you can conclude that forward bias occurs when the external source voltage causes the P-type material to be positive with respect to the N-type material. Under these circumstances, as bias voltage increases from zero, conduction similar to that shown in Figure 2–13 results. Little current flows until the barrier potential of the diode is nearly overcome. Then, there is a rapid rise in current flow. The typical barrier potential for a silicon P-N junction is about 0.7 V. Once you reach this point on the conduction curve, you can see that the voltage across the P-N junction increases very little compared with large increases in forward current.

You have also seen that reverse biasing a P-N junction means we connect the negative terminal of the external source to the P-type material and the external positive terminal to the N-type material. As you increase the amount of applied voltage, only a minute amount of reverse current (leakage current) flows through the semiconductor by means of minority carriers. This reverse current remains very small until the reverse voltage is increased to a point known as the **reverse breakdown voltage** level, Figure 2–14. Before reaching this breakdown point, the leakage current in silicon P-N junctions is on the order of a few microamperes. For germanium junctions, the leakage current is less than a milliampere. The symbol for leakage current is I_{CO}.

An interesting effect occurs when you exceed the reverse breakdown voltage level of a P-N junction. The reverse current increases dramatically, but the voltage across the junction remains fairly constant once you reach the breakdown level.

The reverse breakdown voltage level of a P-N junction is mainly determined at the time of its manufacture. These voltage specifications can be as low as 5 V for certain special applications, are frequently between 100 V and 400 V, and may be as high as 5,000 V for other special applications.

The curves for forward conduction current, Figure 2–13, and reverse leakage and breakdown current, Figure 2–14, are called *I-V curves.* This name is appropriate because the curves show the amount of current flow (I) as you vary the voltage (V) applied to the P-N junction. Forward and reverse I-V curves are usually combined into a single graph, as shown in Figure 2–15. The barrier potential is about 0.7 V (silicon) whereas the reverse breakdown voltage can be over 100 V. The break in the Figure 2–14 graph is used to display the barrier potential and reverse breakdown voltage on one graph.

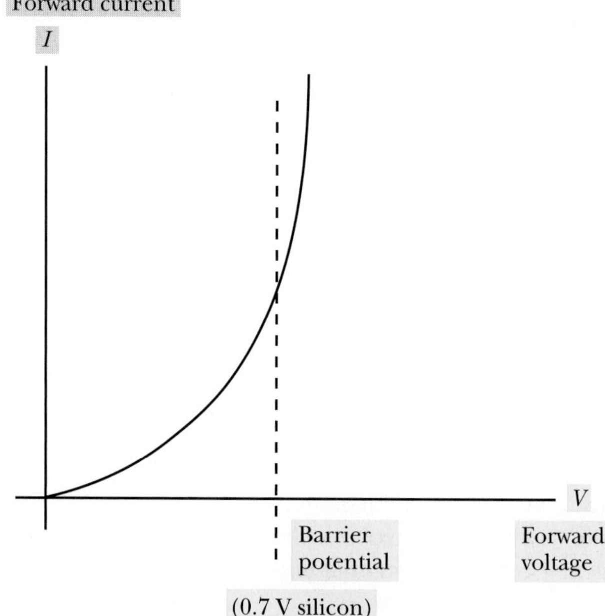

FIGURE 2–13 Forward-bias portion of the I-V curve for a typical P-N junction

Application Problem

General Electronics

EXAMPLES OF SILICON

The silicon semiconductor devices (also called solid-state devices) are commonly used in our daily life. When students are asked in class to list some examples of silicon base items, they point on a computer saying the microprocessor is made of silicon. That is correct. What other silicon items would you find in a classroom or in a household?

Solution

a chip controlling an overhead projector
a window
a glass bottle
a monitor screen or a TV screen
a light bulb

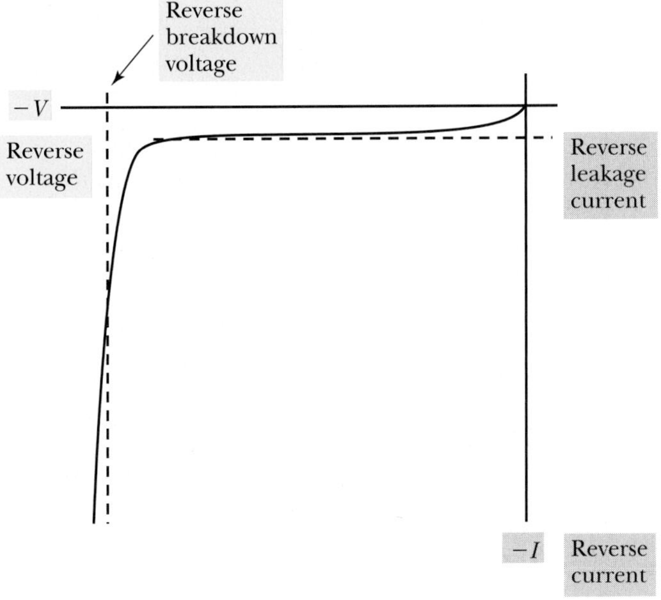

FIGURE 2–14 Reverse-bias portion of the I-V curve for a typical P-N junction

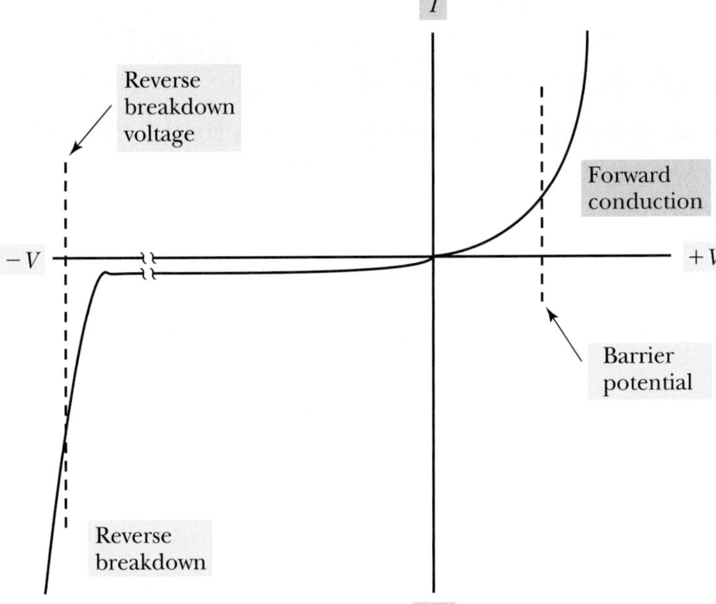

FIGURE 2–15 A single I-V curve showing both the forward- and reverse-bias characteristics of a typical P-N junction

Application Problem

EXAMPLES OF NPN JUNCTIONS

The NPN junction is commonly used as an output stage of computer TTL (transistor-transistor logic) chip. For example, a PC parallel port is TTL compliant, and the state HIGH (one) is 5 V and the state LOW (zero) is 0 V. That is also true for the Motorola based architecture—for example, the MC68HC11/12 microcontrollers. The following figure shows a microprocessor port with graphically depicted NPN junctions. The control voltage is applied to the middle of the junction. What is the voltage between pin D0 and the ground if the control voltage turns the junction ON?

(a) 1 V (b) 0.7 V (c) 5 V (d) 0.2 V

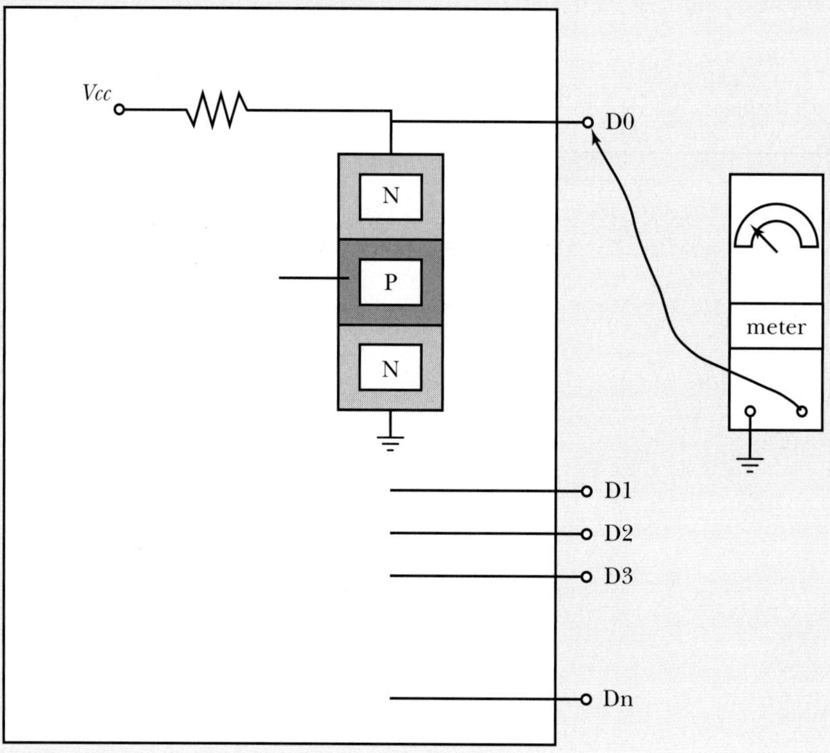

A microprocessor port with graphically depicted NPN junction

Solution Answer (d) is correct. A voltage drop between an emitter and a collector is about 0.15 V to 0.2 V. The exact voltage depends on the output loading.

Summary

- Semiconductor materials have conductivity levels falling in a range between good conductors and good insulators.
- Adding energy (electricity, heat, or light) to an atom causes its electrons to gain energy. In conductors and semiconductors, valence electrons that gain energy will jump up to a conduction band. When conduction-band electrons fall back to their valence bands, they give up the energy they had gained from an outside source. They usually give up their energy in the form of heat, but certain semiconductor materials emit light as well.
- Semiconductor materials such as silicon and germanium have four electrons in their valence shell.

- Introducing selected impurities into the crystal structure of intrinsic semiconductor materials is called *doping*. Doping a tetravalent material with a pentavalent material introduces donor atoms in the structure that donate relatively free electrons. This produces an N-type semiconductor material. Doping a tetravalent material with a trivalent material introduces acceptor atoms into the structure that easily accept electrons to fill the holes where the covalent bonds lack an electron. This produces a P-type semiconductor material.
- The majority carriers in N-type semiconductor materials are electrons. In P-type materials, the majority carriers are holes.
- When P- and N-type materials are formed together, a P-N junction is created at the area of contact. Near the junction, a depletion region is created by electrons from the N-type material moving in to fill holes in the P-type material, and holes moving in the opposite direction (from the P-type material) to combine with available electrons. The depletion region is electrically neutral, but separates the N- and P-type materials, which have a difference in potential called the barrier potential (or junction voltage). This potential is positive on the N-type side and negative on the P-type side of the depletion region.
- Applying an external voltage to a P-N junction causes the semiconductor to conduct current freely (forward bias) or act as a good insulator (reverse bias). This depends on the polarity of the applied voltage. The junction is forward biased when the applied voltage is positive to the P-type material and negative to the N-type material. On the other hand, the junction is reverse biased when the applied voltage is negatively to the P-type material and positive to the N-type material.
- The I-V curve for a P-N junction shows that forward conduction begins when the applied voltage reaches the junctions's barrier potential. The curve also shows that very little current flows through a P-N junctions that is reverse biased until the applied voltage reaches the reverse breakdown voltage level.

Formulas and Sample Calculator Sequences

FORMULA 2–1 $2n^2$

(To find maximum number of electrons in shell n.)

shell number, [x^2], [×], 2, [=]

Review Questions

1. Explain the meaning of the following terms:
   a. Trivalent
   b. Tetravalent
   c. Pentavalent
2. List the number of valence electrons that are found in:
   a. silicon.
   b. arsenic.
   c. gallium.
3. For an N-type semiconductor material, cite:
   a. the name of the majority carrier.
   b. the name of the minority carrier.
4. For a P-type semiconductor material, cite:
   a. the name of the majority carrier.
   b. the name of the minority carrier.
5. Indicate the typical barrier potential for:
   a. a silicon P-N junction.
   b. a germanium P-N junction.
6. Which type of semiconductor material occurs when you dope a tetravalent material with a pentavalent material?
   a. N-type material
   b. P-type material
7. Which type of semiconductor material occurs when you dope a tetravalent material with a trivalent material?
   a. N-type material
   b. P-type material
8. If the electrons in a P-type material are flowing from left to right, the holes in the same material are:
   a. flowing from left to right.
   b. flowing from right to left.
   c. not flowing at all because they are the minority carriers.
9. Draw a P-N junction semiconductor showing an external dc power supply connected for forward biasing.
10. Draw a P-N junction semiconductor showing an external dc power supply connected for reverse biasing.
11. Sketch the forward-bias portion of the I-V curve for a P-N junction. Indicate the barrier potential on the $+V$ (horizontal) axis.
12. Sketch the revere-bias portion of the I-V curve for a P-N junction. Indicate the breakdown voltage level on the $-V$ axis.

Problems

1. What is the maximum number of electrons that can occupy the fourth shell from the nucleus in a normal atom?
2. What is the maximum number of electrons that can occupy the second shell from the nucleus in a normal atom?
3. Which atomic shell has a maximum of 18 electrons?

Analysis Questions

1. Explain the main difference between valence-band and conduction-band electrons.
2. Explain the main difference between intrinsic and extrinsic semiconductor materials.
3. Describe how increasing and decreasing the amount of forward-bias voltage affects the thickness of the depletion region of a P-N junction.
4. Explain why the conductance of a P-N junction is much greater when it is forward biased than when it is reverse biased.
5. Explain why it can be correctly said that holes follow conventional current flow and electrons follow electron flow.
6. Applying energy, such as heat and light, to a semiconductor increases carrier activity. Explain how this accounts for the fact that semiconductors tend to be more conductive as their temperature rises. Compare this effect with the reaction of a normal conductive material such as copper.

OUTLINE

3–1 Diodes
3–2 Diode models
3–3 General diode ratings
3–4 Rectifier diodes
3–5 Switching diodes
3–6 Diode clippers and clampers
3–7 Zener diodes
3–8 Zener diode applications
3–9 The varactor diode
3–10 The light emitting diode (LED)
3–11 Other types of diodes

OBJECTIVES

After studying this chapter, you should be able to:

1. Describe how to connect a dc source to a **diode** for forward bias and for reverse bias
2. Sketch the waveforms found in an ac circuit consisting of a junction diode and resistor
3. Explain the function of diode **clamping** and **clipper circuits**
4. Describe the operation and specifications for **zener diodes**
5. Describe the purpose of **laser diodes, tunnel diodes,** and **varactor diodes**

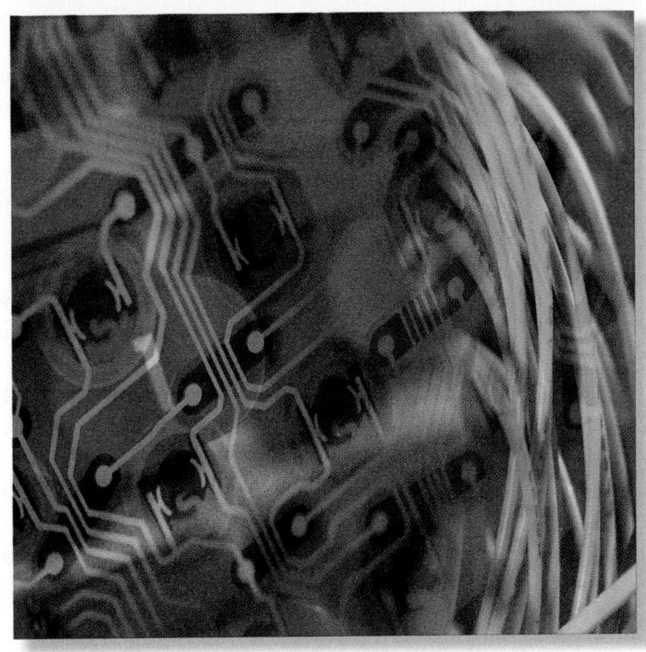

CHAPTER 3

Diodes and Diode Circuits

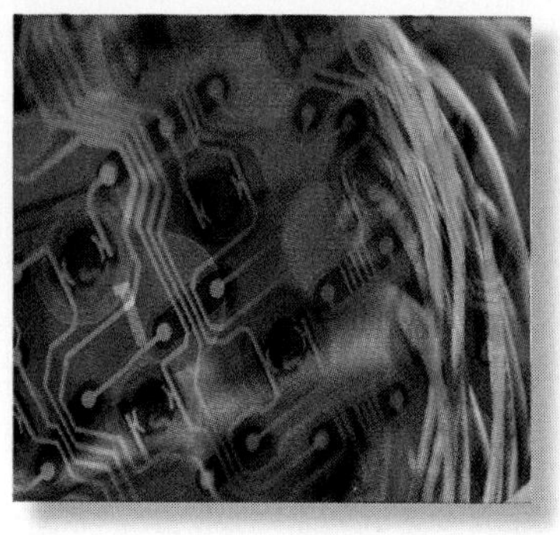

PREVIEW

You have learned that a P-N junction normally allows current to pass in only one direction. This normal direction of current flow occurs when the junction is forward biased. This chapter introduces you to a practical application of P-N junctions in the form of semiconductor diodes.

You will discover that diodes are used for allowing or stopping electron flow and for controlling the direction of current flow through useful electronic circuits. Diode specifications include limitations on the amount of current they can carry and the amount of reverse-bias voltage they can withstand before breaking down. As long as you use diodes properly and within their specified limits, they are very reliable devices.

Rectifier diodes are typically used for power supply applications. Within the power supply, you will see the rectifier diode as an element that converts ac power to dc power.

Switching diodes, on the other hand, have lower current ratings than rectifier diodes, but you will see that switching diodes can function better in high frequency applications and in clipping and clamping operations that deal with short-duration pulse waveforms.

Zener diodes are introduced in this chapter as a special type of P-N junction device. These diodes are very commonly used as voltage-level regulators and protectors against high voltage surges.

Some high frequency diode applications use tunnel diodes (used for producing high frequency oscillators) and varactor diodes (used mainly in high frequency tuning circuitry).

KEY TERMS

Anode
Cathode
Clamping circuit
Clipper circuit
Heat sink
Limiter circuit
Rectifier diode
Switching diode
Tunnel diode
Varactor diode
Zener diode

3–1 Diodes

The diode is constructed using the P-N junction described in Section 2–2. You have seen that current flows easily only in one direction through a P-N junction. Figure 3–1 shows the correlation between the physical construction of the diode, the schematic symbol representation, and the P-N junction. The arrow represents the **anode**—the P-type material. The straight bar or line represents the **cathode**—the N-type material. The size and shape of many diodes are similar to metal film and carbon resistors. Usually there is only one color band near one end of a diode that marks the cathode.

How a Diode Works in a Circuit

The diode operation will be evaluated by using the circuit shown in Figure 3–2a. As the voltage source V_s is increased measurements of the current flowing through the diode and the voltage dropped across the diode are taken. Plotting the diode current with respect to the diode voltage for this experiment provided the results shown in the Figure 3–3a graph. The voltage supply is now reversed (Figure 3–2b). While the voltage source is increased, measurements of the current flowing through the diode and the voltage dropped across the diode are taken. Plotting the diode current with respect to the diode voltage provides the results shown in Figure 3–3b. Combining the diode current and voltage measurements into one graph gives you the diode's characteristic curve (Figure 3–3c).

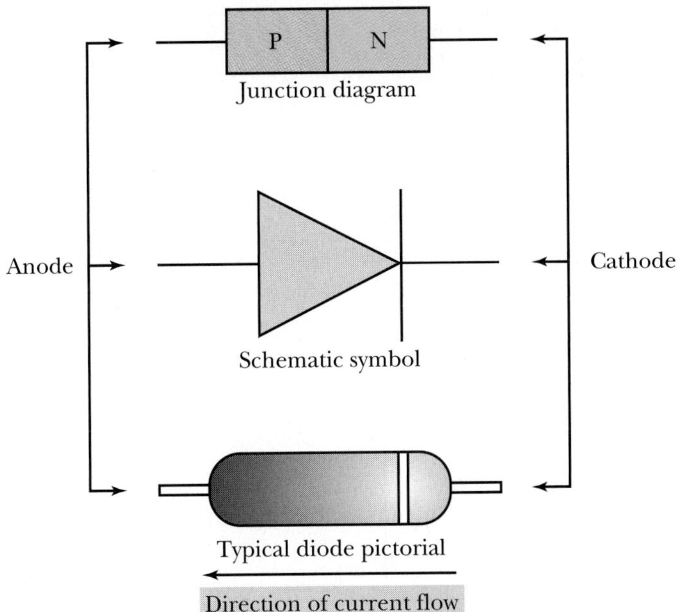

FIGURE 3–1 A diode

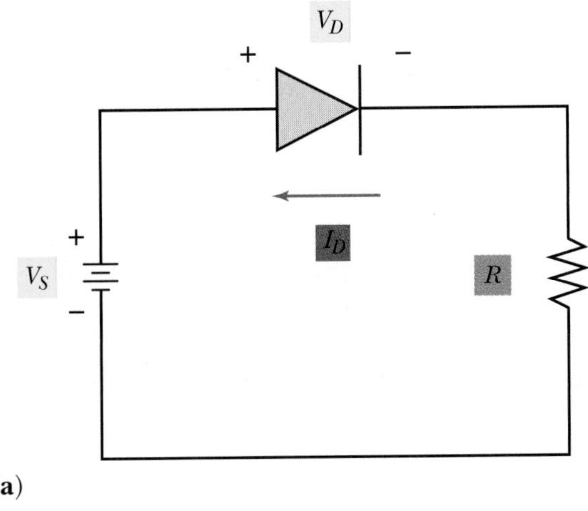

(a)

V_D

\+ −

I_D

− V_S +

R

(b)

FIGURE 3–2 (a) Diode forward-bias circuit; (b) Diode reverse-bias circuit

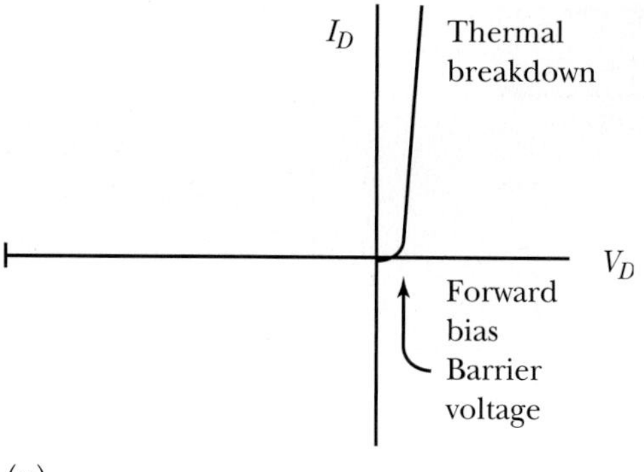

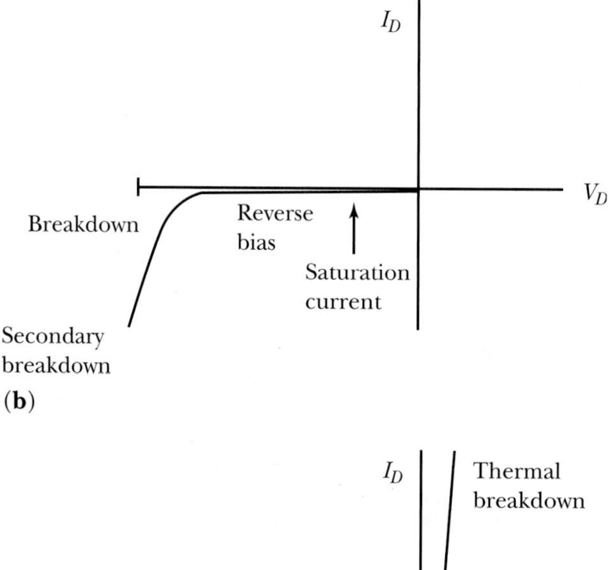

FIGURE 3–3 (a) Diode characteristic curve (forward bias); (b) Diode characteristic curve (reverse bias); (c) Diode characteristic curve (nonlinear relationship)

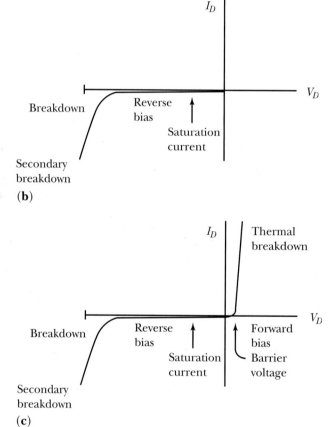

Prior to reflecting on the results of this experiment, we will perform the same experiment using a resistor in place of the diode. This circuit is shown in Figure 3–4a. As the voltage source V_S is increased, measurements of the current flowing through the resistor R_1 and the voltage dropped across the resistor R_1 are taken. Plotting the results of this experiment provided the graph shown in Figure 3–5a. In Figure 3–4b, the voltage supply is now reversed. The voltage supply is increased from 0 V while taking measurements of the current flowing through the resistor R_1 and the voltage dropped across the resistor R_1. Plotting the results produced the Figure 3–5b graph. Combining the resistor and current measurements onto one graph results in the Figure 3–5c resistor characteristic curve. Evaluating this graph, we observe a *linear* relationship between the voltage dropped across the resistor R_1 and the current flowing through the resistor R_1. This relationship is known as Ohm's law (V = IR). A negative voltage dropped across the resistor resulted in the same amount of current flowing through the resistor as a positive voltage dropped across the resistor with the only difference being the direction of the current flow.

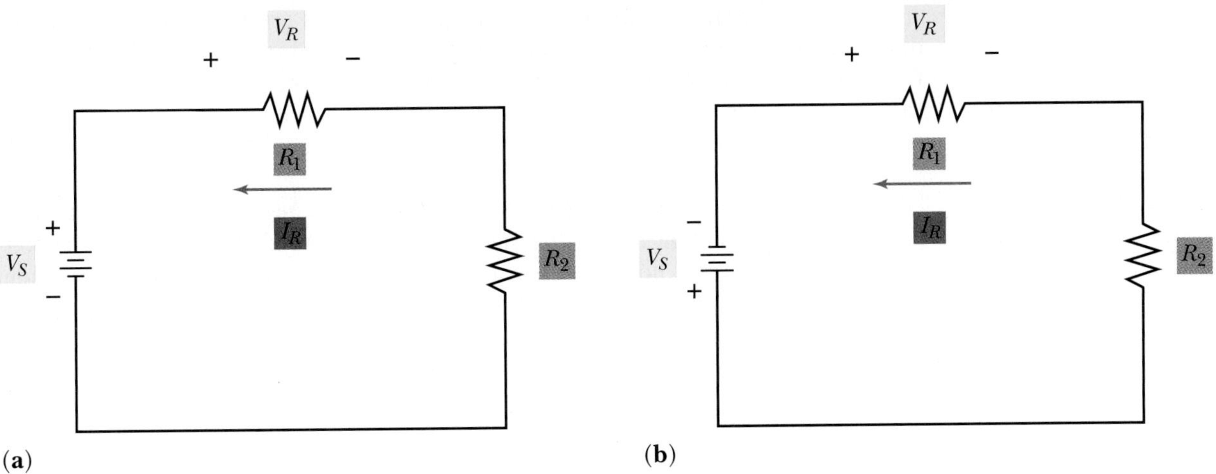

FIGURE 3–4 (a) Resistor positive voltage bias circuit; (b) Resistor negative voltage bias circuit

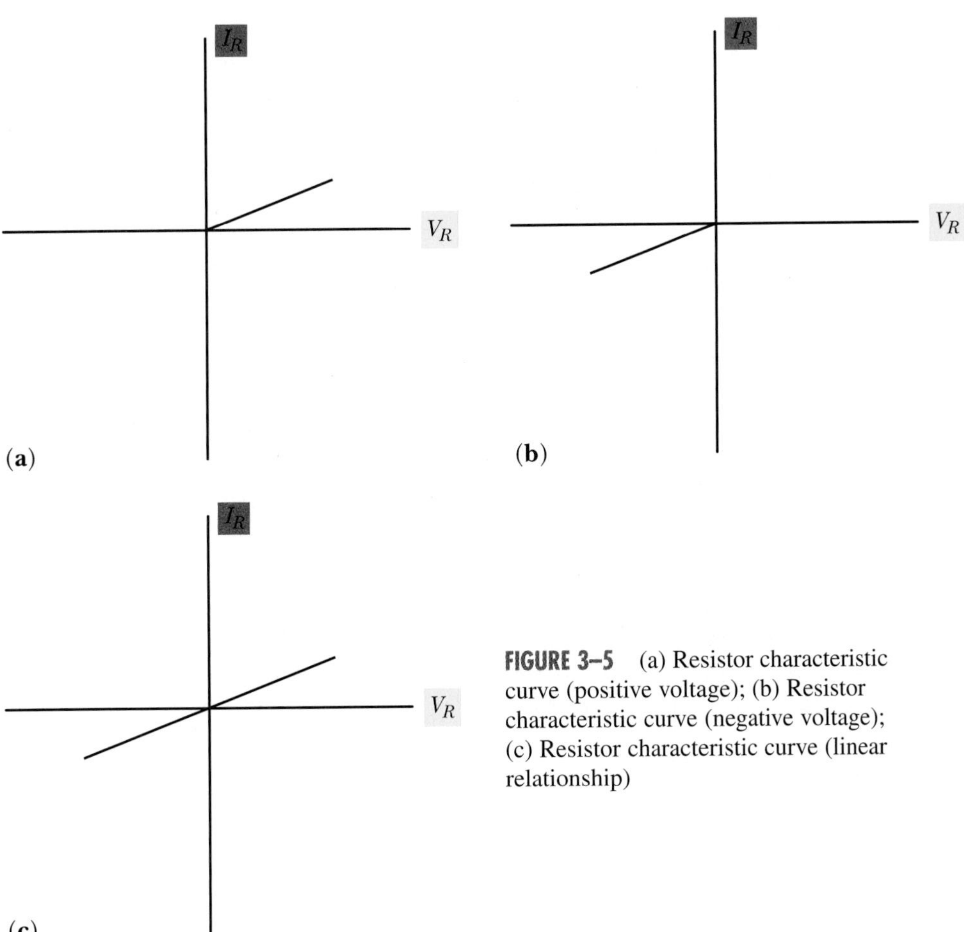

FIGURE 3–5 (a) Resistor characteristic curve (positive voltage); (b) Resistor characteristic curve (negative voltage); (c) Resistor characteristic curve (linear relationship)

Returning to the graph that was obtained for the diode, the relationship is *non-linear.* The magnitude of the current flowing through the diode with positive voltage across the diode is different from the magnitude of the current flowing through the diode with negative voltage across the diode. The two different regions of operation for the diode are called *forward biased* and *reverse biased.* Additional items observed from the graph are *breakdown, secondary breakdown, thermal breakdown, barrier voltage,* and *leakage* (saturation) current.

A diode is forward biased when the supply voltage is greater than or equal to the diode barrier voltage. When the diode is forward biased, electrons will be moving from the cathode to the anode. The voltage dropped across the diode will be greater than or equal to the diode barrier voltage (0.3 V for germanium diodes and 0.7 V for silicon diodes). When the diode is forward biased, current is allowed to flow freely through the circuit.

A diode is reverse biased when the supply voltage is negative. When the diode is reversed biased, electron will not be moving from the cathode to the anode. The voltage dropped across the diode will be equal to the supply voltage. A very small amount of reverse leakage current will flow through this circuit, but for all practical purposes, we can say that there is no current flow.

3–2 Diode Models

To effectively predict or analyze the circuits that contain diodes, an electrical model of the diode must be defined. A *model* is a mathematical or circuit representation of a component, device, or system. Through the device models, we are attempting to match the performance of the device model to the performance of the component. The diode model used in this text is the voltage only model with the barrier voltage for silicon of 0.7 shown in Figure 3–6. Other diode models include the ideal model and the piecewise linear model also known as the voltage resistor model. As you compare each diode model's characteristic curve to the actual diode characteristic curve shown in Figure 3–3 obtained from the Figure 3–2 circuit, you can observe how the models correspond to the actual performance of the diode.

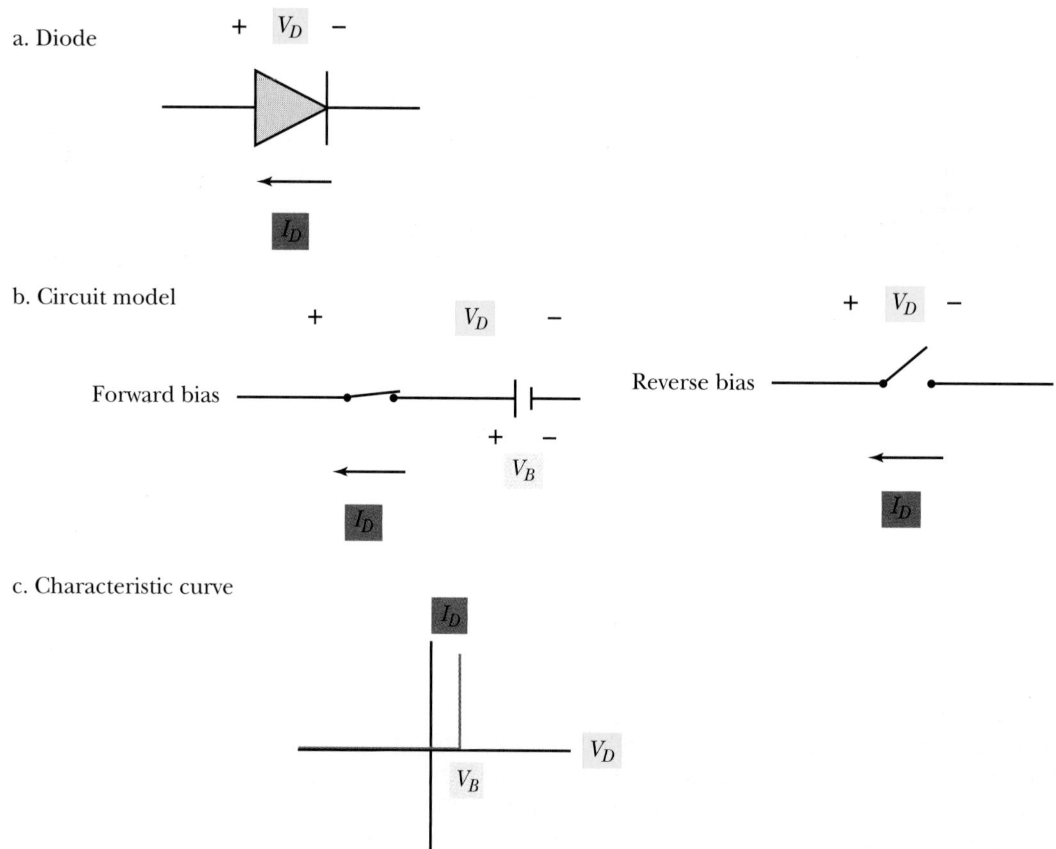

FIGURE 3–6 The voltage only diode model

Appling Kirchhoff's voltage law (KVL) to the Figure 3–2 circuit, we obtain the following equation:

$$V_S - V_D - V_R = 0$$

If the supply voltage V_S and the value of the resistor R were provided, you cannot determine the voltage dropped across the diode V_D nor the voltage dropped across the resistor V_R. Ohm's law provides you with a relationship between the voltage dropped across the resistor, the current flowing through the resistor, and the value of the resistance. Without the diode model, you are unable to calculate the voltage dropped across the diode and the current flowing through the diode.

The Voltage Only Diode Model

In Figure 3–6 the diode is replaced with a closed switch and a 0.7 V battery for the forward biased condition ($V_S > 0.7$ V) and an open switch for the reverse biased condition.

The mathematical equation for this model is:

$$\text{Forward bias } (V_S > 0.7\text{ V}) \quad V_D = 0.7\text{ V}$$
$$I_D > 0\text{ A}$$

$$\text{Reverse bias } (V_S \text{ negative}) \quad V_D = V_S$$
$$I_D = 0\text{ A}$$

You will now analyze the circuit shown in Figure 3–7.

EXAMPLE Using the voltage only diode model, find the voltage drop across the diode, the voltage drop across the resistor, and the current flowing through the diode for the Figure 3–7 circuit. The supply voltage V_S is 10 V and the resistor R is 1000 Ω

Step 1. Write the circuit equation.

$$V_S - V_D - V_R = 0$$

Step 2. Select the diode model.
The voltage only diode model will be used.

Step 3. Document assumption.
You will assume that the diode is forward biased.

Step 4. Calculate the mathematical results.
Based upon the model selected and the condition assumed for the diode.

$$V_D = 0.7\text{ V}$$

Substituting this value into the circuit equation,

$$10\text{ V} - 0.7\text{ V} - V_R = 0$$

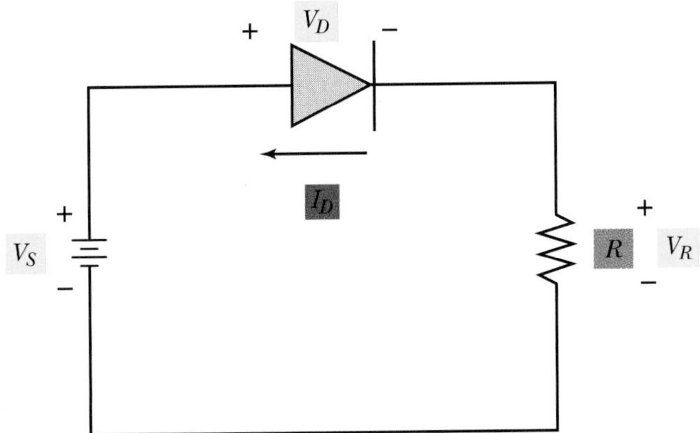

FIGURE 3–7 Example circuit
multiSIM

Solving for the voltage drop across the resistor,

$$V_R = 9.3 \text{ V}$$

The current flowing through the resistor and diode can be found using Ohm's law.

$$I = \frac{V_R}{R}$$

Solving for the current,

$$I = \frac{9.3 \text{ V}}{1000 \ \Omega}$$

Step 5. Evaluate the mathematical results (verify the assumptions).
For the diode to be forward biased, the current flowing through the diode must be greater than 0 mA.

Since the current flowing through the diode is 9.3 mA and 9.3 mA is greater than 0 mA, then the diode is forward biased and your analysis is complete. ◆

This example has shown you the analysis of a circuit using the voltage only model where the diode is forward biased. To calculate the current flowing in a two-resistor circuit, you divided the supply voltage by the addition of the resistance values. For the forward-biased diode circuit, the current flowing in the circuit is calculated by dividing the supply voltage minus the diode voltage drop with the resistance value. Using Formula 3–1, you can solve for the current flowing through the diode and resistor.

FORMULA 3–1 $I = \frac{(V_S - V_D)}{R}$

◆ **EXAMPLE** Using the voltage only diode model, find the voltage drop across the diode, the voltage drop across the resistor, and the current flowing through the diode for the Figure 3–7 circuit. The supply voltage V_S is 10 V and the resistor R is 150 Ω

Step 1. Write the circuit equation.

$$V_S - V_D - V_R = 0$$

Step 2. Select the diode model.
The voltage only diode model will be used.

Step 3. Document assumption.
You will assume that the diode is forward biased.

Step 4. Calculate the mathematical results.
Based upon the model selected and the condition assumed for the diode.

$$V_D = 0.7 \text{ V}$$

You can solve for the current by using Formula 3–1

$$I = \frac{(V_S - V_D)}{R}$$

Substituting the known values into the circuit equation,

$$I = \frac{(10 \text{ V} - 0.7 \text{ V})}{150 \ \Omega}$$

$$I = \frac{9.3 \text{ V}}{150 \ \Omega}$$

Practical Notes

Resistance should always be used in series with the diode. The series resistance limits the current, preventing damage to the diode. Sometimes the resistance is in the form of a resistor.

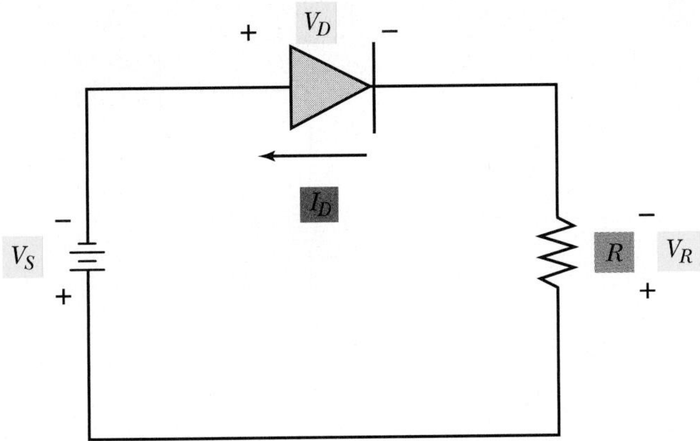

FIGURE 3–8 Example circuit

Step 5. Evaluate the mathematical results (verify the assumptions).
For the diode to be forward biased, the current flowing through the diode must be greater than 0 mA.

Since the current flowing through the diode is 62 mA and 62 mA is greater than 0 mA, then the diode is forward biased and your analysis is complete. ◆

This example has shown you that decreasing the resistance value resulted in an increase of the current flowing through the diode and the resistor. The voltage drop across the resistor and the diode did not change.

You will now analyze the circuit shown in Figure 3–8.

◆ **EXAMPLE** Using the voltage only diode model, find the voltage drop across the diode, the voltage drop across the resistor, and the current flowing through the diode for the Figure 3–8 circuit. The supply voltage V_s is 10 V and the resistor R is 1000 Ω

Step 1. Write the circuit equation.

$$-V_S - V_D - V_R = 0$$

Step 2. Select the diode model.
The voltage only diode model will be used.

Step 3. Document assumption.
You will assume that the diode is reverse biased.

Step 4. Calculate the mathematical results.
Based upon the model selected and the condition assumed for the diode,

$$I_D = 0 \text{ A}$$

Since the Step 1 circuit equation includes only voltages and you do not know the voltage drop across the diode, an equation that relates the current to the voltage must be provided. Using Ohm's law the voltage drop across the resistor can be calculated.

$$V_R = I \times R$$
$$V_R = 0 \text{ A} \times 1000\ \Omega$$
$$V_R = 0 \text{ V}$$

Substituting this value into the circuit equation.

$$-10 \text{ V} - V_D - 0 \text{ V} = 0$$

Solving for the voltage drop across the diode,

$$V_D = -10 \text{ V}$$

FIGURE 3–9

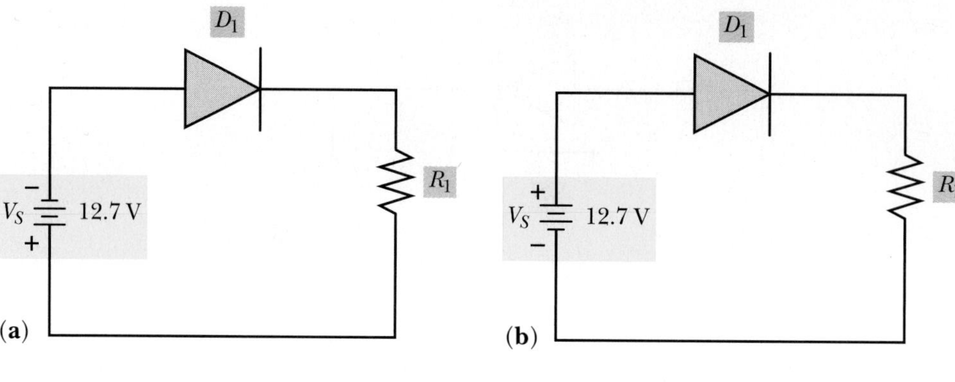

FIGURE 3–10 multiSIM

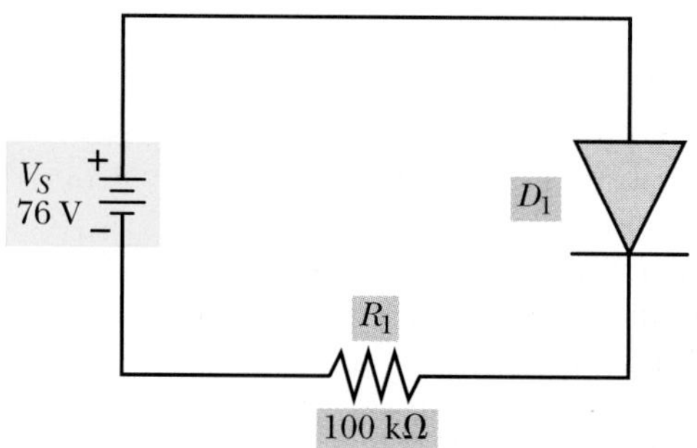

Step 5. Evaluate the mathematical results (verify the assumptions).
For the diode to be reverse biased, the voltage across the diode must be negative.

Since the voltage across the diode is –10 V and –10 V is a negative voltage, then the diode is reverse biased and your analysis is complete. ◆

This example has shown you the analysis of a circuit using the voltage only model where the diode is reverse biased.

If you built these circuits in the lab, you will measure voltages and currents. Your analysis is used to determine how the components will perform in the circuit.

PRACTICE PROBLEMS 1

1. What is the voltage across silicon diode D_1 in Figure 3–9a with $V_s = 12.7$ V?
2. What is the voltage across resistor R_1 in Figure 3–9a?
3. What is the voltage across silicon diode D_1 in Figure 3–9b with $V_s = 12.7$ V?
4. What is the voltage across resistor R_1 in Figure 3–9b?

PRACTICE PROBLEMS 2

Referring to the circuit in Figure 3–10 with $V_s = 76$ VDC and $R_1 = 100$ kΩ;

1. What is the voltage across the silicon diode D_1?
2. What is the voltage across the resistor R_1?
3. How much current is flowing through the resistor R_1?

3–3 General Diode Ratings

Diodes have a number of ratings or specifications. Some ratings are more important than others, most often depending on the application for which the diode is designed. Typically, in most practical cases, if you give careful attention to the most important diode ratings, the ratings of lesser importance automatically fall into line.

There are four diode ratings that apply in one way or another to all types of diodes and applications.

1. *Forward voltage drop, V_F.* As you have seen through earlier discussions, the forward voltage drop (or barrier potential) is the forward-conducting junction voltage level: about 0.3 V for germanium diodes and 0.7 V for silicon. Since germanium diodes start conducting at a voltage less than 0.3 V and silicon diodes start conducting at a voltage less than 0.7 V, it should be noted that engineers might use 0.2 V for germanium diodes and 0.6 for silicon diodes. Experimental results most commonly show a 0.68 V drop for silicon. The actual voltage drop depends on the diode type and the current flowing through the diode.
2. *Average forward current. I_O.* This is the maximum amount of forward conduction that the diode can carry for an indefinite period of time. If the average current exceeds this value, the diode will overheat and eventually destroy itself (thermal breakdown).
3. *Peak reverse voltage, V_R.* The peak reverse voltage (PRV) is sometimes called the *reverse breakdown voltage* or peak inverse voltage (PIV). This is the largest amount of reverse-bias voltage the diode's junction can withstand for an indefinite period of time. If a reverse voltage exceeds this level, the voltage will "punch through" the depletion layer and allow current to flow backwards through the diode (secondary breakdown), which usually destroys the diode. Certain special application diodes (the zener diodes) are designed to permit reverse breakdown conduction until power limits are exceeded.
4. *Maximum power dissipation.* The actual diode power dissipation (Watts) is calculated by multiplying the forward voltage drop and the forward current.

FORMULA 3–2 $P = I \times V_D$

The actual power dissipation must be less than the maximum power dissipation rating of the diode. Exceeding the maximum power dissipation will result in thermal breakdown of the diode, which is catastrophic.

Excessive forward current and exceeding reverse breakdown voltage are the most common causes of diode failure. In both cases, the diode gets very hot, and this heat destroys the P-N junction. Occasional "surges" of voltage or current exceeding these ratings for a very short duration (milliseconds) may not overheat the delicate P-N junction to a point of failure, but repeated "surges" will fatigue the junction and ultimately cause failure. To prevent this from occurring, diodes are selected for a circuit with ratings that are two to three times the expected "surge" values for that circuit.

■ IN-PROCESS LEARNING CHECK 1 Fill in the blanks as appropriate.

1. To cause conduction in a diode, the diode must be _______ biased.
2. To reverse bias a diode, connect the negative source to the _______ and the positive source voltage to the _______.
3. The anode of a diode corresponds to the _______ type material, and the cathode corresponds to the _______ type material.
4. The forward conduction voltage drop across a silicon diode is approximately _______ V.
5. When a diode is reverse biased in a circuit, it acts like a(n) _______ (open, closed) switch.
6. When the diode is _______ biased, there will be no current flow through the diode.
7. When a diode is connected in series with a resistor, the voltage across the resistor is very nearly equal to the dc source voltage when the diode is _______ biased.
8. The four most general diode ratings are _______, _______, _______, and _______. ■

3–4 Rectifier Diodes

Rectifier diodes are used where it is necessary to change an alternating-current power source (generally a low frequency sinusoidal waveform typically, 60 Hz), into a direct-current power source. Rectifier diodes are the most rugged and durable of the semiconductors in the junction diode family. They are especially noted for their large average forward current and reverse breakdown voltage ratings. Figure 3–11 lists some common rectifier diode specifications.

Practical Notes

Part identification numbers for JEDEC registered diodes begin with the designator 1N.JEDEC registration is a formal process allowing multiple manufacturers to produce components in compliance with the registered specification and label the components identically. Therefore, a 1N4004 diode manufactured by ON Semiconductor will have the same specifications as a 1N4004 diode manufactured by Philips Semiconductor. If a part is labeled 1N5243, you can safely assume that it is some sort of diode. You will need a catalog or data sheet to identify the type of diode. Component manufacturers utilize other part number designations for diodes. ON Semiconductor (*www.onsemi.com*) manufactures the MBR150, which is a 1 A 50 V rectifier diode.

JEITA establishes the standards for registered electronic components and devices manufactured in Japan.

JEDEC is an acronym for Joint Electron Device Engineering Council. JEITA is the acronym for Japan Electronics and Information Technology Association.

The physical size of a rectifier diode is a general indication of its current-carrying capacity. Actually, heating is the only factor that limits the forward current-carrying capacity of a diode. So when a diode is made physically larger, it has more surface area for dissipating heat but reduces the diode's switching speed. Also, you can dissipate unwanted diode heat by mounting the diode onto a **heat sink.** A heat sink is an aluminum plate and may include an arrangement of grooves and fins made of alu-

Part Identification Code	Forward Current I_O (A)	Breakdown Voltage V_R (V)	Power Rating P(W)
1N4001	1.0	50	2.5 W
1N4002	1.0	100	2.5 W
1N4003	1.0	200	2.5 W
1N4004	1.0	400	2.5 W
1N4007	1.0	1000	2.5 W
1N5400	3.0	50	6.25 W
1N5401	3.0	100	6.25 W
1N5402	3.0	200	6.25 W
1N5404	3.0	400	6.25 W
1N5408	3.0	1000	6.25 W

FIGURE 3–11 Typical ratings for rectifier diodes

minum that carries away the heat generated by a semiconductor and delivers it efficiently into the surrounding air. Figure 3–12 shows a *stud mounted,* high current diode. (The mounting stud must be bolted to a heat sink.) Figure 3–13 shows the ratings for some stud mounted, high current, rectifier diodes. A heat sink designed for stud mounted rectifier diodes is shown in Figure 3–14.

FIGURE 3–12 High current diode *(Photo courtesy of International Rectifier)*

Part Identification Code	I_O (A)	V_R (V)
1N1199A	12	50
1N1183A	40	50
1N1200A	12	100
1N1184A	40	100
1N1185A	35	150
1N1202A	12	200
1N1186A	40	200
1N1187A	40	300
1N1204A	12	400
1N1188A	40	400
1N1189A	40	500
1N1206A	12	600
1N1190A	40	600

FIGURE 3–13 Typical ratings for high current, stud mounted, rectifier diodes

Practical Notes

1. *The maximum voltage applied to a rectifier diode should not exceed the diode's reverse breakdown voltage level.*
   If a diode is being used in a circuit that is operated from a 120 V_{rms} source, the diode's reverse breakdown rating must exceed the peak value of the applied ac voltage:

$$V_{PK} = 1.414 \times V_{rms}$$
$$V_{PK} = 1.414 \times (120\ V_{rms})$$
$$V_{PK} = 170\ \text{V}$$

   So the reverse breakdown rating of the diode must be greater than 170 V. As noted earlier, a design rule of thumb suggests doubling the calculated reverse voltage value. In this example, the doubled value is 340 V. Because no diodes are rated exactly at 340 V, it is good practice to go to the next higher standard rating: 400 V.

2. *The maximum current through the diode portion of a circuit should not exceed the diode's average forward current rating.*
   A silicon diode is connected to a 120-V_{rms} source through a 100 Ω resistor. After calculating the peak voltage, you can calculate the maximum forward current by using Formula 3–1 (restated here).

$$V_{PK} = 1.414 \times V_{rms}$$
$$V_{PK} = 1.414 \times (120\ V_{rms})$$
$$V_{PK} = 170\ \text{V}$$

$$I = \frac{(V_{pk} - V_D)}{R}$$
$$I = \frac{(170\ \text{V} - 0.7\ \text{V})}{100\ \Omega}$$
$$I = \frac{(169.3\ \text{V})}{100\ \Omega}$$
$$I = 1.693\ \text{A}$$

   The 1.693 A is the peak current for the circuit. This shows that the diode should have an average current rating of at least 1.693 A. Of the common silicon diodes used as power supply rectifiers on the market today, the smallest have I_O ratings of 1.0 A, which is less than the average current in the circuit. From Figure 3–11, you will select a diode with a current rating of 3.0 A. So using a 3 A diode in this circuit would be a practical choice.

3. *The power dissipated by the diode should not exceed the diode's maximum power dissipation rating.*
   A silicon diode is connected (forward biased) to a 120-V_{RMS} source through a 100 Ω resistor. After calculating the peak forward current using. Formula 3–1, you can calculate the peak power dissipation with Formula 3–2 (restated here).

$$P = I \times V_D$$
$$P = 1.693\ \text{A} \times 0.7\ \text{V}$$
$$P = 1.185\ \text{W}$$

   The power level is the peak power that the diode will experience while the diode is forward biased. The average power dissipated by the diode is less than 1.185 W since the value calculated is the peak power and the diode will be reversed biased with 0 W power dissipation for half of the sinusoidal waveform cycle. Notice that all of the diodes listed in Figure 3–11 have power dissipation ratings greater than the 1.185 W.

You can use a VOM ohmmeter or a DMM to test the general condition of a diode that is removed from a circuit. Refer to Figure 3–15a and Figure 3–15b as you take note of the following procedures for the VOM diode check.

1. Set the ohmmeter for a midrange scale: $R \times 100$ for an analog ohmmeter.
2. Connect the ohmmeter leads to the diode: the positive lead is connected to the anode and the negative lead is connected to the cathode. Note the resistance (you don't have to be exact).
3. Reverse the ohmmeter connections to the diode so that the positive lead is now connected to the cathode and the negative lead is connected to the anode. Note the resistance (again, you don't have to be exact).

If the diode is in good working order, you should find a much higher resistance when the diode is reverse biased than when it is forward biased. The diode is defective (usually shorted) when the forward and reverse ohmmeter readings are in the same general range.

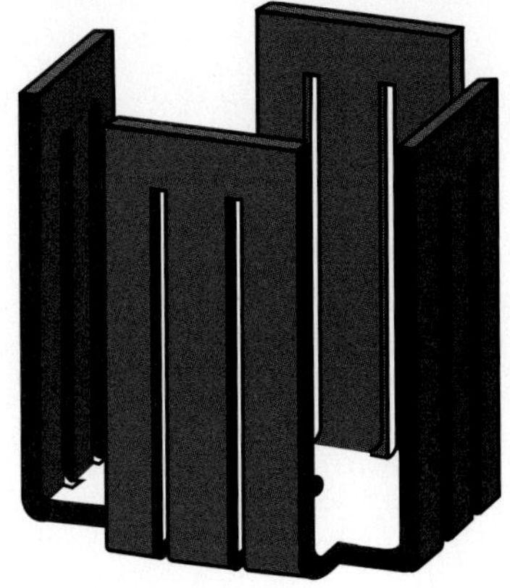

FIGURE 3–14 Heat sink

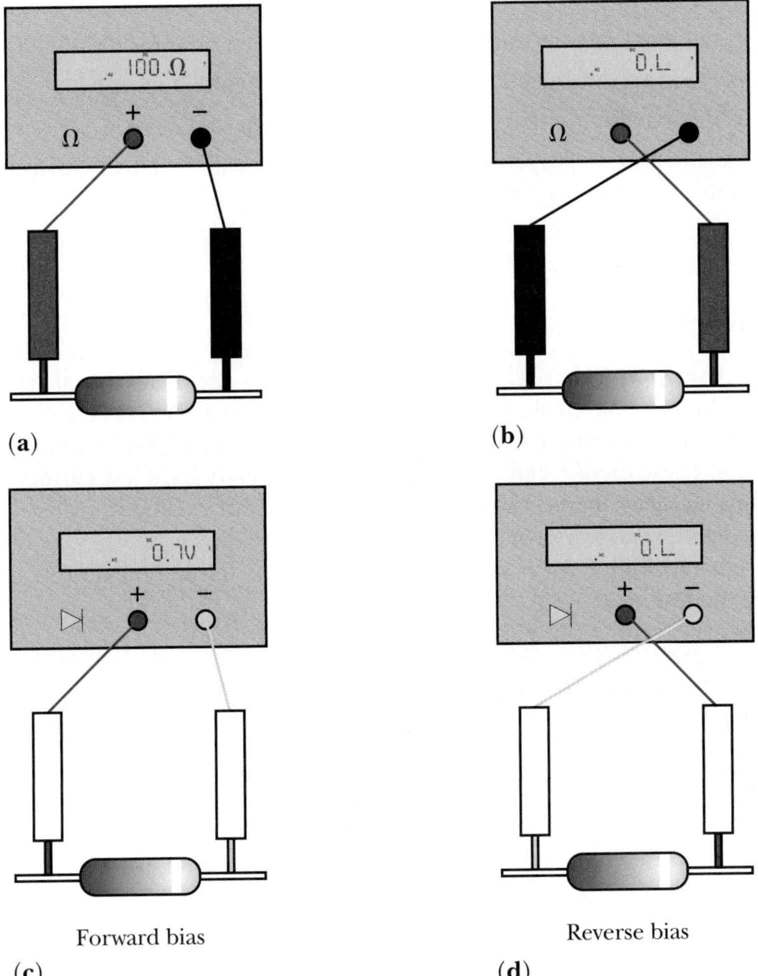

FIGURE 3–15 Diode tests. (a) Ohmmeter forward bias diode test (low forward resistance); (b) Ohmmeter reverse bias diode test (high reverse resistance); (c) DMM forward bias diode test; (d) DMM reverse bias diode test

Some DMMs include a diode test mode (Figures 3–15c and Figures 3–15d). When using a DMM with the diode test feature, you would select the device's "diode test" circuit rather than the ohmmeter function.

1. Set the DMM for the diode test mode.
2. Connect the DMM leads to the diode: the positive lead is connected to the anode and the negative lead is connected to the cathode. Note the reading. Most digital multimeters (DMM) display the forward voltage drop with between 1 to 3 mA of current applied. If the DMM displays 0 V, greater than 1.5 V, OI or OVLD is indicated, then the diode is defective. Some DMMs actually display the diode barrier potential.
3. Reverse the DMM connections to the diode so that the positive lead is now connected to the cathode and the negative lead is connected to the anode. Most DMMs should display an over limits or over voltage (OL) indicating that the diode is reverse biased. If any other reading is indicated, then the diode is defective.

Datasheet 3–1 provides the parameters from ON Semiconductor's 1N4000 series rectifier diodes.

Datasheet Questions

1. What is the maximum average forward voltage drop rating for the 1N4004 diode?
2. What is the average forward current rating for the 1N4004 diode?
3. What is the maximum rated peak reverse voltage for the 1N4004?
4. Calculate the power dissipation for the 1N4004 diode at the maximum average forward voltage drop rating.
5. What name does the manufacturer provide for the 1N4004?

3–5 Switching Diodes

A second type of diode is designed for high frequency and small signal applications such as the circuits commonly found in communications equipment and in computers. They are also used in circuits where diode action has to take place reliably in a very short period. These diodes, called *switching diodes,* usually have low current and voltage ratings, and they are no larger than the smaller rectifier diodes. (Most switching diodes are less than a quarter-inch long, and the surface mount versions are only about 0.1 inch square.) Figure 3–16 shows the specifications for several JEDEC registered switching diodes.

FIGURE 3–16 Typical ratings for switching diodes

Part Identification	V_R (V)	V_F	I_R	t_{rr} (ns)
1N4150	50	0.74 V @ 10 mA	100 nA @ 50 V	4
1N4153	75	0.88 V @ 20 mA	50 nA @ 50 V	2
1N4151	75	1.00 V @ 50 mA	50 nA @ 50 V	2
1N4148	75	1.00 V @ 10 mA	25 nA @ 20 V	4
1N914B	75	0.72 V @ 5 mA	25 nA @ 20 V	4
1N914A	75	1.00 V @ 20 mA	25 nA @ 20 V	4
1N914	75	1.00 V @ 10 mA	25 nA @ 20 V	4
1N4448	75	0.72 V @ 5 mA	25 nA @ 20 V	4

1N4001, 1N4002, 1N4003, 1N4004, 1N4005, 1N4006, 1N4007

1N4004 and 1N4007 are Preferred Devices

ON Semiconductor®

http://onsemi.com

Axial Lead Standard Recovery Rectifiers

This data sheet provides information on subminiature size, axial lead mounted rectifiers for general-purpose low-power applications.

LEAD MOUNTED RECTIFIERS
50-1000 VOLTS
DIFFUSED JUNCTION

CASE 59-10
AXIAL LEAD
PLASTIC

Mechanical Characteristics

- Case: Epoxy, Molded
- Weight: 0.4 gram (approximately)
- Finish: All External Surfaces Corrosion Resistant and Terminal Leads are Readily Solderable
- Lead and Mounting Surface Temperature for Soldering Purposes: 220°C Max. for 10 Seconds, 1/16″ from case
- Shipped in plastic bags, 1000 per bag.
- Available Tape and Reeled, 5000 per reel, by adding a "RL" suffix to the part number
- Available in Fan-Fold Packaging, 3000 per box, by adding a "FF" suffix to the part number
- Polarity: Cathode Indicated by Polarity Band
- Marking: 1N4001, 1N4002, 1N4003, 1N4004, 1N4005, 1N4006, 1N4007

MARKING DIAGRAM

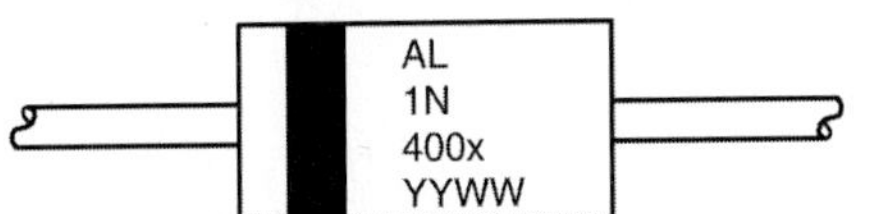

AL = Assembly Location
1N400x = Device Number
x = 1, 2, 3, 4, 5, 6 or 7
YY = Year
WW = Work Week

MAXIMUM RATINGS

Rating	Symbol	1N4001	1N4002	1N4003	1N4004	1N4005	1N4006	1N4007	Unit
*Peak Repetitive Reverse Voltage Working Peak Reverse Voltage DC Blocking Voltage	V_{RRM} V_{RWM} V_R	50	100	200	400	600	800	1000	Volts
*Non-Repetitive Peak Reverse Voltage (halfwave, single phase, 60 Hz)	V_{RSM}	60	120	240	480	720	1000	1200	Volts
*RMS Reverse Voltage	$V_{R(RMS)}$	35	70	140	280	420	560	700	Volts
*Average Rectified Forward Current (single phase, resistive load, 60 Hz, $T_A = 75°C$)	I_O	1.0							Amp
*Non-Repetitive Peak Surge Current (surge applied at rated load conditions)	I_{FSM}	30 (for 1 cycle)							Amp
Operating and Storage Junction Temperature Range	T_J T_{stg}	-65 to +175							°C

*Indicates JEDEC Registered Data

ORDERING INFORMATION

See detailed ordering and shipping information on page 2 of this data sheet.

Preferred devices are recommended choices for future use and best overall value.

January, 2003 - Rev. 8

Publication Order Number:
1N4001/D

Switching diodes are made in such a way that they can respond very quickly to changes in the polarity of the signal applied to them. This *reverse recovery time* (t_{rr}) rating is less than 50 ns (nanoseconds), or 50×10^{-9} s. This means the diode will recover from a reverse-voltage state and begin forward conducting within 50 ns. By contrast, the bulkier construction of rectifier diodes gives them a longer recovery time. A typical reverse recovery time for a rectifier diode is on the order of 30 msec, which is 600 times longer than a typical switching diode.

Datasheet 3–2 provides information from Fairchild Semiconductor's 1N914, 1N9716, 1N4148 and 1N4448 small signal diodes.

Datasheet Questions

1. What is the maximum forward voltage drop rating for the 1N4148 diode at an ambient temperature of 25° C?
2. What is the diode current for the maximum forward voltage drop rating of the 1N4148?
3. What is the breakdown voltage for the 1N4148 with a reverse current of 100µA?
4. Calculate the power dissipation for the 1N4148 diode at the maximum forward voltage rating. What is the maximum power dissipation for the 1N4148?
5. What name does the manufacturer provide for the 1N4148?

3–6 Diode Clippers and Clampers

Diode Clipper Circuits

A diode **clipper circuit** removes the peaks from an input waveform. There are a number of reasons for using such a circuit. For example, a diode clipper can be used in waveshaping circuits to remove unwanted spikes. Diode clippers are also called *limiter* circuits.

There are two basic kinds of diode clipper circuits: series and shunt (parallel) clippers. In a series diode clipper, the diode is connected in series between the circuit's input and output terminals. In a shunt diode clipper, the diode is connected in parallel between the input and output terminals of the circuit.

Negative Series Diode Clipper

Figure 3–17a is the schematic for the negative series diode clipper. The input to the circuit is a sinusoidal waveform. To establish the output signal waveform, the biasing of the diode during the positive half cycle and negative half cycle must be determined.

During the positive half cycle of the input waveform, the diode is forward biased. As soon as the supply voltage applied to the circuit increases above the diode barrier voltage (0.7 V for silicon diodes), the diode will conduct current. The output voltage will then be equal to the supply voltage minus the voltage drop across the diode caused by the diode barrier voltage. When the supply voltage reaches its peak amplitude, the current flowing through the circuit will reach its maximum value (which must be less than the diode maximum current rating) and the output voltage will be at its peak amplitude. Once the supply voltage drops below the diode barrier voltage, the diode no longer conducts current. The output voltage will be 0 V.

During the negative half cycle of the input waveform, the diode is reverse biased. No current will be conducted through the diode. The output voltage will then be equal to 0 V. The supply voltage will appear across the diode (the diode PIV rating must be greater than the peak supply voltage).

1N/FDLL 914/A/B / 916/A/B / 4148 / 4448

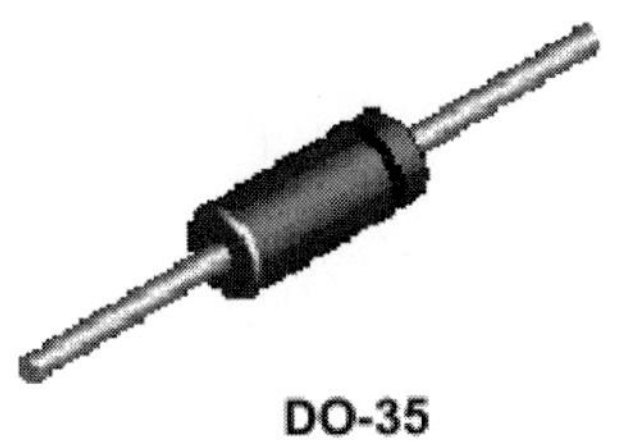

DO-35

LL-34

THE PLACEMENT OF THE EXPANSION GAP HAS NO RELATIONSHIP TO THE LOCATION OF THE CATHODE TERMINAL

COLOR BAND MARKING

DEVICE	1ST BAND	2ND BAND
FDLL914	BLACK	BROWN
FDLL914A	BLACK	GRAY
FDLL914B	BROWN	BLACK
FDLL916	BLACK	RED
FDLL916A	BLACK	WHITE
FDLL916B	BROWN	BROWN
FDLL4148	BLACK	BROWN
FDLL4448	BROWN	BLACK

Small Signal Diode

Absolute Maximum Ratings*

T_A = 25°C unless otherwise noted

Symbol	Parameter	Value	Units
V_{RRM}	Maximum Repetitive Reverse Voltage	100	V
$I_{F(AV)}$	Average Rectified Forward Current	200	mA
I_{FSM}	Non-repetitive Peak Forward Surge Current Pulse Width = 1.0 second Pulse Width = 1.0 microsecond	 1.0 4.0	 A A
T_{stg}	Storage Temperature Range	-65 to +200	°C
T_J	Operating Junction Temperature	175	°C

*These ratings are limiting values above which the serviceability of any semiconductor device may be impaired.

NOTES:
1) These ratings are based on a maximum junction temperature of 200 degrees C.
2) These are steady state limits. The factory should be consulted on applications involving pulsed or low duty cycle operations.

Thermal Characteristics

Symbol	Characteristic	Max 1N/FDLL 914/A/B / 4148 / 4448	Units
P_D	Power Dissipation	500	mW
$R_{\theta JA}$	Thermal Resistance, Junction to Ambient	300	°C/W

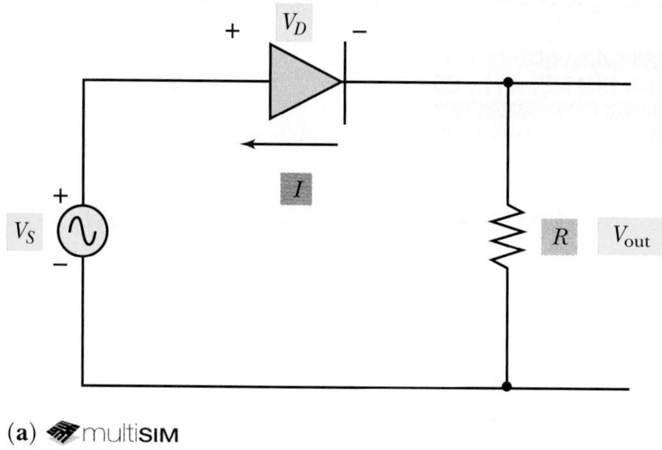

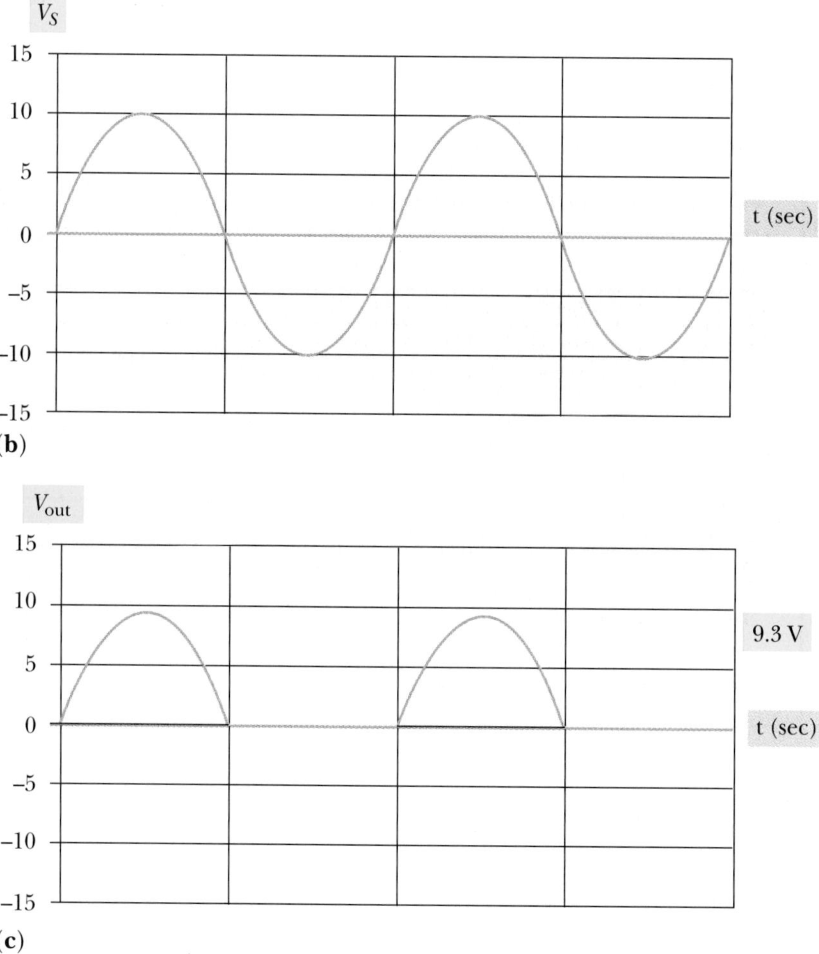

FIGURE 3–17 Negative clipper (series): (a) schematic; (b) input waveform; (c) output waveform

◆ **EXAMPLE** A 20 V_{PP} sine wave signal is applied to the negative series diode clipper circuit (Figure 3–17a) consisting of a silicon diode and a 1 kΩ resistor. Sketch the supply voltage waveform and the output signal waveform indicating the peak voltages.

Answer

The silicon diode barrier voltage is 0.7 V. When the supply voltage V_S is positive, the diode is forward biased. For V_S less than the diode barrier voltage of 0.7 V, no current is flowing in the circuit and the output voltage which appears across the resistor is 0 V ($V_{out} = I \times R$ where $I = 0$ A). Once the supply voltage V_S is greater than 0.7 V diode barrier voltage, the diode will be conducting current. As a result, the output voltage is calculated as

$$V_{out} = V_S - V_f$$

When V_S reaches the positive peak of 10 V, the output voltage is

$$V_{out} = 10\text{ V} - 0.7\text{ V}$$
$$V_{out} = 9.3\text{ V}$$

When V_S is less than 0.7 V, the current will not be conducting through the diode. With no current flow through the circuit, the output voltage is 0 V ($V_{out} = I \times R$ where $I = 0$ A).

When the supply voltage V_S is negative, the diode is reverse biased. No current will be flowing in the circuit. As a result, the output voltage is calculated as

$$V_{out} = I \times R$$
$$V_{out} = 0\text{ A} \times 1\text{ k}\Omega$$
$$V_{out} = 0\text{ V}$$

Figure 3–17b is the sketch of the supply voltage waveform. The sketch of the output waveform is shown in Figure 3–17c. Observe that the negative portion of the output waveform has been clipped off with only the positive sections of the waveform remaining. The positive cycle amplitude of the output waveform is 0.7 V less than the positive cycle amplitude of the input waveform. ◆

Positive Series Diode Clipper

Figure 3–18a is the schematic for the positive series diode clipper. The input to the circuit is a sinusoidal waveform. To establish the output signal waveform, the biasing of the diode during the positive half cycle and negative half cycle must be determined.

During the positive half cycle of the input waveform, the diode is reverse biased. No current will be conducted through the diode. The output voltage will then be equal to 0 V. The supply voltage will appear across the diode (the diode PIV rating must be greater than the peak supply voltage).

During the negative half cycle of the input waveform, the diode is forward biased. As soon as the supply voltage applied to the circuit exceeds the diode barrier voltage (0.7 V for silicon diodes), the diode will conduct current. The output voltage will then be equal to the supply voltage minus the voltage drop across the diode caused by the diode barrier voltage. When the supply voltage reaches its peak amplitude, the current flowing through the circuit will reach its maximum value (which must be less than the diode maximum current rating) and the output voltage will be at its peak amplitude. Once the supply voltage is below the diode barrier voltage, the diode no longer conducts current. The output voltage will be 0 V.

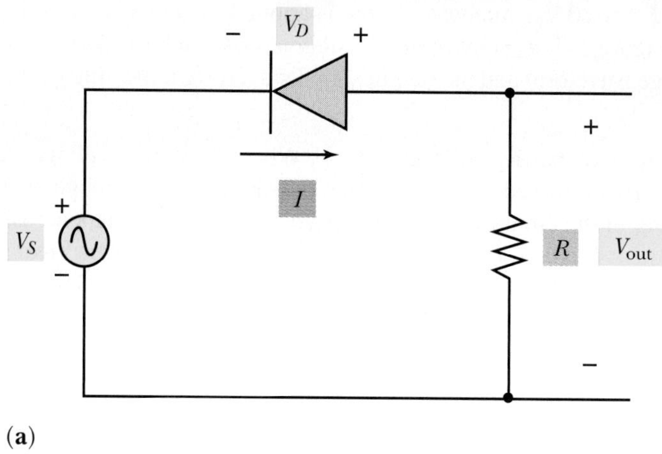

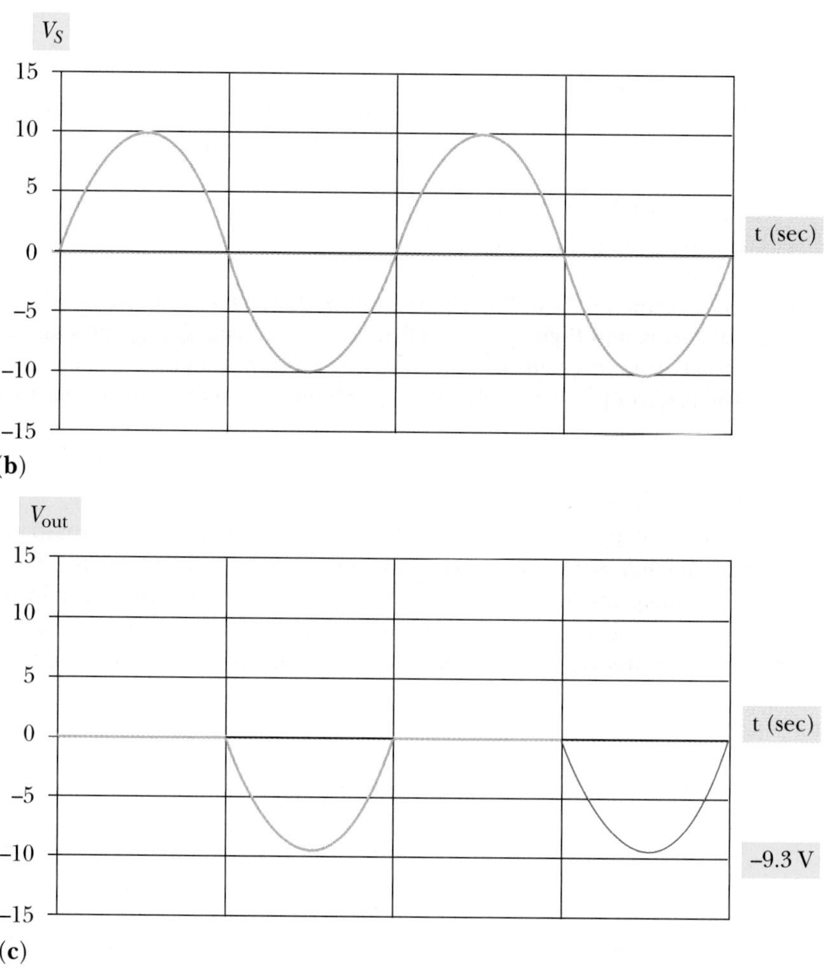

FIGURE 3–18 Positive clipper (series): (a) schematic; (b) input waveform; (c) output waveform

◆ **EXAMPLE** A 20 V_{PP} sine wave signal is applied to the positive series diode clipper circuit (Figure 3–18a) consisting of a silicon diode and a 1 kΩ resistor. Sketch the supply voltage waveform and the output signal waveform indicating the peak voltages.

Answer

The silicon diode barrier voltage is 0.7 V. When the supply voltage V_S is positive, the diode is reverse biased. No current will be flowing in the circuit. As a result, the output voltage is calculated as

$$V_{out} = I \times R$$
$$V_{out} = 0\text{ A} \times 1\text{ k}\Omega$$
$$V_{out} = 0\text{ V}$$

When the supply voltage V_S is negative, the diode is forward biased. For V_S below the diode barrier voltage of 0.7 V, no current is flowing in the circuit and the output voltage which appears across the resistor is 0 V ($V_{out} = I \times R$ where $I = 0$ A). Once the supply voltage V_S exceeds the 0.7 V diode barrier voltage, the diode will be conducting current. As a result, the output voltage is calculated as

$$V_{out} = V_S + V_f$$

When V_S reaches the negative peak of 10 V, the output voltage is

$$V_{out} = -\ 10\text{ V} + 0.7\text{ V}$$
$$V_{out} = -\ 9.3\text{ V}$$

When V_S is below 0.7 V, the current will not be conducting through the diode. With no current flow through the circuit, the output voltage is 0 V ($V_{out} = I \times R$ where $I = 0$ A).

Figure 3–18b is the sketch of the supply voltage waveform. The sketch of the output waveform is shown in Figure 3–18c. Observe that the positive portion of the output waveform has been clipped off with only the negative sections of the waveform remaining. The negative cycle amplitude of the output waveform is 0.7 V less than the negative cycle amplitude of the input waveform. ◆

Series Diode Clipper

Figure 3–17 and Figure 3–18 show a pair of series diode clippers. The only difference between the two is that the circuit in Figure 3–17 clips off the negative portion of the input waveform, while the Figure 3–18 circuit clips off the positive portion. The clipping action of these series clippers works according to the principle that the diode will conduct current only when it is forward biased. In Figure 3–17, the diode is forward biased on the positive portion of the input waveform when the generator voltage exceeds the diode barrier voltage (0.7 V for silicon diodes), thereby allowing the positive portion of the waveform to be duplicated across the output resistor. When the generator voltage is less than the diode barrier voltage, which includes the waveform's negative half-cycle, the diode is reverse biased and cannot pass that part of the input waveform to the output resistor. In Figure 3–18, the diode is turned around so that it is forward biased when the negative portion of the input waveform is greater than the diode barrier voltage.

Positive Shunt Diode Clipper

Figure 3–19a is the schematic for the positive shunt diode clipper. The input to the circuit is a sinusoidal waveform. To establish the output signal waveform, the biasing of the diode during the positive half cycle and negative half cycle must be determined.

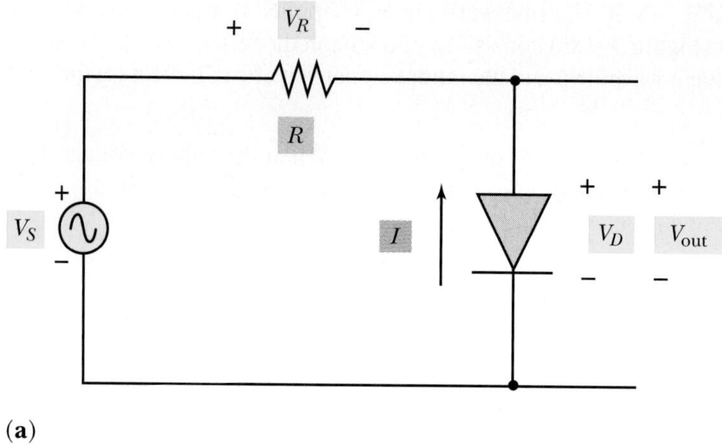

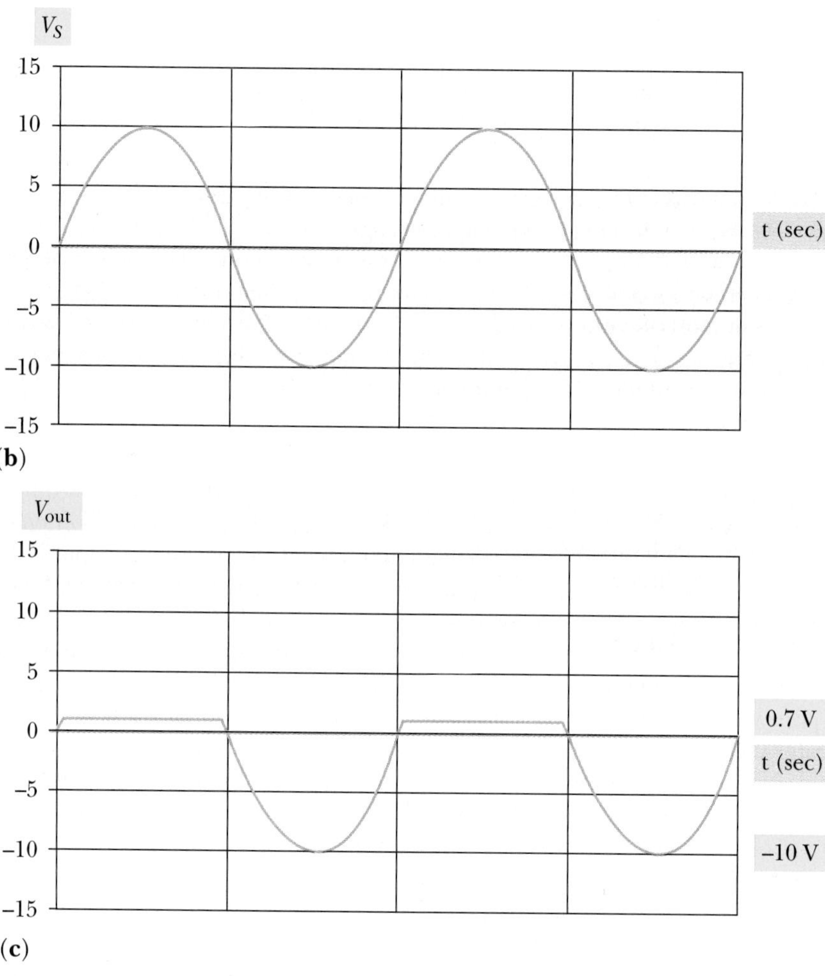

FIGURE 3–19 Positive diode clipper (shunt): (a) schematic; (b) input waveform; (c) output waveform multiSIM

During the positive half cycle of the input waveform, the diode is forward biased. As soon as the supply voltage applied to the circuit increases above the diode barrier voltage (0.7 V for silicon diodes), the diode will conduct current. The output voltage will then be equal to the voltage drop across the diode caused by the diode barrier voltage. When the supply voltage reaches its peak amplitude, the current flowing through the circuit will reach its maximum value (which must be less than the diode maximum current rating) and the output voltage will remain at the diode barrier voltage level. Once the supply voltage drops below the diode barrier voltage, the diode no longer conducts current. The output voltage will be the same as the supply voltage.

During the negative half cycle of the input waveform, the diode is reverse biased. No current will be conducted through the diode. The output voltage will then be equal to the supply voltage. The supply voltage will appear across the diode (the diode PIV rating must be greater than the peak supply voltage).

◆ **EXAMPLE** A 20 V_{PP} sine wave signal is applied to the positive shunt diode clipper circuit (Figure 3–19a) consisting of a silicon diode and a 1 kΩ resistor. Sketch the supply voltage waveform and the output signal waveform indicating the peak voltages.

Answer

The silicon diode barrier voltage is 0.7 V. When the supply voltage V_S is positive, the diode is forward biased. For V_S less than the diode barrier voltage of 0.7 V, no current is flowing in the circuit and the output voltage, which appears across the diode, is identical to the supply voltage. Once the supply voltage V_S is greater than the 0.7 V diode barrier voltage, the diode will be conducting current. As a result, the output voltage will be the 0.7 V diode barrier voltage.

When V_S reaches the positive peak of 10 V, the output voltage will remain at the 0.7 V level. When V_S is less than 0.7 V, the current will not be conducting through the diode. With no current flow through the circuit, the output voltage is identical to the supply voltage.

When the supply voltage V_S is negative, the diode is reverse biased. No current will be flowing in the circuit. As a result, the output voltage is identical to the supply voltage.

Figure 3–19b is the sketch of the supply voltage waveform. The sketch of the output waveform is shown in Figure 3–19c. Observe that the positive portion of the output waveform has been clipped off with only the negative sections of the waveform remaining. The positive cycle amplitude of the output waveform is 0.7 V while the negative cycle amplitude is 10 V. ◆

Negative Shunt Diode Clipper

Figure 3–20a is the schematic for the negative shunt diode clipper. The input to the circuit is a sinusoidal waveform. To establish the output signal waveform, the biasing of the diode during the positive half cycle and negative half cycle must be determined.

During the positive half cycle of the input waveform, the diode is reverse biased. No current will be conducted through the diode. The output voltage will then be equal to the supply voltage. The supply voltage will appear across the diode (the diode PIV rating must be greater than the peak supply voltage).

During the negative half cycle of the input waveform, the diode is forward biased. As soon as the supply voltage applied to the circuit exceeds the diode barrier voltage (0.7 V for silicon diodes), the diode will conduct current. The output voltage will then be equal to the voltage drop across the diode caused by the diode barrier voltage. When the supply voltage reaches its peak amplitude, the current flowing through the circuit will reach its maximum value (which must be less than the diode maximum current rating) and the output voltage will remain at the diode barrier voltage. Once the supply voltage is below the diode barrier voltage, the diode no longer conducts current. The output voltage will be identical to the supply voltage.

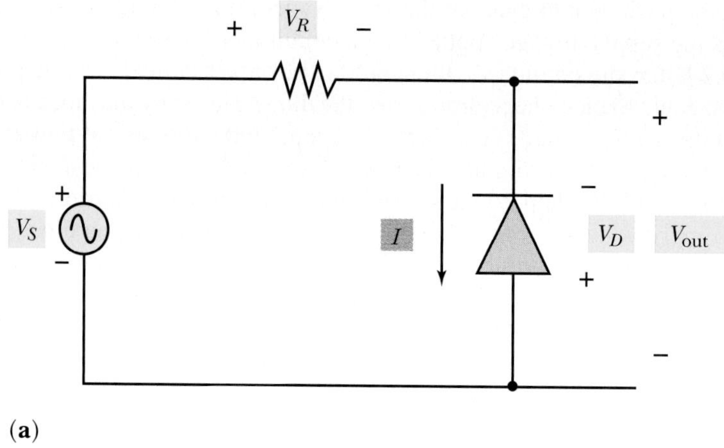

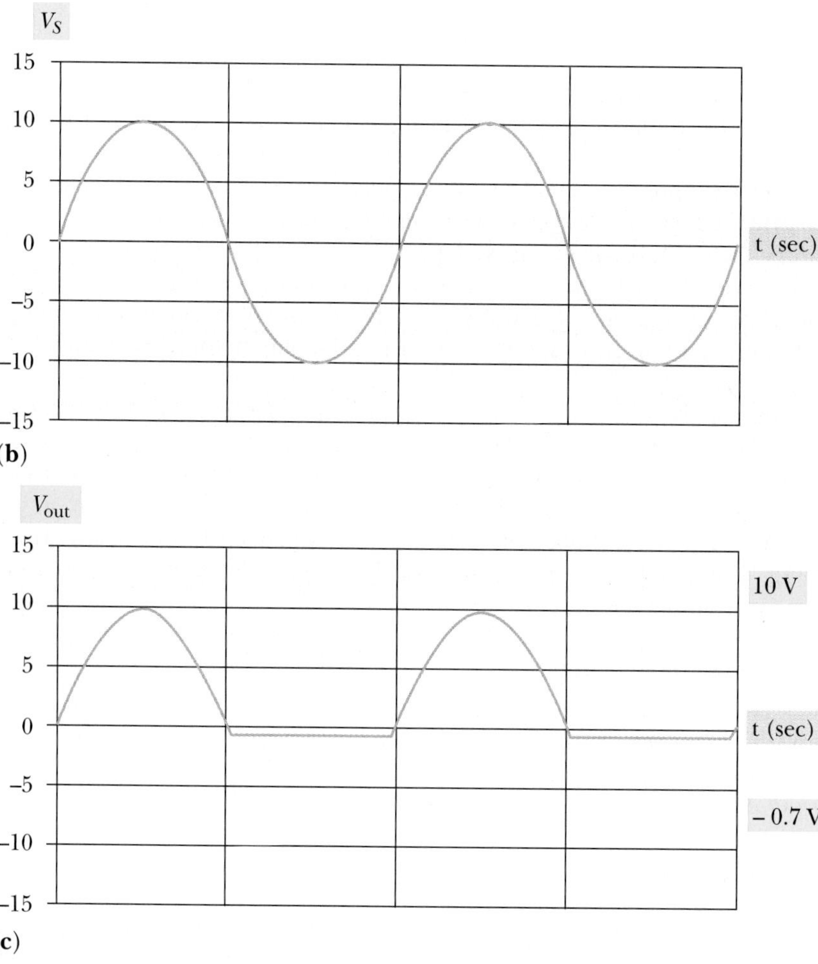

FIGURE 3–20 Negative diode clipper (shunt): (a) schematic; (b) input waveform; (c) output waveform

◆ **EXAMPLE** A 20 V_{PP} sine wave signal is applied to the negative shunt diode clipper circuit (Figure 3–20a) consisting of a silicon diode and a 1 kΩ resistor. Sketch the supply voltage waveform and the output signal waveform indicating the peak voltages.

Answer

The silicon diode barrier voltage is 0.7 V. When the supply voltage V_S is positive, the diode is reverse biased. No current will be flowing in the circuit. As a result, the output voltage is identical to the supply voltage.

When the supply voltage V_S is negative, the diode is forward biased. For V_S below the diode barrier voltage of 0.7 V, no current is flowing in the circuit and the output voltage is the same as the supply voltage. Once the supply voltage V_S exceeds the 0.7 V diode barrier voltage, the diode will be conducting current. As a result, the output voltage will be equal to the 0.7 V diode barrier voltage. When V_S reaches the negative peak of 10 V, the output voltage remains at the 0.7 V diode barrier voltage. When V_S is below 0.7 V, the current will not be conducting through the diode. With no current flow through the circuit, the output voltage is identical to the supply voltage.

Figure 3–20b is the sketch of the supply voltage waveform. The sketch of the output waveform is shown in Figure 3–20c. Observe that the negative portion of the output waveform has been clipped off at –0.7 V with only the positive sections of the waveform remaining. The positive cycle amplitude of the output waveform is 10 V while the negative cycle amplitude is 0.7 V. ◆

Shunt Diode Clipper

Figure 3–19 and Figure 3–20 show examples of shunt clippers. Note that the diodes are in parallel with this output. That is why they are called shunt clippers. In Figure 3–19, the diode conducts while the input waveform is greater than the diode barrier voltage. In this circuit when the diode is forward biased, the voltage across the diode remains fairly close to the diode's barrier potential. When a silicon switching diode is used, this voltage is about 0.7 V. So while the input waveform is greater than the diode barrier voltage in Figure 3–19, the output is fixed rather close to a steady 0.7-V level. (NOTE: Using the voltage only diode model, the diode voltage is fixed at 0.7 V. Using the voltage resistor diode model, the diode voltage will increase slightly from the 0.7 V.) When the input waveform goes below the diode barrier potential voltage, the diode is then reverse biased and the entire negative half-cycle appears across the diode and output of the circuit. So in this case, the positive half-cycle is effectively clipped from the input waveform.

The simple shunt clipper in Figure 3–20 is identical with the Figure 3–19 circuit except that the diode is reversed in its place across the output. The input waveform is clipped, or limited, during the negative portion of the input cycle. The diode is reverse biased during the positive half-cycle, so that is the portion of the input waveform that appears at the output terminals of the circuit. The addition of a load resistance or the input impedance of another component will change the output waveform.

Biased Diode Clippers

An interesting variation of the basic diode clipping action is to bias the diodes purposely so they limit the waveform to a value other than zero. Examples of *biased diode clippers* are shown in Figure 3–21 and in Figure 3–22.

In Figure 3–21, the circuit is biased at 50 Vdc and the positive peak input voltage is 170 V. You can see that the positive limiter is a +50.7 V clipper. Everything above that bias level is clipped off. The negative half of the input waveform is passed to the output without change.

In Figure 3–22, the circuit is biased at 50 Vdc. The clipping action is of the opposite polarity. Here, the negative swing is clipped at –50.7 V, and the positive half-cycle is unchanged.

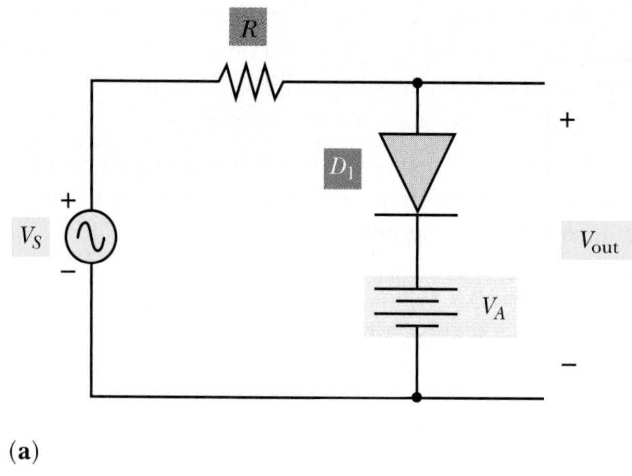

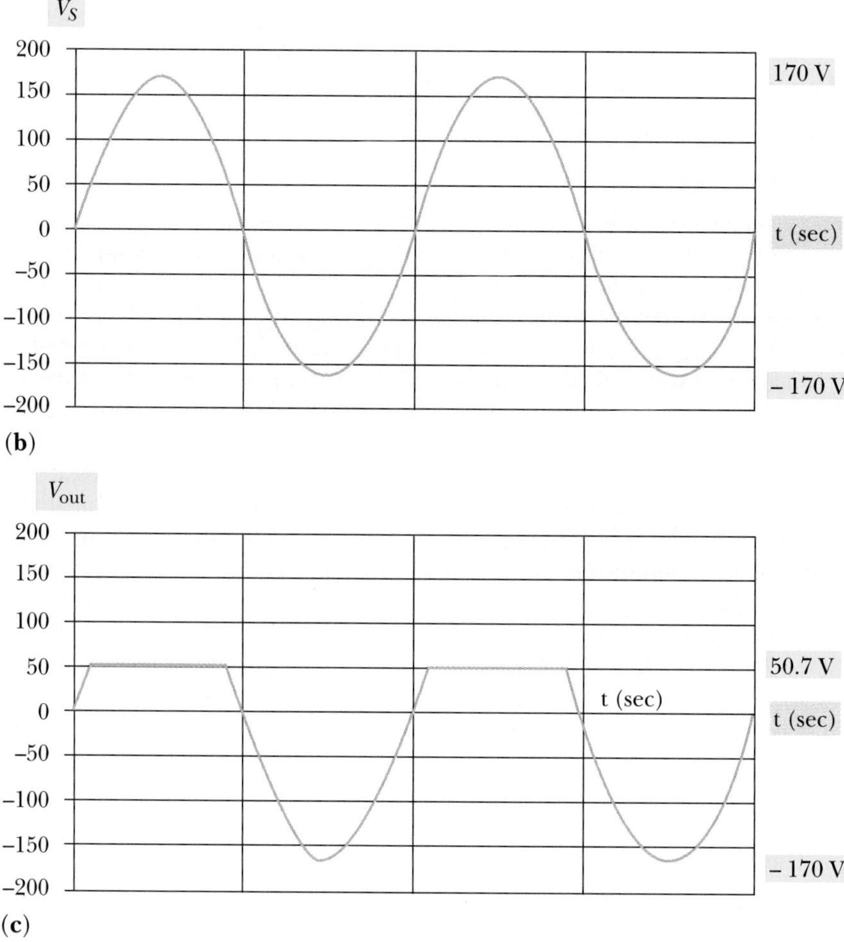

FIGURE 3–21 Biased positive shunt diode clipper: (a) schematic; (b) input waveform); (c) output waveform multiSIM

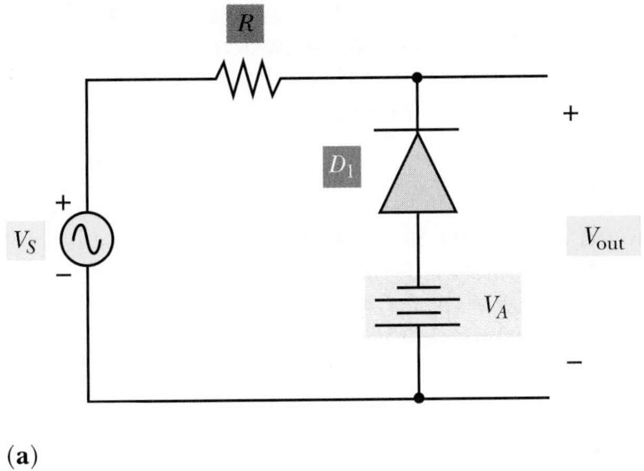

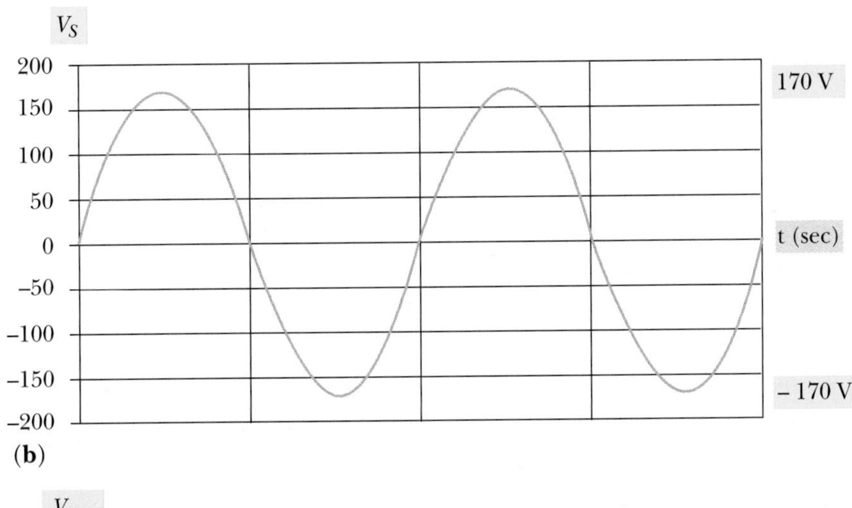

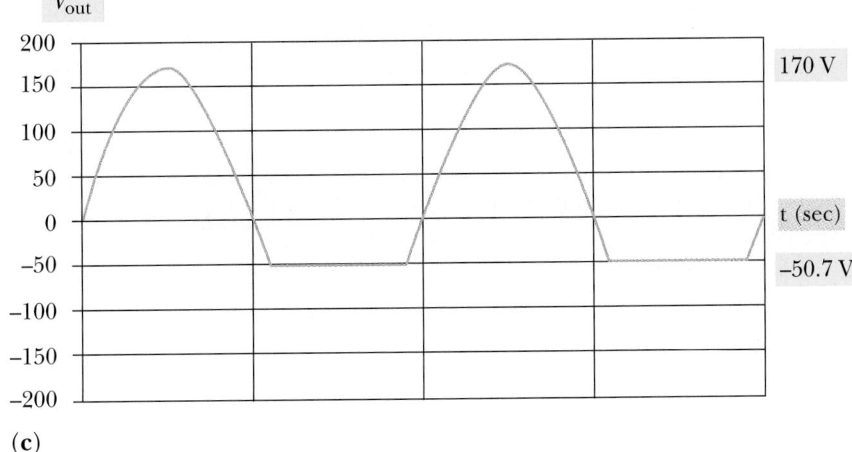

FIGURE 3–22 Biased negative shunt diode clipper: (a) schematic; (b) input waveform; (c) output waveform

The dc source that is used for setting the bias level for these clipper circuits can be a battery, but you will find that a zener diode (which you will study very soon) is simpler and less expensive. Most electronic circuits that require clipping action of any kind now accomplish the job with operational amplifiers.

Diode Clamping Circuits

A **clamping circuit** or **clamper** is one that changes the baseline voltage level of a waveform. Unless stated otherwise, you can safely assume that the baseline level for a waveform is 0 V. A 160-V (peak) sine waveform, for instance, normally goes to + 160 V and swings down through 0 V on its way to –160 V. The 0-V level is the exact middle of that voltage swing. It is the baseline voltage. Sometimes, however, it is desirable to change the baseline level.

Figure 3–23 and Figure 3–24 show two different diode clamping circuits. In both cases, the peak input waveform is shown to be 10 V. Because of the combined action of the capacitor, resistor, and diode, the baseline of the output is radically shifted—positive in Figure 3–23 and negative in Figure 3—24.

The output voltage levels of these diode clamping circuits depend on the values of C_1 and R_1, and the frequency of the input waveform. We will not be calculating those values in this chapter, but you should be able to recognize this kind of circuit when you see it on a schematic diagram.

The baseline voltage value equals the voltage across the capacitor which is equal to the supply voltage V_s reduced by the forward barrier voltage V_f. The polarity of the base line voltage depends on the orientation of the diode. If the diode's cathode is connected to the capacitor, it is a positive (+) baseline value. If the diode's anode is connected to the capacitor, it is a negative (–) baseline value.

■ IN-PROCESS LEARNING CHECK 2 Fill in the blanks as appropriate

1. Rectifier diodes are used where it is necessary to change _______ current power to _______ current power.
2. Where additional cooling is necessary, a rectifier diode can be connected to a(n) _______ to dissipate heat more efficiently.
3. When an ac waveform is applied to a rectifier diode, the diode's _______ rating must be greater than the peak voltage level.
4. The main current specification for rectifier diodes is _______.
5. To test the forward conduction of a diode with an ohmmeter, connect the _______ lead of the meter to the cathode and the _______ lead to the anode.
6. The forward resistance of a good diode should be much _______ than its reverse resistance.
7. Switching diodes have a(n) _______ rating that is hundreds of times less than most rectifier diodes.
8. The type of diode circuit that removes the peaks from an input waveform is called a(n) _______ circuit.
9. The type of diode circuit that changes the baseline level of an input waveform is called a(n) _______ circuit.

■

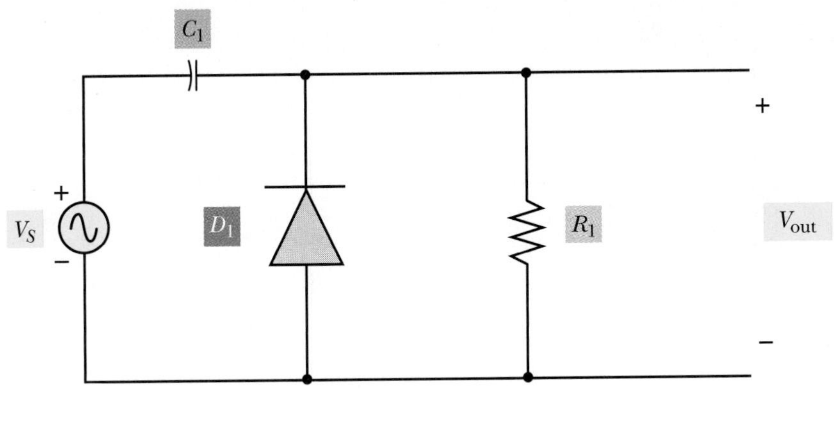

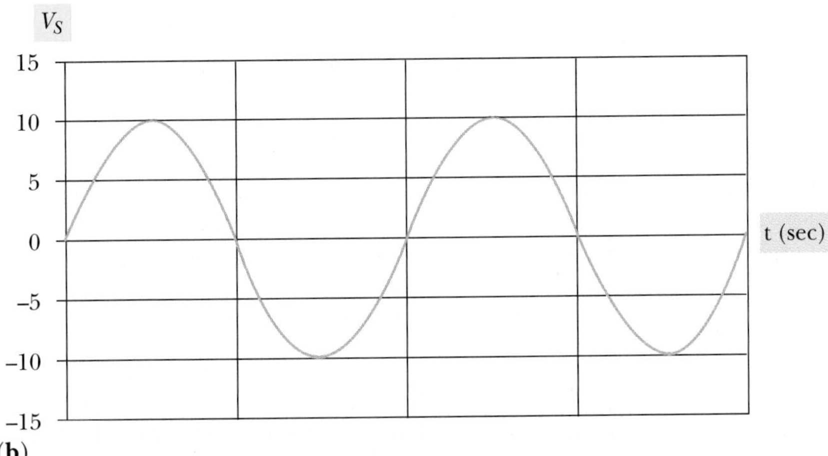

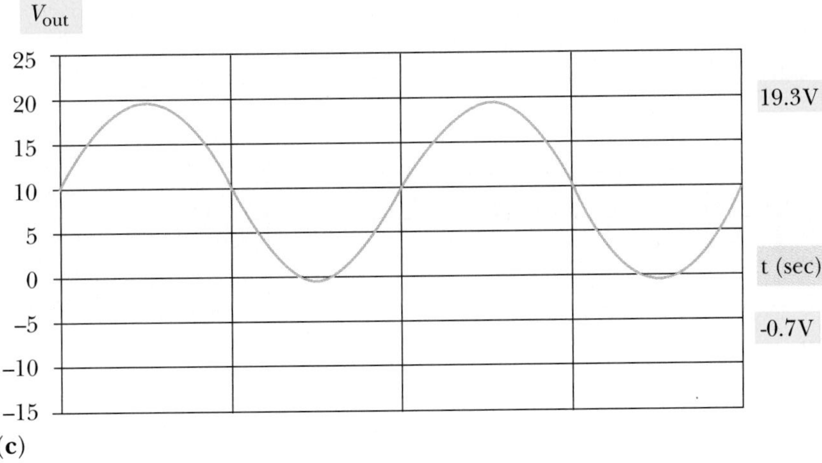

FIGURE 3–23 Positive diode clamping circuit: (a) schematic; (b) input waveform; (c) output waveform multiSIM

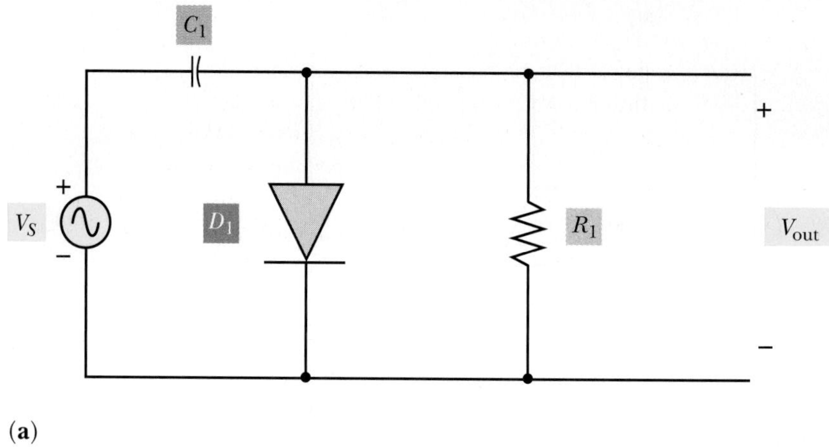

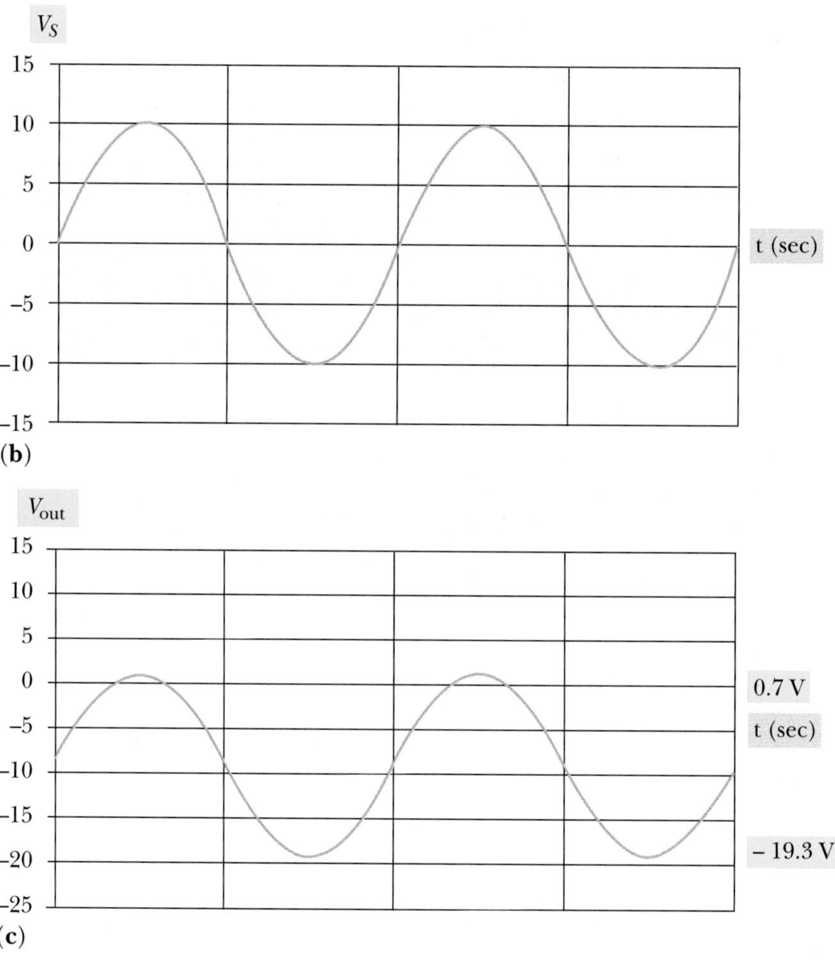

FIGURE 3–24 Negative diode clamping circuits: (a) schematic; (b) input waveform; (c) output waveform

3–7 Zener Diodes

You have already learned that every diode has a certain reverse breakdown voltage specification. For rectifier and switching diodes, it is essential to avoid circuit conditions that approach their reverse breakdown ratings. The zener diode is different. It is a diode that is designed to operate normally in its reverse-breakdown mode.

The zener diode operation is evaluated by using the Figure 3–25a and Figure 3–25b circuits. Figure 3–26 shows the I-V curve for a zener diode. It is important that you notice the part of the curve that represents the reverse-breakdown current. Once the breakdown voltage is reached, the zener will conduct current. In other diodes, this "reverse break-down" would cause the diode to fail (open). The zener is specially constructed to permit current flow at the reverse breakdown voltage. Once the PIV rating is reached, the zener diode conducts, allowing current to flow through the balance of the circuit (similar to a forward biased diode). The unique action of the zener diode causes its REVERSE DROP to stay at this breakdown voltage value. Should the voltage applied to the zener circuit exceed this voltage (called the "zener" voltage, labeled V_Z), all excess voltage will be passed on to the balance of the circuit. This is similar to the 0.7 V drop seen in the forward biased diode (all else is passed on to the circuit).

Practical Notes

Zener diodes look exactly like rectifier diodes, a stripe near one end marks the cathode terminal, and the JEDEC registered zener diodes have part designations that begin with 1N. This means that you cannot distinguish between a standard diode and a zener diode without checking the specifications in a data book, catalog, or online resource (such as www.FreeTradeZone.com and www.digikey.com).

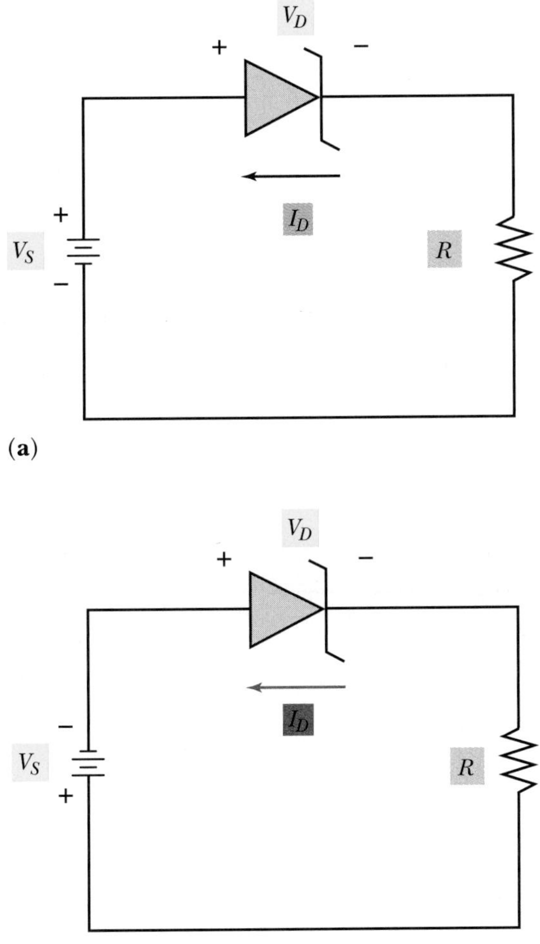

FIGURE 3–25 (a) Zener diode test circuit (positive supply voltage); (b) Zener diode test circuit (negative supply voltage)

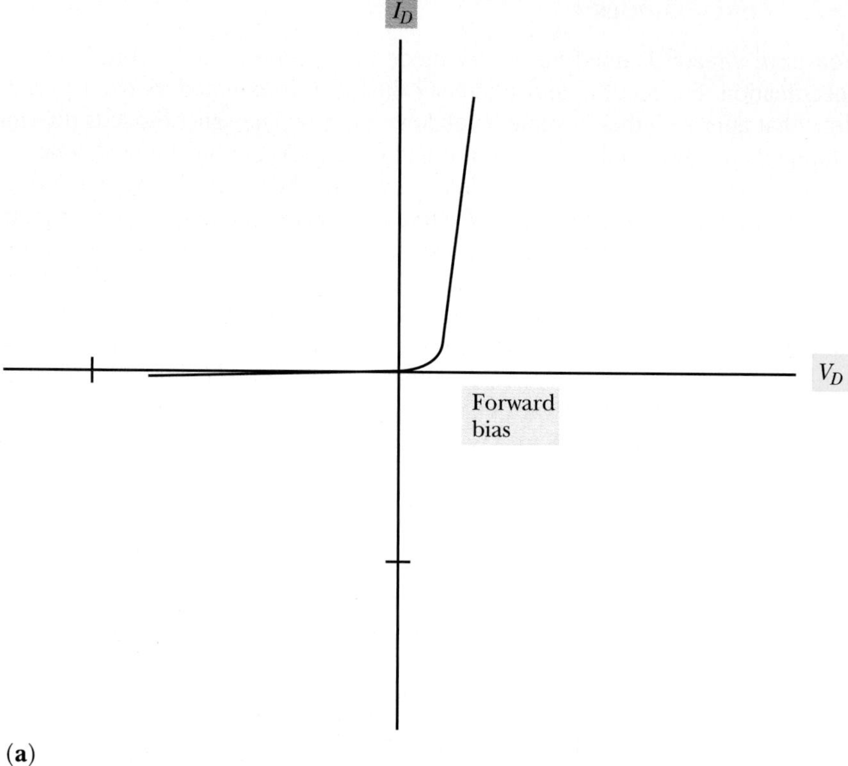

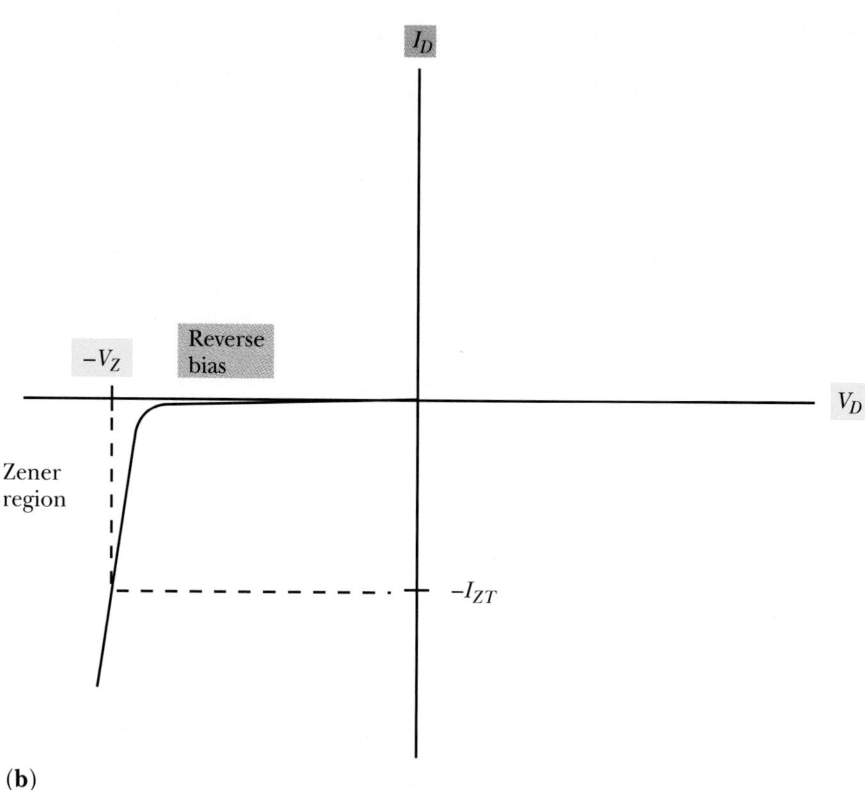

FIGURE 3–26 (a) Zener diode characteristic curve (positive voltage); (b) Zener diode characteristic curve (negative voltage)

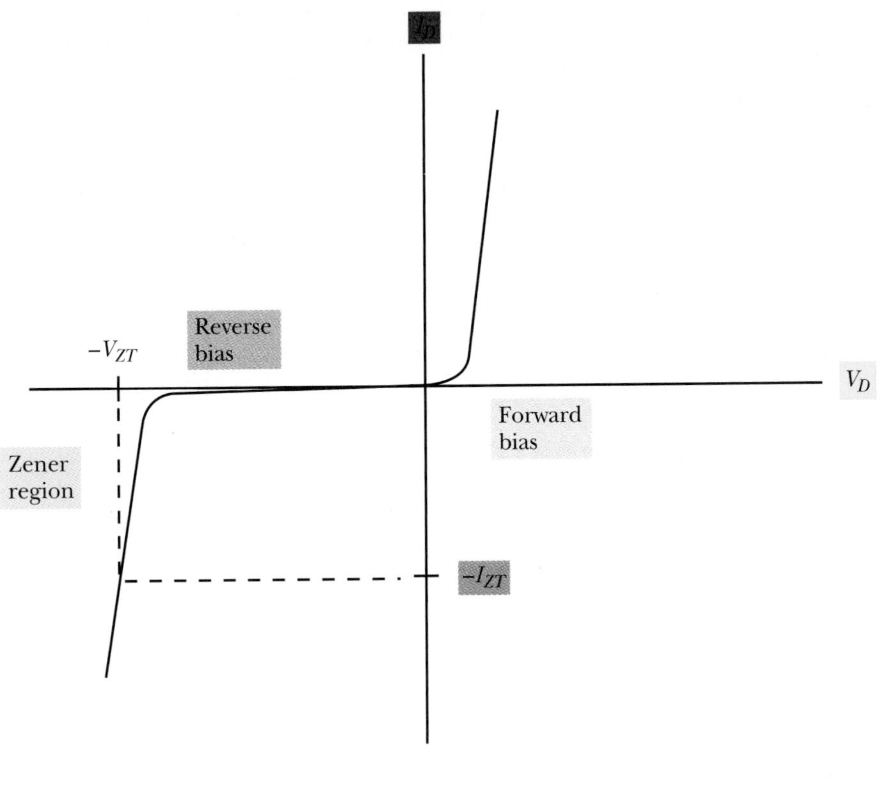

FIGURE 3–26 (c) Zener diode characteristic curve

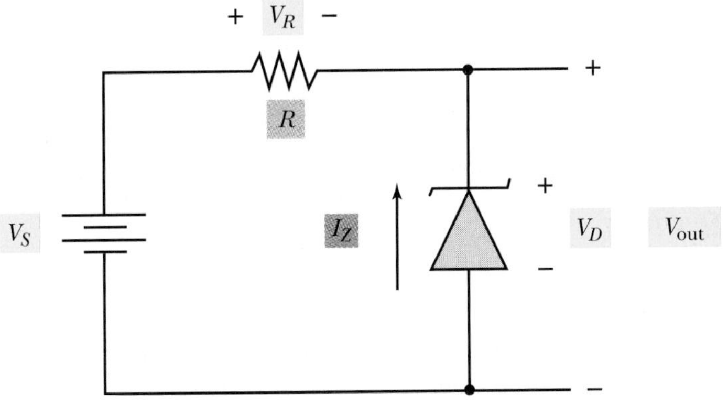

FIGURE 3–27 Zener diode circuit

Zener diodes are available with a wide variety of zener voltages ranging typically from 3 V to 400 V. The zener diode is often used as a *voltage regulator.* A voltage regulator is a circuit or device that will maintain a constant voltage output given varying voltage input. For example, the voltage produced from the alternator in your car varies its output voltage between 14 V and 18 V with engine speed. Zener diodes are used to stabilize this voltage at about 12 V. Thus the headlights don't get brighter and dimmer as the car speeds up and slows down.

The diagram in Figure 3–27 is a simple demonstration circuit. Notice that the symbol for a zener diode looks like an ordinary diode, but with the cathode-marker bar being bent on the ends. Manufacturers specify the zener diode based on the zener voltage (V_Z) at the zener test current (I_{ZT}) using positive values. Another critical rating for zener diodes is their maximum power dissipation. It is very important that the power

Part Identification	Zener Voltage V_Z (V)	Zener Test Current I_{ZT} (mA)	Max. power (W)
1N4728A	3.3	76	1.0
1N4731A	4.7	58	1.0
1N4733A	5.1	45	1.0
1N5233B	6.0	20	0.5
1N4739A	9.1	28	1.0
1N5346B	9.1	150	5.0
1N5347B	10.0	125	5.0
1N4742A	12.0	21	1.0
1N5252B	24.0	5.2	0.5

FIGURE 3–28 Typical ratings for zener diodes

dissipation of an operating zener diode does not exceed its power rating. The actual power dissipation of a zener diode is found using Formula 3–3 by multiplying the voltage dropped across the zener diode by the current flowing through it.

FORMULA 3–3 $P = I_Z \times V_Z$

Figure 3–28 lists some zener diode part designations along with their zener voltage and power ratings. Datasheet 3–3 provides information on Microsemi Corporation's 1N4728A through 1N4764A zener diodes.

Datasheet Questions

1. What is the Zener voltage rating for the 1N4733A?
2. What is the test current (I_{ZT}) for the 1N4733A?
3. Calculate the power dissipated by the 1N4733A at the rated zener voltage and zener test current. What is the maximum power dissipation rating for the 1N4733A?
4. What name does the manufactures provide for the 1N4733A?

To effectively analyze the circuits that contain zener diodes, an electrical model of the zener diode must be defined. The voltage only model emphasizing the zener region shown in Figure 3–29 will be used as the zener diode model.

The Voltage Only Zener Diode Model

In Figure 3–29 the zener diode is replaced with an open switch when the supply voltage is less than the zener test voltage ($0 \leq V_S < V_Z$), and a closed switch with a battery of V_Z volts for the zener region ($V_S > V_Z$).

The mathematical equation for this model is:

Non-conduction	$0 \leq V_S < V_Z$ Volts	$V_D = V_S$ $I_Z = 0$ A
Zener region	$V_S > V_Z$ Volts	$V_D = V_Z$ $I_Z > 0$ A

You will now analyze the circuit shown in Figure 3–27.

SCOTTSDALE, AZ
For more information call:
(602) 941-6300

1N4764A DO-41 GLASS

SILICON 1 WATT ZENER DIODES

FEATURES

- 3.3 THRU 100 VOLTS
- HERMETIC GLASS PACKAGE
- CONSULT FACTORY FOR VOLTAGES OVER 100 V

MAXIMUM RATINGS

Junction and Storage Temperature: −65°C to +200°C
Power Dissipation at T_L 100°C; 1.0 Watt
Power Derating from 100°C; 10 mW/°C
T_L = lead temperature at 3/8″ from body

*ELECTRICAL CHARACTERISTICS

(at +25°C ambient.)

Maximum forward voltage 1.2 volts at 200 mA

JEDEC TYPE NUMBER (Note 1)	ZENER VOLTAGE (V_Z) (NOTE 4)	TEST CURRENT (I_{ZT})	MAXIMUM DYNAMIC IMPEDANCE (Z_{ZT} @ I_{ZT}) (Note 2)	MAXIMUM REVERSE CURRENT (I_R @ V_R)	TEST VOLTAGE (V_R)	MAXIMUM REGULATOR CURRENT (I_{ZM}) T_A = 50°C	MAXIMUM KNEE IMPEDANCE (Z_{ZK} @ I_{ZK}) (Note 2)	TEST CURRENT (I_{ZK})	MAXIMUM (SURGE) CURRENT (I_S) (Note 3)
	VOLTS	mA	OHMS	μA	VOLTS	mA	OHMS	mA	mA
1N4728A	3.3	76	10	100	1	276	400	1.0	1380
1N4729A	3.6	69	10	100	1	252	400	1.0	1260
1N4730A	3.9	64	9	50	1	234	400	1.0	1190
1N4731A	4.3	58	9	10	1	217	400	1.0	1070
1N4732A	4.7	53	8	10	1	193	500	1.0	970
1N4733A	5.1	49	7	10	1	178	550	1.0	890
1N4734A	5.6	45	5	10	2	162	600	1.0	810
1N4735A	6.2	41	2	10	3	146	700	1.0	730
1N4736A	6.8	37	3.5	10	4	133	700	1.0	660
1N4737A	7.5	34	4.0	10	5	121	700	0.5	605
1N4738A	8.2	31	4.5	10	6	110	700	0.5	550
1N4739A	9.1	28	5.0	10	7	100	700	0.5	500
1N4740A	10	25	7	10	7.6	91	700	0.25	454
1N4741A	11	23	8	5	8.4	83	700	0.25	414
1N4742A	12	21	9	5	9.1	76	700	0.25	380
1N4743A	13	19	10	5	9.9	69	700	0.25	344
1N4744A	15	17	14	5	11.4	61	700	0.25	304
1N4745A	16	15.5	16	5	12.2	57	700	0.25	285
1N4746A	18	14	20	5	13.7	50	750	0.25	250
1N4747A	20	12.5	22	5	15.2	45	750	0.25	225
1N4748A	22	11.5	23	5	16.7	41	750	0.25	205
1N4749A	24	10.5	25	5	18.2	38	750	0.25	190
1N4750A	27	9.5	35	5	20.6	34	750	0.25	170
1N4751A	30	8.5	40	5	22.8	30	1000	0.25	150
1N4752A	33	7.5	45	5	25.1	27	1000	0.25	135
1N4753A	36	7.0	50	5	27.4	25	1000	0.25	125
1N4754A	39	6.5	60	5	29.7	23	1000	0.25	115
1N4755A	43	6.0	70	5	32.7	22	1500	0.25	110
1N4756A	47	5.5	80	5	35.8	19	1500	0.25	95
1N4757A	51	5.0	95	5	38.8	18	1500	0.25	90
1N4758A	56	4.5	110	5	42.6	16	2000	0.25	80
1N4759A	62	4.0	125	5	47.1	14	2000	0.25	70
1N4760A	68	3.7	150	5	51.7	13	2000	0.25	65
1N4761A	75	3.3	175	5	56.0	12	2000	0.25	60
1N4762A	82	3.0	200	5	62.2	11	3000	0.25	55
1N4763A	91	2.8	250	5	69.2	10	3000	0.25	50
1N4764A	100	2.5	350	5	76.0	9	3000	0.25	45

*JEDEC Registered Data

0.107 / 2.718 MAX.
0.032 ± 0.002 / 0.813 ± 0.051
1.10 / 27.940 MIN.
0.205 / 5.207 MAX.
1.10 / 27.940 MIN.
All dimensions in INCH / m.m.

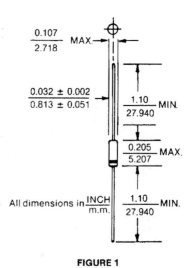

FIGURE 1

MECHANICAL CHARACTERISTICS

CASE: Hermetically sealed glass case. DO-41.

FINISH: All external surfaces are corrosion resistant and leads solderable.

THERMAL RESISTANCE: Less than 100°C / Watt junction to lead at 0.375-inches from body.

POLARITY: Banded end is cathode.

WEIGHT: 0.378 grams (Typical).

a. Zener diode schematic symbol

+ V_D −

I_Z

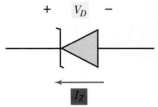

FIGURE 3–29 The zener diode voltage only model

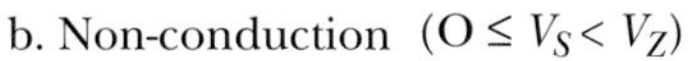

b. Non-conduction $(O \leq V_S < V_Z)$

c. Zener region $(V_S > V_Z)$

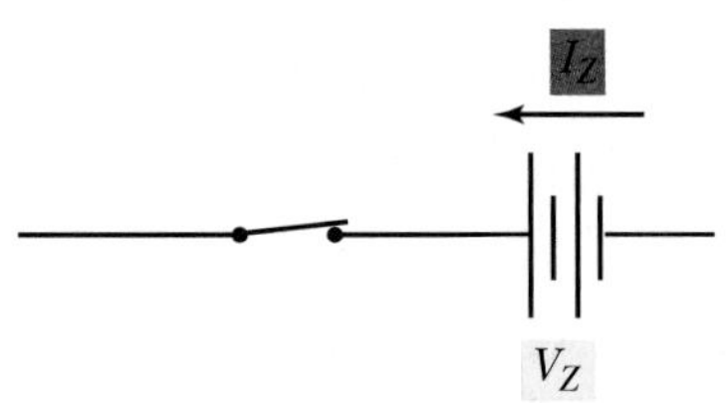

Practical Notes

The actual zener diode voltage V_z changes as a function of the zener current I_z. To account for this characteristic, the zener diode model would consist of a voltage source V_{zk} (the zener knee voltage) and a series resistance (Z_{zt}).

◆ **EXAMPLE** Analyze the Figure 3–27 circuit. A power supply provides 13 V to the circuit. The resistor is 100 Ω. The zener diode has a rating of 12 V. Using the voltage only zener diode model, calculate the voltage across the 12-V zener diode, the voltage across the resistor, the current flowing through the zener diode, and the power dissipated by the zener diode.

Step 1. Write the circuit equation.

$$V_S - V_R - V_{out} = 0$$
$$\text{where } V_{out} = V_D$$

Step 2. Select the zener diode model.
The voltage only zener diode model will be used.

Step 3. Document assumption.
You will assume that the diode is the zener region.

Step 4. Calculate the mathematical results.
Based upon the model selected and the condition assumed for the diode.

$$V_D = V_Z \text{ where } V_Z = 12 \text{ V}$$

Substitute this value into the circuit equation.

$$13 \text{ V} - V_R - 12 \text{ V} = 0$$

Solve for the voltage drop across the resistor.

$$V_R = 1.0 \text{ V}$$

The current flowing through the resistor and diode can be found using Ohm's law.

$$I_Z = \frac{V_R}{R}$$

Solve for the current

$$I_Z = \frac{1.0 \text{ V}}{100\ \Omega}$$
$$I_Z = 10 \text{ mA}$$

Step 5. Evaluate the mathematical results (verify the assumptions).
For the zener diode to be in the zener region, the zener current flowing through the diode must be greater than 0 mA.

Since the zener current flowing through the diode is 10 mA and 10 mA is greater than 0 mA, then the zener diode is operating in the zener region and your analysis is complete.

You can calculate the power dissipated by the zener diode using Formula 3–3 (restated here).

$$P = I_Z \times V_Z$$
$$P = 10 \text{ mA} \times 12 \text{ V}$$
$$P = 120.0 \text{ mW}$$ ◆

This example has shown you the analysis of a circuit using the voltage only model where the zener diode is operating in the zener region.

◆ **EXAMPLE** Analyze the Figure 3–27 circuit. A power supply provides 18 V to the circuit. The resistor is 100 Ω. The zener diode has a rating of 12 V. Using the voltage only zener diode model, calculate the voltage across the 12-V zener diode, the voltage across the resistor, the current flowing through the zener diode, and the power dissipated by the zener diode.

Step 1. Write the circuit equation.

$$V_S - V_R - V_{out} = 0$$
$$\text{where } V_{out} = V_D$$

Step 2. Select the zener diode model.
The voltage only zener diode model will be used.

Step 3. Document assumption.
You will assume that the diode is the zener region.

Step 4. Calculate the mathematical results.
Based upon the model selected and the condition assumed for the diode.

$$V_D = V_Z \text{ where } V_Z = 12 \text{ V}$$

Solve for the voltage drop across the resistor.

$$18 \text{ V} - V_R - 12 \text{ V} = 0$$
$$V_R = 6.0 \text{ V}$$

Solve for the current flowing through the resistor and diode.

$$I_Z = \frac{V_R}{R}$$
$$I_Z = \frac{6.0 \text{ V}}{100 \,\Omega}$$
$$I_Z = 60 \text{ mA}$$

Step 5. Evaluate the mathematical results (verify the assumptions).
Since the zener current is 60 mA, then the zener diode is operating in the zener region.

The power dissipated by the zener diode is calculated using Formula 3–3 (restated here).

$$P = I_Z \times V_Z$$
$$P = 60 \text{ mA} \times 12 \text{ V}$$
$$P = 720.0 \text{ mW}$$ ◆

From these two examples, you can observe that the output voltage remained at 12 V even though the input voltage changed from 13 V to 18 V. The voltage drop across the resistor increased, as did the zener current.

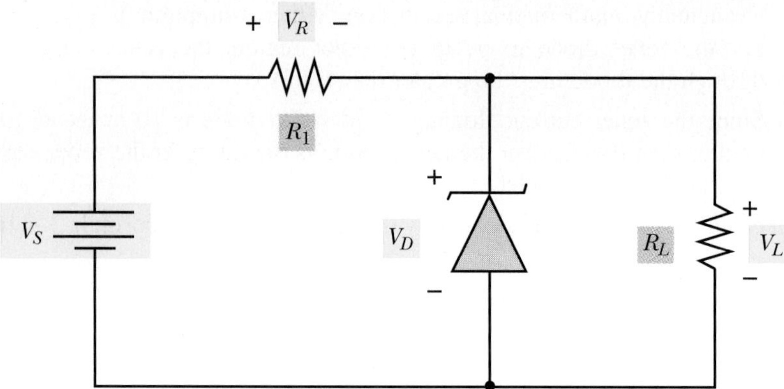

FIGURE 3–30 Zener diode voltage regulator circuit
multiSIM

Figure 3–30 illustrates a practical voltage regulator circuit. The function of this circuit is to maintain a steady voltage across the load resistor R_L while the input voltage varies. Whenever the supply voltage increases, the voltage across the zener diode and the load resistor R_L will remain fixed at the zener diode rated voltage. The voltage drop across the resistor R varies along with the supply voltage. In addition, the current flowing through the series resistor R and the zener diode vary accordingly. The load (R_L) voltage and current will remain regulated.

Whenever the output voltage with the load resistor is less than the zener diode rated voltage, the zener diode is not conducting current and will no longer regulate the voltage for the load. The supply voltage for the regulation transition point can be calculated using Formula 3–4. If the supply voltage is less than this voltage, the zener diode is not conducting current and you can calculate the output voltage by Formula 3–5. If the supply voltage is greater than the regulation transition point supply voltage, the output voltage is regulated by the zener diode rated voltage.

FORMULA 3–4 $V_S = V_z \times \frac{(R_1 + R_L)}{R_L}$

FORMULA 3–5 $V_L = V_S \times \frac{R_L}{(R_1 + R_L)}$

EXAMPLE Analyze the Figure 3–30 circuit. The power supply varies from 12 V to 20 V. The load resistor R_L is 750 Ω and the resistor R_1 is 100 Ω. The 1N5346B zener diode is used. Calculate the regulation transition point supply voltage using Formula 3–4. The zener test voltage rating V_Z for the zener diode 1N5346B is 9.1 V.

$$V_S = V_Z \times \frac{(R_1 + R_L)}{R_L}$$

$$V_S = 9.1\text{ V} \times \frac{(100\ \Omega + 750\ \Omega)}{750\ \Omega}$$

$$V_S = 10.31\text{ V}$$

Calculate the load voltage, and zener current when the supply is 12 V. Since the supply voltage is greater than 10.31 V, the voltage that will appear at the load will be limited to 9.1 V since the load resistor is in parallel with the zener diode.

The current flowing through the load resistor is calculated by

$$I_L = \frac{V_L}{R_L}$$

$$I_L = \frac{9.1\text{ V}}{750\ \Omega}$$

$$I_L = 12.13\text{ mA}$$

To calculate the current flowing through the zener diode, you must calculate the current flowing through the resistor R_1. This current is calculated by

$$I_1 = \frac{(V_S - V_R)}{R_1}$$
$$I_1 = \frac{(12\text{ V} - 9.1\text{ V})}{100\ \Omega}$$
$$I_1 = 29.0\text{ mA}$$

The current flowing through the zener diode is calculated using Kirchhoff's current law.

$$I_Z = I_1 - I_L$$
$$I_Z = 29.0\text{ mA} - 12.13\text{ mA}$$
$$I_Z = 16.87\text{ mA}$$

Now, calculate the load voltage, the load current, and the zener current when the supply is 20 V.

Since the supply voltage is greater than 10.31 V, the voltage that will appear at the load will be limited to 9.1 V.

The current flowing through the load resistor is calculated by

$$I_L = \frac{V_L}{R_L}$$
$$I_L = \frac{9.1\text{ V}}{750\ \Omega}$$
$$I_L = 12.13\text{ mA}$$

To calculate the current flowing through the zener diode, you must calculate the current flowing through the resistor R_1. This current is calculated by

$$I_1 = \frac{(V_S - V_D)}{R_1}$$
$$I_1 = \frac{(20\text{ V} - 9.1\text{ V})}{100\ \Omega}$$
$$I_1 = 109.0\text{ mA}$$

The current flowing through the zener diode is calculated using Kirchhoff's current law.

$$I_Z = I_1 - I_L$$
$$I_Z = 109.0\text{ mA} - 12.13\text{ mA}$$
$$I_Z = 96.87\text{ mA}$$ ◆

This example has shown you the analysis of a circuit where the zener diode is used as a voltage regulator. The voltage across the load will drop below the zener rated value of 9.1 V if the supply voltage is less than 10.31 V. When this occurs the zener diode is no longer conducting current and is no longer regulating the load voltage.

PRACTICE PROBLEMS 3

Use the zener circuit in Figure 3–30 where $R_1 = 1.0\text{ k}\Omega$ $R_L = 1.0\text{ k}\Omega$ with a 1N5233B zener diode.

1. What is the zener diode voltage rating?
2. Calculate the regulation transition point supply voltage.
3. How much voltage is dropped across the resistor R_1 when $V_{in} = 10$ Volts? When $V_{in} = 20$ Volts?
4. What is the voltage at V_{out} when $V_{in} = 10$ Volts? When $V_{in} = 20$ Volts?

Application Problem

ZENER DIODE BANK VOLTAGE REGULATOR

A CMOS broadcast amplifier requires a constant 12 V to function properly. The dc power supply output voltage varies from 12 to 14 V. Figure 3–28 lists a zener diode regulator that we ordered five 1N4742A from a Texas electronics supplier. However, three 1N5233B voltage regulators have arrived. How can we temporarily solve this matter if there is no time for receiving correct shipment?

Solution A quick fix is to connect two 6 V regulators in series. But we have to be careful that only the very same kind are connected together. The same part number made by the same company should be used to avoid slight process variations.

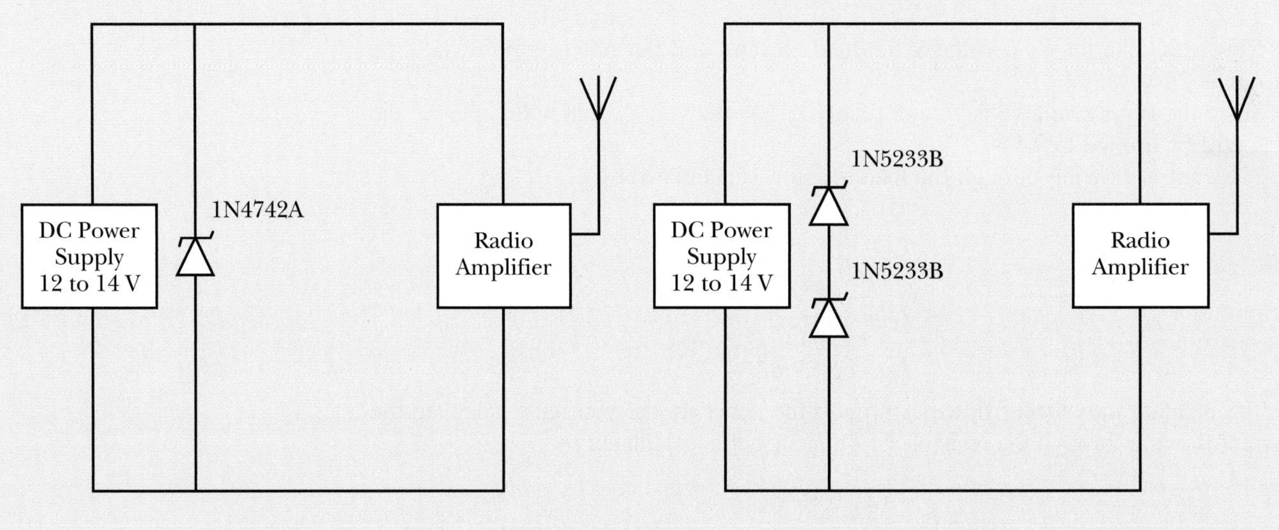

5. What is the current flowing through R_L when V_{in} = 10 Volts? When V_{in} = 20 Volts?
6. What is the current flowing through R_1 when V_{in} = 10 Volts? When V_{in} = 20 Volts?
7. What is the current flowing through the zener diode when V_{in} = 10 Volts? When V_{in} = 20 Volts?
8. What is the power dissipation of the zener diode when V_{in} = 10 Volts. When V_{in} = 20 Volts?

3–8 Zener Diode Applications

Zener Diode Limiter Circuits

A diode **limiter circuit** removes the peaks from an input waveform. Figure 3–31 and Figure 3–32 show examples of shunt limiters using zener diodes. Note that the zener diodes are parallel with this output. In Figure 3–31, the zener diode conducts while the input waveform is greater than the zener voltage and less than the zener diode barrier voltage. In this circuit the output voltage levels will be limited to between –0.7 V and the zener voltage rating.

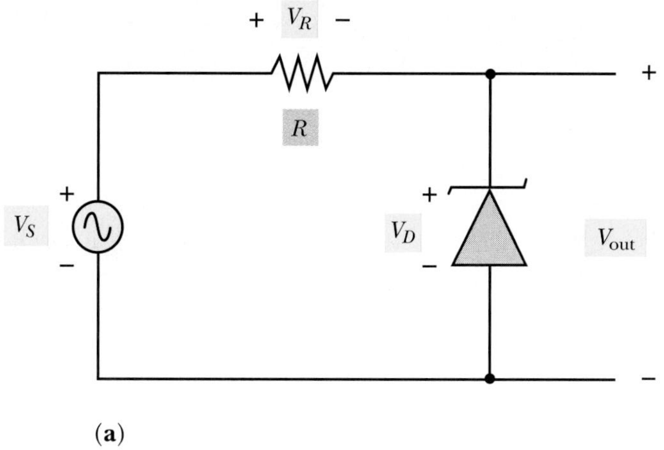

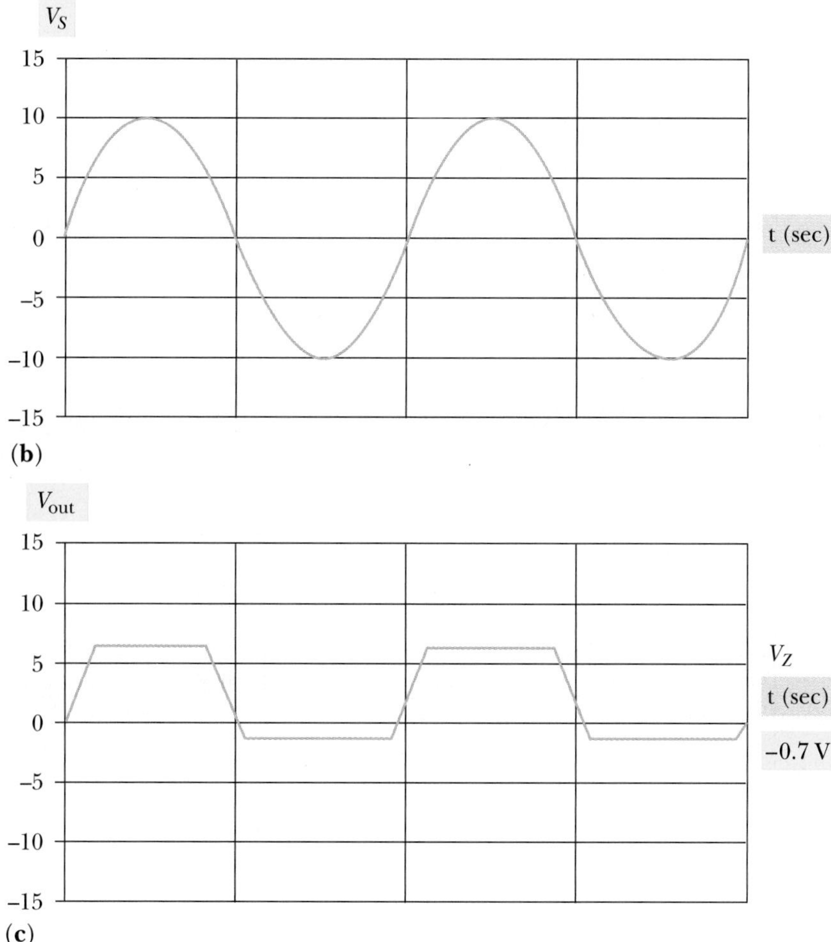

FIGURE 3–31 (a) Zener diode limiter circuit; (b) input waveform; (c) output waveform

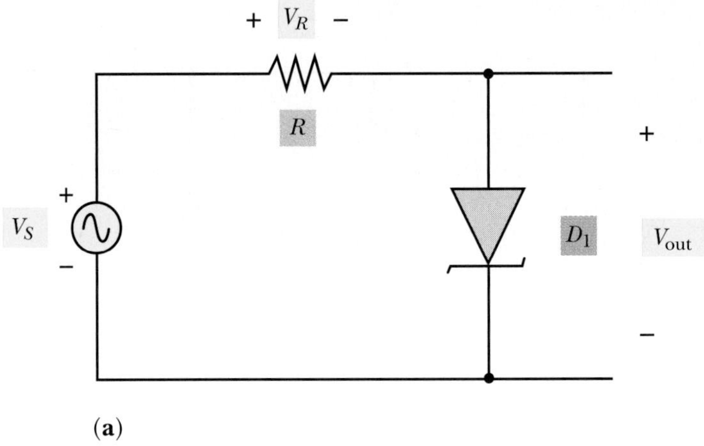

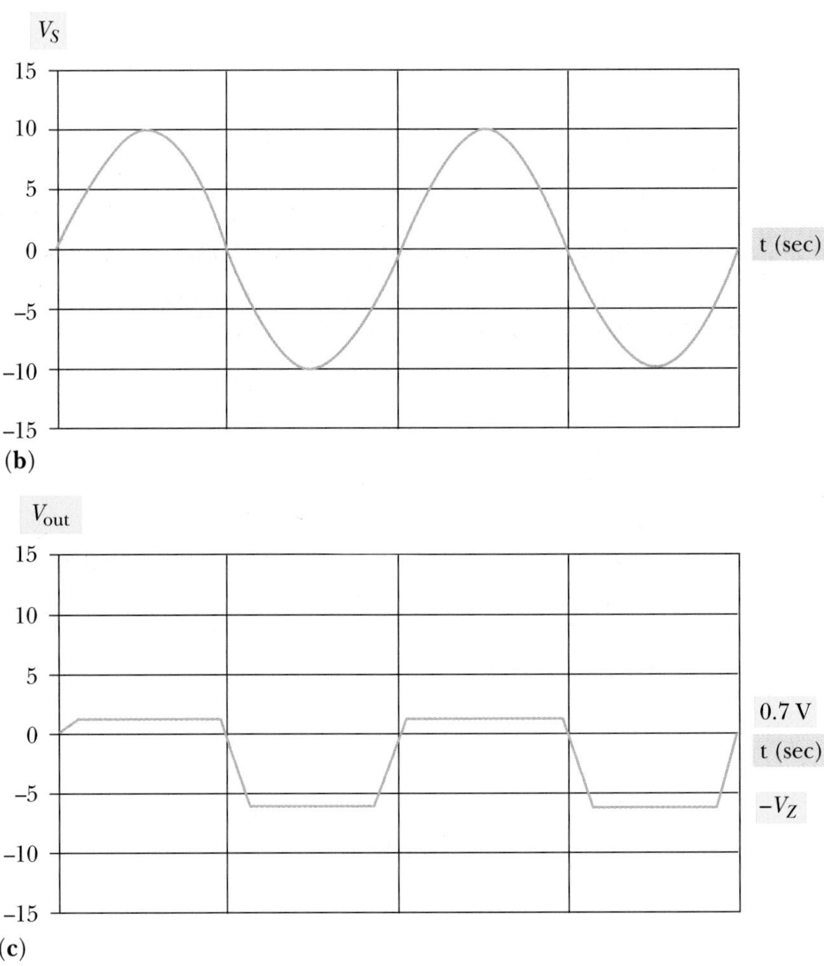

FIGURE 3–32 (a) Zener diode limiter circuit; (b) input waveform; (c) output waveform

The simple shunt limiter in Figure 3–32 is identical with the Figure 3–31 circuit except that the zener diode is reversed in its place across the output. The input waveform is limited, resulting in the output waveform from the negative zener rated voltage to +0.7 V. The addition of a load resistance or the input impedance of another component will change the output waveform.

A device similar to the zener diode was developed for high speed, high energy voltage limiting. This classification of devices is referred to as *transient suppressors.* These devices are available as unidirectional (DC applications) and bidirectional (AC applications). Transient suppressors are used to protect electronic equipment from short duration power surges from lightning strikes. The metal oxide varistor (MOV) has a slower response time but is designed to handle longer duration transient power surges.

3–9 The Varactor Diode

Varactor diodes act like small variable capacitors. By changing the amount of reverse bias across their P-N junction, their capacitance can be varied. The greater the amount of reverse bias, the wider the depletion region becomes. Also, the greater the effective distance (depletion region) between the P- and N-type sections, the smaller the capacitance is. On the other hand, the smaller the amount of reverse bias, the narrower the depletion region and the greater the amount of capacitance produced. In effect, the varactor is a voltage-variable capacitance compared with a variable capacitor whose physical plates must be moved apart or closer together to change capacitance. The devices are designed with capacitance values ranging from 1 pF to 500 pF. The varactor symbol is shown in Figure 3–33.

As an example the varactor diode is used in an "electronically tuned" radio, the type that uses a digital display. A precise reverse bias voltage is applied to the varactor, which changes the capacitance. The change in capacitance changes the resonant circuit used to tune in the radio station's broadcast frequency.

3–10 The Light Emitting Diode (LED)

The LED is a diode that emits light whenever it is forward biased and conducting current. Refer to the simple circuit in Figure 3–34. If you are using a gallium arsenide LED, the barrier potential (forward voltage drop) is 1.5 V, and you need about 15 mA of forward current to generate a useful amount of light. The value of the current limiting resistor connected in series with the LED depends on the amount of supply voltage. Formula 3–6 lets you determine the amount of voltage that has to be dropped across the resistor (V_R) in terms of the source voltage (V_S) and the forward voltage drop across the diode (V_D).

FORMULA 3–6 $V_R = V_S - V_D$

FIGURE 3–33 Varactor diode schematic symbol

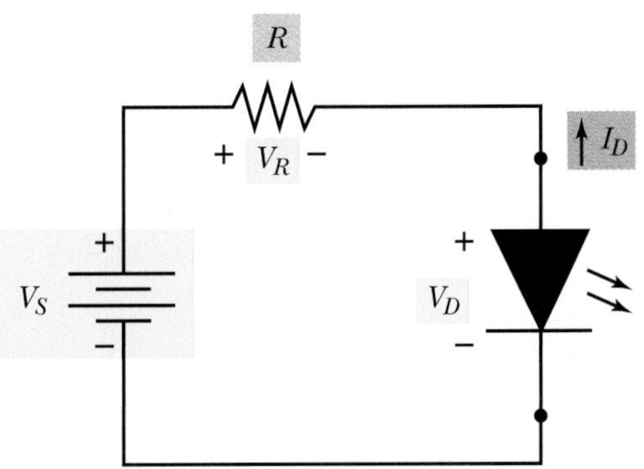

FIGURE 3–34 A practical LED circuit multiSIM

Application Problem

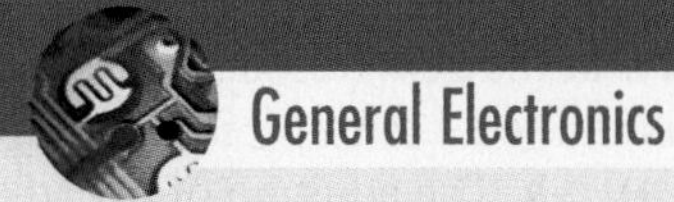

DIMMERS

A dimmer controls illumination in an automobile dashboard. The dashboard is illuminated by three light bulbs. All three bulbs are rated for 12 V, and a current through one bulb is 300 mA. When daylight is striking a photodiode, its resistance is 150 Ω. The resistance of a photodiode is quite high when it is dark (on the order of megaohms ~ 5 MΩ. such as)

(a) Calculate the transistor base voltage, V_B.

(b) Calculate the power consumed by three light bulbs.

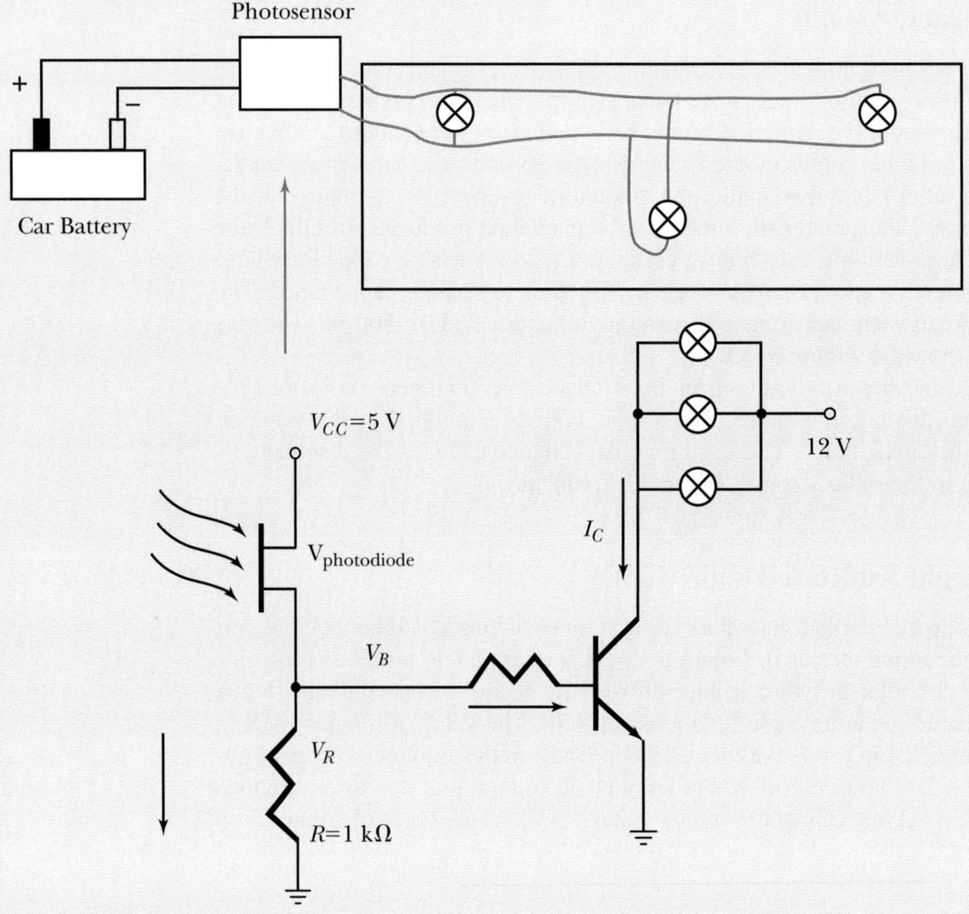

Solution

(a) The photodiode and the 1kΩ resistor form a voltage divider. When the photodiode is ON

$$V_{B(\text{on})} = \frac{R}{R_{\text{photodiode-on}} + R} V_{CC} = \frac{1000}{150 + 1000} 5 = 4.35 \text{ V}$$

When the photodiode is OFF

$$V_{B(\text{off})} = \frac{R}{R_{\text{photodiode-off}} + R} V_{CC} = \frac{1000}{5 \times 10^6 + 1000} 5 = 0.001 \text{ V}$$

(b) When the light bulbs are ON, the consumed power is

$$P = 3 \times VI = 3 \times 12 \times 0.3 = 10.8 \text{ W}$$

Application Problem

Industrial

PN JUNCTION DIODES—LED CIRCUIT LIMIT

An indication LED (light emitting-diode) is widely used throughout the industry—for example, kitchen appliances have ON/OFF indication, and also, computers, videos, TVs, and satellite receivers have several LEDs indicating different states. The following figure shows an LED circuit. The current limiting resistor placed in series with the LED is extremely important; otherwise the LED will burn out. If the desired current flow through an LED is 20 mA and a bias voltage is 5 V, calculate the resistance of such a resistor.

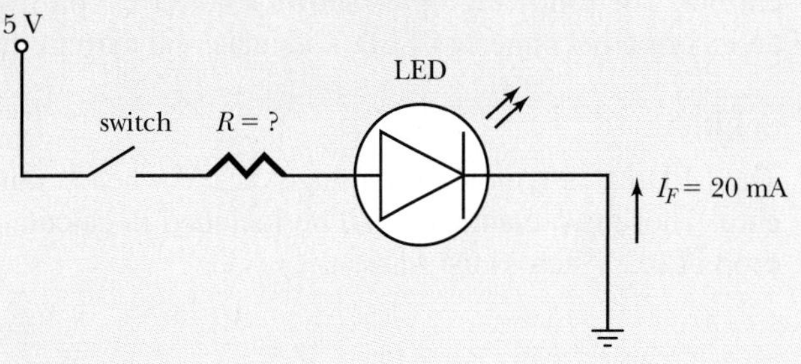

LED Circuit

Solution

An LED has a typical ON voltage of 2 V when conducting a forward current of 20 mA.

$$R = \frac{V_{CC} - 2.0\text{ V}}{I_F} = \frac{5\text{ V} - 2.0\text{ V}}{0.020} = 150\ \Omega$$

Choose the next resistor commercially available to increase a life expectancy of the LED.

$$R = 220\ \Omega$$

The resistance is 220 Ω

EXAMPLE Referring to the circuit in Figure 3–34, let VS = 12 V, and assume that the forward voltage drop for the LED (V_D) is 1.5 V and the LED current is 20 mA. Calculate the voltage across the current limiting resistor (V_R) and find the resistance of the resistor.

From Formula 3–6:

$$V_R = V_S - V_D$$
$$V_R = 12\text{ V} - 1.5\text{ V} = 10.5\text{ V}$$

Using Ohm's law, the resistance of the resistor can be determined.

$$R = \frac{V_R}{I}$$
$$R = \frac{10.5\text{ V}}{20\text{ mA}}$$
$$R = 525\ \Omega$$

In this example, the voltage across the 525-Ω current limiting resistor is 10.5 V.

3–11 Other Types of Diodes

The Tunnel Diode

The **tunnel diode** is a two-terminal semiconductor that is manufactured with a high proportion of the doping material and an especially thin depletion region. Because of the heavy doping, the barrier potential is relatively high, yet because of the thin depletion region, electrons can be forced to "tunnel" through the depletion region. At the

Application Problem

INCORRECT LED INTERFACING

An indication LED (light emitting-diode) is incorrectly connected to a computer buffer. The buffer's driving capability is up to 20 mA when "sinking"; however, connection in the other direction, "sourcing," provides much lower current. The following figure shows a buffer circuit taken from a Motorola or TI data sheet. The next figure shows an incorrectly connected LED. Calculate the current through LED.

Solution

The LED has a typical ON voltage of 2 V when conducting. There is one 130 Ω resistor and one diode inside the chip. Those two elements shall be included in calculation. (To keep things simple, we will not include the voltage drop of 0.2 V across Q4 transistor.)

$$I_F = \frac{V_{CC} - V_{LED} - V_{DIODE}}{R_{LIMIT} + R_{CHIP}} = \frac{5 - 2.0 - 0.7}{220 + 130} = 6.6 \text{ mA}$$

The current will be significantly lower; the LED will still glow but its intensity will be lower. Furthermore, the 130 Ω resistor is not designed to handle much overload and the chip might overheat and fail.

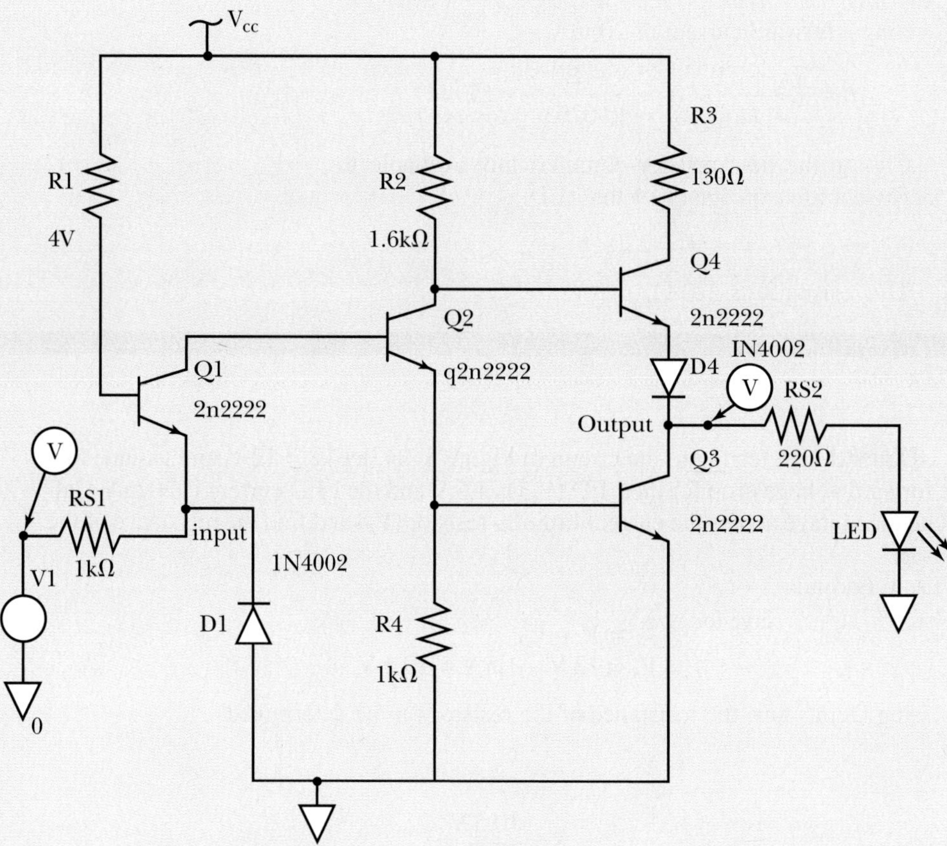

Schematic of 74LS04 buffer-inverter

LED

$R = 220\Omega$

74LS04

$I_F = ?$ mA

Incorrectly connected LED circuit

biasing voltage levels, which cause this tunneling effect, there is a unique feature exhibited by the diode, called the *negative-resistance* region of the I-V curve (see Figure 3–35). This means there is a portion of the characteristic curve where an increase in forward voltage actually causes a decrease in forward current.

Negative resistance effects are advantageous in circuits known as *oscillators,* which you will learn about later. These oscillator circuits are commonly used to generate high frequency ac signals. The tunnel diode is also sometimes used as an electronic switch. Its key feature in this application is its high switching speed and low power consumption. Figure 3–36 shows the different schematic symbols for tunnel diodes.

Schottky Diode

The Schottky diode, or Schottky barrier diode (SBD), is specially constructed and doped for ultra-fast switching speeds. Ordinary junction diodes have difficulty keeping up with changes in polarity greater than 1 MHz. It isn't unusual, however, to find Schottky diodes operating comfortably at 20 MHz and higher. There is an entire family of digital IC devices that uses diodes and transistors built according to Schottky's hot-carrier junction effect. Figure 3–37 shows the schematic symbol for a Schottky diode.

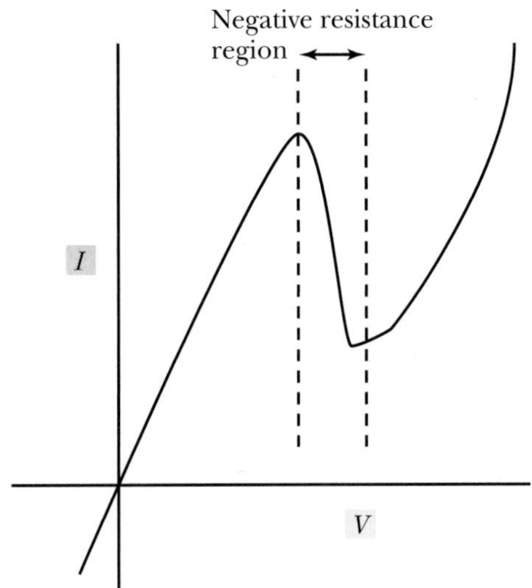

FIGURE 3–35 I-V curve for a tunnel diode

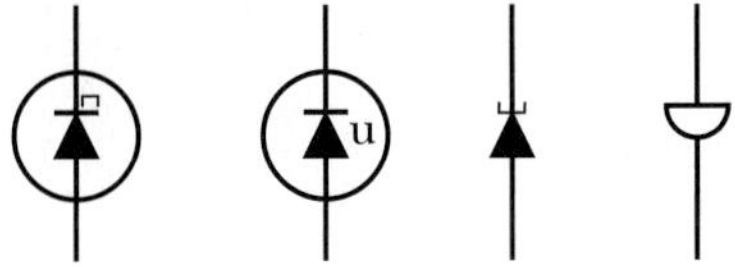

FIGURE 3–36 Tunnel diode schematic symbols

FIGURE 3–37 Schematic symbol for a Schottky diode

USING TECHNOLOGY: DIODES USED IN A SYSTEM

In this and following chapters, you will study the fundamental concepts of different electronic active devices. Where appropriate, how the device is used in a system will be described in detail.

A system is defined as a collection of components that are interconnected from an input to an output to accomplish a prescribed end. A system can be self-contained or composed of a variety of subsystems integrated to accomplish the required operation. From the outside you may not be able to observe how the components are used to make the system perform the desired tasks. The system block diagram and the associated schematic are the tools to communicate the function of the internal circuitry. As you evaluate the consumer and business electronic products, you can observe a vast variety of electronic devices that provide the different end user functions. Today alone, you probably used your digital watch, cell phone, pocket calculator, computer, microwave oven, television, and CD and DVD players. These are but a few of the products we use every day that contain electronics. How do they perform the task for which you use them? What components are inside of these products? Are there any active devices, such as diodes? If there are any such devices inside, how are they used?

The simple diode is designed to control the direction of current flow. Some of the primary uses for the diode are to protect devices, regulate voltages, and rectify signals and certain digital operations. Any one or more of these functions might be required for the product that you are using. All products need a power source. Is it a battery, solar cell, dc input, or ac input? Do the internal components require a single or multiple voltage levels to operate? What inputs are required to cause the system to perform the necessary tasks?

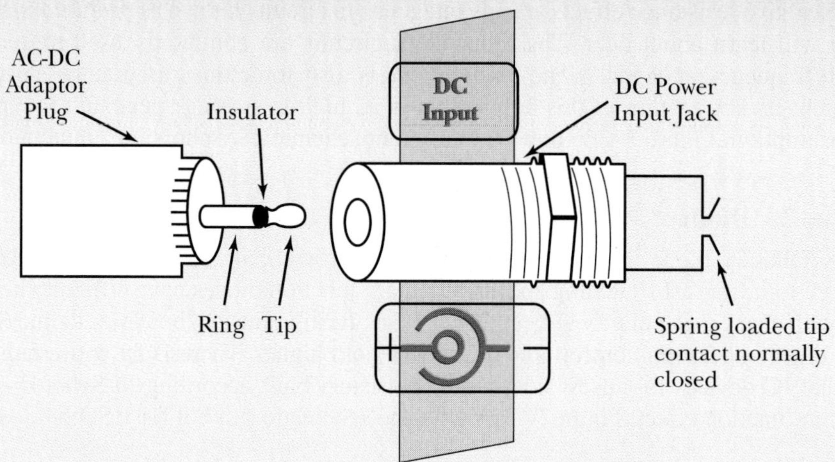

FIGURE 1 Pictorial

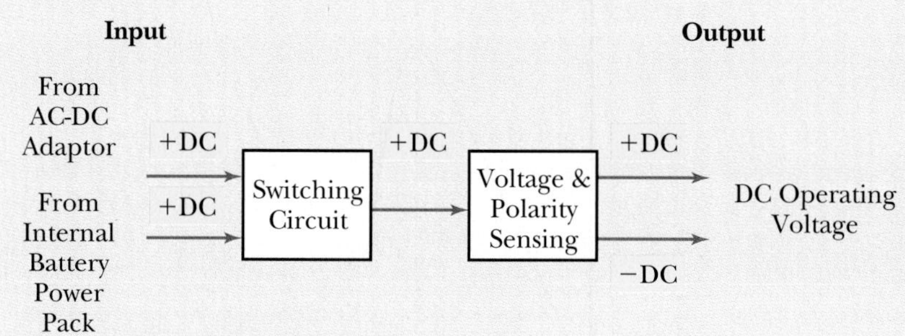

FIGURE 2 Block diagram

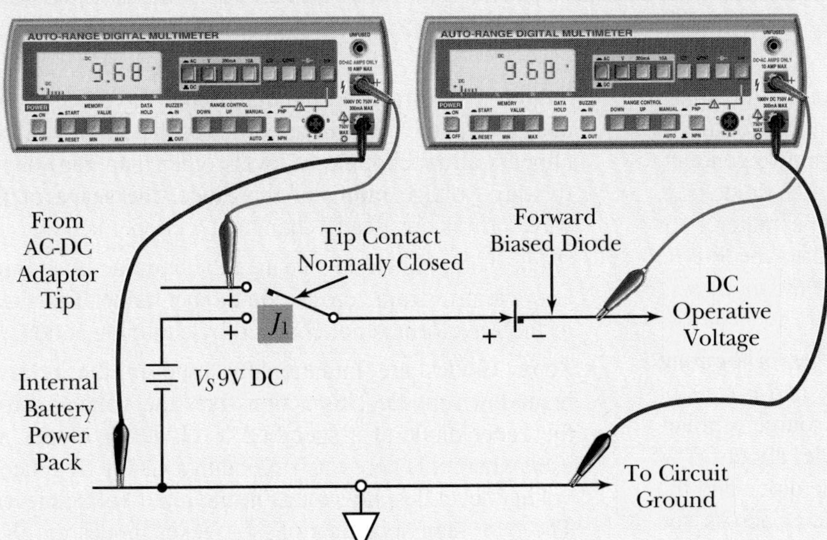

FIGURE 3 Circuit schematic

Will the output be a signal, a display, or an action such as the movement of a mechanical device?

One of the most common uses for a simple diode is voltage and polarity sensing. Many of the battery-operated products mentioned previously incorporate a DC IN jack. This jack is usually not identified with the required dc voltage but does usually have a symbol for identifying the polarity of the required dc voltages. The larger outer circle indicates the ring of the ac-dc adaptor jack; the smaller inner circle or dot represents the tip of the jack as shown in Figure 1. The DC IN jack incorporates a spring loaded tip contact usually in the normally closed position.

The system application block diagram in Figure 2 shows system inputs and outputs along with the switching circuit and the voltage and polarity sensing circuit. The switching circuit is activated by the inserting of the ac-dc adaptor plug into the DC IN jack. The selected dc voltage from either the ac-dc adaptor if inserted or from the internal battery power pack is connected by the switching circuit to the voltage and polarity sensing circuit. Only the presence of the dc voltage of the proper polarity, in this case a positive, is passed through to the battery-operated product.

The schematic in Figure 3 shows the details of the ciruit. The spring loaded tip contact is normally closed. This allows the internal battery pack to supply the proper voltage to the device. Inserting the ac-dc adaptor plug opens the normally closed movable contact and connects it mechanically to the tip of the plug. This first opens the power circuit from the internal battery power pack, then reconnects the power circuit to the ac-dc adaptor.

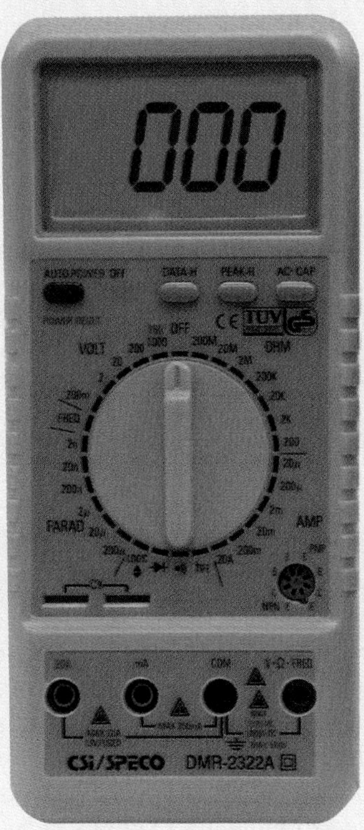

Summary

- The P-N junction creates a diode. (The prefix "di" indicates two parts.) By applying an external bias voltage to the diode, it can be made to conduct or not to conduct. Conduction is caused by forward biasing the diode (i.e., negative voltage connected to the N-type material or cathode, and positive voltage connected to the P-type material or anode). When reverse biased, the diode will not conduct and acts as an open switch.
- There is a definite distribution of voltages in a circuit consisting of a dc power source, diode, and a resistor connected in series. When the dc power source is polarized such that it forward biases the diode, about 0.7 V (forward voltage drop) appears across the diode and the remainder of the source voltage appears across the series resistor. The current through the forward-biased circuit can be calculated by dividing the resistor value into the source voltage minus the forward drop across the diode. When the power source is connected so as to reverse bias the diode, there is virtually no current flow through the circuit; the full source voltage can be found across the diode, and zero voltage across the resistor.
- The three important diode specifications (or ratings) are forward voltage drop (V_F), average forward current (I_O), and peak reverse voltage (V_R). The forward voltage drop, also known as the barrier potential, is the voltage found across the diode when it is forward biased and fully conducting. This value is about 0.7 V for silicon diodes. Average forward current is the maximum amount of forward-bias current a diode can carry indefinitely. Operating a diode outside these specifications for extended periods will eventually destroy it.
- Rectifier diodes are designed for converting ac power to dc power. They can have relatively high current ratings (e.g., 10A) and heat sinks are often used to cool the diodes where this is required.
- Rectifier and switching diodes can be tested with an ohmmeter. A diode in good working order will show a much higher resistance when it is reverse biased by the ohmmeter connected than when it is forward biased. A diode is defective whenever you find the forward and reverse ohmmeter readings in the same general range.
- Switching diodes usually have lower current and voltage ratings than rectifier diodes. However, they have much faster reverse recovery times. This means switching diodes recover much faster from reverse-bias conditions than rectifier diodes do.
- Diode clipper circuits (also called limiters) remove all or a portion of one polarity from an input waveform. In a series clipper, the diode is connected in series with the output of the circuit. In a shunt clipper, the diode is connected in parallel with the output of the circuit. Biased clippers allow clipping at levels other than zero or the forward voltage rating of the diode. the shape of the waveform is drastically changed by clipper action.
- Diode clamping circuits change the baseline of an input ac waveform from zero to some other value. The shape of the waveform is not changed by clamping action.
- Zener diodes are intentionally used in the reverse-breakdown mode. This action fixes the voltage across the zener diode at a specified level. Zener diodes are used where it is necessary to obtain a steady (regulated) voltage level despite swings in the input voltage levels. The two main specifications for zener diodes are their reverse voltage drop (V_Z) and their maximum continuous power.
- A tunnel diode has a negative-resistance I-V curve that makes it useful for generating high frequency waveforms.
- The varactor diode acts like a small capacitor that is variable by means of changing the amount of reverse dc bias applied to it. The larger the reverse-bias voltage, the smaller the amount of capacitance.
- The Schottky diode is specially constructed for ultra-fast switching times. It is found in high frequency circuits and digital IC devices.

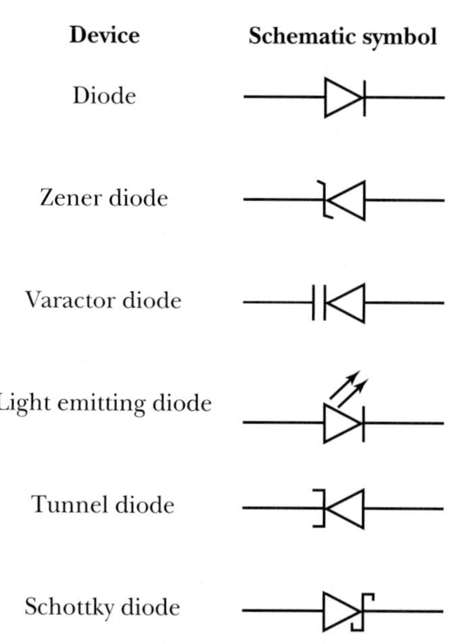

FIGURE 3–38 Diode summary

CIRCUIT SUMMARY

Description	Circuit	Waveform/Formula
Forward bias diode circuit	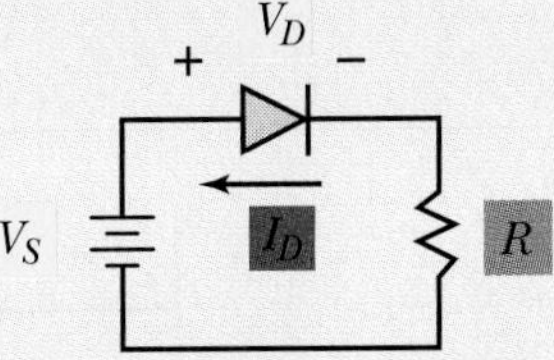	For $V_S > 0.7\text{ V}$ $V_D = 0.7\text{ V}$ $I_D = \frac{V_S - V_D}{R}$ $P_D = I_D \times V_D$
Reverse bias diode circuit	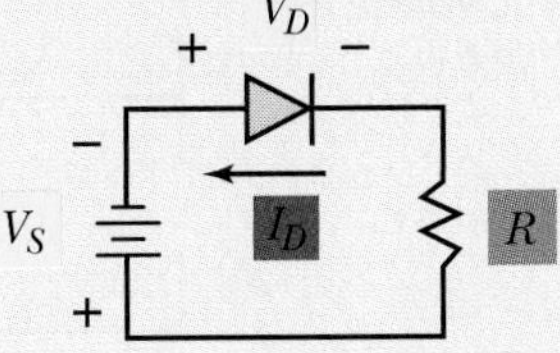	$I_D = 0\text{A}$ $V_D = -V_S$ $P_D = 0\text{W}$
Negative diode clipper series circuit	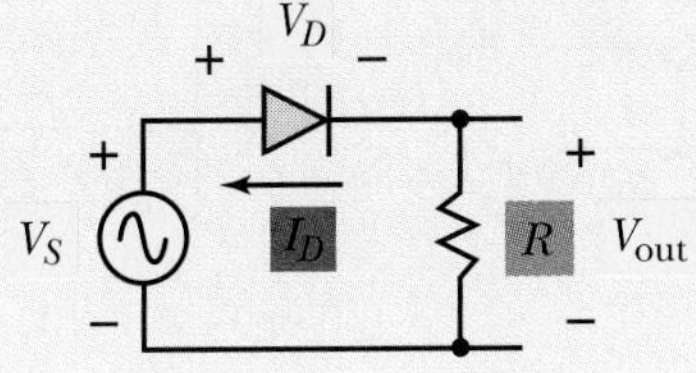	V_S: V_{pk}, $-V_{pk}$ V_{out}: $V_{pk} - 0.7\text{ V}$
Positive diode clipper series circuit	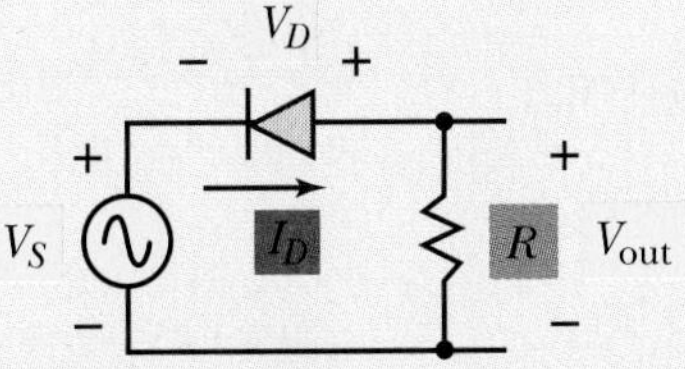	V_S: V_{pk}, $-V_{pk}$ V_{out}:  $-V_{pk} + 0.7\text{ V}$
Positive diode clipper (shunt) circuit	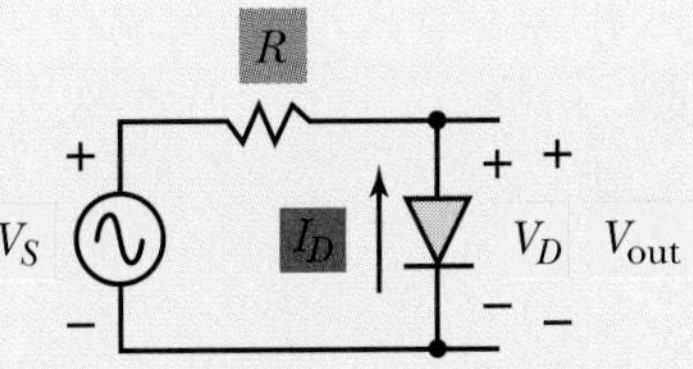	V_S: V_{pk}, $-V_{pk}$ V_{out}: 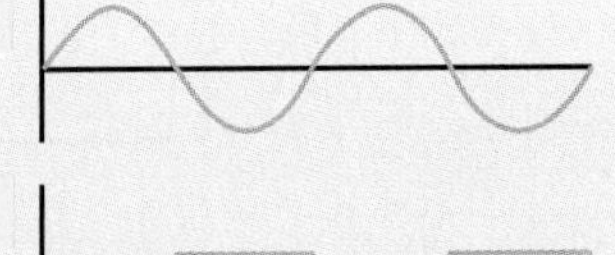 0.7 V, $-V_{pk}$
Negative diode clipper (shunt) circuit	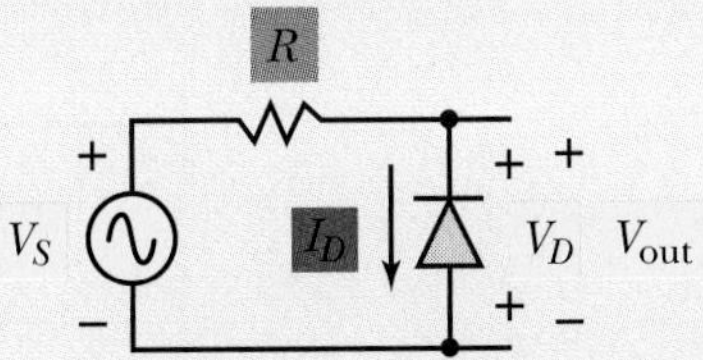	V_S: V_{pk}, $-V_{pk}$ V_{out}: V_{pk}, -0.7 V
Biased positive shunt diode clipper circuit	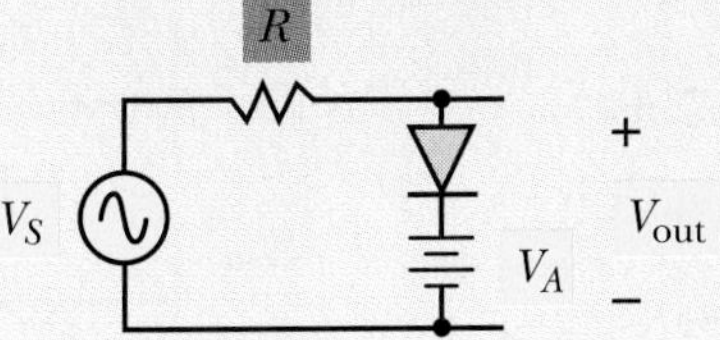	V_S: V_{pk}, $-V_{pk}$ V_{out}: 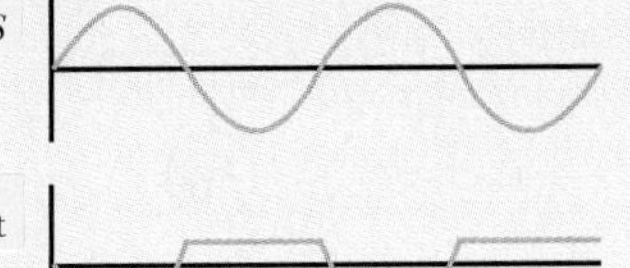 $V_A + 0.7\text{ V}$, $-V_{pk}$
Biased negative shunt diode clipper circuit	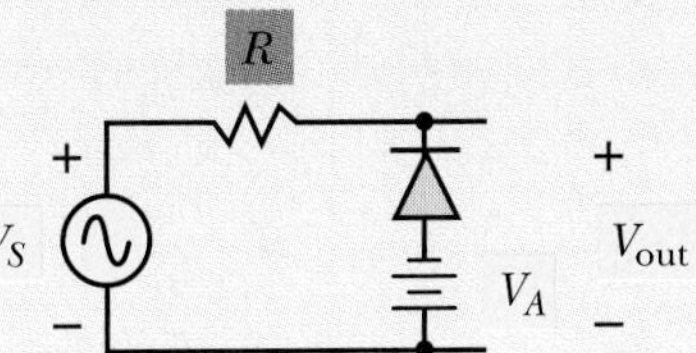	V_S: V_{pk}, $-V_{pk}$ V_{out}: 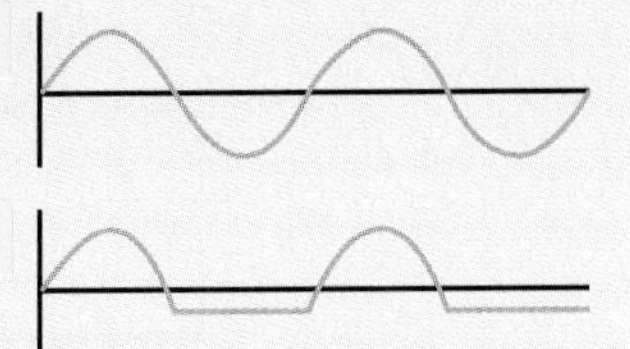 V_{pk}, $-V_A - 0.7\text{ V}$

FIGURE 3–39 Diode circuit summary

CIRCUIT SUMMARY

Description	*Circuit*	*Waveform/Formula*
Positive diode clamping circuit	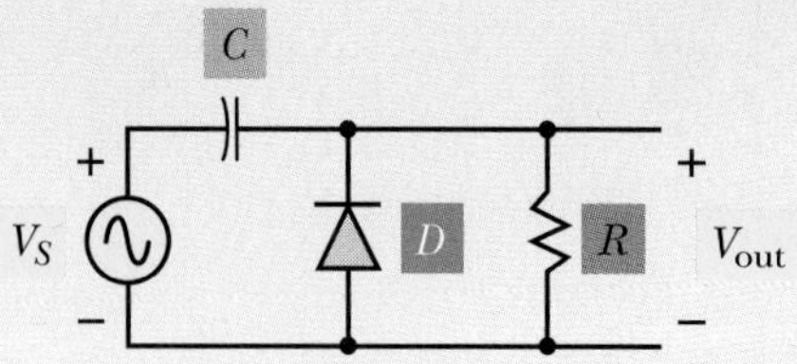 C, D, R, V_S, V_{out}	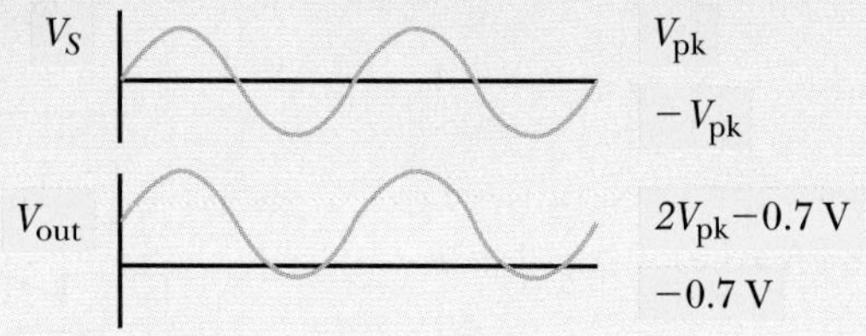 V_S: V_{pk}, $-V_{pk}$; V_{out}: $2V_{pk} - 0.7\text{ V}$, -0.7 V
Negative diode clamping circuit	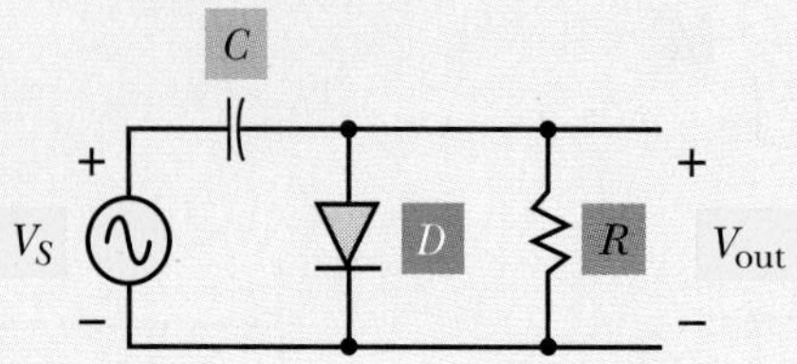 C, D, R, V_S, V_{out}	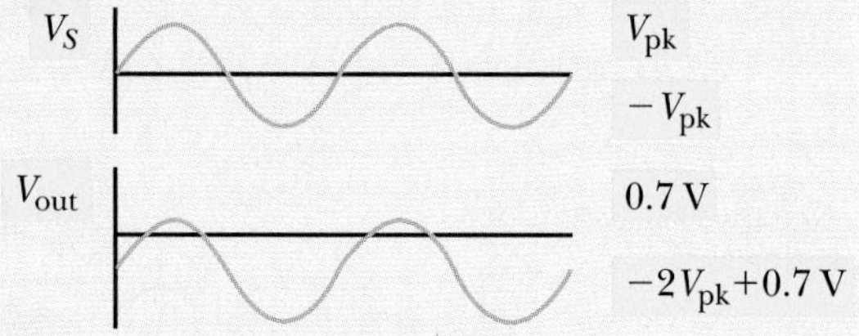 V_S: V_{pk}, $-V_{pk}$; V_{out}: 0.7 V, $-2V_{pk} + 0.7\text{ V}$
Zener diode circuit	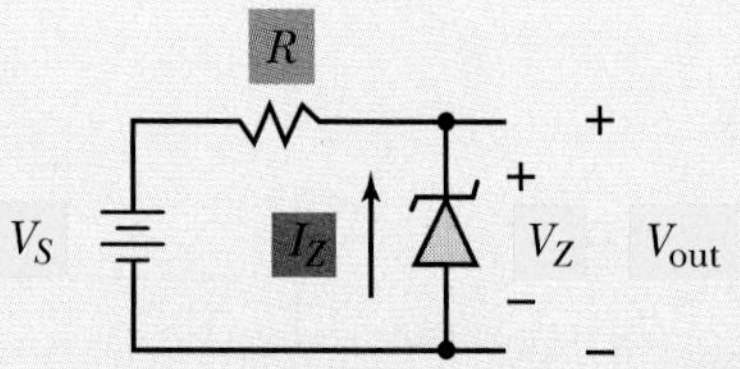 R, V_S, I_Z, V_Z, V_{out}	For $V_S < V_Z$, $V_{out} = V_S$ For $V_S \geq V_Z$, $V_{out} = V_Z$ $P = I_Z \times V_Z$ Formula 3 − 3
Zener diode voltage regulator circuit	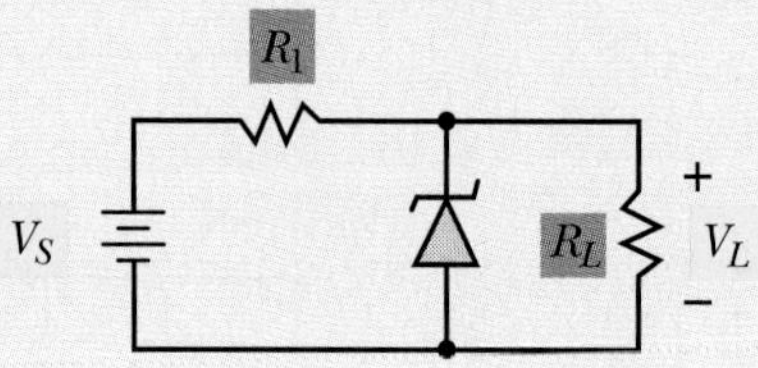 R_1, V_S, R_L, V_L	For $V_S < V_Z \times \frac{R_1 + R_L}{R_L}$ $V_L = V_S \times \frac{R_L}{R_L + R_1}$ Formula 3 − 4 For $V_S \geq V_Z \times \frac{R_1 + R_L}{R_L}$ $V_L = V_Z$
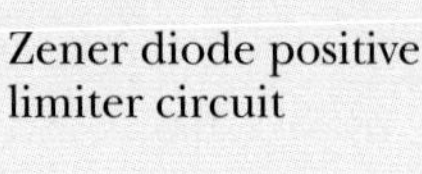 Zener diode positive limiter circuit	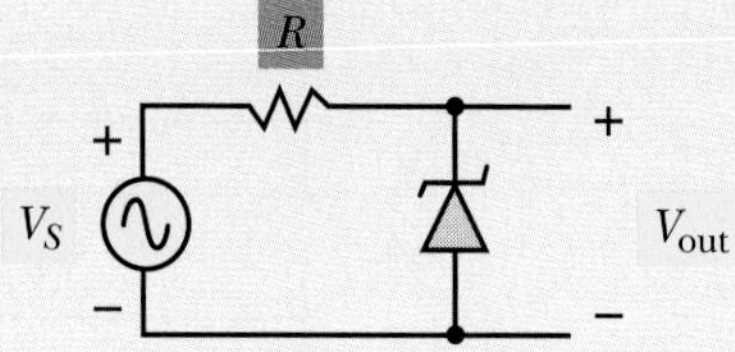	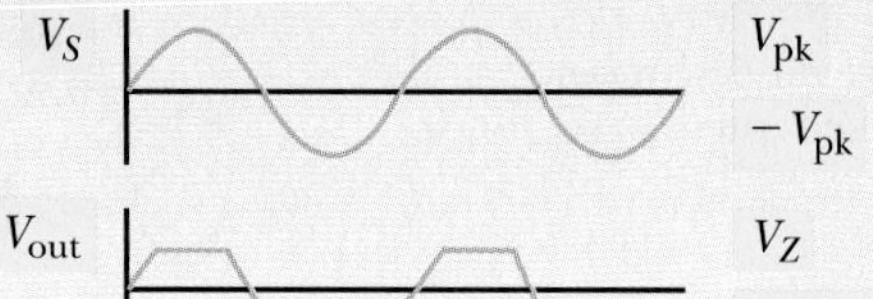
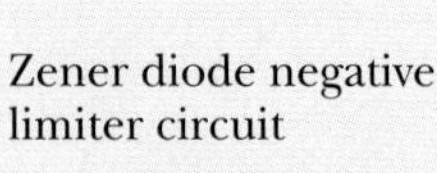 Zener diode negative limiter circuit	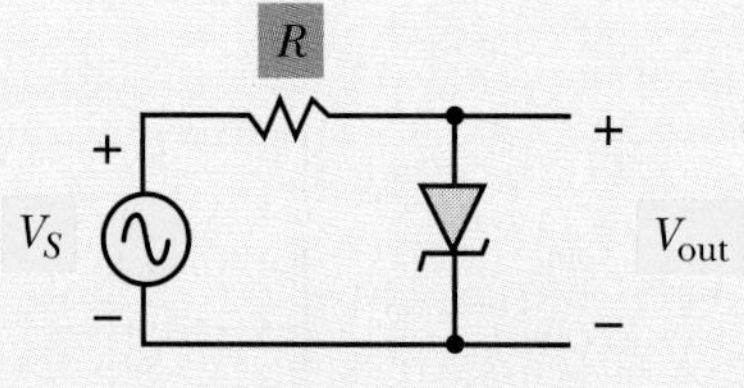	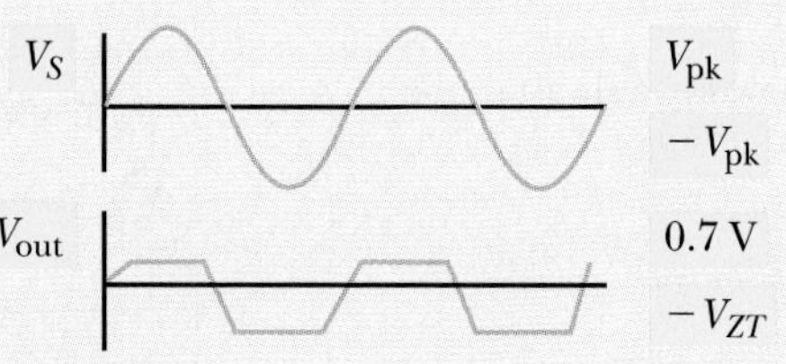
 LED circuit	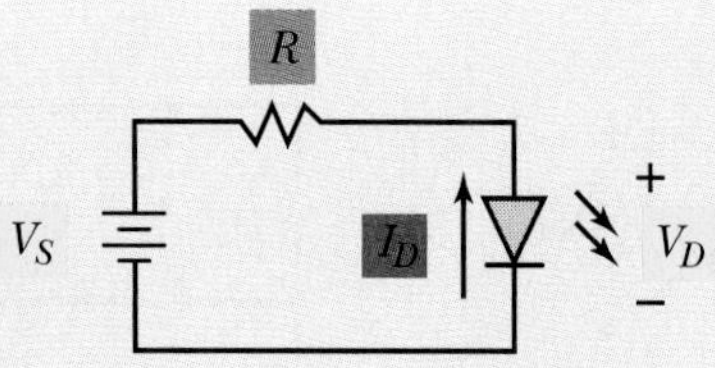	$V_R = V_s - V_D$ Formula 3 − 6 $R = \frac{V_R}{I_D}$

FIGURE 3–39 (Continued)

Formulas and Sample Calculator Sequences

FORMULA 3–1
(To find series resistor and diode current)

$$I = \frac{(V_S - V_D)}{R}$$

(, V_S value, [−], V_D value,), [÷], R value, [=]

FORMULA 3–2
(To find diode power)

$P = I \times V_D$

I value, [×], V_D value, [=]

FORMULA 3–3
(To find zener diode power)

$P = I_Z \times V_Z$

I_Z value, [×], V_Z value, [=]

FORMULA 3–4
(To find regulation transition point supply voltage)

$$V_S = V_Z \times \frac{(R_1 + R_L)}{R_L}$$

V_Z value, [×], (, R_1 value, [+] R_L value,), [÷], R_L value, [=]

FORMULA 3–5
(To fine unregulated load voltage)

$$V_L = V_S \times \frac{R_L}{(R_1 + R_L)}$$

V_S value, [×], R_L value, [÷], (, R_1 value, [+], R_L value,), [=]

FORMULA 3–6
(To find series resistor voltage drop)

$V_R = V_S - V_D$ Voltage drop across the series resistor

V_S value, [−], V_D value, [=]

Using Excel

Diode Formulas

(Excel file reference: FOE3_01.xls)

DON'T FORGET! It is NOT necessary to retype formulas, once they are entered on the worksheet! Just input new parameters data for each new problem using that formula, as needed.

- Use the Formula 3–1 spreadsheet sample and the parameters given for Chapter Problems #1 question **a.** Solve for the total current flowing through the circuit. Check your answer against the answer for this question in the Appendix.
- Use the Formula 3–2 spreadsheet sample and the parameters given for Chapter Problems #1. Solve for the power dissipated by the diode.

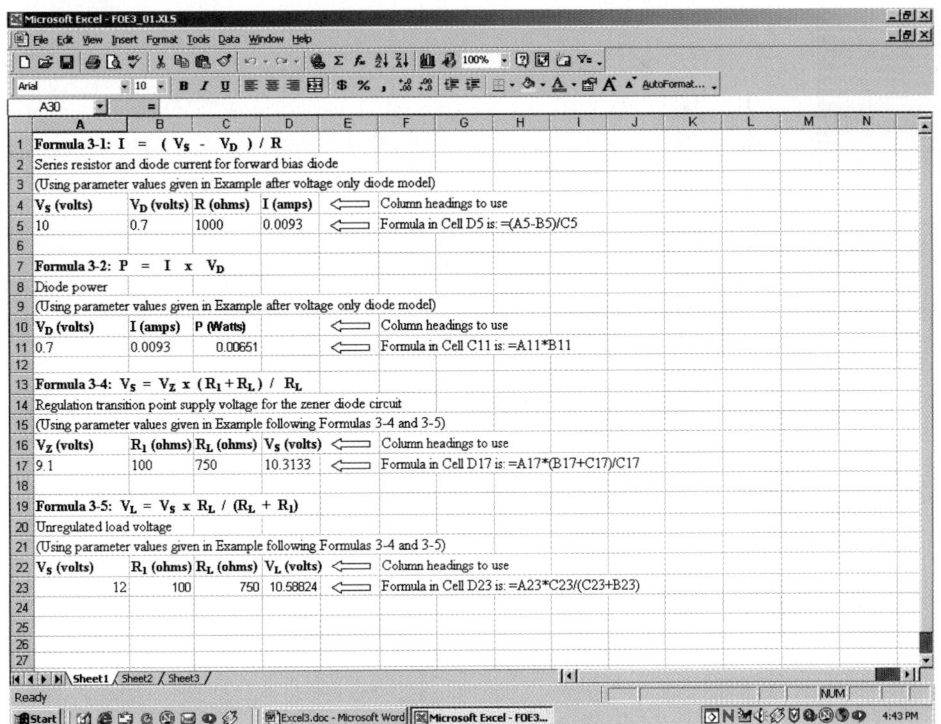

Review Questions

1. Draw two diode circuits, one showing the diode as it is reverse biased and the second as it is forward biased. Label the drawings as appropriate.
2. Indicate the polarity of the voltages dropped across the resistor and diode in Figure 3–40.
3. Modify the circuit in Figure 3–40 by reversing the polarity of the dc source. Indicate on your drawing the polarity and amount of voltage across the diode and resistor. Show the amount of current flowing through the circuit.
4. Sketch the waveform found across the resistor in Figure 3–41. What is the maximum amount of reverse voltage found across the diode?
5. Sketch a circuit for a series diode clipper that clips the positive levels from the input ac waveform.
6. Sketch a circuit for a shunt diode clipper that clips the negative levels from the input ac waveform.
7. Describe how a diode clamp circuit affects an ac input waveform.
8. Explain the difference between the action of a biased diode clipper and a diode clamping circuit.
9. Explain the following diode specifications:
   - **a.** Reverse breakdown voltage
   - **b.** Forward conduction voltage
   - **c.** Average forward current
   - **d.** Reverse recovery time
10. Describe the condition necessary for a zener diode to hold its rated V_Z level.
11. Suppose the source voltage in Figure 3–42 increases. What will happen to the voltage across the zener diode? The resistor? How will this increase in applied voltage affect the current through the circuit?
12. What is the peculiar feature of the I-V curve for tunnel diodes?
13. Draw at least one schematic symbol for a tunnel diode.
14. Name two important applications of tunnel diodes.
15. Describe how a varactor changes capacitance with changing levels of reverse bias.
16. Describe how Schottky diodes are different from most other kinds of semiconductor diodes.

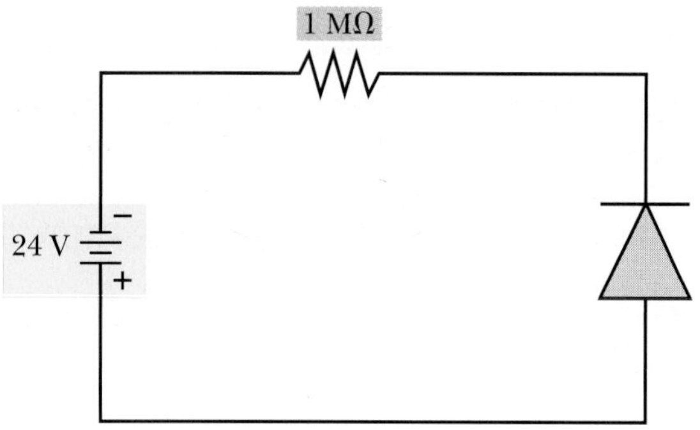

FIGURE 3–40

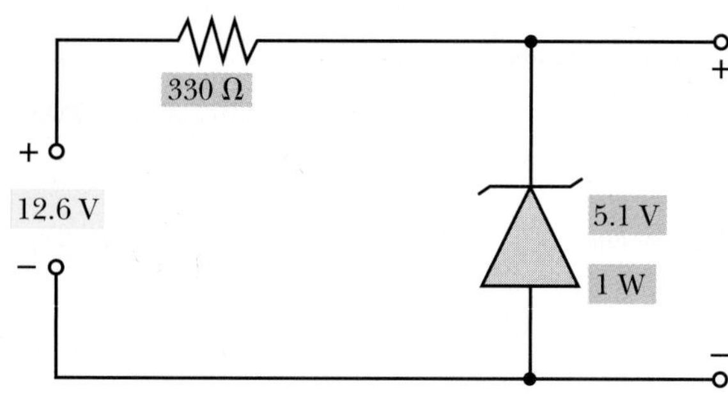

FIGURE 3–42

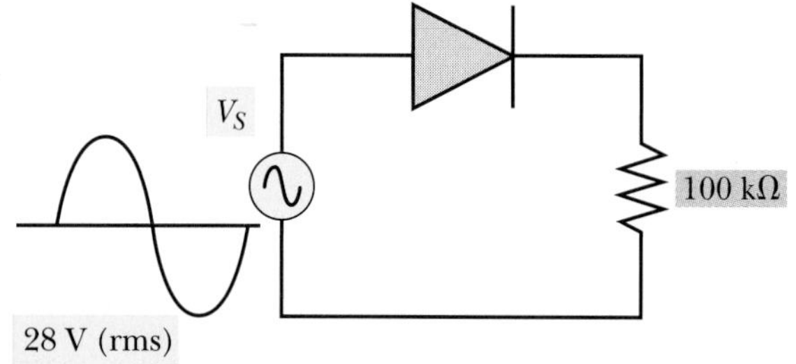

FIGURE 3–41

Problems

1. Refer to the circuit in Figure 3–43. $V_S = 6$ V, $R_S = 1$ kΩ, and the diode is a silicon diode. Determine:
   a. Is the diode forward biased or reverse biased?
   b. The total current flowing through the circuit.
   c. The voltage drop across the resistor R.
   d. The forward current through the diode.
2. Refer to the circuit in Figure 3–44. $V_S = 9$ V, $R = 10$ kΩ, and $V_F = 0.7$ V. Determine:
   a. Is the diode forward biased or reverse biased?
   b. The total current flowing through the circuit.
   c. The voltage across the resistor R.
   d. The voltage across the diode.
3. Refer to the circuit in Figure 3–45. $V_S = 12$ V, $R = 200$ Ω, and D_1 is a germanium diode. Determine:
   a. Is the diode forward biased or reverse biased?
   b. The total current flowing through the circuit.
   c. The voltage across the resistor R.
   d. The voltage across the diode.
4. Refer to the circuit in Figure 3–46. $V_S = 10$ V, $R = 200$ Ω, and $V_F = 0.3$ V. Determine:
   a. Is the diode forward biased or reverse biased?
   b. The total current flowing through the circuit.
   c. The voltage across the resistor R.
   d. The voltage across the diode
5. A silicon diode is used in the circuit in Figure 3–43. The supply voltage V_S is 12 V. Determine:
   a. The value for R that will limit the diode current to 500 mA.
   b. The power dissipation of R.
6. A germanium diode is used in the circuit in Figure 3–46. The supply voltage V_S is 5 V. Determine:
   a. The value for R that will limit the diode current to 500 mA.
   b. The power dissipation of R.

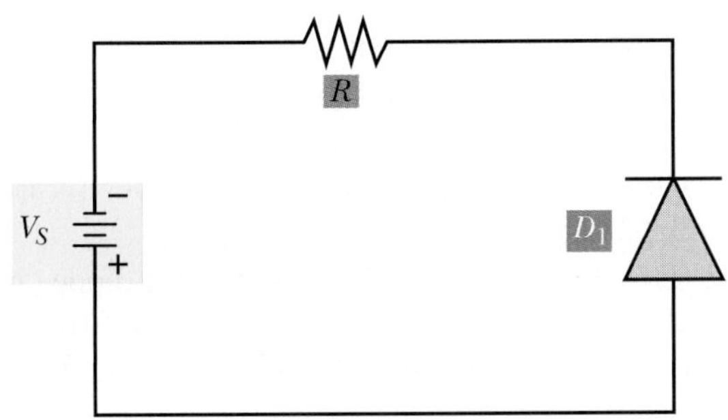

FIGURE 3–43

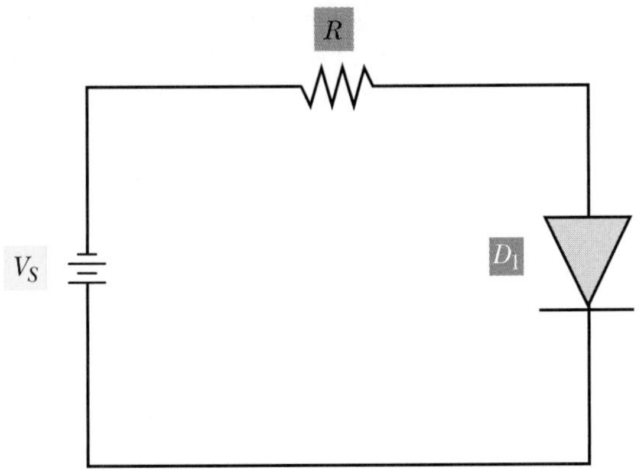

FIGURE 3–44

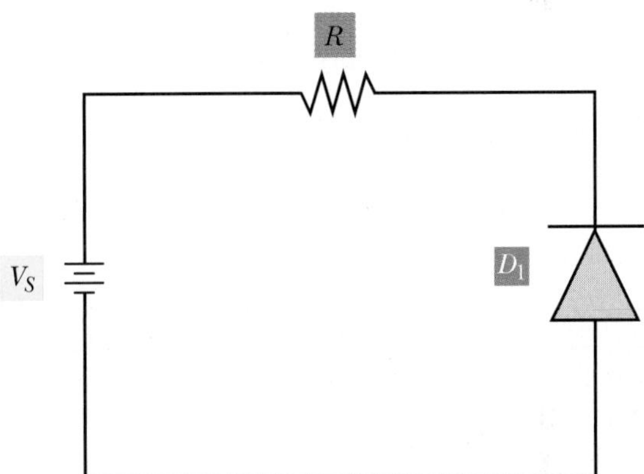

FIGURE 3–45

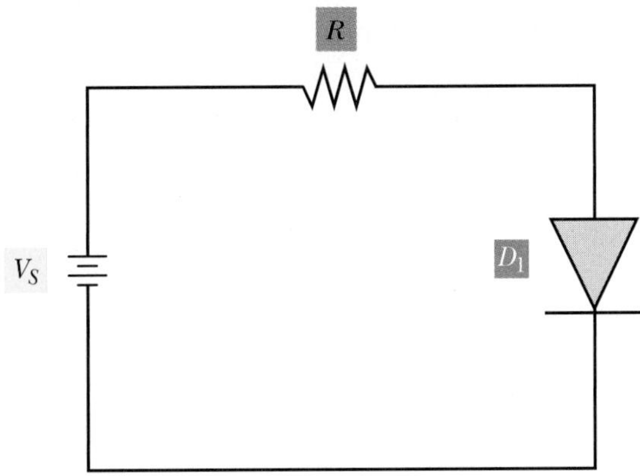

FIGURE 3–46

7. The maximum current rating for the diode used in the circuit in Figure 3–43 is 1 A. The resistor R is 100 Ω. Calculate the maximum supply voltage V_S that can be applied to the circuit without exceeding the diode's current rating.

8. A silicon switching diode has a maximum current rating of 80 mA. The supply voltage V_S is 5 V.

a. Calculate the lowest value resistor that can be used without exceeding the diode's current rating.

b. Select a standard value resistor that can be used.

c. Calculate the power dissipated by the standard value resistor.

9. A 10-V_{pp} sine wave is applied to the circuit in Figure 3–17a. A 1-kΩ resistor is used with the silicon diode. Sketch the supply voltage and output voltage waveforms.

10. A 10-V_{pp} sine wave is applied to the circuit in Figure 3–18a. A 10-kΩ resistor is used with the silicon diode. Sketch the supply voltage and output voltage waveforms.

11. A 5-V_{pp} sine wave is applied to the circuit in Figure 3–19a. A 2-kΩ resistor is used with the silicon diode. Sketch the supply voltage and output voltage waveforms.

12. A 5-V_{pp} sine wave is applied to the circuit in Figure 3–20a. A 470-Ω resistor is used with the silicon diode. Sketch the supply voltage and output voltage waveforms.

13. A 16-V_{pp} sine wave is applied to the circuit in Figure 3–21a. A 6-V battery is connected in the circuit, which contains a silicon diode and a 1-kΩ resistor. Sketch the supply voltage waveform and the output signal waveform.

14. A 12-V_{pp} sine wave is applied to the circuit in Figure 3–22a. A 3-V voltage source is connected in the circuit, which contains a silicon diode and a 330-Ω resistor. Sketch the supply voltage waveform and the output signal waveform.

15. A 10-V_{pp} sine wave is applied to the circuit in Figure 3–23a, which contains a silicon diode and a 2-kΩ resistor. Sketch the supply voltage waveform and the output signal waveform.

16. A 8-V_{pp} sine wave is applied to the circuit in Figure 3–24a, which includes a silicon diode and a 1-kΩ resistor. Sketch the supply voltage waveform and the output signal waveform.

17. The 1N4740A zener diode is used in the circuit in Figure 3–47. The resistor has a value of 390 Ω.

a. When the supply voltage V_S is 8 V, what are the values for the output voltage and zener current?

b. Find the output voltage, zener current, and zener power dissipation when the supply voltage V_S is 20 V.

18. The 1N4735A zener diode is used in the circuit in Figure 3–27. The resistor has a value of 150 Ω. The supply voltage V_S is 12 V. Calculate the output voltage, zener current, and zener power dissipation.

19. Determine the following value for the zener diode circuit in Figure 3–47 where $V_S = 20$ V, $R = 100$ Ω, and D_1 = IN4742A:

a. voltage drop across the zener diode

b. voltage drop across the resistor

c. current through the circuit

d. the power dissipation of D_1.

20. For the circuit in Figure 3–47, let $V_S = 12$ V, $R = 120$ Ω, and $V_Z = 9$ V. Determine:

a. the voltage across R.

b. the current through the circuit.

c. the power dissipation of R.

d. the power dissipation of D_1.

21. Referring to the circuit in Figure 3–47, let $V_S = 12$V and $V_Z = 9$ V. If the current through the circuit is to be set at 1 A, determine:

a. the value of R.

b. the voltage across R.

c. the power dissipation of R.

d. the power dissipation of D_1.

22. Suppose the zener diode in Figure 3–47 is rated at 6.3 V, the dc power source is 9 V, and the value of $R = 680$ Ω. Determine the amount of power dissipated by the zener diode.

23. Suppose the dc source voltage in a zener diode circuit (such as in Figure 3–43) increases by 2 V (from 12 V to 14 V). How much will the voltage across the resistor increase? How much will the voltage across the zener diode increase?

24. The circuit in Figure 3–34 has a supply voltage V_S of 5 V with a 200-Ω resistor and an LED forward voltage drop of 2.0 V. Calculate the voltage across the current limiting resistor (V_R), the current flowing in the circuit, and the power dissipated by the resistor.

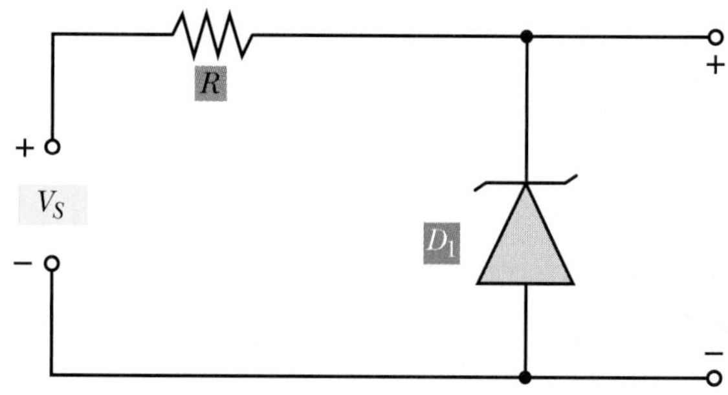

FIGURE 3–47

Analysis Questions

1. Only one of the diodes in the circuit in Figure 3-48 will conduct. The conducting diode will carry the full current load, and the other will carry no current. Explain why this happens. (*Hint:* There are slight variations in the forward conduction voltage.) Explain why connecting a 5-Ω resistor in series with each diode will allow both to conduct, each carrying about half the current load.
2. Use an electronics supply catalog to locate the part number for the rectifier diode that has the highest peak reverse voltage. Also locate the diode having the highest average current rating.
3. Consult an electronics supply website or catalog for the highest and lowest values for zener diode voltage and power.
4. Sketch the waveform you might see across the zener diode in Figure 3–47 if the zener is rated at 6 V, and the de voltage source is replaced with an 8-V peak-to-peak ac source.
5. Research the basic circuits and range of operating frequencies for tunnel diode oscillators.
6. Use an electronics supply website or catalog to determine the specifications that are important for selecting Schottky diodes.
7. Use a dictionary or glossary of technical terms to describe the difference between discrete and integrated electronic devices.
8. When the switch is closed in the circuit in Figure 3–49, the motor is conducting 0.5 A current. Evaluate and explain what happens when the switch opens for the following situations:
   a. No diode is connected across the motor.
   b. The 1N4001 diode is connected across the motor.
   c. A 10-Ω resistor is connected in series with the 1N4001 diode across the motor.

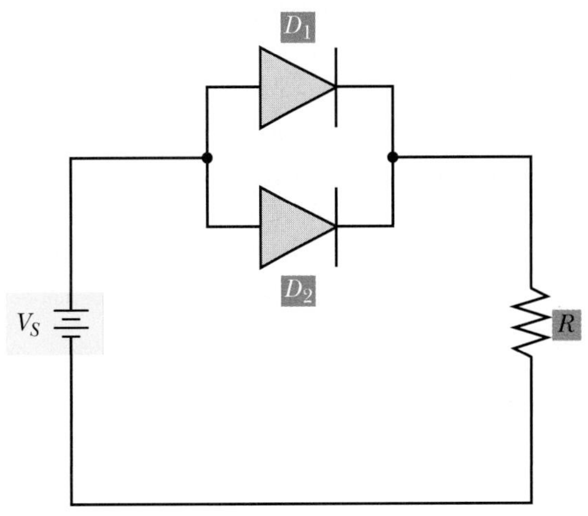

FIGURE 3–48

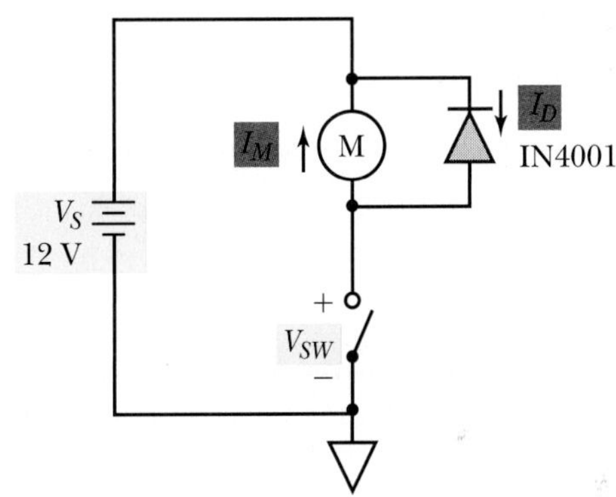

FIGURE 3–49 Motor control circuit

MultiSIM Exercise for Resistor-Diode *Series Circuits*

1. Use the MultiSIM program and utilize the circuit shown in Figure 3–17a.
2. Set the function generator to a 12 V peak sine wave at 1 kHz.
3. Measure and record the following parameters:
   a. Input waveform
   b. Output waveform
4. Compare your MultiSIM results with the values given in the Instructor Guide for this circuit. Were the input and output waveforms made with MultiSIM reasonably close to the Figure 3–17b and Figure 3–17c waveforms?

Datasheet Challenge Problems

1. List at least three manufacturers of the 1N4148 diode.
   Approach Options:
   1) Start with *www.digikey.com*
      Enter component part number
   2) Start with *www.freetradezone.com*
      Enter component part number
2. List at least three manufacturers of the 1N4004 diode.
3. List at least three manufacturers of the 1N4733A zener diode.
4. Compare the 1N4148 datasheets from two different manufacturers. (Focus on forward voltage drop, forward current, breakdown voltage, and power dissipation.)
5. Compare the 1N4004 datasheets from two different manufacturers.
6. Compare the 1N4733A datasheets from two different manufacturers.
7. Visit *www.diodes.com.* List and compare at least three different switching diodes.
8. Visit *www.diodes.com.* List and compare at least three different rectifier diodes.
9. Obtain part number and datasheet information for a 6.2-V zener diode. (Include manufacturer and data source.)
10. Research and describe the following; manufacturer, sales or manufacturer representative, and distributor.

Performance Projects Correlation Chart

Labs are available on the accompanying Lab Source CD.

Chapter Topic	Performance Project	Project Number
P-N Junction Diodes	Forward and Reverse Bias, and *I* vs. *V*	1
Diode Clipper Circuits	Diode Clipper Circuits	2
Zener Diodes	Zener Diodes	3

NOTE: It is suggested that after completing the above projects, the student should be required to answer the questions in the "Summary" at the end of this section of projects in the Laboratory Manual.

Troubleshooting Challenge

CHALLENGE CIRCUIT 1

(Follow the SIMPLER sequence by referring to inside front cover.)

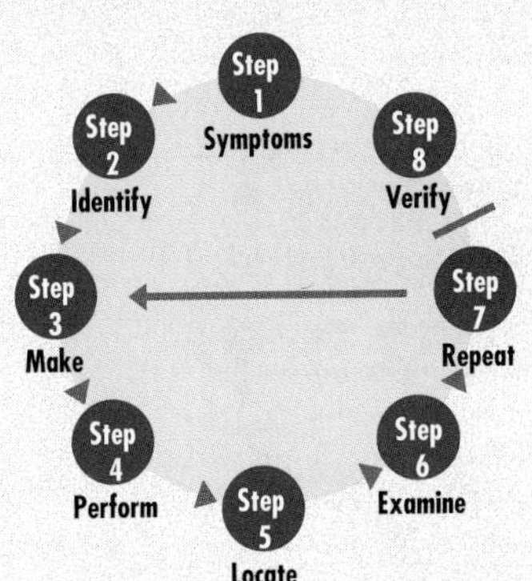

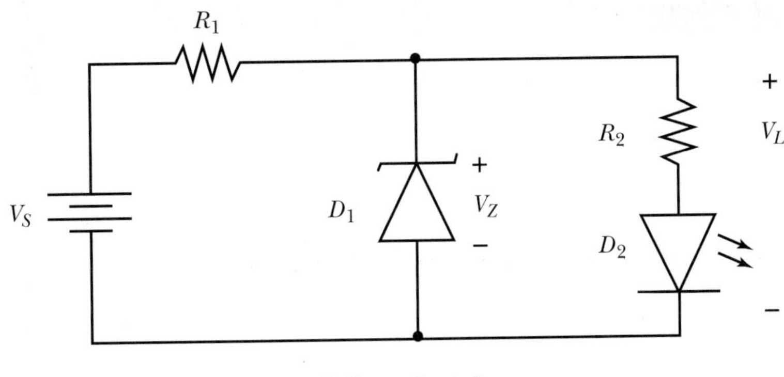

Challenge Circuit 1

STARTING POINT INFORMATION

1. Circuit diagram ($R_1 = 100\ \Omega$, $R_2 = 470\ \Omega$)
2. $V_S = 25$ V
3. Zener diode 1N5347B ($V_Z = 10$ V)
4. Measured $I_T = 150$ mA
5. LED off

TEST	Results in Appendix C
I_T	....................(118)
V_L	....................(22)
V_D	....................(44)
R_S	....................(58)
I_L	....................(119)

CHALLENGE CIRCUIT 1

STEP [1]

SYMPTOMS The total current is correct. The LED is not lit.

STEP [2]

IDENTIFY initial suspect area: The LED is not powered or is burned out.

STEP [3]

MAKE test decision: **1st Test:** Check the load voltage.

STEP [4]

PERFORM Look up the test result. V_l is 10 V, which is correct.

STEP [5]

LOCATE new suspect area: The LED.

STEP [6]

EXAMINE available data.

STEP [7]

REPEAT Analysis and testing: New suspect area: Resistor R_2.
2nd Test: Use the DMM diode test to measure the LED forward voltage drop, which is normal.
3rd Test: Measure the resistance of R_2. R_2 is infinite ohms (open).

STEP [8]

VERIFY **4th Test:** Replace R_2 and note the circuit parameters. When this is done, the circuit operates properly.

2nd Test

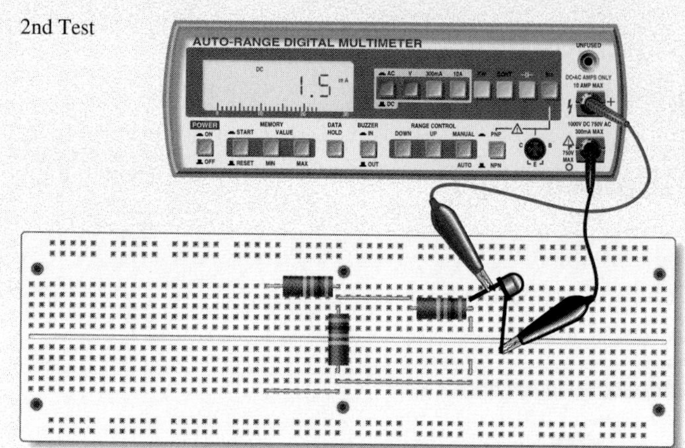

3rd Test

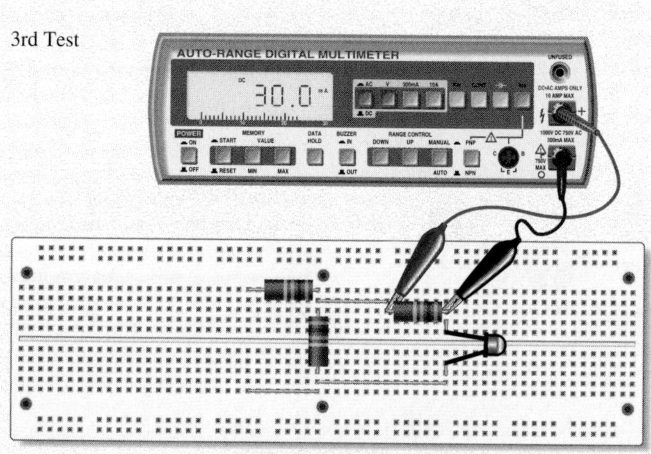

4th Test

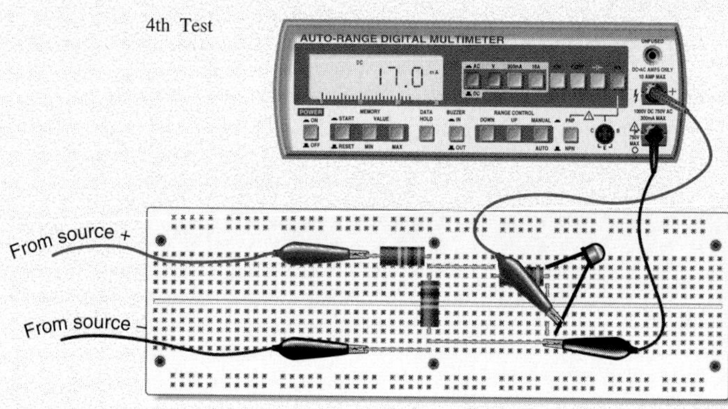

1st Test

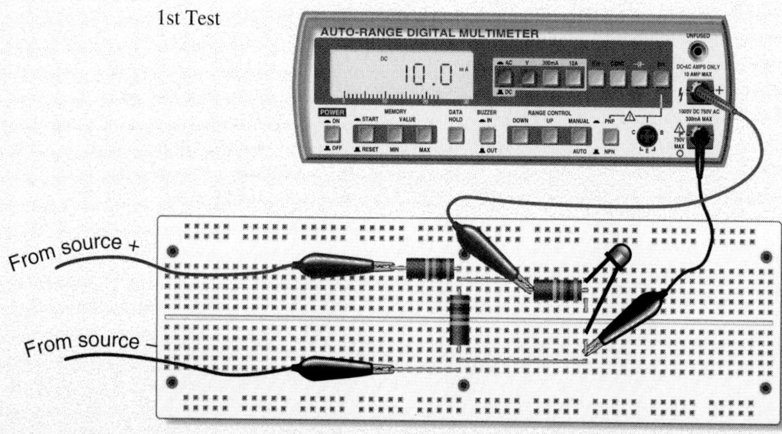

Troubleshooting Challenge

CHALLENGE CIRCUIT 2

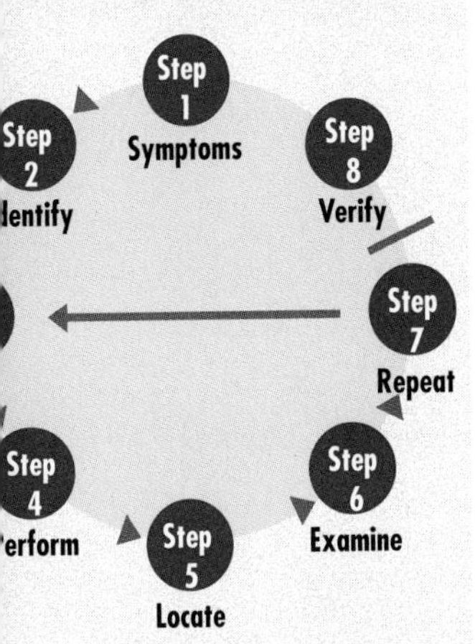

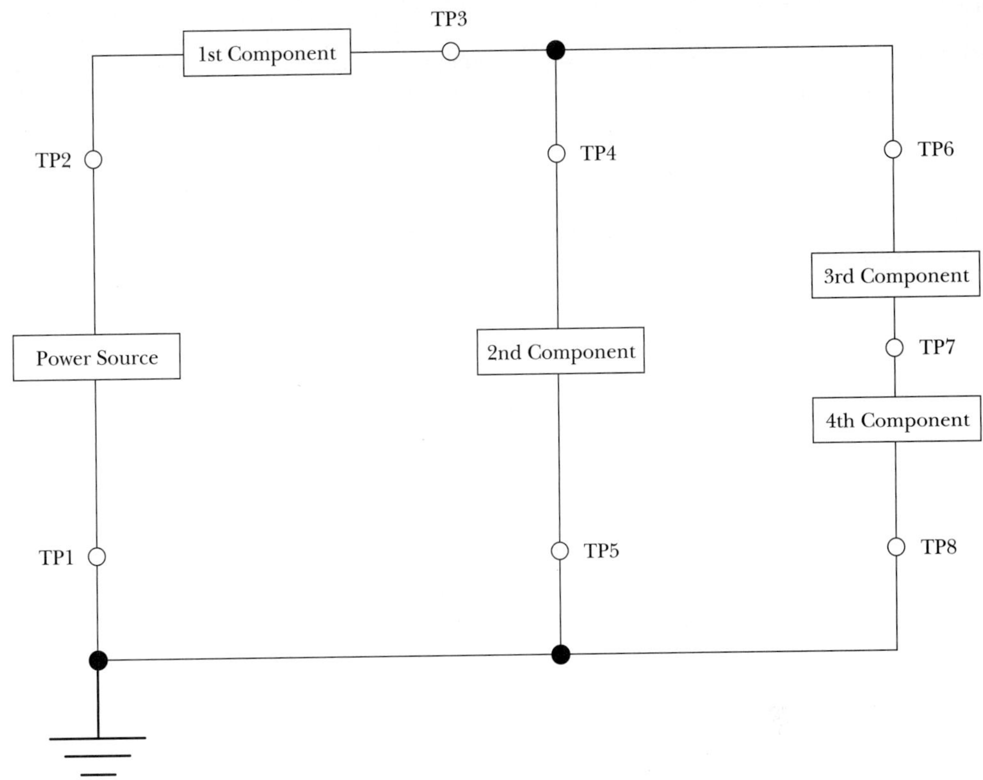

Challenge Circuit 2

General Testing Instructions:

Measurement Assumptions:
Signal = from **TP** to ground
I = at the **TP**
V = from **TP** to ground
R = from **TP** to ground (with the power source disconnected from the circuit)

Possible Tests & Results:
Signal (normal, abnormal, none)
Current (high, low, normal)
Voltage (high, low, normal)
Resistance (high, low, normal)

Starting Point Information	Test Points	Test Results in Appendix C *V*	*I*	*R*	*Signal*
At TP1:	TP2	(229	230	NA	NA)
	TP3	(231	232	NA	NA)
V = normal	TP4	(233	234	NA	NA)
I = low	TP5	(235	236	NA	NA)
R = high	TP6	(237	238	NA	NA)
S = NA	TP7	(239	240	NA	NA)
	TP8	(241	242	NA	NA)

STEP [1]

SYMPTOMS At TP1, voltage is normal, current is normal, and with the power source disconnected, resistance to ground measures normal. The LED is not lit.

STEP [2]

IDENTIFY initial suspect area. Since the supply voltage and total current are correct, the voltage at TP6 could be incorrect.

STEP [3]

MAKE test decision is based on the symptoms information. Let's start with the voltage measurements at strategic test points.

STEP [4]

PERFORM **1st Test:** Voltage value at TP6. Voltage measures normal. This indicates that the zener diode is regulating the voltage.

STEP [5]

LOCATE new suspect area. Components 3 and 4 are suspect.

STEP [6]

EXAMINE available data.

STEP [7]

REPEAT analysis and testing.

2nd Test: Isolate Component Block #4 from Component Block #3. Using the DMM diode test, measure the LED forward voltage (with power off). The LED voltage value is normal.

3rd Test: Isolate Component Block #3 from Component Block #4. Measure the resistance between TP6 and TP7 (with power off). The resistance measurement is higher than normal.

STEP [8]

VERIFY **4th Test:**When the 3rd Component Block is replaced with a new module that meets specifications, all circuit parameters come back to their norms.

Troubleshooting Challenge

CHALLENGE CIRCUIT 3
FIND THE FAULT

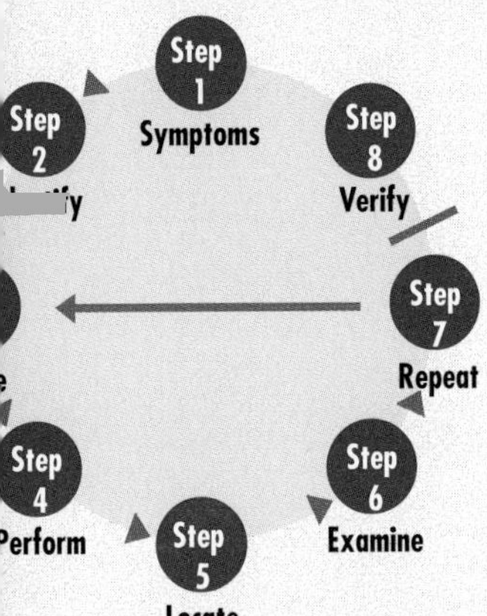

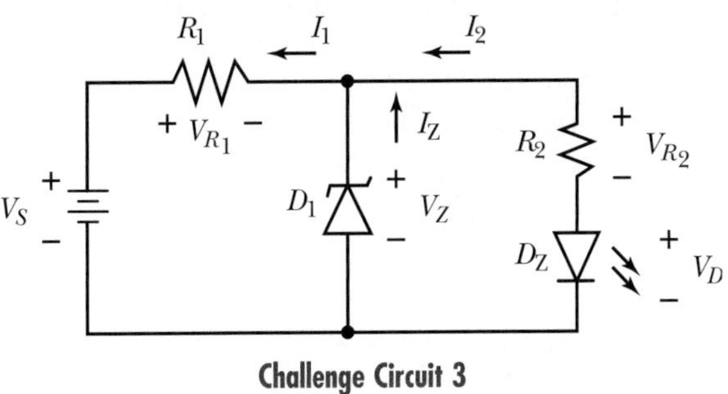

Challenge Circuit 3

STARTING POINT INFORMATION

1. Circuit diagram ($R_1 = 200\ \Omega$, $R_2 = 330\ \Omega$)
2. $V_S = 18$ V
3. Zener diode 1N4733A
4. LED voltage rating 1.5 V
5. LED is very bright

Test	Circuit Measurements	Theoretical Expected Value
V_S	18 V	______
V_{R_1}	6.23 V	______
V_{R_2}	10.27 V	______
V_D	1.5 V	______
V_Z	11.77 V	______
I_1	31.1 mA	______
I_2	31.1 mA	______
I_Z	0 mA	______
R_1	....	______
R_2	....	______

STEP [1]

You will analyze the circuit to calculate the expected voltages, currents, and resistances for a properly functioning circuit. Record these values beside each test listed in the table provided (Refer to Analysis Techniques)

STEP [2]

Starting with the symptom, document your steps for troubleshooting this circuit using the SIMPLER troubleshooting technique or instructions provided by your instructor.

STEP [3]

For each circuit measurement, you will indicate the test to be accomplished.

STEP [4]

Using the previous table, you will find the voltage and current measurements taken for the faulted circuit.

STEP [5]

You will now compare the faulted circuit measurement to the expected measurement for a properly functioning circuit. Are these values the same? (Refer to General Testing Instructions)

STEP [6]

You will repeat steps 2 through 5 until you have located the fault.

STEP [7]

When you have located the fault, you will identify the fault and the characteristics of the failure (Refer to Types of Failures)

ANALYSIS TECHNIQUES (CHECK WITH YOUR INSTRUCTOR)

METHOD [1]

Using circuit theory and algebra, evaluate the circuit.

METHOD [2]

Using an Excel spreadsheet for this circuit, evaluate the circuit.

METHOD [3]

Using MultiSIM, build the circuit schematic and generate the results.

METHOD [4]

Assemble this circuit on a circuit board. Apply power and make the appropriate measurements using your DMM and oscilloscope.

Troubleshooting Challenge

CHALLENGE CIRCUIT 3 (continued)

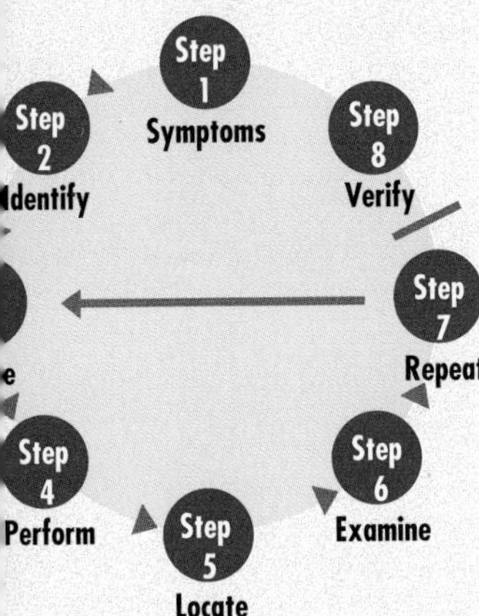

General Testing Instructions:

Measurement
Assumptions:
Signal = from **TP** to ground
I = at the **TP**
V = from **TP** to ground
R = from **TP** to ground (with the power source disconnected from the circuit)

Possible Tests & Results:
Signal (normal, abnormal, none)
Current (high, low, normal)
Voltage (high, low, normal)
Resistance (high, low, normal)

Types of Failures

Component failures
- wrong part value
- part shorted
- part open

Supply voltage—incorrect voltage
Ground—floating

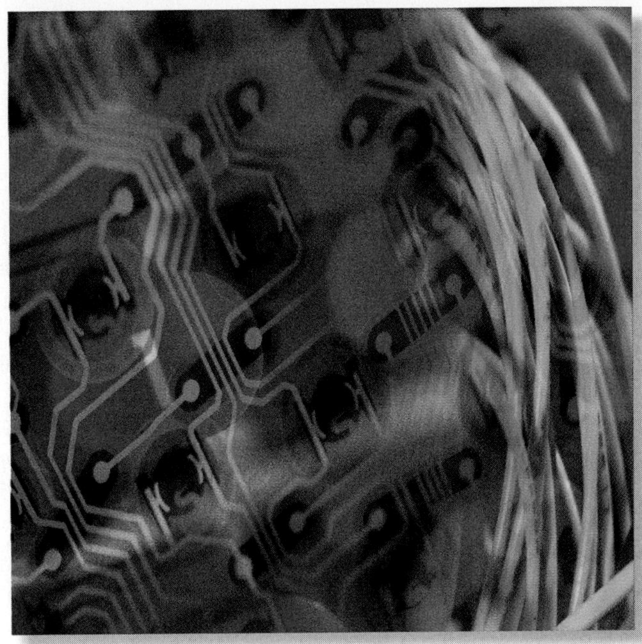

CHAPTER 4

Power Supply Circuits

OUTLINE

4–1 Elements of power supply circuits
4–2 Rectifier circuits
4–3 Basic power supply filters
4–4 Simple power supply regulators
4–5 Basic voltage multiplier circuits
4–6 Advanced power supply topics
4–7 Shunt regulators (zener diodes)
4–8 Series regulators
4–9 Switching regulators
4–10 Troubleshooting power supplies

OBJECTIVES

After studying this chapter, you should be able to:

1. List the basic elements of a power supply system
2. Draw the three basic types of rectifier circuits: **half-wave, center-tapped full-wave,** and **bridge circuits**
3. Explain the paths for current flow through the three basic types of rectifier circuits
4. Describe the waveforms found across the diode(s) and at the output of the three basic types of rectifier circuits
5. Determine the unfiltered dc output voltage of specified rectifier circuits
6. Briefly describe power supply **filter action**
7. Identify power supply filter configurations
8. Explain the purpose of a power supply **voltage regulator**
9. Recognize two **voltage multiplier** circuits
10. Briefly describe the most common types of power supply troubles

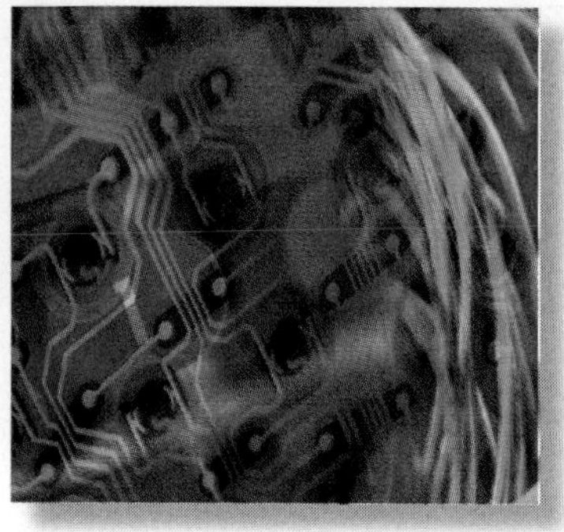

PREVIEW

The voltage supplied by power companies to our homes and businesses is ac voltage. As you learned earlier, this is because of the higher efficiency of transporting power from the power company to its customers by ac rather than dc. This efficiency is due to the step-up or step-down ability for ac voltages by using transformers, which minimize line losses.

Power supplies, which change ac source voltages to appropriate dc voltages, are found in most electronic equipment. Because of this, circuits associated with power supplies are a critical area that a technician should know. In this chapter, you will study each of the basic parts of a power supply. You will have a chance to apply directly what you have already learned about rectifier diodes as well as R, C, and L networks.

More complex power supplies have provisions for regulating their output voltage levels against swings in the input voltage. Also, there are instances where it is important to rectify and step up voltage levels with a voltage multiplier power supply.

KEY TERMS

Bleeder resistor
Bridge rectifier
Filter
Full-wave rectifier
Half-wave rectifier
Power supply
Pulsating dc
Rectifier circuit
Regulator

4–1 Elements of Power Supply Circuits

The purpose of a **power supply** circuit is to convert available ac power into usable dc power. Most of the circuits inside an ordinary TV receiver, for example, require dc voltage levels such as +12 V and +48 V. The source of power for the TV is 120 Vrms from a standard wall outlet. A power supply inside the TV receiver is responsible for converting electrical power from the 120-Vac source into the lower dc level required for operating most of the circuits. Power supplies are required in any electronic device that requires dc and is operated from ac power sources.

Figure 4–1 shows the main elements of a complete power supply system:

1. A transformer is used to step up or step down the ac voltage level and isolate the remainder of the electronic system from the ac power source.
2. A **rectifier circuit** is used to change the ac power into pulsating dc.
3. A **filter** removes the ac components from the pulsating dc waveform and provides a smooth dc power level.
4. An optional **bleeder resistor** or resistor divider circuit is used to aid filtering and discharge any capacitors in the power supply that might otherwise remain charged when ac power is removed.
5. An optional **regulator** provides an especially smooth and constant output voltage or current level.

You will learn more about these elements as the various power supply configurations are discussed in this chapter. You will also learn about power supply specifications, including:

- output voltage levels;
- output current levels;
- amount of voltage regulation; and
- ripple voltage levels.

And you will see how these specifications are met by the selection of components for the various sections in the power supply system (e.g., transformer, rectifier, filter, and regulator).

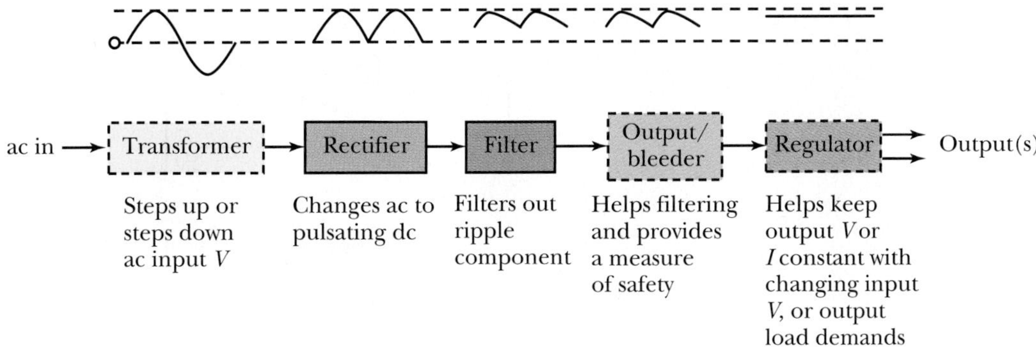

FIGURE 4–1 Power supply system block diagram

4–2 Rectifier Circuits

Rectifier circuits transform ac waveforms into pulsating dc waveforms. This is an important first step in the process of obtaining useful, smooth, dc sources from commercial ac power sources. The most important part of a rectifier circuit is its diodes (the rectifier diodes you studied in Chapter 3).

Half-Wave Rectifier Circuits

The **half-wave rectifier** introduced in this section is an ac version of the simple junction diode circuit described in the previous chapter. In Figure 4–2, note that the ac source is supplying power to a diode and resistor that are connected in series. The diode conducts current whenever the applied voltage forward biases the diode, and the diode blocks current flow through the circuit when the applied voltage reverse biases the diode. To put it another way, you will find the applied voltage dropped across resistor R_L whenever the diode conducts and zero voltage across the resistor whenever the diode is not conducting.

Refer to the circuit in Figure 4–2 while you consider the following points:

- The direction of electron flow through the diode is from cathode to anode (against the direction of the arrow in the diode symbol).
- The diode (D_1) conducts only on each ac alternation when its anode is positive with respect to the cathode. When this is the case, virtually all of the ac voltage is found across the load resistor (R_L).
- On ac input alternations when the diode is reverse biased, it cannot conduct because its anode is negative with respect to its cathode. When this happens, all of the ac voltage is found across the diode and virtually none across the load resistor.

The waveforms in Figure 4–2 show the operation of this circuit through two full cycles of applied ac voltage. During the first half-cycle, the diode is forward biased. As a result, only the forward conduction voltage appears across the diode, and most of the waveform is applied to the resistor at the output. On the second half-cycle (the negative half-cycle), the diode is reverse biased. Under this condition, current cannot flow through the circuit and all of the input voltage appears across the diode. (NOTE: This is indicated by the flat, 0 V, part of the R_L volts or output waveform.)

Because there is no filtering connected to this rectifier circuit, the output waveform is called **pulsating dc.** It is considered a pulsating waveform because the current flows in half-cycle bursts. It is a dc waveform because the output voltage is of just one polarity.

Recall for ac voltage waveforms, V_{rms} equals 0.707 times V_{pk} and that V_{avg} for each alternation = 0.637 times V_{pk}. Since 0.637 is about nine-tenths of 0.707, V_{avg} equaled 0.9 times V_{rms}.

Since a half-wave rectifier circuit is producing an output pulse only during one-half cycle, it is logical that the average dc output is only half the average ac voltage. That

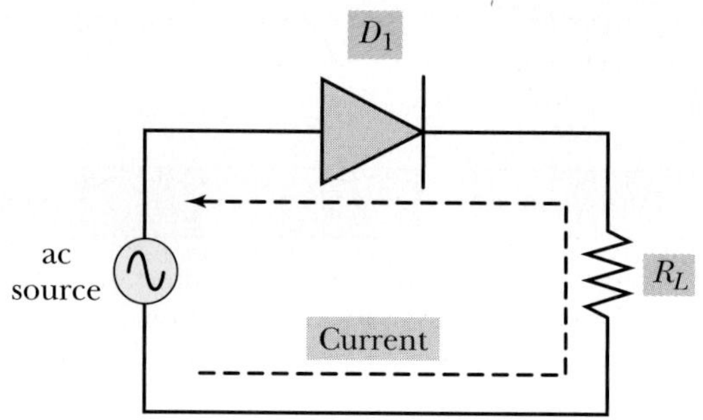

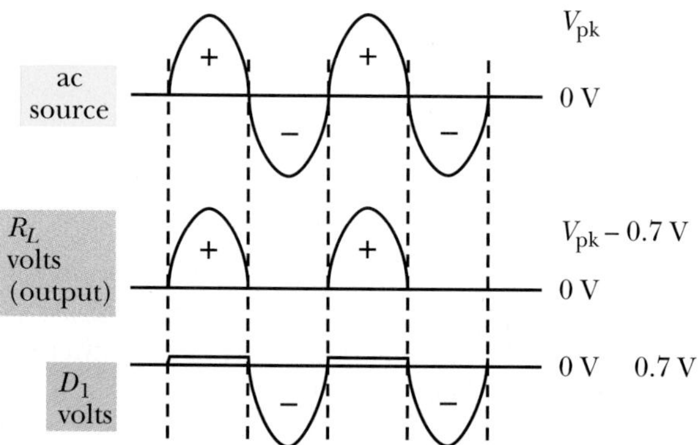

FIGURE 4–2 Half-wave rectifier circuit and waveforms

is, half of 0.9 = 0.45. Hence, the average dc output voltage of the half-wave rectifier circuit equals 0.45 times V_{rms}. Of course you can also compute the dc output as half the average ac voltage. That is, half of 0.637 times V_{pk}. This means the unfiltered dc output (half-wave rectifier) is calculated from either of the following formulas:

FORMULA 4–1 $V_{dc} = 0.45 \times V_{rms}$

$$V_{dc} = \frac{V_{pk}}{\pi}$$

FORMULA 4–2 $V_{dc} = 0.318 \times V_{pk}$

Formulas 4–1 and 4–2 do not account for the signal reduction due to the diode barrier voltage.

◆ **EXAMPLE** What is the unfiltered dc output voltage of a half-wave rectifier circuit if the ac input voltage is 100 V peak?

Answer:

$$V_{dc} = 0.318 \times V_{pk} = 0.318 \times 100 \text{ V} = 31.8 \text{ V}_{dc}$$ ◆

PRACTICE PROBLEMS 1

What is the unfiltered dc output voltage of a half-wave rectifier circuit if the ac input is 200 V rms?

You already know that the voltage drop across a conducting silicon diode is about 0.7 V. And that's the voltage you will find across the diode in a half-wave rectifier circuit on the half-cycle when the diode is conducting. But what about the voltage across the diode during the half-cycle when it is reverse biased? More specifically, what is the maximum voltage across the diode when it is not conducting? This value is called the *peak inverse voltage,* PIV.

Application Problem

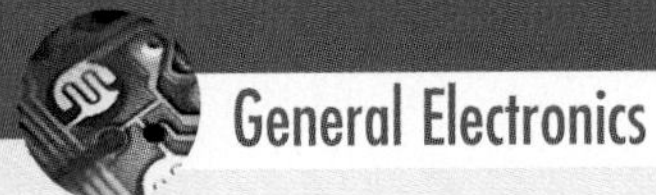

POWER SUPPLY CIRCUITS AND BATTERIES

A half-wave rectifier power supply containing an ideal diode is used to charge a lead acid battery. Batteries are used, for example, as a backup power supply in a telephone distribution station, or in a house alarm system, if there is a power failure. It is a simple process to charge a battery with the dc current; however, we do not want to cause any damage to battery electrodes. A half-wave rectification allows the battery to degas during the zero cycle, increasing its life. This process is shown in the figure below:

A typical power supply arrangement has a diode in series with the circuit (shown below). We wish to calculate the dc value of the output voltage ($V_{battery}dc$).

Solution

First we need to convert the rms power supply voltage to its peak voltage value (amplitude) as follows:

$$V_m = \sqrt{2}\, V_{rms} = \sqrt{2} \times 24 = 33.94 \text{ V}$$

Then we determine the battery dc voltage.

$$V_{battery}\,dc = \frac{V_m}{\pi} = 10.80 \text{ V}$$

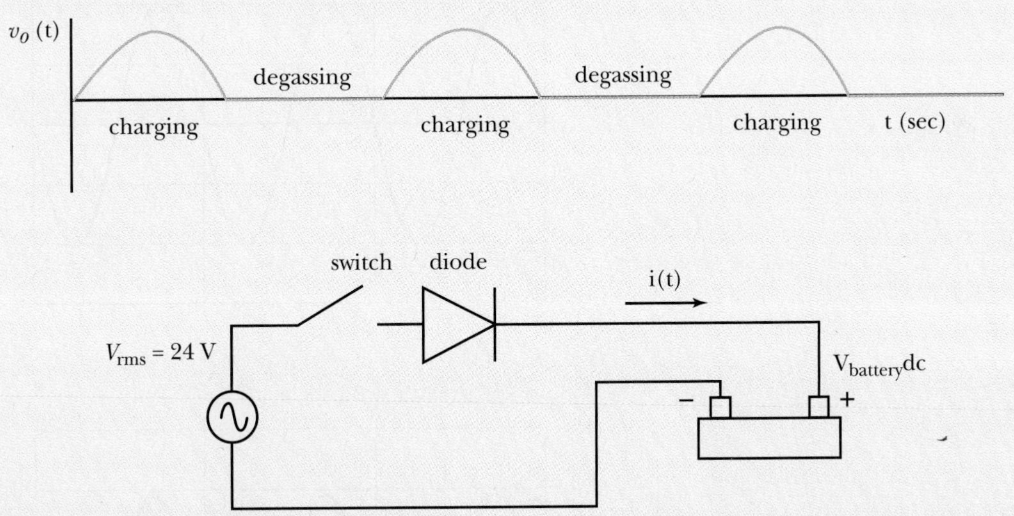

The PIV appears across the diode when the applied ac voltage has a polarity that reverse biases the diode. Because there is no current flowing through the diode in Figure 4–2, there can be no current flowing through the load resistor, R_L. The diode is acting like an open switch, so you will find the full amount of applied voltage dropped across it. The peak value of this reverse voltage waveform is PIV = 1.414 × V_{rms}.

FORMULA 4–3 PIV (H-W) = 1.414 × V_{rms}

where (H-W) means Half-Wave.

As you have seen in Figure 4–2, there is one ripple in the output for each ac cycle of input.

FORMULA 4–4 Ripple Frequency (H-W) = ac input frequency

If the ac input frequency to a half-wave rectifier circuit is 60 Hz, the ripple frequency of the output resistor R_L will also be 60 Hz.

The circuits in Figure 4–3 are both half-wave rectifiers. They are identical except for the direction the rectifier diode is connected. In Figure 4–3a, the diode is connected so that the output waveform across R_L is positive with respect to ground. In Figure 4–3b, the diode is connected in such a way that the output waveform is negative with respect to ground.

> ***Practical Notes***
>
> The reason for showing you Formula 4–1 is that you can measure the rms ac input voltage with any DMM or VOM—then easily compute average dc output. Not all meters can measure peak values directly. (Of course, a scope will!)
>
> The V_{dc} formulas shown don't take into account the very small dc voltage dropped across the diode, since it does not appreciably change the output voltage.

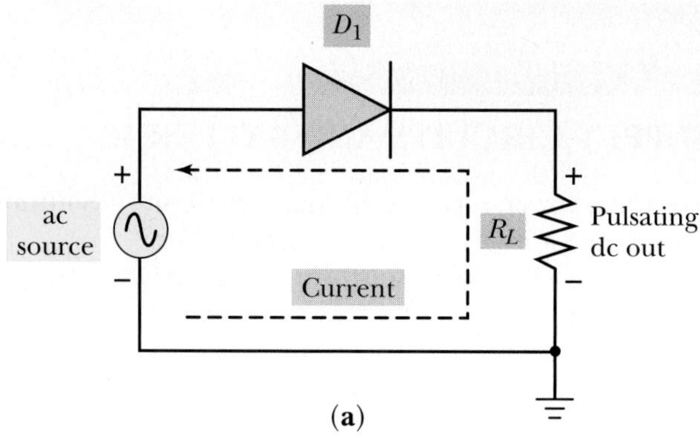

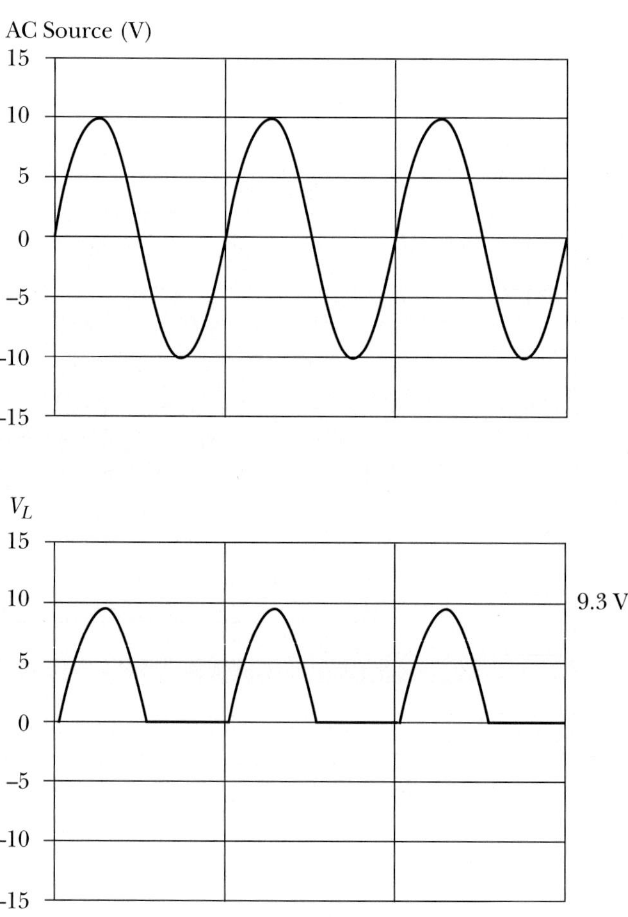

FIGURE 4–3 Changing output polarity from a half-wave rectifier: (a) circuit with positive output; (b) circuit with negative output

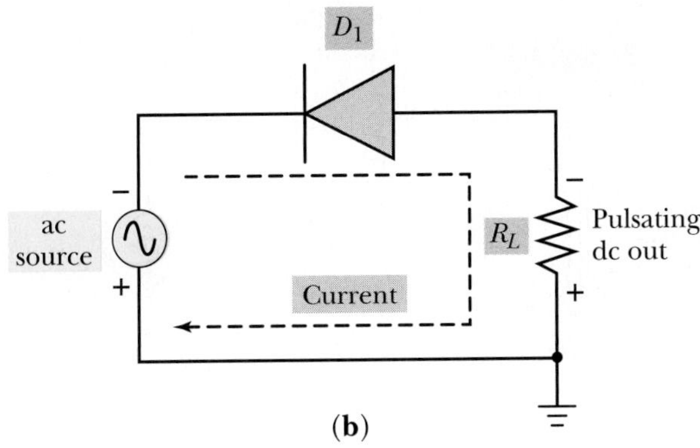

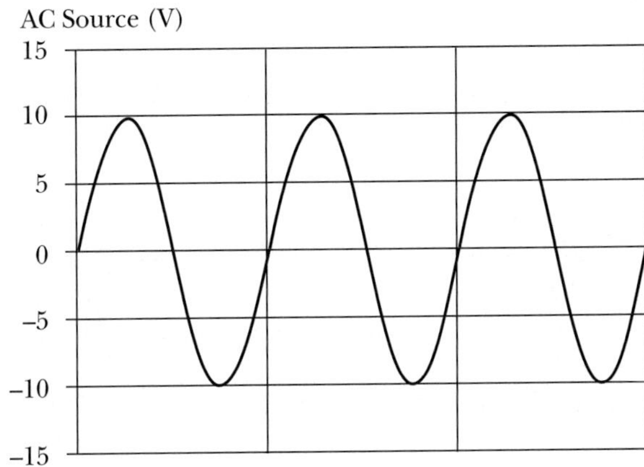

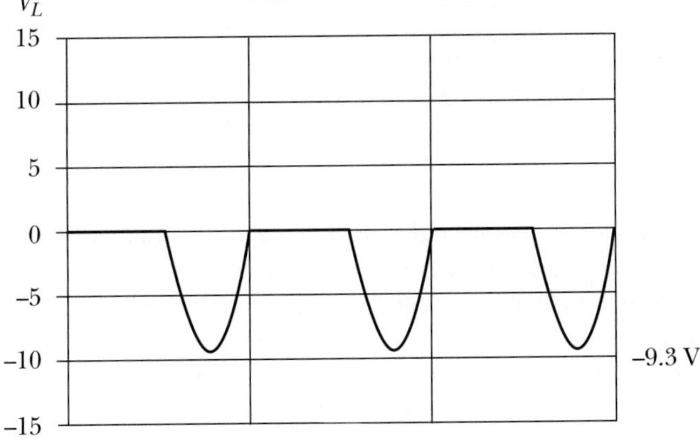

FIGURE 4–3 Continued

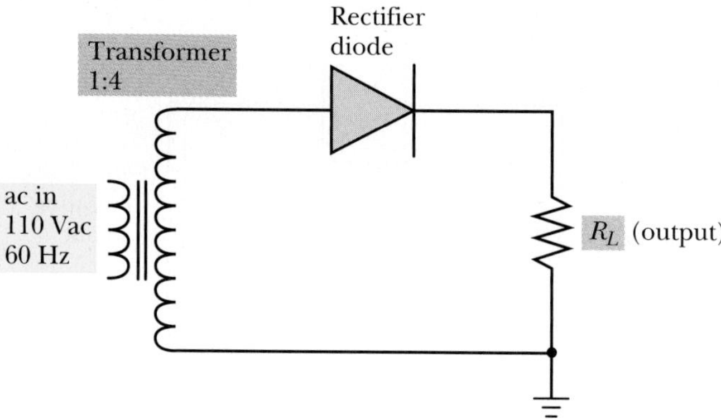

FIGURE 4–4 multiSIM

PRACTICE PROBLEMS 2

Refer to Figure 4–4 and find the dc output voltage, PIV, output ripple frequency, and output polarity with respect to ground.

Full-Wave Rectifier Circuits

Refer to the rectifier circuit in Figure 4–5 while you study the following discussion of the basic operation of a **full-wave rectifier** circuit. Notice that the secondary of the transformer is *center tapped.* The center-tapped transformer is absolutely necessary for the operation of this full-wave rectifier circuit.

For a transformer with a center-tapped secondary winding, when the top of the center-tapped secondary is positive, the bottom of the secondary winding is negative. It is important to note also that when the top of the tapped secondary is positive, the center tap is negative with respect to the top of the winding, and positive with respect to the bottom of the secondary winding. Of course, on the next alteration of ac input, when the top of the secondary is negative, the bottom of the full secondary is positive. In this case, the center tap is positive with respect to the top of the secondary, and negative with respect to the bottom of the secondary. It is obvious that from center tap to either end of the secondary, the voltage equals only half of the full secondary voltage. Keep these thoughts in mind as you follow the discussion of the operation of the full-wave rectifier circuit.

The rectifier diodes are connected in such a way that D_1 conducts on the first half-cycle of the input ac, when the top half of the center-tapped secondary is forward-biasing it. The current path is from center tap, up through R_L through D_1, back to the top of the secondary winding. D_2 conducts on the opposite half-cycle of the input ac, when the bottom half of the secondary causes D_2 to be forward biased. The two diodes thus conduct alternately, D_1 on one half-cycle of input ac, and D_2 on the other half-cycle.

The following four formulas allow you to calculate the average dc voltage from a full-wave rectifier circuit that uses a center-tapped transformer. The first two yield the output voltage when you know the rms voltage that is applied.

Formula 4–5 lets you calculate V_{dc} on the basis of the rms voltage (V_{rms}) measured between the center tap and one end of the transformer secondary (as shown in Figure 4–6a):

FORMULA 4–5 $V_{dc} = 0.9 \times V_{rms}$ of half the secondary

Formula 4–6 also lets you calculate V_{dc}, but after measuring the rms voltage across the full transformer secondary (see Figure 4–6b):

FORMULA 4–6 $V_{dc} = 0.45 \times V_{rms}$ of the full secondary

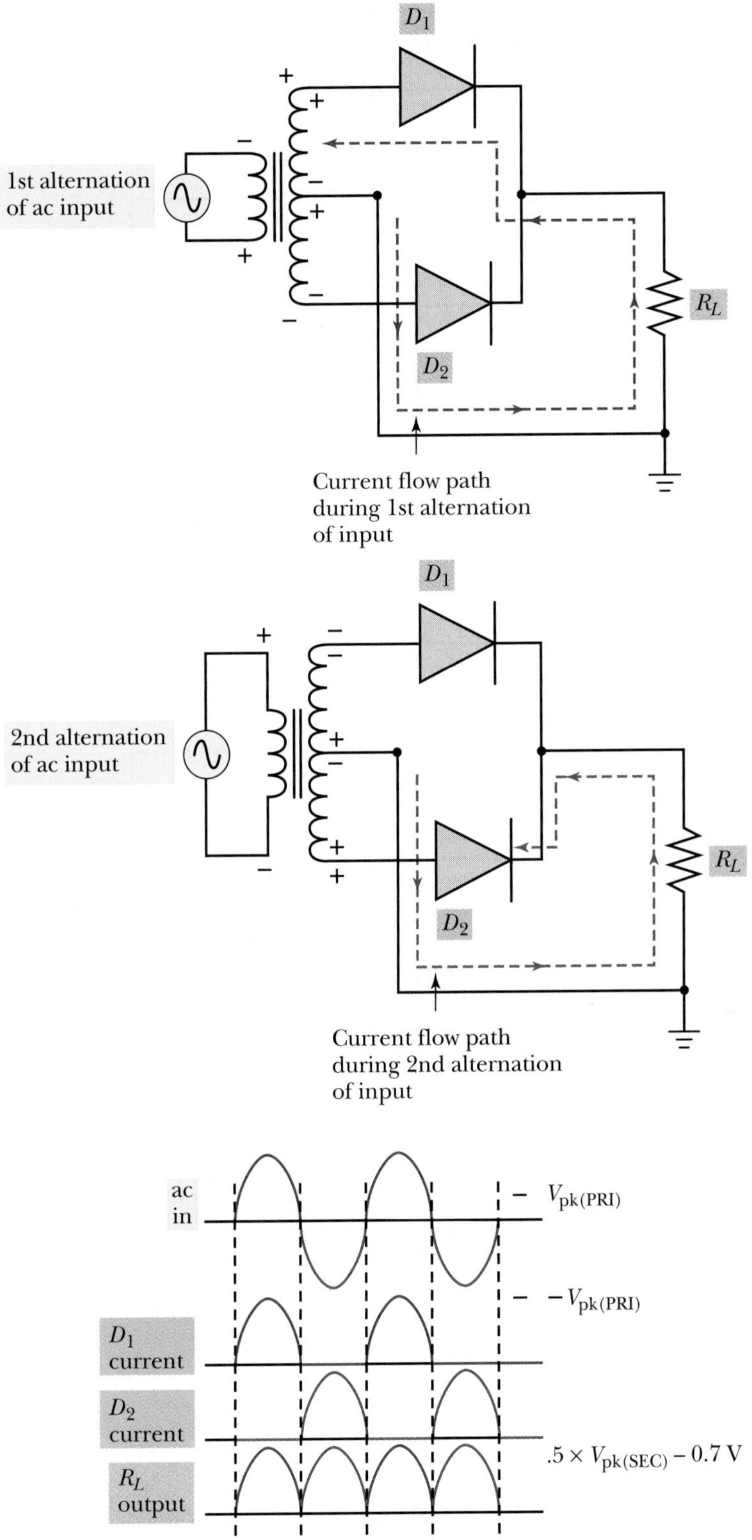

FIGURE 4–5 Full-wave rectifier circuit and waveforms

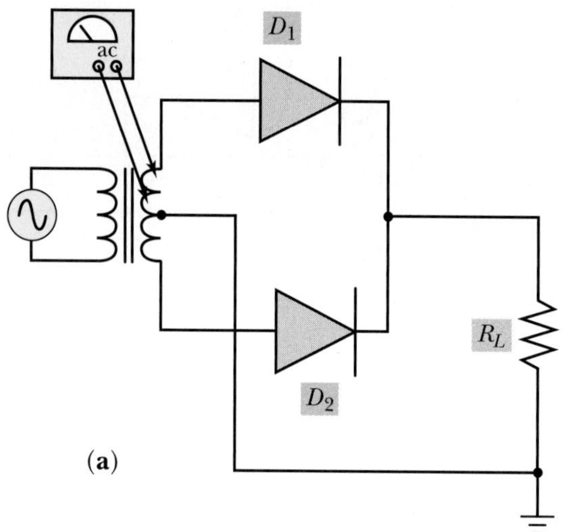

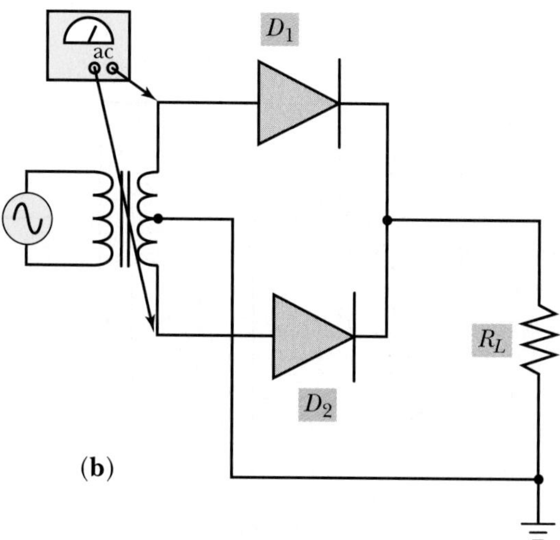

FIGURE 4–6 Center-tapped transformer voltages: (a) center tap to one end; (b) end-to-end.

Formulas 4–7 and 4–8 allow you to calculate the average output dc voltage level on the basis of peak voltage measurements (V_{pk}) as measured from the center tap or across the full secondary:

$$Wave = \frac{2V_{pk}}{\pi}$$

FORMULA 4–7 $V_{dc} = 0.637 \times V_{pk}$ of half the secondary

FORMULA 4–8 $V_{dc} = 0.318 \times V_{pk}$ of the full secondary

The peak inverse voltage (PIV) across EACH diode in the full-wave rectifier circuit equals the peak voltage measured across the FULL secondary of the transformer (end-to-end).

FORMULA 4–9 PIV (F–W) $= 1.414 \times V_{rms}$ of full secondary

NOTE: Formulas 4–5 through 4–8 inclusive do not account for the signal reduction due to the diode barrier voltage.

◆ **EXAMPLE** The end-to-end (full secondary) voltage on the secondary of a center-tapped transformer is 260 V. What is the minimum reverse voltage rating for the diodes if they are used in a full-wave rectifier?

Answer:

From Formula 4–9:

$$\text{PIV} = 1.414 \times V_{\text{rms}} \text{ of full secondary}$$
$$\text{PIV} = 1.414 \times 260 = 368 \text{ V}$$

So the reverse voltage rating of the diodes has to be at least 368 V. You could select diodes having a 400-V peak reverse rating, but a 600-V version would be even better.

As far as the ripple frequency is concerned, a full-wave rectifier produces an output ripple for each half-cycle of ac input. This means the output ripple frequency equals two times the frequency of the ac input.

FORMULA 4–10 Ripple Frequency = 2 × ac input frequency ◆

PRACTICE PROBLEMS 3

Assume a full-wave rectifier circuit is connected to a transformer having a full-secondary rms voltage of 300 V. Further assume that the frequency of the ac source feeding the transformer is 400 Hz. Determine the dc output, ripple frequency, and PIV for each diode.

The output polarity of a full-wave rectifier with center-tapped transformer, can be reversed by reversing the cathode and anode connections for *both* diodes.

Caution: Both diodes must be reversed! If only one is reversed, the transformer winding will be shorted when both diodes are forward biased during one of the half-cycles of applied power. Compare the circuits in Figure 4–7. It is also possible to reverse the output polarity by taking the output from the center tap and grounding the common-connection point of the two diodes.

The Bridge Rectifier

A **bridge rectifier** is a type of full wave rectifier circuit, but one that does not require a transformer having a center-tapped secondary. As shown in Figure 4–8, a bridge rectifier uses four rectifier diodes.

You can see from the diagrams that two of the diodes are forward biased on one half-cycle of the input ac waveform and the other two are forward biased on the opposite half-cycle. When the input waveform has the polarity shown in Figure 4–8a, you can see that diodes D_1 and D_3 conduct. To trace the series path for electron flow during this half-cycle of applied voltage:

- start from the negative end of the ac power source;
- pass through diode D_1 (against the symbol arrow);
- pass through the output load resistor in the direction indicated on the diagram;
- pass through diode D_3; and
- return to the positive side of the ac power source.

Diodes D_2 and D_4 do not conduct on this half-cycle because they are reverse biased.

On the second half-cycle, Figure 4–8b, D_2 and D_4 are allowed to conduct, while D_1 and D_3 are reverse biased. To trace the series path for electron flow during this second half-cycle:

- start from the negative end of the ac power source;
- pass through diode D_4;

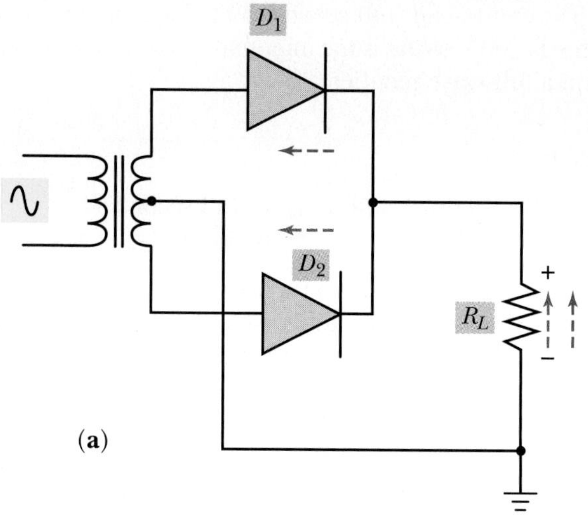

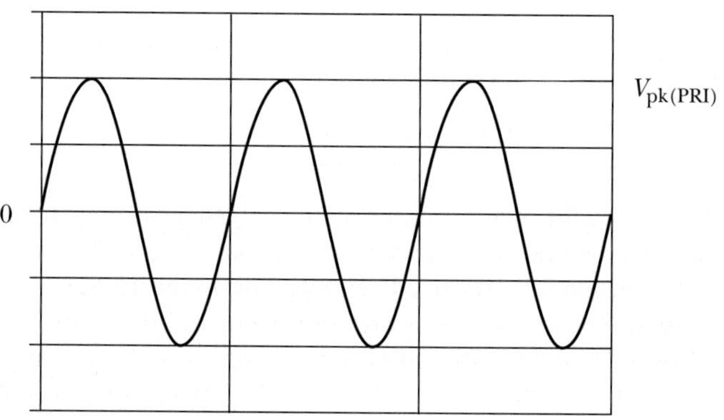

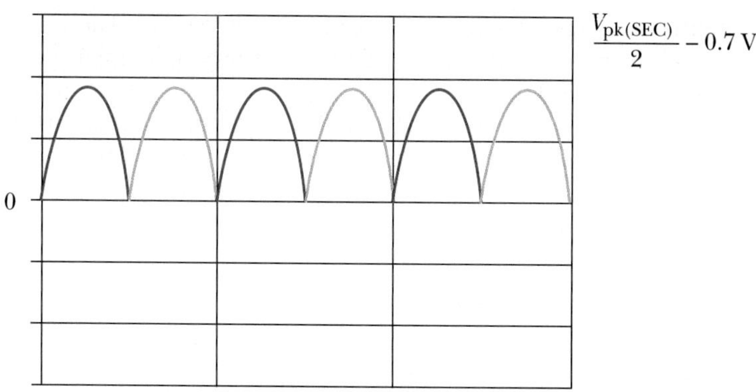

FIGURE 4–7 Diode direction affects output polarity: (a) positive output; (b) negative output

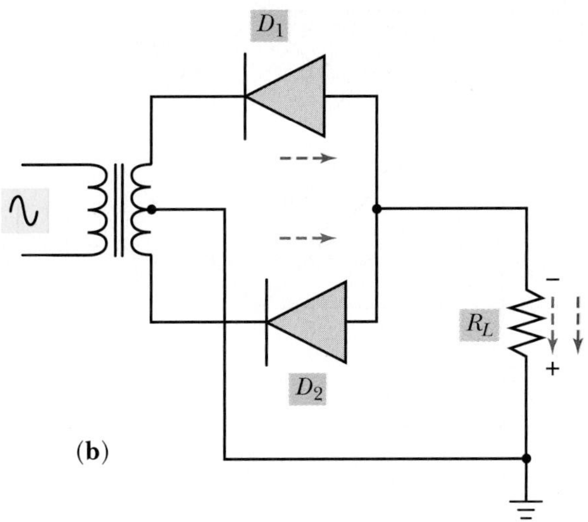

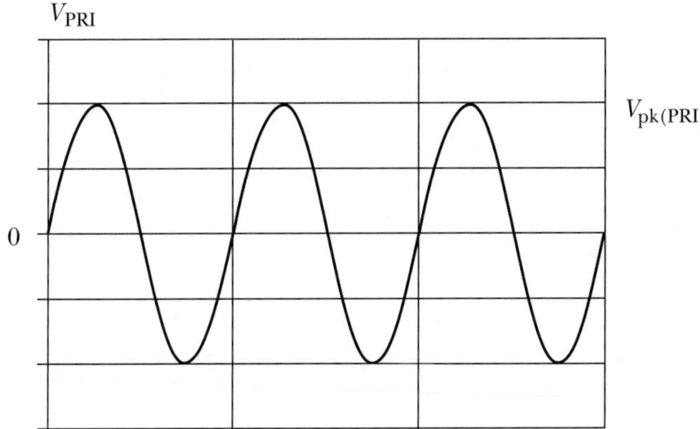

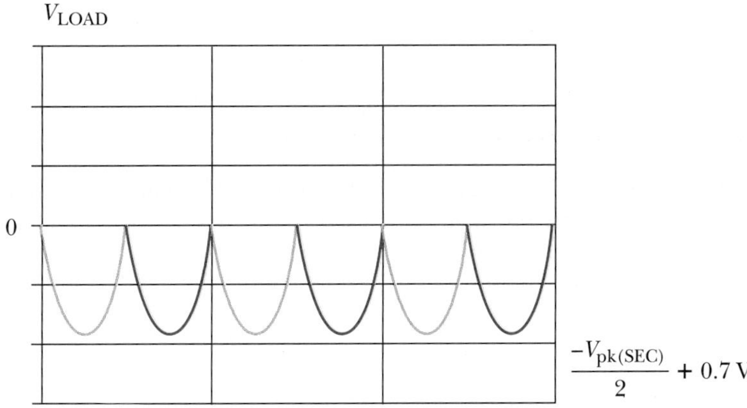

FIGURE 4–7 Continued

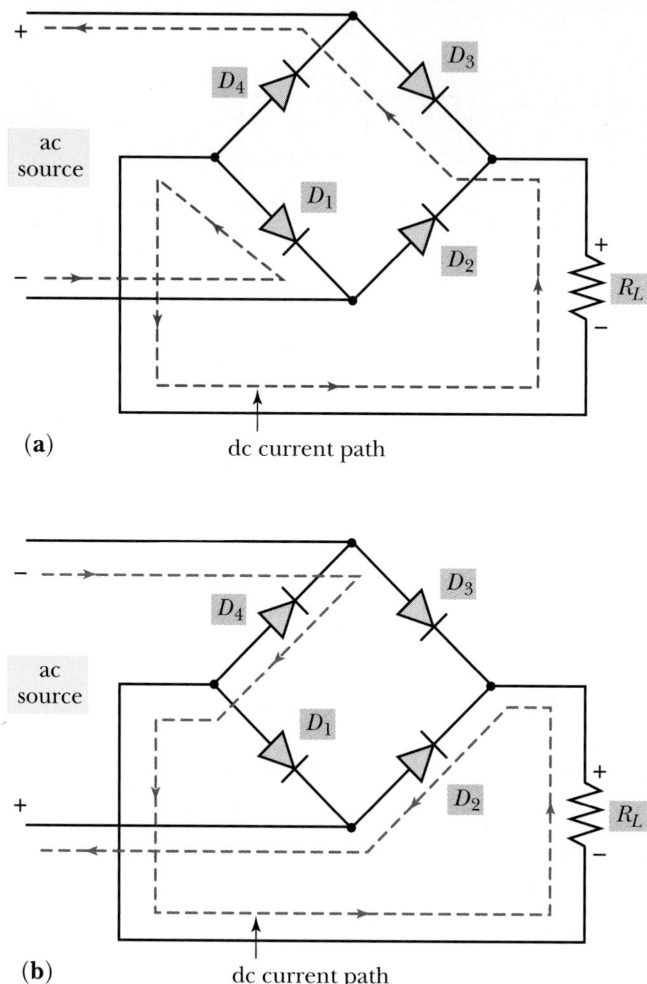

FIGURE 4–8 Bridge rectifier circuit: (a) first half-cycle; (b) second half-cycle

- pass through the output load resistor in the direction indicated on the diagram;
- pass through diode D_2; and
- return to the positive side of the ac power source.

Since there is a pulse of output current during each half-cycle of the ac input waveform, the frequency of the pulsating dc waveform at the output is twice that of the input frequency.

Formulas 4–11 and 4–12 allow you to calculate the average dc voltage level at the output on the basis of the input voltage. If you know the rms voltage (V_{rms}) at the input, you can use Formula 4–11 to calculate the average dc output. But if you know the peak voltage level at the input (V_{pk}), use Formula 4–12.

FORMULA 4–11 $V_{dc} = 0.9 \times V_{rms}$ input

FORMULA 4–12 $V_{dc} = 0.637 \times V_{pk}$ input

NOTE: Formula 4–11 and 4–12 do not account for the signal reduction due to the diode barrier voltage.

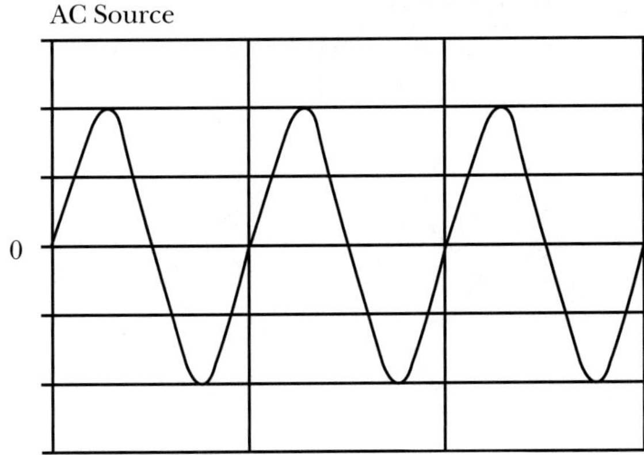

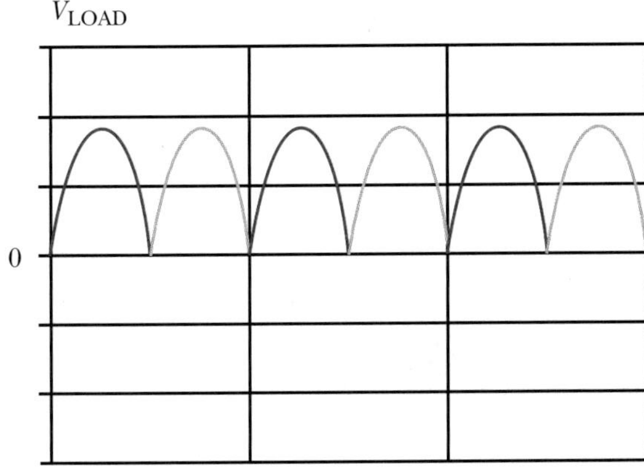

FIGURE 4–8 Continued

EXAMPLE What is the average dc voltage level at the output of the bridge rectifier circuit in Figure 4–9 if you measure 12.6 V (rms) across the secondary of the transformer?

Answer:

Using Formula 4–11:

$$V_{dc} = 0.9 \times V_{rms} \text{ input}$$
$$V_{dc} = 0.9 \times 12.6 \text{ V} = 11.34 \text{ V}$$

The peak inverse voltage across each diode of an unfiltered bridge rectifier circuit equals the peak value of the ac input waveform. You can use Formula 4–13 to calculate the peak value, given the rms voltage present at the input:

FOMULA 4–13 PIV (Bridge) = $1.414 \times V_{rms}$ input

As seen earlier in this chapter, the ripple frequency at the output of an unfiltered bridge rectifier circuit is twice the frequency of the input waveform:

FORMULA 4–14 Ripple Frequency (Bridge) = 2 × ac input frequency

You have seen that a bridge rectifier requires four separate diodes. This arrangement is so commonly used, however, that semiconductor manufacturers simplify the

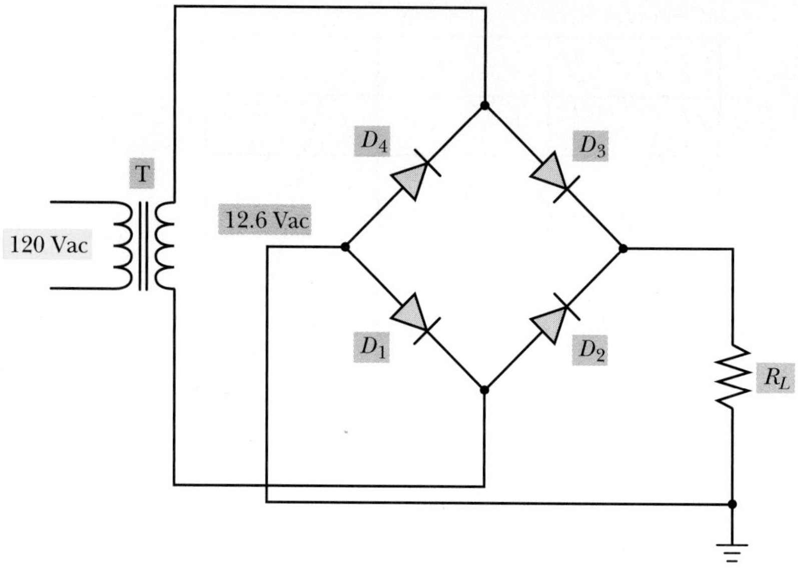

FIGURE 4–9 multiSIM

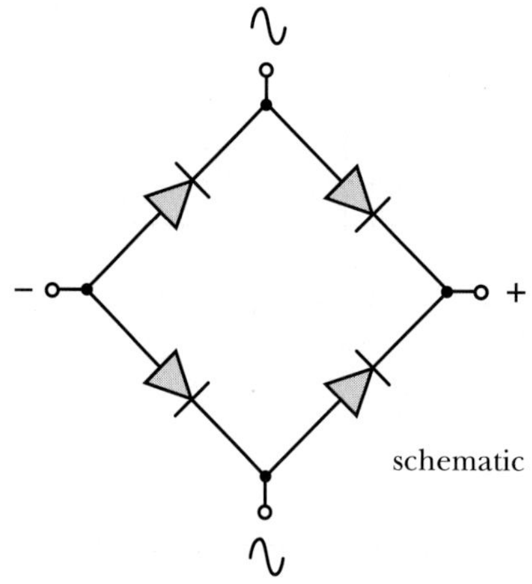

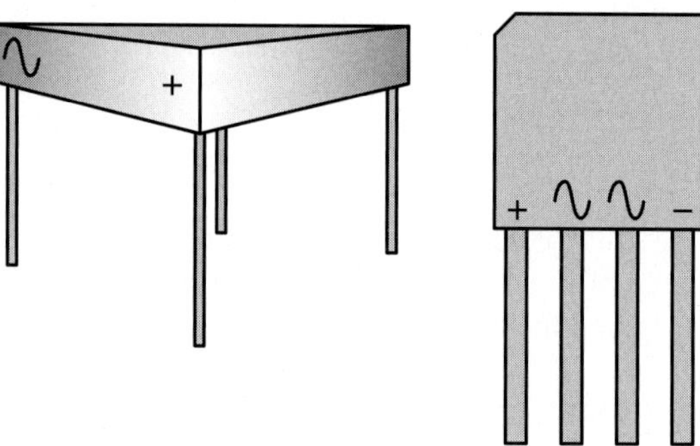

FIGURE 4–10 Bridge rectifier assemblies

actual circuitry by providing the four-diode bridge assembly in a single package. Figure 4–10 shows two popular packages for bridge rectifier assemblies. The ac inputs are marked with a small sine-wave symbol, and the two dc outputs are labeled plus (+) and minus (–).

Bridge rectifier assemblies simplify the procedures for building, testing, and repairing power supplies. These assemblies are rated much like single rectifier diodes—by average forward current and maximum reverse breakdown voltage.

Application Problem

Computers

POWER SUPPLY BRIDGE CIRCUITS

A full-wave rectifier power supply containing a bridge rectifier shown in Figure 4–9 is used to deliver power to a microprocessor. The V_{rms} of the bridge rectifier is 12.6 V. A large capacitor is used to change pulsing output to nearly dc. What will happen to the microprocessor input voltage? Determine the value of the voltage.

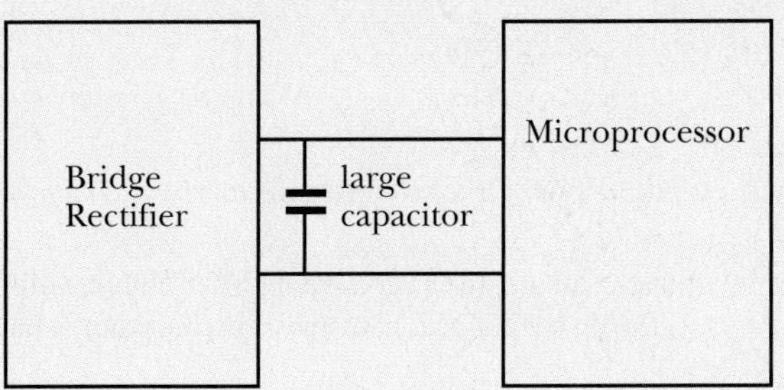

Solution A large capacitor will change the V_{rms} to the peak voltage, V_{peak}. The process is shown in the figure below.

First we need to convert the rms power supply voltage to its peak voltage value (amplitude) as follows:

$$V_{peak} = \sqrt{2}\ V_{rms} = \sqrt{2} \times 12.6 = 17.81\ \text{V}$$

The microprocessor will see 17.81 V dc and most likely will get damaged because it normally uses 5 volts.

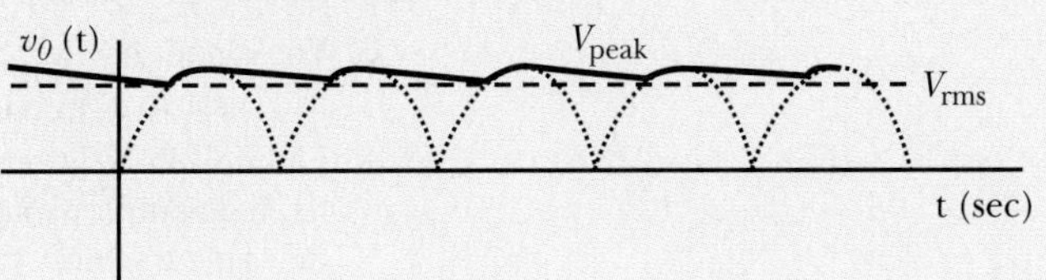

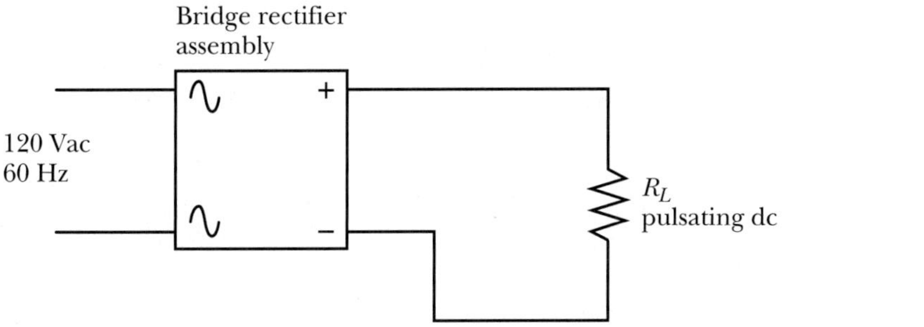

FIGURE 4–11

PRACTICE PROBLEMS 4

Refer to the circuit in Figure 4–11.

1. If the input rms voltage level is 120 V as shown in the figure, what is the average dc output voltage level?
2. If the load (R_L) is 50 Ω, what is the average dc current through the load? (*Hint:* Use Ohm's law, $I_{dc} = V_{dc}/R_L$.)
3. What is the minimum reverse breakdown voltage specification for the diodes?
4. What is the ripple frequency across the load?

Summary of Basic Rectifier Circuit Features

Consider the three basic kinds of rectifier circuits you have just studied.

- A half-wave rectifier requires only a single diode in addition to the power source and output load. It is the simplest of the three basic kinds of rectifier circuits. Its main drawback is that there is no power delivered to the load during one-half of each input ac cycle.
- A center-tapped full-wave rectifier circuit requires two diodes; but more important, it also requires the use of a center-tapped power transformer. The full-wave rectifier circuit provides output power to the load during both alternations of ac input, and that is its advantage over a half-wave rectifier.

- The full-wave bridge circuit requires four diodes, but has the power delivering capability of full-wave rectifiers and does not require the use of a center-tapped power transformer. Furthermore, for a given transformer, the bridge rectifier circuit delivers twice the voltage output, when compared with either the half-wave, or center-tapped transformer full-wave circuit. Although the need for four diodes might seem to be a disadvantage, the low cost and reliability of suitable rectifier diodes (especially bridge rectifier assemblies) make this circuit the first choice for most practical applications.

■ IN-PROCESS LEARNING CHECK 1

Fill in the blanks as appropriate.

1. The simplest rectifier circuit is the _______ circuit.
2. The ripple frequency of a bridge rectifier circuit is _______ the ripple frequency of a half-wave rectifier.
3. The circuit having the highest output voltage for a given transformer secondary voltage is the _______ rectifier.
4. For a given full transformer secondary voltage, the full-wave rectifier circuit unfiltered dc output voltage is _______ the dc output of a half-wave rectifier that is using the same transformer.
5. To find the average dc output (unfiltered) of a center-tapped, full-wave rectifier, multiply the full secondary rms voltage by _______.

4–3 Basic Power Supply Filters

The purpose of a power supply filter is to keep the ripple component from appearing in the output. It is designed to convert pulsating dc from rectifier circuits into a suitably smooth dc level.

Capacitance Filters

Figure 4–12 shows a simple capacitance (C-type) filter. You can see that the output of the rectifier circuit is connected in parallel with a filter capacitor and the load resistance. Figure 4–12b shows how the C-type filter helps smooth out the valleys in the waveform from a half-wave rectifier, and Figure 4–12c shows the filtering when using a full-wave rectifier.

The capacitor charges to the peak value of the rectified signal through the forward bias diode. The capacitor will not follow the normal decrease of the rectified signal as its voltage reverse biases the initially conducting diode shutting it off. The capacitor now slowly discharges through the load until the next cycle when the diode becomes forward biased again and recharges the capacitor.

Figure 4–13 illustrates an example of a C-type filter that is used with a simple half-wave rectifier network. This is the simplest and most economical sort of power supply. It is used when simplicity and economy are important and when a relatively large amount of ripple does not seriously affect the operation of electronic devices that operate from it.

Inductance Filters

Figure 4–14 shows an inductance (L-type) power supply filter. In this case, the output of a half- or full-wave rectifier (including a bridge rectifier) is connected to the load through an inductor, or choke. Figure 4–14b and 4–14c show how an L-type filter circuit affects the waveforms from half- and full-wave rectifier circuits, respectively.

During the time a rectifier is supplying power through the inductor in an L-type filter, energy is stored in the form of a magnetic field. As the circuit current tries to decrease, the inductor's energy is used to keep current flowing; thus, providing a "smoothing" action on the output waveform. The quality of the filtering depends on the value of the inductance and the amount of current required to operate the load.

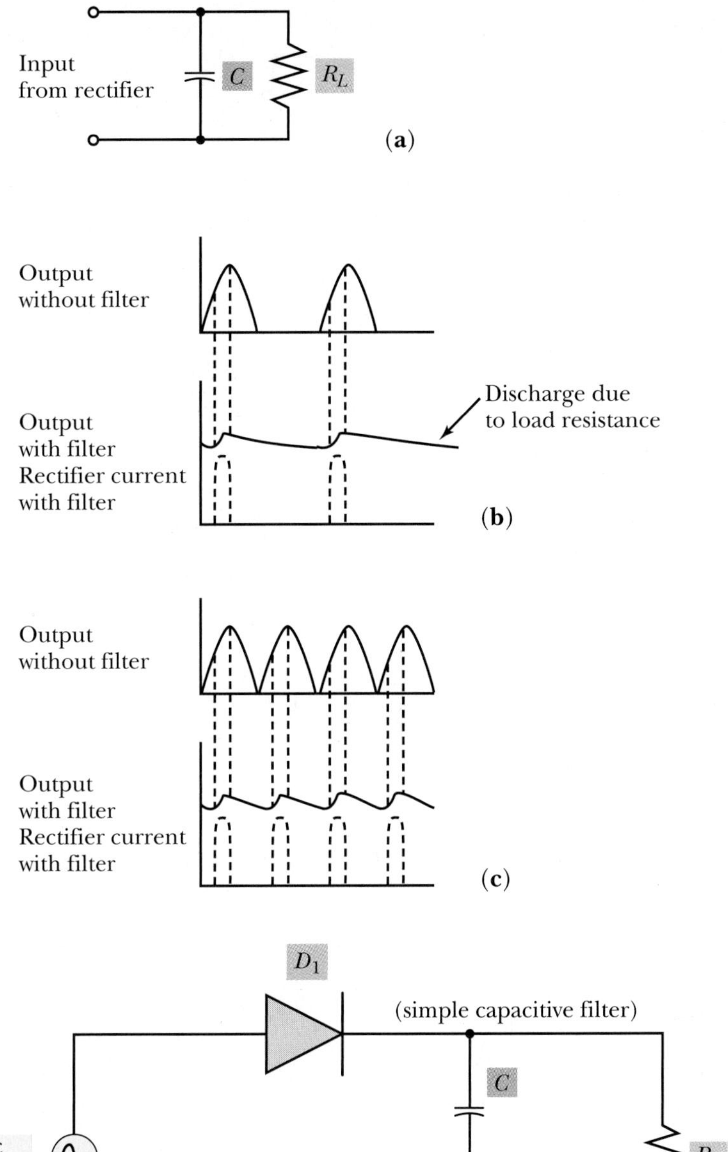

FIGURE 4–12 Simple C-type filter circuit: (a) circuit diagram; (b) half-wave rectification before and after C-type filter action; (c) full-wave rectification before and after C-type filter action

FIGURE 4–13 Half-wave power supply with C-type filter

Figure 4–15 is an example of an L-type filter that is driven by a center-tapped full-wave rectifier. This is a relatively expensive circuit because of the need for electromagnetic components (namely the center-tapped power transformer and filter choke).

Filter Networks

You have seen how capacitors and inductors, used separately, can filter pulsating dc waveforms. The best kinds of power supply filter circuits use combinations of L, C, and R. The main considerations in the selection and quality of a power supply filter network are:

1. allowable output ripple;
2. allowable minimum regulation;
3. rectifier peak current limits;
4. load current; and
5. output voltage.

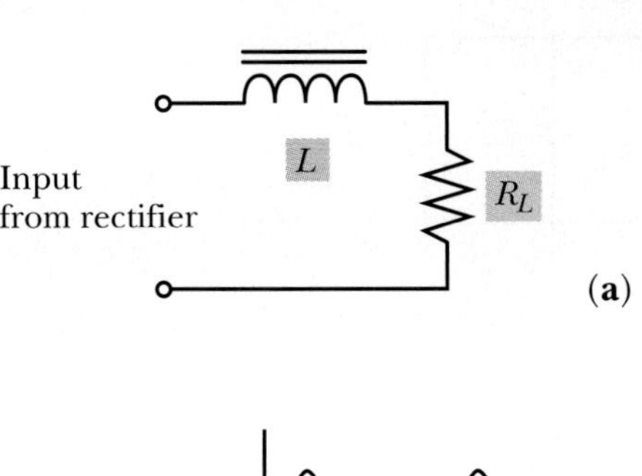

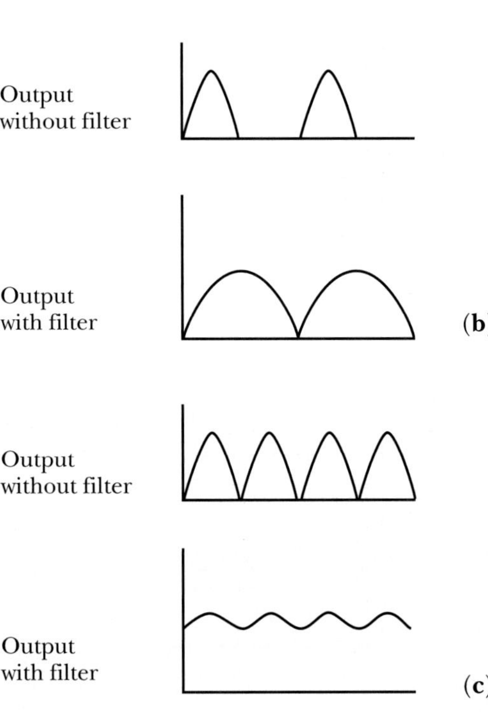

FIGURE 4–14 Simple L-type filter circuit: (a) circuit diagram; (b) half-wave rectification before and after L-type filter action; (c) full-wave rectification before and after L-type filter action

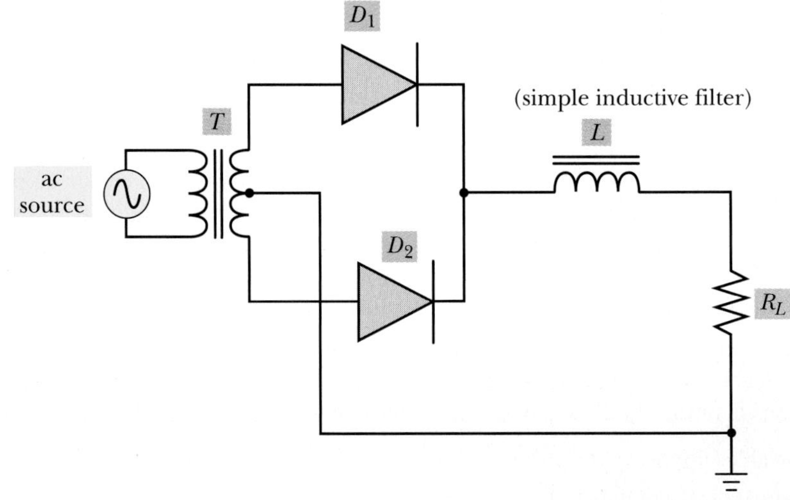

FIGURE 4–15 Full-wave power supply with L filter

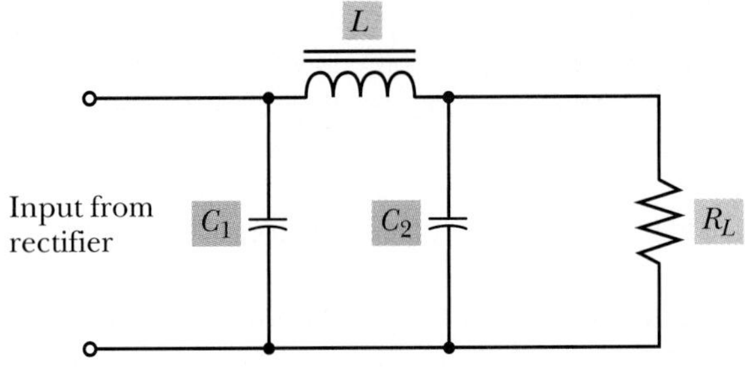

FIGURE 4–16 Capacitor-input pi-type LC filter

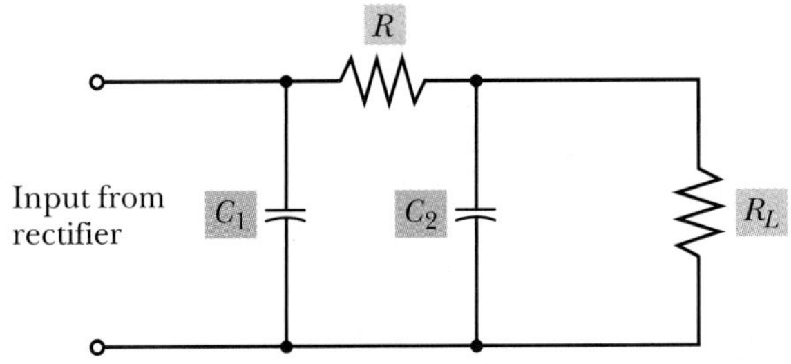

FIGURE 4–17 Capacitor-input pi-type RC filter

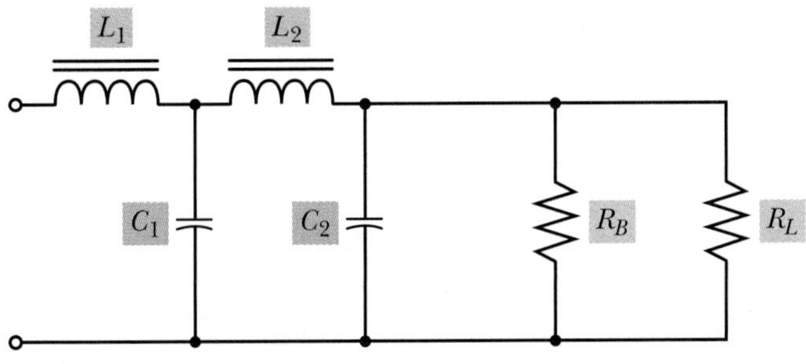

FIGURE 4–18 Choke-input, two-section L filter

The filter circuit in Figure 4–16 is one of the most popular in equipment that requires a very high amount of filtering (low ripple) and good regulation. It is called a *capacitor input pi filter.* It is called "capacitor input" because the output from the rectifier sees a shunt capacitor first. It is called a pi filter because the formation of the two capacitors and inductor on a schematic diagram resembles the Greek letter pi (π).

A slightly different capacitor-input pi filter is shown in Figure 4–17. The inductor is replaced with a resistor. The ripple and regulation qualities for this circuit are not as good as for the LC version, but the RC version is less expensive and good for many less-demanding applications.

Figure 4–18 shows the best kind of LCR power supply filter circuit. This is a *choke-input, two-section L filter.* It is a "choke-input" filter because inductor L_1 is the first component seen by the pulsating dc output from the rectifiers. It is called an "L" filter because the schematic drawing of L_1 and C_1 looks like an inverted letter L (as does L_2 and C_2). It is a two-section filter because there are two L sections connected one after the other.

This type of two-section L filter is by far the best kind of power supply filter network, both in terms of extremely low ripple and excellent regulation. It is very expensive, however, so it is used only in the most demanding kinds of equipment such as radar, medical instruments, and certain kinds of communications gear.

Power Supply Ripple

You have seen how filter circuits go a long way toward converting pulsating dc from a rectifier circuit to a reasonably steady dc level. In actual practice, no filter circuit is perfect—there is always some slight amount of variation left over. The part of the original ac waveform that remains at the output of a filtered rectifier circuit is called the *ripple*. You can see the ripple on the output waveforms for the circuits in Figures 4–12 and 4–14.

The higher the quality of a power supply, the smaller the amount of ripple exists in its output voltage. The amount of ripple, however, is not always the same value. Drawing more current from a power supply, for example, tends to increase the amount of ripple. The specifications for a power supply ought to cite the amount of ripple (usually measured in ac volts or millivolts) at a certain level of current.

A useful specification for power supply ripple is the percent of ripple. As shown in Formula 4–15, you can determine the percent of ripple of a power supply by using an oscilloscope to measure the ac ripple (peak-to-peak) and the amount of dc voltage below the ripple. You then divide the ac amount by the dc amount, and multiply by 100.

FORMULA 4–15 $$\%\text{Ripple} = \frac{\text{ac}}{\text{dc}} \times 100$$

For example, if the ripple is found to be 1 $V_{p\text{-}p}$ and the dc level is 48 V, the percent of ripple is 1/48 × 100 = 2.08%. That is a pretty good power supply, considering that many allow up to 10% ripple.

4–4 Simple Power Supply Regulators

The purpose of a **power supply voltage regulator** is to provide a very steady, or well-regulated, dc output. Two different factors commonly cause the voltage from a rectifier circuit to swing away from the desired output level. One factor is the changes in the ac input voltage. Power supplies that are operated from 120-Vac, 60-Hz utility sources can be affected by fluctuations in that so-called 120-Vac level. This ac voltage can drop as low as 110 Vac at times and rise to 128 Vac or so at other times. Without the help of a voltage regulator, the output of a power supply will vary according to such changes in the ac input level.

Another factor that causes the dc output from a power supply to vary is the amount of current that is required to operate the load at any given moment. This is not ordinarily a problem with very simple electronic devices, but the more complex an electronic system becomes, the more the current requirement varies as the equipment performs different tasks. The voltage from a power supply tends to decrease when the load "pulls" more current from it. Then when the load lightens, the power supply voltage tends to rise.

Electronic power supply regulators eliminate dc voltage swings caused by variations in input voltage and output loading. Whenever the dc output from a regulated power supply tries to change, the regulator instantly makes an internal adjustment that causes the dc output to remain steady. Modern regulators respond so rapidly that they can do much of the ripple filtering as well. So the output from a good power supply regulator is not only very steady, but virtually ripple free.

Some regulation is offered by the type of filter network that is used in a power supply. Traditionally, these filters were made up of passive components that have limited capabilities. In recent years, however, nearly ideal voltage regulation has

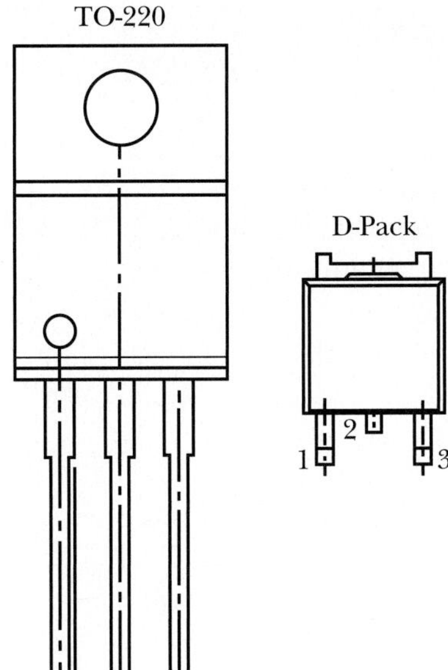

FIGURE 4–19 A typical power supply regulator IC device

become possible at a low cost with the help of small, integrated-circuit voltage regulator components.

Integrated-circuit (IC) components are made up of many semiconductors, resistances, and interconnections that are all included on a single chip of silicon. IC power supply regulators may have just three terminals on them. Figure 4–19 shows a typical IC voltage regulator used with a bridge rectifier circuit. The three terminals are for:

- unregulated dc input;
- regulated dc output; and
- common ground connection.

Voltage regulator ICs are rated according to their output regulation voltage and maximum current capability. Typical regulators are rated at +5 V at 1 A, +12 V at 1 A, and +24 V at 1/2 A.

The quality of a regulator is expressed in terms of percent of regulation. This is a figure that expresses how well a power supply retains its dc output level under changes in the amount of current being drawn from its output called the output load regulation. A power supply is said to have *no load* when no current is being drawn from it, and *full load* when the maximum rated current is being drawn from it. Formula 4–16 shows that the percent of regulation depends on:

- V_{FL}, the output of dc voltage when maximum current is being drawn from it, the full load voltage; and
- ΔV, the difference in dc output voltage when the supply is operated at full load and no load.

V_{NL}-No load voltage

$$\frac{V_{NL} - V_{FL}}{V_{FL}} \times 100\%$$

FORMULA 4–16 $\%\text{Regulation} = \dfrac{\Delta V}{V_{FL}} \times 100$

IN-PROCESS LEARNING CHECK 2

Fill in the blanks as appropriate.

1. In a C-type filter, the output of the rectifier circuit is connected in _______ with a filter _______ and the load resistance.
2. In an L-type filter, the output of the rectifier circuit is connected in _______ with a filter _______ and the load resistance.
3. In an L-type filter, energy is stored in the form of a _______ field.
4. The five main considerations in the selection and quality of a power supply filter network are: _______, _______, _______, _______, and _______.
5. The two main causes of swings in the output voltage of a power supply are changes in the _______ voltage level and changes in the _______ current demand.
6. The three terminals on an integrated circuit voltage regulator are the unregulated _______, regulated _______, and _______.
7. The (higher, lower) _______ the percent of ripple, the higher the quality of the power supply.
8. Full-load output voltage is taken when the current from the power supply is (maximum, minimum) _______.
9. The (higher, lower) _______ the percent of regulation, the higher the quality of the power supply.

EXAMPLE You are working with a power supply that has a dc output of 12.2 V at no load and 11.8 V at full load. What is the percent of regulation?

Answer:

In this instance, V_{FL} = 11.8 V.

$$\Delta V = V_{FL} - V_{NL}$$
$$\Delta V = 12.2\text{ V} - 11.8\text{ V}$$
$$\Delta V = 0.4\text{ V}$$

Using Formula 4–16, the percent of regulation can be calculated.

$$\%\text{ Regulation} = \frac{\Delta V}{V_{FL}} \times 100\%$$
$$\%\text{ Regulation} = \frac{0.4\text{ V}}{11.8\text{ V}} \times 100\%$$
$$\%\text{ Regulation} = 3.39\%$$

The regulation for this power supply is 3.39%.

4–5 Basic Voltage Multiplier Circuits

A **voltage doubler** circuit is a type of rectifier circuit that not only changes an ac input to a dc output, but doubles the voltage in the process. An ideal voltage doubler circuit provides an output voltage level that is equal to two times the peak value of the ac input voltage. There are no ideal voltage doubler circuits, so you can expect the output voltage to be slightly lower than twice the peak input voltage, especially under heavy current demands from the load circuit.

EXAMPLE A voltage doubler power supply is used with the normal household 120-Vac input. What is the ideal dc voltage level at the output of the doubler?

Answer:

The 120-Vac rating is an rms value. The peak value is 1.414 × 120 V = 169.7 V. The output of the multiplier is 2 × 169.7 V = 339.4 Vdc.

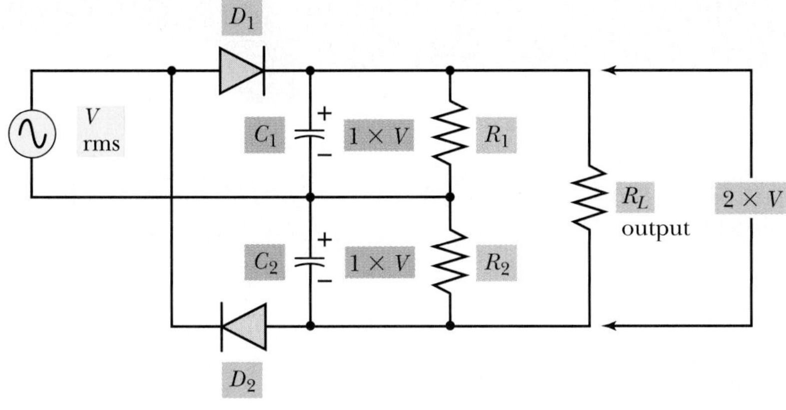

FIGURE 4–20 Full-wave voltage doubler

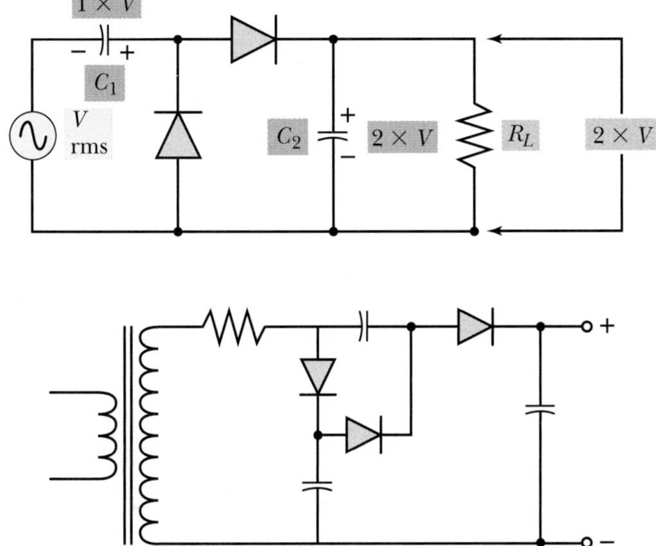

FIGURE 4–21 Half-wave cascaded voltage doubler

FIGURE 4–22 Voltage tripler circuit

The circuit in Figure 4–20 is a full-wave voltage double circuit. The rectified input voltage charges the capacitors series-aiding, with each one charging to the input voltage. Therefore, the output voltage is about twice the input voltage (under light load). This is called a "full-wave" doubler because there are two output pulses for every ac input cycle.

Figure 4–21 shows a different type of voltage doubler that is known as a half-wave cascaded voltage doubler. Capacitor C_2 is charged close to twice the ac input voltage. In this circuit, C_2 gets a pulse of charging current once during each ac input cycle, thus the "half-wave" designation. The reason C_2 charges to twice the input voltage is because the charge on C_1 is series-aiding with the source in the charge path for C_2. Thus, C_2 charges to the voltage across C_1 plus the peak value of the ac source.

Voltage doubler circuits are used where it is desirable to obtain a higher dc voltage without the use of a step-up transformer.

Figures 4–22 and 4–23 are the basic circuits for voltage tripler and quadrupler power supplies, respectively. The dc output from the tripler is about three times the peak input voltage, and the output from the quadrupler is approximately four times the peak input voltage (under light load).

Although it is possible to connect combinations of doublers and triplers to create voltage multiplication greater than four, it isn't practical to do so. The main problem is that the full-load current capability and percent of regulation drop off dramatically as you increase the number of times the input voltage is multiplied. Once you get past the quadrupler, voltage multiplier circuits have virtually no practical applications.

Application Problem

Industrial

Stereo Amplifiers

A stereo amplifier system used in automotive industry is shown in the following figure. The amplifier system drives two 8-Ω speakers per channel in a car. The amplifier produces the maximum voltage to speakers of 14 V. Determine a vehicle power supply size by calculating the power requirements of the amplifier.

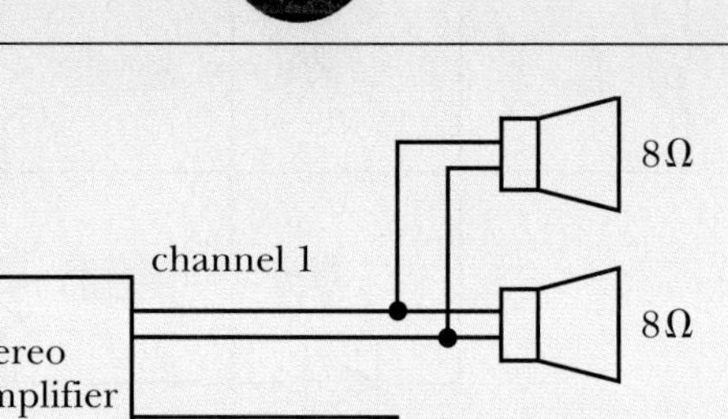

Stereo amplifier assembly

Solution Each channel has two speakers connected in parallel. We can calculate the required power for one speaker as follows:

$$P_{\text{speaker}} = \frac{V^2}{R} = \frac{14^2}{8} = 24.5 \text{ W}$$

The required power per channel is

$$P_{\text{channel}} = P_{\text{speaker}} \times 2 = 24.5 \times 2 = 49 \text{ W}$$

The stereo amplifier has 2 channels, thus the total power that the stereo amplifier must deliver is

$$P_{\text{amplifier}} = P_{\text{channel}} \times 2 = 49 \times 2 = 98 \text{ W}$$

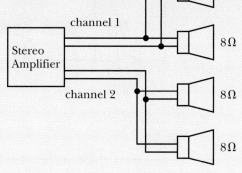

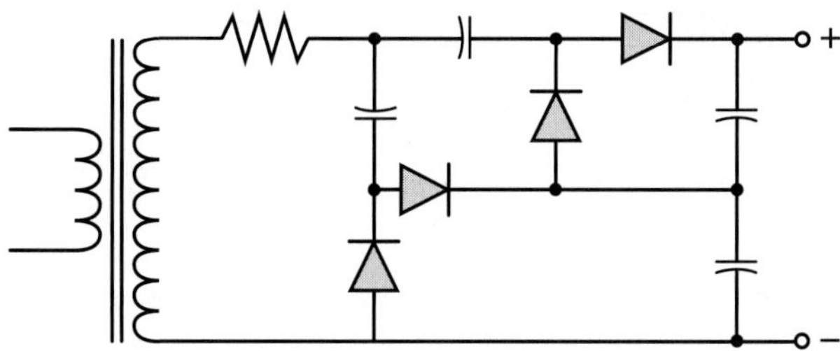

FIGURE 4–23 Voltage quadrupler circuit

4–6 Advanced Power Supply Topics

A power supply is known as both a power converter and a power conditioner; it is a device that converts power with one set of parameters into power with a different set of parameters to meet specific system requirements. The power supply's primary purpose is to convert input power into a stable voltage and/or current to operate electronic equipment.

There are four classifications of power conversion circuits: ac-ac converters (example: frequency changers); ac-dc converters (example; rectifiers); dc-ac converters (example: uninterruptible power supplies—PC UPS); and dc-dc converters (also called converters). The term "converter" is often used for all four classifications, with its true meaning being determined from the context.

Up to this point, our emphasis has been on converting an ac input signal into dc output voltages. A few examples of consumer electronic equipment that use the ac-dc converter are desktop personal computers, televisions, and home stereo systems. The output of the ac-dc converter might be a single voltage or multiple voltages. The desktop personal computer power supply is designed for an ac input of 115 V_{rms} and dc outputs of +5 V, +12 V and –12 V.

The category of emphasis for this section is the dc-dc converter. Several consumer electronic systems that use dc-dc converters are boom boxes, laptop computers, and camcorders. Some power supplies use the ac-dc converter to produce a dc voltage and the dc-dc converter to produce additional dc voltages. In some power supplies circuits, the ac-dc converter produces a raw dc voltage, and the dc-dc converter produces a regulated output voltage. Another use of dc-dc converters is to convert a battery's dc voltage into regulated dc voltages to power the electronic circuitry.

A critical element of both the ac-dc converter and the dc-dc converter is the regulator. The regulator is used to maintain a fixed dc output voltage. Three primary types of regulators are the shunt regulator, the series regulator, and the switching regulator.

4–7 Shunt Regulators (Zener Diodes)

One example of the shunt regulator is the zener diode regulator circuit shown in Figure 4–24. Ideally, the output voltage (V_{out}) of the zener diode regulator circuit is equal to the rating of the zener diode (V_Z) for all loads (R_L). In other words, the ideal zener diode regulator would have a voltage regulation of 0%. The desired intent of the circuit is to provide a stable output voltage to the load as variations in the input voltage and/or load resistance occur. To maintain the constant output voltage, the zener current will increase and decrease as the input voltage and/or load resistance changes. For the practical zener diode regulator circuit, the zener voltage will increase as the zener current increases, and the zener voltage will decrease as the zener current decreases. The output voltage appearing across the load is always equal to the zener voltage.

Several items must be considered when selecting components for the zener diode voltage regulator. The input voltage range, the output voltage requirement, and the full load resistance must be established. With this information, the resistance value of R_1 and the type of zener diode can be determined. The efficiency of the power supply regulator circuit is calculated by using Formula 4–17.

FORMULA 4–17 $\eta = \frac{P_L}{P_T} \times 100$

EXAMPLE A zener diode regulator circuit shown in Figure 4–24 consists of a 1N751A zener diode and 180 Ω R_1 resistor. The input voltage will vary from 15 V to 18 V. The output voltage must be 5 V ± 0.5 V. The full load resistance is 100 Ω.

Calculate the load current (I_L).

Calculate the power dissipated by the load (P_L).

When V_{in} is 15 V, calculate:

the current flowing through the R_1 resistor (I_T).

the power dissipated by the R_1 resistor (P_1).

the current flowing through the zener diode (I_Z).

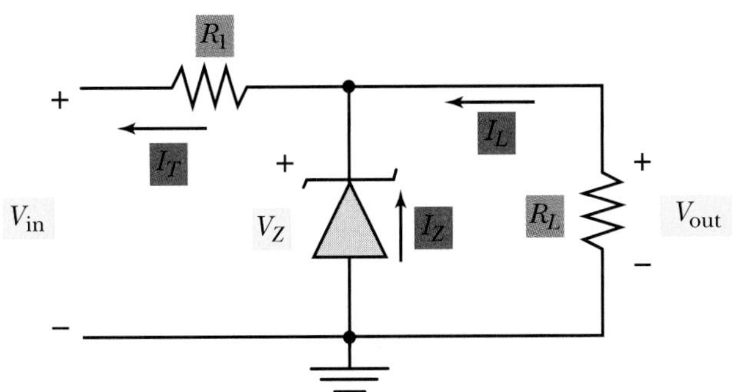

FIGURE 4–24 Shunt regulator (zener diode)

the power dissipated by the zener diode (P_Z).

the power supplied to the regulator circuit (P_T).

the efficiency (η) of the zener diode voltage regulator circuit.

When V_{in} is 18 V, calculate:

the current flowing through the R_1 resistor (I_T).

the power dissipated by the R_1 resistor (P_1).

the current flowing through the zener diode (I_Z).

the power dissipated by the zener diode (P_Z).

the power supplied to the regulator circuit (P_T).

the efficiency (η) of the zener diode voltage regulator circuit.

Select the power rating for the R_1 resistor.

Answer

The 1N751A is a 5.1 V zener diode (test current of 20 mA) with a power rating of 0.5 W. For the zener diode shunt regulator circuit, the output voltage is equal to the zener diode voltage. The output voltage for the regulator circuit is 5.1 V, which is within the 4.5 V to 5.5 V specified requirement. The load current can now be calculated.

$$I_L = \frac{V_L}{R_L}$$

$$I_L = \frac{5.1\text{ V}}{100\ \Omega}$$

$$I_L = 51\text{ mA}$$

The analysis here will evaluate only the ideal zener diode condition where the zener voltage will not change as the zener current changes. As the input voltage changes from 15 V to 18 V, the zener voltage remains constant causing the output voltage and load current to remain unchanged. The power dissipated by the load resistor is

$$P_L = I_L \times V_{out}$$

$$P_L = 51\text{ mA} \times 5.1\text{ V}$$

$$P_L = 260.1\text{ mW}$$

When V_{IN} is 15 V:

Using KVL, the voltage across the resistor R_1 is

$$V_1 = V_S - V_Z$$

$$V_1 = 15\text{ V} - 5.1\text{ V}$$

$$V_1 = 9.9\text{ V}$$

Using Ohm's law, the current flowing through the resistor R_1 is

$$I_T = \frac{V_1}{R_1}$$

$$I_T = \frac{9.9\text{ V}}{180\ \Omega}$$

$$I_T = 55.0\text{ mA}$$

The power dissipated by the resistor R_1 is

$$P_1 = V_1 \times I_T$$

$$P_1 = 9.9\text{ V} \times 55.0\text{ mA}$$

$$P_1 = 544.5\text{ mW}$$

Using KCL, the current flowing through the zener diode is

$$I_Z = I_T - I_L$$
$$I_Z = 55.0 \text{ mA} - 51.0 \text{ mA}$$
$$I_Z = 4.0 \text{ mA}$$

The power dissipated by the zener diode is

$$P_Z = V_Z \times I_Z$$
$$P_Z = 5.1 \text{ V} \times 4.0 \text{ mA}$$
$$P_Z = 20.4 \text{ mW}$$

The power supplied to the regulator circuit is

$$P_T = V_{in} \times I_T$$
$$P_T = 15 \text{ V} \times 55.0 \text{ mA}$$
$$P_T = 825 \text{ mW}$$

The efficiency of the zener diode voltage regulator circuit is calculated using Formula 4–17.

$$\eta = \frac{P_L}{P_T} \times 100$$
$$\eta = \frac{260.1 \text{ mW}}{825 \text{ mW}} \times 100$$
$$\eta = 31.5\%$$

When V_{in} is 18 V:
Using KVL, the voltage across the resistor R_1 is

$$V_1 = V_S - V_Z$$
$$V_1 = 18 \text{ V} - 5.1 \text{ V}$$
$$V_1 = 12.9 \text{ V}$$

Using Ohm's law, the current flowing through the resistor R_1 is

$$I_T = \frac{V_1}{R_1}$$
$$I_T = \frac{12.9 \text{ V}}{180 \, \Omega}$$
$$I_T = 71.7 \text{ mA}$$

The power dissipated by the resistor R_1 is

$$P_1 = V_1 \times I_T$$
$$P_1 = 12.9 \text{ V} \times 71.7 \text{ mA}$$
$$P_1 = 924.9 \text{ mW}$$

Using KCL, the current flowing through the zener diode is

$$I_Z = I_T - I_L$$
$$I_Z = 71.7 \text{ mA} - 51.0 \text{ mA}$$
$$I_Z = 20.7 \text{ mA}$$

The power dissipated by the zener diode is

$$P_Z = V_Z \times I_Z$$
$$P_Z = 5.1 \text{ V} \times 20.7 \text{ mA}$$
$$P_Z = 105.6 \text{ mW}$$

The power supplied to the regulator circuit is

$$P_T = V_{in} \times I_T$$
$$P_T = 18 \text{ V} \times 71.7 \text{ mA}$$
$$P_T = 1291 \text{ mW}$$

The efficiency of the zener diode voltage regulator circuit is calculated using Formula 4–17.

$$\eta = \frac{P_L}{P_T} \times 100$$

$$\eta = \frac{260.1\text{ mW}}{1291\text{ mW}} \times 100$$

$$\eta = 20.2\%$$

The minimum required power rating for the resistor R_1 is 1 W. It is recommended that a 2 W resistor be used to limit the temperature to be dissipated by the resistor.

Note that if the maximum zener diode power dissipation exceeds the 500 mW manufacturer's rating, a higher wattage zener diode would be required. ◆

Shunt regulators are also available in linear integrated circuit (IC) modules.

4–8 Series Regulators

Fixed Positive-Voltage Series Regulator

An example of the positive voltage output IC series regulator is the 78XX family of devices (7805, 7812, and 7815). These devices include internal circuitry that provides current-limiting and thermal-shutdown making them indestructible to overload conditions. If the voltage regulator output experiences a very small resistance or a short to ground, the current-limiting feature of the voltage regulator will control the output current preventing damage to the voltage regulator. When the internal junction temperature of the voltage regulator exceed 150° C, the voltage regulator shuts down and will no longer provide an output voltage until the internal junction temperature cools down. The high junction temperature can be caused by high power dissipation by the voltage regulator and/or a high temperature in the air where the voltage regulator is operating.

Some important parameters of the series voltage regulator are provided in Figure 4–25. It should be noted that the minimum input voltage for the voltage regulators must be at least 2 to 2.5 V greater than the specified output voltage to operate properly. The maximum input voltage should not be exceeded for the device to function correctly. Exceeding the absolute maximum input voltage (a parameter not included in this table) has the potential of destroying the device. The bias current is used to provide power to the internal circuitry of the regulator. The 78XX family of devices is available with different package options. The packaging of the device strongly affects the power dissipation capability of the voltage regulator. As indicated in the table, the TO 220 package has a power rating of 2 W. The voltage regulator can experience thermal-shutdown even though the package is rated to handle 2 W of power. Remember—the thermal-shutdown is caused by the internal junction temperature exceeding

Device	V_{in}min V	V_{in}max V	Bias current I_B mA	V_{out} V	Power rating P_{REG} W	Package
7805	7	25	4.2	5	2	TO 220
7812	14.5	30	4.3	12	2	TO 220
7815	17.5	30	4.4	15	2	TO 220
7905	–7	–25	1	–5	2	TO 220
7912	–14.5	–30	1.5	–12	2	TO 220
7915	–17.5	–30	1.5	–15	2	TO 220

Typical values

$T_{AMBIENT}$ = 25° C with no heat sink

Note: Additional package options are available.
Bias current and power rating may be different.

FIGURE 4–25 Series voltage regulator parameters

150° C, which is influenced by the power dissipated within the device and the air temperature around the device. Additional parameters are available by obtaining a copy of the component's datasheet.

Figure 4–26 shows us a circuit using the series voltage regulator. The power dissipated by the linear IC regulator can be calculated using Formula 4–18. The efficiency of the regulator is found using Formula 4–17.

FORMULA 4–18 $P_{REG} = I_B \times V_{in} + I_L \times (V_{in} - V_{out})$

◆ **EXAMPLE** The 7805 voltage regulator is used in the circuit shown in Figure 4–26. The input voltage will vary from 15 V to 18 V. The full load resistance is 100 Ω.

Calculate the load current (I_L).

Calculate the power dissipated by the load (P_L).

When V_{in} is 15 V, calculate:

the regulator input current (I_T).

the power dissipated by the voltage regulator (P_{REG}).

the power supplied to the regulator circuit (P_T).

the efficiency (η) of the voltage regulator circuit.

When V_{in} is 18 V, calculate:

the regulator input current (I_T).

the power dissipated by the voltage regulator (P_{REG}).

the power supplied to the regulator circuit (P_T).

the efficiency (η) of the voltage regulator circuit.

Answer

The 7805 has an output voltage of 5 V. The load current is found using Ohm's law.

$$I_L = \frac{V_L}{R_L}$$

$$I_L = \frac{5\text{ V}}{100\ \Omega}$$

$$I_L = 50\text{ mA}$$

The power dissipated by the load resistor is

$$P_L = I_L \times V_{out}$$

$$P_L = 50\text{ mA} \times 5\text{ V}$$

$$P_L = 250\text{ mW}$$

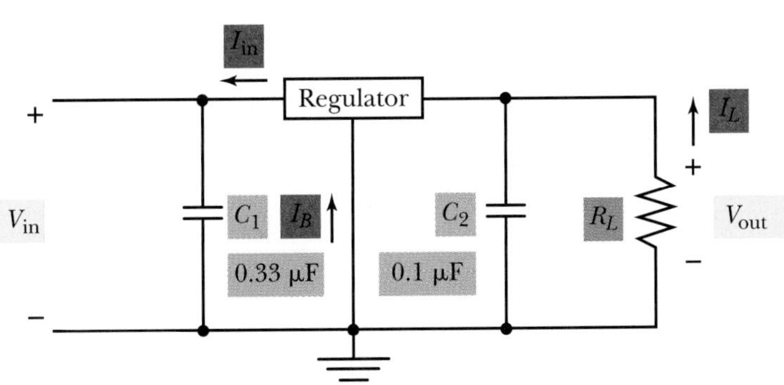

FIGURE 4–26 Fixed voltage series voltage regulator

When V_{in} is 15 V:
The current flowing into the voltage regulator is

$$I_T = I_L + I_B$$
$$I_T = 50.0 \text{ mA} + 4.2 \text{ mA}$$
$$I_T = 54.2 \text{ mA}$$

Using Formula 4–18 the power dissipated by the regulator is

$$P_{REG} = I_B \times V_{in} + I_L \times (V_{in} - V_{out})$$
$$P_{REG} = 4.2 \text{ mA} \times 15 \text{ V} + 50.0 \text{ mA} \times (15 \text{ V} - 5 \text{ V})$$
$$P_{REG} = 563 \text{ mW}$$

The power supplied to the voltage regulator is

$$P_T = I_T \times V_{in}$$
$$P_T = 54.2 \text{ mA} \times 15 \text{ V}$$
$$P_T = 813 \text{ mW}$$

The efficiency of the voltage regulator circuit is calculated using Formula 4–17.

$$\eta = \frac{P_L}{P_T} \times 100$$
$$\eta = \frac{250 \text{ mW}}{813 \text{ mW}} \times 100$$
$$\eta = 30.75\%$$

When V_{IN} is 18 V:
The current flowing into the voltage regulator is

$$I_T = I_L + I_B$$
$$I_T = 50.0 \text{ mA} + 4.2 \text{ mA}$$
$$I_T = 54.2 \text{ mA}$$

Using Formula 4–18 the power dissipated by the regulator is

$$P_{REG} = I_B \times V_{in} + I_L \times (V_{in} - V_{out})$$
$$P_{REG} = 4.2 \text{ mA} \times 18 \text{ V} + 50.0 \text{ mA} \times (18 \text{ V} - 5 \text{ V})$$
$$P_{REG} = 725.6 \text{ mW}$$

The power supplied to the voltage regulator is

$$P_T = I_T \times V_{in}$$
$$P_T = 54.2 \text{ mA} \times 18 \text{ V}$$
$$P_T = 975.6 \text{ mW}$$

The efficiency of the voltage regulator circuit is calculated using Formula 4–17.

$$\eta = \frac{P_L}{P_T} \times 100$$
$$\eta = \frac{250 \text{ mW}}{975.6 \text{ mW}} \times 100$$
$$\eta = 25.6\%$$

◆

PRACTICE PROBLEM 5

Figure 4–26 voltage regulator uses the 7815. The input voltage is 24 V and the load resistance is 100 Ω. Calculate the power dissipated by the voltage regulator and the efficiency of the regulator circuit.

Fixed Negative-Voltage Series Regulator

The 79XX family of devices (7905, 7912, and 7915) is a negative voltage output IC series regulator. The 79XX device can be used as the regulator in the Figure 4–26 schematic. The input to these devices must be a negative voltage as defined in Fig-

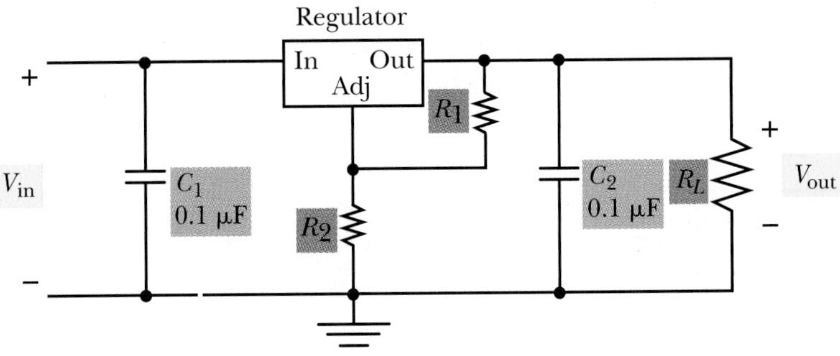

FIGURE 4–27 Adjustable series voltage regulator

ure 4–25. Formula 4–18 and Formula 4–19 will be used to calculate the power dissipated by the voltage regulator and the efficiency of the voltage regulator. The 79XX family of devices is provided with current-limiting and thermal-shutdown protection features.

Adjustable Voltage Regulator

The positive adjustable series regulator (LM317) and the negative adjustable series regulator (LM337) allow the user to design the desired output voltage. Figure 4–27 is a common circuit configuration for the LM317. The LM317 requires a positive dc input voltage resulting in a regulated positive dc output voltage. The LM337 requires a negative dc input voltage resulting in a regulated negative dc output voltage. The regulated output voltage for the circuit is calculated using Formula 4–19.

FORMULA 4–19 $$V_{out} = V_{REF} \times \left(\frac{1 + R_2}{R_1} \right)$$

where $V_{REF} = 1.25$ V for LM317 or $V_{REF} = -1.25$ V for LM337 and $R_1 = 240\ \Omega$

4–9 Switching Regulators

The low efficiency of the shunt and the series regulators is a disadvantage. The power dissipated by the series and shunt elements is high and is an energy loss to the system. More efficient regulation occurs utilizing the switching regulator. This improvement is accomplished by turning on and off a pass transistor (a BJT or MOSFET component). The three configurations of switching regulators are the Buck, Boost, and Flyback.

The Buck regulator (Figure 4–28) is a step-down regulator. The unregulated dc input is always larger than the regulated dc output voltage. The output voltage of the Buck regulator is calculated using Formula 4–20 and Formula 4–21.

FORMULA 4–20 $$D = \frac{T_{on}}{T_{on} + T_{off}}$$

FORMULA 4–21 $$V_{out} = V_{in} \times D$$

The Boost regulator (Figure 4–29) is a step-up regulator. The unregulated dc input is always less than the regulated dc output voltage. The output voltage of the Boost regulator is calculated using Formula 4–20 and Formula 4–22.

FORMULA 4–22 $$V_{out} = \frac{V_{in}}{1 - D}$$

Practical Notes

Some power supplies require a minimum load connected to them to provide the stable regulated output voltage.

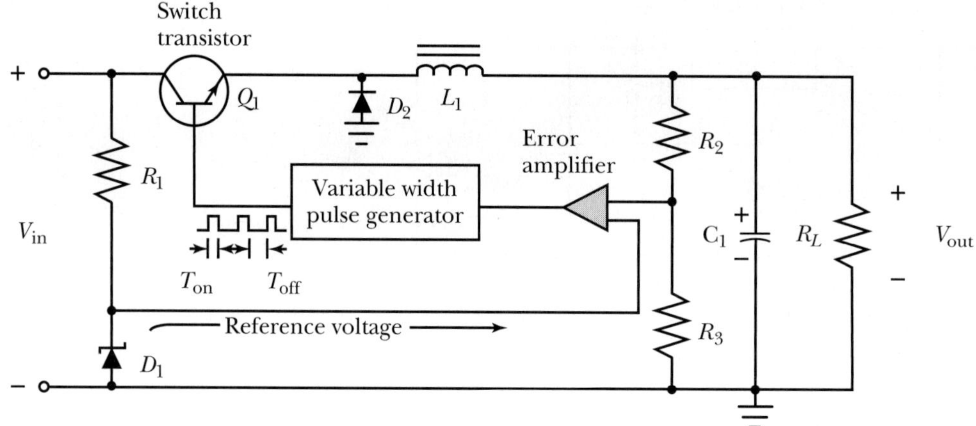

FIGURE 4–28 Buck switching voltage regulator

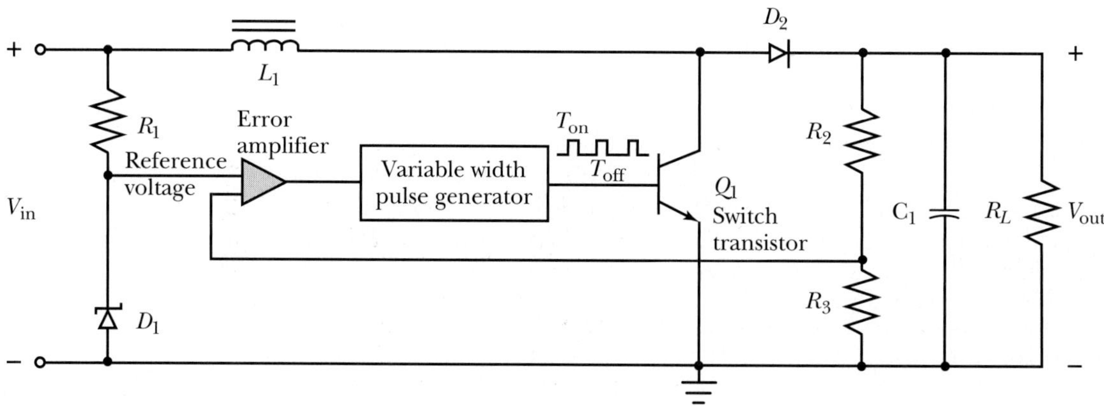

FIGURE 4–29 Boost switching voltage regulator

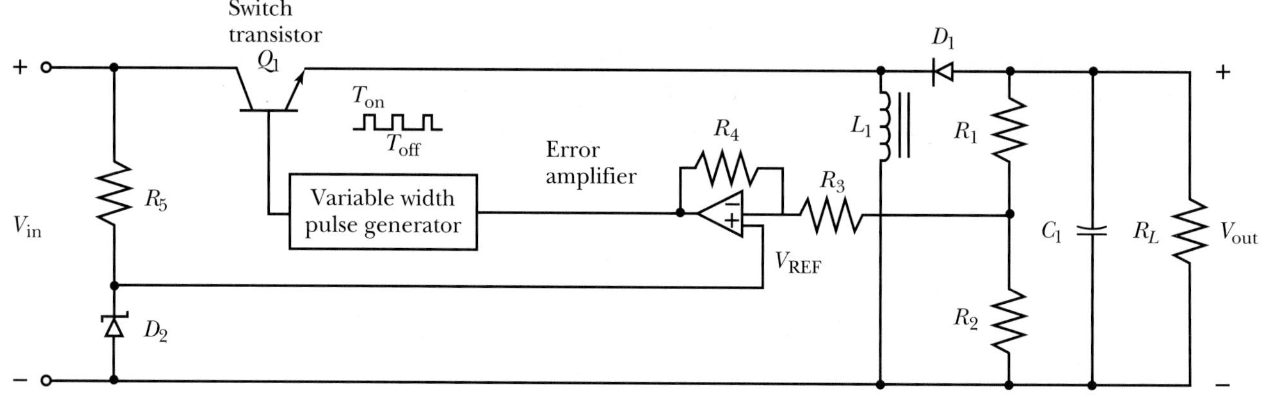

FIGURE 4–30 Flyback switching voltage regulator

The Flyback regulator (Figure 4–30) allows the dc regulated output voltage to be a smaller or a larger value than the unregulated dc input voltage. However, the output is always opposite in polarity. The output voltage of the Flyback regulator is calculated using Formula 4–20 and Formula 4–23.

FORMULA 4–23 $V_{out} = \dfrac{-V_{in} \times D}{(1 - D)}$

Switching power supplies operate at high frequencies (20 kHz to 1 MHz), whereas linear power supplies operate at 60 Hz. Magnetic components (transformers and inductors) operating at high frequencies are physically smaller. Since very little power is dissipated by the switching transistor and control circuitry, the output power is approximately the same as the input power ($P_{out} \approx P_{in}$) resulting in the high efficiency of the switching regulator. With the high operating frequency and the high efficiency, the switching power supplies are more compact than linear power supplies.

Even though you could build each of the power supply configurations using discrete components, the power supplies are readily available in prepackaged assemblies. Due to the special packaging concerns of switching regulators, it is advantageous to use a prepackaged assembly complying with your electronic system requirements. Additional features common included within power supplies are surge and over voltage protection, overload protection, and thermal shutdown.

Practical Notes

Even though the power supply load is constantly changing and the input voltage can be fluctuating, the power supply output must maintain an extremely stable dc voltage.

Practical Notes

It is not unusual for power transformers to operate at high temperatures that seem a bit uncomfortable to the touch. Whenever a transformer contains shorted windings (a definite trouble that must be fixed), it runs much hotter than normal. Often, a strong odor of melted wax or burned insulation is given off, and/or some smoke, and areas showing discoloration may be seen.

4–10 Troubleshooting Power Supplies

A common trouble with rectifiers is short circuits. A shorted diode places a short across the ac input to the power supply, often blowing a fuse and overheating a transformer that might be connected at the input to the rectifier portion of the circuit. The voltage is lower than normal at the dc output of the power supply.

The best way to check a rectifier while it remains in the circuit is to measure the ac voltage across each diode. If the diode is shorted, the ac voltage reading across it is virtually zero. If the diode is open, the ac voltage across it is much higher than normal. You can also check a rectifier by removing it from the circuit and measuring its forward and reverse resistance with an ohmmeter. The diode is defective if the forward and reverse resistance readings are not vastly different.

Often, electrolytic capacitors are used in the filter section of power supplies. When a capacitor fails, an internal short circuit is usually the problem. Short circuits in capacitors are usually caused by voltage spikes that exceed the capacitor's rated value. Shorted capacitors quickly overheat, sometimes exploding or popping open with a sharp cracking sound. Capacitors in this condition are quite easy to locate by visual inspection. (They may give off a strong acrid odor as well.)

To check for open filter capacitors in an operating circuit, measure the *ac voltage* (the ripple level) across each capacitor. If the capacitor is good, the ripple level will be within tolerance for the power supply. If it is open, the ripple voltage is a large percentage of the rated dc output level.

If the power supply uses filter inductors, or chokes, they are usually connected in series with the load current. Excessive load current causes too much current to flow through the filter choke. Overheating is the result. Shorted windings within a filter choke also cause it to overheat. The symptoms are virtually identical with those you have seen earlier for shorted power transformers.

An open filter choke is easy to diagnose because (1) it will have full rectifier voltage at the input end and no voltage at the output end, and (2) an ohmmeter check will show infinite resistance across the choke when it is disconnected from the circuit.

Troubles in voltage regulators are easily detected by comparing the input and output voltages for the regulator circuit. If the voltage at the input of the regulator is within its normal range, but the output of the regulator is not, the trouble is most likely with the regulator.

Practical Notes

Some integrated-circuit regulators have a built-in circuit protection feature that greatly reduces the output voltage in the event of excessive load current. The fact that the output of the regulator is lower than normal might indicate a short-circuit condition in the load rather than a faulty regulator circuit.

Application Problem

Communications

INDUCTION COIL FOR AN ANTENNA SYSTEM

A broadcast antenna coil has an unknown coupling coefficient. We can very simply calculate the coil coupling using the circuit shown in the following figure. We can vary the capacitance so that there is no deflection on the galvanometer while cycling the switch ON-OFF. If the capacitor value is 0.88 μF, calculate the coupling coefficient.

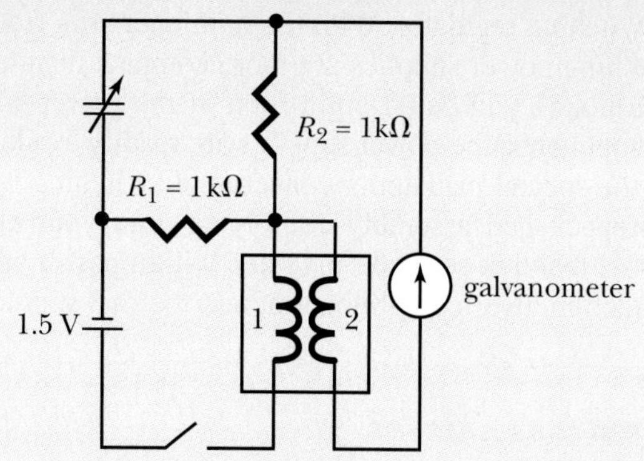

Solution

$L_{12} = R_1 R_2 C = 1000 \times 1000 \times 0.88 \times 10^{-6} = 0.88$

The coupling coefficient between two coils is 88%.

Troubleshooting Hints

First, isolate a power supply problem to the general section of trouble by carefully considering the symptoms. Recall the main sections of a typical power supply are the transformer, the rectifier(s), the filter network, and possibly a regulator circuit. Excellent tools to help you with symptoms are the oscilloscope to look at waveforms, and/or a volunteer to make voltage measurements. An ohmmeter can also bc helpful for checking rectifier diodes as described in the previous chapter. Your eyes become a tool as you can see burned parts, blown fuses, and so forth, and you can use them to analyze the readings and scope patterns. Your nose is a great symptom locator to find overheated parts. If the system you are troubleshooting involves an audio output (such as a radio receiver or cassette tape player) your ears are a very helpful tool, since you can hear hum and can probably tell if it is 60 Hz or 120 Hz, and you can hear "motor-boating."

By noting whether there is excessive ripple in the output and zero or too low an output voltage, you can isolate the section you might want to troubleshoot.

Troubles in the transformer section of a power supply can include power sources feeding the primary of the transformer (loose or broken connections), fuses in the primary and/or secondary, and the transformer itself. Transformers usually don't go bad by themselves. Usually some external circuit or component connected to the transformer causes too much current through the transformer circuit that either blows a fuse or damages the transformer. The best way to check a transformer is to measure the voltage on the primary and secondary windings and compare the results with known norms.

USING TECHNOLOGY: DIODES USED IN A SYSTEM

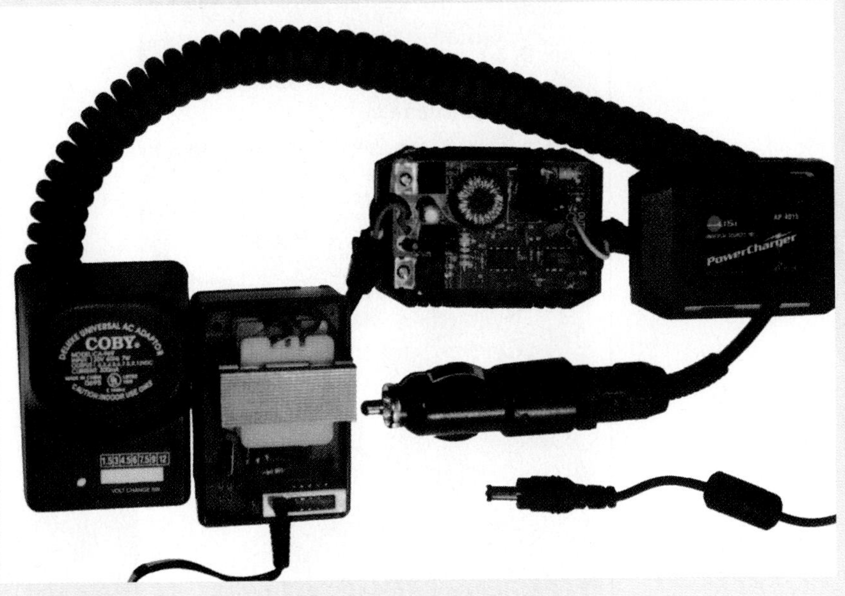

All electronic systems require a source of power to function. The power may be provided from a battery pack allowing the product to be portable, a solar cell allowing the system to be installed in a remote location, such as a highway emergency phone, or the most commonly connected to an ac power source. Throughout the chapter you have been provided with fundamental concepts of power supply circuits. Some of the power supply circuits are designed specifically to convert the ac power source into the dc voltages required to operate a system. Other power supply circuits are configured to convert a dc power source into another dc voltage required by a system, or from a dc power source to an ac voltage.

One of the most common uses for a diode is as a rectifier in a system application. Almost every portable battery-operated product is equipped with a DC IN jack to allow the use of an ac-dc adaptor. The operation of such a dc power pack was described previously in Chapter 3. We will now turn our attention to one of the most common electronic products available, the ac to dc adaptor.

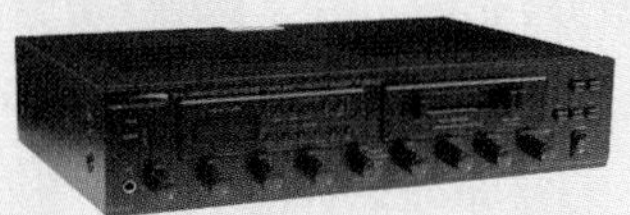

Figure 4a shows a picture of this type of ac to dc adaptor. You will note it consists of a front panel multivoltage selector switch and includes six color-coded snap-on dc adaptor plugs. The color-coded tips make it easy to identify the correct plug to use with the individual DC IN jack on the electronic product to be powered with the ac-dc adaptor. These color-coded jacks can be inserted in the plug from the ac-dc adaptor in either of two ways, allowing for a quick change in the polarity of the supplied dc voltage. Six adaptor jacks are supplied, each one classified by style, either tip or ring, and diameter: 3.5 mm tip, 1.5 mm ring, 1.7 mm ring, 2.1 mm ring, 2.5 mm ring and 3.5 mm ring.

The rectifier diode is designed to control the direction of current flow. The primary usage of the rectifier diode is to convert an incoming ac voltage from the local electric utility to a dc voltage for supplying the required voltages and currents to operate the electronic product. Rectifier diodes are specifically designed to handle large currents, withstand high peak inverse voltages, and dissipate significant heat.

The system application block diagram for the common ac-dc adaptor is shown in Figure 6. This shows the ac input, step-down transformer, voltage selector switch, bridge rectifier, filter circuit, and dc output. The incoming ac is applied to the step-down transformer and to the selector switch which can choose any of seven different low ac voltages. The selected ac voltage is rectified by the full wave bridge rectifier to a pulsating dc voltage. The pulses are applied to an electrolytic capacitor which filters the ac

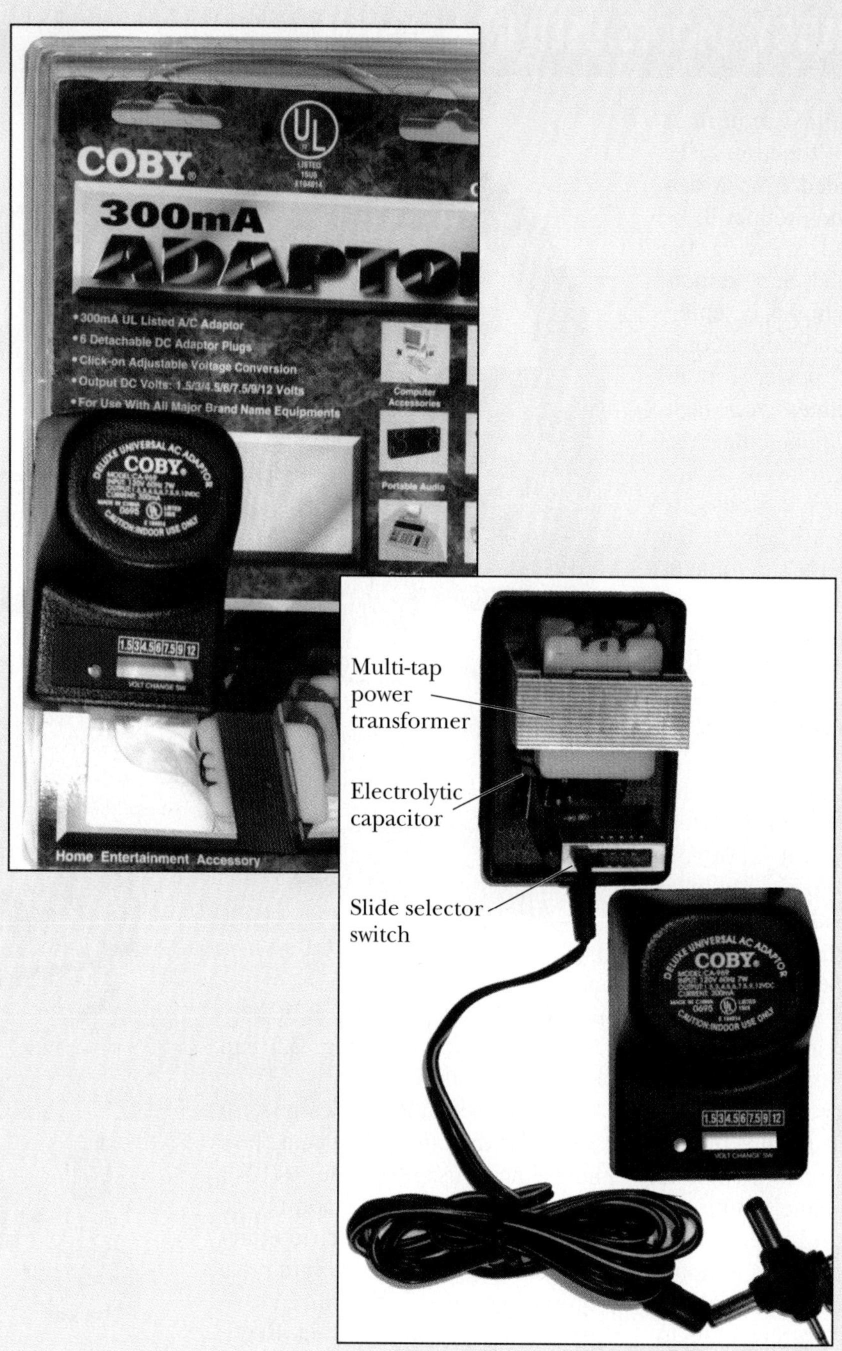

ripple portion of the voltage to supply a fixed dc voltage with a much reduced ac ripple voltage.

The circuit schematic in Figure 7 shows the details of the circuit. The input circuit consists of a polarized ac line plug. The wide blade sets up an earth ground for the ac-dc adaptor. Capacitor C1 is a 0.1 uf ceramic disc connected from the hot wire to the earth ground to filter out ac line noise. The auto reset circuit breaker, F1, protects the transformer and other components from excessive line voltages, but most importantly, from excessive current demands from the load.

The multivoltage line transformer steps down the incoming 117 vac from the plug to seven preset voltages from preselected taps on the secondary winding of the transformer. A slide style selector switch S1 allows a choice of seven ac voltages: 1.5 V, 3 V, 4.5 V, 6 V, 7.5 V,

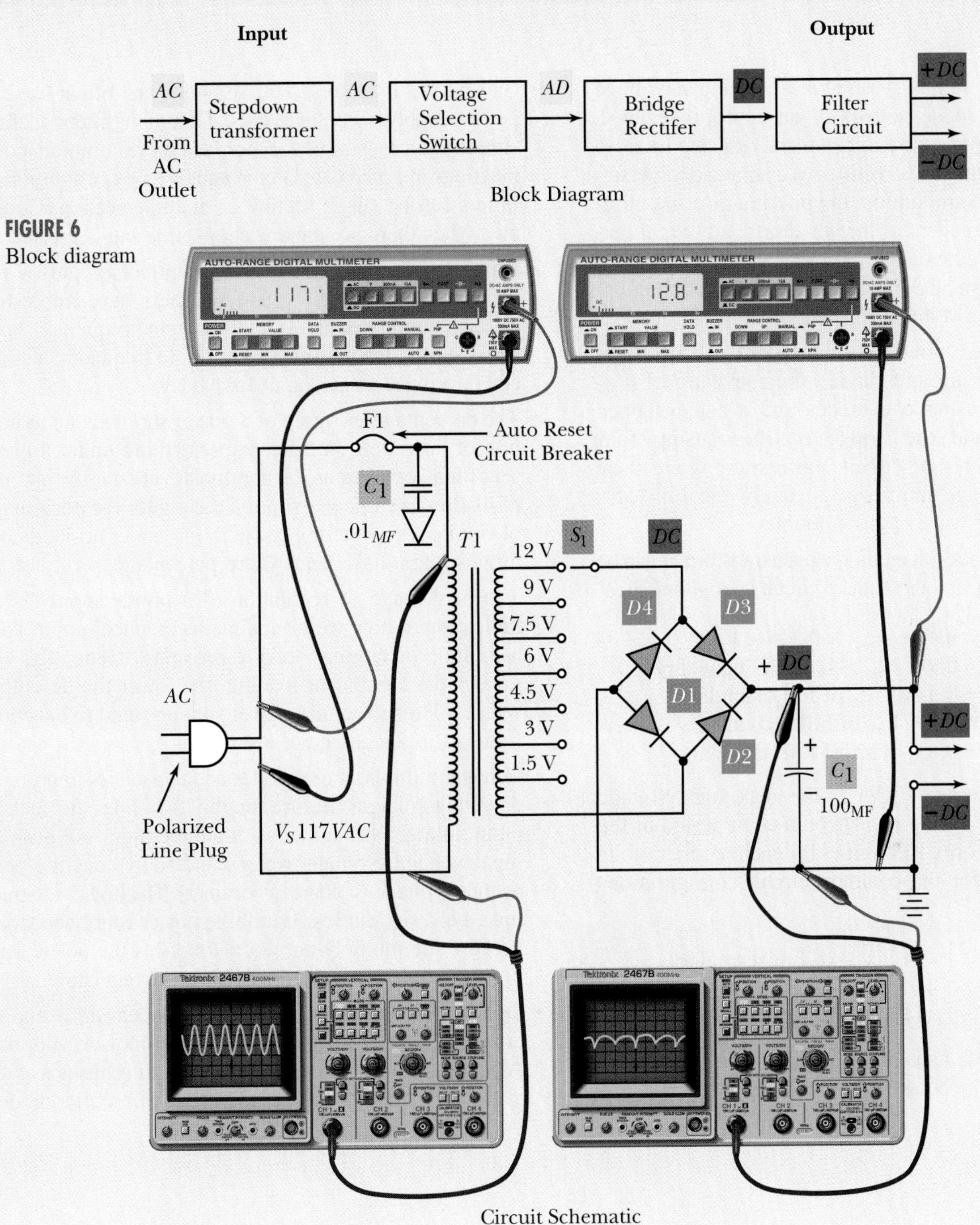

FIGURE 6 Block diagram

FIGURE 7 Circuit Schematic

9 V or 12 V. This ac voltage is applied to the opposite sides of a full wave bridge rectifier. The bridge rectifier converts the ac to a pulsating dc voltage. A 100 uF electrolytic capacitor charges up to the selected dc voltage and between the pulses supplies a current to smooth or filter out the variations in the pulsating dc voltage. The positive and negative dc voltages appear at the top and bottom contacts of the ac-dc adaptor plug. By switching the way the proper adaptor jack is inserted into this plug, the polarity appearing at the inner tip and at the outer ring of the adaptor jack can be reversed. The most common polarity required by most portable products is for the inner tip to be positive and the outer ring to be negative. This follows the common rule that the product's chassis is grounded not energized.

Summary

- A typical power supply system takes existing ac voltage and converts it to dc voltage of an appropriate level. Many power supplies use a transformer to step up or to step down ac voltages, a rectifier to change ac to pulsating dc, a filter to smooth out the pulsating dc and eliminate the ac component of the waveform, an output network (in some cases) that can be a voltage divider, a bleeder, and so on, and, in some cases, a voltage regulator that keeps the voltage constant with varying loads.
- Common rectifier circuit configurations are the half-wave rectifier (using one diode), the conventional full-wave rectifier (using two diodes and a center-tapped transformer), and the bridge rectifier (using four diodes). Other special circuit configurations are used that act as voltage multipliers, namely the full-wave doubler and half-wave cascade doubler.
- Unfiltered dc output voltages for a given transformer can be computed for the various standard circuit configurations:

$$V_{dc}\text{ (H-W)} = 0.45 \times V_{rms} \text{ or } 0.318 \times V_{pk}$$
$$V_{dc}\text{ (F-W)} = 0.9 \times V_{rms} \text{ of half the secondary}$$
$$= 0.318 \times V_{pk} \text{ of full secondary}$$
$$V_{dc}\text{ (Bridge)} = 0.9 \times V_{rms} \text{ of full secondary}$$
$$= 0.637 \times V_{pk} \text{ of full secondary}$$

- Peak inverse voltage (PIV) is the maximum voltage across the diode during its nonconduction period of the applied ac waveform. PIV values for each rectifier diode (without filter) for the common circuit configurations are as follows:

$$\text{PIV (H-W)} = 1.414 \times V_{rms} \text{ of full secondary}$$
$$\text{PIV (F-W)} = 1.414 \times V_{rms} \text{ of full secondary}$$
$$\text{PIV (Bridge)} = 1.414 \times V_{rms} \text{ of full secondary}$$

(NOTE: With capacitor filters, peak inverse voltage can be as high as $2.828 \times V_{rms}$.)

- Capacitors, inductors, and resistors are often used in power supply filter networks. Common filters include simple C types, simple L types, and more common, combinations of L and C in L-type and pi-type configurations. Filters can be single section or multiple section. Generally, when there are more sections, filtering is greater.
- Ripple frequency relates to the number of pulses per cycle of ac input. Half-wave rectifiers have ripple frequency outputs equal to the ac input frequency. Full-wave and bridge rectifiers have ripple frequency outputs equal to twice the ac input frequency.
- The percentage of ripple of a power supply is an indication of how well the filter is performing under a given set of load conditions. Generally, the greater the amount of loading on a power supply, the higher the percentage of ripple. A power supply operating under no-load conditions might have nearly zero percentage of ripple.
- The percentage of regulation of a power supply is an indication of how well it maintains its rated output voltage under changing load conditions. Generally, the greater the amount of loading, the lower the dc output voltage. Higher-quality power supplies tend to have low figures for percentage of regulation.
- Voltage multipliers use diodes and capacitors to produce dc output voltages that are roughly two times the peak ac input voltage (voltage doubler), three times the peak ac input voltage (voltage tripler), and four times the peak ac input voltage (voltage quadrupler). The higher the multiple, however, the less useful the power supply becomes because the multiplying process reduces the power supply's current capacity and percentage of regulation.
- Oscilloscope waveforms and voltage measurements are commonly used for troubleshooting power supplies. Also, the ohmmeter is useful for checking fuses as well as the forward and reverse resistances of rectifier diodes.

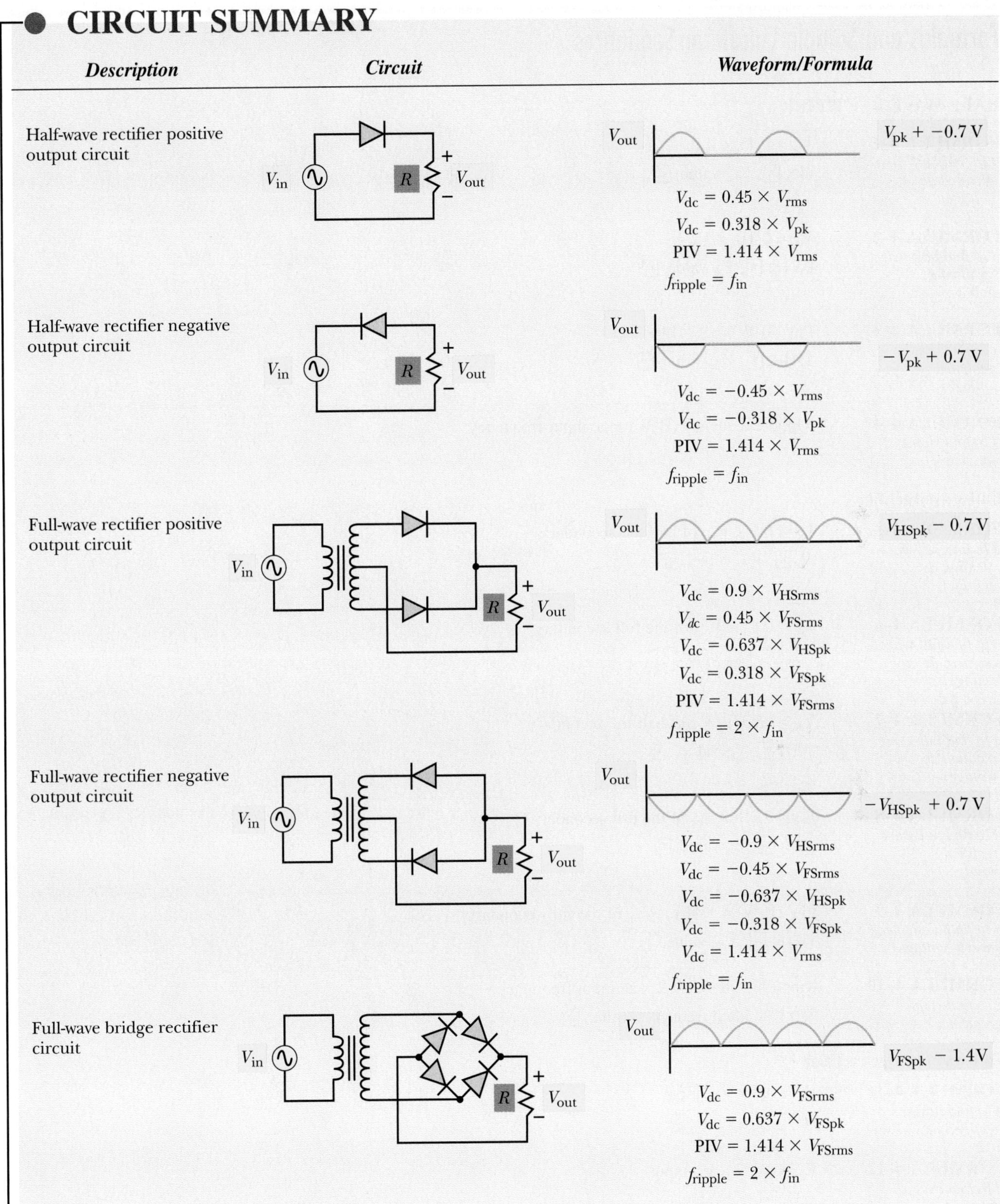

FIGURE 4–31 Rectifier diode circuit summary

Formulas and Sample Calculator Sequences

HALF-WAVE RECTIFIER

FORMULA 4–1
(To find half-wave rectified dc voltage)

$V_{dc} = 0.45 \times V_{rms}$

.45, $\boxed{\times}$, V_{rms} value, $\boxed{=}$

FORMULA 4–2
(To find half-wave rectified dc voltage)

$V_{dc} = 0.318 \times V_{pk}$

0.318, $\boxed{\times}$, V_{pk} value, $\boxed{=}$

FORMULA 4–3
(To find diode peak inverse voltage)

PIV (II-W) = $1.1414 \times V_{rms}$

1.414, $\boxed{\times}$, V_{rms} value, $\boxed{=}$

FORMULA 4–4
(To find ripple frequency)

Ripple Frequency (H-W) = ac input frequency

Full-wave rectifier

FORMULA 4–5
(To find full-wave rectified dc voltage)

$V_{dc} = 0.9 \times V_{rms}$ of half the secondary

.9, $\boxed{\times}$, V_{rms} value, $\boxed{=}$

FORMULA 4–6
(To find full-wave rectified dc voltage)

$V_{dc} = 0.45 \times V_{rms}$ of the full secondary

.45, $\boxed{\times}$, V_{rms} value, $\boxed{=}$

FORMULA 4–7
(To find full-wave rectified dc voltage)

$V_{dc} = 0.637 \times V_{pk}$ of half the secondary

.637, $\boxed{\times}$, V_{pk} value, $\boxed{=}$

FORMULA 4–8
(To find full-wave rectified dc voltage)

$V_{dc} = 0.318 \times V_{pk}$ of the full secondary

.318, $\boxed{\times}$, V_{pk} value, $\boxed{=}$

FORMULA 4–9
(To find diode peak inverse voltage)

PIV (F-W) = $1.414 \times V_{rms}$ of the full secondary

1.414, $\boxed{\times}$, V_{rms} value, $\boxed{=}$

FORMULA 4–10
(To find ripple frequency)

Ripple Frequency = 2 × ac input frequency

2, $\boxed{\times}$, ac input frequency value, $\boxed{=}$

Full-wave bridge rectifier

FORMULA 4–11
(To find bridge rectified dc voltage)

$V_{dc} = 0.9 \times V_{rms}$ input

.9, $\boxed{\times}$, V_{rms} value, $\boxed{=}$

FORMULA 4–12
(To find bridge rectified dc voltage)

$V_{dc} = 0.637 \times V_{pk}$ input

.637, $\boxed{\times}$, V_{pk} value, $\boxed{=}$

FORMULA 4–13
(To find diode peak inverse voltage)

PIV (Bridge) = $1.414 \times V_{rms}$ input

1.414, $\boxed{\times}$, V_{rms} value, $\boxed{=}$

FORMULA 4–14
(To find ripple frequency)

Ripple Frequency (Bridge) = 2 × ac input frequency

2, $\boxed{\times}$, ac input frequency value, $\boxed{=}$

FORMULA 4–15
(To find power supply ripple percentage)

% Ripple = ac/dc × 100

ac ripple, [÷], dc voltage, [×], 100 [=]

FORMULA 4–16
(To find percentage of regulation)

% Regulation = $((V_{NL} - V_{FL}) / V_{FL}) \times 100$

[(], V_{NL} value, [−], V_{FL} value, [)], [×], 100, [÷], V_{FL} value, [=]

FORMULA 4–17
(To find efficiency)

$$\eta = \frac{P_L}{P_T} \times 100$$

P_L value, [÷], P_T value, [×], 100, [=]

FORMULA 4–18
(To find regulatory power)

$P_{REG} = I_R \times V_{in} + I_L \times (V_{in} - V_{out})$

I_B value, [×], V_{in} value, [+], [(], I_L value, [×], [(], V_{in} value, [−], V_{out} value, [)], [)], [=]

FORMULA 4–19
(To find LM317 or LM337 output voltage)

$V_{out} = V_{REF} \times (1 + R_2/R_1)$ where $V_{REF} = 1.25$ V for LM317 or $V_{REF} = -1.25$ V for LM337 and $R_1 = 240\ \Omega$

V_{REF} value, [×], [(], 1, [+], [(], R_2 value, [÷], R_1 value, [)], [)], [=]

FORMULA 4–20
(To find duty cycle)

$D = (T_{on} / (T_{on} + T_{off}))$

T_{on} value, [÷], [(], T_{on} value, [+], T_{off} value, [)], [=]

FORMULA 4–21
(To find Back regulator output voltage)

$V_{out} = V_{in} \times D$

V_{in} value, [×], D value, [=]

FORMULA 4–22
(To find Boost regulator output voltage)

$V_{out} = V_{in} / (1 - D)$

V_{in} value, [÷], [(], 1, [−], D value, [)], [=]

FORMULA 4–23
(To find Flyback regulator output voltage)

$V_{out} = -V_{in} \times D / (1 - D)$

V_{in} value, ±, [×], D value, [÷], [(], 1, [−], D value, [)], [=]

Using Excel

Power Supply Formulas

(Excel file reference: FOE4_01.xls)

DON'T FORGET! It is NOT necessary to retype formulas, once they are entered on the worksheet! Just input new parameters data for each new problem using that formula, as needed.

- Use the Formula 4–2 spreadsheet sample and the parameters given for Practice Problems 2. Solve for the half wave rectifier dc output voltage. Check your answer against the answer for this question in the Appendix.
- Use the Formula 4–16 spreadsheet sample and the parameters given for Chapter Problems #9. Solve for the power supply percent of regulation. Check your answer against the answer for this question in the Appendix.

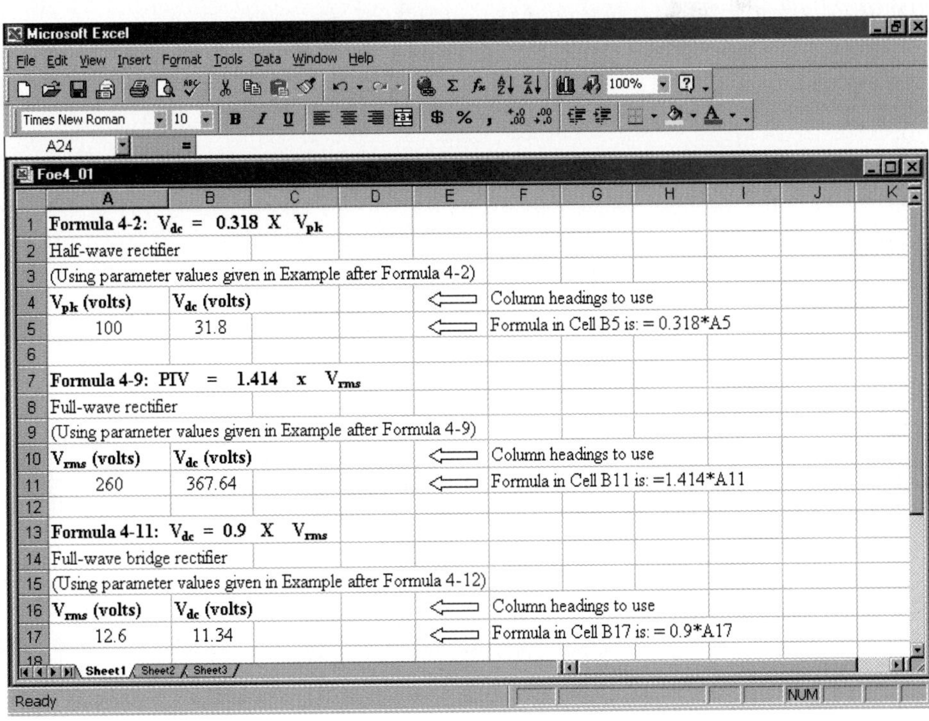

Review Questions

1. Draw the basic diagram for a half-wave rectifier.
2. Sketch a diagram that compares the ac input for a half-wave rectifier with its output when: (a) there is no output filtering, (b) when there is a C-type output filter.
3. What happens to the output polarity of a half-wave rectifier if you turn around the rectifier diode? Switch around the connections to the ac source?
4. If the input ac frequency is 60 Hz, what is the frequency (in pulses per second, pps) of the ripple from a half-wave rectifier circuit?
5. Draw the basic diagram for a typical full-wave rectifier circuit that uses a center-tapped transformer.
6. Sketch a diagram that compares the ac input (transformer primary) with the output of a full-wave, center-tapped transformer rectifier circuit when: (a) there is no output filtering, (b) when there is a C-type output filter.
7. If the input ac frequency is 60 Hz, what is the frequency (in pps) of the ripple from a full-wave, center-tapped, transformer rectifier circuit?
8. Draw the basic diagram for a bridge rectifier circuit.
9. Referring to your drawing in question 8, draw a circle around two diodes that conduct at the same time on the first half-cycle, then draw a square around the two diodes that conduct on the second half-cycle.
10. Sketch a diagram that compares the ac input for a bridge rectifier circuit with its output when: (a) there is no output filtering, (b) when there is a C-type output filter.
11. If the input ac frequency is 400 Hz, what is the frequency (in pps) of the ripple from a bridge rectifier circuit?
12. Draw the circuit for a full-wave power supply that fits the following description:
    - **a.** Center-tapped power transformer
    - **b.** *RC*, capacitor-input, pi-type filter
    - **c.** Resistor load of 100-kΩ

    Is the dc output of this circuit equal to, greater than, or less than the same rectifier without filtering?
13. If the ripple output of a rectifier-filter system is excessive, is the problem most likely in the transformer, rectifier(s), or filter network? Indicate which component(s) could cause the problem.
14. Briefly describe the purpose of a voltage regulator in a power supply.
15. Sketch a block diagram of a complete dc power supply that includes a power transformer, bridge rectifier, integrated-circuit voltage regulator, C-type filter, and a bleeder resistance across the dc output.
16. Describe the difference between percent of ripple and percent of regulation.
17. Explain the difference between no-load and full-load conditions for a power supply.
18. What is the dc output voltage of a voltage tripler that has an ac input of 12.6 V_{rms}?
19. What rms voltage level should be applied to a voltage quadrupler if the dc output voltage is to be 1,200 V?
20. The percent regulation for a quadrupler voltage multiplier is (better than, worse than, the same as) _______ the regulation for a tripler multiplier.

Problems

1. The ac input to a half-wave, unfiltered rectifier is 120 Vac (rms). Determine:
    - **a.** the dc output voltage.
    - **b.** peak inverse voltage applied to the rectifier diode.
    - **c.** frequency of the output ripple.
2. What is the PIV of the diode in a half-wave rectifier circuit that uses a capacitor filter, if the transformer feeding the rectifier circuit is a 1:3 step-up transformer supplied by a 150-Vac source?
3. If the peak output voltage from a half-wave rectifier is 17 V, what is the peak output current when the load resistance is 10Ω?
4. What is the unfiltered dc output of a full-wave rectifier circuit fed by a center-tapped, 1:2 step up transformer that is connected to a primary voltage source of 200 V rms?
5. The output of an unfiltered full-wave rectifier is supposed to be 5.5 V peak dc. If the rectifier is fed by a center-tapped transformer, what is the correct transformer ratio if the primary voltage is 120 Vac (rms)?
6. What is unfiltered dc output of a bridge rectifier circuit fed by 120 Vac (rms)?
7. Calculate the percent of ripple when an oscilloscope shows a 500-mV peak-to-peak ac waveform riding on top of a 5-V dc level.

8. The percent of ripple for a high-quality power supply is supposed to be 5%. If it is a 12-V dc power supply, what is the maximum amount of ripple you should expect to see at the output?
9. Determine the percent of regulation of a power supply that has a no-load output of 26 V and a full-load output of 23.5 V.
10. The percent of regulation for a certain power supply is rated at 6%. If the no-load voltage is 5.5 V, what is the minimum full-load voltage?
11. A zener diode regulator shown in Figure 4–24 consists of a IN4742A zener diode and a 200 Ω R_1 resistor. The input voltage is 20 V. The full load resistance is 560 Ω. Calculate
    - **a.** the load current
    - **b.** the power dissipated by the load
    - **c.** the current flowing through the R_1 resistor
    - **d.** the power dissipated by the R_1 resistor
    - **e.** the current flowing through the zener diode
    - **f.** the power dissipated by the zener diode
    - **g.** the power supplied to the regulator circuit
    - **h.** the efficiency of the regulator circuit
12. A zener diode regulator shown in Figure 4–24 consists of a IN4742A zener diode and a 200 Ω R_1 resistor. The input voltage is 18 V. The full load resistance is 560 Ω. Calculate
    - **a.** the load current
    - **b.** the power dissipated by the load
    - **c.** the current flowing through the R_1 resistor
    - **d.** the power dissipated by the R_1 resistor
    - **e.** the current flowing through the zener diode
    - **f.** the power dissipated by the zener diode
    - **g.** the power supplied to the regulator circuit
    - **h.** the efficiency of the regulator circuit
13. The 7812 voltage regulator is used in the circuit shown in Figure 4–26. The input voltage is 20 V. The full load resistance is 56 Ω. Calculate
    - **a.** the load current
    - **b.** the power dissipated by the load
    - **c.** the regulator input current
    - **d.** the power dissipated by the voltage regulator
    - **e.** the power supplied to the regulator circuit
    - **f.** the efficiency of the voltage regulator circuit
14. The 7812 voltage regulator is used in the circuit shown in Figure 4–26. The input voltage is 20 V. The full load resistance is 42 Ω. Calculate
    - **a.** the load current
    - **b.** the power dissipated by the load
    - **c.** the regulator input current
    - **d.** the power dissipated by the voltage regulator
    - **e.** the power supplied to the regulator circuit
    - **f.** the efficiency of the voltage regulator circuit
15. The 7915 voltage regulator is used in the circuit shown in Figure 4–26. The input voltage is –20 V. The full load resistance is 56 Ω. Calculate
    - **a.** the load current
    - **b.** the power dissipated by the load
    - **c.** the regulator input current
    - **d.** the power dissipated by the voltage regulator
    - **e.** the power supplied to the regulator circuit
    - **f.** the efficiency of the voltage regulator circuit
16. The 7915 voltage regulator is used in the circuit shown in Figure 4–26. The input voltage is –20 V. The full load resistance is 42 Ω. Calculate
    - **a.** the load current
    - **b.** the power dissipated by the load
    - **c.** the regulator input current
    - **d.** the power dissipated by the voltage regulator
    - **e.** the power supplied to the regulator circuit
    - **f.** the efficiency of the voltage regulator circuit
17. The input voltage to the LM317 positive adjustable voltage regulator is 16 V. R_1 = 240Ω and R_2 = 1k Ω. The load resistor is 100 Ω. Calculate
    - **a.** the output voltage
    - **b.** the load current
18. The input voltage to the LM317 positive adjustable voltage regulator is –16 V. R_1 = 240Ω and R_2 = 2 KΩ. The load resistor is 100 Ω. Calculate
    - **a.** the output voltage
    - **b.** the load current

Datasheet Problems

1. Obtain a datasheet for the 7805. What is the output voltage regulation (indicate the test conditions)? Calculate the voltage regulator's regulation percent for the maximum manufacturer specification for the output voltage regulation using the typical output voltage.

Analysis Questions

1. Draw the circuit for a half-wave rectifier that provides a negative output voltage.
2. Research how a center-tapped power transformer can be used with a bridge rectifier to produce a power supply that provides both a positive and a negative dc voltage.
3. Explain why using a larger filter capacitor improves the percent ripple and regulation for a power supply.
4. Research and write a paragraph that defines the term *critical inductance.*
5. Research and write a paragraph defining the term *optimum inductance.*
6. Explain why the voltage rating of a filter capacitor must be larger than the peak voltage level from the rectifier.
7. Explain why a shorted diode in a rectifier circuit will eventually destroy the power supply's electrolytic filter capacitor.
8. Describe how a leaky filter capacitor will affect the ripple and regulation of a power supply circuit.

MultiSIM Exercise for Power Supply Circuits

1. Use the MultiSIM program and utilize the circuit for Practice Problem 2.
2. Measure and record the output voltage.
3. Compare your MultiSIM results with the values given in the Example for this circuit. Was the output voltage that you made with MultiSIM reasonably close to the results of the Example?

Performance Projects Correlation Chart

All labs are available on the LabSource CD.

Chapter Topic	Performance Project	Project Number
Half-Wave Rectifier Circuits Capacitance Filters	Half-Wave Rectifier	5
The Bridge Rectifier Capacitance Filters	Bridge Rectifier	6
Basic Voltage Multiplier Circuits	Voltage Multiplier	7
	Story Behind the Numbers: The Voltage Regulator	

NOTE: It is suggested that after completing the above projects, the student should be required to answer the questions in the "Summary" at the end of this section of projects in the Laboratory Manual.

Troubleshooting Challenge

CHALLENGE CIRCUIT 4

(Follow the SIMPLER sequence by referring to inside front cover.)

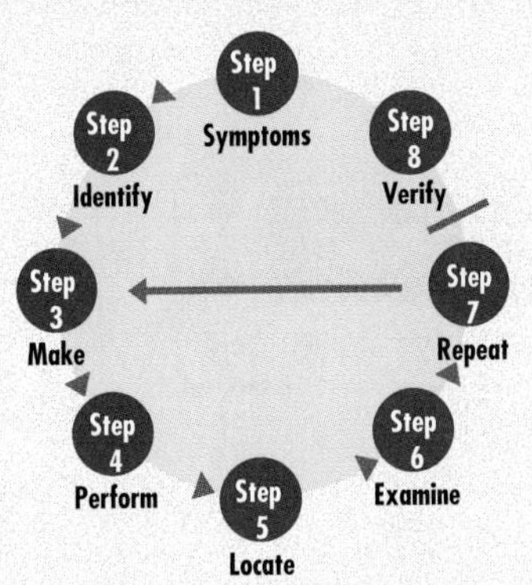

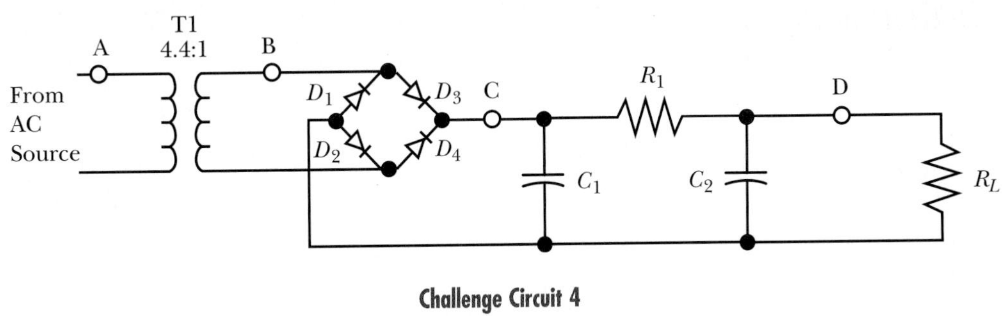

Challenge Circuit 4

STARTING POINT INFORMATION

1. Circuit diagram
2. The output voltage at point D is lower than it should be.

TEST	Results in Appendix C
Waveform at point A shows	(7)
Waveform at point B shows	(53)
Waveform at point C shows	(21)
DC voltage at point D measures	(65)

CHALLENGE CIRCUIT 4

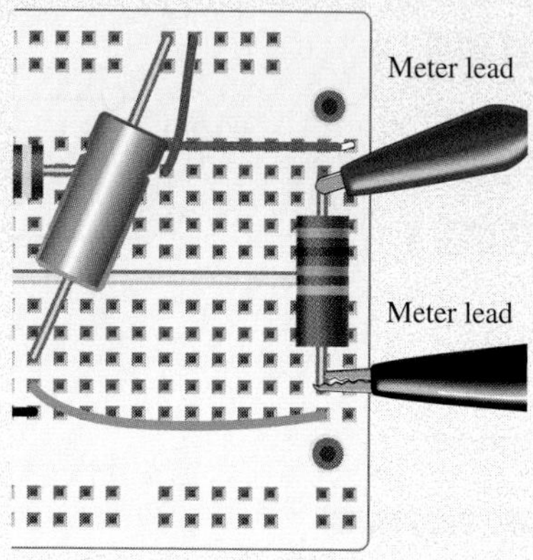

STEP 1

SYMPTOMS The dc output voltage is considerably lower than it should be.

STEP 2

IDENTIFY Initial suspect area: The low voltage can be caused by a transformer problem, a rectifier problem, a filter problem, or an excessively high drain by the load. For now, let's keep all of these possibilities in the initial suspect area.

STEP 3

MAKE Test decision: The magnitude of change in the output voltage is large, and we would probably see smoke if the transformer were that faulty or if the load were that heavy. Therefore we'll make our first check at the midpoint between the rectifier circuit and the filter circuit. That is at point C or the rectifier output.

Symptoms

Transformer

Bridge Rectifier

Filter

Load

From AC source

STEP 4

PERFORM **1st Test:** Using a scope, check the waveform at point C. The result is only one pulse per ac input cycle. Half of the full-wave rectifier output is missing.

STEP 5

LOCATE New suspect area: It looks like we have a rectifier problem. But, using our "input-output" technique, we'll have to verify that the ac input into the rectifier is normal before we exclude the transformer. Thus, the new suspect area includes the transformer and rectifier circuits.

STEP 6

EXAMINE Available data.

STEP 7

REPEAT Analysis and testing: The next test is to look at the ac input to the rectifier circuit at point B.

2nd Test: Check the waveform at point B. The scope check indicates a sine-wave voltage with peak voltage of about 39 V. This is what it should be. So we have isolated the trouble to the rectifier block.

3rd Test: Isolate each diode. Using the digital multimeter in the diode test mode, check the forward bias and reverse bias conduction of each diode. Diodes D1 and D4 are open. Diodes D2 and D3 test normal.

Note: This meter shows 30.00 MΩ when measuring an open (or an R value from 30 MΩ to ∝).

STEP 8

VERIFY **4th Test:** Replace diodes D1 and D4 and note the circuit parameters. When this is done, the circuit operates properly.

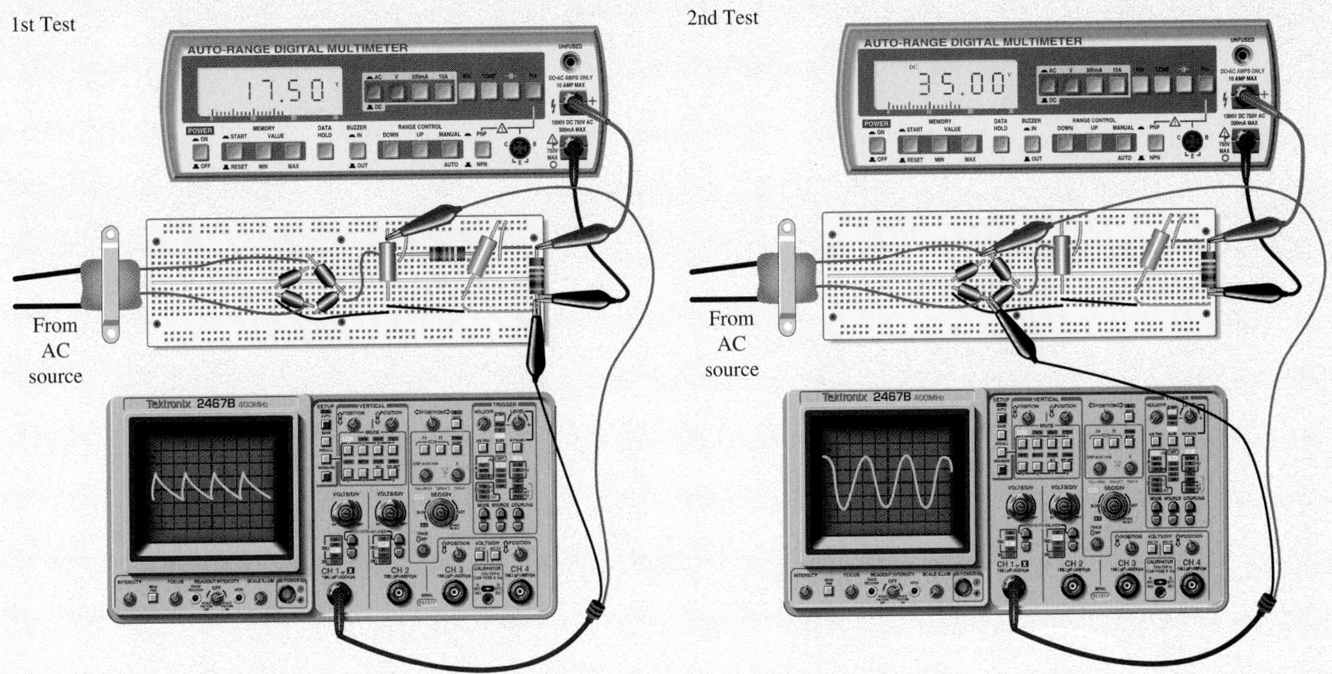
1st Test
AUTO-RANGE DIGITAL MULTIMETER
17.50
From
AC
source
Tektronix 2467B 400MHz
2nd Test
AUTO-RANGE DIGITAL MULTIMETER
35.00
From
AC
source
Tektronix 2467B 400MHz

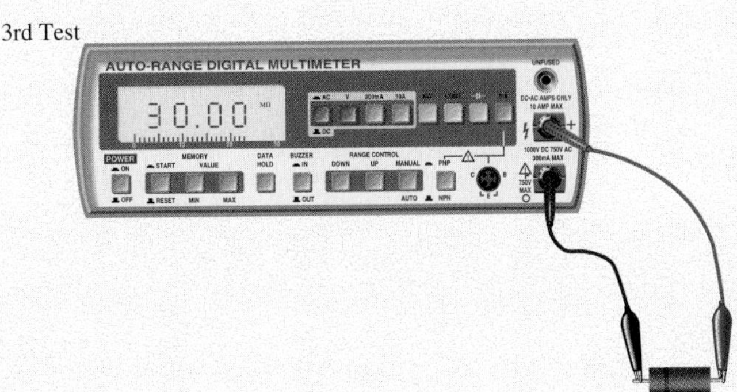
3rd Test
AUTO-RANGE DIGITAL MULTIMETER
30.00

Troubleshooting Challenge

CHALLENGE CIRCUIT 5 (Block Diagram)

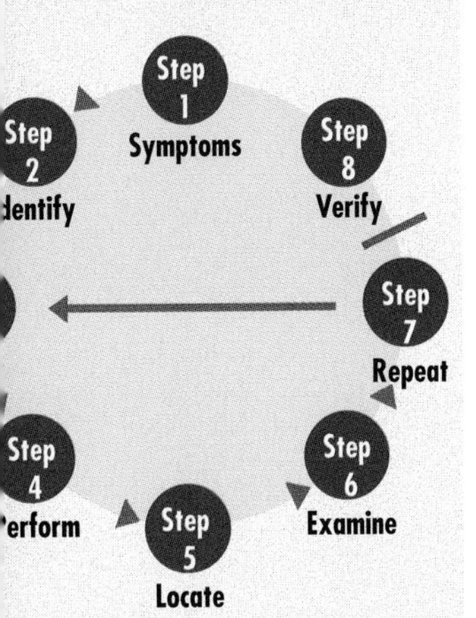

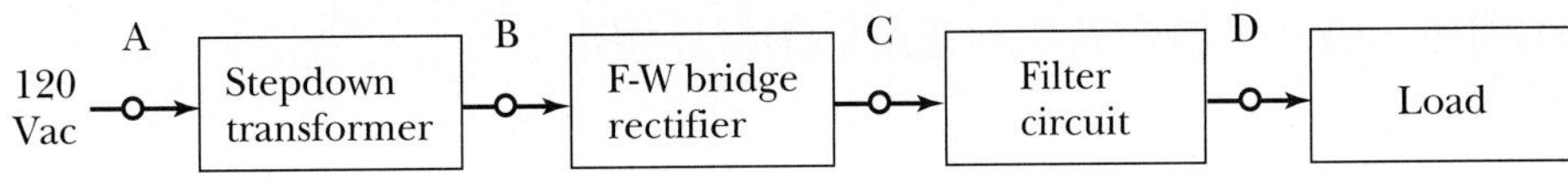

Power Supply Block Diagram

General Testing Instructions:

Measurement
Assumptions:
Signal = from **TP** to ground
I = at the **TP**
V = from **TP** to ground
R = from **TP** to ground (with the power source disconnected from the circuit)

Possible Tests & Results:
Signal (normal, abnormal, none)
Current (high, low, normal)
Voltage (high, low, normal)
Resistance (high, low, normal)

Starting Point Information	Test Points	Test Results in Appendix C			
		V	*I*	*R*	*Signal*
At Point A:	Point B	(243	NA	NA	244)
	Point C	(245	246	247	248)
V = normal	Point D	(249	250	250	252)
I =NA					
R = NA					
Signal = Normal					

POWER SUPPLY BLOCK DIAGRAM CHALLENGE PROBLEM

STEP [1]

SYMPTOMS At point D, voltage is lower than it should be, current is lower than normal, and with the power source disconnected, resistance to ground measures normal.

STEP [2]

IDENTIFY initial suspect area. Since the supply voltage is normal, the voltage waveform at point C could be incorrect.

STEP [3]

MAKE test decision is based on the symptoms information. Let's start with the voltage measurements at strategic test points.

STEP [4]

PERFORM **1st Test:** Voltage waveform at point C. Waveform shows only one pulse per ac input cycle. This indicates that the bridge rectifier is not functioning properly.

STEP [5]

LOCATE new suspect area. The bridge rectifier is suspect.

STEP [6]

EXAMINE available data.

STEP [7]

REPEAT analysis and testing.
2nd Test: Check the waveform at point B. The scope check indicates that the waveform is normal.
3rd Test: Remove the bridge rectifier from the circuit. Using the digital mulitmeter in the diode test mode, test the bridge rectifier. The bridge rectifier fails the test.

STEP [8]

VERIFY **4th Test:** When the bridge rectifier is replaced with a new module that meets specifications, all circuit parameters come back to their norms.

Troubleshooting Challenge

CHALLENGE CIRCUIT 6

(Follow the SIMPLER sequence by referring to inside front cover.)

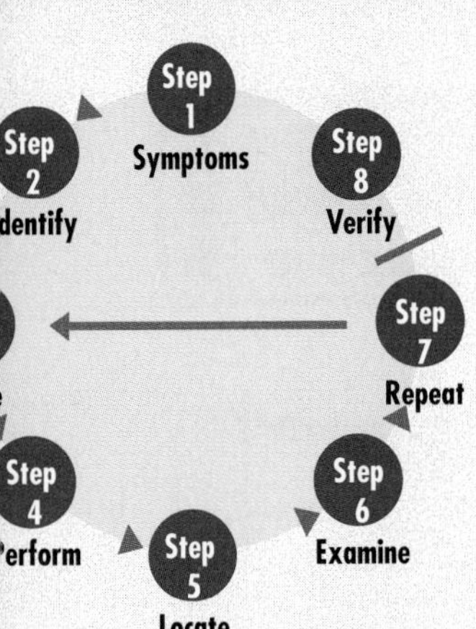

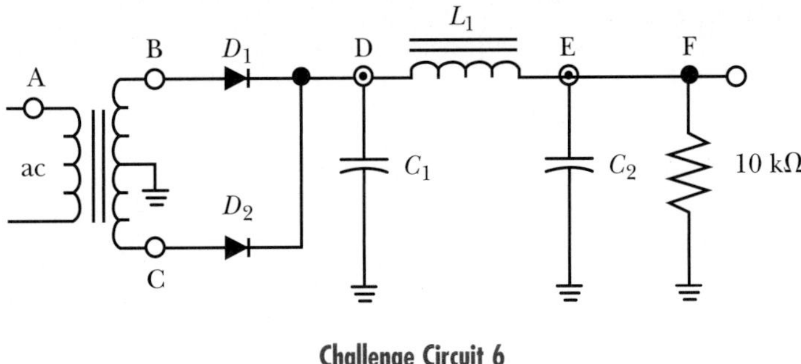

Challenge Circuit 6

STARTING POINT INFORMATION

1. Circuit diagram
2. The output voltage point F is lower than normal. A scope check also indicates higher than normal ac ripple present.

Test	Results in Appendix C
DC voltage at point F	(98)
AC voltage at point F	(124)
DC voltage at point E	(69)
AC voltage at point E	(26)
DC voltage at point D	(76)
AC voltage at point D	(35)
Waveform at point C	(70)
Waveform at point B	(2)
Waveform at point A	(113)
R of L_1	(92)
C_1 R measurements	(45)
C_2 R measurements	(117)

CHALLENGE CIRCUIT 6

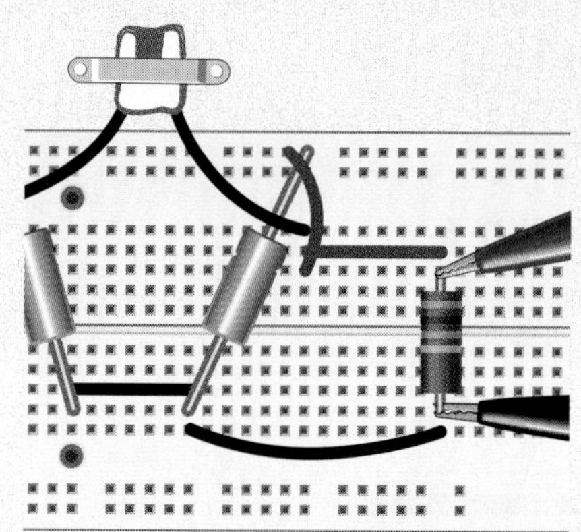

STEP [1]

SYMPTOMS The dc output voltage seems low, and excessive ac ripple is present.

STEP [2]

IDENTIFY Initial suspect area: The ripple symptom indicates the filter components are involved. If the inductor were shorted, the dc output might be slightly higher than normal. If the inductor were open, there would be no output. So this narrows our initial suspect area to the two filter capacitors.

STEP [3]

MAKE Test decision: If either capacitor were shorted, the output voltage would be drastically lower than normal (virtually zero). So we need to see if a capacitor might be open, which would agree with our symptom information. Typically, an aging electrolytic capacitor that has dried out and become virtually open will show much more ac than it would if it were normal. First, let's check the ac voltage across the output filter capacitor (C_2).

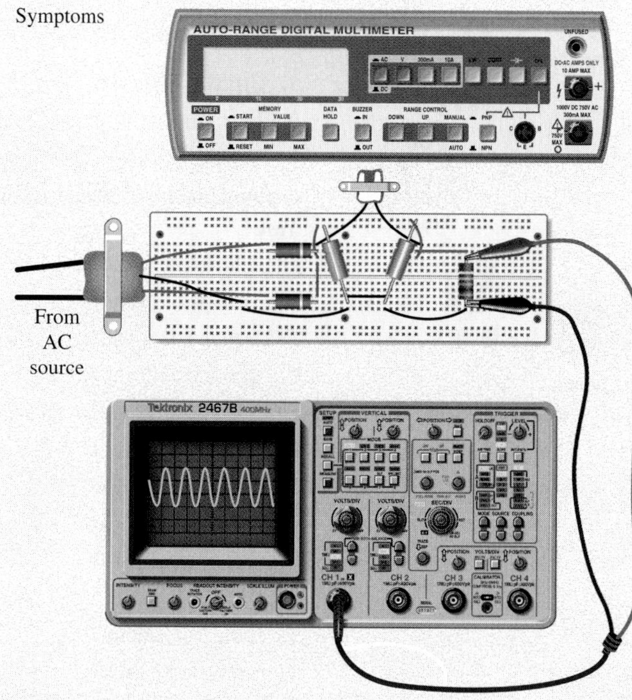

STEP [4]

PERFORM **1st Test:** Check the ac voltage across C_2. It shows higher than the normal ac component. However, this could be caused by the input filter capacitor (C_1) not feeding a normal signal to the inductor and this capacitor.

STEP [5]

LOCATE New suspect area: The first test is not conclusive; therefore the suspect area remains the same at this point.

STEP [6]

EXAMINE Available data.

STEP [7]

REPEAT Analysis and testing:
2nd Test: Measure ac voltage across C_1. The result shows a much higher than normal ac component.
3rd Test: Make a front-and-back resistance check of capacitor C_1, with power removed. Since the capacitor checks open, we have probably found our problem—an open filter capacitor (C_1).

STEP [8]

VERIFY **4th Test:** Replace C_1 with a capacitor of appropriate ratings and restore normal operation.

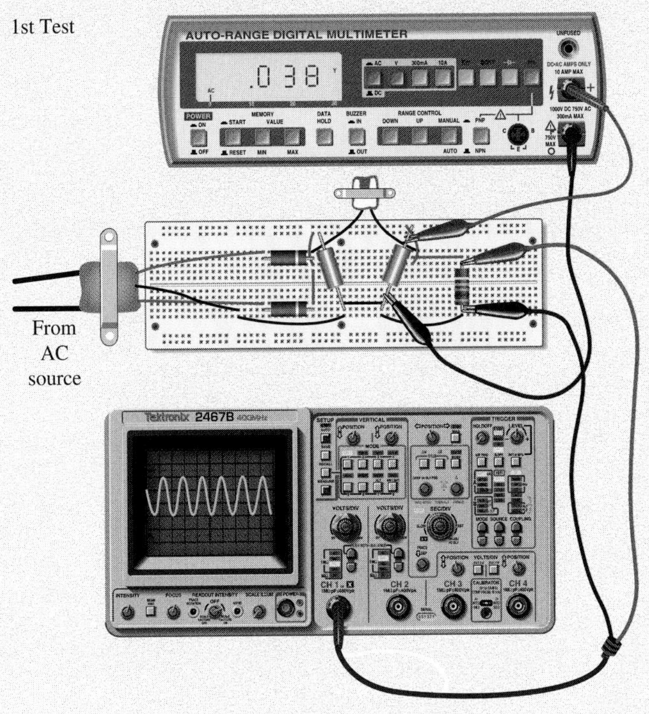
1st Test
AUTO-RANGE DIGITAL MULTIMETER
.038
From
AC
source
Tektronix 2467B

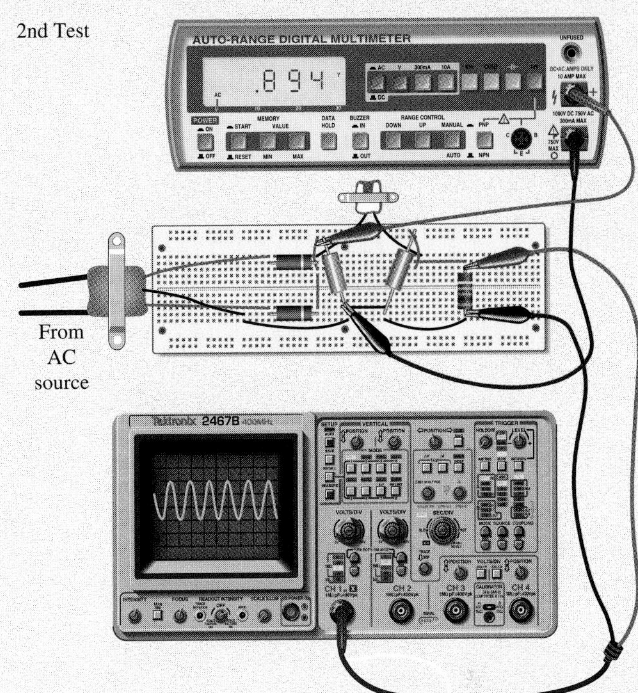
2nd Test
AUTO-RANGE DIGITAL MULTIMETER
.894
From
AC
source
Tektronix 2467B

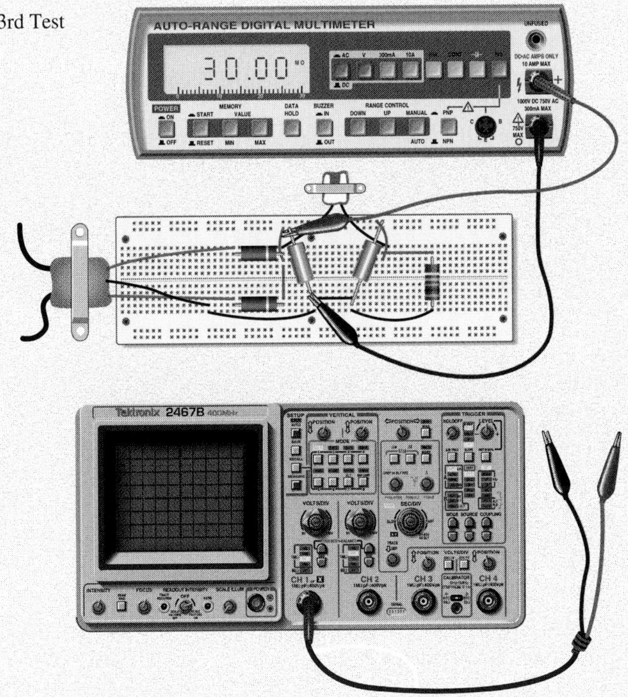
3rd Test
AUTO-RANGE DIGITAL MULTIMETER
30.00
Tektronix 2467B

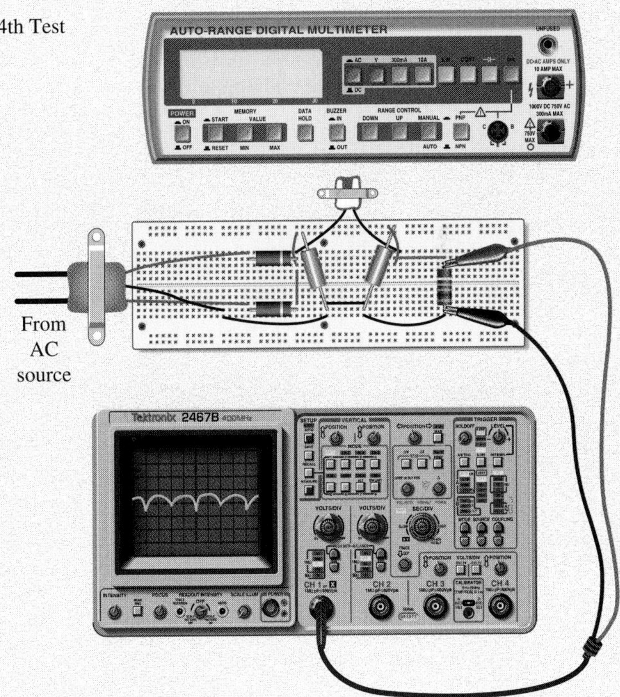
4th Test
AUTO-RANGE DIGITAL MULTIMETER
From
AC
source
Tektronix 2467B

Troubleshooting Challenge

CHALLENGE CIRCUIT 7 (Block Diagram)

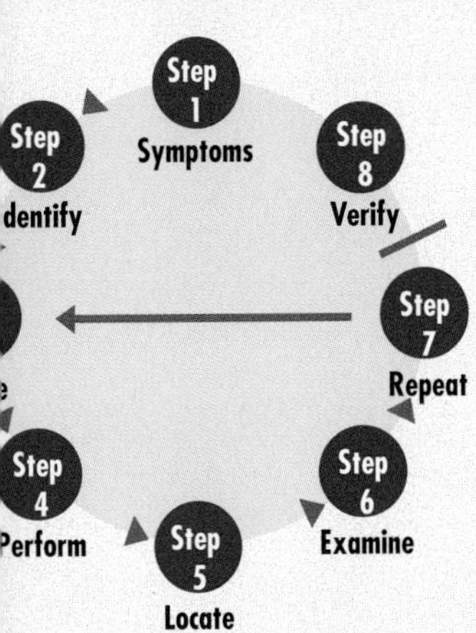

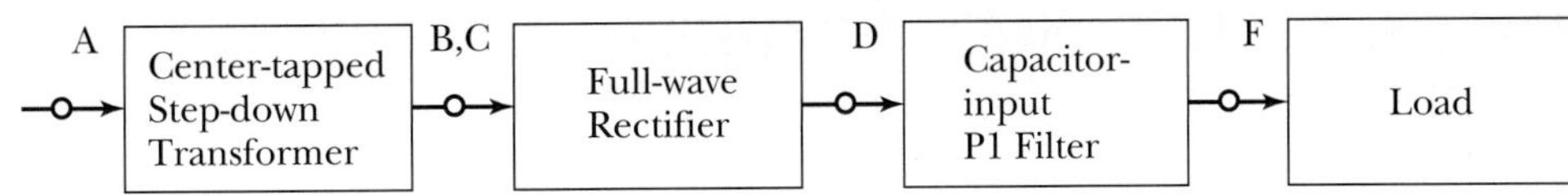

Power Supply Block Diagram

General Testing Instructions:

Measurement Assumptions:
Signal = from **TP** to ground
I = at the **TP**
V = from **TP** to ground
R = from **TP** to ground (with the power source disconnected from the circuit)

Possible Tests & Results:
Signal (normal, abnormal, none)
Current (high, low, normal)
Voltage (high, low, normal)
Resistance (high, low, normal)

Starting Point Information	Test Points	Test Results in Appendix C			
		V	*I*	*R*	*Signal*
At Point A:	Point B	(253	NA	NA	254)
	Point C	(225	NA	NA	256)
V = normal	Point D	(257	NA	NA	258)
I = NA	Point F	(259	260	261	262)
R = NA					
Signal = normal					

POWER SUPPLY BLOCK DIAGRAM CHALLENGE PROBLEM

STEP [1]

SYMPTOMS At point F, voltage is lower than is should be, current is lower than normal, the ac ripple is higher than normal and with the power source disconnected, resistance to ground measures normal.

STEP [2]

IDENTIFY Initial suspect area. Since the supply voltage (point A) is normal, the voltage waveform at points B, C, and D could be incorrect.

STEP [3]

MAKE test decision is based on the symptoms information. Let's start with the voltage measurements at strategic test points.

STEP [4]

PERFORM **1st Test:** Voltage at point D. DC voltage is low and the ac voltage is high. This indicates that the filter, the rectifiers, or the transformer are not functioning properly.

STEP [5]

LOCATE new suspect area. The filter is suspect.

STEP [6]

EXAMINE available data.

STEP [7]

REPEAT analysis and testing.
2nd Test: Check the waveform at point D. The scope check indicates that the waveform has two pulses per ac input cycle. This implies that the rectifiers are okay. It appears that the filter is faulty.

STEP [8]

VERIFY **3rd Test:** When the filter module is replaced with a new module that meets specifications, all circuit parameters come back to their norms.

Troubleshooting Challenge

CHALLENGE CIRCUIT 8
Find the Fault

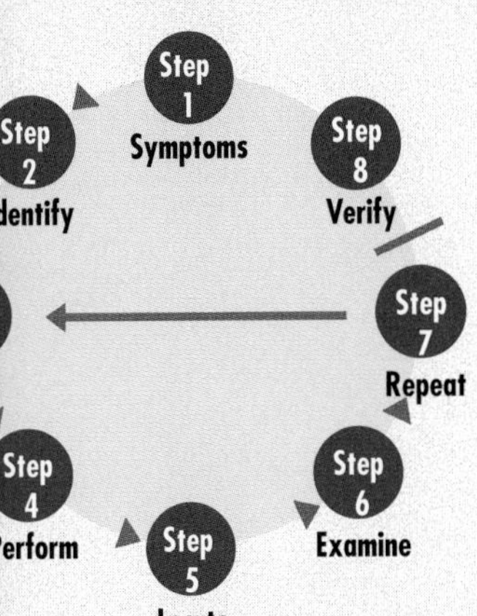

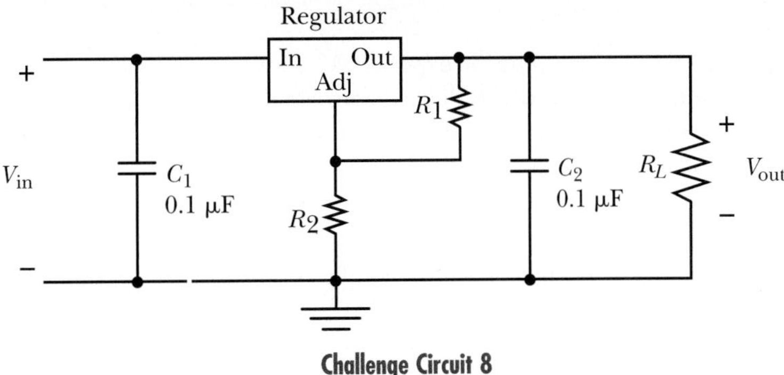

Challenge Circuit 8

STARTING POINT INFORMATION

1. Circuit diagram (R_1 = 200 Ω, R_2 = 1.0 kΩ, R_L = 20 Ω)
2. V_{in} = 25 V V_{REF} = 1.25 V
3. LM317 adjustable voltage regulator
4. V_{out} is a voltage lower than expected

Test	Circuit Measurements	Theoretical Expected Value
V_{in}	25 V	____
V_{out}	1.77 V	____
V_{R_2}	0.52 V	____
I_L	88.5 mA	____
R_1	240 Ω	____
R_2	100 Ω	____
R_L	20 Ω	____

STEP [1]

You will analyze the circuit to calculate the expected voltages, currents, and resistances for a properly functioning circuit. Record these values beside each test listed in the table provided. (Refer to Analysis Techniques.)

STEP [2]

Starting with the symptom, document your steps for troubleshooting this circuit using the SIMPLER troubleshooting technique or instructions provided by your instructor.

STEP [3]

For each circuit measurement, you will indicate the test to be accomplished.

STEP [4]

Using the previous table, you will find the voltage and current measurements taken for the faulted circuit.

STEP [5]

You will now compare the faulted circuit measurement to the expected measurement for a properly functioning circuit. Are these values the same? (Refer to General Testing Instructions.)

STEP [6]

You will repeat steps 2 through 5 until you have located the fault.

STEP [7]

When you have located the fault, you will identify the fault and the characteristics of the failure. (Refer to Types of Failures.)

ANALYSIS TECHNIQUES (Check with your instructor)

METHOD [1]

Using circuit theory and algebra, evaluate the circuit.

METHOD [2]

Using an Excel spreadsheet for this circuit, evaluate the circuit.

METHOD [3]

Using MultiSIM, build the circuit schematic and generate the results.

METHOD [4]

Assemble this circuit on a circuit board. Apply power and make the appropriate measurements using your DMM and oscilloscope.

oubleshooting
hallenge

HALLENGE CIRCUIT 8
(ontinued)

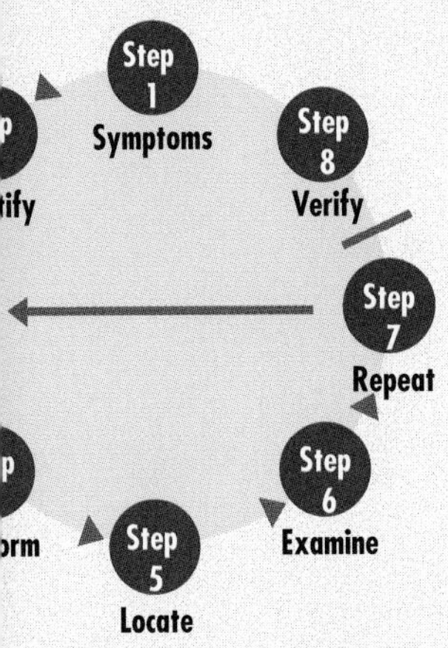

General Testing Instructions:

Measurement
Assumptions:
Signal = from **TP** to ground
I = at the **TP**
V = from **TP** to ground
R = from **TP** to ground (with the power source disconnected from the circuit)

Possible Tests & Results:
Signal (normal, abnormal, none)
Current (high, low, normal)
Voltage (high, low, normal)
Resistance (high, low, normal)

Types of Failures

Component failures
- wrong part value
- part shorted
- part open

Supply voltage—incorrect voltage
Ground—floating

OUTLINE

5–1 Basic types of transistors
5–2 BJT bias voltages and currents
5–3 BJT operation
5–4 BJT characteristic curves
5–5 BJT ratings
5–6 An NPN transistor circuit model
5–7 Information for the technician
5–8 Basic uses of the BJT
5–9 Basic concepts of amplifiers
5–10 Transistor biasing circuits
5–11 Amplifier circuit configurations
5–12 The common-emitter (CE) amplifier
5–13 The common-collector (CC) amplifier
5–14 The common-base (CB) amplifier
5–15 Classification by signal levels
5–16 Classification by class of operation
5–17 Analysis of BJT amplifiers

CHAPTER 5

Bipolar Junction Transistors

OBJECTIVES

After studying this chapter, you should be able to:

1. Draw the symbols and identify the **emitter, base,** and **collector** leads for NPN and PNP transistors
2. Draw the symbols for **NPN** and **PNP transistors** and show the proper **voltage polarities** for the base-emitter terminals and for the collector-base terminals.
3. Explain the meaning and cite the mathematical symbols for **emitter current, base current, collector current, base-emitter voltage,** and **collector-emitter voltage**
4. Describe how increasing the **forward-bias base current** in a **BJT amplifier** decreases the voltage between the collector and emitter
5. Describe the operation of a BJT when applied as a switch
6. Explain the meaning of the curves shown on a family of **collector characteristic curves**
7. Describe the meaning of the **maximum voltage, current,** and **power ratings** listed in BJT data sheets
8. Explain the basic **transistor amplification process.**
9. Describe the input and output characteristics of **common transistor amplifier stages**
10. List the advantages of each common type of transistor amplifier stage
11. Describe the differences between **small-signal** and **power amplifier circuits**
12. Classify amplifiers by **class of operation**
13. Describe the classification of amplifiers and their operations from their **load lines**
14. List typical applications for each classification of amplifier
15. Perform a basic **analysis** of a common-emitter Class A BJT that uses voltage divider biasing

PREVIEW

The transistor was invented in the late 1940s. Credit for its invention is given to three Bell Laboratories scientists, John Bardeen, Walter Brattain, and William Shockley. The term *Transistor* is really a contraction of "transfer" and "resistor." This name suggests that transistors can be thought of as making use of an input current to control an output voltage.

In this chapter you will learn more about the basics of transistors. You will find out exactly what transistors are, you will see some of the things they can do, and you will find out how they work. You will also learn about the transistor specifications that are found in transistor data sheets.

Amplifier circuits are found in numerous systems and subsystems in electronics. For example, the ability to amplify enables applications such as building a very small signal from a microphone or CD player into a large signal that drives the speakers in a stereo amplifier system.

An overview of some key operating features of BJT amplifier circuits is presented. You will learn classifications of amplifiers relative to their circuit configuration, signal levels, and classes of operation. These circuits will provide information to help you in further studies of these important topics, including amplifiers built around field-effect transistors as summarized for you in a later study.

KEY TERMS

Ac beta, β_{ac}
Amplifier
Base
Base current, I_B
Base-emitter voltage, V_{BE}
Bipolar junction transistor (BJT)
Class A amplifiers
Class B amplifiers
Class C amplifiers
Collector
Collector-base voltage, V_{CB}
Collector characteristic curves
Collector current, I_C
Collector-emitter voltage, V_{CE}
Common-base amplifier (CB)
Common-collector amplifier (CC)
Common-emitter amplifier (CE)
Dc beta, β_{dc}
Emitter
Emitter current, I_E
Emitter follower
Gain
Large-signal amplifiers
Load line
NPN transistor
PNP transistor
Push-pull amplifier
Q point
Small-signal amplifiers
Unity gain analog buffer
Voltage follower

5–1 Basic Types of Transistors

The **bipolar junction transistor** (BJT, commonly called transistor) is constructed with three doped semiconductor regions. The physical representation and the schematic symbol for each device are shown in Figure 5–1. The **NPN transistor** consists of two N regions separated by a P region (Figure 5–1a). The **PNP transistor** consists of two P regions separated by an N region (Figure 5–1b). Terminals are connected to each region. The terminals are called the **collector** (C), the **base** (B), and the **emitter** (E). The operational characteristics of the transistor are determined by the structure of the transistor. Both NPN and PNP structures are valid and useful transistor types. The schematic symbols differ only by the direction the arrow points in the emitter part of the symbol. The arrow direction correlates with the P-N junction (remember P = anode and N = cathode). For an NPN transistor, the arrow points outward and away from the base (remember that NPN is "*Not*-Pointing-i*N*"). For a PNP transistor, the emitter arrow points in toward the base (remember that PNP is "*P*ointing-i*N*").

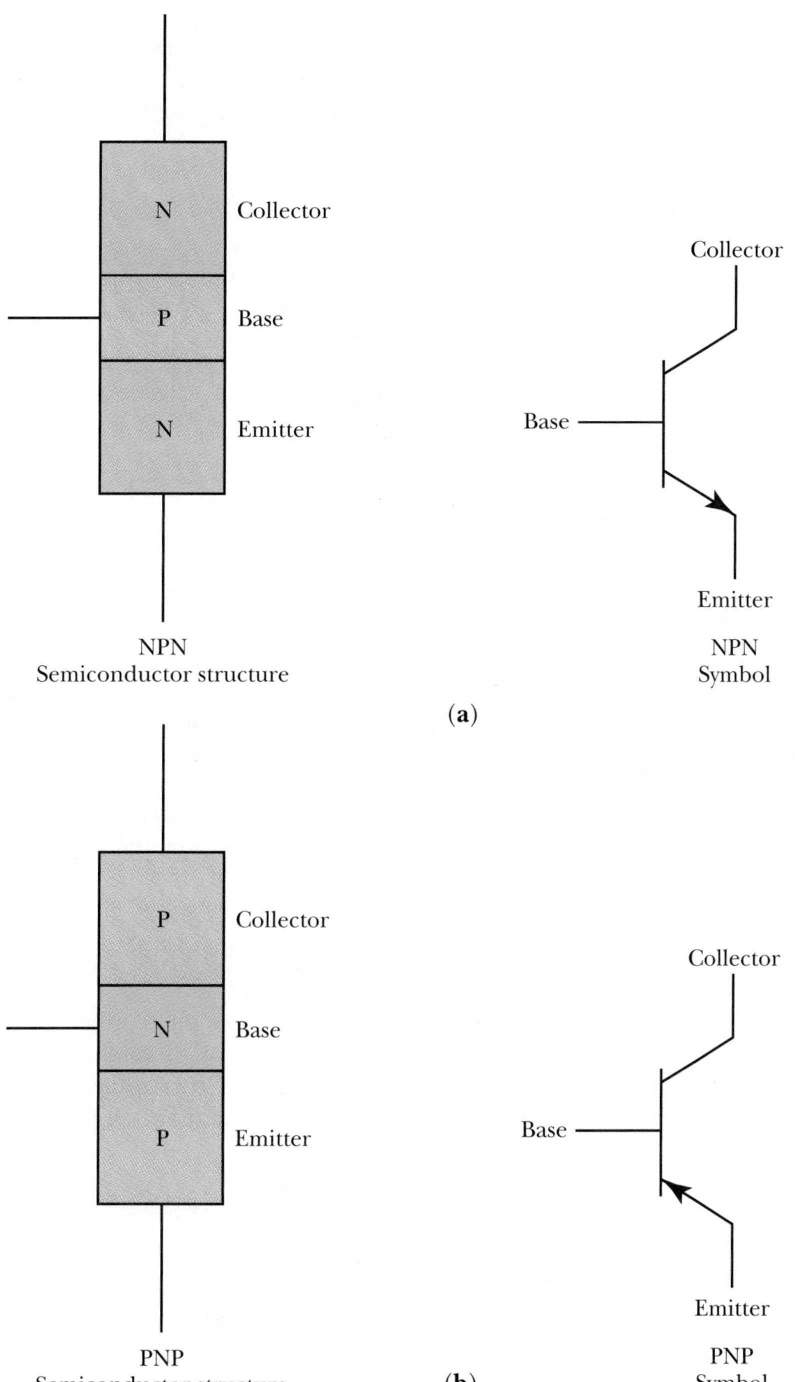

FIGURE 5–1 Representations, symbols, and basing or electrode terminals for BJTs: (a) NPN transistor; (b) PNP transistor

Some Common BJT Packages

In Figure 5–2 you see some common transistor lead base layouts (i.e., bottom view of transistor package). The topology outline numbers (such as TO-3 and SOT-23) identify each specific packaging configuration used by manufacturers. Figure 5–3 shows BJTs mounted on heat sinks. The reason that some transistors are mounted on heat sinks is discussed later in this chapter.

BJTs are listed according to device identification codes. Recall that JEDEC registered diodes are listed according to a code that begins with 1N (1N4001, for instance).

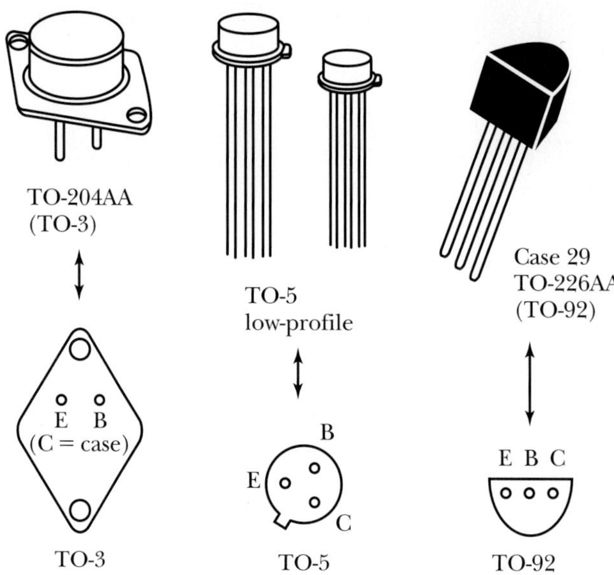

FIGURE 5–2 A sample of some common packaging and basic schemes

FIGURE 5–3 High-power BJTs mounted on heat sinks *(Photo courtesy of Thermalloy Inc.)*

JEDEC registered BJTs are listed according to a code that usually begins with 2N. Some actual NPN transistors are 2N697, 2N2222A, 2N4401, and BSR17A. Some actual PNPs are 2N2904, 2N4030, 2N5400, and BSR18A. There is no way to tell the difference between an NPN and PNP transistor just by looking at the part identification code. You should check the manufacturers' data sheets, a transistor data book, or a parts catalog in order to discover the specifications for a given transistor part identification number.

5–2 BJT Bias Voltages and Currents

Figure 5–4 is a circuit that is intended to show the voltage and currents for an NPN BJT. It also introduces some labels you will find in nearly all discussions and descriptions of BJTs. First note that this is an NPN transistor. Observe the polarity of the power supplies and the direction of the current flowing in the circuit. Now consider the following:

V_{BB} is the *base-bias voltage source.*

V_{CC} is the *collector-bias voltage source.*

V_{EE} is the *emitter-bias voltage source.*

V_{CB} is the *collector to base* voltage drop.

V_{CE} is the *collector to emitter* voltage drop.

V_{BE} is the *base to emitter* voltage drop.

I_B is the ***base current,*** the current at the base terminal of the transistor.

I_C is the ***collector current,*** the current at the collector terminal of the transistor.

I_E is the ***emitter current,*** the current at the emitter terminal of the transistor.

R_B is the resistor connected to the transistor base terminal.

R_C is the resistor connected to the transistor collector terminal.

R_E is the resistor connected to the transistor emitter terminal.

As labeled, all the voltages and currents will be positive for the normal operation of the NPN transistor.

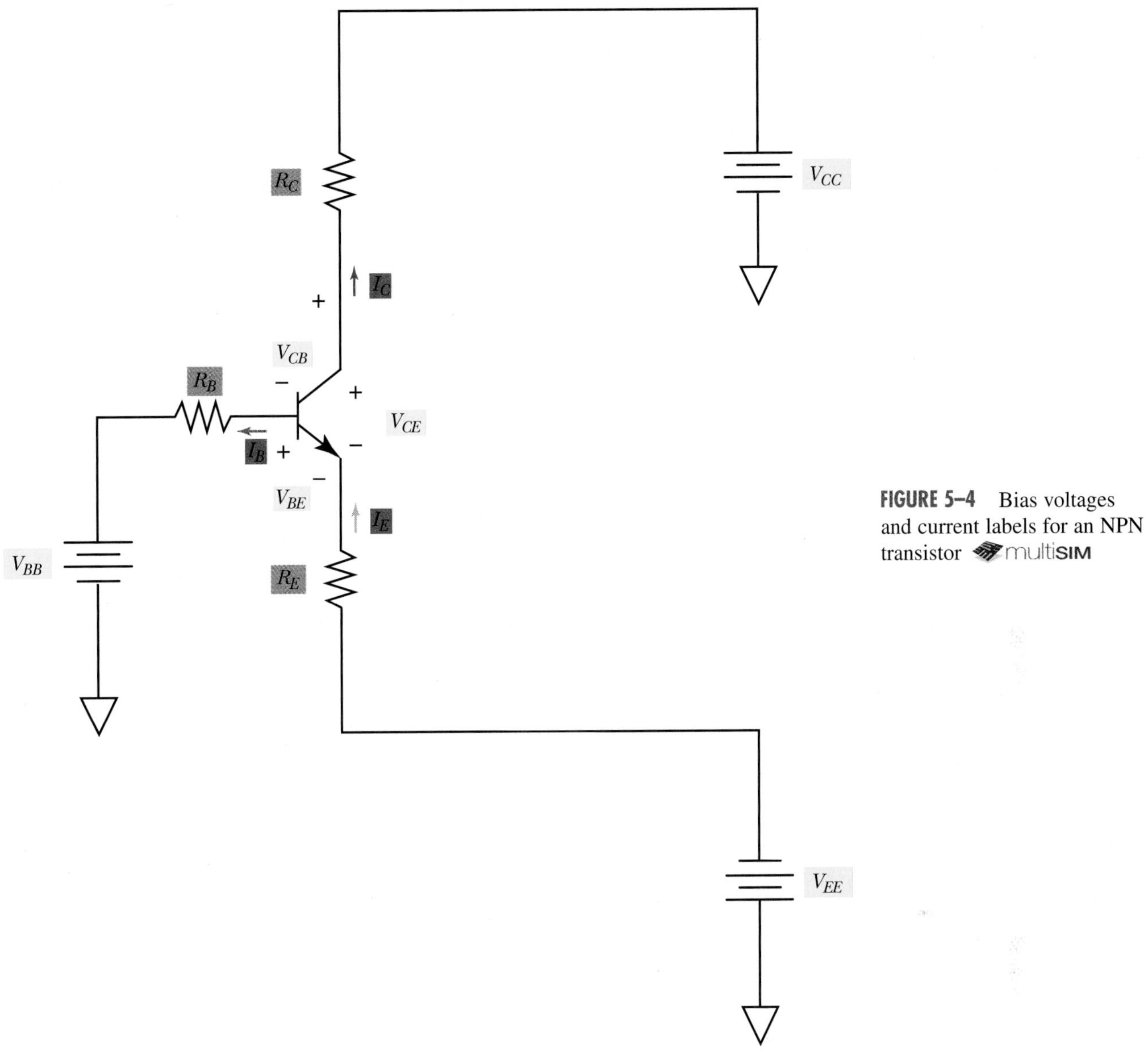

FIGURE 5–4 Bias voltages and current labels for an NPN transistor multiSIM

Figure 5–5 is a circuit that shows how currents flow through a PNP BJT. It also introduces some labels you will find in nearly all discussions and descriptions of BJTs. First note that this is a PNP transistor. Observe the polarity of the power supplies and the direction of the current flowing in the circuit. Notice also the change in the labels for the voltages around the transistor.

V_{BC} is the *base to collector* voltage drop.

V_{EC} is the *emitter to collector* voltage drop.

V_{EB} is the *emitter to base* voltage drop.

As labeled, all the voltages and currents will be positive for the normal operation of the PNP transistor. You will observe that not all of the voltage sources as shown in Figure 5–4 and Figure 5–5 will be required to operate the transistors in actual applications.

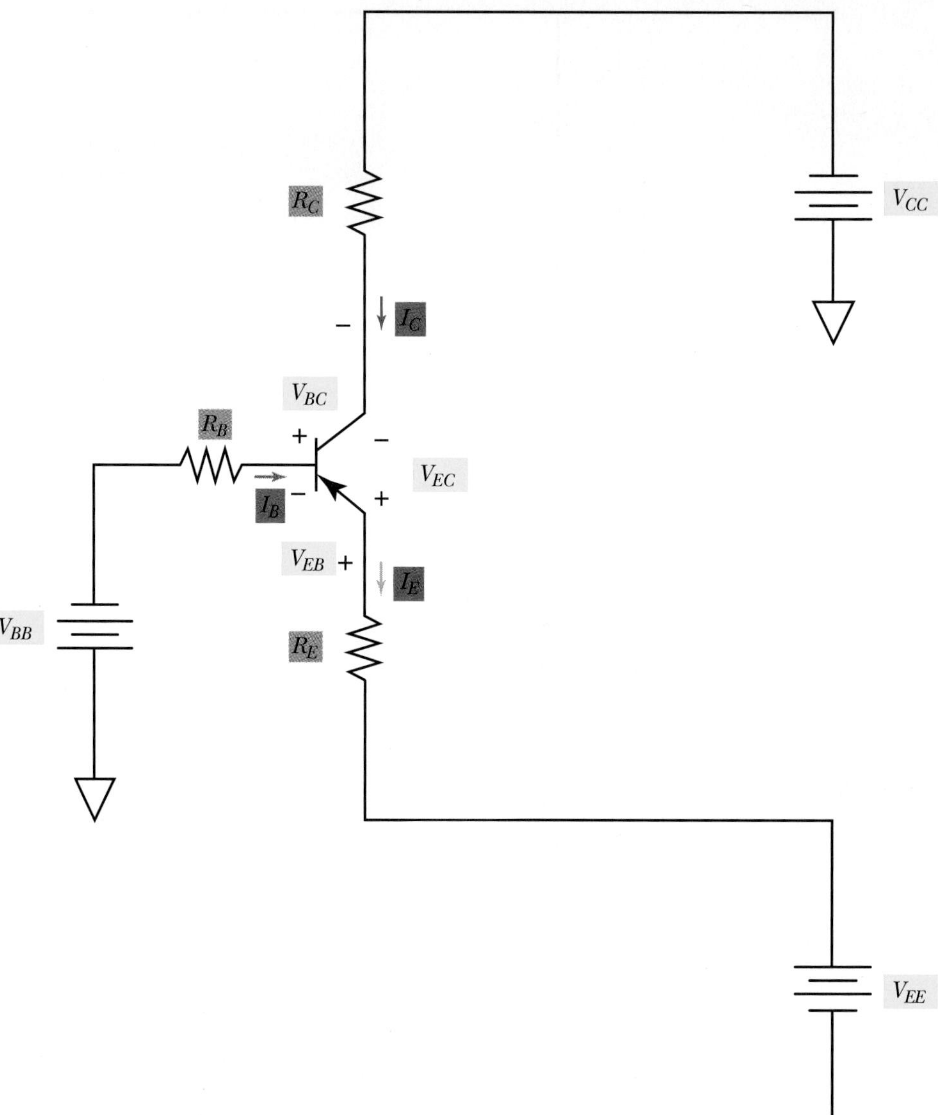

FIGURE 5–5 Bias voltages and current labels for a PNP transistor multiSIM

IN-PROCESS LEARNING CHECK 1 Fill in the blanks as appropriate.

1. The symbol for an NPN transistor shows the emitter element pointing _______ the base element.
2. The symbol for an PNP transistor shows the emitter element pointing _______ the base element.
3. For BJTs:
   V_{BE} is the _______ voltage. V_{CB} is the _______ voltage.
   V_{CE} is the _______ voltage.
4. For a typical BJT circuit:
   I_C stands for the _______ current. I_B is the _______ current.
   I_E is the _______ current.

Application Problem

BJT PRECISION BIAS

A precision voltage is needed for a cell phone chip. The following figure shows an example of a precision biasing with 2N3053 transistor. The microprocessor requires a constant 25 mA current at 3.5 V to function properly. Determine the base current (I_B) that the op-amp needs to supply. Assume transistor beta of 100. Would it work?

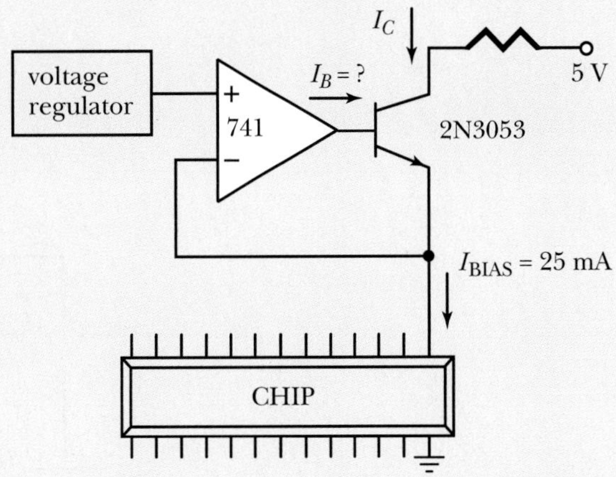

Solution

Knowing that $I_C = I_{BIAS}$, and the transistor beta is

$$\beta = \frac{I_C}{I_B}$$

We now calculate the transistor base current as follows.

$$I_B = \frac{25\text{ mA}}{100} = 0.25\text{ mA}$$

The calculated value is below the 5 mA capacity of a 741C op-amp.

5–3 BJT Operation

Figure 5–6 is a demonstration circuit that is intended to show how currents flow through an NPN BJT. The base-emitter junction of a BJT behaves very much like the P-N junction of an ordinary diode. For instance, you have to forward bias the base-emitter junction in order to allow any current to flow through it. You can see from the diagram that the forward-bias current will indeed flow from the negative terminal of V_{BB}, through the transistor base-emitter junction to the positive terminal of V_{BB}. Increasing the value of V_{BB} increases the value of I_B much the same way that increasing the voltage applied to a forward-biased diode causes the diode current to increase. Base currents for BJTs, however, are usually limited to the microampere range. The forward-bias junction potential for the base-emitter junction of a BJT is identical with that of a diode, namely about 0.3 V for germanium transistors and 0.7 for silicon transistors.

You will confuse yourself, however, if you try to compare the collector-base junction of a BJT with the P-N junction of an ordinary diode. Under normal operating conditions, the collector-base junction is reverse biased. Yet, the internal current flowing from the emitter through the base region to the collector actually is far more current than the forward-biased base-emitter junction. How can this be? Since the base region in a BJT is extremely thin and very lightly doped, the collector *collects* most of the electrons provided or *emitted* by the emitter. This accounts for why base currents are usually quite small. When the base-emitter junction is forward biased, the base current controls the amount of collector current that is flowing within the transistor.

Let's take a closer look at the emitter-collector circuit. Note from the diagram that the dc source V_{CC} is connected in series with the collector resistor R_C. The total voltage applied between the emitter and collector terminals (**collector-emitter voltage, V_{CE}**) of

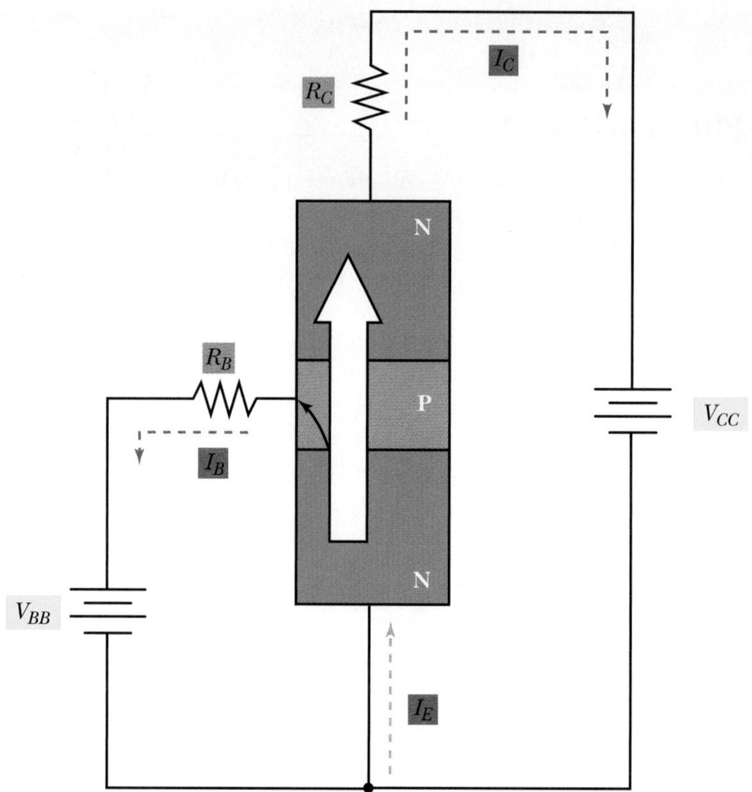

FIGURE 5–6 Bias voltages and currents in an NPN transistor

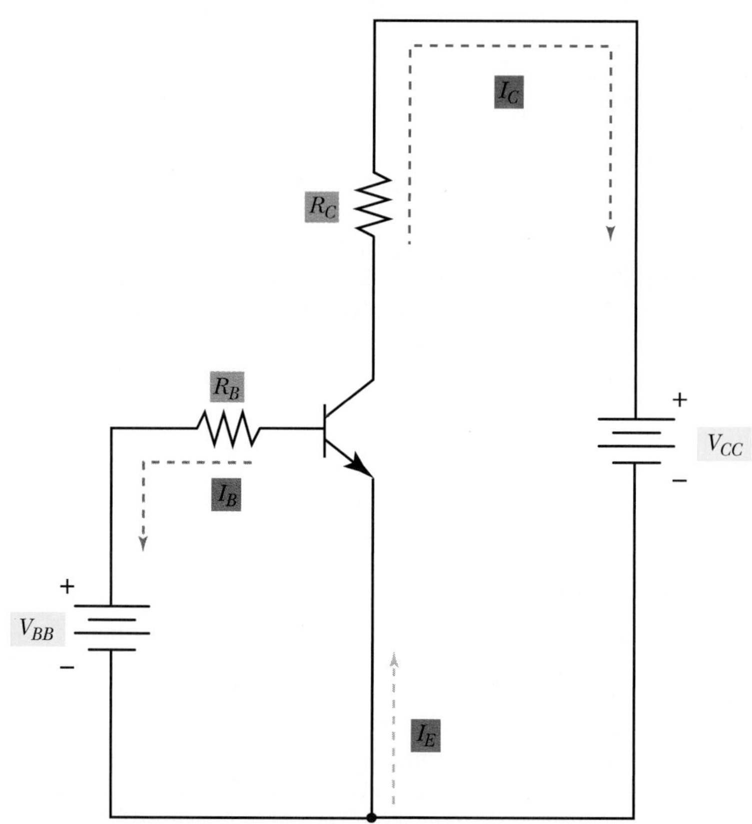

the BJT is equal to the dc voltage source minus the voltage drop across the collector resistor: $V_{CE} = V_{CC} - I_C R_C$. If the base-emitter junction is forward biased and V_{CC} is sufficiently high, there will be a complete path for current flow that begins from the negative terminal of V_{CC}, goes to the emitter, passes through the extremely thin base region of the BJT, goes to the collector terminal, and returns to the positive terminal of V_{CC} through the collector resistor. One of the vital characteristics of a BJT is that the collector current (I_C) is always much larger than the base current (I_B) that controls it.

The normal operating conditions for a BJT can be summarized this way:

- The base-emitter junction must be forward biased with a little bit of base current, I_B.
- The collector-base junction must be reverse biased.

The following occur as a result of setting up these conditions:

- The emitter current, I_E, is equal to the sum of the base and collector currents.
- The collector current, I_C, is always much larger than the base current.

Formula 5–1 expresses the fact that the emitter current is exactly equal to the sum of the collector and the base currents:

FORMULA 5–1 $I_E = I_C + I_B$

The fact that I_C is always much greater than I_B is expressed in terms of a ratio. Typically, I_C is between 40 and 300 times greater than I_B. This ratio is expressed as the β_{dc} **(dc beta)** or sometimes h_{FE}. (We will use the β_{dc} notation in our discussions, but you should expect to see the "h convention" often used in other places.) Formula 5–2 shows how you can calculate the value of dc beta when you know the values of the collector and base currents.

FORMULA 5–2 $\beta_{dc} = \dfrac{I_C}{I_B}$

Used in this way, β_{dc} represents the current gain of the transistor.

◆ **EXAMPLE** In a test setup for a BJT, you find that the collector current is 1.2 mA when the base current is 24 μA. What is the current gain, β_{dc}, of the transistor?

Answer:

From Formula 5–2:

$$\beta_{dc} = \frac{I_C}{I_B}$$

$$\beta_{dc} = \frac{1.2\text{ mA}}{24\ \mu\text{A}}$$

$$\beta_{dc} = 50$$

The current gain of this BJT IS 50. ◆

Formulas 5–3 and 5–4 show how you can determine the collector current and the base current when you know the dc beta and one of the other terms.

FORMULA 5–3 $I_C = \beta_{dc} \times I_B$

FORMULA 5–4 $I_B = \dfrac{I_C}{\beta_{dc}}$

◆ **EXAMPLE** The β_{dc} of a certain BJT is known to be approximately 180. How much base current is required in order to have 18 mA of collector current?

Answer:

From Formula 5–4:

$$I_B = \frac{I_C}{\beta_{dc}}$$
$$I_B = \frac{18 \text{ mA}}{180}$$
$$I_B = 100\ \mu\text{A}$$ ◆

There is another BJT specification that is important when you are working with certain transistor configurations. This is the α (alpha) ratio—the ratio of collector current to emitter current.

FORMULA 5–5 $\alpha = \frac{I_C}{I_E}$

Values of α for typical BJTs are always less than 1, but usually greater than 0.9.

◆ **EXAMPLE** The collector current for a certain BJT is 9.8 mA when the emitter current is 10 mA. What is the BJT's α rating?

Answer:

From Formula 5–5:

$$\alpha = \frac{I_C}{I_E}$$
$$\alpha = \frac{9.8 \text{ mA}}{10 \text{ mA}}$$
$$\alpha = 0.98$$

The α for this BJT is 0.98. ◆

All of the BJT specification described so far apply to the experimental setup for an NPN transistor in Figure 5–6. You might be pleased to know that the same kind of thinking, expressions, and formulas apply equally well to PNP transistors. See Figure 5–7. The only difference is the polarity of the applied bias voltages.

You can see that the base-emitter junction is forward biased by V_{BB}, and that the collector-base junction is reverse biased by V_{CC}. This is the same set of conditions required for operating an NPN transistor.

PRACTICE PROBLEMS 1

1. How much emitter current flows in a BJT that has a base current of 100 μA and a collector current of 10 mA?
2. Suppose you measure the base current for a certain BJT and find it is 15 μA. The corresponding collector current is 2 mA. What is the dc current gain, β_{dc}, of the BJT if it is an NPN transistor? If it is a PNP transistor?
3. A certain BJT has a β_{dc} of 100. Calculate the amount of collector current, amount of emitter current, and the value of α when the base current is 10 μA.

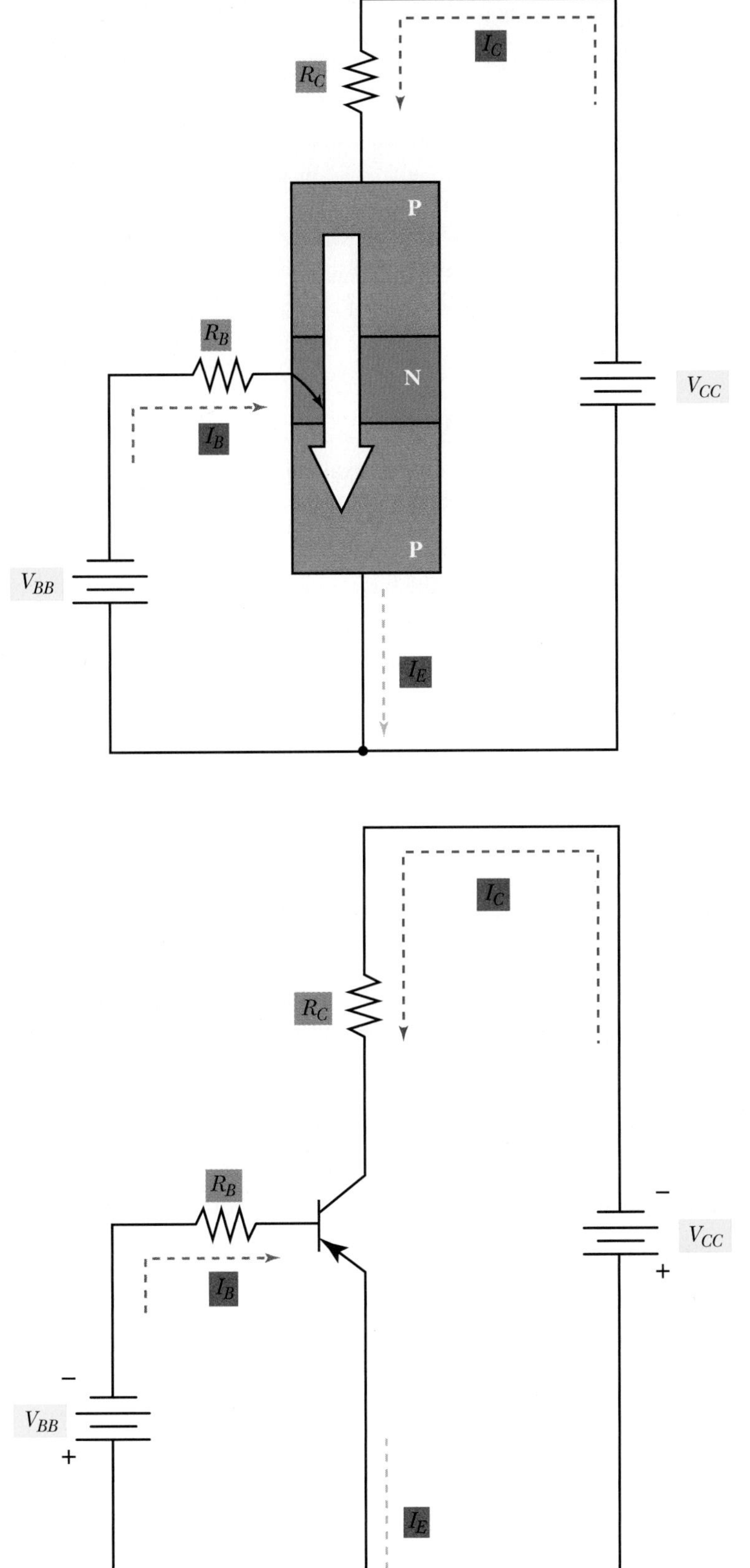

FIGURE 5–7 Bias voltages and currents in a PNP transistor

■ **IN-PROCESS LEARNING CHECK 2**

Fill in the blanks as appropriate.

1. Under normal operating conditions, the base-emitter junction of a BJT must be _______ biased, while the collector-base junction is _______ biased.
2. Current must be flowing through the _______ junction of a BJT before current can flow between the emitter and collector.
3. Stated in words, the dc beta of a BJT is the ratio of _______ divided by _______.
4. Stated in words, the alpha of a BJT is the ratio of _______ divided by _______.
5. The α of a BJT is _______ than 1, while the β_{dc} is _______ than 1. ■

5–4 BJT Characteristic Curves

Using the circuit of Figure 5–8a, you can evaluate how the NPN transistor operates. The evaluation will consist of the repetition of the following steps.

1. Adjust the base-bias supply voltage V_{BB} to 0 V.
   Measure the voltage drop across the base-emitter junction V_{BE}.
   Measure the base current flowing in the circuit I_B.
   a. Adjust the collector bias supply voltage V_{CC} to a fixed voltage of 0 V.
      Measure the voltage drop across the collector-emitter junction V_{CE}.
      Measure the collector current flowing in the circuit I_C.
      Measure the emitter current flowing in the circuit I_E.
   b. Repeat step a, increasing the collector bias supply voltage V_{CC} in small increments up to about 30 V.
2. Repeat part 1 steps a and b increasing the base-bias supply voltage V_{BB} starting at 1.7 V in 1-V increments up to about 10.7 V.

Figure 5–8b provides a graph of the collector current I_C versus the collector-emitter voltage drop V_{CE} using the data collected from this experiment. The graph of this data is called a **collector characteristic curve.** All NPN transistors will result in similar graphs. The collector current can be different.

As you evaluate the circuit and the graph, you can observe the following:

when the base supply $V_{BB} < 0.7$ V:

the base-emitter voltage drop $V_{BE} < 0.7$ V
the base current $I_B = 0$ A
the collector current $I_C = 0$ A
the collector-emitter voltage drop $V_{CE} = V_{CC}$
This region of operation is referred to as the *cutoff* region.

when the base supply $V_{BB} > 0.7$ V and $V_{CE} \leq 0.3$ V:

the base-emitter voltage drop $V_{BE} = 0.7$ V
the base current $I_B > 0$ A
the collector current $I_C > 0$ A, I_C increases sharply
This region of operation is referred to as the *saturation* region.

when the base supply $V_{BB} > 0.7$ V and $V_{CE} > 0.3$ V:

the base-emitter voltage drop $V_{BE} = 0.7$ V
the base current $I_B > 0$ A
the collector current $I_C > 0$ A, the increase in I_C tapers off
This region of operation is referred to as the *active* or *linear* region.

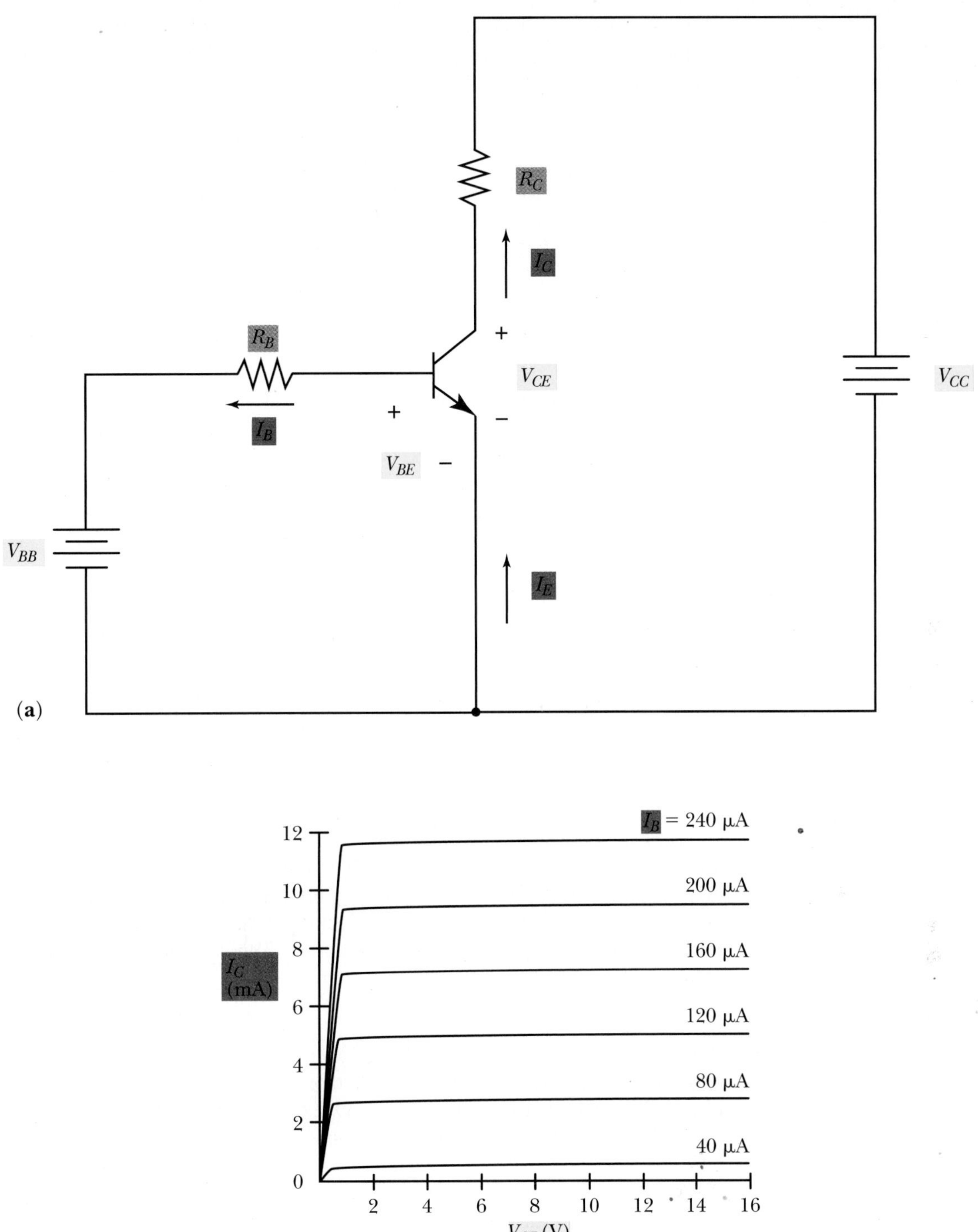

FIGURE 5–8 (a) Transistor circuit used to evaluate transistor characteristics; (b) NPN family of collector characteristic curves

Practical Notes

All BJTs have a frequency limitation often stated as the Current-Gain—Bandwidth Product. At a frequency of $100MH_Z$, the 2N3904 would have a current gain of 3.

The family of collector characteristic curves for a PNP transistor is similar to the characteristic curves for an NPN transistor except the PNP values for voltage and current are negative values.

Collector characteristic curves are very nearly the same for a given type of BJT. The 2N697 is a popular general-purpose NPN transistor. Every 2N697 that is tested should have nearly identical families of collector characteristic curves. The 2N2222A is another popular NPN transistor, and each of them should have the same set of collector characteristic curves. But the family of collector characteristic curves for a 2N697 is different from that of a 2N2222A. Transistor manufacturers publish the collector characteristic curves for each of their BJT products.

5–5 BJT Ratings

The ON Semiconductor (formerly Motorola) data sheet for the 2N3904 general purpose transistor is provided in Figure 5–9. The part number is located in the top right of the data sheet. The device type is identified in the top left. The package configuration is shown below the part number with the pin assignment shown on the diagram and the provided schematic symbol. The second page of the data sheet contains values for the dc current gain (h_{FE}), the collector-emitter saturation voltage ($V_{CE(sat)}$), and the base-emitter saturation voltage ($V_{BE(sat)}$) under the On Characteristics category. Additional parameters of interest are presented on the data sheet.

Maximum Collector-Emitter Voltage ($V_{(BR)CEO}$)

Recall there must be a voltage applied between the emitter and collector in order to operate a BJT properly. What's more, the polarity is important and depends on whether the BJT is an NPN or PNP type. This is the collector-emitter voltage, V_{CE}. There is a definite limit to the amount of collector-emitter voltage that a BJT can withstand without breaking down and destroying itself with an excess of heat-producing current flow between the emitter and collector. This upper voltage limit is called the maximum collector-emitter voltage rating, and it applies when the transistor is turned off—when $I_B = 0$ A (Figure 5–10). The symbol is $V_{(BR)CEO}$. The $V_{(BR)}$ stands for "breakdown voltage" and CEO stands for "collector emitter" voltage when the remaining terminal, the base, is "open" (no connection to it). Typical values of $V_{(BR)CEO}$ for BJTs are between 25 and 100 V.

Maximum Collector-Base Voltage ($V_{(BR)CBO}$)

You have already learned that the collector-base junction of a BJT must be reverse biased. This voltage is specified as V_{CB}. One of the important ratings, or specifications, for a BJT is the maximum allowable amount of **collector-base voltage.** If this reverse-bias value is exceeded, the voltage will break down the collector-base junction and destroy the transistor. The symbol for the maximum collector-base voltage is $V_{(BR)CBO}$. This signifies a breakdown voltage between the collector and base when the emitter is open. See Figure 5–11.

Values for $V_{(BR)CBO}$ run about 50 to 100% higher than the $V_{(BR)CEO}$ rating of the same transistor. So if a certain BJT has a collector-emitter breakdown voltage rating of 80 V, you can expect the collector-base rating to be on the order of 120 V to 160 V.

Maximum Base-Emitter Voltage ($V_{(BR)EBO}$)

The **base-emitter voltage** should either be 0 V (transistor turned off) or some forward-biasing value (transistor conducting). This P-N junction is rarely reverse biased. One reason is that the base portion of the BJT is extremely thin, which means it cannot withstand very much reverse-bias voltage without breaking down and destroying the junction.

2N3903, 2N3904

2N3903 is a Preferred Device

General Purpose Transistors

NPN Silicon

ON Semiconductor

http://onsemi.com

Features

- Pb–Free Package May be Available. The G–Suffix Denotes a Pb–Free Lead Finish

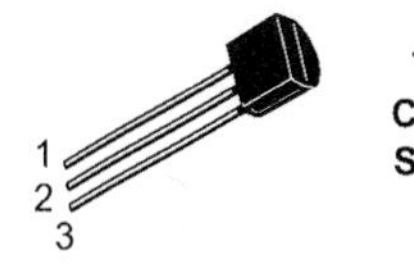

TO–92
CASE 29
STYLE 1

MAXIMUM RATINGS

Rating	Symbol	Value	Unit
Collector–Emitter Voltage	V_{CEO}	40	Vdc
Collector–Base Voltage	V_{CBO}	60	Vdc
Emitter–Base Voltage	V_{EBO}	6.0	Vdc
Collector Current – Continuous	I_C	200	mAdc
Total Device Dissipation @ $T_A = 25°C$ Derate above 25°C	P_D	625 5.0	mW mW/°C
Total Device Dissipation @ $T_C = 25°C$ Derate above 25°C	P_D	1.5 12	W mW/°C
Operating and Storage Junction Temperature Range	T_J, T_{stg}	−55 to +150	°C

THERMAL CHARACTERISTICS (Note 1)

Characteristic	Symbol	Max	Unit
Thermal Resistance, Junction-to-Ambient	$R_{\theta JA}$	200	°C/W
Thermal Resistance, Junction-to-Case	$R_{\theta JC}$	83.3	°C/W

1. Indicates Data in addition to JEDEC Requirements.

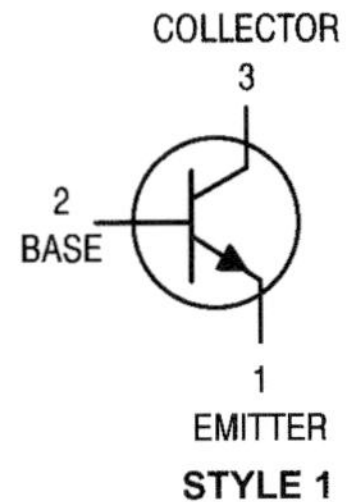

STYLE 1

MARKING DIAGRAMS

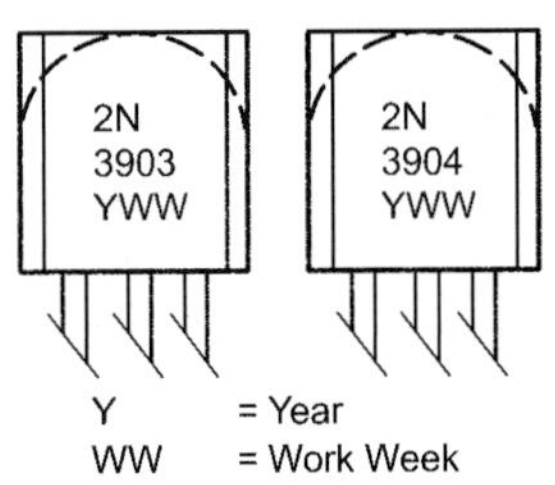

Y = Year
WW = Work Week

ORDERING INFORMATION

Device	Package	Shipping†
2N3903	TO–92	5000 Units/Box
2N3903RLRM	TO–92	2000/Ammo Pack
2N3904	TO–92	5000 Units/Box
2N3904RLRA	TO–92	2000/Tape & Reel
2N3904RLRE	TO–92	2000/Tape & Reel
2N3904RLRM	TO–92	2000/Ammo Pack
2N3904RLRMG	TO–92	2000/Ammo Pack
2N3904RLRP	TO–92	2000/Ammo Pack
2N3904RL1	TO–92	2000/Tape & Reel
2N3904ZL1	TO–92	2000/Ammo Pack

†For information on tape and reel specifications, including part orientation and tape sizes, please refer to our Tape and Reel Packaging Specifications Brochure, BRD8011/D.

*For additional information on our Pb–Free strategy and soldering details, please download the ON Semiconductor Soldering and Mounting Techniques Reference Manual, SOLDERRM/D.

Preferred devices are recommended choices for future use and best overall value.

December, 2003 – Rev. 4

Publication Order Number:
2N3903/D

FIGURE 5–9

2N3903 2N3904

ELECTRICAL CHARACTERISTICS (T_A = 25°C unless otherwise noted) (Continued)

Characteristic		Symbol	Min	Max	Unit
ON CHARACTERISTICS					
DC Current Gain[1]		h_{FE}			—
(I_C = 0.1 mAdc, V_{CE} = 1.0 Vdc)	2N3903		20	—	
	2N3904		40	—	
(I_C = 1.0 mAdc, V_{CE} = 1.0 Vdc)	2N3903		35	—	
	2N3904		70	—	
(I_C = 10 mAdc, V_{CE} = 1.0 Vdc)	2N3903		50	150	
	2N3904		100	300	
(I_C = 50 mAdc, V_{CE} = 1.0 Vdc)	2N3903		30	—	
	2N3904		60	—	
(I_C = 100 mAdc, V_{CE} = 1.0 Vdc)	2N3903		15	—	
	2N3904		30	—	
Collector–Emitter Saturation Voltage[1]		$V_{CE(sat)}$			Vdc
(I_C = 10 mAdc, I_B = 1.0 mAdc)			—	0.2	
(I_C = 50 mAdc, I_B = 5.0 mAdc			—	0.3	
Base–Emitter Saturation Voltage[1]		$V_{BE(sat)}$			Vdc
(I_C = 10 mAdc, I_B = 1.0 mAdc)			0.65	0.85	
(I_C = 50 mAdc, I_B = 5.0 mAdc)			—	0.95	
SMALL–SIGNAL CHARACTERISTICS					
Current–Gain — Bandwidth Product		f_T			MHz
(I_C = 10 mAdc, V_{CE} = 20 Vdc, f = 100 MHz)	2N3903		250	—	
	2N3904		300	—	
Output Capacitance (V_{CB} = 5.0 Vdc, I_E = 0, f = 1.0 MHz)		C_{obo}	—	4.0	pF
Input Capacitance (V_{EB} = 0.5 Vdc, I_C = 0, f = 1.0 MHz)		C_{ibo}	—	8.0	pF
Input Impedance		h_{ie}			k Ω
(I_C = 1.0 mAdc, V_{CE} = 10 Vdc, f = 1.0 kHz)	2N3903		1.0	8.0	
	2N3904		1.0	10	
Voltage Feedback Ratio		h_{re}			X 10^{-4}
(I_C = 1.0 mAdc, V_{CE} = 10 Vdc, f = 1.0 kHz)	2N3903		0.1	5.0	
	2N3904		0.5	8.0	
Small–Signal Current Gain		h_{fe}			—
(I_C = 1.0 mAdc, V_{CE} = 10 Vdc, f = 1.0 kHz)	2N3903		50	200	
	2N3904		100	400	
Output Admittance (I_C = 1.0 mAdc, V_{CE} = 10 Vdc, f = 1.0 kHz)		h_{oe}	1.0	40	μmhos
Noise Figure		NF			dB
(I_C = 100 μAdc, V_{CE} = 5.0 Vdc, R_S = 1.0 k Ω, f = 1.0 kHz)	2N3903		—	6.0	
	2N3904		—	5.0	
SWITCHING CHARACTERISTICS					
Delay Time	(V_{CC} = 3.0 Vdc, V_{BE} = 0.5 Vdc, I_C = 10 mAdc, I_{B1} = 1.0 mAdc)	t_d	—	35	ns
Rise Time		t_r	—	35	ns
Storage Time	(V_{CC} = 3.0 Vdc, I_C = 10 mAdc, I_{B1} = I_{B2} = 1.0 mAdc) 2N3903	t_s	—	175	ns
	2N3904		—	200	
Fall Time		t_f	—	50	ns

1. Pulse Test: Pulse Width ≤ 300 μs; Duty Cycle ≤ 2.0%.

FIGURE 5–9 Continued

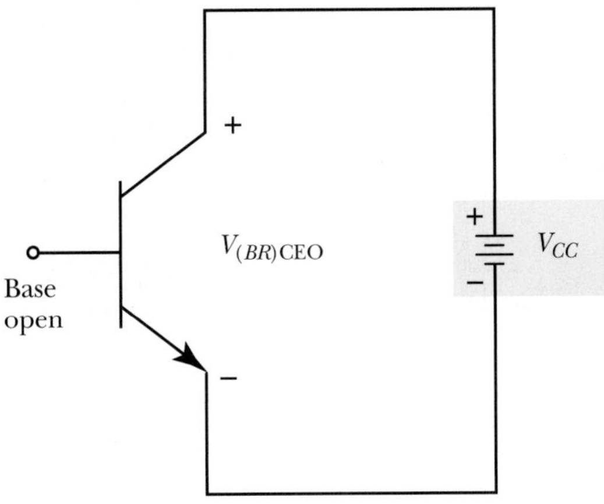

FIGURE 5–10 Diagram of maximum collector-emitter voltage

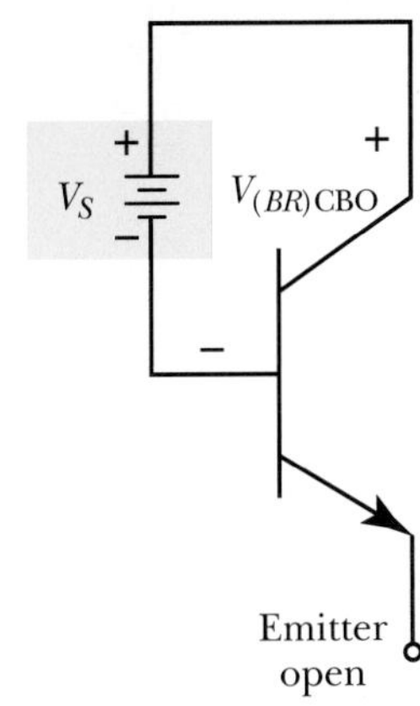

FIGURE 5–11 Diagram of maximum collector-base voltage

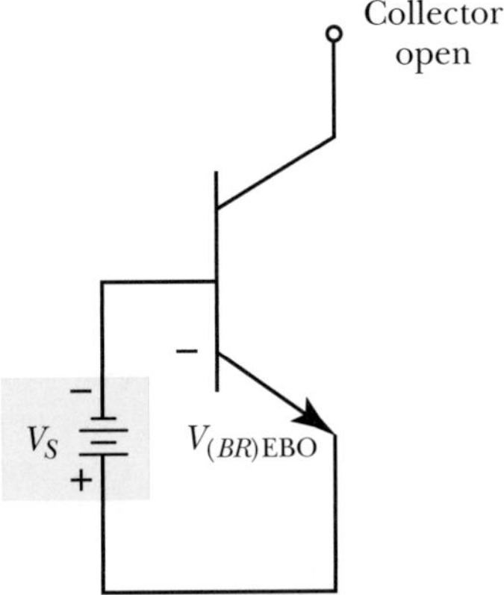

FIGURE 5–12 Diagram of maximum base emitter-voltage

The voltage between emitter and base, Figure 5–12, is usually shown as V_{BE}. The reverse breakdown voltage level for this parameter is shown as $V_{(BR)EBO}$. This is the breakdown voltage (reverse) between the emitter and base terminals when the collector terminal is open. Typical values for $V_{(BR)EBO}$ are between 5 V and 10 V.

Practical Notes

Be careful when applying reverse dc levels and ac waveforms directly between the emitter and base of a BJT. The P-N junction between the emitter and base of a BJT is easily destroyed by seemingly harmless voltage levels.

Maximum Collector Current (I_C) and Power Dissipation (P_d)

The collector current for a BJT is usually denoted at I_C. The maximum allowable amount of continuous collector current is shown under maximum ratings as I_C. For BJTs that are designed for small-signal and light-duty applications, the maximum I_C is less than 1 A. For power BJTs, the maximum I_C is 1 A or more. Larger BJTs also have a maximum surge current rating. BJTs can withstand surges of current that are two or three times greater than their continuous current rating. The data sheet will list the allowable duration of the surges and their maximum duty cycle. The power dissipation (P_d) of a BJT is figured by multiplying the collector-emitter voltage by the collector current. Power dissipation is related to the amount of heating the BJT will undergo. Whenever the power dissipation of a BJT exceeds the maximum power rating (P_d), the transistor's own heat will cause it to destroy itself. However, you can effectively increase the maximum power dissipation of a transistor by mounting it to a heat sink assembly (see Figure 5–3).

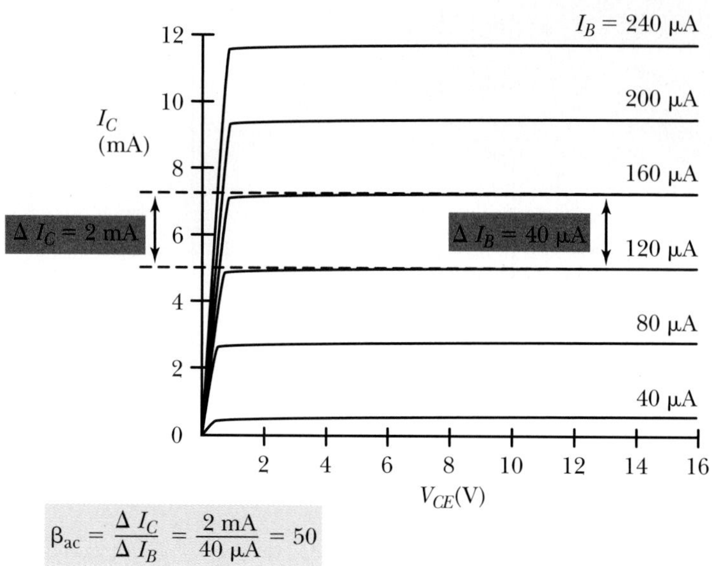

$$\beta_{ac} = \frac{\Delta I_C}{\Delta I_B} = \frac{2\text{ mA}}{40\text{ µA}} = 50$$

FIGURE 5–13 Calculating β_{ac} from a family of collector characteristic curves

AC Current Gain (β_{ac} or h_{FE})

You have learned that a BJT has a certain amount of dc current gain that is called dc beta, β_{dc}. This is found in the transistor data sheets (sometimes as h_{FE}), and it is defined as the ratio of collector current to base current. There is a second version of current gain that appears on transistor data sheets under small-signal characteristics. This is the **ac beta, β_{ac}** (sometimes shown as h_{FE}). Figure 5–13 will help you understand how the value of β_{ac} can be determined.

Notice that a change in base current between 120 µA and 160 µA (40 µA change) causes the collector current to change from 5.5 mA to 7.5 mA (2 mA change). Ac beta is calculated as the ratio of the change in collector current (2 mA in this example) to the change in base current (40 µA) in this example. This comes out to provide β_{ac} = 2 mA/40 µA = 50. In this formula, Δ means " a change in."

FORMULA 5–6 $\beta_{ac} = \dfrac{\Delta I_C}{\Delta I_B}$

> *Practical Notes*
>
> Some transistor troubles are caused by physical factors. An example is when a solder connection fails to make contact between a transistor electrode and the conductive foil on a circuit board.

5–6 An NPN Transistor Circuit Model

Figure 5–14b shows a dc model of the NPN transistor. The model considers how the transistor operates in the three regions (cutoff, saturation, and linear). Figure 5–14a models the collector-emitter junction of the transistor as a switch. If the voltage drop across the base-emitter junction is less than 0.7 V, then the switch is open (cutoff

Application Problem

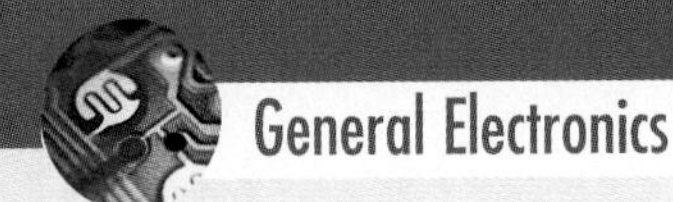

POWER CONSIDERATIONS OF A TRANSISTOR

A medium-power NPN silicon transistor (2N4877) has the following maximum ratings taken from the Motorola, Inc. data sheet.

*** Maximum Ratings**

Rating	Symbol	Value	Unit
Collector-Emitter Voltage	V_{CEO}	60	V_{DC}
Collector-Base Voltage	V_{CB}	70	V_{DC}
Emitter-Base Voltage	V_{EB}	5.0	V_{DC}
Collector Current–Continuous	I_C	4.0	A_{DC}
Base Current	I_B	1.0	A_{DC}
Total Device Dissipation @ $T_C = 25°C$	P_D	10	Watts

Calculate the voltage drop (V_{CE}) across the fully saturated transistor.

Solution We know $I_C = 4.0$ A from the Motorola data sheet, and the maximum transistor power dissipation is 10 W. Using the power equation:

$$P = V \times I$$

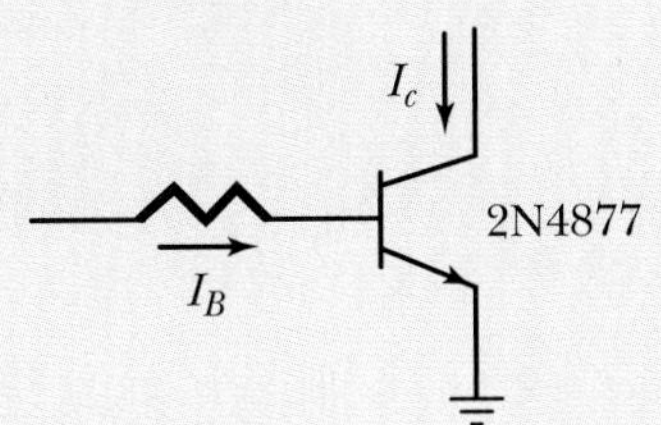

Precision voltage bias

We now calculate the transistor voltage drop (V_{CE}) as follows.

$$V_{CE} = \frac{P_D}{I_C} = \frac{10\text{ W}}{4.0\text{ A}} = 2.5\text{ V}$$

Note: Using the Collector-Emitter voltage of 60 V would not be correct.

region). If the voltage drop across the base-emitter junction equals 0.7 V, then the switch is closed (saturation region). Figure 5–14b models the collector-emitter junction of the transistor as a current-controlled current source (the base current controls the amount of collector current flowing in the circuit). Figure 5–15 shows an ac model of the NPN transistor operating in the linear region. The dynamic resistance r'_e is located within the emitter of the transistor.

5–7 Information for the Technician

Many problems occur in transistors when their electrical ratings are exceeded. Damage results from excessive current through or excessive heating of the materials in the device. (You can damage a semiconductor device when you are soldering it to the circuit board if you do not appropriately heat-sink the leads that you are soldering.) Exceeding the voltage or current parameters for such devices is easy to do, and they "die silently" with no warning, that is, there are no outward signs of damage. Also, wrong voltage polarities applied to the leads will cause problems.

To discover if the transistor is operational, static or dynamic testing can be done. Static testing involves the removal of the transistor from the circuit and checking the transistor characteristics using a digital multimeter (DMM). Dynamic testing is more frequently used in the field where the transistor is tested with power applied to it. Sophisticated *in-circuit testers* and *out-of-circuit testers* are available to test transistors.

The static test can be performed using the DMM. When using the DMM for transistor testing, you will select the DMM diode test mode. Transistor failures are found by checking the forward- and reverse-biased conditions between terminals. Many

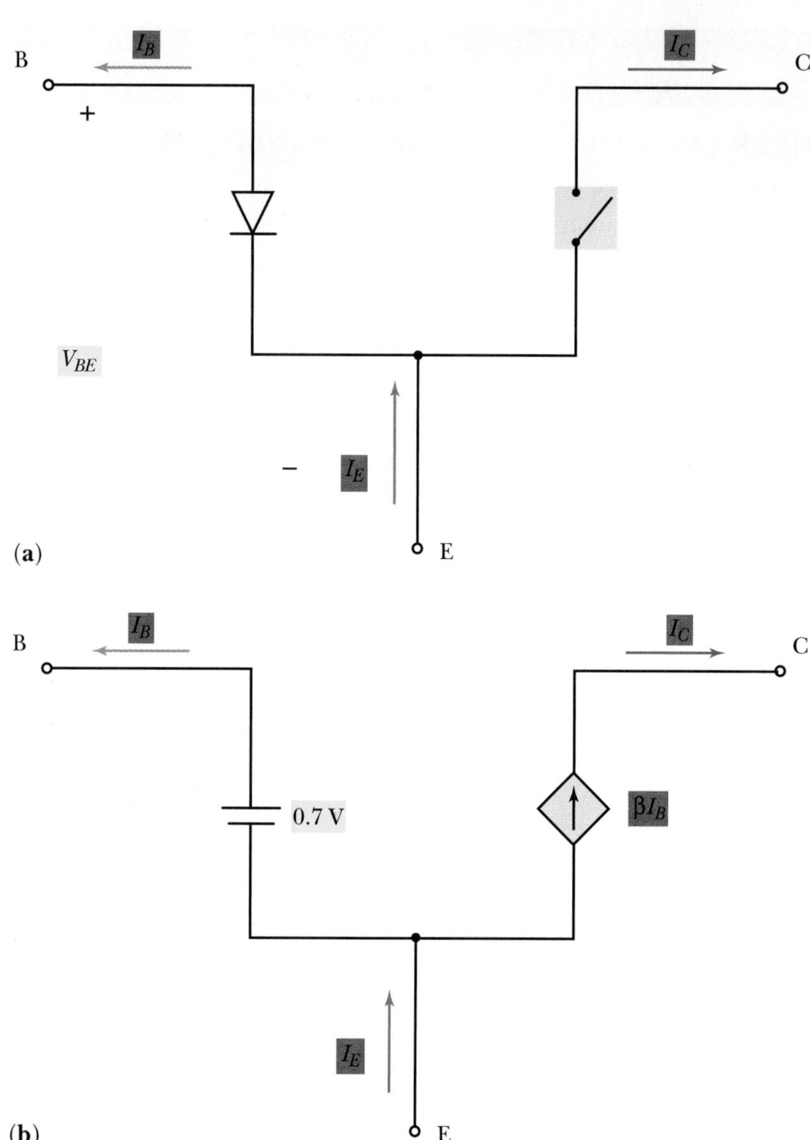

FIGURE 5–14 (a) NPN BJT switch circuit model: $V_{BE} < 0.7$ V switch open cutoff region; $V_{BE} = 0.7$ V switch closed saturation region; (b) NPN BJT dc circuit model: linear (active) region

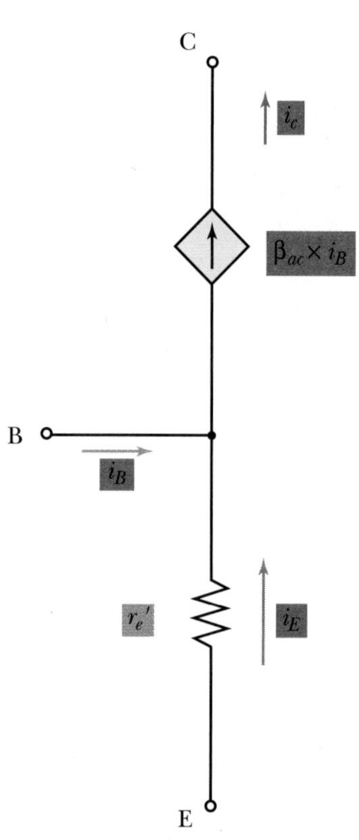

FIGURE 5–15 Ac transistor model (r – parameter): $r_e' = 25 \text{ mV}/I_E$

times, the problem is an internal short or open. Of course, the DMM supplies the battery voltage for these tests. (NOTE: It is important to know the voltage polarity of the test probes when in the diode test mode.) The DMM diode test current is typically between 1 to 2 mA. Refer to Figure 5–16 for some examples of how DMM tests are used for the NPN and PNP transistors. Note that the values shown are only relative and can differ from device to device.

In the dynamic test, the circuit is either operated in its normal system (box) or is placed in a "test-jig" (simulates the function of the rest of the system). The circuit has power applied and the technician proceeds to measure voltage levels, either dc levels for bias or ac levels for the signal flowing through the system. When a device (BJT, resistor, capacitor, etc.) fails, voltage levels around the defective device will not be the values expected. The technician then evaluates the test results to locate the defective component(s) and possible source of the fault (what caused the part(s) to fail). Since semiconductor devices are usually most sensitive to other system faults, they are often the first parts to "break" in a system. Often though, the reason the BJT failed is not because it was, in itself, defective, but as a reaction to some other circuit fault. (Much like when a fuse blows to disconnect power from a circuit, it is not because the fuse was defective. However, some other element within the circuit failed, which caused the fuse to blow.)

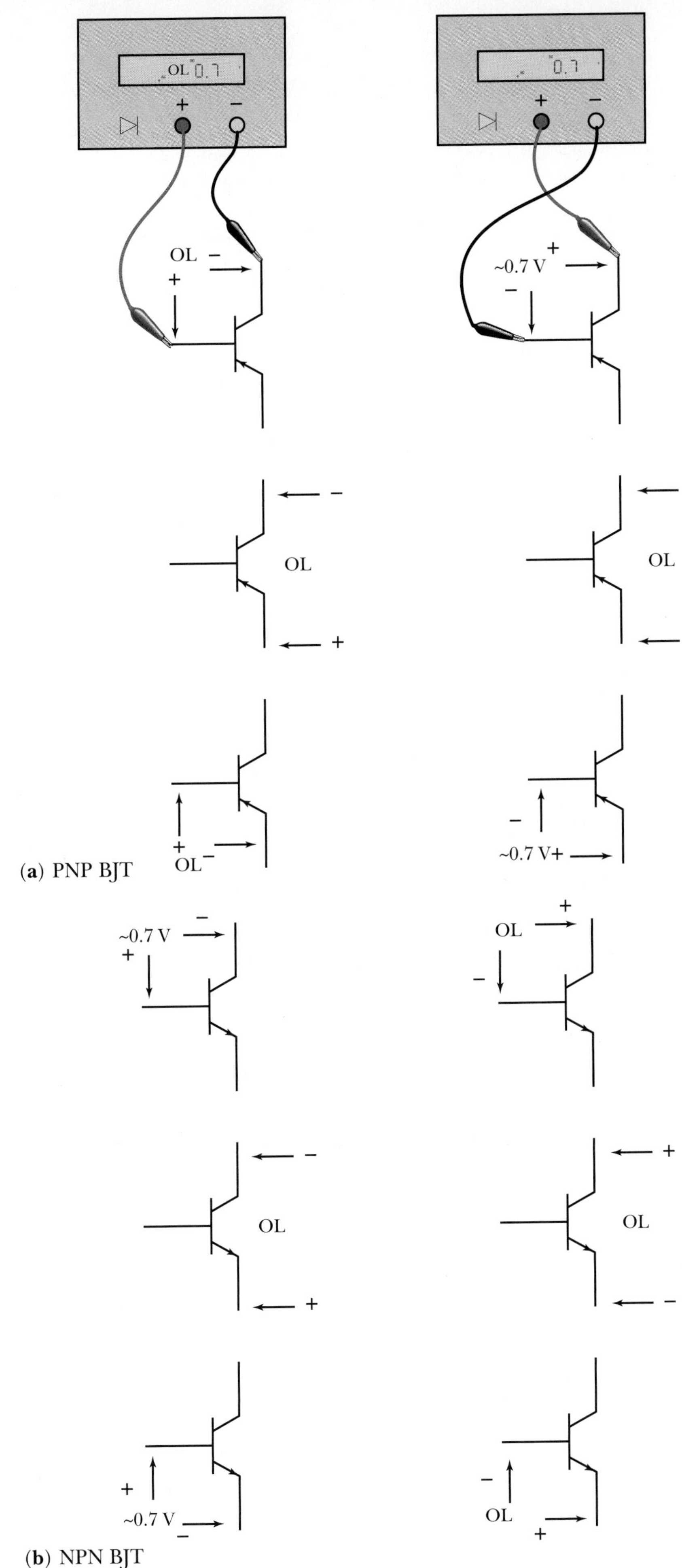

FIGURE 5–16 DMM testing of silicon BJT with diode test function; (a) PNP BJT; (b) NPN BJT

Practical Notes

Some transistor troubles are caused by physical factors. An example is when a solder connection fails to make contact between a transistor electrode and the conductive foil on a circuit board.

5–8 Basic Uses of the BJT

Before we look at a few more details about the operation of BJTs, it is important to be able to describe basically how transistors are used. Transistors are controllers of current. The BJT uses a small amount of base current to control a larger amount of collector current. BJTs are basically used in two ways: as switches and as amplifiers. Often, these two applications are combined in one circuit.

The BJT as a Switch and Current Amplifier

Figure 5–17 shows a typical circuit where an NPN transistor is used as a switch and as an *amplifier* of current. The rectangular waveform at the input switches between 0 V and +5 V. The LED has a forward voltage drop of 1.5 V when conducting. Resistor values are $R_B = 33\ \text{k}\Omega$ and $R_C = 200\ \Omega$. The NPN transistor that is used in this circuit is a 2N3904. Let's examine how the BJT responds to the 0 V input and to the +5 V input.

1. When the input is at 0 V, the transistor is in cutoff.
   - there is no base current flowing in the BJT ($I_B = 0$ A)
   - there is no collector current in the BJT ($I_C = 0$ A)
   - the collector voltage is equal to the V_{CC} level ($V_C = V_{CE} = V_{CC} = +5$ V)

     (NOTE: If you measured the voltage at the collector, it may read less than 5 V since the digital multimeter provides a current conduction path and some voltage may be dropped across the LED.)
   - the LED does not light

   The circuit is switched off.
2. When the input is at +5 V, the transistor is in saturation.
   - base current is flowing in the BJT ($I_B > 0$ A)

$$I_B = \frac{(V_{BB} - V_{BE})}{R_B}$$

$$I_B = \frac{(5\text{ V} - 0.7\text{ V})}{33\ \Omega}$$

$$I_B = 130\ \mu\text{A}$$

   - the collector voltage is equal to the saturation voltage of the BJT ($V_C \leq +0.3$ V)
   - the collector current is flowing in the BJT ($I_C > 0$ mA)

$$I_C = \frac{(V_{CC} - V_D - V_{CE\,(\text{sat})})}{R_C}$$

$$I_C = \frac{(5\text{ V} - 1.5\text{ V} - 0.3\text{ V})}{200\ \Omega}$$

$$I_C = 16\text{ mA}$$

   - the LED is turned on

   The circuit is switched on.

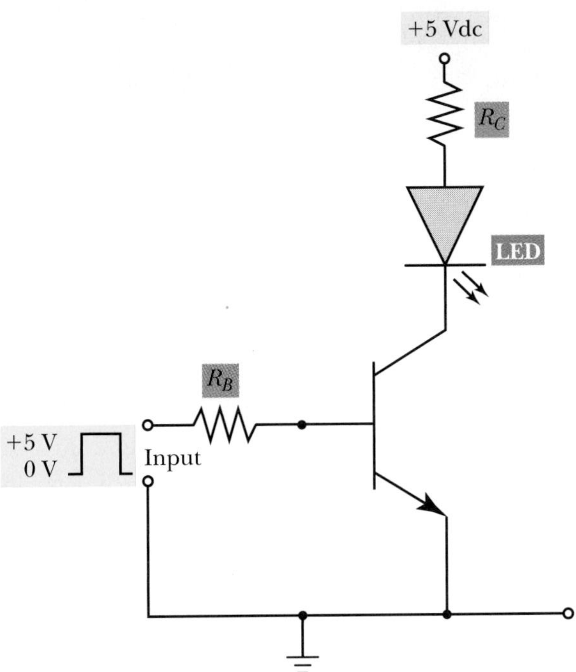

FIGURE 5–17 BJT used as a switch and current amplifier
multiSIM

In summary, the LED lights whenever the input waveform switches to +5 V, and the LED goes out whenever the input waveform switches to its 0 V level. The LED switches on and off at the frequency of the input waveform. At frequencies above 60 Hz, the LED visually appears to be on at all times. However, using the scope, you can observe that the LED is still turning on and off.

But why do we go to the trouble of using a BJT? Why not connect the input waveform direct to the LED and simplify the circuit? Well, the reason is that the input waveform probably comes from a source that cannot provide the 16 mA current level that is required for operating most LEDs. Maybe the waveform source can provide a maximum of 1 mA. The BJT circuit draws only 130 μA to provide the 16 mA current for the LED. In this case, the BJT serves as a current amplifier as well as a switch.

The BJT as a Voltage Amplifier

The circuit in Figure 5–18 is a voltage amplifier. The input waveform in this example is a sinusoidal waveform that measures 2 V peak-to-peak. The values of the resistors at the base connection are selected so that the transistor is operating in the linear region ($V_B = 2.2$ V and $V_C = 6$ V). There is some base-bias current flowing, even when the input signal is at 0 V. Let's study how the BJT reacts to one complete cycle of the waveform at 1/4-cycle intervals.

1. At 0°, the input is at 0 V.

   Due to the base-bias circuit, the base voltage is at 2.2 V_{dc}. The base current flowing is due to only the base-bias current. The BJT is conducting, so the collector voltage is somewhere between 0 V and the $+V_{CC}$ level of +12 V. Let's suppose that the base bias is adjusted so that the collector voltage is +6 V.

2. At 90°, the input is at +1 V.

   The positive polarity of the input signal adds to the base voltage. The base voltage is now at 3.2 V. The increased base voltage causes a higher base current resulting in a larger collector current. This means the collector voltage (from collector to ground) drops to a lower level due to the $I \times R$ drop across the collector resistor. Thus, less voltage is available to be dropped by the transistor. (Remember Kirchhoff's voltage law?) Let's say we find it drops down to +2 V.

Application Problem

Computers

BJT PC PARALLEL PORT DRIVER

A PC is used to control a small dc motor. A technician initiates commands through the keyboard to turn the electric motor ON/OFF. The actual switching circuit is a BJT transistor that turns the dc motor ON and OFF. The dc motor's current is 800 mA when it is running. The following figure 1 shows a PC parallel port connected to the induction motor interface.

(a) Calculate the transistor base current needed to operate the induction motor. (assume β of 100)

(b) Calculate a correct value for resistor R. Assume that the buffer chip (74ACT04) output is 5 V when HIGH.

Solution

(a) We know the dc motor requires $I_C = 800$ mA, and the transistor beta is 100. Using standard transistor parameter equation:

$$\beta = \frac{I_C}{I_B}$$

We now calculate the transistor base current as follows.

$$I_B = \frac{I_C}{\beta} = \frac{800 \text{ mA}}{100} = 8 \text{ mA}$$

The calculated value of 8 mA is needed to properly turn the transistor ON.

(b) Next we calculate a correct resistor value. We know that the voltage drop between the base and the emitter is 0.7 V. Using Ohm's Law and assuming the HIGH state of 74ACT04 output,

$$R = \frac{5 \text{ V} - \Delta V_{BE}}{I_C} = \frac{5 \text{ V} - 0.7}{8 \text{ mA}} = 537.5 \ \Omega$$

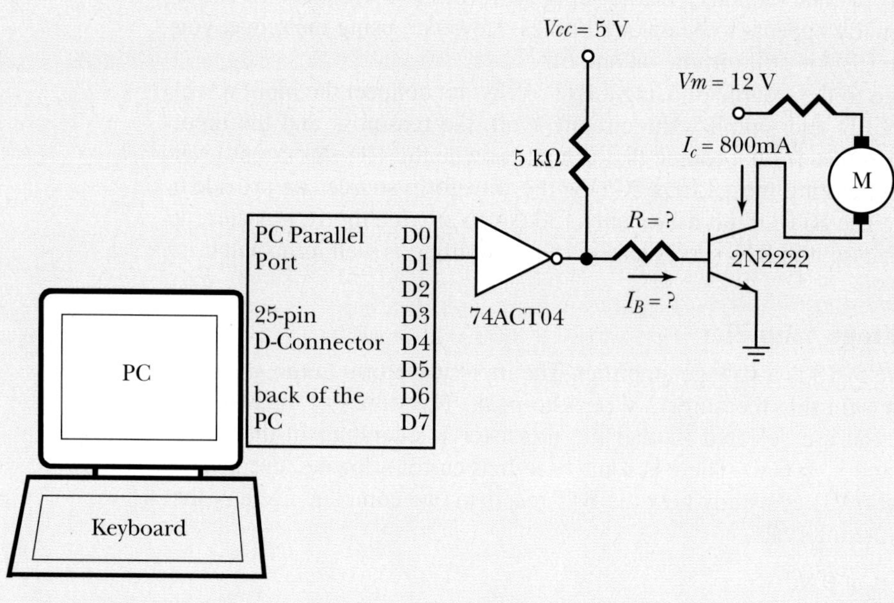

A PC to small motor driver circuit

3. At 180°, the input is at 0 V.

   Again the base current flowing is due to only the base-bias current. There is less base current than while the input waveform was positive; there is less collector current, so there is a larger collector voltage. The collector voltage returns to +6 V.

4. At 270°, the input is at –1 V.

   The negative polarity of the input signal subtracts from the base-bias current set by resistors R_1 and R_2. The reduction in base current causes the base current I_B to be

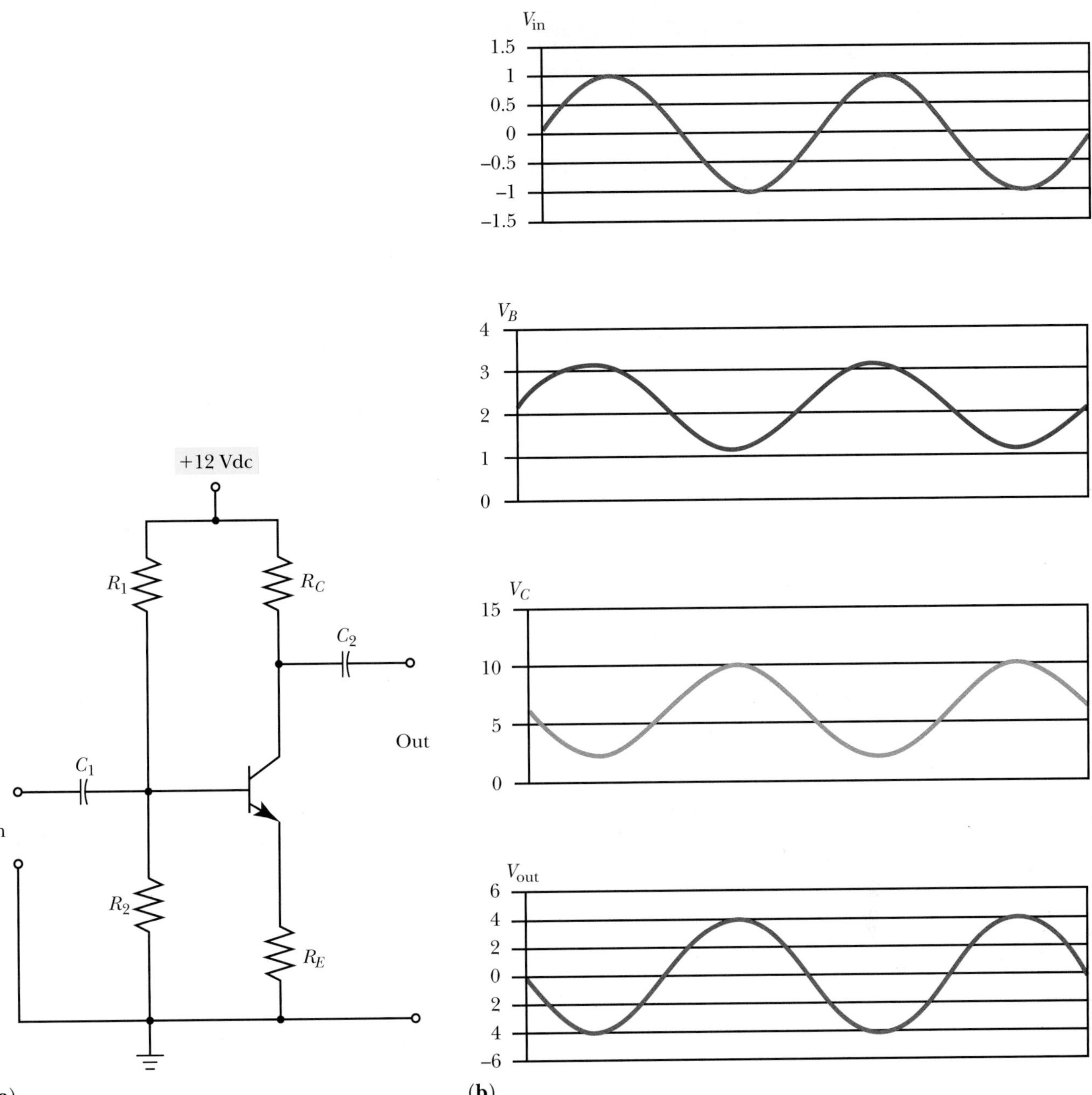

FIGURE 5–18 (a) BJT used as a voltage amplifier; (b) amplifier waveforms

lower. Subsequently, the collector current is lower resulting in an increase in the collector voltage. The collector voltage rises to a higher level, say, +10 V.

You can see from the resulting plot of the output waveform that it is a sinusoidal waveform that is 8 V peak-to-peak. This circuit is a voltage amplifier that increases the input signal from 2 $V_{p\text{-}p}$ to 8 $V_{p\text{-}p}$. It is said that this amplifier has a voltage gain of 4 (8 $V_{p\text{-}p}$ divided by 2 $V_{p\text{-}p}$). Also notice that this type of BJT amplifier inverts the input waveform or, in other words, shifts the waveform by 180°. Even though the BJT is still a current amplifying device (I_B is amplified to I_C), the surrounding circuit converted the input voltage into a current flow and used the output current (I_C) to create an output voltage. Thus making the *system* a voltage amplifier.

■ **IN-PROCESS LEARNING CHECK 3**

Fill in the blanks as appropriate.

1. BJTs use a small amount of base _______ to control a larger amount of collector _______. It can be said that BJTs are _______ controllers.
2. When the base current in a BJT is zero, the collector current is _______.
3. BJTs are basically used as _______ and _______.
4. When a BJT is being used as a switch (as in Figure 5–17, V_{CE} is maximum when I_B is _______, and V_{CE} is minimum when I_B is _______. In the same circuit, I_C is maximum when I_B is _______, and I_C is minimum when I_B is _______.
5. When a BJT is being used as a voltage amplifier (as in Figure 5–18, an increase in base current causes a(n) _______ in V_{CE}.
6. On a family of collector characteristic curves, the horizontal axis represents the _______ and the vertical axis represents the _______. Each curve in the family represents a different level of _______.

5–9 Basic Concepts of Amplifiers

An **amplifier** is a circuit that can increase the peak-to-peak voltage, current, or power of a signal. The change in the signal is called **gain.** Gain is the ratio of the signal output to the signal input. Since both the input signal and the output signal have the same units, gain has no units. The change in the voltage, current, and power caused by the amplifier is respectively known as voltage gain, current gain, and power gain. A triangle is used as a block diagram symbol for an amplifier (Figure 5–19).

Gain Calculations

Voltage gain, or voltage amplification (A_V), can be calculated by comparing the output voltage amplitude to the input voltage amplitude (Formula 5–7).

FORMULA 5–7 $A_V = \frac{V_{out}}{V_{in}}$

Current gain (A_I) can be computed in a similar fashion—that is, the value of the output current divided by the value of the input current (Formula 5–8)

FORMULA 5–8 $A_I = \frac{I_{out}}{I_{in}}$

Power gain (A_P) is calculated as the voltage gain times the current gain (Formula 5–9).

FORMULA 5–9 $A_p = |A_V \times A_I|$

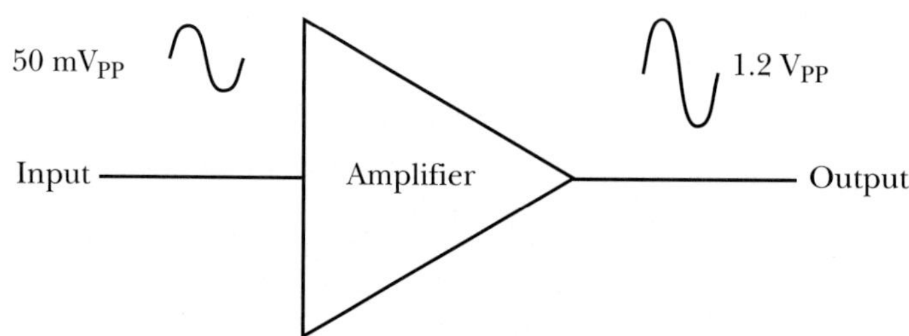

FIGURE 5–19 Voltage amplification

◆ **EXAMPLE** In Figure 5–19, the input voltage applied to an amplifier is 50 mV_{PP}. The output voltage of the amplifier is 1.2 V_{PP}. The input current is 0.5 mA_{PP} and the output current is 10 mA_{PP}. What is the voltage gain? The current gain? The power gain?

Answers:

From Formula 5–7, the voltage gain is:

$$A_V = \frac{V_{out}}{V_{in}}$$

$$A_V = \frac{1.2\ V_{PP}}{50\ mV_{PP}}$$

$$A_V = 24$$

From Formula 5–8, the current gain is:

$$A_I = \frac{I_{out}}{I_{in}}$$

$$A_I = \frac{10\ mA_{PP}}{0.5\ mA_{PP}}$$

$$A_I = 20$$

From Formula 5–9, the power gain is:

$$A_P = |A_V \times A_I|$$

$$A_P = |24 \times 20|$$

$$A_P = 480$$ ◆

IN-PROCESS LEARNING CHECK 4

1. The input signal to an amplifier is 4 mV_{PP} and the output signal is 1 V_{PP}. What is the voltage gain of the amplifier?
2. The input current amplitude is 0.03 mA. The output current amplitude is 1.5 mA. What is the current gain of the amplifier? _______
3. An amplifier has a voltage gain of 60 and a current gain of 30. What is the power gain? _______

You already discovered that the collector current of bipolar junction transistors is larger than the base current. This characteristic is due to the transistor's current gain (β). For the transistor to function as an amplifier, the transistor must be *biased* (the application of dc voltage and current). With proper biasing, ac signals can be applied to the transistor input resulting in an amplified nondistorted output signal.

5–10 Transistor Biasing Circuits

The biasing circuit determines the dc operating point for the transistor. The bias circuit is composed of the dc power source(s), the transistor, and resistors. The dc voltages and currents in the circuit establish the dc operating point. The dc operating point is also known as the **quiescent point** or **Q point.** The Q point establishes the dc voltages and currents around which amplification will take place for the transistor. There are several bias circuits. The following are some very common bias circuits:

1. Fixed base biasing
2. Voltage divider with emitter stabilized biasing

Fixed Base Biasing

Figure 5–20 shows the schematic for the fixed base biasing circuit. To determine the Q point of this circuit, you must first evaluate the base-emitter loop. Writing the KVL (Kirchholf's voltage law) around the closed loop, you get:

$$V_{BB} - I_B \times R_B - V_{BE} = 0$$

If $V_{BB} > 0.7$ V, then $V_{BE} = 0.7$ V and the transistor is operating in either the linear or saturation regions. Using the following formula, you will calculate the base current I_B.

$$I_B = \frac{(V_{BB} - V_{BE})}{R_B}$$

Assuming that the transistor is operating in the linear region, you can calculate the collector current using the dc current gain (β) of the transistor.

$$I_C = \beta \times I_B$$

By writing the KVL for the collector-emitter loop, you can calculate the collector-emitter voltage drop.

$$V_{CC} - I_{CC} \times R_C - V_{CE} = 0$$
$$V_{CE} = V_{CC} - I_C \times R_C$$

For the transistor to be operating in the linear region, the collector-emitter voltage V_{CE} must be greater than 0.3 V. If V_{CE} is not greater than 0.3 V, then the transistor is in saturation. You then can solve for the collector current using $V_{CE} \leq 0.3$ V. Upon completion of your calculations, you now have determined the Q point for this circuit. As you observe, the operating point for this circuit depends on the transistor's dc current

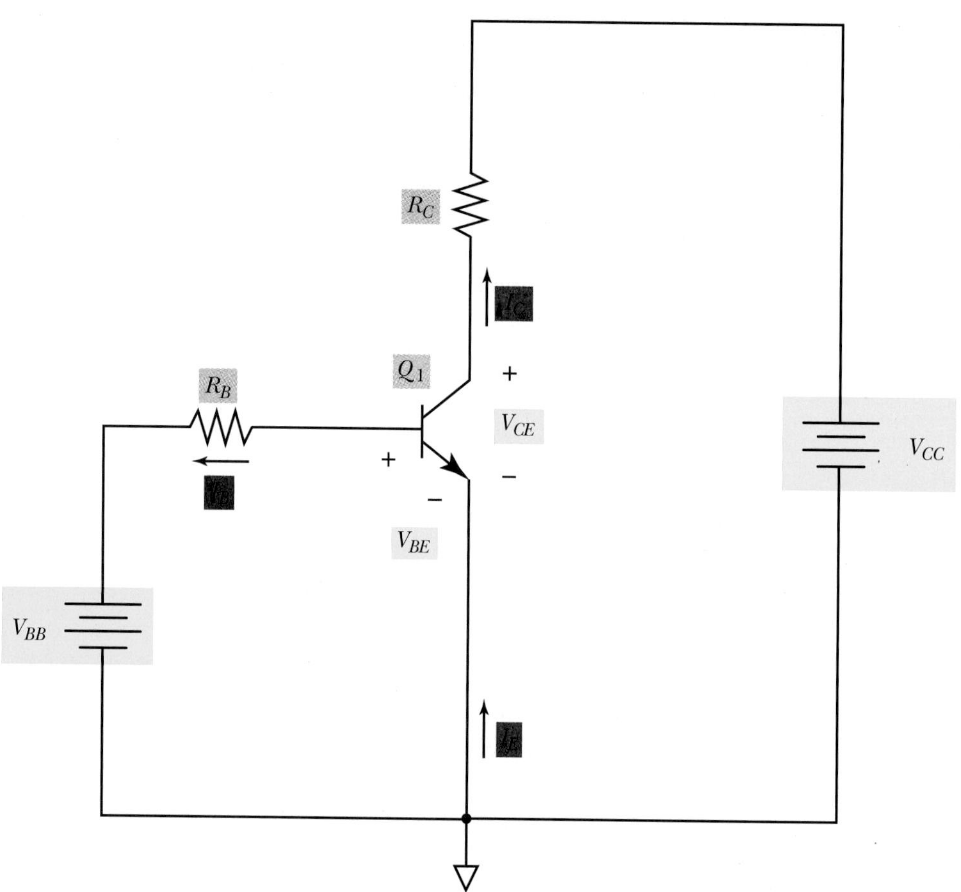

FIGURE 5–20 Fixed base bias circuit

gain (β). Typical transistor current gains β can exist between 40 and 300. The collector current and the collector-emitter voltage drop will have different values as the value of β is changed.

Voltage Divider with Emitter Stabilized Biasing

Figure 5–21 shows the schematic for the voltage divider with emitter stabilized biasing circuit. To determine the Q point of this circuit, you must evaluate the base-emitter loop. To accomplish this, you will first Thevenize the base circuit V_{CC}, R_1 and R_2.

$$V_{TH} = V_{CC} \times \frac{R_2}{(R_1 + R_2)}$$

$$R_{TH} = \frac{R_1 \times R_2}{(R_1 + R_2)}$$

Figure 5–22 shows the schematic for the Thevenin equivalent circuit of Figure 5–21. If $V_{TH} > 0.7$ V, then $V_{BE} = 0.7$ V and the transistor is operating in either the linear or saturation regions. Assuming that the transistor is operating in the linear region and due to the very small voltage drop across the Thevenin resistor (R_{TH}), the base voltage is approximately equal to the Thevenin voltage (Formula 5–10).

FORMULA 5–10 $V_B \approx V_{TH} \times \frac{R_2}{(R_1 + R_2)}$

Writing the KVL (Kirchhoff's voltage law) around the base emitter closed loop, you get:

$$V_B - V_{BE} - I_E \times R_E = 0$$

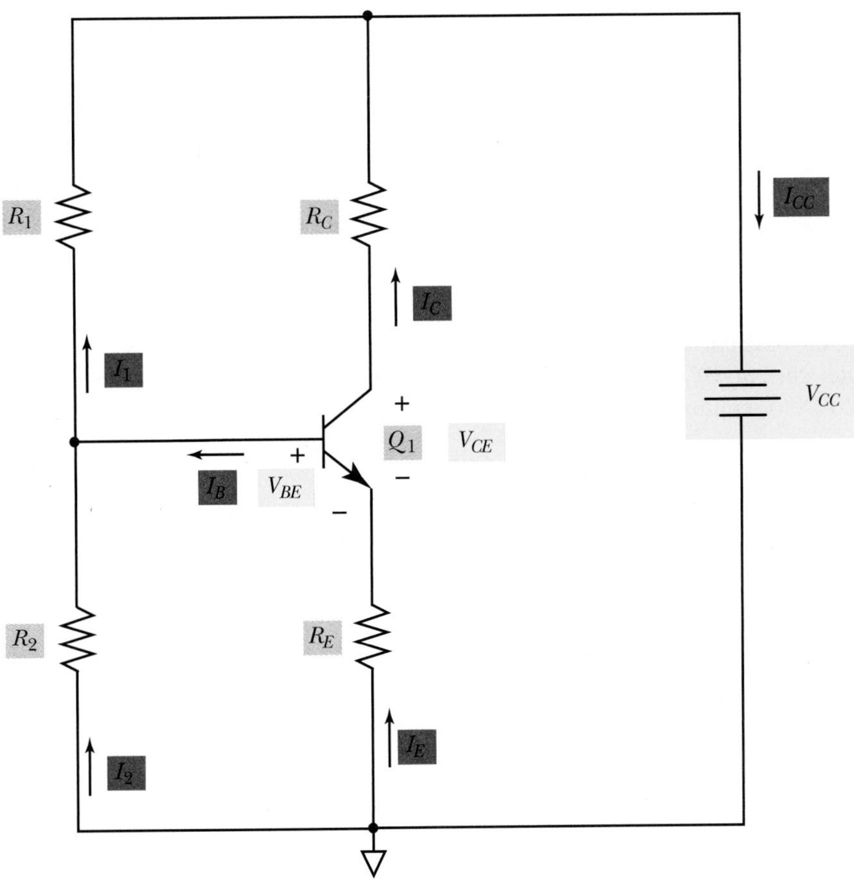

FIGURE 5–21 Voltage divider biasing circuit with emitter stabilized

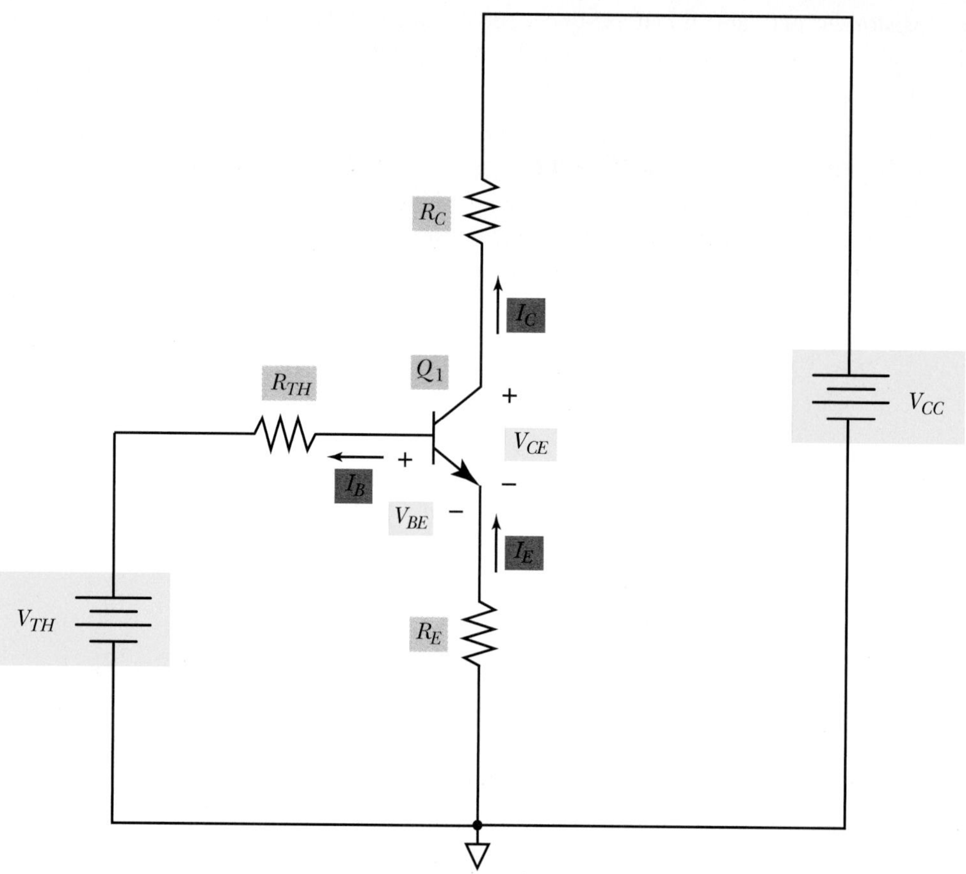

FIGURE 5–22 The Thevenin equivalent circuit for the voltage divider bias circuit

As long as $V_B > 0.7$ V, then $V_{BE} = 0.7$ V and the transistor is operating in either the linear or saturation regions. Assuming that the transistor is operating in the linear region, you can calculate the emitter current I_E using Formula 5–11.

FORMULA 5–11 $I_E = \dfrac{V_B - V_{BE}}{R_E}$

Since the base current is very small, the collector current is approximately equal to the emitter current. By writing the KVL for the collector-emitter loop, you can calculate the collector-emitter voltage drop.

$$V_{CC} - I_C \times R_C - V_{CE} - I_E \times R_E = 0$$

Since $I_C \approx I_E$, you can solve for the collector-emitter voltage.

$$V_{CE} = V_{CC} - I_C \times R_C - I_C \times R_E$$

FORMULA 5–12 $V_{CE} = V_{CC} - I_C \times (R_C + R_E)$

For the transistor to be operating in the linear region, the collector-emitter voltage V_{CE} must be greater than 0.3 V. If V_{CE} is not greater than 0.3 V, then you will assume the transistor is in saturation and solve for the collector current using $V_{CE} \leq 0.3$ V. Upon completion of your calculations, you now have determined the Q point for this circuit.

As you observe, the operating point for this circuit does not depend on the transistor's dc current gain (β), due to the resistor values used for R_1, R_2, and R_E. Therefore, the collector current and the collector-emitter voltage drop have very small changes as the value of β is changed. The Q point is stabilized by the emitter resistor for variations in β.

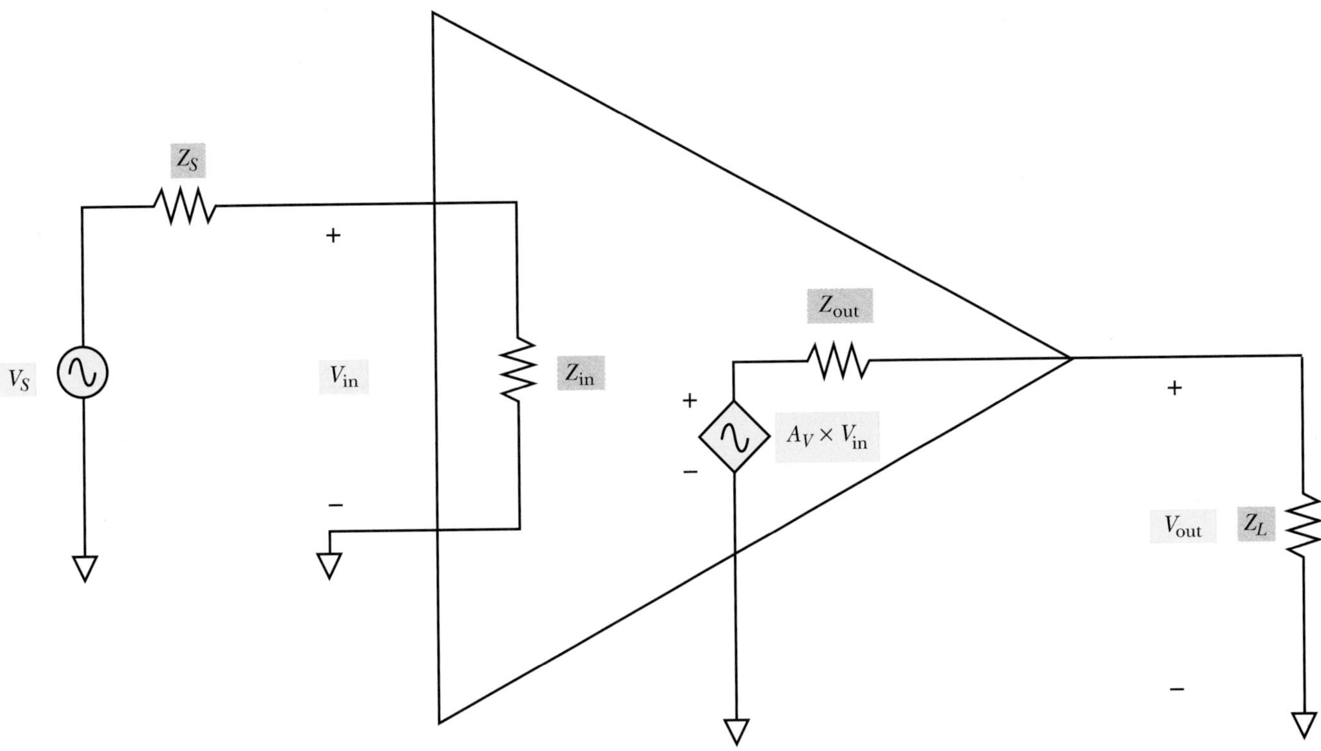

FIGURE 5–23 A schematic model of any amplifier

5–11 Amplifier Circuit Configurations

There are several common ways to classify amplifiers:

1. By circuit configuration
2. By signal levels involved
3. By class of operation
4. By the type of transistor used

There are three basic configurations for BJT amplifier circuits. Any amplifier circuit found in electronics is one of these three basic configurations or, in some special instances, a combination of two of them. It is important that you understand and recognize these basic circuit configurations. Without this understanding, you will become confused when attempting to analyze and troubleshoot more complex transistor amplifier circuits. Figure 5–23 shows a generic circuit model for any amplifier with some of the parameters that you will need to know.

5–12 The Common-Emitter (CE) Amplifier

Figure 5–24 shows the common-emitter amplifier, one of the most commonly used BJT amplifier circuits. Notice that the input signal is applied between the BJT base terminal and ground, and that the output signal is taken between the collector terminal and ground. (NOTE: In evaluating the ac signal operation, the transistor dc sources are bypassed (i.e., considered as shorts) and capacitors act as very low impedances as far as the ac signals are concerned. C_1 and C_2 are referred to as coupling capacitors and C_E is called a bypass capacitor.) Thus, the input signal is effectively between base and ground [emitter], and the output signal is effectively between the collector and ground [emitter]. What BJT terminal is used for both the input and output connections? The emitter. For this reason, this circuit configuration is called the **common-emitter,** or

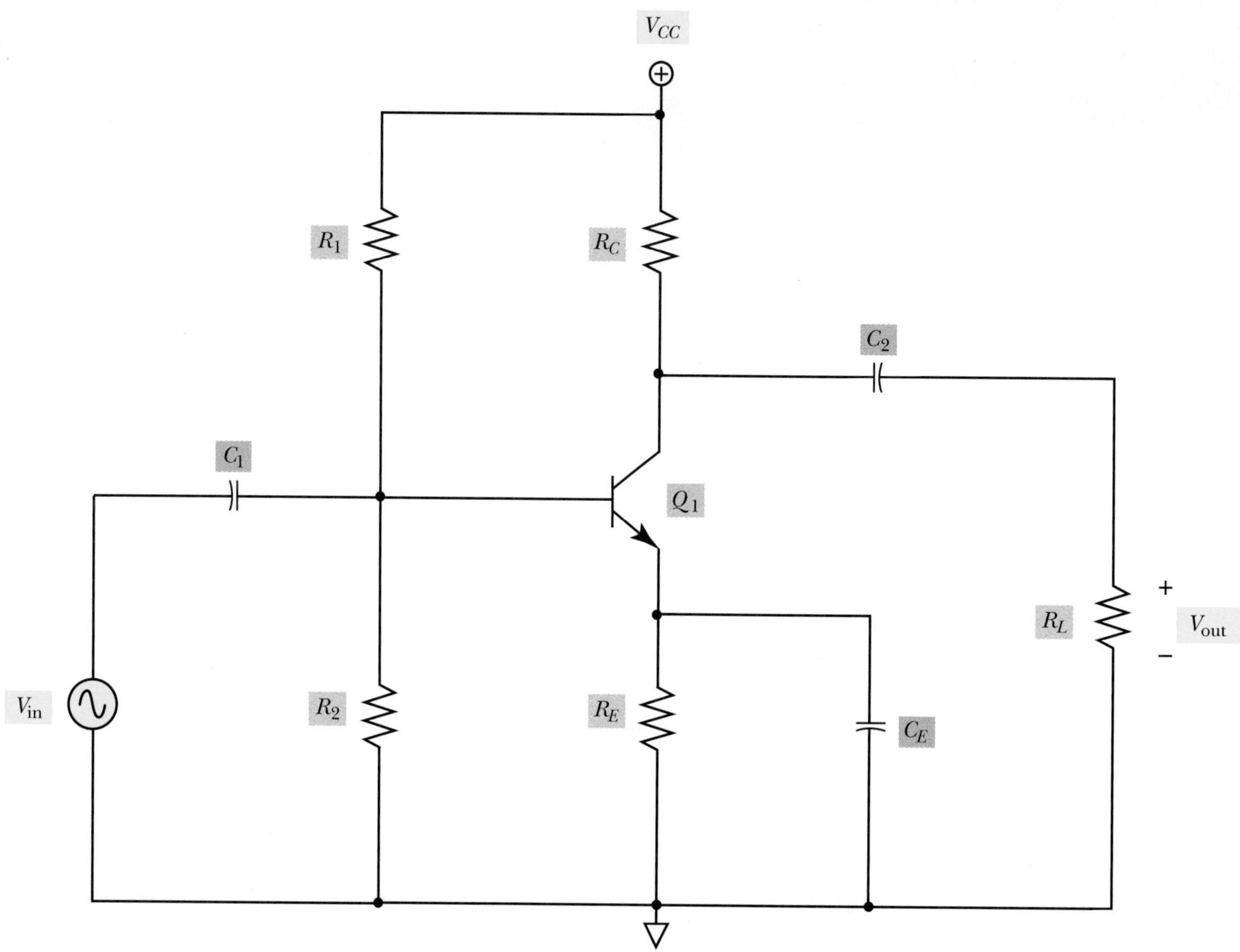

FIGURE 5–24 Common-emitter amplifier multiSIM

CE, amplifier configuration. Let's study several things that happen whenever the input signal level is changed.

Referring to Figure 5–24, suppose the input signal changes and causes the base current to increase. The small increase in the base current causes an increase in the collector current resulting in a decrease in the collector-emitter voltage drop. Notice that an increase in the base current at the input of this circuit causes a decrease in the output voltage. On the other hand, causing the base current to decrease results in an increase in the output voltage.

Figure 5–25 highlights the bias circuit for the common-emitter amplifier circuit. For dc voltages, the capacitors appear as open circuits. The bias circuit used for the common-emitter amplifier is the voltage divider emitter stabilized biasing circuit. In the previous section, you developed the equations to calculate the Q point for this bias circuit. The bias circuit establishes the base current, collector current, and collector-emitter voltage around which the ac signal will be varied. The positioning of the Q point is critical to prevent clipping of the output waveform due to driving the transistor into the cutoff or saturation regions.

Several parameters of importance for amplifiers are their input impedance, output impedance, voltage gain, current gain, and power gain. The equations to calculate these parameters are (NOTE: Equations do not include the effects of the load resistor):

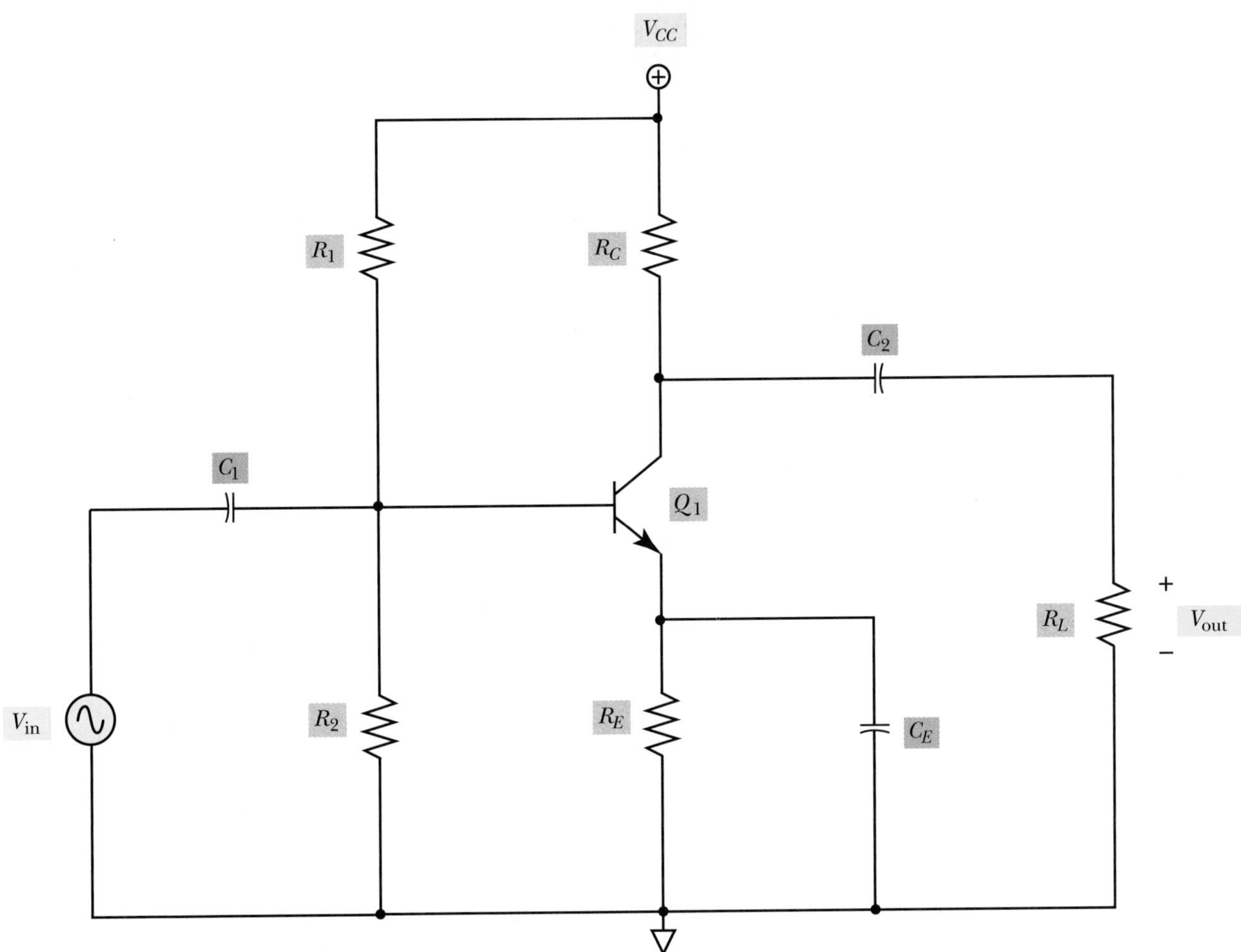

FIGURE 5–25 Common-emitter amplifier biasing

FORMULA 5–13 $r'_e = \dfrac{25\text{ mV}}{I_E}$ dynamic emitter resistance

FORMULA 5–14 $Z_{in} = R_1 \| R_2 \| (\beta + 1)r'_e$ input impedance

FORMULA 5–15 $Z_{out} = R_C$ output impedance

FORMULA 5–16 $A_V \approx \dfrac{-R_C}{r'_e}$ voltage gain

FORMULA 5–17 $A_i = \dfrac{\beta(R_1 \| R_2)}{(R_1 \| R_2) + (\beta + 1)r'_e}$ current gain

Practical Notes

In actual practice, amplifier distortion occurs whenever the input signal is larger than the circuit design intends. Distortion (usually in the form of signal clipping) is also evident under a number of different types of circuit faults.

Features of CE Amplifiers

Here is a summary of the basic features of CE amplifiers:

1. The input signal is introduced into the base circuit, and the output is taken from the collector (the emitter is connected to ground through the emitter bypass capacitor and common to the input and output circuits).
2. The input circuit has medium impedance (because of the forward-biased base-emitter junction). Typically, the input impedance is about 1 kΩ.
3. The output circuit has medium impedance (approximately 5 kΩ).
4. The circuit provides voltage, current, and power gain. Power gains range as high as 10,000. Current gain is approximately equal to β_{ac}.
5. There is a 180° phase reversal between the input and output signals (the voltage gain has a negative value).

The Swamped Common-Emitter (CE) Amplifier

Figure 5–26 and Figure 5–27 are variations of the common-emitter amplifier. Notice that the input signal is applied between the BJT base terminal and ground, and that the output signal is taken between the collector terminal and ground. Observe the differences between these circuits and the common-emitter amplifier circuit. In Figure 5–26 the emitter resistor is separated into two resistors, R_{E_1} and R_{E_2}. The emitter bypass capacitor is located across R_{E_2}. In Figure 5–27 there is no emitter bypass capacitor. The voltage and current gain for the Figure 5–26 and Figure 5–27 amplifier circuits is reduced (swamped).

The equations to calculate the parameters for the swamped common-emitter amplifiers are slightly different. (NOTE: The equations do not include the effects of the load resistor.)

FORMULA 5–18 $Z_{in} = R_1 \| R_2 \| ((\beta + 1)(r'_e + R_{E_1}))$

FORMULA 5–19 $Z_{out} = R_C$

FORMULA 5–20 $A_V \approx \dfrac{-R_C}{(r'_e + R_{E_1})}$

FORMULA 5–21 $A_i = \dfrac{\beta(R_1 \| R_2)}{(R_1 \| R_2) + (\beta + 1)(r'_e + R_{E_1})}$

Features of Swamped CE Amplifiers

Here is a summary of the basic features of swamped CE amplifiers with negative feedback:

1. The input signal is introduced into the base circuit, and the output is taken from the collector (the emitter is connected to signal ground through the emitter resistor R_{E1} and the emitter bypass capacitor).
2. The input circuit has slightly high medium impedance (because of the forward-biased base-emitter junction and negative feedback). Typically, the input impedance is about 5 kΩ.
3. The output circuit has medium impedance (approximately 5 kΩ).
4. The circuit provides generally lower voltage, current, and power gain; with negative feedback through R_{E_1}.

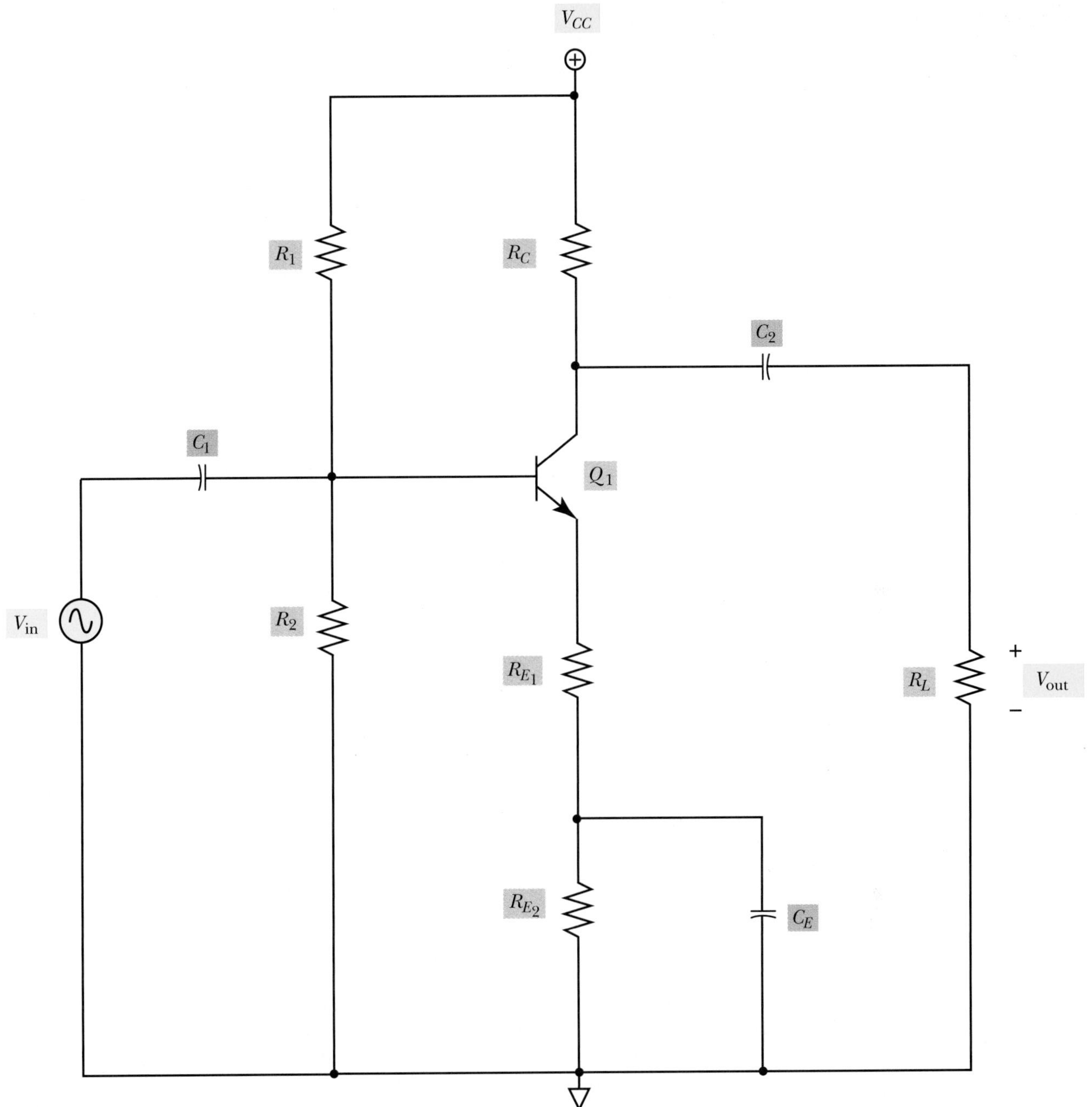

FIGURE 5–26 Swamped common-emitter amplifier

5. There is a 180° phase reversal between the input and output signals.
6. There is more stable operation under varying conditions of current flow, temperature changes, and transistor specifications.

Multiple-Stage CE Amplifier Circuits

Figure 5–28 shows two separate CE amplifier circuits connected one after the other. This is a *two-stage amplifier* circuit. The original input signal is applied at the base circuit of the first stage, transistor Q_1. An amplified version is taken from the collector circuit of Q_1 and coupled by capacitor C_2 to the input of the second amplifier stage, transistor Q_2. So the signal is amplified once again at Q_2 and the output taken from the collector of that stage. The signal is thus amplified twice.

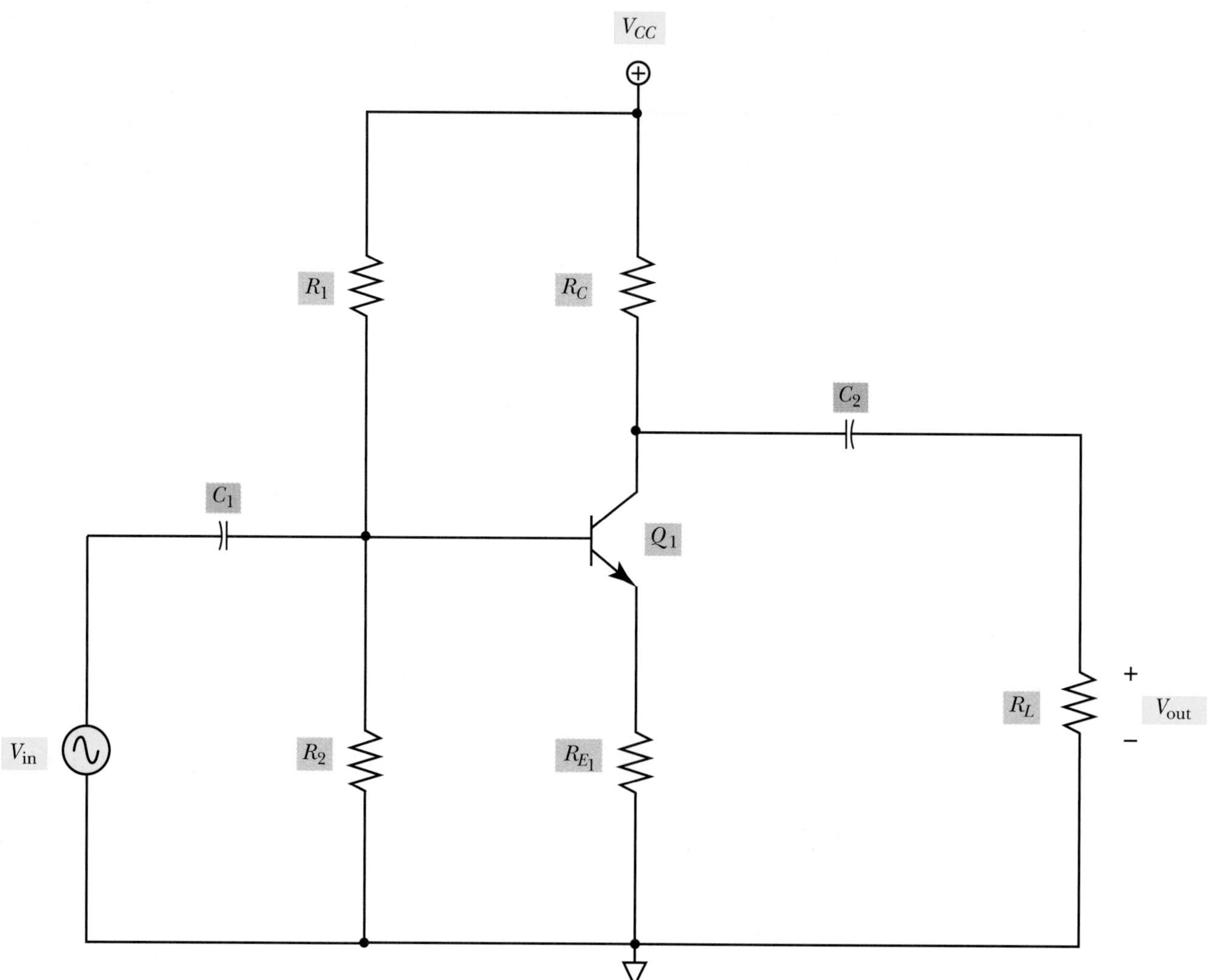

FIGURE 5–27 Swamped common-emitter amplifier

> *Practical Notes*
>
> A disadvantage of using capacitor coupling between the stages of transistor amplifiers used as audio amplifiers is that under these circumstances the capacitors act as high pass filters that block the low frequency response that is desirable in modern audio equipment. For this reason, many audio amplifiers are direct coupled (no C components between stages.)

The two-stage circuit in Figure 5–28 is also considered *capacitor coupled* because the signal is introduced into the circuit through a capacitor (C_1), the signal is passed (ac-coupled) from the first to the second stage through a capacitor (C_2), and the signal leaves from the output through yet another capacitor (C_3). Recall that capacitors pass only ac signals. So in a multistage transistor circuit, gradual fluctuations in dc bias levels for the transistors are not amplified. Additionally, the different dc bias levels do not affect the quality of the ac portion of the signal that is supposed to be amplified, as long as each stage is properly biased.

A two-stage amplifier circuit is shown in Figure 5–29 using a triangle amplifier symbol, identifying the input and output impedances and the voltage gain of each stage. In an amplifier that uses more than one stage, the overall gain multiplies from one stage to the next. So if the voltage gain of the input signal at the first stage is –10 and the voltage gain of the second stage is –6, then the overall gain is (–10) × (–6), or 60. In each stage of a CE amplifier the signal is inverted (indicated by the negative gain). This means that the output signal from a two-stage CE amplifier is in phase with the input signal. Due to the effects of the load impedance on the second amplifier output impedance, the second amplifier input impedance on the first amplifier output impedance, and the first amplifier input impedance on the source impedance of the sine wave, the overall gain will be less than 60. You can calculate the overall gain of the amplifier using Formula 5–22.

FORMULA 5–22 $A_V = A_{V_1} \times A_{V_2} \times A_{V_3} \times A_{V_4} \times A_{V_5}$

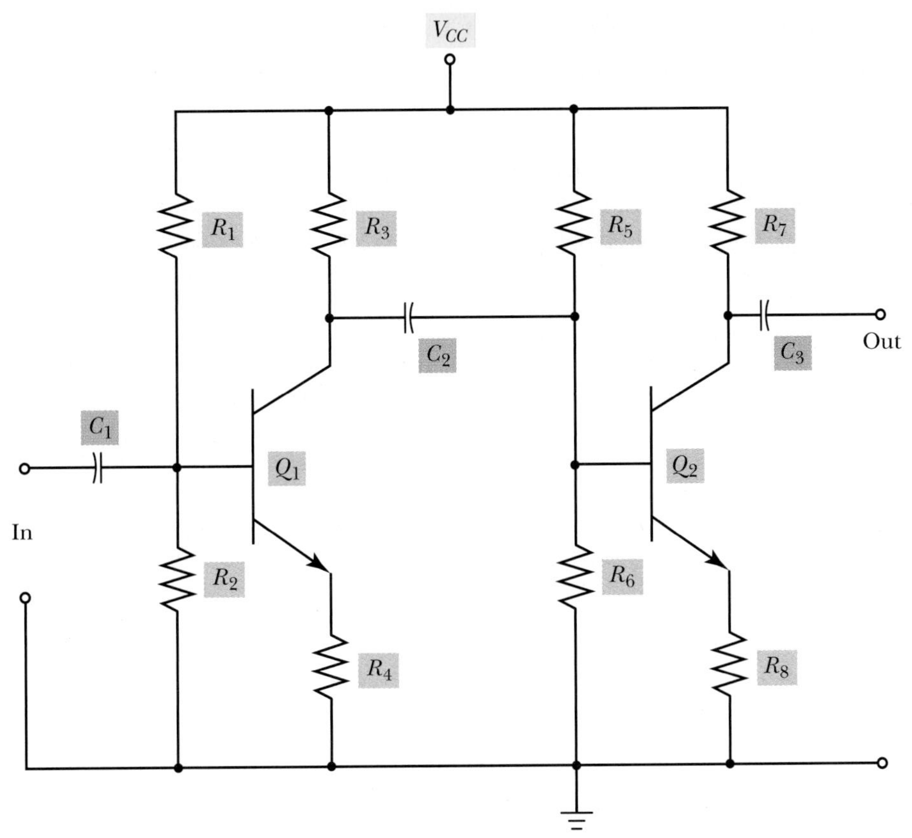

FIGURE 5–28 Two-stage, ac-coupled, common-emitter amplifier circuit

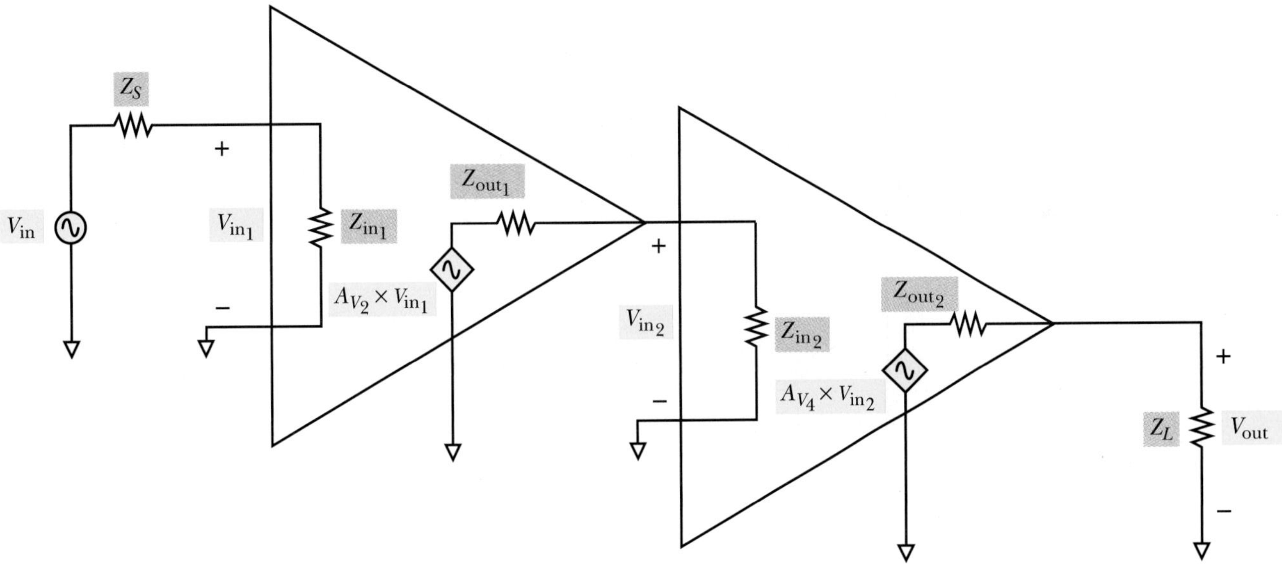

FIGURE 5–29

where:

$$A_{V_1} = \frac{Z_{in_1}}{Z_{in_1} + Z_S}$$

A_{V_2} is the gain of the first amplifier

$$A_{V_3} = \frac{Z_{in_2}}{Z_{in_2} + Z_{out_1}}$$

A_{V_4} is the gain of the second amplifier

$$A_{V_5} = \frac{Z_L}{Z_L + Z_{out_2}}$$

5–13 The Common-Collector (CC) Amplifier

Figure 5–30 is a schematic diagram of an amplifier configuration that has the input signal applied to the base terminal, and output taken from emitter terminal. If the input is applied to the base and the output taken from the emitter, it is apparent that the collector terminal is common to both the input and output. This circuit configuration is called the **common-collector amplifier (CC)** configuration. It is also known as an **emitter follower** and **unity gain analog buffer,** and works well for impedance matching.

It is important to notice that the BJT is still biased to meet the basic operating conditions outlined in earlier discussions. First, you can see that the dc source V_{CC} reverse biases the BJT's base-collector junction. Second, the voltage divider effect of V_{CC}, R_1, and R_2 forward biases the base-emitter junction. No matter what BJT amplifier configuration you are using these two important conditions will always be met.

Upon establishing the dc operating point of the Figure 5–30 amplifier and knowing that for ac signals the coupling capacitors C_1 and C_2 will appear as "short circuits," you can gain a clearer impression of how the circuit operates. As the input signal goes positive, the base current I_B will increase. As a result (1) the voltage drop between the collector and emitter V_{CE} decreases, (2) the collector current I_C and emitter current I_E increase, and (3) the voltage across R_E increases. Note that an increase in base current at the input of this circuit causes an increase in voltage at the output.

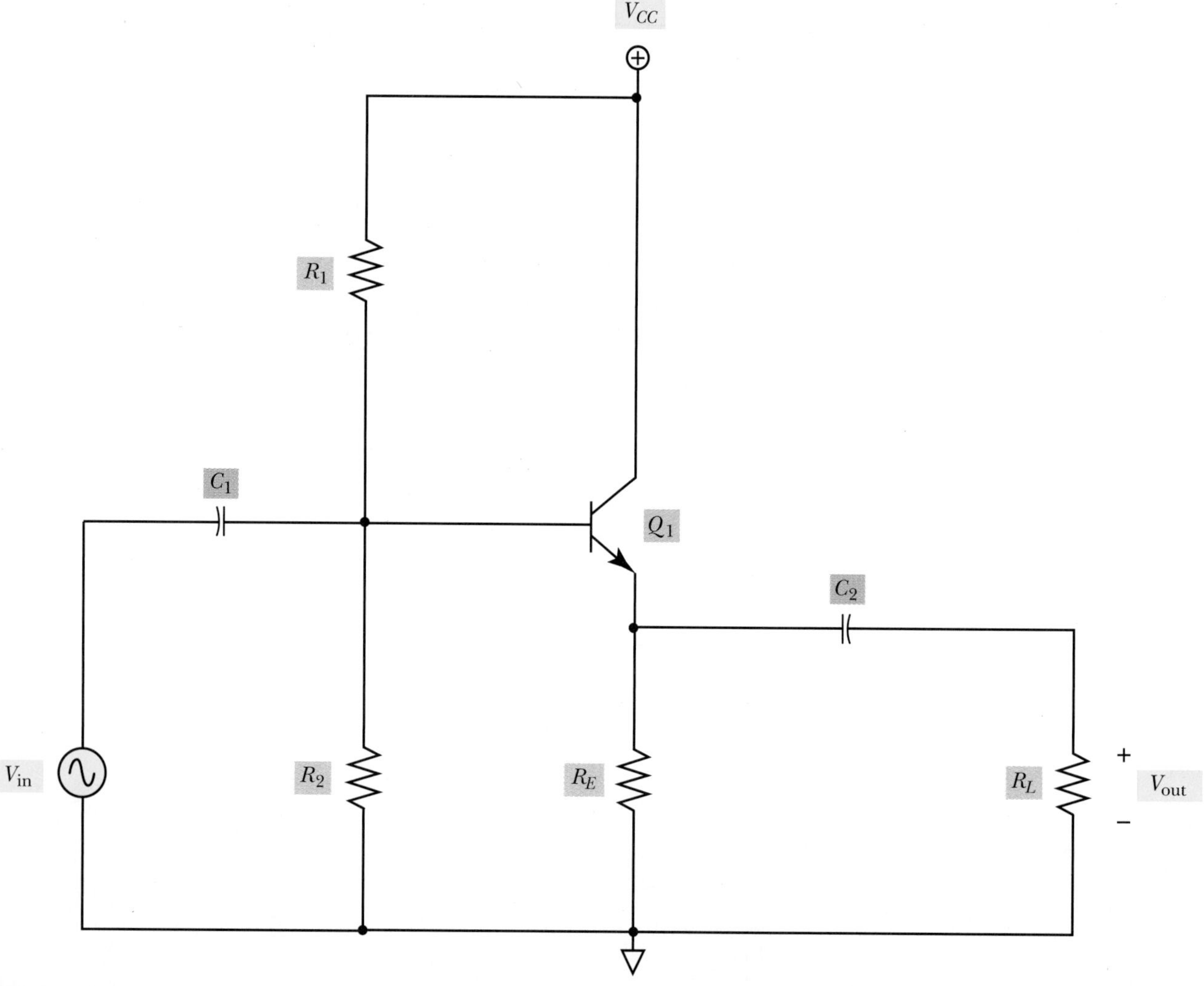

FIGURE 5–30 Common-collector amplifier multiSIM

When the input signal goes negative, the amount of base current I_B will decrease. As a result (1) the voltage drop between the collector and emitter V_{CE} increases. (2) the collector current I_C and emitter current I_E decrease, and (3) the voltage across R_E decreases. Note that a decrease in base current at the input of this circuit causes a decrease in voltage at the output.

Carefully notice that the output voltage "follows" the input voltage for a CC amplifier. As the input voltage goes more positive, so does the output; and as the input voltage goes less positive, so does the output. In other words, the input and output signals for a CC amplifier are in phase. For this reason a CC amplifier is often called a **voltage follower** amplifier.

If you study the circuit composed only of the input signal source, the base-biasing source, output resistor R_E, and the BJT's base-emitter junction, you will discover that the voltage dropped across the output resistor can never exceed the sum of the input signal and biasing sources. In fact, the voltage across R_E will always be less than the total input voltage (signal plus bias) because of the portion of the dc bias that will be dropped across the base-emitter junction. Since the ac output voltage is approximately equal to the ac input voltage, the amplifier voltage gain is one ($A_V = 1$). The amplifier is also known as a *unity gain analog buffer.*

The equations to calculate the parameters for the common-collector amplifier are (NOTE: Equations do not include the effects of the load resistor):

FORMULA 5–23 $Z_{\text{in}} = R_1 \| R_2 \| ((\beta + 1)(r_e' + R_E))$

FORMULA 5–24 $Z_{\text{out}} = R_E \| r_e'$

FORMULA 5–25
$$A_V = \frac{R_E}{(r_e' + R_E)}$$
$$A_V \approx 1$$

FORMULA 5–26
$$A_i = \frac{-(\beta+1)(R_1 \| R_2)}{((R_1 \| R_2) + (\beta+1)(r_e' + R_E))}$$
$$A_i \approx -\beta$$

Features of CC Amplifiers

Here is a summary of the basic features of practical CC amplifiers:

1. The input signal is introduced into the base circuit, and the output is taken from the emitter circuit (the collector is common to both the input and output).
2. The input circuit can have very high impedance.
3. The output impedance is relatively low.
4. The circuit provides voltage gain approximately equal to 1.
5. The circuit provides good current gain, but less power gain than the common emitter amplifier configurations.
6. There is no phase reversal between the input and output voltage waveforms.

Application of the CC Amplifier

You have seen that the input impedance of a CE voltage amplifier is relatively low. We often want to take advantage of the superior voltage-amplifying characteristic of a

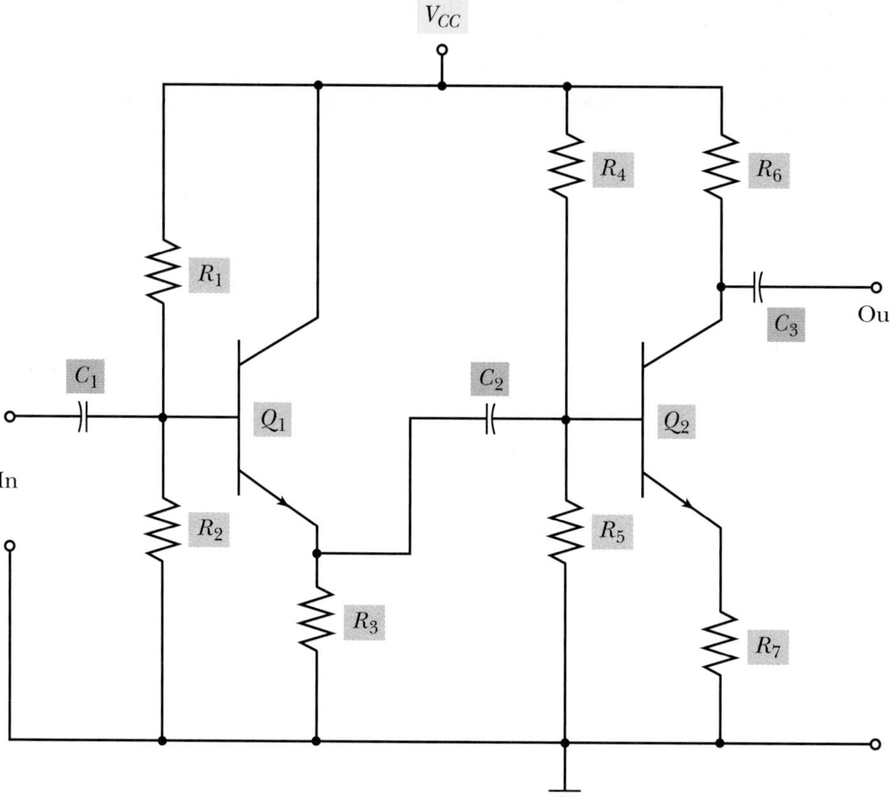

FIGURE 5–31 A CC amplifier followed by a CE amplifier

CE amplifier, but there is sometimes a requirement for a high impedance input. A two-stage amplifying process as shown in Figure 5–31 readily solves this problem. The first stage uses a CC amplifier, which is noted for high input impedance and a voltage gain of 1. By the time the signal passes through the CC stage, there is no longer a need for high input impedance, so the second stage can be a CE amplifier.

By using these two amplifiers in cascade, we get a result that cannot be achieved by either one alone. This principle is used frequently in modern electronic technology. If two CC amplifiers are connected together as shown in Figure 5–32, the result is a circuit that has extremely high input impedance. The input impedance is created by the multiplication of the β_{ac} for Q_1, the β_{ac} for Q_2, and the value of R_E. It is not unusual for such a circuit to have input impedance rated in megohms. This circuit is very commonly used in the input stages of sensitive measuring equipment. The circuit is known in the industry as a *Darlington amplifier* or Darlington pair.

Besides very high input impedances, Darlington amplifiers are noted for high current gain (the product of the individual betas), but have a voltage gain less than 1. The circuit is so commonly used today that it is made available in integrated circuit packages that include both BJTs and their interconnections.

5–14 The Common-Base (CB) Amplifier

Figure 5–33 and Figure 5–34 are **common-base (CB) amplifier** circuits. These circuits have the input signal applied to the emitter terminal and the output signal coming from the collector terminal. Both biasing requirements for BJT are met:

1. The base-emitter junction is forward biased.
2. The base-collector junction is reverse biased.

The voltage-divider model in Figure 5–34 is useful for stressing how changes in the base current of the BJT affect the output voltage level. The analysis is very nearly the

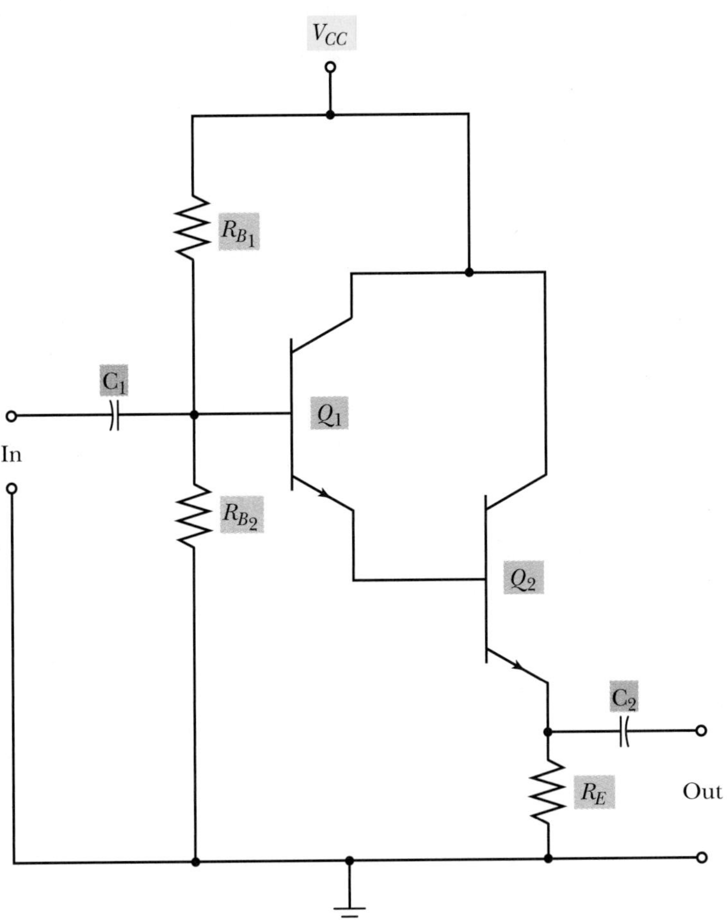

FIGURE 5–32 A Darlington common-collector amplifier circuit

same as for a CE amplifier. In fact, the bias circuit of the Figure 5–34 CB amplifier is the same as the CE amplifier shown in Figure 5–25. As the input signal goes more positive at the emitter, the emitter current I_E will decrease. You can then expect (1) the collector current I_C to decrease, (2) the voltage across R_C to decrease, and (3) the voltage drop between the collector and emitter V_{CE} to increase. Notice that a positive-going input signal causes a positive-going output signal. If you then consider the case where the input signal is negative, you will find that the output voltage follows along with the input signal. For a CB amplifier, the output signal is in phase with the input signal.

The equations to calculate the parameters for the common-base amplifier are (NOTE: Equations do not include the effects of the load resistor):

FORMULA 5–27 $Z_{in} = R_E \| r'_e$

FORMULA 5–28 $Z_{out} = R_C$

FORMULA 5–29 $A_V = \dfrac{R_C}{r'_e}$

FORMULA 5–30 $A_i \approx -1$

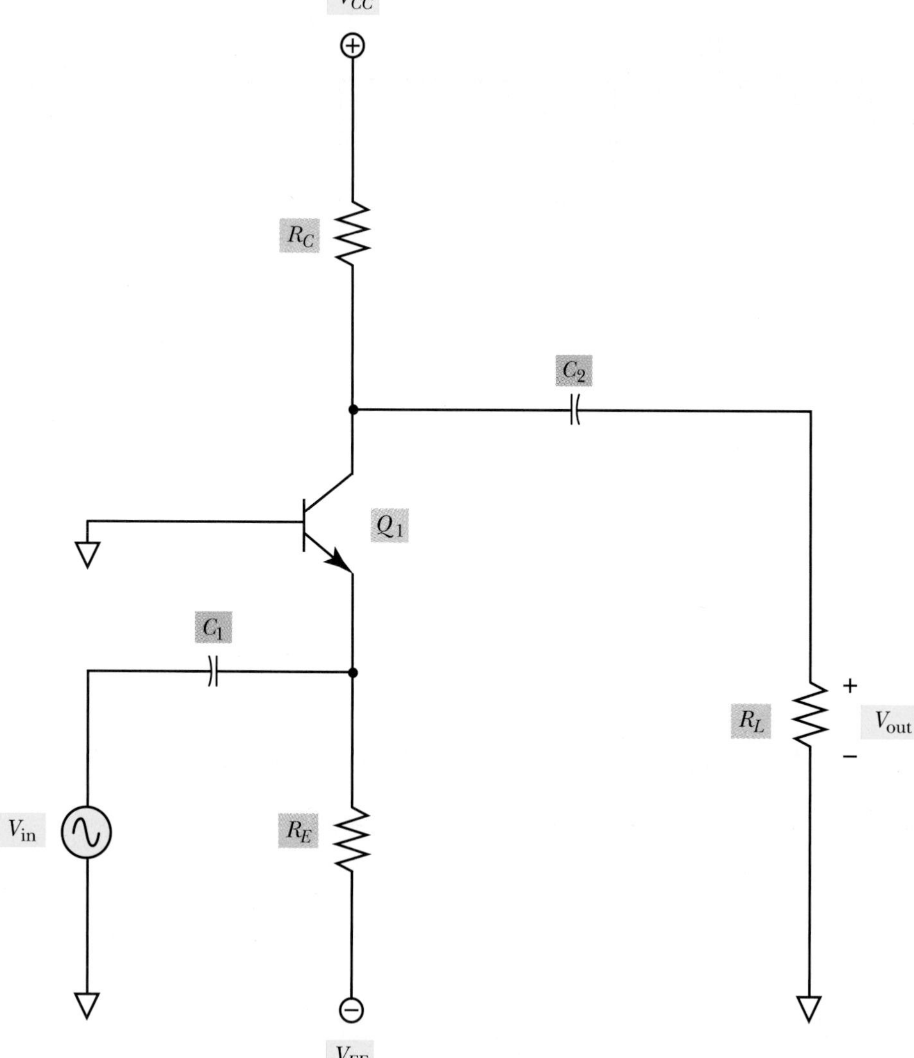

FIGURE 5–33 Common-base amplifier (dual supply)
multiSIM

Features of CB Amplifiers That Use BJTs

Here is a summary of the basic features of practical CB amplifiers:

1. The input signal is introduced into the emitter, and the output is taken from the collector circuit (the base is common to the input and output).
2. The input circuit has very low impedance (usually between 1 Ω and 50 Ω).
3. The output circuit has medium impedance (about 1 kΩ).
4. The circuit provides good voltage and power gain. The magnitude of the current gain is always less than 1 (because collector current is always less than the emitter current).
5. There is no phase reversal between the input and output voltage waveforms.

Applications of CB Amplifier Circuits

The very low input impedance of CB amplifier circuits seriously limits their use. In fact, this is the least common of the three basic amplifier configurations. Even so, CB amplifiers are consistently found in one application that requires low input impedances, namely the sensitive input stages of high-frequency communications receivers.

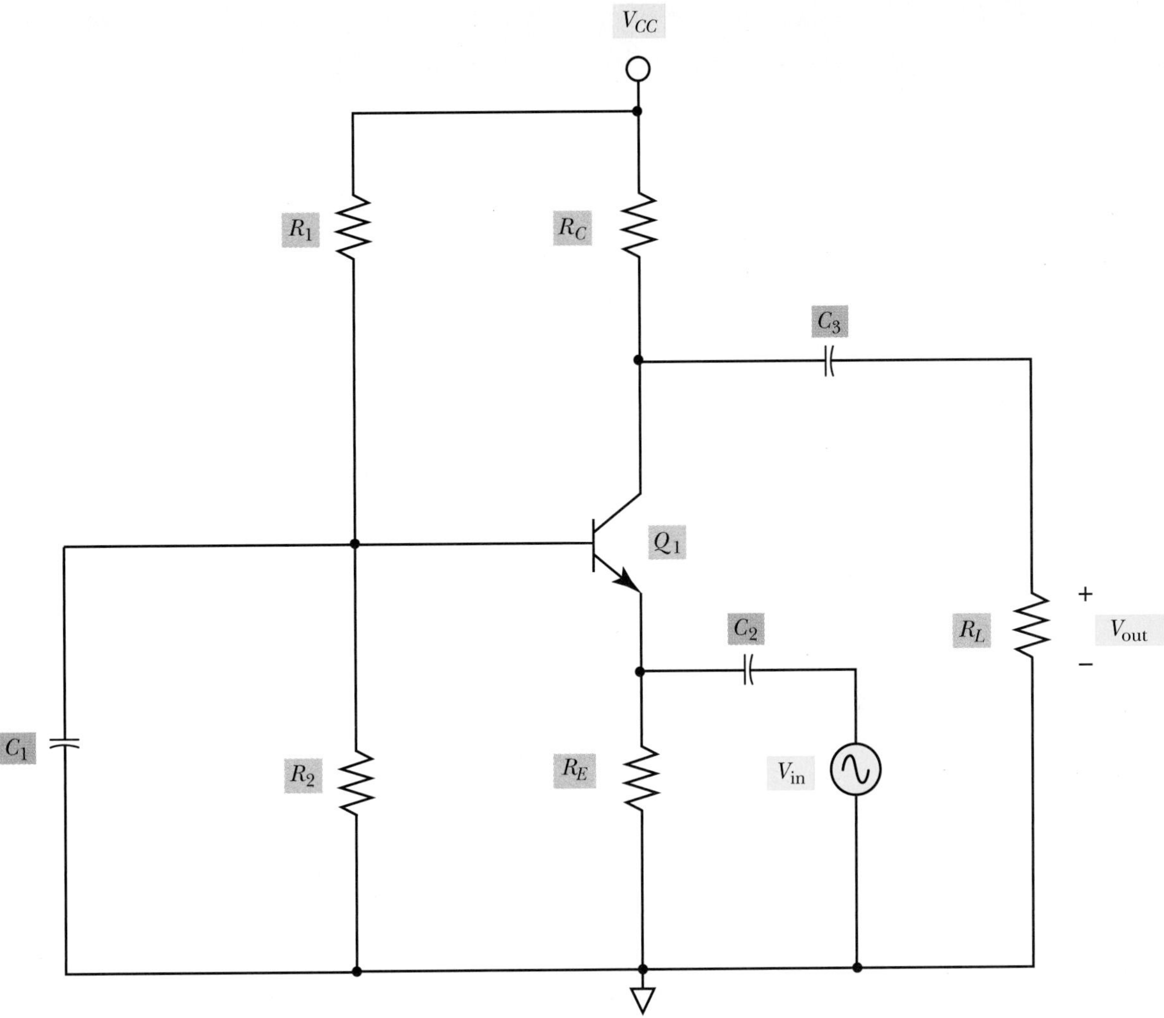

FIGURE 5–34 Common-base amplifier (single supply)

■ IN-PROCESS LEARNING CHECK 5

1. The BJT amplifier circuit having the base terminal common to both the input and output circuits is the _______________________ amplifier.
2. The BJT amplifier circuit providing the magnitude of both voltage and current gain greater than 1 is the _______________________ amplifier.
3. The BJT amplifier circuit having 180° phase difference between input and output voltage waveforms is the _______________________ amplifier.
4. The BJT amplifier circuit having the magnitude of the voltage gain less than 1 is the _______________________ amplifier.
5. The BJT amplifier circuit having an input to the base and output from the emitter is called a common-________ amplifier.
6. In an amplifier that uses more than one stage, the overall gain is found by ________ the gains of the individual stages.
7. The voltage-follower amplifier is another name for a common-_______ amplifier.
8. The Darlington amplifier is made up of two common-_______ amplifiers.

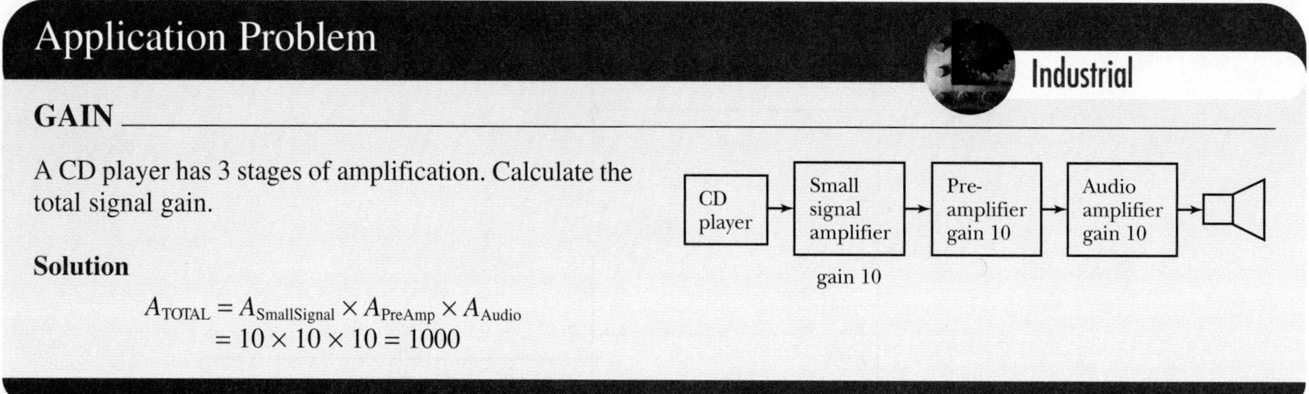

Application Problem

Industrial

GAIN

A CD player has 3 stages of amplification. Calculate the total signal gain.

Solution

$$A_{\text{TOTAL}} = A_{\text{SmallSignal}} \times A_{\text{PreAmp}} \times A_{\text{Audio}}$$
$$= 10 \times 10 \times 10 = 1000$$

5–15 Classification by Signal Levels

As its name implies, a **small-signal amplifier** is designed to operate effectively with small ac input signals and at power levels less than 1 W. What is a small signal? For our purposes, a small signal is one whose peak-to-peak ac current value is less than 0.1 times the amplifier's input bias current.

Examples of small-signal amplifiers are the first stages in radio and TV receivers. The small signals coming from the antennas are usually in the range of microvolts or millivolts. See the block diagram in Figure 5–35.

Large-signal amplifiers properly handle signals that have significantly larger input current levels and output power levels. The final *power amplifier* stages that drive the loudspeakers in an audio system are examples of large-signal amplifiers. See the example in the block diagram in Figure 5–36.

5–16 Classification by Class of Operation

The family of collector characteristic curves, which you studied in an earlier section, plays an important part in your understanding of classes of amplifier operation. By using these curves, we can explain the operating points and determine operating values for any class of amplifier operation.

One of the important elements in this procedure is a **load line.** The load line describes the voltage drop across the collector-emitter terminals and the current flowing from the emitter to the collector terminals as the characteristics of the element

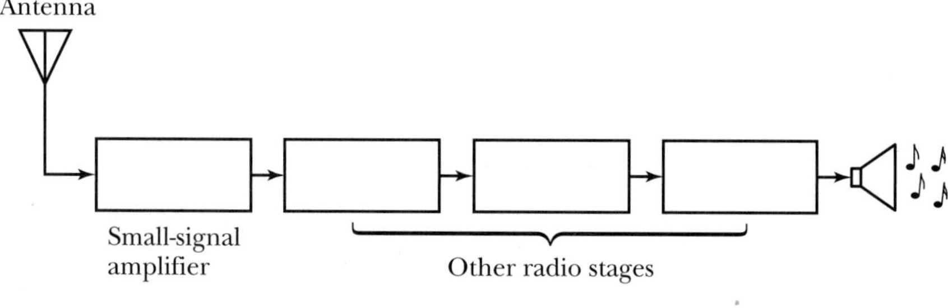

FIGURE 5–35 Small-signal amplifier fed by an antenna signal

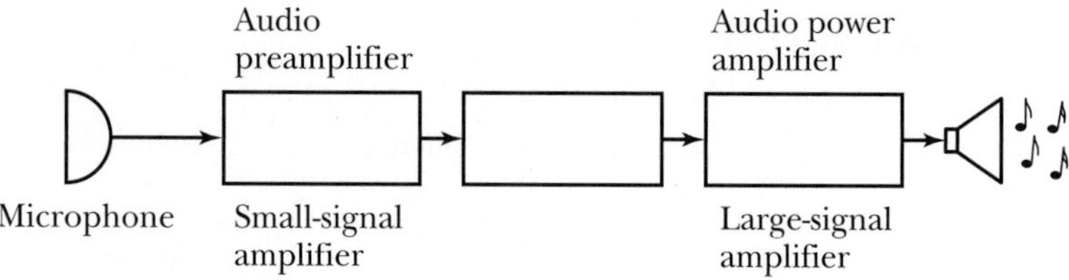

FIGURE 5–36 Large-signal amplifier feeding a loudspeaker

between these terminals is varied from one extreme (short = 0 Ω) to the other extreme (open = ∞ Ω). You can observe this concept in Figure 5–37 where the transistor is replaced with a variable resistor. When the load line is drawn onto a set of collector characteristic curves, the load line will describe in detail how the BJT will respond in a particular circuit. To construct a load line on a set of curves, you need to know two extreme values: (1) the collector-emitter voltage when the BJT is biased to its cutoff point (resistor open), and (2) the amount of collector current when the BJT is at its saturation point (resistor shorted). These values can always be determined from the amount of supply voltage and the values of resistors that are in series with the collector-emitter circuit.

When the base current for a BJT is zero, the collector current is newly zero with a very small amount of collector leakage current. This condition occurs when the base-emitter junction is less than 0.7 V (reverse biased), and when the base-collector junction is reverse biased (as it usually is). Under these conditions, the collector-emitter voltage drop (V_{CE}) is maximum because the BJT acts as an open circuit (very high impedance). At that point, V_{CE} virtually equals the source voltage (V_{CC}). This is the cutoff point on the collector characteristic curves. See Figure 5–38.

FORMULA 5–31 $V_{CE} = V_{CC}$

The opposite extreme operational condition occurs when I_C is maximum or the transistor operates at a point called *saturation*. This state exists when the forward

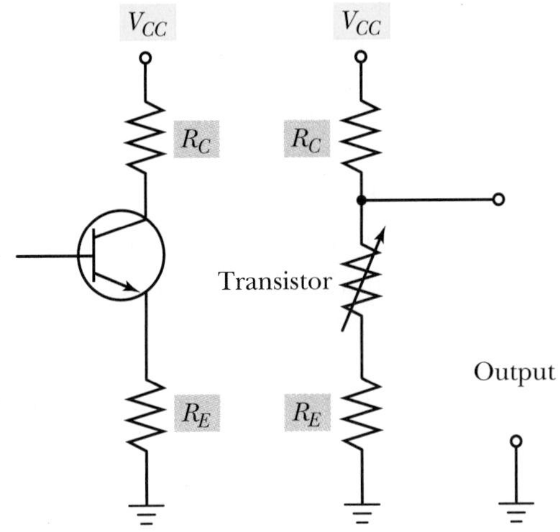

FIGURE 5–37 Analogy of a transistor as a variable resistor

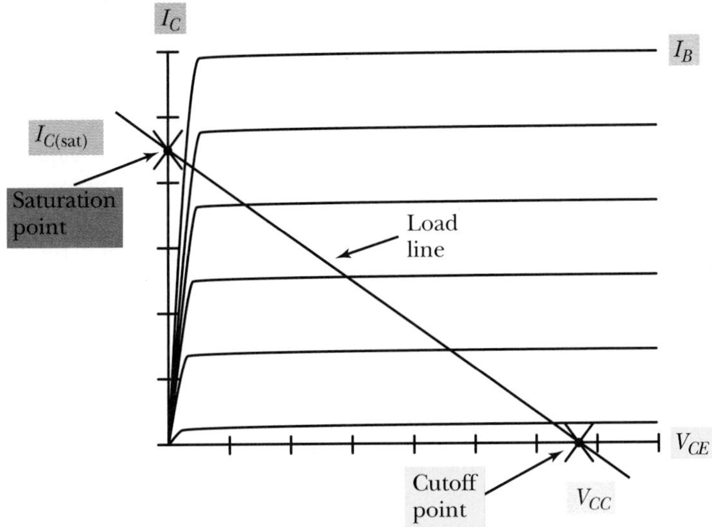

FIGURE 5–38 Plotting the load line on the collector characteristic curves

base current increases to a point where a further increase in base current causes no additional increase in collector current. At this point, the collector-emitter voltage is minimum and is taken as virtually zero. See this saturation point plotted on the family of curves in Figure 5–38. The symbol for the amount of collector saturation current is $I_{C(sat)}$.

Referring to the circuit in Figure 5–25, you can calculate $I_{C(sat)}$ with the following formula:

FORMULA 5–32 $I_{C(sat)} = \dfrac{V_{CC}}{(R_C + R_E)}$

◆ **EXAMPLE** Referring to the circuit in Figure 5–25, let $V_{CC} = 12$ V, $R_C = 10$ kΩ, and $R_E = 2.2$ kΩ. Determine the cutoff point and saturation current.

Answer:

The cutoff voltage is the same as V_{CC}, 12 V.

From Formula 5–32

$$I_{C(sat)} = \frac{V_{CC}}{(R_C + R_E)}$$

$$I_{C(sat)} = \frac{12\text{ V}}{(10\text{ k}\Omega + 2.2\text{ k}\Omega)} = 984\ \mu\text{A}$$ ◆

By drawing a straight line from ($V_{CE} = 0$ V, $I_C = I_{C(sat)}$) to ($V_{CE} = V_{CC}$, $I_C = 0$ A), you have constructed the dc load line for the circuit, Figure 5–38. From the dc load line, you can determine the voltage across the transistor and the amount of collector current for a given value of base current. As you will soon see, by setting the base-bias value, you set the operating point for the transistor.

To summarize this load line procedure (refer to Figure 5–38):

1. Mark the cutoff point on the collector where $V_{CE} = V_{CC}$.
2. Determine the saturation, or maximum I_C, by calculating the collector current required to cause the total V_{CC} voltage value to be dropped across R_E and R_C ($V_{CE} = 0$ V).
3. Draw the dc load line by connecting the saturation and cutoff points on the characteristics curves with a straight line.

PRACTICE PROBLEMS 2

Referring to the circuit in Figure 5–25, Let $V_{CC} = 5$ V, $R_C = 470$ Ω, and $R_E = 47$ Ω. Determine the cutoff point and saturation current.

The Q point, or *quiescent operating point,* of a transistor is determined by the value of dc base bias. The intersection of the base-bias (I_B) line and the load line is called the Q point. By drawing a line down to the V_{CE} axis, we can see the V_{CE} value for that circuit operating at that particular bias point. By projecting a line horizontally from the Q point to the I_C axis, we can determine the collector current (I_C) value for that circuit operating with that base bias and the circuit's component values. See, for example, the load line and Q point in Figure 5–39.

Class A Amplifier Operation

When the Q point is set at the mid-point of the load line, Figure 5–39, the amplifier stage is operating at or near the definition for a **Class A amplifier.** By definition, a Class A amplifier is one where the collector current flows throughout the entire input cycle. Under these conditions, the maximum applied base current does not drive the BJT into saturation, and the minimum base current does not drive the BJT into cutoff.

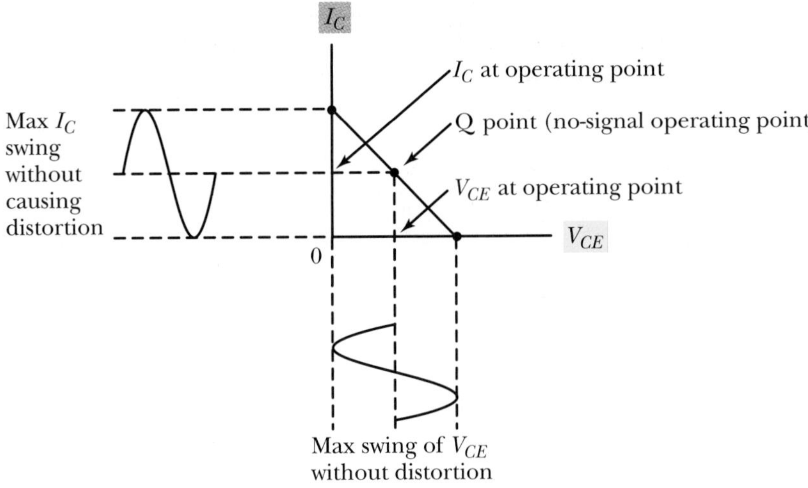

FIGURE 5–39 Q point set at mid-point on the load line for Class A amplifier operation

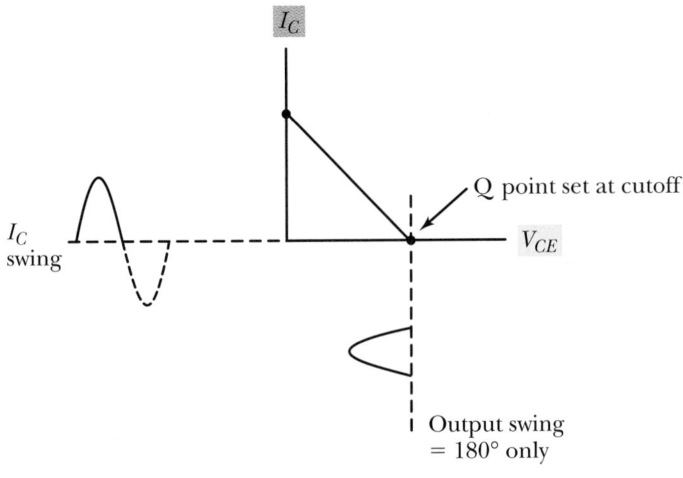

FIGURE 5–40 Q point set at cutoff for Class B amplifier operation

So the transistor is always conducting above and below the Q point, but never hitting either extreme, staying within the linear operating region.

Because the output signal is not distorted by driving the amplifier into cutoff or saturation. Class A amplifiers are used where it is desirable to keep signal distortion as low as possible. This is especially important in the small-signal amplifiers for radio, TV, and audio equipment. The signal swing uses little of the available load line to maintain linearity.

Class A amplifiers provide minimum signal distortion, buy they are also known for their low operating efficiency. (The amplifier efficiency is calculated by dividing the output power by the sum of the input power and supply power.) The low efficiency occurs due to a significant amount of dc bias current flowing even when there is no signal applied to the amplifier.

Class B Amplifier Operation

When the BJT amplifier is biased so that its Q point is at cutoff, Figure 5–40, the stage is biased as a **Class B amplifier.** Collector current flows only as long as the input signal is positive. This is usually just 180° of a sinusoidal input waveform. Because of being biased at cutoff, half of the input cycle biases the transistor stage below cutoff. The output waveform resembles the half-wave rectified signal you studied in an earlier chapter. Class B amplifiers have higher efficiency than Class A stages because the Class B transistors are in cutoff through half the input waveform. When no signal is applied, the transistor is turned off (recall that a Class A amplifier is conducting even

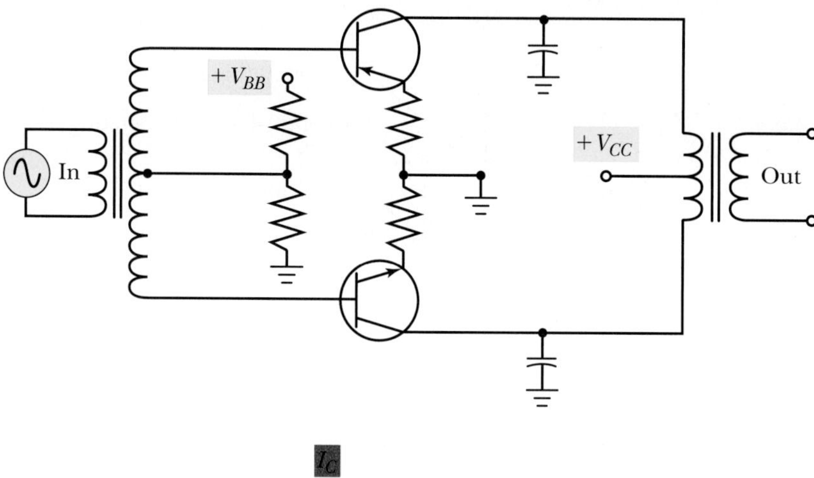

FIGURE 5–41 Two Class B amplifiers used in a push-pull amplifier circuit

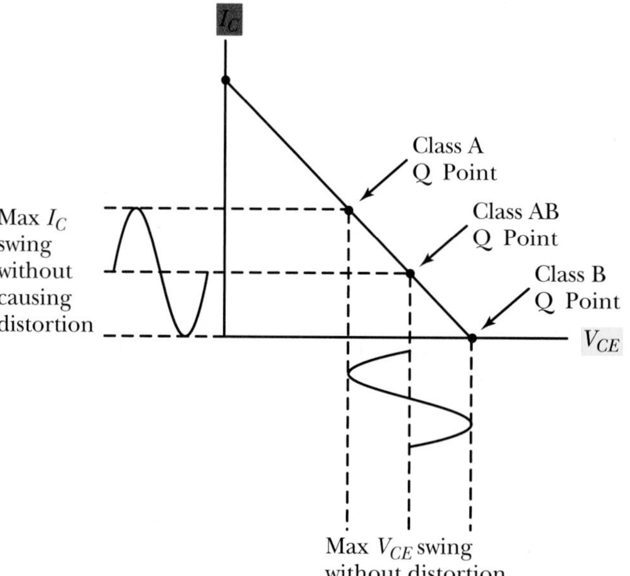

FIGURE 5–42 Class AB amplifier operation

when no signal is applied). However, Class B amplifiers distort the signal by clipping off any part of the signal that drives the transistor into cutoff (Class A amplifiers do not distort the signal). This class of amplifier is used in radio frequency circuits and can be used in audio amplifiers by using a special two-transistor circuit known as a **push-pull amplifier,** Figure 5–41. The transistors conduct on opposite half-cycles, thus providing a full 360° signal at the output.

Class AB Amplifier Operation

The Class AB amplifier is a compromise between Class A (low distortion, low efficiency) and Class B (high distortion, high efficiency). As long as the input signal isn't too large, a Class AB amplifier shows no signal distortion and it is more efficient than a Class A amplifier, but not quite as efficient as the Class B. Class AB amplifiers commonly avoid distortion in audio amplifier circuits.

Figure 5–42 shows that the Q point for a Class AB amplifier is about halfway between the Q points for the Class A and Class B amplifiers. When Class AB amplifiers are used in the push-pull amplifier configuration, the crossover distortion observed in the Class B amplifier is eliminated.

Class C Amplifier Operation

Figure 5–43 is a schematic of a Class C amplifier. When a transistor amplifier stage is biased below cutoff value, Figure 5–44, collector current flows for less than 180° of the input cycle. Typical conduction time for **Class C amplifier** operation is about 120°.

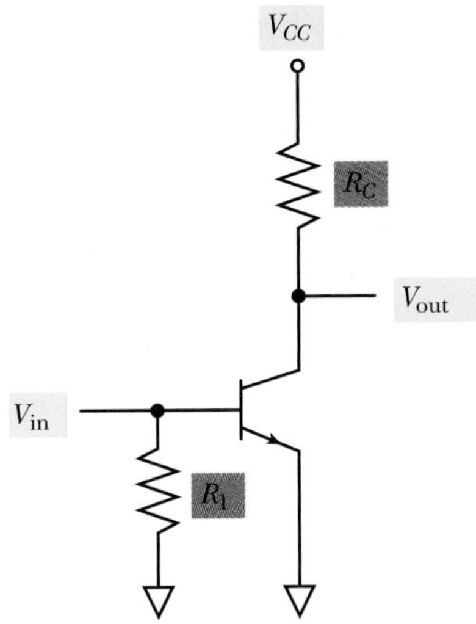

FIGURE 5–43 Class C amplifier schematic

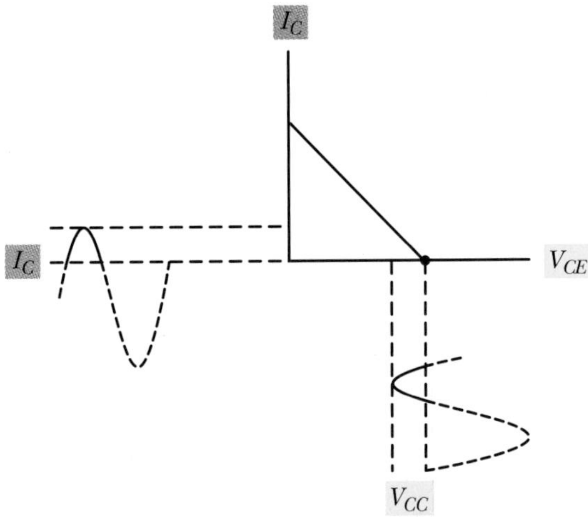

FIGURE 5–44 Q point set below cutoff for class C amplifier operation

Because Class C operation can be used in radio-frequency (RF) circuits where the other circuit components (such as a tuned LC circuit) complete the missing part of the output waveform, you frequently find this operation used in RF power amplifiers. Class C operation has high efficiency and large-signal handling capability.

■ IN-PROCESS LEARNING CHECK 6

1. Class A amplifiers conduct during _______° of the input cycle; Class B amplifiers conduct for _______° of the input cycle; and Class C amplifiers conduct for approximately _______° of the input cycle.
2. The dc load line shows all operating points of a given amplifier from _______ to _______.
3. What is the amount of saturation current for a Class A CE amplifier when the collector resistance is 1 kΩ, the emitter resistance is zero (this resistor is not used), and the supply voltage is 12 V? What is the value of the cutoff voltage for this circuit?
4. The Q point for a Class A amplifier is located at the _______ of the dc load line.
5. The Class _______ amplifier has the Q point located below the cutoff.
6. Collector current flows in a Class _______ amplifier, even when there is no signal applied to the input.

5–17 Analysis of BJT Amplifiers

Analysis of BJT amplifier circuits can be crucial to understanding and troubleshooting electronic equipment. This section introduces a method for analyzing BJT amplifiers in a simple way using suitable approximations. The examples are all for an NPN swamped common-emitter amplifier, but you can use them easily in other kinds of circuits that you have studied.

Voltage and Current Calculations

Now, let's analyze the swamped CE circuit in Figure 5–45. Some assumptions will be made for an easier analysis. We are able to use these assumptions because they introduce only a small percentage of error.

First the dc bias circuit must be evaluated. The voltage divider emitter stabilized biasing (Figure 5–25) is utilized in this amplifier.

Using Formula 5–10, you can calculate the Thevenin voltage for the bias circuit. The base voltage will always be less than or equal to the Thevenin voltage.

Assuming that the transistor is operating in the linear region, Formula 5–11 can be used to calculate the emitter current.

Assuming that the collector current is equal to the emitter current (if the transistor is operating in the linear region, this assumption is practical), Formula 5–12 will be used to calculate the collector-emitter voltage drop. The collector-emitter voltage should be greater than 0.3 V and less than the supply voltage. If the collector-emitter voltage is not within this voltage range, then either a mis-

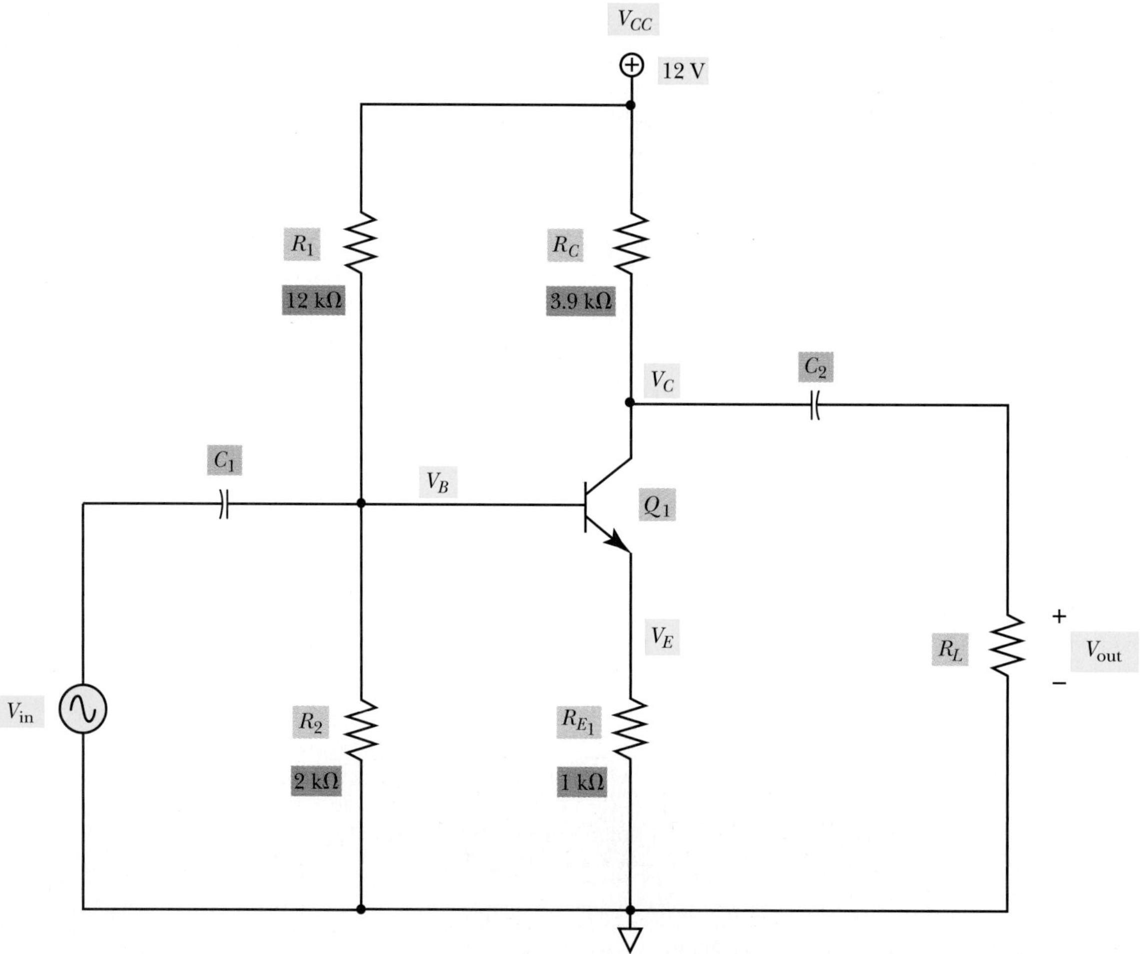

FIGURE 5–45 Swamped common-emitter amplifier

take was made in our calculations, or one or more of our assumptions are wrong. (How to analyze the circuit for different assumptions is beyond the scope of this text.)

You have now calculated the Q point for the swamped CE amplifier. Traditionally, as a rule of thumb, the Q point should be mid-point on the load line. The "mid-point" Q point allows the amplifier to have the largest nondistorted ac output signal.

After observing that the Q point is located in the linear region of the transistor curves, you can proceed with the ac circuit analysis. Using the equations from Formulas 5–18 through 5–21 and Formulas 5–7 through 5–9, the input impedance, output impedance, voltage gain, current gain, power gain, and output voltage can all be calculated.

◆ **EXAMPLE** Using the above information, let's analyze the circuit parameters in Figure 5–45 with $\beta = 100$, $V_{in} = 0.5\ V_{PP}$, and $R_L = 100\ k\Omega$ so you can see how easy such analysis is.

DC bias circuit: voltage divider emitter stabilized

1. Thevenin voltage (base voltage) using Formula 5–10:

$$V_B \approx V_{TH} = \frac{V_{CC} \times R_2}{(R_1 + R_2)}$$
$$V_B \approx \frac{12\ V \times 2\ k\Omega}{(12\ k\Omega + 2\ k\Omega)}$$
$$V_B \approx 1.714\ V$$

2. Emitter current using Formula 5–11:

$$I_E = \frac{(V_B - V_{BE})}{R_E}$$
$$I_E = \frac{(1.714\ V - 0.7\ V)}{1\ k\Omega}$$
$$I_E = 1.014\ mA$$

3. Collector-emitter voltage using Formula 5–12 (note: $I_C = I_E$):

$$V_{CE} = V_{CC} - I_C \times (R_C + R_E)$$
$$V_{CE} = 12\ V\ 1.014\ mA \times (3.9\ k\Omega + 1\ k\Omega)$$
$$V_{CE} = 7.03\ V$$

AC circuit

4. Dynamic emitter resistance using Formula 5–13:

$$r_e' = \frac{25\ mV}{I_E}$$
$$r_e' = \frac{25\ mV}{1.014\ mA}$$
$$r_e' = 24.65\ \Omega$$

5. Input impedance using Formula 5–18:

$$Z_{in} = R_1 \| R_2 \| ((\beta + 1)(r_e' + R_{E1}))$$
$$Z_{in} = \frac{1}{\left(\frac{1}{(12\ k\Omega)} + \frac{1}{(2\ k\Omega)} + \frac{1}{((100 + 1)(24.65\ \Omega + 1\ k\Omega))}\right)}$$
$$Z_{in} = 1.686\ k\Omega$$

6. Output impedance using Formula 5–19:

$$Z_{out} = R_C$$
$$Z_{out} = 3900\ \Omega$$

7. Voltage gain using Formula 5–20:

$$A_V \approx \frac{-R_C}{(r_e' + R_{E1})}$$

$$A_V \approx \frac{-3.9\ \text{k}\Omega}{(24.65\ \Omega + 1\text{k}\Omega)}$$

$$A_V \approx -3.806$$

8. Current gain using Formula 5–21:

$$A_i \approx \frac{\beta(R_1 \| R_2)}{((R_1 \| R_2) + (\beta+1)(r' + R_{E1}))}$$

$$A_i \approx 100 \times \frac{\left(\dfrac{12\ \text{k}\Omega \times 2\ \text{k}\Omega}{(12\ \text{k}\Omega + 2\ \text{k}\Omega)}\right)}{\left(\left(\dfrac{(12\ \text{k}\Omega \times 2\ \text{k}\Omega)}{(12\ \text{k}\Omega + 2\ \text{k}\Omega)}\right) + (100+1)(24.65\ \Omega + 1\ \text{k}\Omega)\right)}$$

$$A_i \approx 1.63$$

9. Power gain using Formula 5–9:

$$A_P = |A_V \times A_i|$$

$$A_P = |-3.806 \times 1.63|$$

$$A_P \approx 6.204$$

10. Output voltage using Formula 5–7:

$$V_{\text{out}} = A_V \times V_{\text{in}}$$

$$V_{\text{out}} = -3.806 \times 0.5\ \text{V}_{\text{PP}}$$

$$V_{\text{out}} = -1.903\ \text{V}_{\text{PP}}$$ ◆

The negative sign for the output voltage indicates that the output signal has a 180° phase shift with respect to the input signal. Figure 5–46 shows the waveforms throughout the swamped common emitter amplifier.

PRACTICE PROBLEMS 3

Assume that the circuit in Figure 5–45 has $R_1 = 10\ \text{k}\Omega$, $R_2 = 2\ \text{k}\Omega$, $R_E = 1\ \text{k}\Omega$, $R_C = 4.7\ \text{k}\Omega$, and $V_{CC} = 10$ V with $\beta = 100$, $V_{\text{in}} = 0.2\ \text{V}_{\text{PP}}$, and $R_L = 100\ \text{k}\Omega$. Calculate the base voltage, emitter current, collector-emitter voltage, dynamic emitter resistance, input impedance, output impedance, voltage gain, current gain, power gain, and output voltage. Sketch the waveforms throughout the amplifier indicating the dc and ac voltage values.

Practical Notes

The critical parameters for the swamped common emitter amplifier transistor stage are calculated with three previously stated formulas.

1. $V_B \approx V_{TH} = \dfrac{V_{CC} \times R_2}{(R_1 + R_2)}$

2. $I_E = \dfrac{(V_{TH} - V_{BE})}{R_E}$

3. $V_{CE} = V_{CC} - I_E \times (R_C + R_E)$

Using your approximations and knowing these parameters allows you to determine easily all the other parameters of interest.

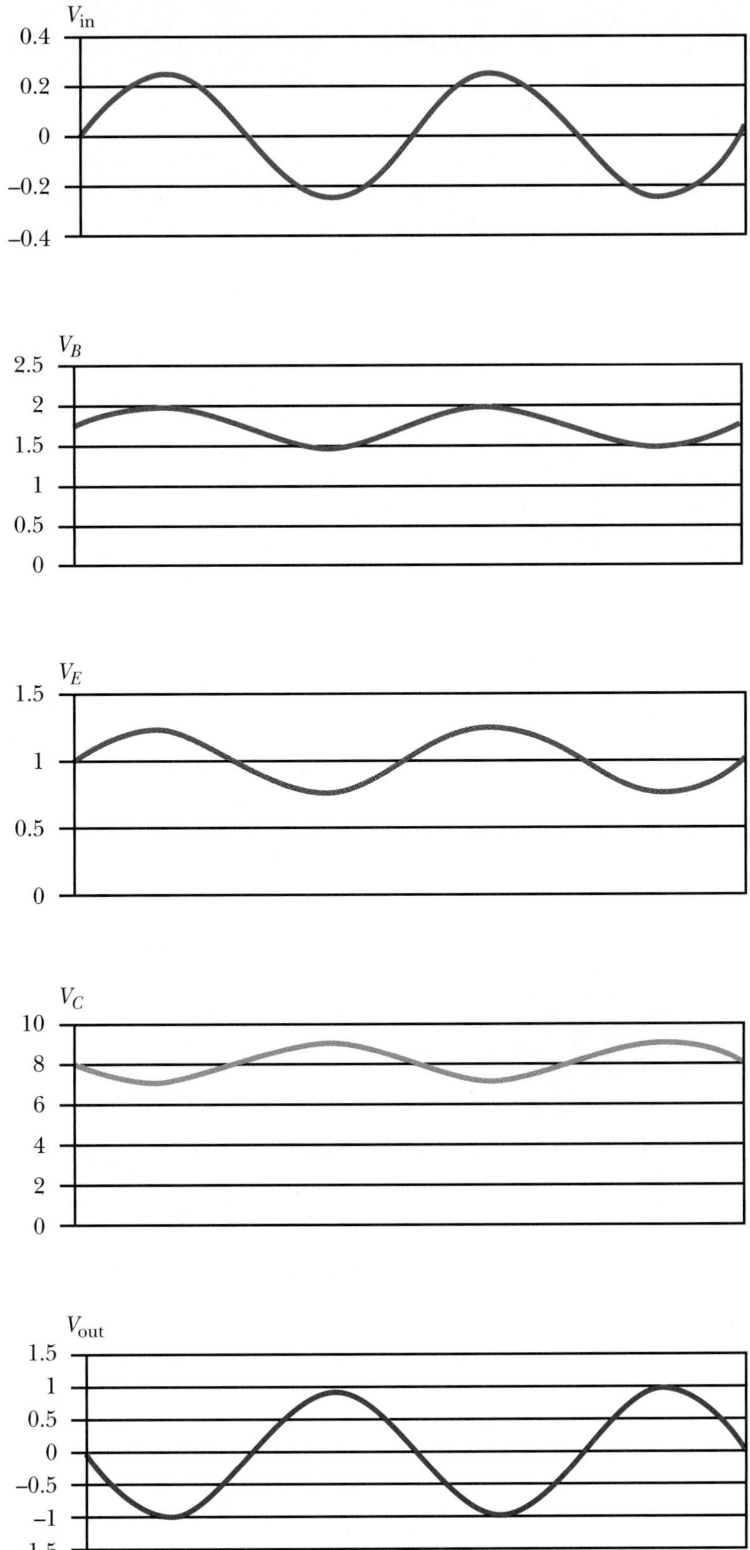

FIGURE 5–46 Amplifier waveforms

USING TECHNOLOGY: BJTs USED IN A SYSTEM

Every electronic system requires a source of dc power to function. The power is usually provided from an ac-dc supply. This power must be maintained at a constant voltage and current under varying load conditions. Bipolar junction transistors, which can be fabricated as discrete devices, are capable of handling such heavy demands, either as a current amplifier or acting as a variable load. In addition, by carefully choosing the semiconductor materials used and the amount of doping added, the BJT can be used in high voltage power supply applications where both high voltages and heavy current demands need to be satisfied.

Reviewing the ideal characteristics of the BJT, it can be seen that this device can be packaged as a discrete device with an attached heat sink. Such devices possess high voltage, current and power gain abilities, stable operation under varying conditions of temperature and operating voltages, and/or currents and uniform response over a wide range of frequencies. Such devices are shown in Figure 8.

One of the most common, almost universally used applications of the BJT is in the horizontal drive portion of every television. To display information on the face of a television picture tube, streams of highly focused electrons, in essence a beam of electrons, must paint the picture elements in terms of intensity, hue, and tint on the screen. This beam must move from left to right across the screen in direct synchronization with the horizontal scan of the television camera back at the broadcasting station studio. This signal is called the horizontal drive signal, since it drives the internal circuitry of the television receiver.

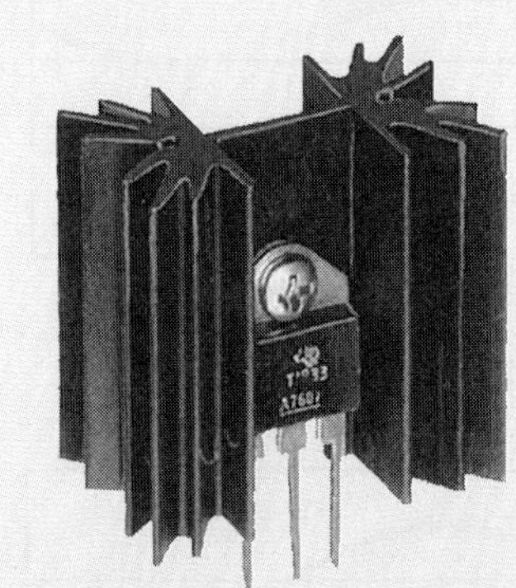

FIGURE 8 High-power BJTs mounted on heat sinks *(Photo courtesy of Thermalloy Inc.)*

The system application block diagram for the horizontal drive circuit from input to output is shown in Figure 9. Discrete BJTs are utilized in each stage starting with the horizontal buffer which reduces coupling between IC U1001, the T chip, and the horizontal drive amplifier. The horizontal drive amplifier increases the power of the horizontal drive

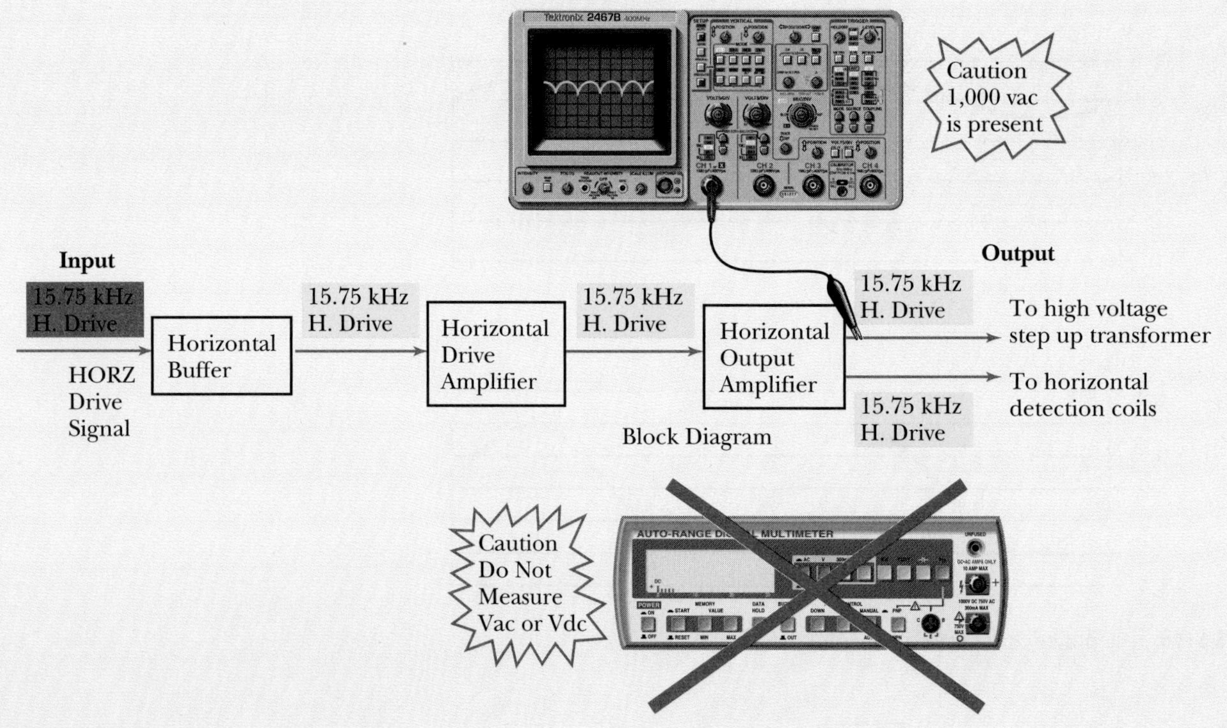

FIGURE 9 Block Diagram

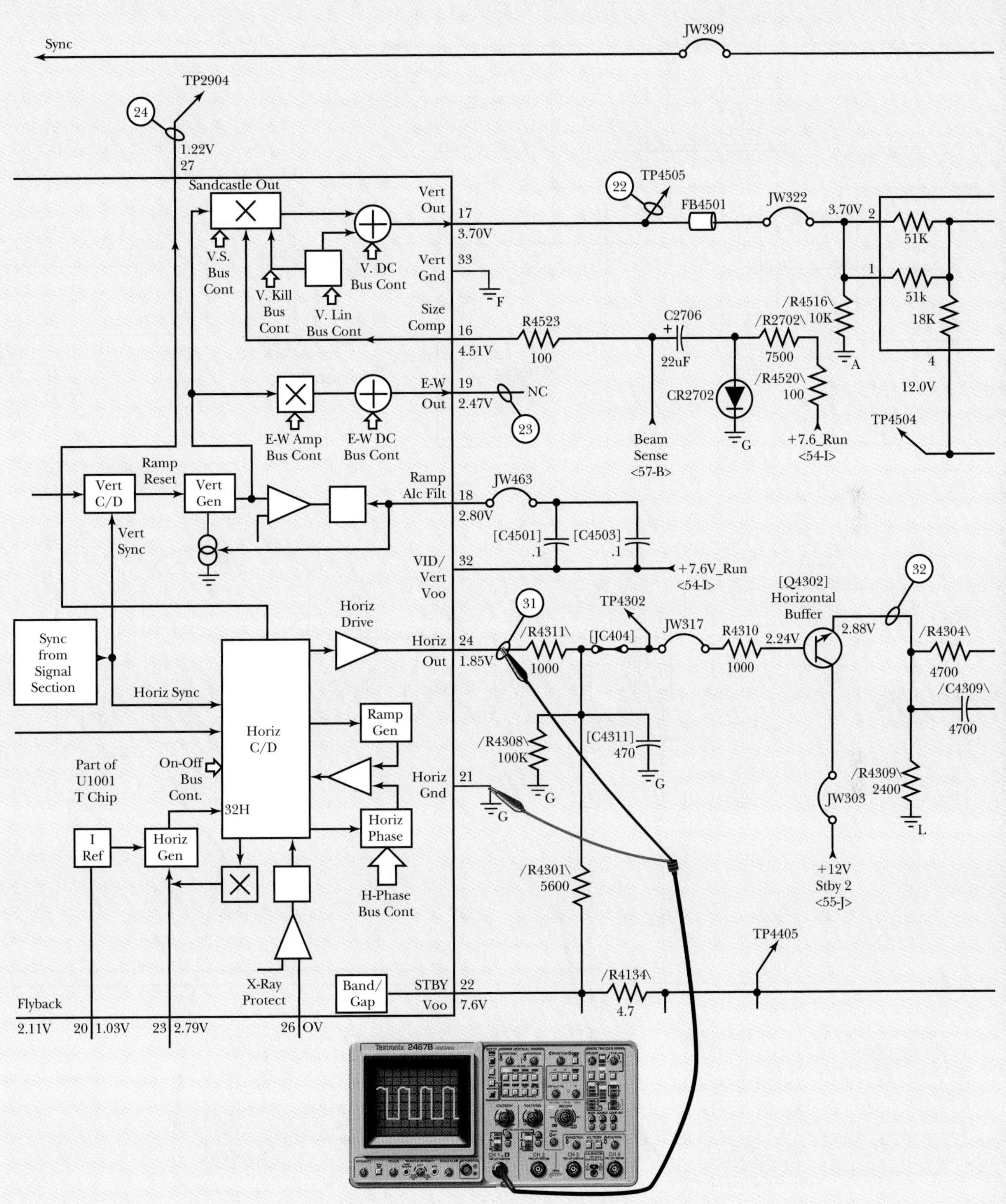

FIGURE 10 Circuit schmatic

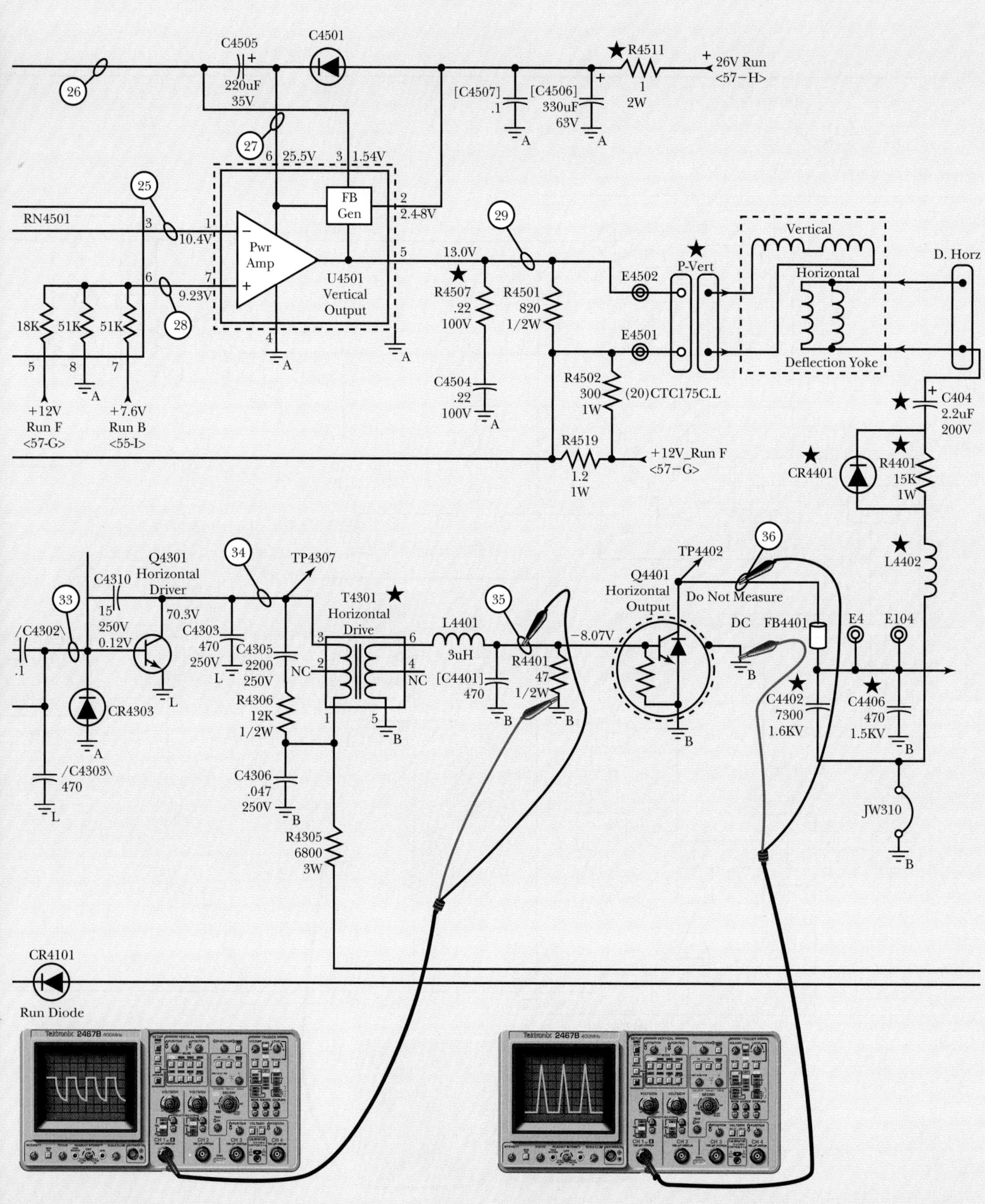

FIGURE 10 Continued

signals and converts them to a significant drive current which it supplies to the horizontal output amplifier. The amplified high voltage horizontal drive signal now exceeds 1,000 volts peak-to-peak. This signal is coupled to the horizontal deflection coils and also to the high voltage transformer, where it is stepped up to 24,800 volts for use on the anode and focusing electrodes of the picture tube.

The circuit schematic in Figure 10 shows the details of the circuit. This shows the H SYNC pulse controlling the pulse repetition rate of the horizontal constant drive circuit. This square wave with the horizontal sync pulse in position (Figure 11) is fed to BJT Q4302, operating as a buffer amplifier. The signal is passed to BJT Q4301, operating as a high gain horizontal drive amplifier. This fully formed signal (see Figure 12) is applied through a high frequency pulse transformer as a drive current to BJT Q4401, a specialized type with high voltage handling capabilities. This 1000-volt (1 KV ac) horizontal pulse signal (see Figure 13) is fed to the horizontal deflection coils to literally snap the scanning beam horizontally across the face of the picture tube from left to right drawing one single horizontal line of the picture. At the end of the trace, the sync pulse, now a spike, forces an even faster retrace of the beam back across to its starting point at the left side of the screen, to begin tracing the next scanning line. The horizontal drive pulse is also fed to the high voltage step up transformer. The 15.750 kHz signal voltage is now 24,800 volts. This and a slightly lower stepped up voltage are applied to the anode, focusing electrodes and metal screen of the television picture tube. The high voltage step up transformer and the horizontal drive transistor are shown in Figure 14.

BJT's used in a CCTV Convert Camera

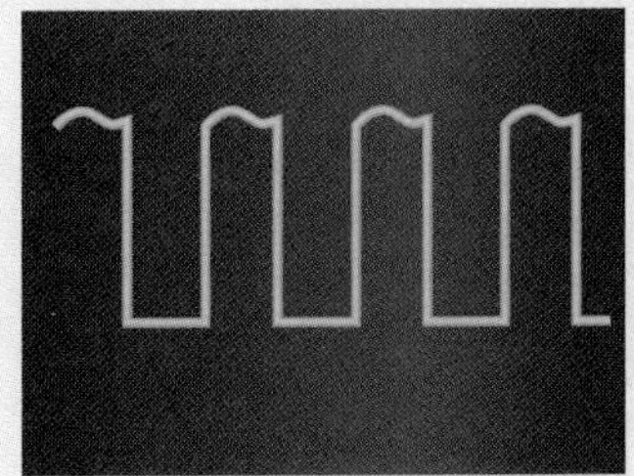

7.0 Vp-p 20μS per Div.

FIGURE 11 Horizontal drive signal

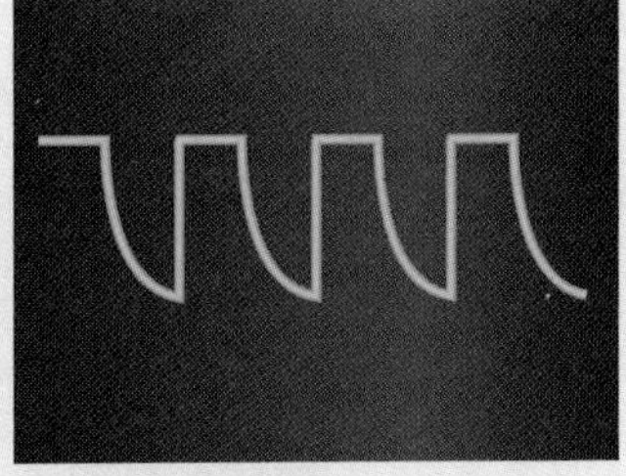

5.0 Vp-p 20 μS per Division

FIGURE 12 Horizontal current drive signal

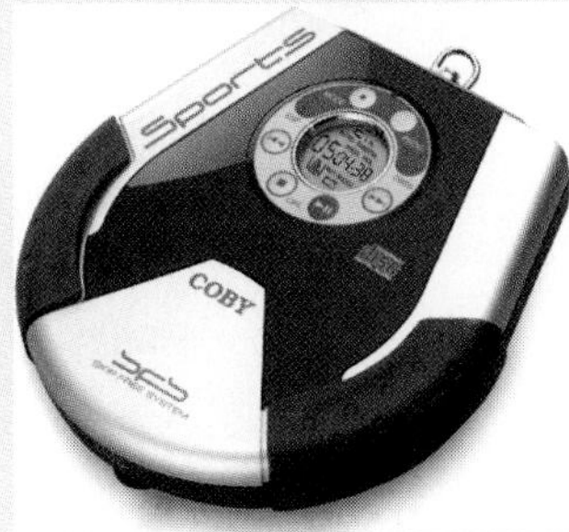

BJT's used in "skip-free" all weather sports CD player.

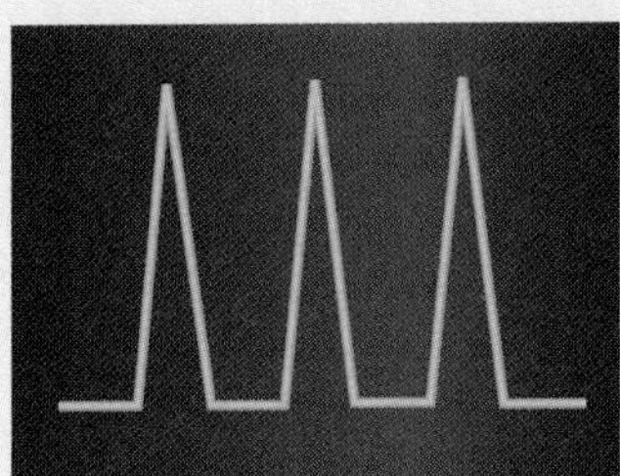

1000 Vp-p 20 μS per Div.
(bright flat segment is
horz scan, peak is retrace)

FIGURE 13 Horizontal pulse signal

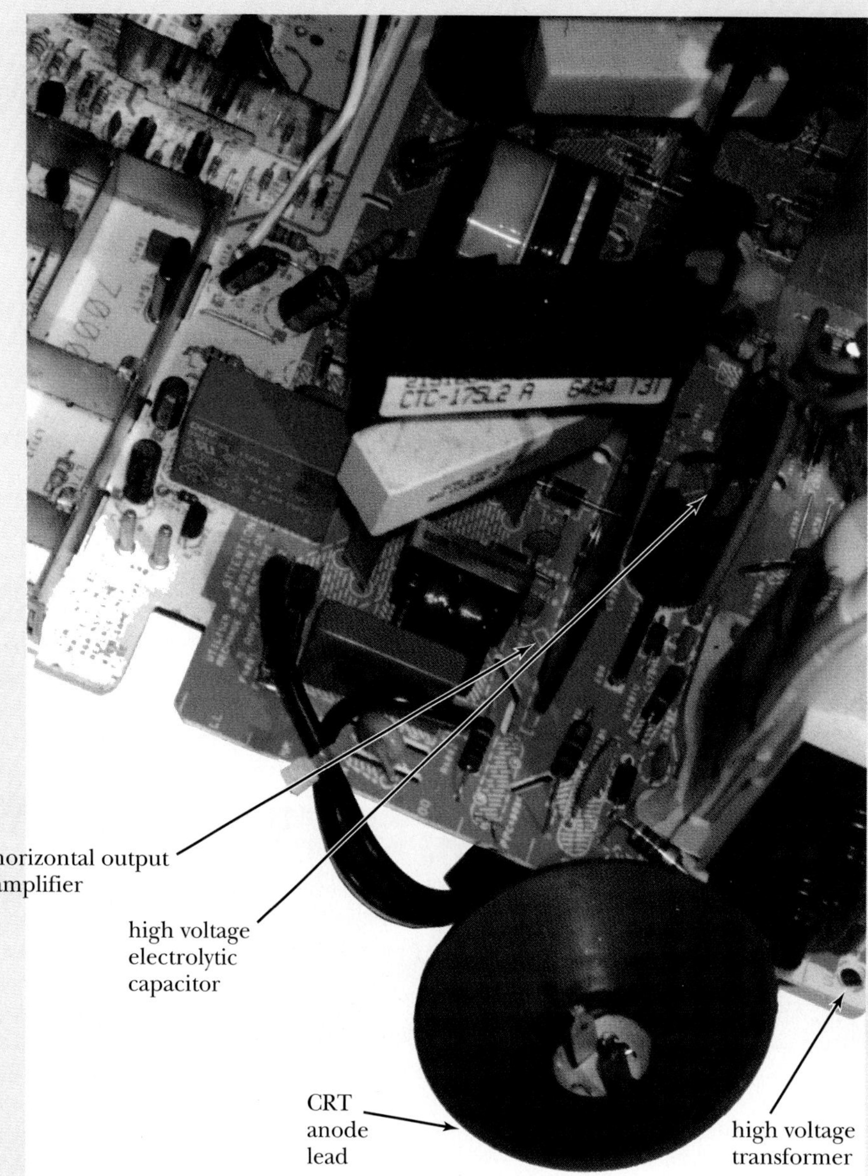

FIGURE 14 High voltage transformer

Summary

- *Transistor* is derived from the words *transfer* and *resistor.*
- A transistor consists of two P-N junctions. Typically, one P-N junction is forward biased and is a low resistance path for current; the other junction is reverse biased and is a high resistance path for current.
- Because a transistor transfers signals from its low resistance input circuit to its high resistance output circuit, it exhibits power gain.
- Two basic BJTs are the NPN and PNP. The letters N and P designate the type of semiconductor material in the three sections. In the NPN device, P-type material is sandwiched between two areas of N-type material. In the PNP device, N-type material is placed between two outside areas of P-type material.
- Three regions of a BJT are the emitter, base, and collector.
- Forward bias is a voltage applied to a P-N junction so that positive polarity is connected to the P-type material and the negative polarity voltage is connected to the N-type material. In a BJT, the base-emitter junction is forward biased by V_{BB}.
- Reverse bias exists when a source's negative polarity voltage is connected to the P-type material, and the positive polarity voltage is connected to the N-type material. In a BJT, the collector-base junction is reverse biased by V_{CC}.
- Currents at the connections of a BJT are:

  emitter current I_E;

  base current I_B; and

  collector current I_C.
- The base region in a BJT is extremely thin and very lightly doped. This is why it is possible to force current to flow between the emitter and collector whenever the base emitter junction is forward biased.
- DC beta β_{dc} is an indication of the current gain of the BJT; it is the ratio of collector current to base current. In the linear region, the dc beta is typically between 40 and 300, depending on the design of the transistor. This specification is often shown as h_{FE} as well as β_{dc}.
- The alpha (α) of a BJT is the ratio of collector current to emitter current. In the linear region, alpha is always less than 1.
- A BJT is a current controlling device. It uses a relatively small amount of base current to control a larger amount of collector current. When there is no base current flowing in a BJT, the transistor is in cutoff and there is no collector current. As the forward-bias base current increases, the collector current increases and the transistor is in either saturation or linear operation.
- Even though a BJT is essentially a current-controlled device, it can be used as a voltage amplifier. Used in this way, changes in base current are said to cause changes in collector voltage.
- BJT's are basically used as switches and as amplifiers, and often as both at the same time.
- An *I-V* curve for a BJT shows how collector current increases with increasing collector-emitter voltage at first, but then levels off so that further increases in collector-emitter voltage cause only insignificant increases in collector current. A set of such curves for several different levels of base current is called a family of collector characteristic curves.
- Maximum ratings for BJTs are maximum collector-emitter voltage ($V_{(BR)CEO}$), maximum collector-base voltage ($V_{(BR)CBO}$), maximum emitter-base voltage ($V_{(BR)EBO}$), maximum collector current (I_C), and maximum power dissipation (P_d).
- The power dissipation (P_d) of a BJT can be determined by the power formula as it applies to the collector current and voltage drop, i.e., $P_d = I_C \times V_{CE} + I_B \times V_{BE} \approx I_C \times V_{CE}$.
- Ac beta, β_{ac}, is a measure of dynamic current gain. This means the ratio of a change in collector current to the related change in base current that caused it. Ac beta can be determined from the BJT's family of collector characteristic curves. This same specification is also indicated by h_{FE}.
- Semiconductors are sensitive to overheating. Therefore, proper caution must be used when soldering and desoldering them.
- BJTs are tested in various ways. Test equipment is available to test them in circuit and out of circuit. Also, the ohmmeter and DMM, if properly used, are good tools for general testing. Bias voltages are also good indicators of normal or abnormal circuit operation.
- The three basic transistor amplifier circuit configurations are the common-emitter (CE), common-collector (CC), and common-base (CB) circuits. Each has special features that are used in specific applications.
- For the common-emitter amplifier, the input is to the base and the output is from the collector. For the common-collector amplifier, the input is to the base and the output is from the emitter. For the common-base amplifier, the input is to the emitter, and the output is from the collector.

- The common-emitter circuit provides both current and voltage gain, and the phase of the amplified signal is 180° different from the input signal. That is, the signal is inverted using the amplification process. The common-base amplifier output is in phase with the input signal. The current gain of the common-base amplifier is approximately 1. The common-collector circuit features include a voltage gain of less than 1 and high current gain. It has a high input impedance and low output impedance, causing it to be used frequently for impedance matching or circuit isolation purposes.
- Regardless of the type of BJT amplifier configuration that is used (CE, CC, or CB), the basic requirements for biasing a BJT are strictly followed: the base-emitter P-N junction is forward biased and the collector-base P-N junction is reverse biased.
- Amplifier circuits can be composed of more than one stage connected one after the other. This enables the use of advantages of the three basic circuit configurations. A CC amplifier followed by a CE amplifier, for example, provides high input impedance (CC advantage) and high voltage gain (CE advantage).
- A Darlington common collector amplifier is composed of two stages of direct-coupled common-collector amplifiers. The circuit has extremely high input impedance (the two beta values multiplied together), but a voltage gain that is less than unity.
- Small-signal amplifiers are designed to operate at small ac input signals (peak-to-peak ac current value is less than 0.1 times the amplifier's input bias current) and at power levels less than 1 W. Small-signal amplifiers are used where it is required to boost low voltage, low power signals.
- Large-signal amplifiers are designed for higher input current levels and output power levels (such as the output stages of an audio amplifier system).
- Four classes of operation for BJT amplifiers are Class A, Class B, Class AB, and Class C. In Class A amplifiers, collector current flows at all times, or during 360° of the input cycle. In Class AB amplifiers, collector current flows for more than 180° of the input cycle, but less than 360°. In Class B operation, current flows during half the input signal cycle (180°). And in Class C operation, collector current flows for approximately 120° of the input cycle.
- A dc load line is constructed on collector characteristic curves for a given transistor and circuit by drawing a line between the extremes of cutoff and saturation. In other words, the dc load line illustrates all possible operating points between those two extremes.
- The Q point, or quiescent operating point, of a transistor amplifier circuit is identified on the load line at the junction of the base current and the dc load line for the given circuit. The "no-signal" values (quiescent dc values) for base current, collector current, and collector-to-emitter voltage are read from the collector curves/load line representation.
- If the Q point for a BJT amplifier stage is halfway between cutoff and saturation, the stage is operating as a Class A amplifier. If the Q point is set at cutoff, the stage is operating Class B. When the Q point is located somewhere between Class A and B, the circuit is said to be operating Class AB. And if the Q point is set below cutoff, the stage is operating Class C.
- A summary of BJT amplifier circuits is provided in Figure 5–47.

CIRCUIT SUMMARY

Description	*Formula*
Voltage divider bias circuit	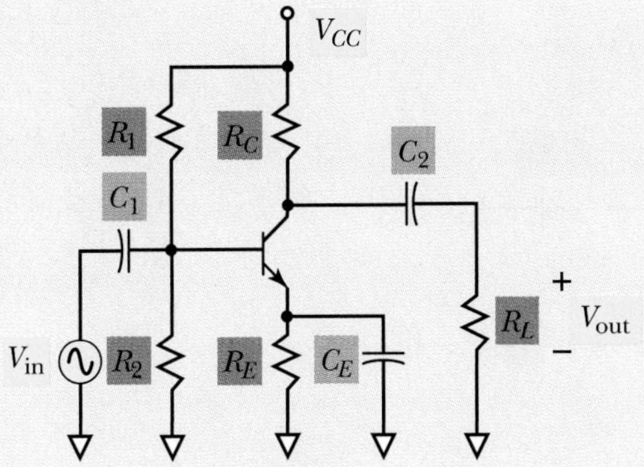 DC operating point (Q point) $V_B \approx V_{CC} \times \dfrac{R_2}{R_1 + R_2}$ (5 – 10) $I_E \approx \dfrac{V_B - 0.7\text{ V}}{R_E}$ (5 – 11) $I_C \approx I_E$ $I_B = \dfrac{I_C}{\beta}$ $V_{CE} = V_{CC} - I_E(R_C + R_E)$ (5 – 12) $r_e' = \dfrac{25\text{ mV}}{I_E}$ (5 – 13)
Common Emitter Amplifier	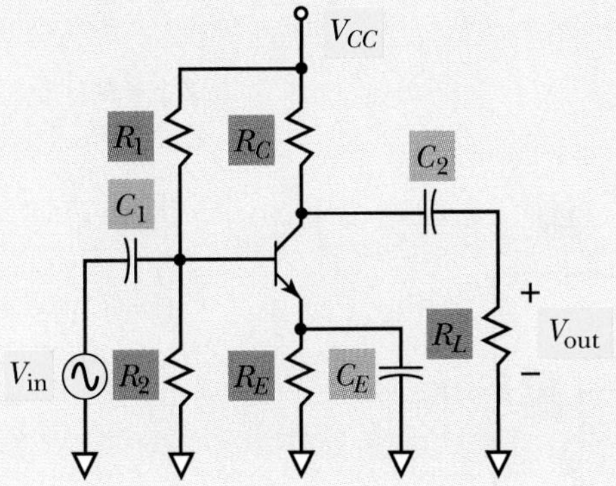 $Z_{in} = R_1 \parallel R_2 \parallel (\beta + 1)r_e'$ (5 – 14) $Z_{out} = R_c$ (5 – 15) $A_V = -\dfrac{R_c}{r_e'}$ (5 – 16) $A_i = \beta \dfrac{R_1 \parallel R_2}{R_1 \parallel R_2 + (\beta + 1)r_e'}$ (5 – 17) $A_p = \lvert A_V \times A_i \rvert$ (5 – 9) $V_{out} = A_V \times V_{in}$ (5 – 7)
Swamped Common Emitter Amplifier	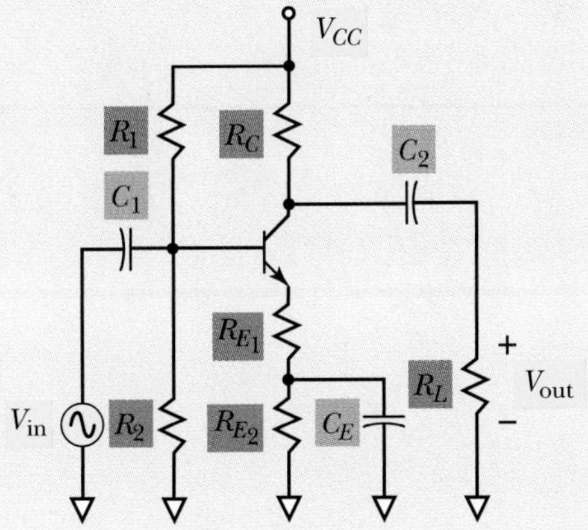 $Z_{in} = R_1 \parallel R_2 \parallel ((\beta + 1)(r_e' + R_{E_1}))$ (5 – 18) $Z_{out} = R_C$ (5 – 19) $A_V = -\dfrac{R_C}{(r_e' + R_{E_1})}$ (5 – 20) $A_i = \beta \dfrac{(R_1 \parallel R_2)}{(R_1 \parallel R_2) + (\beta + 1)(r_e' + R_{E_1})}$ (5 – 21) $A_p = \lvert A_V \times A_i \rvert$ (5 – 8) $V_{out} = A_V \times V_{in}$ (5 – 7)
Common Collector (Voltage Follower) Amplifier	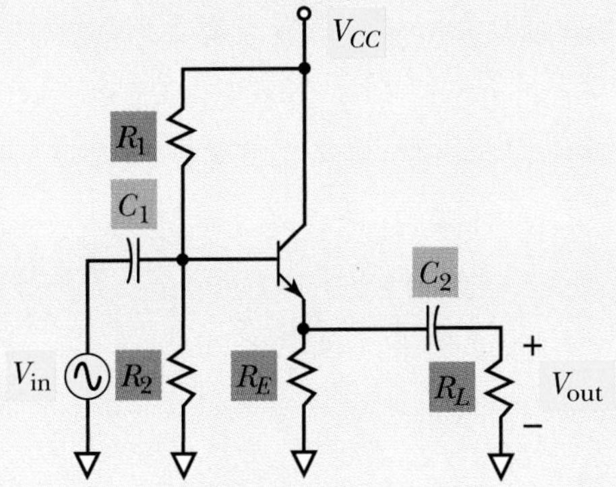 $Z_{in} = R_1 \parallel R_2 \parallel ((\beta + 1)(r_e' + R_E))$ (5 – 23) $Z_{out} = R_E \parallel r_e'$ (5 – 24) $A_V \approx +1$ (5 – 25) $A_i \approx -\beta$ (5 – 26) $A_p = \lvert A_V \times A_i \rvert$ (5 – 9) $V_{out} = A_V \times V_{in}$ (5 – 7)

FIGURE 5–47 BJT amplifier circuit summary

CIRCUIT SUMMARY

Description	*Circuit*	*Formula*
Common-base (dual supply) amplifier		$Z_{in} = R_E \parallel r_e'$ (5 – 27) $Z_{out} = R_C$ (5 – 28) $A_V = \dfrac{R_C}{r_e'}$ (5 – 29) $A_i \approx -1$ (5 – 30) $A_p = \lvert A_V \times A_i \rvert$ (5 – 9) $V_{out} = A_V \times V_{in}$ (5 – 7)
Common-base (single supply) amplifier		$Z_{in} \approx R_E \parallel r_e'$ (5 – 27) $Z_{out} = R_C$ (5 – 28) $A_V \approx \dfrac{R_C}{r_e'}$ (5 – 29) $A_i \approx -1$ (5 – 30) $A_p = \lvert A_V \times A_i \rvert$ (5 – 9) $V_{out} = A_V \times V_{in}$ (5 – 7)

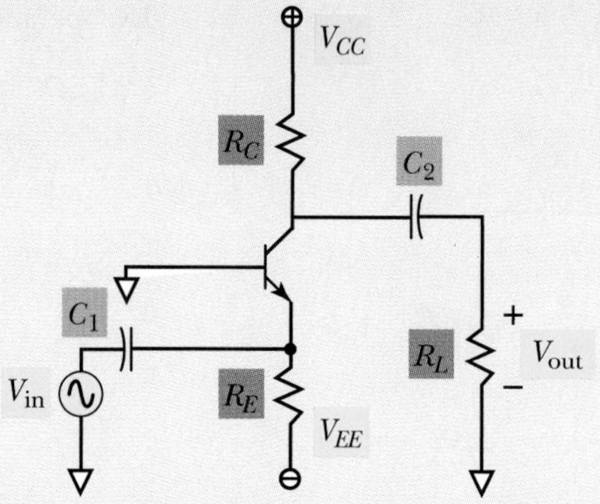

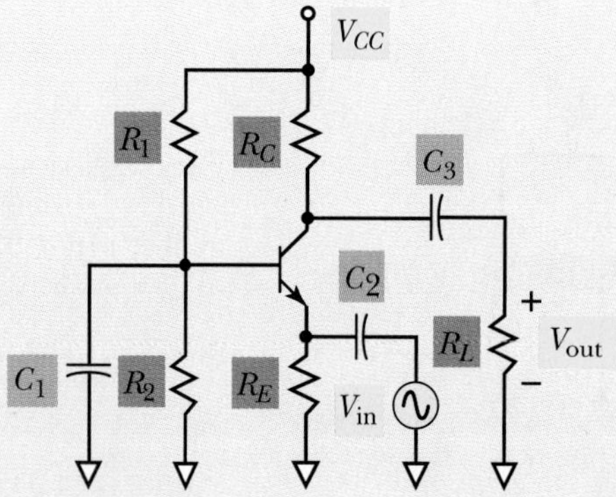

FIGURE 5–47 Continued

Formulas and Sample Calculator Sequences

FORMULA 5–1
(To find emitter current)

$I_E = I_C + I_B$

I_C value, [+], I_B value, =

FORMULA 5–2
(To find dc current gain)

$\beta_{dc} = \dfrac{I_C}{I_B}$

I_C value, [÷], I_B value, =

FORMULA 5–3
(To find collector current)

$I_C = \beta_{dc} \times I_B$

β_{dc} value, [×], I_B value, =

FORMULA 5–4
(To find base current)

$I_B = \dfrac{I_C}{\beta_{dc}}$

I_C value, [÷], β_{dc} value, =

FORMULA 5–5
(To find α)

$\alpha = \dfrac{I_C}{I_E}$

I_C value, [÷], I_E value, =

FORMULA 5–6 *(To find ac current gain)*

$\beta_{ac} = \dfrac{\Delta I_C}{\Delta I_B}$

ΔI_C value, ÷, ΔI_B value, =

Gain calculations

FORMULA 5–7 *(To find voltage gain)*

$A_V = \dfrac{V_{out}}{V_{in}}$

V_{out} value, ÷, V_{in} value, =,

FORMULA 5–8 *(To find current gain)*

$A_I = \dfrac{I_{out}}{I_{in}}$

I_{out} value, ÷, I_{in} value, =

FORMULA 5–9 *(To find power gain)*

$A_P = |A_V \times A_I|$

A_V value, ×, A_I value, =

Voltage divider with emitter stabilized biasing

FORMULA 5–10 *(To find base voltage)*

$V_B \approx V_{TH} = \dfrac{V_{CC} \times R_2}{(R_1 + R_2)}$

V_{CC} value, ×, R_2 value, ÷, (, R_1 value, +, R_2 value,), =

FORMULA 5–11 *(To find emitter current)*

$I_E = \dfrac{(V_B - V_{BE})}{R_E}$

(, V_B value, −, V_{BE} value,), ÷, R_E value, =

FORMULA 5–12 *(To find collector-emitter voltage)*

$V_{CE} = V_{CC} - I_C \times (R_C + R_E)$

V_{CC} value, −, I_C value, ×, (R_C value, +, R_E value,), =

The common-emitter (CE) amplifier

FORMULA 5–13 *(To find dynamic emitter resistance)*

$r_e' = \dfrac{25\text{ mV}}{I_E}$

0.025, ÷, I_E value, =

FORMULA 5–14 *(To find input impedance)*

$Z_{in} = R_1 \parallel R_2 \parallel (\beta + 1)\, r_e'$

1, ÷, (, (, 1, ÷, R_1 value,), +, (, 1, ÷, R_2 value,), +, (, 1, ÷, (, (, β value, +, 1,), ×, r_e' value,),),), =

FORMULA 5–15 *(To find output impedence)*

$Z_{out} = R_C$

FORMULA 5–16 *(To find voltage gain)*

$A_V \approx -\dfrac{R_C}{r_e'}$

R_C value, +, ÷, r_e' value, =

FORMULA 5–17 *(To find current gain)*

$A_i = \beta\,(R_1 \parallel R_2)/((R_1 \parallel R_2) + (\beta + 1)\, r_e')$

The swamped common-emitter (CE) amplifier

FORMULA 5–18 *(To find input impedance)*

$Z_{in} = R_1 \parallel R_2 \parallel ((\beta + 1)\,(r_e' + R_{E_1}))$

FORMULA 5–19 *(To find output impedance)*

$Z_{out} = R_C$

FORMULA 5–20 $A_V = \frac{R_C}{r_e'}$

(To find voltage gain)

FORMULA 5–21 $A_i \approx \frac{\beta(R_1 \| R_2)}{((R_1 \| R_2) + (\beta+1)(r_e' + R_{E1}))}$

(To find current gain)

Multiple-stage CE amplifier circuits

FORMULA 5–22 $A_V = A_{V_1} \times A_{V_2} \times A_{V_3} \times A_{V_4} \times A_{V_5}$

(To find voltage gain)

where: $A_{V_1} = \frac{Z_{\text{in}_1}}{(Z_{\text{in}_1} + Z_S)}$

A_{V_2} is the gain of the first amplifier

$A_{V_3} = \frac{Z_{\text{in}_2}}{(Z_{\text{in}_2} + Z_{\text{out}_1})}$

A_{V_4} is the gain of the second amplifier

$A_{V_5} = \frac{Z_L}{(Z_L + Z_{\text{out}_2})}$

The common-collector (CC) amplifier

FORMULA 5–23 $Z_{\text{in}} = R_1 \| R_2 \| ((\beta + 1)(r_e' + R_E))$

(To find input impedance)

FORMULA 5–24 $Z_{\text{out}} = R_E \| r_e'$

(To find output impedance)

FORMULA 5–25 $A_V \approx 1$

(To find voltage gain)

FORMULA 5–26 $A_i \approx -\beta$

(To find current gain)

The common-base (CB) amplifier

FORMULA 5–27 $Z_{\text{in}} = R_E \| r_e'$

(To find input impedance)

FORMULA 5–28 $Z_{\text{out}} = R_C$

(To find output impedance)

FORMULA 5–29 $A_V = \frac{R_C}{r_e'}$

(To find voltage gain)

FORMULA 5–30 $A_i \approx -1$

(To find current gain)

DC Load Line

FORMULA 5–31 $V_{CE} = V_{CC}$

(To find maximum collector-emitter voltage)

FORMULA 5–32 $I_{C(\text{sat})} = \frac{V_{CC}}{(R_C + R_E)}$

(To find saturation current)

Using Excel *Bipolar Junction Transistor* Formulas

(Excel file reference: FOE5_01.xls)

DON'T FORGET! It is NOT necessary to retype formulas, once they are entered on the worksheet! Just input new parameters data for each new problem using that formula, as needed.

- Use the Formula 5–1 spreadsheet sample and the parameters given for Practice Problems 1 question 1. Solve for the emitter current. Check your answer against the answer for this question in the Appendix.
- Use the Formula 5–2 spreadsheet sample and the parameters given for Practice Problems 1 question 2. Solve for the dc current gain. Check your answer against the answer for this question in the Appendix.

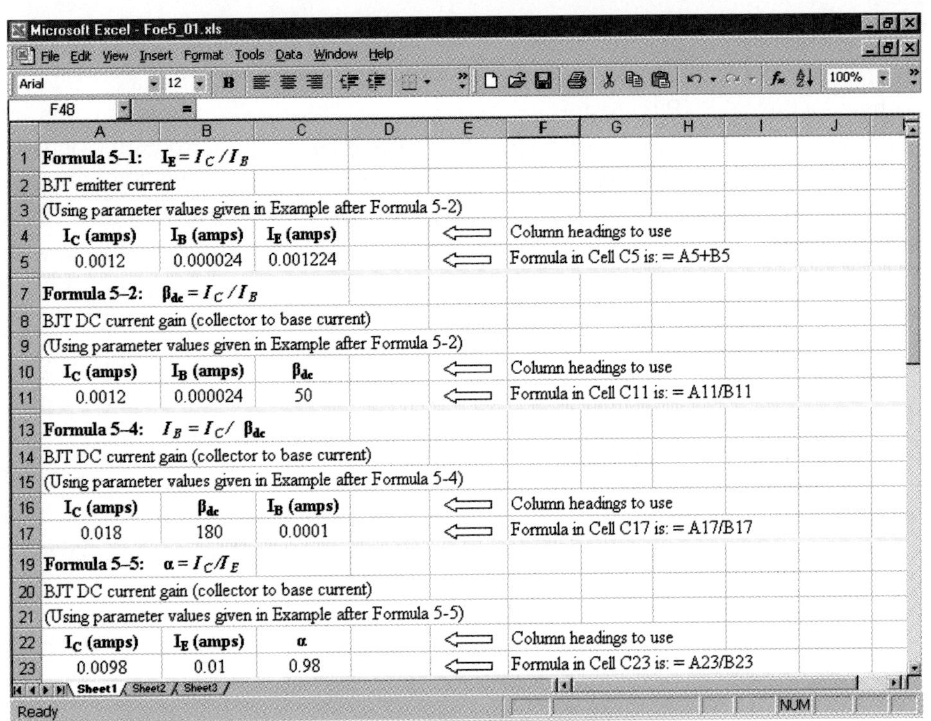

Microsoft Excel - Foe5_01.xls

	A	B	C	D	E	F
1	Formula 5–1: $I_E = I_C / I_B$					
2	BJT emitter current					
3	(Using parameter values given in Example after Formula 5-2)					
4	I_C (amps)	I_B (amps)	I_E (amps)		⟸	Column headings to use
5	0.0012	0.000024	0.001224		⟸	Formula in Cell C5 is: = A5+B5
7	Formula 5–2: $\beta_{dc} = I_C / I_B$					
8	BJT DC current gain (collector to base current)					
9	(Using parameter values given in Example after Formula 5-2)					
10	I_C (amps)	I_B (amps)	β_{dc}		⟸	Column headings to use
11	0.0012	0.000024	50		⟸	Formula in Cell C11 is: = A11/B11
13	Formula 5–4: $I_B = I_C / \beta_{dc}$					
14	BJT DC current gain (collector to base current)					
15	(Using parameter values given in Example after Formula 5-4)					
16	I_C (amps)	β_{dc}	I_B (amps)		⟸	Column headings to use
17	0.018	180	0.0001		⟸	Formula in Cell C17 is: = A17/B17
19	Formula 5–5: $\alpha = I_C / I_E$					
20	BJT DC current gain (collector to base current)					
21	(Using parameter values given in Example after Formula 5-5)					
22	I_C (amps)	I_E (amps)	α		⟸	Column headings to use
23	0.0098	0.01	0.98		⟸	Formula in Cell C23 is: = A23/B23

(Excel file reference: FOE5_01.xls)

DON'T FORGET! It is NOT necessary to retype formulas, once they are entered on the worksheet! Just input new parameters data for each new problem using that formula, as needed.

- Use the Formula 5–32 spreadsheet sample and the parameters given for Practice Problems 2. Solve for the saturation current. Check your answer against the answer for this question in the Appendix.
- Use the Formula 5–7 spreadsheet sample and the parameters given for Chapter Problems 19. Solve for the amplifier voltage gain. Check your answer against the answer for this question in the Appendix.

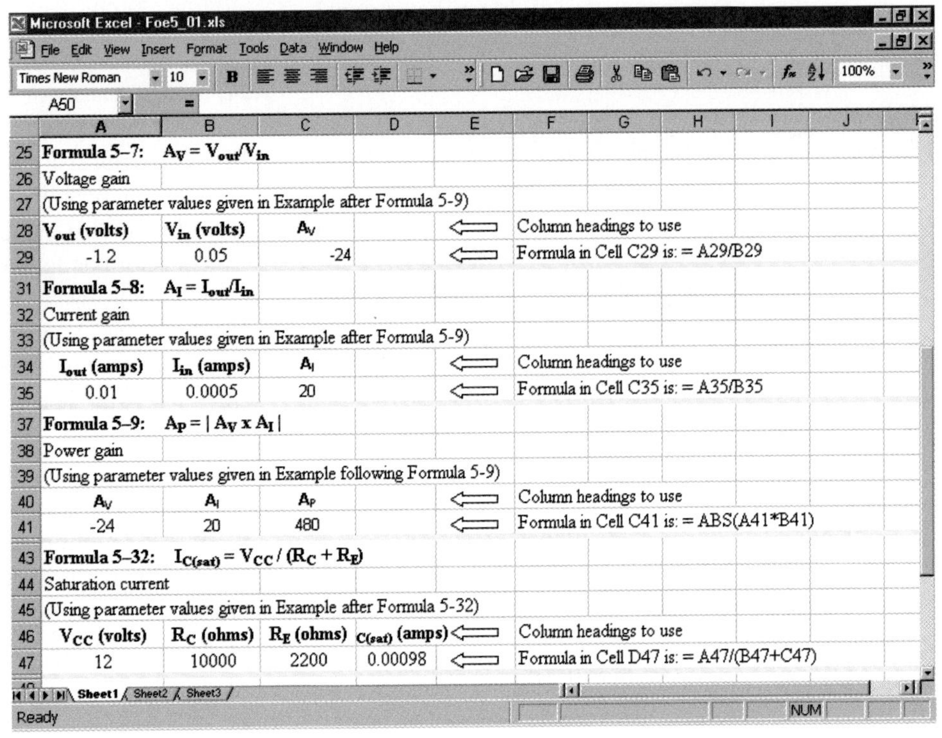

Microsoft Excel - Foe5_01.xls

	A	B	C	D	E	F
25	Formula 5–7: $A_V = V_{out}/V_{in}$					
26	Voltage gain					
27	(Using parameter values given in Example after Formula 5-9)					
28	V_{out} (volts)	V_{in} (volts)	A_V		⟸	Column headings to use
29	-1.2	0.05	-24		⟸	Formula in Cell C29 is: = A29/B29
31	Formula 5–8: $A_I = I_{out}/I_{in}$					
32	Current gain					
33	(Using parameter values given in Example after Formula 5-9)					
34	I_{out} (amps)	I_{in} (amps)	A_I		⟸	Column headings to use
35	0.01	0.0005	20		⟸	Formula in Cell C35 is: = A35/B35
37	Formula 5–9: $A_P = \lvert A_V \times A_I \rvert$					
38	Power gain					
39	(Using parameter values given in Example following Formula 5-9)					
40	A_V	A_I	A_P		⟸	Column headings to use
41	-24	20	480		⟸	Formula in Cell C41 is: = ABS(A41*B41)
43	Formula 5–32: $I_{C(sat)} = V_{CC} / (R_C + R_E)$					
44	Saturation current					
45	(Using parameter values given in Example after Formula 5-32)					
46	V_{CC} (volts)	R_C (ohms)	R_E (ohms)	$I_{C(sat)}$ (amps)	⟸	Column headings to use
47	12	10000	2200	0.00098	⟸	Formula in Cell D47 is: = A47/(B47+C47)

Review Questions

1. How many P-N junctions does a typical NP transistor have? a typical PNP transistor?
2. What semiconductor material is used in the construction of the base for an NPN transistor? a PNP transistor?
3. What semiconductor material is used in the construction of the collector for an NPN transistor? a PNP transistor?
4. For the NPN transistor to operate in the linear region, what polarity of voltage is applied to the base-emitter junction? to the collector-base junction?
5. For the PNP transistor to operate in the linear region, what polarity of voltage is applied to the base-emitter junction? to the collector-base junction?
6. Draw the schematic symbols for the NPN and PNP transistors. Label all three terminals.
7. Approximately what percentage of emitter current appears in the collector circuit of a typical silicon BJT operating in the linear region.
8. Explain forward biasing with respect to a P-N junction.
9. Explain reverse biasing with respect to a P-N junction.
10. For normal operation, the base-emitter circuit of a BJT is (forward, reverse) __________ biased.
11. For normal operation, the base-collector circuit of a BJT is (forward, reverse) __________ biased.
12. Is the base current in a BJT operating in the linear region a high or low percentage of the emitter current?
13. The emitter current and collector current in the BJT linear region have vastly different values. (True/False)
14. Explain the term *basing diagram.*
15. Describe how increasing the amount of base current affects the collector current in a BJT amplifier.
16. Explain why the voltage drop between emitter and collector of a BJT being used as a switch will be at its minimum level when the base current is at its maximum level.
17. Describe the relationship between forward-biasing current and collector-emitter voltage drop in a typical BJT voltage amplifier.
18. Define the following ratings for a BJT:
    a. Maximum collector-emitter voltage ($V_{(BR)\mathrm{CEO}}$)
    b. Maximum collector-base voltage ($V_{(BR)\mathrm{CBO}}$)
    c. Maximum emitter-base voltage ($V_{(BR)\mathrm{EBO}}$)
19. Define the following ratings for a BJT:
    a. Maximum collector current (I_C)
    b. Maximum power dissipation (P_d)
20. Explain the difference between dc beta and ac beta.
21. In the common-emitter amplifier configuration, what transistor terminal is common to both input and output circuits?
22. Indicate between which junctions or electrodes the input signal is introduced, and between which electrodes the output signal is taken for the following circuit configurations:
    a. Common-emitter amplifier:
       input is between _____ and _____
       output is between _____ and _____
    b. Common-base amplifier:
       input is between _____ and _____
       output is between _____ and _____
    c. Common-collector amplifier:
       input is between _____ and _____
       output is between _____ and _____
23. For the following transistor amplifier configurations, what are the typical input and output impedance characteristics? (NOTE: Use "high," "low," or "medium" to describe them.)
    a. Common-emitter amplifier:
       input Z = _____
       output Z = _____
    b. Common-base amplifier:
       input Z = _____
       output Z = _____
    c. Common-collector amplifier:
       input Z = _____
       output Z = _____
24. Which transistor amplifier circuit configuration provides both voltage and current gain?
25. Which transistor amplifier circuit configuration cause 180° phase shift from input to output?
26. What is the name for a two-transistor, common-collector configuration that is noted for extremely high input impedance, but a voltage gain of less than 1?
27. Explain why a circuit would have a common-collector amplifier stage followed by a common-emitter stage.
28. What is the least popular of the three basic amplifier configurations?
29. Name one application for small-signal and one application for large-signal transistor amplifier stages.
30. Which amplifier class has the lowest efficiency?
31. Which amplifier class has the least distortion?
32. Which amplifier class has the highest efficiency?
33. Where is the Q point set to achieve Class A operation?
34. Where is the Q point set to achieve Class B operation?
35. A common-emitter circuit has a V_{CC} of 20 V. What is the approximate quiescent value of V_{CE} if the circuit operates as a Class A amplifier?

36. Briefly list four assumptions frequently used when analyzing a common-emitter circuit that uses voltage-divider bias.

37. For a voltage-divider bias circuit, V_{CC} equals 15 V, R_C equals 10 kΩ, and R_E equals 1 kΩ. (Assume R_1 = 9.1 kΩ and R_2 = 1 kΩ in the bias divider and β = 150.) Find the following values:
   a. Divider current
   b. Base voltage
   c. Emitter voltage (assuming a silicon transistor)
   d. Collector voltage
   e. Collector-to-emitter voltage

38. If you need a transistor amplifier to provide 180° phase difference between the input and output signals, which circuit configuration would you choose?

39. A transistor amplifier circuit is needed that matches a high impedance at its input side to a low impedance connected to its output side. Which circuit configuration would you choose?

40. Which class of amplifier is used in a push-pull circuit?

Problems

1. For a typical BJT amplifier circuit, the collector current is 40 mA and the base current is 500 μA. Determine the following:
   a. The value of the emitter current
   b. The value of β_{dc}

2. When I_E = 50 mA and I_B = 120 μA, determine the following:
   a. The value of I_C
   b. The value of β_{dc}

3. What is the dc beta of a BJT if the collector current is 10 mA when applying 20 μA of base current?

4. When β_{dc} = 100 and I_C = 800 mA, determine the following:
   a. The value of I_B
   b. The value of I_E

5. What is the value of I_C for a BJT that has a β_{dc} of 80 and an I_B of 20 μA?

6. What is the alpha of a BJT if you find 20 mA of emitter current and 19.5 mA of collector current?

7. Determine α_{dc} when I_C = 200 mA and I_E = 198 mA.

8. For a certain BJT circuit, α_{dc} = 0.95 and I_C = 400 mA. Determine the following.:
   a. I_E
   b. I_B
   c. β_{dc}

9. For a certain BJT circuit, α_{dc} = 0.96 and I_E = 40 mA. Determine the following:
   a. I_C
   b. I_B
   c. β_{dc}

10. For a certain BJT circuit, β_{dc} = 200 and I_B = 40 μA. Determine the following:
   a. I_C
   b. I_E
   c. α_{dc}

11. A change of 200 μA at the base causes a change of 20 mA at the collector. What is the β_{ac} for this transistor?

12. If the β_{ac} for a BJT is 400, how much change in collector current can you expect to see when the base current changes by 10 μA?

13. What is the power dissipation of a BJT where V_{CE} = 4 V and I_C = 120 mA?

14. If $P_{d(max)}$ for a BJT = 600 mW, what is the maximum allowable collector current when V_{CE} = 1.2 V?

15. Determine the β_{ac} for ΔI_B = 80 μA as shown on the family of collector characteristic curves in Figure 5–48.

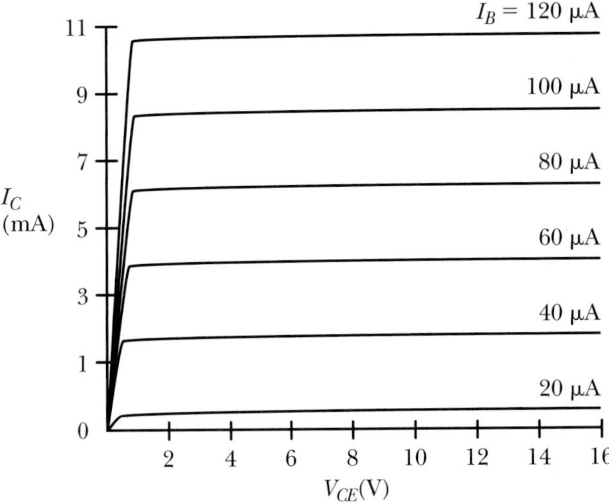

FIGURE 5–48

16. Referring to the circuit in Figure 5–49, let $+V_{CC} = 12$ V, $R_1 = 10$ kΩ, $R_2 = 8.2$ kΩ, $R_E = 1$kΩ, and $\beta = 100$. Determine the following:

a. base voltage

b. emitter current

17. For the circuit in Figure 5–49, let $+V_{CC} = 12$ V, $R_1 =$ 8.2 kΩ, $R_2 = 10$ kΩ $R_E = 1$ kΩ, and $\beta = 100$. Determine the following.

a. base voltage

b. emitter current

18. In Figure 5–49, let $+V_{CC} = 9$ V, $R_1 = 10$ kΩ, $R_2 =$ 10 kΩ, $R_E = 120$ Ω, $R_C = 120$ Ω, and $\beta = 100$.

a. $V_B =$ _____

b. $I_E =$ _____

c. $V_{CE} =$ _____

d. V_{BE} (voltage measured between base and emitter)

19. If $+V_{CC} = 6$ V, $R_C = 1$ kΩ, and $R_E = 100$ Ω in Figure 5–9, determine the following:

a. The saturation current for the circuit

b. The cutoff voltage for the circuit

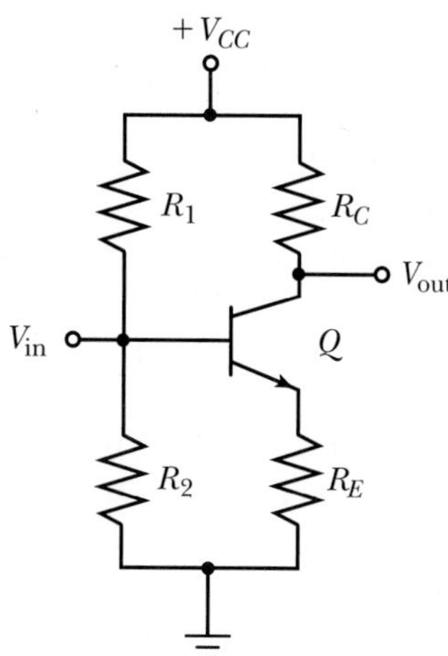

FIGURE 5–49

20. Referring to the circuit and values assigned in problem 18, determine the following:

a. The saturation current for the circuit

b. The cutoff voltage for the circuit

21. In a common-emitter BJT amplifier, such as the one shown in Figure 5–49, suppose $+V_{CC} = 10$ V and resistor R_1 is removed from the circuit. Determine the following:

a. The current through R_2

b. The voltage across R_E

c. The voltage between the emitter and collector of the transistor

d. The voltage across R_C

e. The transistor's emitter current

f. The class of operation

g. The cutoff voltage

22. In the C_E amplifier of Figure 5–49, suppose $V_{CC} = +18$ V and resistor R_2 is shorted to common with $R_1 = 10$ kΩ, R_E and $\beta = 100$. Determine the following:

a. The current through R_1

b. The voltage across R_1

23. The input signal to a Class A amplifier is 120 mV, and the output signal level is 2.2 V. What is the voltage gain of the circuit?

24. There is a need for an audio amplifier that boosts the 50 mV signal from a crystal microphone to 7.1 V. What is the required voltage gain for the amplifier?

25. The voltage gain of a certain common-collector amplifier is 0.9, and its current gain is 2.2. What is the power gain?

Analysis Questions

1. Look up and draw the transistor outlines for the following case/basing configurations: TO-3, TO-5, TO-92, TO-220, and SOT-89. Label each electrode on your drawings.
2. Look up the part identification codes for five NPN and five PNP transistors.
3. Research and explain the features and use of heat sinks associated with power transistors.
4. Locate data sheets for several BJTs and find the ratings for β_{ac} (or h_{FE}), BV_{CEO}, BV_{CBO}, BV_{EBO}, $I_{C(max)}$, and $P_{d(max)}$
5. Use Formulas 5–1 and 5–5 to prove that α is always a value that is less than 1.
6. Explain how the circuit in Figure 5–50 operates.
7. Figure 5–51 shows a family of collector characteristic curves and a load line for three different values of R_C: low resistance, medium resistance, and high resistance. Does this drawing indicate that to change the resistance value of R_C is to change the voltage gain of the amplifier? Explain your answer.
8. From the curves in Figure 5–52, determine the collector current value for a base current of 60 μA and a V_{CE} of 15 V.
9. What are the V_{CE} and I_C values at cutoff and saturation if a collector resistor with a value of 1 kΩ and an emitter resistor with a value of 220 Ω were present? (Use Figure 5–49 and assume V_{CC} = 15 V.)
10. Using the information from question 9, what is the I_E value when the Q point is set for Class A operation? What is the V_E value?
11. For the conditions described in question 10, what are the divider resistor values that will set the Q point for Class A operation?

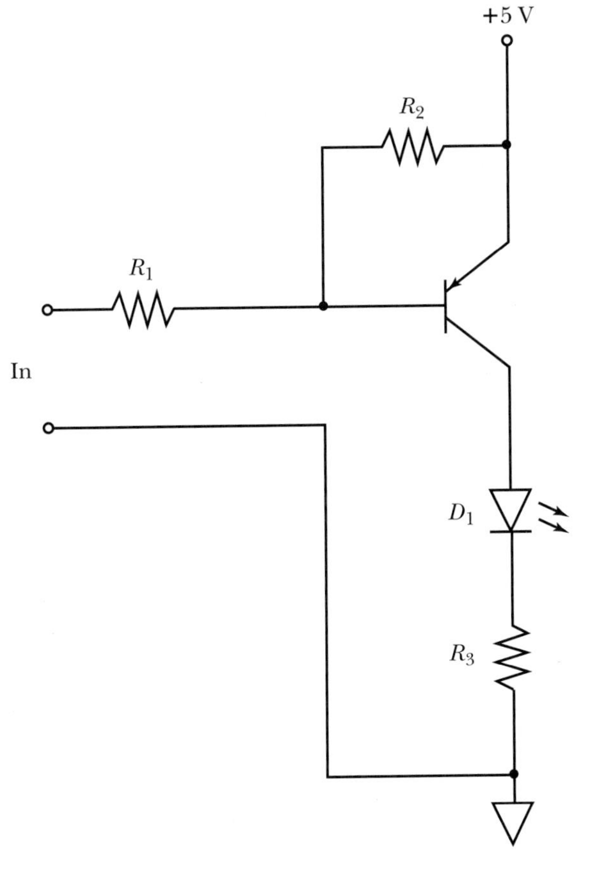

FIGURE 5–50

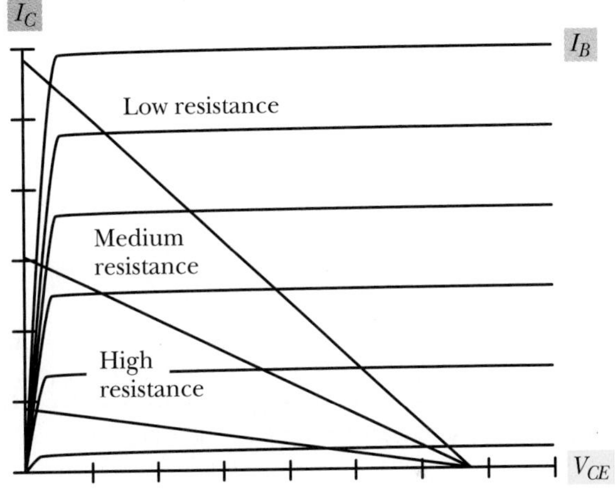

FIGURE 5–51

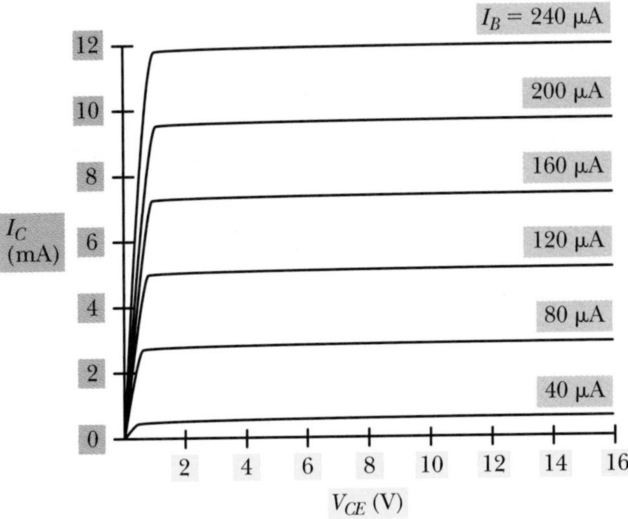

FIGURE 5–52

MultiSIM Exercise for *BJT Amplifier Circuits*

1. Use the MultiSIM program and utilize the circuit for Practice Problem 3. (Capacitors C_1 and C_2 are both 10 μF.)
2. Set the function generator to a 0.1 V peak sine wave at 1 kHz.
3. Measure and record the following parameters:
   - **a.** Input waveform
   - **b.** Output waveform
4. If allowed, compare your MultiSIM results with the answers given in the Instructor Guide for this circuit. Did the input and output waveforms agree with the Instructor Guide?

Performance Projects Correlation Chart

Chapter Topic	Performance Project	Project Number
BJT Bias Voltages and Currents	BJT Biasing	8
	Story Behind the Numbers: BJT Transistor Characteristics	
The Common-Emitter (CE) Amplifier	Common-Emitter Amplifier	9
The Common-Collector Amplifier	Common-Collector Amplifier	10
Classification by Class of Operation	BJT Class A Amplifier	11
Class A Amplifier Operation		
Analysis of BJT Amplifiers		
Class B Amplifier Operation	BJT Class B Amplifier	12
Class C Amplifier Operation	BJT Class C Amplifier	13

NOTE: It is suggested that after completing the above projects, the student should be required to answer the questions in the "Summary" at the end of this section of projects in the Laboratory Manual.

Troubleshooting Challenge

CHALLENGE CIRCUIT 9

(Follow the SIMPLER sequence by referring to inside front cover.)

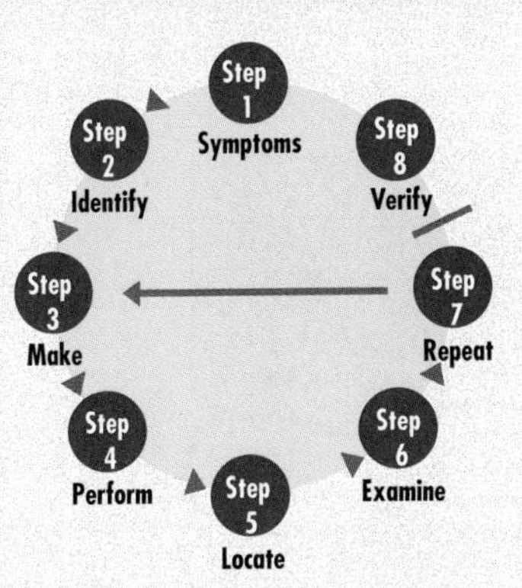

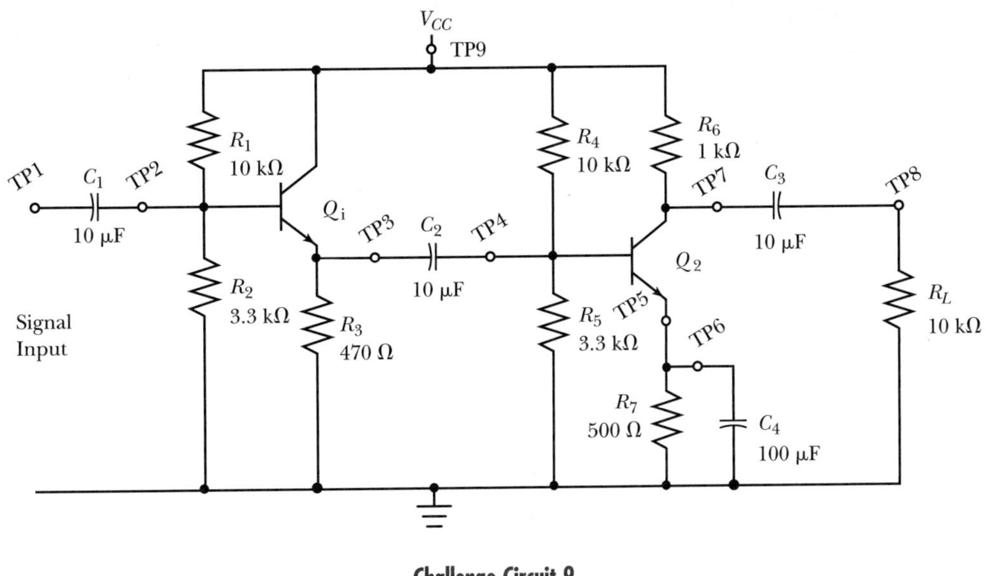

Challenge Circuit 9

STARTING POINT INFORMATION

1. Circuit diagram
2. The output waveform at TP8 is distorted.

TEST	Results in Appendix C
Waveform at TP1 shows	(2)
DC voltage at TP2	(92)
Waveform at TP3 shows	(57)
DC voltage at TP4	(76)
DC voltage at TP5	(27)
Resistance at TP6	(46)
Waveform at TP7 shows	(122)
Waveform at TP8 shows	(121)

CHALLENGE CIRCUIT 9

STEP [1]

SYMPTOMS The output signal is distorted.

STEP [2]

IDENTIFY initial suspect area. The low voltage can be caused by a failure of several components in the amplifier circuits, the coupling capacitors, and the load resistor. We will begin with an evaluation of the swamped common-emitter amplifier.

STEP [3]

MAKE test decision. We will start by looking at the signal at TP7 (the collector of Q_2).

STEP [4]

PERFORM **First Test:** Using a scope, check the waveform at TP7. The result is a distorted sine wave riding on a low dc voltage. The lower portion of the sine wave is clipped.

STEP [5]

LOCATE new suspect area. It looks like we might have a transistor-biasing problem. To evaluate the transistor biasing, we will remove the ac input. Thus, the new suspect area includes the swamped common-emitter amplifier bias components.

STEP [6]

EXAMINE available data.

STEP [7]

REPEAT analysis and testing.
The next test is to look at the dc voltage at TP7.
Second Test: Remove the signal source from the input. Check the voltage at TP7. The check indicates a voltage lower than normal.
Third Test: With the signal source removed from the input, let us now measure the voltage at TP5. This voltage is 0 V. The voltage reading indicates that either the capacitor C_4 is shorted or the resistor R_8 is shorted. At it is more common that capacitors fail, we will suspect that the capacitor is damaged.
Fourth Test: Next, we measure the capacitor resistance at TP6 will the capacitor removed from the circuit. This resistance is 0 Ω. The reading indicates that the capacitor C_4 is shorted.

STEP [8]

VERIFY **Fifth Test:** Replace capacitor C_4 and note the circuit parameters. When this is done, the circuit operates properly.

Symptoms

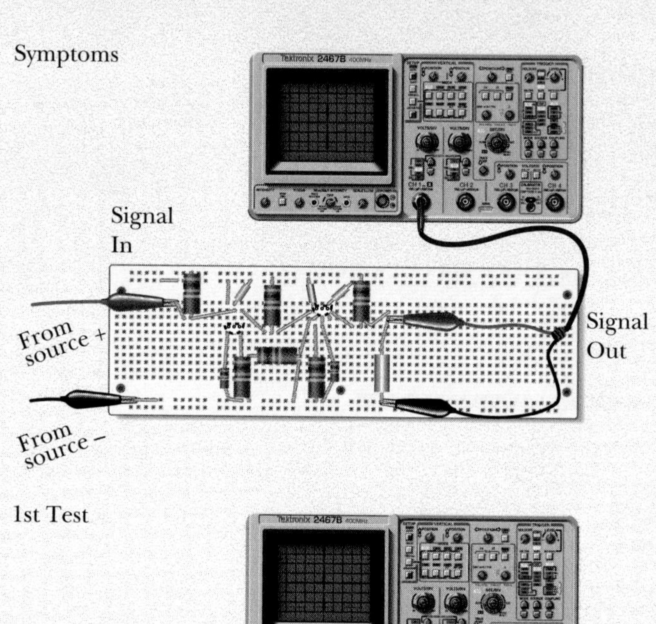

1st Test

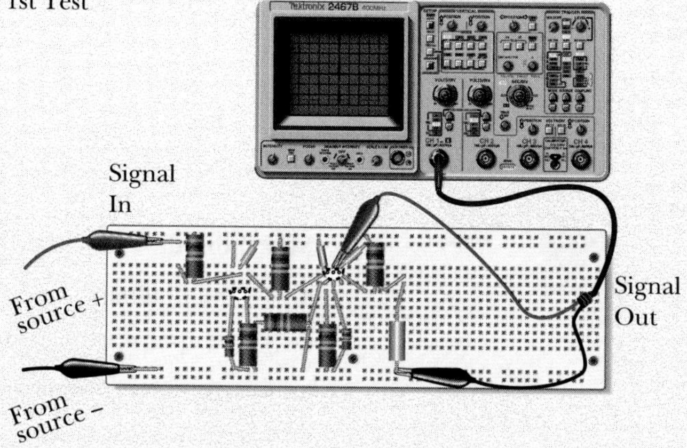

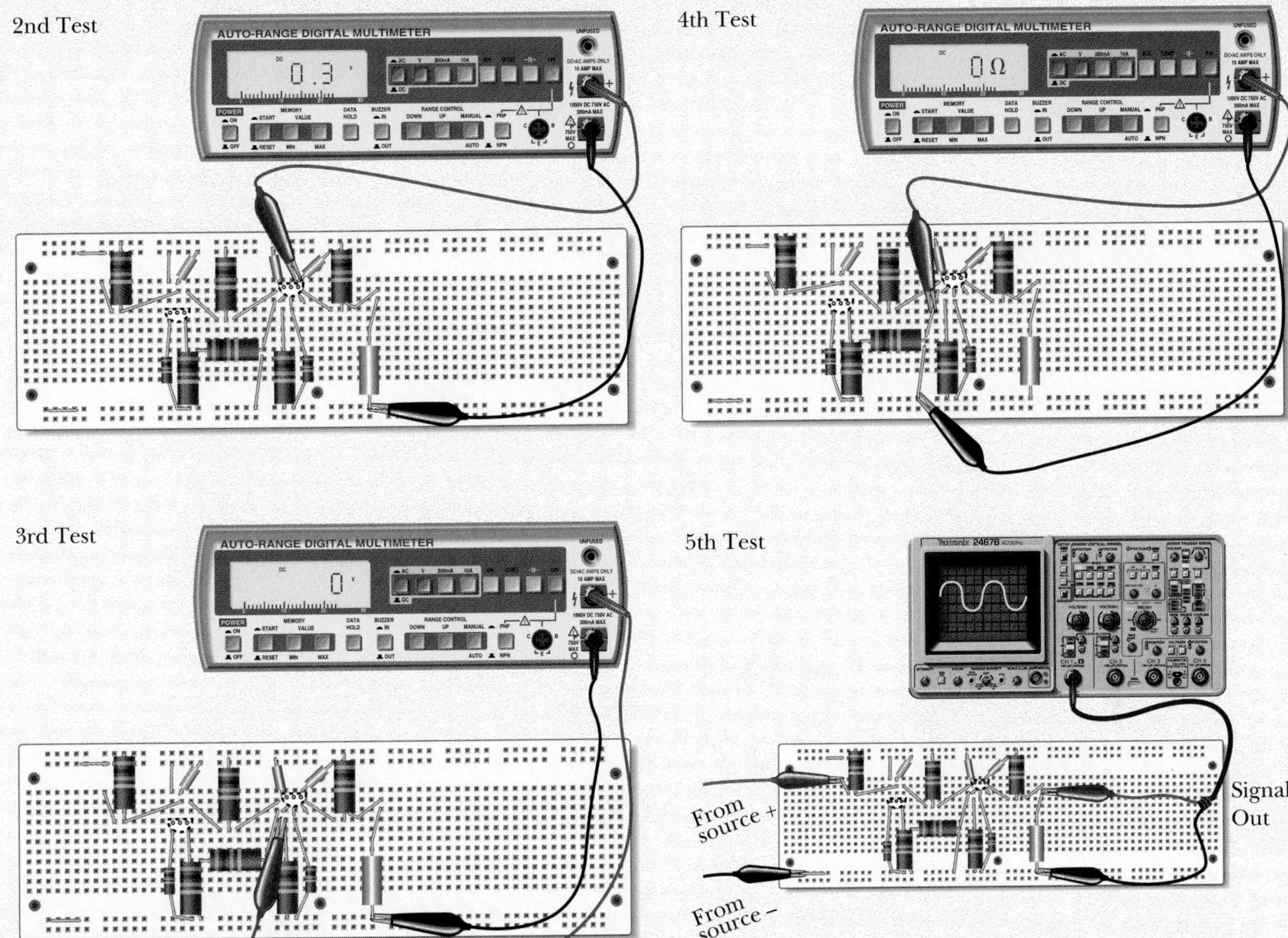
2nd Test
AUTO-RANGE DIGITAL MULTIMETER
0.3
4th Test
AUTO-RANGE DIGITAL MULTIMETER
0 Ω
3rd Test
AUTO-RANGE DIGITAL MULTIMETER
0
5th Test
From source +
From source –
Signal Out

Troubleshooting Challenge

CHALLENGE CIRCUIT 10 (Block Diagram)

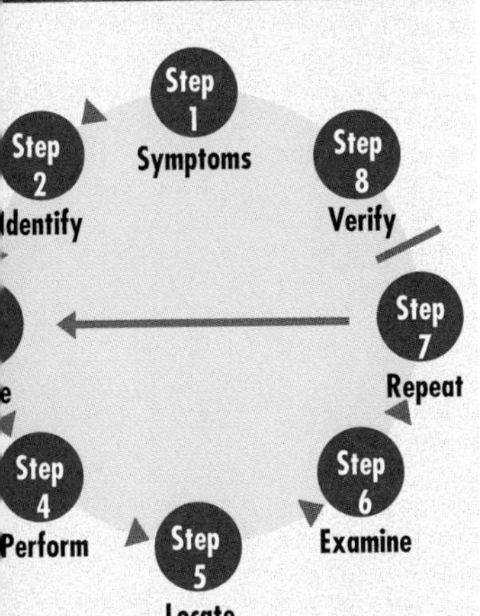

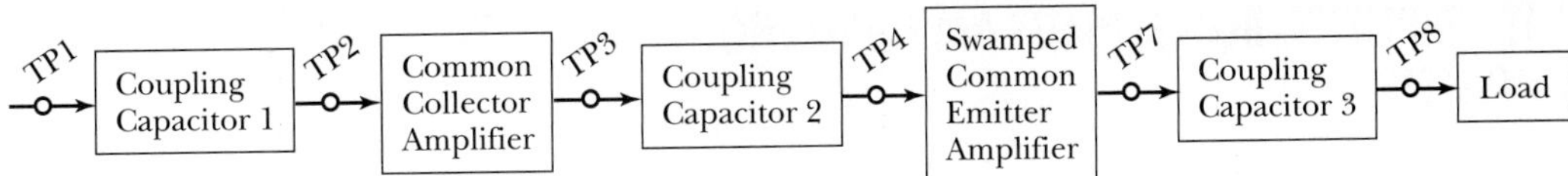

BIT Amplifier Block Diagram

General Testing Instructions:

Measurement
Assumptions:
Signal = from **TP** to ground
I = at the **TP**
V = from **TP** to ground
R = from **TP** to ground (with the power source disconnected from the circuit)

Possible Tests & Results:
Signal (normal, abnormal, none)
Current (high, low, normal)
Voltage (high, low, normal)
Resistance (high, low, normal)

Starting Point Information	Test Points	Test Results in Appendix C *V*	*I*	*R*	*Signal*
At TP_1	TP2	(263	NA	NA	264)
	TP3	(265	NA	NA	266)
V = normal	TP4	(267	NA	NA	268)
I = NA	TP7	(269	270	NA	271)
R = NA	TP8	(272	273	274	275)
Signal = Normal					

BJT AMPLIFIER BLOCK DIAGRAM CHALLENGE PROBLEM

STEP [1]

SYMPTOMS At TP8, output signal is distorted, and with the power source disconnected, resistance to ground measures normal.

STEP [2]

IDENTITY initial suspect area. Since the supply voltage is normal, the problem could be caused by a failure of the coupling capacitors, the common collector amplifier, or the swamped common-emitter amplifier.

STEP [3]

MAKE test decision based on the symptoms information. Let's start with the voltage measurements at strategic test points.

STEP [4]

PERFORM First Test: Voltage waveform at point TP7. Waveform shows a clipped sine wave riding on a low dc voltage. This indicates that the biasing circuit for the swamped common-emitter amplifier is not functioning properly.

STEP [5]

LOCATE new suspect area. The swamped common-emitter amplifier is suspect.

STEP [6]

EXAMINE available data.

STEP [7]

REPEAT analysis and testing.
Second Test: Check the waveform at TP4. The waveform shows a nondistorted sine wave riding on a dc voltage slightly below normal.
Third Test: Remove the swamped common-emitter amplifier. Check the waveform at TP3. The signal is normal. The swamped common emitter contains a faulty component.

STEP [8]

VERIFY Fourth Test: After replacing the swamped common-emitter amplifier with a new module that meets specifications, all circuit parameters come back to their norms.

Troubleshooting Challenge

CHALLENGE CIRCUIT 11
Find the Fault

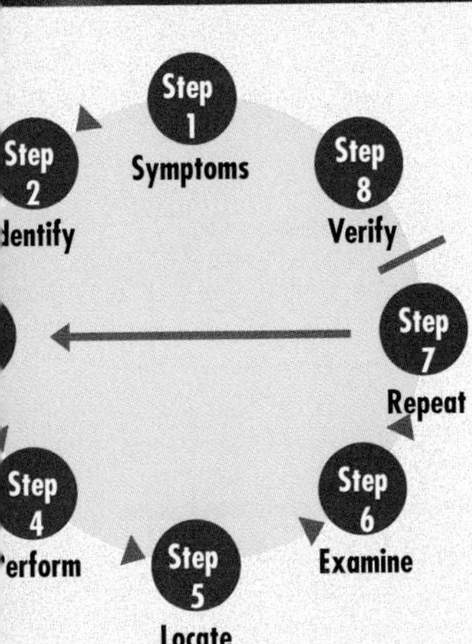

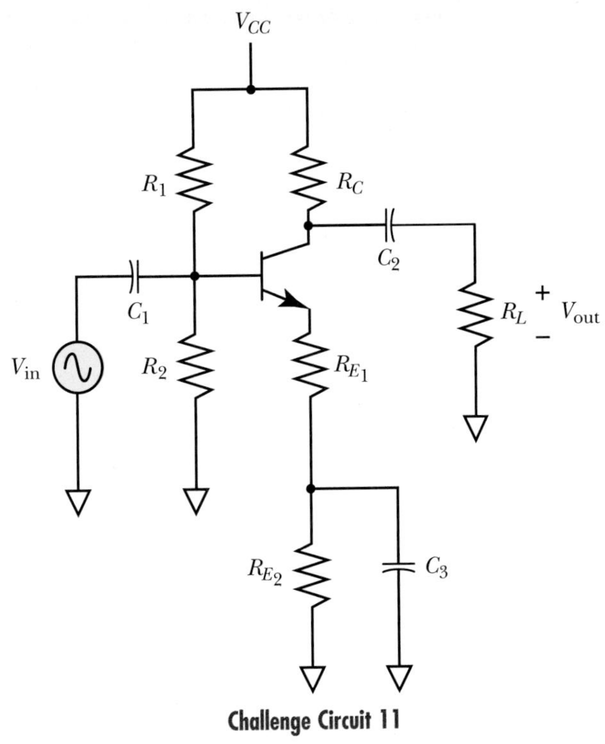

Challenge Circuit 11

STARTING POINT INFORMATION

1. Circuit diagram
   ($R_1 = 10$ kΩ, $R_2 = 3.3$ kΩ, $R_C = 2$ kΩ, $R_{E_1} = 200$ Ω, $R_{E_2} = 1$ kΩ, $R_L = 100$ kΩ)
   ($C_1 = 2.2$ μF, $C_2 = 2.2$ μF, $C_3 = 10$ μF)
2. $V_{CC} = 12$ V
3. $V_{in} = 0.4\ V_{pp}\ f = 1.0$ kHZ
4. Transistor 2N3904 ($100 < \beta < 400$)
5. Output voltage amplitude is too low.

Test	Circuit Measurements	Theoretical Expected Value
DC Voltages:		
V_{CC}	12 V	______
V_{in}	0 V	______
V_{out}	0 V	______
V_B	2.98 V	______
V_C	8.2 V	______
V_E	2.28 V	______
V_{RE_2}	1.9 V	______
V_{BE}	0.7 V	______

Test	Circuit Measurements	Theoretical Expected Value
V_{CE}	5.92 V	____
I_{CC}	2.8 mA	____
I_{R_1}	0.9 mA	____
I_{R_2}	0.9 mA	____
I_C	1.9 mA	____
I_E	1.9 mA	____
R_1	10 kΩ	____
R_2	3.3 kΩ	____
R_C	2.0 kΩ	____
R_{E_1}	200 Ω	____
R_{E_2}	1.0 kΩ	____
R_L	100 kΩ	____
C_1	2.2 µf	____
C_2	2.2 µf	____
C_3	open	____

Test	Circuit Measurements	Theoretical Expected Value
AC Voltages:		
V_{in}	0.4 V_{PP}	____
V_{out}	0.66 V_{PP}	____
V_{B}	0.4 V_{PP}	____
V_{C}	0.66 V_{PP}	____

STEP [1]

You will analyze the circuit to calculate the expected voltages, currents, and resistances for a properly functioning circuit. Record these values beside each test listed in the table provided. (Refer to Analysis Techniques)

STEP [2]

Starting with the symptom, document your steps for troubleshooting this circuit using the SIMPLER troubleshooting technique or instructions provided by your instructor.

STEP [3]

For each circuit measurement, you will indicate the test to be accomplished.

STEP [4]

Using the previous table, you will find the voltage and current measurements taken for the faulted circuit.

STEP [5]

You will now compare the faulted circuit measurement to the expected measurement for a properly functioning circuit. Are these values the same? (Refer to General Testing Instructions)

STEP [6]

You will repeat steps 2 through 5 until you have located the fault.

STEP [7]

When you have located the fault, you will identify the fault and the characteristics of the failure. (Refer to Types of Failures).

ANALYSIS TECHNIQUES (Check with your instructor)

METHOD [1]

Using circuit theory and algebra, evaluate the circuit.

METHOD [2]

Using an Excel spreadsheet for this circuit, evaluate the circuit.

METHOD [3]

Using MultiSIM, build the circuit schematic and generate the results.

METHOD [4]

Assemble this circuit on a circuit board. Apply power and make the appropriate measurements using your DMM and oscilloscope.

General Testing Instructions:

Measurement
Assumptions:
Signal = from **TP** to ground
I = at the **TP**
V = from **TP** to ground
R = from **TP** to ground
(with the power
source disconnected
from the circuit)

Possible Tests & Results:
Signal (normal, abnormal, none)
Current (high, low, normal)
Voltage (high, low, normal)
Resistance (high, low, normal)

Types of Failures

Component failures
- wrong part value
- part shorted
- part open

Supply voltage—incorrect voltage

Ground—floating

OUTLINE

OBJECTIVES

After studying this chapter, you should be able to:

1. Describe the **semiconductor structure** and identify the schematic symbols for N- and P-channel **JFETs, D-MOSFETs,** and **E-MOSFETs**
2. Determine the proper voltage polarities for operating **N-** and **P-channel FETs**
3. Explain the difference between depletion and enhancement modes of operation for FETs
4. Identify and explain the operation of **common-source, common-drain,** and *common-gate FET amplifier* circuits
5. Name some common practices for storing and handling MOSFET devices that ensure they are not destroyed by static electricity

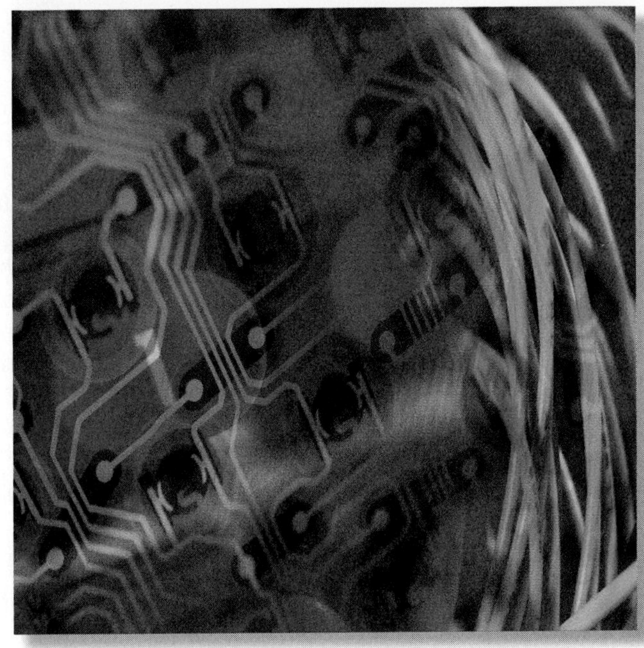

CHAPTER 6

Field-Effect Transistors and Circuits

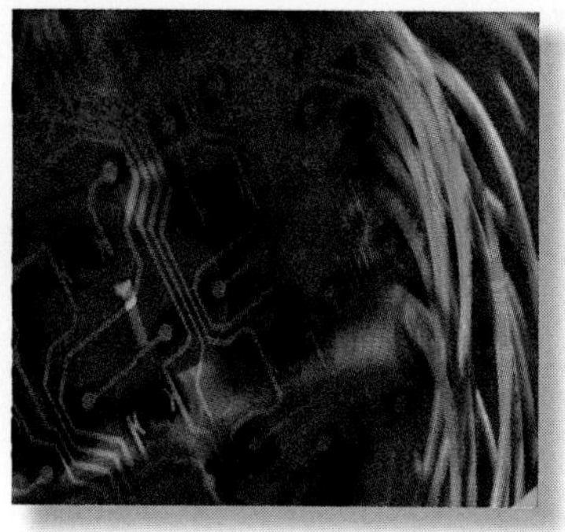

PREVIEW

The bipolar junction transistors (BJTs) you learned about earlier depend on current flow by means of two different charge carriers, namely, holes and electrons. The need for two different charge carriers is why such devices are called *bipolar.* By contrast, the transistor devices introduced in this lesson are *unipolar.* That is, current through the device is carried by just one type of charge carrier—either holes or electrons, depending on the type of semiconductor being used.

You will see how these unipolar devices control electron flow through them by means of applied electrical fields. Recall that BJTs control current flow by means of another path for current flow. Unipolar semiconductor devices that control their current flow by means of electrical fields are called **field-effect transistors** (abbreviated FETs).

You will learn about two important families of FETs: junction field-effect transistors and MOS field-effect transistors. You will see how each type is biased for use in basic types of amplifier circuits, and you will learn how the circuits operate.

KEY TERMS

Common-drain amplifier
Common-gate amplifier
Common-source amplifier
Depletion mode
Drain
Enhancement mode
Field-effect transistor (FET)
Gate
Junction field-effect transistor (JFET)
Metal-oxide semiconductor FET (MOSFET)
Pinch-off voltage, V_P
Self-bias
Source
Transconductance, g_m

6–1 Junction Field-Effect Transistors

Figure 6–1 shows a simplified drawing of the semiconductor structure and symbols for a type of FET known as a **junction field-effect transistor,** or JFET. It is called a junction device because it contains a single P-N junction (which is always reverse biased under normal operating conditions).

The JFET has three external connections that are called the **gate** (G), **source** (S), and **drain** (D).

The source connection on a JFET is the connection that provides the source of charge carriers for the channel current. The drain connection acts as the place where the charge carriers are removed (or "drained") from the device. It doesn't matter whether you are working with an N-channel or a P-channel JFET, the charge carriers always flow from the source connection to the drain connection. Now, let's consider this important point in greater detail.

Remember from your earlier lessons on semiconductor materials that current flows through a solid block of N or P materials. The sample is neither a good conductor nor an insulator—it is a *semi*conductor. This is true for JFETs. And you can see from the structures in Figure 6–1 that the channel running between the source and drain connections are simply solid pieces of N or P semiconductor material. So if you just apply a negative voltage to the source terminal of an N-type JFET and a positive voltage to its drain terminal, the device will conduct current by means of electron carriers that flow from source to drain, Figure 6–2a. Or if you are working with a P-channel FET and you apply a positive voltage to the source connection and a negative voltage to the drain terminal, you will get current flow through the channel. In this case, holes carry the current charges, and the holes flow from source to drain, Figure 6–2b. The whole purpose behind the operation of JFETs is to control the flow of charge carriers through the channel material.

Figure 6–3 shows how an N-channel JFET is to be biased. Notice in the structure diagram that a dc voltage source, V_{DD}, is connected between the source and drain terminals. V_{DD} is connected so that its positive polarity is to the JFET's drain connection and its negative polarity is to the source connection. This lines up with what you just

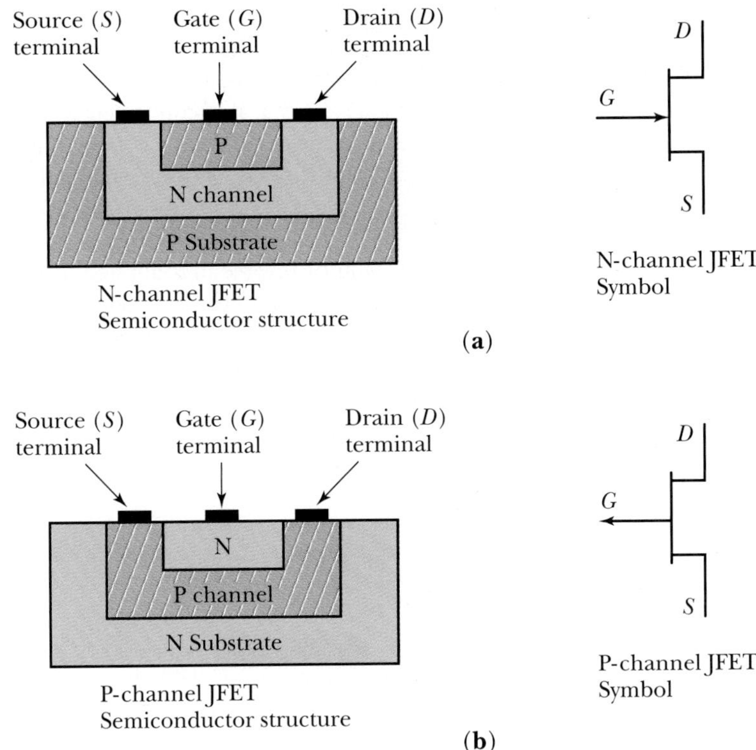

FIGURE 6–1 JFET semiconductor structures and symbols: (a) N-channel FET; (b) P-channel FET

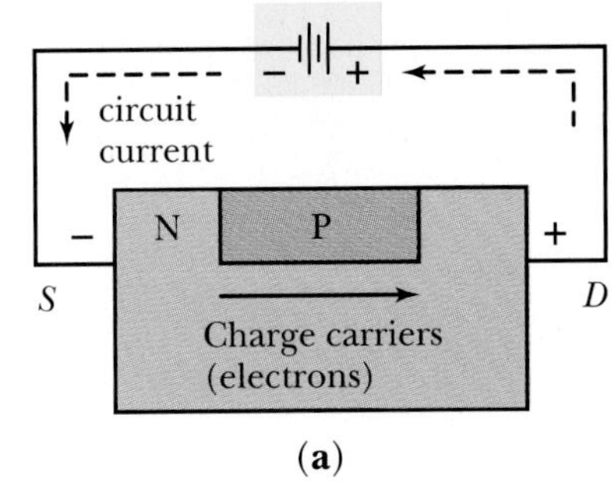

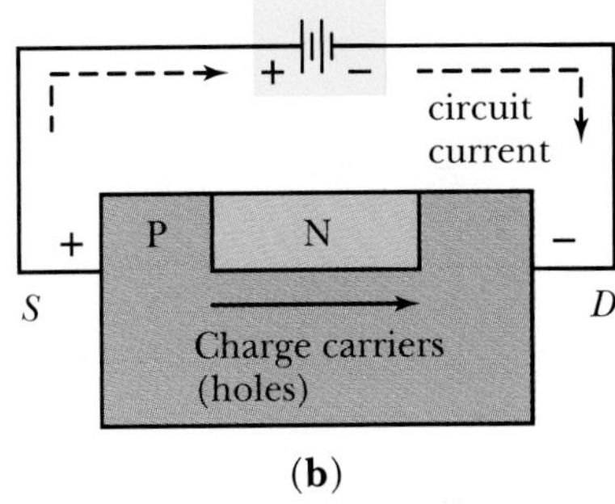

FIGURE 6–2 Charge carrier flow through JFET channels: (a) electron carriers in an N channel; (b) hole carriers in a P channel

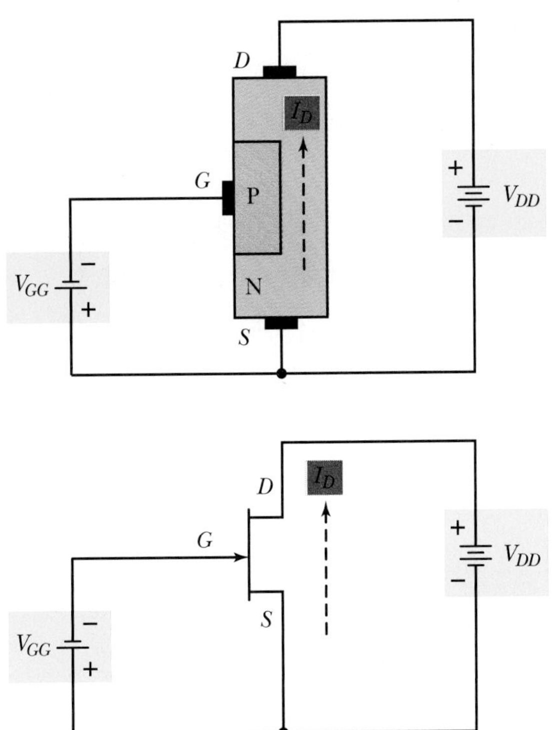

FIGURE 6–3 Bias voltages for an N-channel JFET

learned about the polarity and direction of current through an N-channel device. The second voltage source (V_{GG} connected between the source and gate terminals) is the one that controls the amount of current that flows through the channel.

One of the most important things to notice about V_{GG} is that it is connected with a polarity that reverse biases the P-N junction between the gate and source connections. This means there is nothing but reverse leakage current flowing between the source and gate (we usually consider there is no reverse current at all). More important, however, is the effect that the P-N depletion region has on the flow of charge carriers through the channel, Figure 6–4. The larger the amount of reverse bias applied to the gate-source junction, the farther the depletion region extends into the channel; and the farther the depletion region extends into the channel, the higher the resistance to current flow through the channel. The current flowing through the channel is called the *drain current,* and its symbol is I_D.

When there is no voltage applied to the gate-source junction, drain current is maximum. As V_{GG} increases, the effective width of the channel decreases until V_{GG} reaches a point where it stops the drain current altogether.

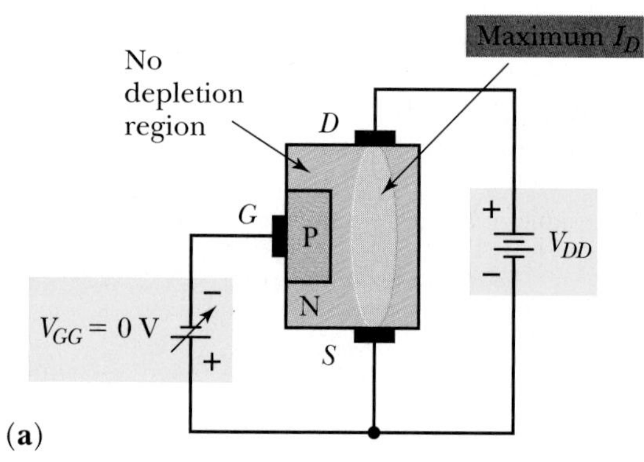

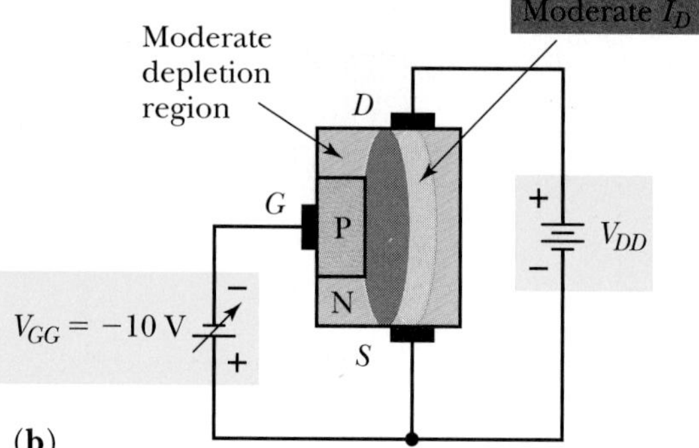

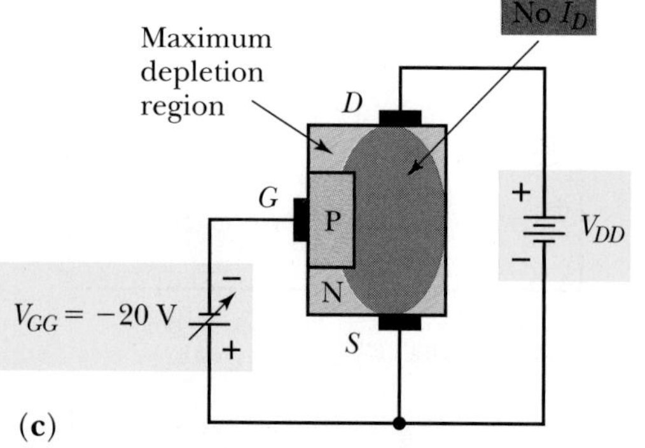

FIGURE 6–4 Depletion region in an N-channel JFET

Practical Notes

There are some important differences to notice between BJTs and JFETs:

1. BJTs use a current (I_B) to control a current (I_C), but JFETs use a voltage (V_{GG}) to control a current (I_D).
2. BJTs are turned off when there is zero base current ($I_B = 0$ A), but JFETs are turned on and maximum drain current flows when there is zero gate voltage ($V_{GG} = 0$).

In summary, BJTs are turn-on devices and JFETs are turn-off devices. BJTs and JFETs, however, are identical in appearance. They have three leads, they are provided in plastic and metal cases, and they may have part designations that begin with 2N. To distinguish BJETs from JFETs, you must have access to a catalog, data book, or other listing of devices that specifies which 2N designations belong to BJTs and which belong to JFETs.

6–2 JFET Characteristic Curves and Ratings

Figure 6–5 is a family of drain characteristic curves for an N-channel JFET. The JFET drain characteristic curves are similar to the BJT collector characteristic curves. Selecting any one of the drain characteristic curves, you can see that I_D increases with V_{DS} up to a certain point, then there is no more increase in I_D. What happens is that the channel becomes saturated with charge carriers, and the rate of flow changes very little as you increase or decrease the drain-source voltage. The value of V_{DS} where drain current goes into saturation is called the **pinch-off voltage, I_P**.

Notice that the amount of drain current (I_D) becomes less as the gate-source voltage (V_{GS}) increases from 0 V to –4 V. The maximum amount of drain current flows through the channel when V_{GS} is at 0 V. This particular drain current level has a symbol of its own, I_{DSS} (known as drain-source saturation current). The amount of gate-source voltage that is required for completely turning off the JFET in this example is about –4 V. This turn-off gate-source voltage level is denoted as $V_{GS_{(off)}}$.

Figure 6–6 shows a simple JFET amplifier. If the gate bias voltage (V_{GG}) is set to 0 V, the transistor will be normally conducting to its fullest extent. This means V_{DS}

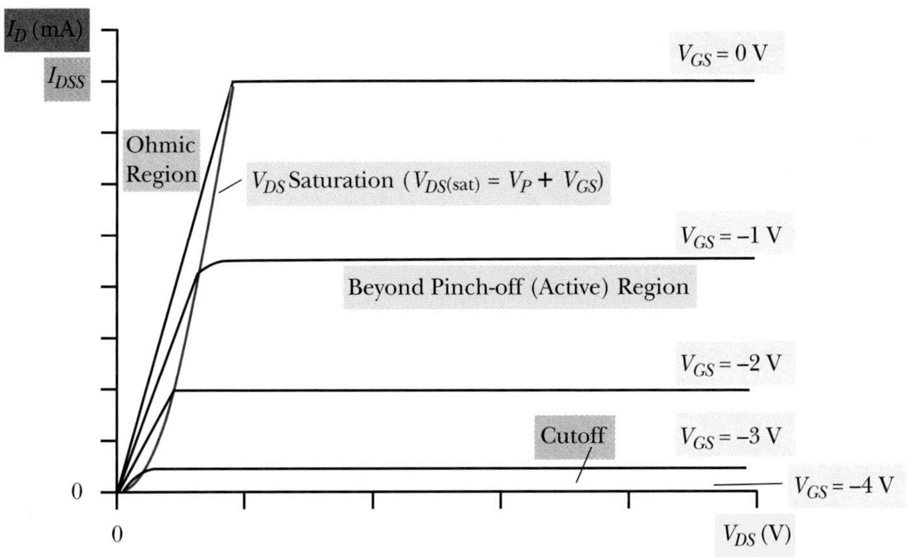

FIGURE 6–5 Drain characteristic curves for an N-channel JFET

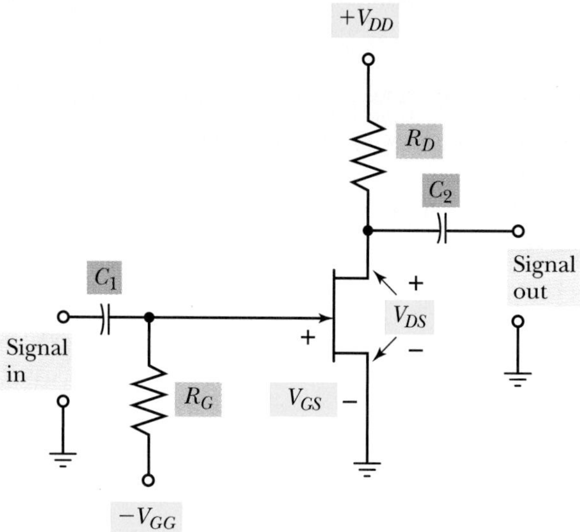

FIGURE 6–6 A simple N-channel JFET amplifier

will be at its lowest voltage level, and I_D at its highest. A negative-going input signal, however, will tend to decrease the conduction of the transistor, thereby increasing the V_{DS} voltage. This is not a very efficient way to operate an amplifier because drain current is maximum when there is no input signal. A better approach is to set the bias level, V_{GG}, to a negative value that will tend to hold down the normal conduction of the FET. An input signal can then aid or oppose the V_{GG} bias level and thereby control the amount of drain current accordingly. In this circuit, the more negative the input, the more positive the output. (You will learn more about FET amplifier circuits later.)

You may notice that we have used N-channel JFETs in most of the examples. Generally speaking, N-channel versions are more commonly in use. You should have no trouble learning about and working with P-channel versions as long as you remember that the only difference is the polarity of the supply and bias voltages.

Although drain characteristic curves are useful for explaining the operation of a JFET, a transconductance curve shows some of the important features even more clearly. The sample of a transconductance curve in Figure 6–7 shows how the drain current (I_D) decreases as the gate voltage (V_{GS}) is made more negative. The curve also shows maximum drain current (I_{DSS}) when V_{GS} is zero, and how drain current is stopped when V_{GS} reaches the $V_{GS(\text{off})}$ level. Shockley's equation (Formula 6–1) is the equation for the transconductance curve relating the drain current to the applied gate to source voltage.

FORMULA 6–1 $$I_D = I_{DSS}\left(1-\left(\frac{V_{GS}}{V_{GS(\text{off})}}\right)\right)^2$$ Shockley' s equation

Another important specification for an FET is its **transconductance** (g_m). This value is based on the FET's transconductance curve and provides an indication of how much it amplifies signals. Transconductance for an FET serves much the same function as beta for BJT. Transconductance is not a simple ratio like beta is, however. Formula 6–2 shows that transconductance for an FET is the ratio of a change in drain cur-

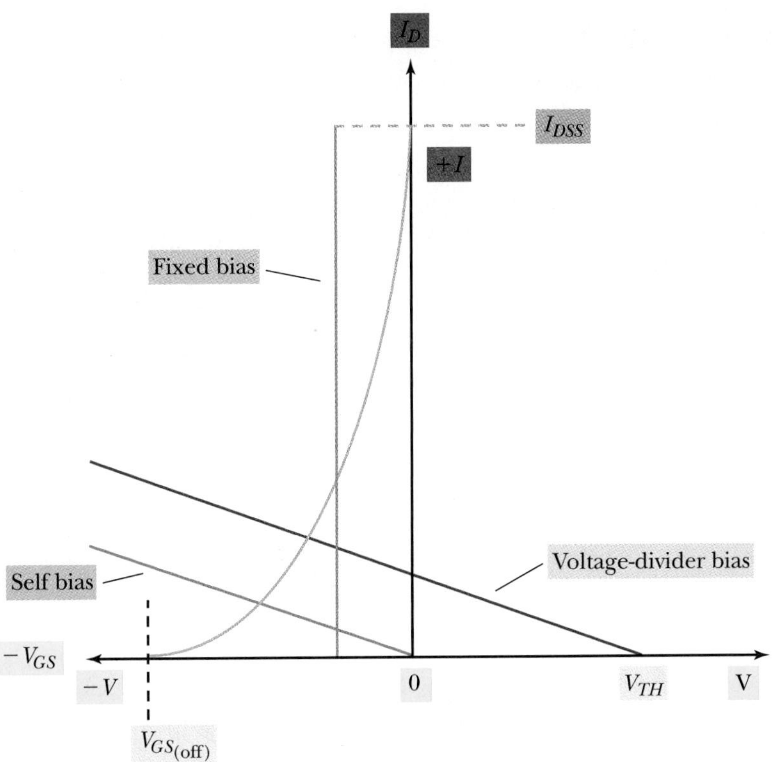

FIGURE 6–7 A typical JFET transconductance curve (active region)

rent to the corresponding change in gate-source voltage. (In mathematics, Δ symbolizes "a change in.") The unit of measurement for transconductance is the siemens. This is the same unit of measure for conductance you studied in an earlier lesson. And like conductance, the unit letter symbol for transconductance is *S* (Siemens) or ℧ (mhos).

FORMULA 6–2 $$g_m = \frac{\Delta I_D}{\Delta V_{GS}}$$

The maximum ratings for JFETs enable the engineer to design circuits that will not allow the device to be destroyed by reverse voltage, excessive voltage, or excessive current flow. The most critical maximum JFET ratings follow:

- Maximum gate-source voltage (V_{GS}). This is the maximum reverse-bias voltage that can be applied continuously to the P-N junction. Typical values are between 20 V and 50 V. (Recall that the gate-source junction is normally operated with reverse bias.)
- Maximum gate-drain voltage (V_{GD}). This is the maximum reverse-bias voltage between the gate and drain sections that can be applied continuously. The value is usually identical with the maximum gate-source voltage rating, typically 20 V to 50 V.
- Maximum drain-source voltage (V_{DS}). This is the maximum forward voltage that can be controlled. Exceeding this rating causes the channel to conduct fully, regardless of the amount of applied gate voltage. This basically states the largest amount of supply voltage to be used with the device, and it is on the same order as the other maximum voltage levels, between 20 V and 50 V.

■ **IN-PROCESS LEARNING CHECK 1**

Fill in the blanks as appropriate.

1. *JFET* is the abbreviation for ________.
2. The three terminals on a JFET are called _____, _____, and _____.
3. The arrow in the symbol for an N-channel JFET points _____ the source-drain connections, while the arrow for a P-channel JFET points ________ source-drain connections.
4. In an N-channel JFET, the supply-drain voltage must be connected so that the positive polarity is applied to the _____ terminal of the JFET and the negative polarity is applied to the _____ terminal.
5. In a P-channel JFET, the drain supply voltage must be connected so that the positive polarity is applied to the _____ terminal of the JFET and the negative polarity is applied to the _____ terminal.
6. The charge carriers in the channel material of a JFET always flow from the _____ terminal to the _____ terminal.
7. In an N-channel JFET, the gate supply voltage must be connected so that the positive polarity is applied to the _____ terminal of the JFET and the negative polarity is applied to the _____ terminal.
8. In a P-channel JFET, the gate-supply voltage must be connected so that the positive polarity is applied to the _____ terminal of the JFET and the negative polarity is applied to the _____ terminal.
9. V_{DS} stands for _____.

   V_{GS} stands for _____.

   I_D stands for _____.

   g_m stands for _____.
10. The greatest amount of current flows through the channel of a JFET when V_{GS} is at its _____ level, while the least amount of drain current flows when V_{GS} is at its _____ level.

6–3 Biasing JFET Amplifiers

JFETS are somewhat more difficult to bias than BJTs are. A BJT is normally nonconducting, and a base bias of the same polarity as the collector supply voltage is applied in order to increase conduction. A JFET, however, is normally conducting, and a gate bias having the opposite polarity from the drain supply voltage is required to achieve biasing effects. The need for a biasing potential that has a polarity opposite the polarity required for the drain-current circuit complicates JFET biasing circuits.

Bear in mind the following facts and assumptions while analyzing the JFET biasing circuits:

1. As long as the gate terminal has the proper polarity, there is no current flowing in the gate circuit.
2. Source current and drain current are the same.

Gate Bias

The most direct procedure for biasing a JFET amplifier is to apply a voltage to the gate that is opposite in polarity from the voltage applied to the drain. The drain supply voltage for an N-channel JFET is positive, for instance: so the amplifier can be biased by applying a negative voltage to the gate. This method is called the *gate bias* method. See the example in Figure 6–8.

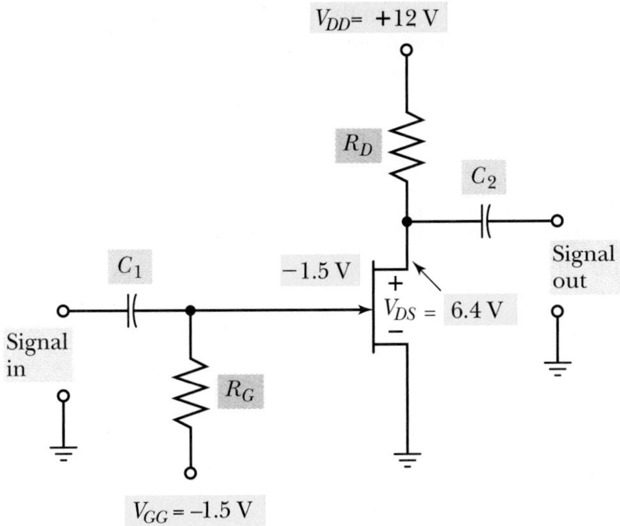

FIGURE 6–8 Gate-biased JFET amplifier

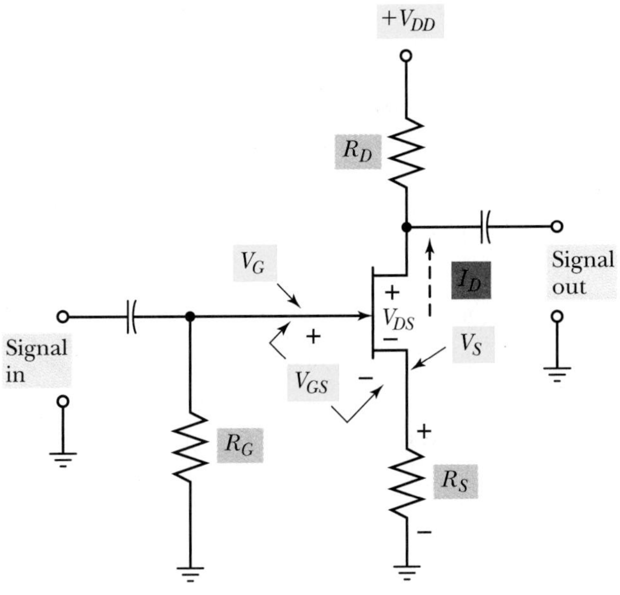

FIGURE 6–9 Self-biased JFET amplifier

The incoming ac signal passes through the input capacitor, where its voltage level is shifted by the value of V_{GG}, which increases and decreases the total voltage applied to the gate. The more positive the input signal goes, the more current the JFET conducts; and of course the more negative the input signal goes, the less the JFET conducts.

FORMULA 6–3 $V_{GS} = V_{GG}$

FORMULA 6–4 $V_{DD} - I_D R_D - V_{DS} = 0$

Self-Bias

Figure 6–9 shows a JFET with the gate grounded through resistor R_G and the source grounded through resistor R_S. With no signal applied, the voltage at V_G will be zero (because of its connection to ground through R_G), and the drain current will be maximum. The voltage at the source terminal will be determined by the amount of drain current and the value of R_S. In any event, there will be a voltage at the source terminal of the JFET that is more positive than the ground potential. There will be a difference in potential between the source and gate: the gate voltage will be zero and the source voltage some positive level. The gate, in other words, will be less positive (more negative) than the source; and that is exactly what is required for biasing an N-channel JFET amplifier. Because this kind of bias is generated by the FET's own current, it is called **self-bias.**

One of the important advantages of using self-bias is that it does not require a separate negative-voltage power supply as does the gate-biasing method. A disadvantage of self-bias is that the amount of bias varies with the amount of drain current.

FORMULA 6–5 $V_{GS} = -I_D R_S$

FORMULA 6–6 $V_{DD} - I_D R_D - V_{DS} - I_D R_S = 0$

Voltage-Divider Bias

The voltage-divider method of bias is illustrated in Figure 6–10. Resistors R_1 and R_2 make up a voltage divider that applies a bias voltage to the gate that has the same polarity as the drain supply. Recall this is not the proper polarity for biasing a JFET. However, the voltage-divider potential is combined with the self-bias potential found across the source resistor (R_S), which has the proper polarity for biasing. This type of circuit is designed so that the self-bias feature is balanced by the voltage-divider feature. The result is a stable biasing method that does not require a second power source that has a polarity opposite from V_{DD}.

FORMULA 6–7 $V_G = V_{TH} = V_{DD} \times \frac{R_2}{R_1 + R_2}$

FORMULA 6–8 $V_{GS} = V_G = I_D R_S$

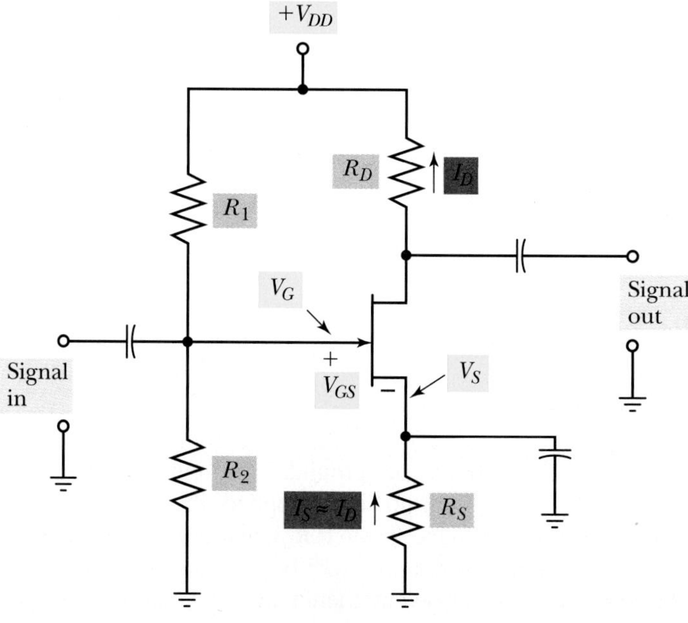

FIGURE 6–10 Voltage-divider biasing for JFET amplifiers

Application Problem

BIAS

A voltage-divider biasing technique is used in the circuit board. Figure 6–10 shows the circuit with $R_1 = 1\ \text{k}\Omega$ and $R_2 = 100\ \Omega$, and $V_{DD} = 12$ V. Calculate the power losses in the bias circuit.

Solution

$$P = \frac{V_{DD}^2}{R_1 + R_2} = \frac{12^2}{1000 + 100} = 131\ \text{mW}$$

That is quite a high number and we should choose larger resistors.

6–4 JFET Amplifier Circuits

The basic configurations for JFET amplifiers are the same as for BJTs. The names are slightly different, but the features are generally the same; **common-source,** *common drain,* and *common-gate.*

All of the JFET circuits shown to this point have been common-source (CS) JFET amplifier circuits. The basic features of CS JFET amplifiers correspond to the basic features you have studied for the common-emitter BJT amplifier:

1. The input signal is introduced into the gate circuit, and the output is taken from the drain circuit (the source is common to the input and output).
2. The input circuit can have a high impedance (because of the reverse-biased source-drain P-N junction). The actual input impedance of the circuit is determined by the values of the resistors used in the gate biasing.
3. The output circuit is a high impedance one, determined by the value of the drain resistor.
4. The circuit provides voltage, current, and power gain. Current and power gains are extremely high.
5. There is a 180° phase reversal between the input and output signals.

Figure 6–11 shows the JFET version of a common-collector amplifier, which is called a **common-drain** (CD) amplifier. The common drain is also called a source follower. The basic features of practical CD amplifiers include:

1. The input signal is introduced into the gate, and the output is taken from the source (the drain is common to both the input and output).
2. The input circuit can have a high impedance, but the actual input impedance of the circuit is determined by the values of the resistors used in the gate biasing.
3. The output circuit is a low impedance one, determined by the value of the source resistor and the g_m of the FET.
4. The circuit provides voltages gain less than 1.
5. The circuit provides high current and power gains.
6. There is no phase reversal between the input and output signals.

The practical **common-gate** (CG) Amplifier in Figure 6–12 has the following features:

1. The input signal is introduced into the source, and the output is taken from the drain (the gate is common to the input and output).
2. the input impedance can be fairly low as determined by the values of R_S, R_D and g_m.
3. The output circuit has high impedance.
4. The circuit provides good voltage and high power gains. Current gain is always less than 1.
5. There is no phase reversal between the input and output signals.

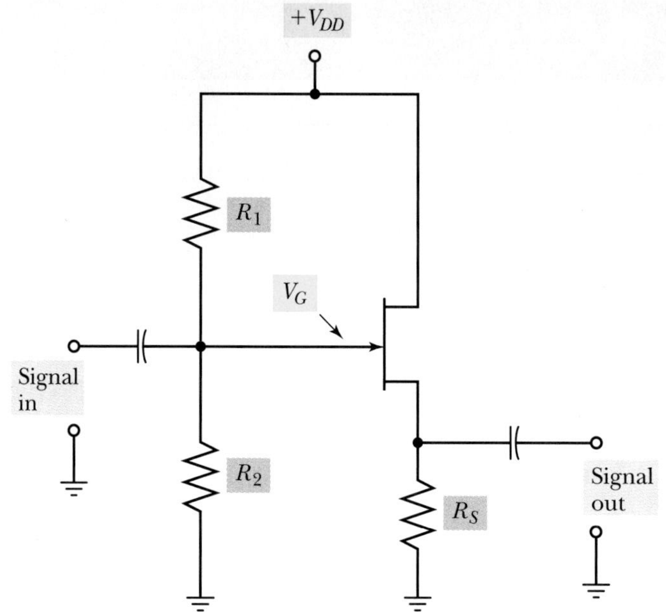

FIGURE 6–11 A common-drain JFET amplifier

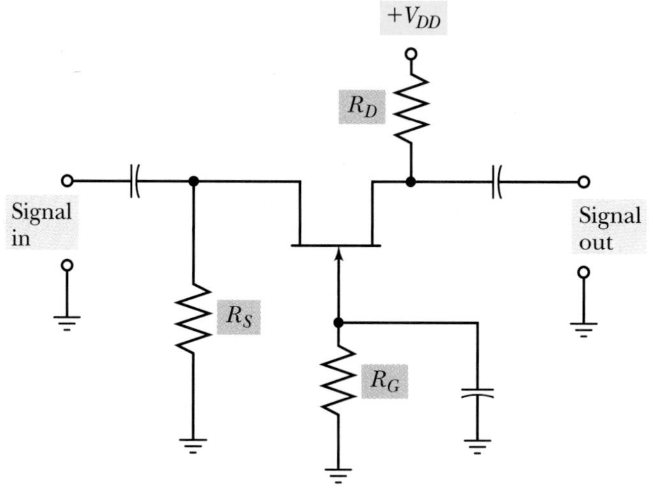

FIGURE 6–12 A common-gate JFET amplifier

■ IN-PROCESS LEARNING CHECK 2

Fill in the blanks as appropriate.

1. When a polarity opposite the drain polarity is applied to the gate of a JFET through a large-value resistor, the bias method is called _______ bias.
2. When the gate and source of JFET are both connected to ground through resistors, the bias method being used is called _______.
3. When the gate of a JFET is connected to a voltage divider and the source is connected through a resistor to ground, the bias method being used is called _______ bias.
4. The common _______ FET amplifier has the input signal applied to the source terminal and the signal taken from the drain.
5. The common _______ FET amplifier has the signal applied to the gate terminal and the output signal taken from the drain.
6. The common _______ FET amplifier has the signal applied to the gate terminal and taken from the source terminal.

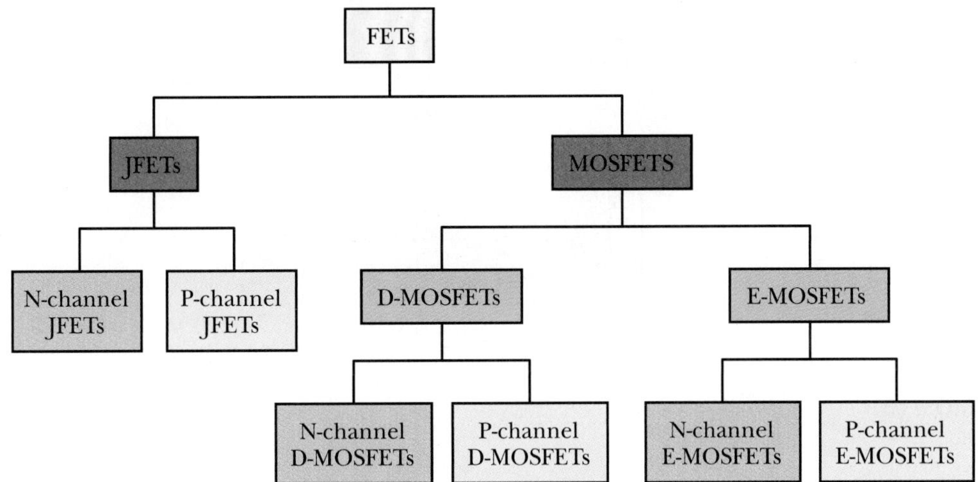

FIGURE 6–13 The "family tree" of FETs

6–5 MOS Field-Effect Transistors

You have just learned how JFETs use the voltage of a reverse-biased P-N junction to control a current flow through a solid piece of semiconductor material. There is a second kind of FET technology that also uses a voltage to control a current. This is MOS technology. MOS stands for metal-oxide semiconductor. Field-effect transistors that operate according to this technology are called **metal-oxide semiconductor FETs,** or MOSFETs. (This device technology has also been called IGFET—insulated gate field effect transistor.)

There are two types of MOSFETs. These two types of MOSFETs differ according to their structure and according to the way their controlling voltage affects the flow of current. The two types of MOSFETs are known as *enchancement type* (E-MOSFET) and *depletion type* (D-MOSFET). You will be studying both types of MOSFETs in this chapter. But first, let's look at a diagram that will aid your understanding of these families of MOSFET devices as you further your study, Figure 6–13.

The devices described in this chapter are all field-effect transistors, or FETs. The diagram shows that there are two different classifications of FETs: JFETs (which you have already studied) and MOSFETs (which you are about to study). Then the JFET family is further classified into N-channel and P-channel devices. The MOSFET family has an additional "generation." MOSFETs are basically divided into E-MOSFETs and D-MOSFETs—of which there are both N- and P-channel versions.

There are other types of MOSFET devices available today, but they work by the same principles shown in this chapter. Once you know how the "family tree" of FETs work, you will have no trouble learning about the more specialized types that are not shown here.

6–6 D-MOSFETs

Figure 6–14 shows the elemental structure and symbol for depletion-type MOSFETs. Notice that the source and drain terminals are connected to a "straight-through" piece of semiconductor material. In the case of an N-type D-MOSFET, the charges are carried by electrons in the N-type material. In the case of a P-type D-MOSFET, the charges are carried by holes through a P-type material.

The main difference between these MOSFETs and a typical JFET is the manner in which the gate terminal is separated from the channel. In a MOSFET, gate-channel separation is enabled by a very thin layer of a metal oxide known as silicon dioxide. Silicon dioxide is a extremely good insulator material. Thus, you can see that there

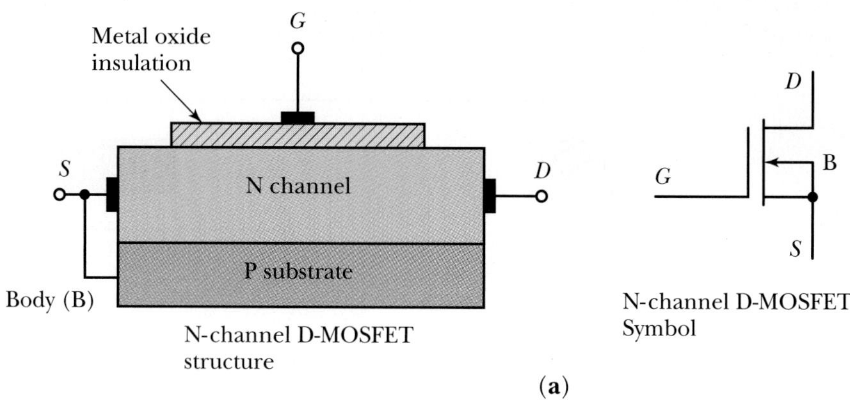

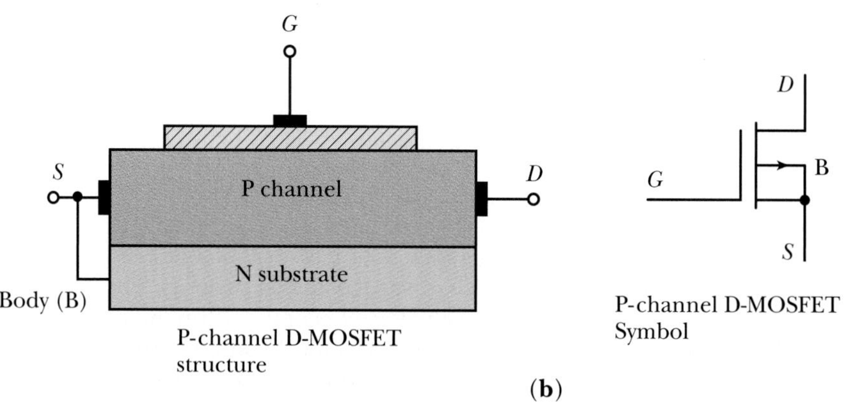

FIGURE 6–14 D-MOSFET structures and symbols

will be no current flow between the gate terminal and the channel. However, as in a JFET, a voltage applied to the gate of a MOSFET greatly influences the amount of current flowing through the channel.

The electronic symbols for D-MOSFETs show the gate symbol entirely separated from the channel part of the symbol. This emphasizes that there is no possibility of current flow between the gate and the channel.

The arrow part of the symbols for D-MOSFETs indicates the nature of the base material, or substrate. You can see from the structure diagrams in Figure 6–14 that the substrate is not electrically connected to any of the terminals. It mainly serves as a foundation material for the channel, and it is always of the opposite type. For the N-type D-MOSFET, for instance, the substrate is a P-type material. The inward-pointing arrow in the symbol for an N-channel D-MOSFET indicates that the substrate is a P-type material (remember "*P*ointing in = *P*-type"). An outward-pointing arrow signifies a P-channel MOSFET that is built upon an N-type substrate (remember "*N*ot pointing in = *N*-type").

The channel portion of the device is indicated on the symbol as an unbroken line between the source and drain connections. This reminds us that the D-MOSFET uses an uninterrupted piece of one type of semiconductor material between the gate and drain.

Now let's consider how a D-MOSFET operates and where the "D" designation comes from.

D-Mode of Operation

D-MOSFETs are specifically designed to operate in the **depletion mode.** Figure 6–15 shows what is meant by this. Notice the polarity of the biasing voltages:

1. The V_{DS} polarity is determined by the fact that the device in this example has an N-type material for a channel. In order to satisfy the rule that charge carriers must flow from source to drain (and since the carriers in N-type semiconductors are elec-

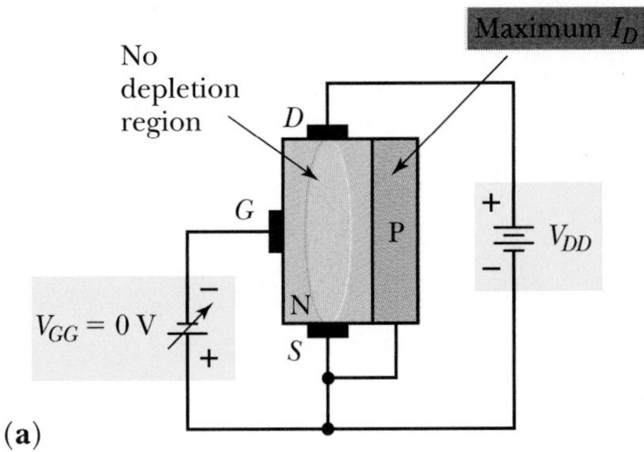

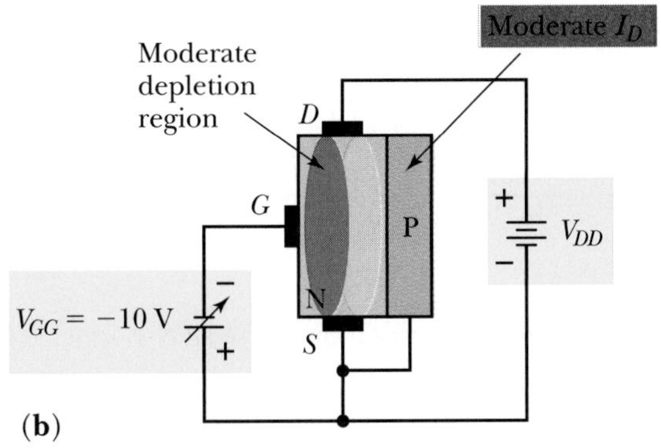

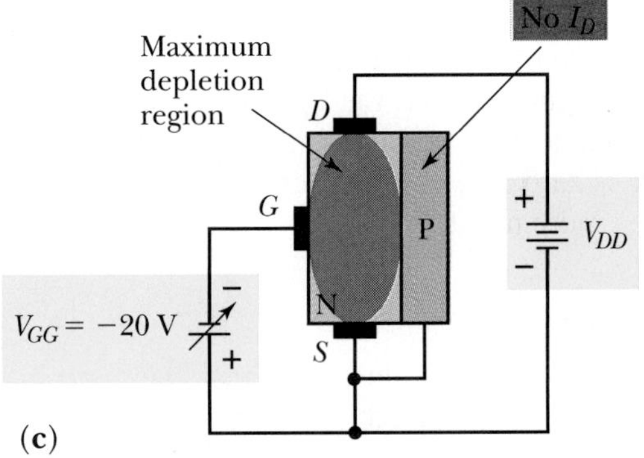

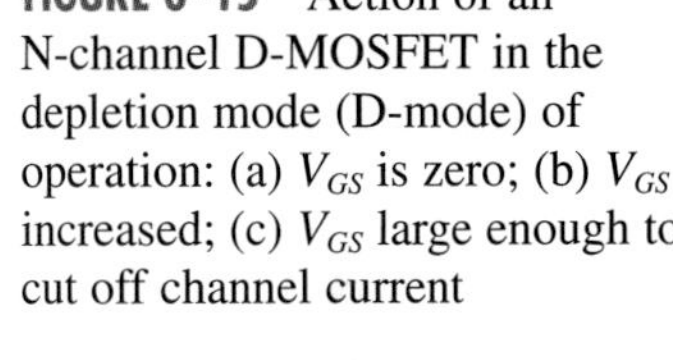
FIGURE 6–15 Action of an N-channel D-MOSFET in the depletion mode (D-mode) of operation: (a) V_{GS} is zero; (b) V_{GS} increased; (c) V_{GS} large enough to cut off channel current

trons), it follows that the positive supply voltage is connected to the drain terminal and the negative to the source terminal.

2. The V_{GS} polarity is shown here as positive voltage to the source terminal and negative to the gate terminal. This is what determines the depletion mode of operation.

While V_{GG} is at 0 V, Figure 6–15a, the charge carriers have a clear path through the channel, and drain current (I_D), flows freely. As V_{GG} is increased, however, electron carriers in the channel are forced away from the gate region, creating a depletion region of increasing size, Figure 6–15b. And as the size of this depletion region increases, the size of the charge-carrying channel decreases. Current through the channel can be completely cut off by making V_{GG} sufficiently large. Figure 6–15c. This is called the depletion mode of operation because V_{GG} controls channel current by depleting the number of available charge carriers.

Figure 6–16 shows the family of drain characteristic curves for an N-channel D-MOSFET operating in the depletion mode. The curves are virtually the same except for their drain current levels. Pick any one of them, and you can see that drain current increases rapidly with drain-source voltage at first. But the drain current soon reaches

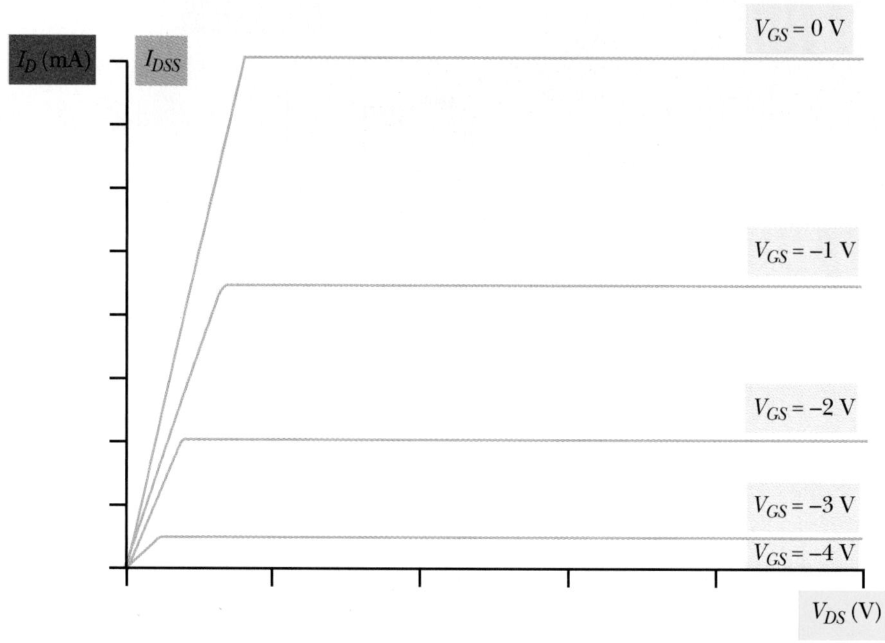

FIGURE 6–16 Family of drain characteristic curves for an N-channel D-MOSFET operating in its depletion mode

a saturation point where the only way to get more current flowing is to make the gate voltage less negative.

It is this depletion mode of operation that accounts for the "D" in the name D-MOSFET. And can you see that the operation of a D-MOSFET in its depletion mode works much like a JFET? Notice that drain current flows readily through the channel when there is zero gate voltage, but the current drops off dramatically as the gate voltage is taken more negative.

E-Mode of Operation

D-MOSFETs are very special in the sense that they can also be operated in an **enhancement mode.** Let's see what this means.

Figure 6–17 shows an N-channel D-MOSFET that looks like the basic circuit in Figure 6–15, but has the polarity of V_{GG} reversed. Instead of restricting current flow through the channel, V_{GG} in the enhancement mode actually increases the effective size of the channel and *enhances* the flow of current through the channel. The positive gate voltage actually allows the channel to widen into the substrate material by forcing substrate carriers (holes in the case of a P-type substrate) away from the channel region. Figure 6–18 shows the enhancement-mode portion of the drain characteristic curves.

Figure 6—19 is a complete set of drain curves for an N-channel D-MOSFET. Note that the range of gate voltages is from +2 V (maximum drain current) to –4 V (cutoff). None of the other FETs discussed can be used in both a depletion and an enhancement mode. JFETs cannot be used in an enhancement mode because attempting to do so would simply forward bias the P-N junction and would not enable a voltage field that could control drain current. You will soon see that the structure of E-MOSFETs makes depletion-mode operation impossible. In this sense, the D-MOSFET is an extremely useful device, especially when you consider it can amplify true ac waveforms without the need for clamping the signal all positive or all negative.

The transconductance curve, Figure 6–20, for an N-channel D-MOSFET is very distinctive because the curve covers the – V_{GS} part of the axis as well as the + V_{GS} part of the axis. This emphasizes that a D-MOSFET can be operated in both the depletion (– V_{GS} part of the axis) and enhancement (+ V_{GS} part of the axis) modes.

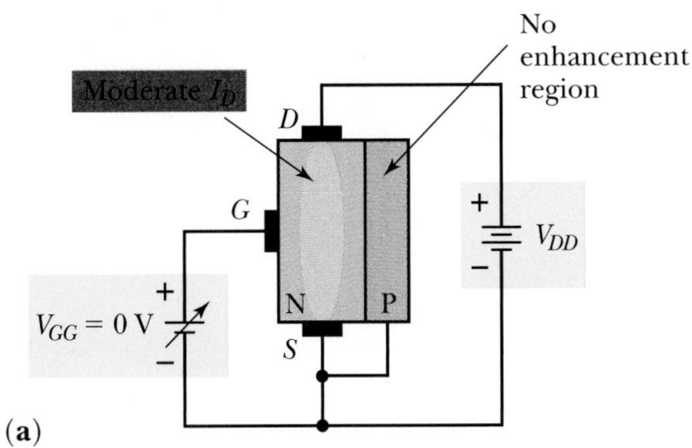

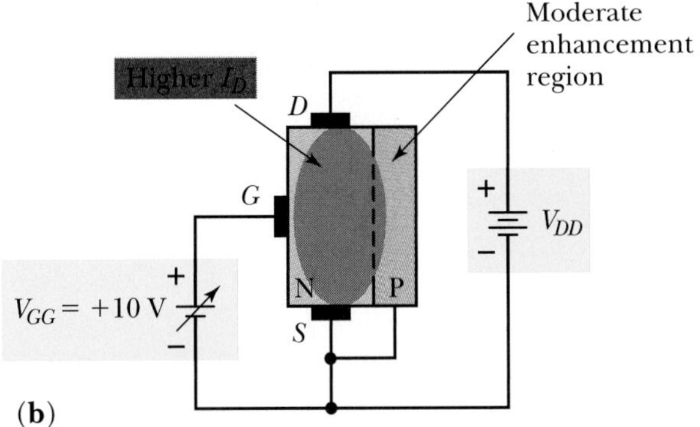

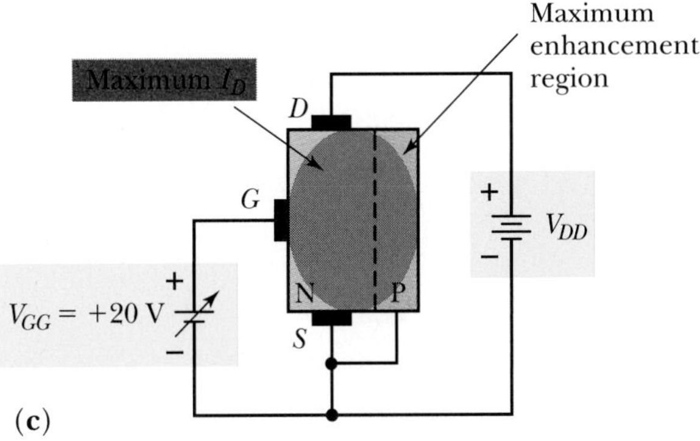

FIGURE 6–17 Action of an N-channel D-MOSFET in the enhancement mode (E-mode) of operation: (a) V_{GG} is zero; (b) V_{GG} increased; (c) V_{GG} further increased

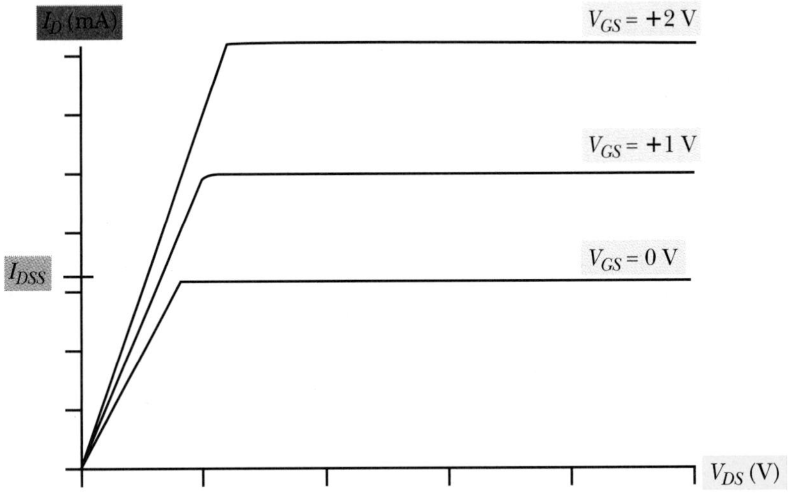

FIGURE 6–18 Family of drain characteristic curves for an N-channel D-MOSFET operating in its enhancement mode

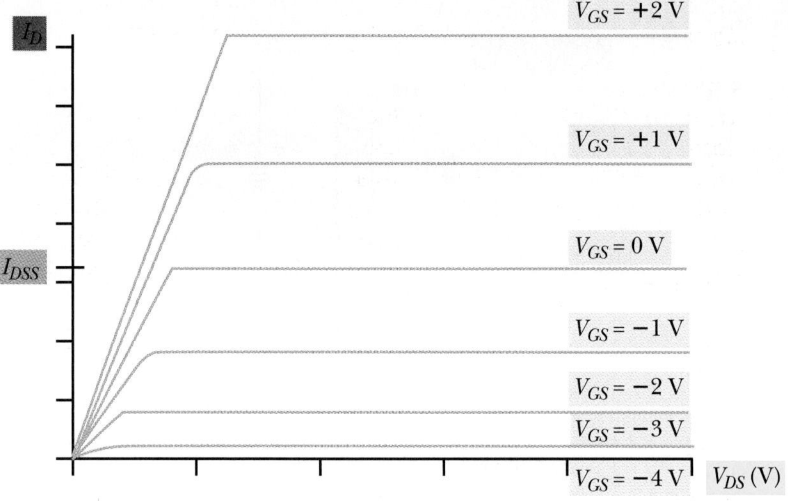

FIGURE 6–19 Complete family of drain characteristic curves for an N-channel D-MOSFET

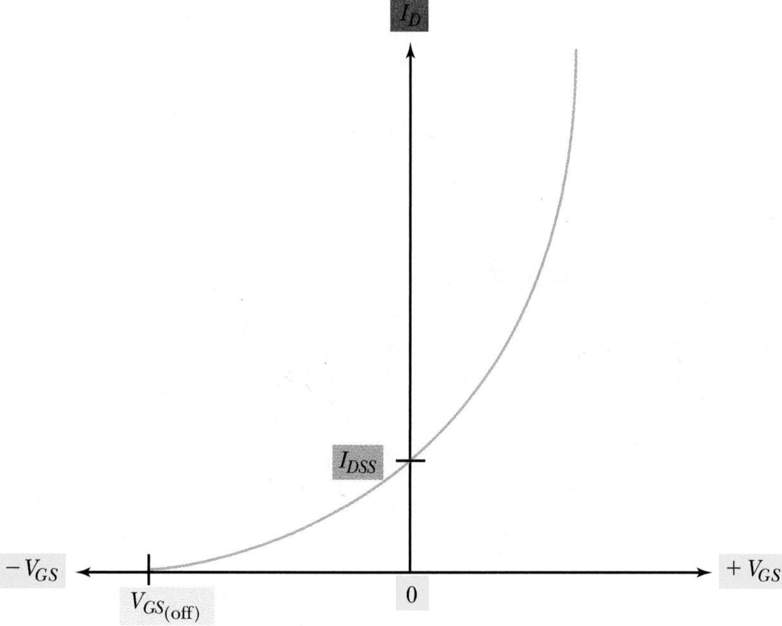

FIGURE 6–20 A typical N-channel D-MOSFET transconductance curve (active region)

A simple amplifier circuit using an N-channel D-MOSFET is shown in Figure 6–21. By having no dc gate bias, the transistor is set up for Class A operation. Recall this means the entire input waveform (both positive and negative phases) is amplified without distortion. In this amplifier, the positive portion of the input signal causes an increase in enhancement-mode drain current and a corresponding decrease in V_{DS} voltage. So as the input signal goes more positive, the output signal goes less positive. And when a negative portion of the input signal occurs, the depletion mode goes into effect and decreases the channel conduction. The result is an increase in V_{DS} voltage. As the input signal goes more negative, the output signal goes more positive.

Common-drain and common-gate D-MOSFET amplifiers are possible, and they have the same general characteristics as the JFET amplifiers you studied earlier. Also, the examples used in this section are all of N-channel D-MOSFETS. P-channel D-MOSFETs operate the same way as the N-channel versions, but with the polarities of V_{DS} and V_{GS} reversed.

Application Problem

General Electronics

MOTORCYLE REGULATOR

A motorcyle uses a 6 V electrical system. The alternator generates a voltage that depends on the RPM of a motorcycle engine. That voltage varies from 5.5 V to 7.1 V. The motorcycle voltage regulator is a bank of several MOSFETs. The transistor drain characteristics are shown in Figure 6–18 on page 269. Assume that the transistors operate at 70 to 90% of linear region, V_{DS} is equal to the generator voltage, and the gate voltage is 2 V. One day we turn the ignition key ON and we realize we have a dead battery and we connect a battery charger to the battery leads. But we forget two things: to disconnect one wire between the battery and the motorcycle circuit, and to flip the switch from 12 V to 6 V on the battery charger panel. Estimate a current increase through the transistors. What will happen to the transistors?

Solution The figure shows that the transistor operates in its linear range. The transistor saturates when the charger voltage is 13 V. That temporarily protects the circuit, but longer exposure would increase heat dissipation inside the transistor and the transistor would fail.

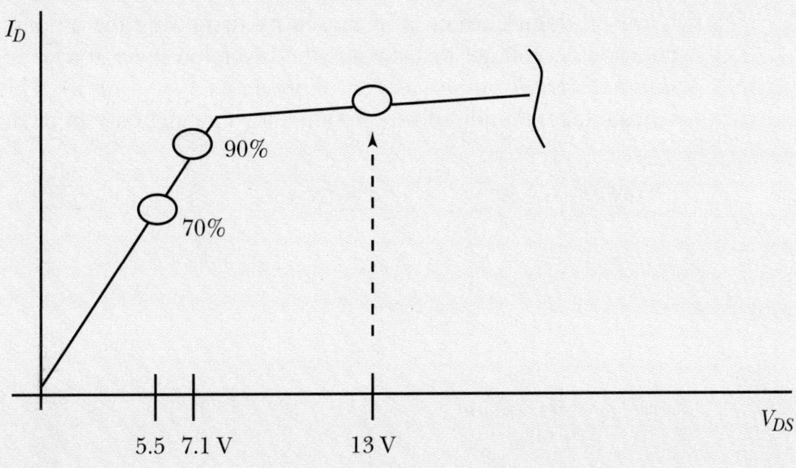

Drain transistor characteristic

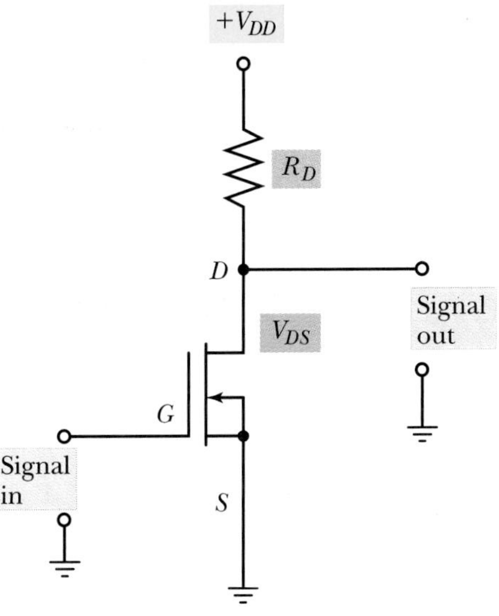

FIGURE 6–21 A simple Class A amplifier using an N-channel D-MOSFET

6–7 E-MOSFETs

Figure 6–22 shows the basic structure and symbol for enhancement-type ("E") MOSFETs. The channel between the source and drain terminals is interrupted by a section of the substrate material. This interruption is noted on the schematic symbols by showing the channel broken into three segments: one for the source, one for the substrate (arrow), and one for the drain.

Current cannot possibly flow between source and drain on this device when there is no gate voltage applied. Figure 6–23a. There are two P-N junctions, and the middle one is far too wide to punch carriers through it (as is done with BJTs). But as shown in Figure 6–23b, a V_{GG} that is positive to the gate terminal causes holes to move away from the gate area and/or electrons to move into the gate area. This creates an *inversion layer* that completes a conductive channel between the source and drain. The more positive the gate becomes, the wider the inversion layer and the more conductive the channel. Figure 6–23c. The E-MOSFET is the only FET that is nonconducting (switched off) when there is zero voltage applied to the gate.

Figure 6–24 is a complete set of drain curves for an N-channel E-MOSFET. Each of the curves has the same general shape as other FET curves. In the case of E-MOSFETS, however, drain current is increased by increasing the amount of V_{GG}. The minimum gate-source voltage that creates the inversion layer for an E-MOSFET is called the *threshold gate-source voltage,* abbreviated $V_{GS_{(TH)}}$ or V_T. This is the amount of gate voltage that is required for allowing any channel current to flow.

FORMULA 6–9 $I_D = \mathrm{k}\,(V_{GS} - V_T)^2$

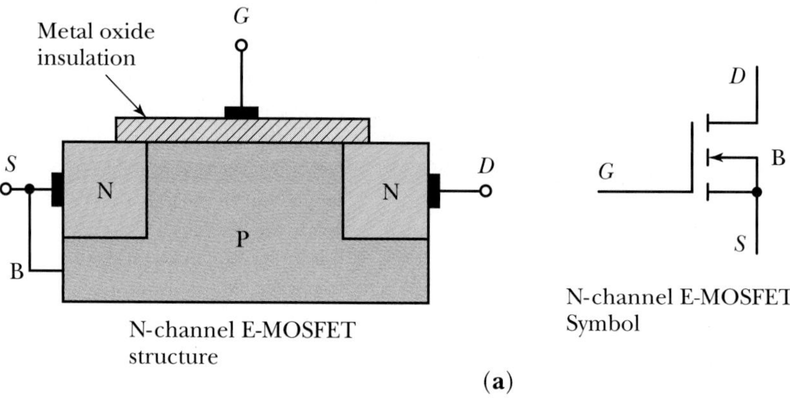

FIGURE 6–22 E-MOSFET structures and symbols

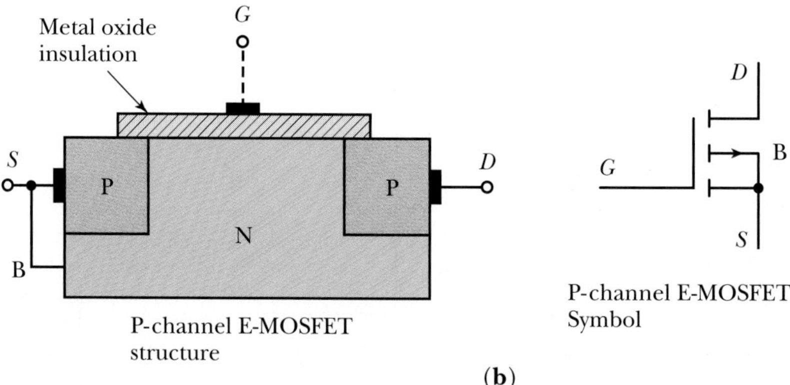

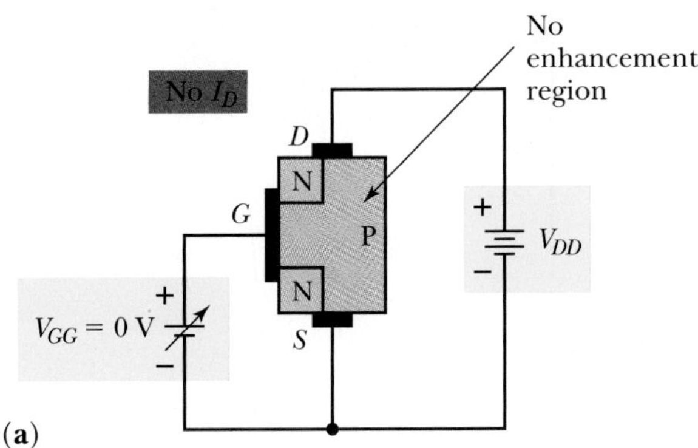

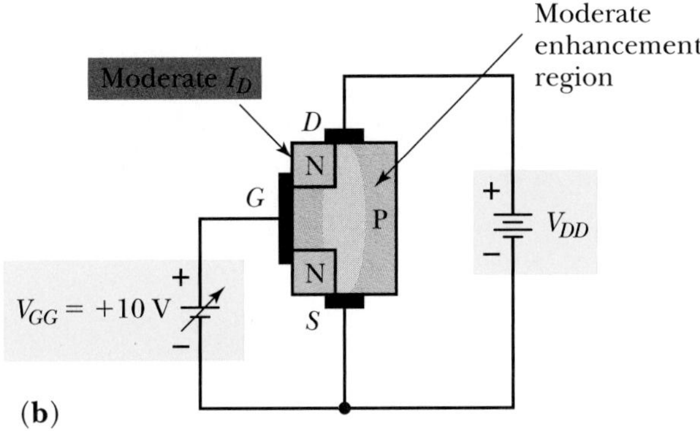

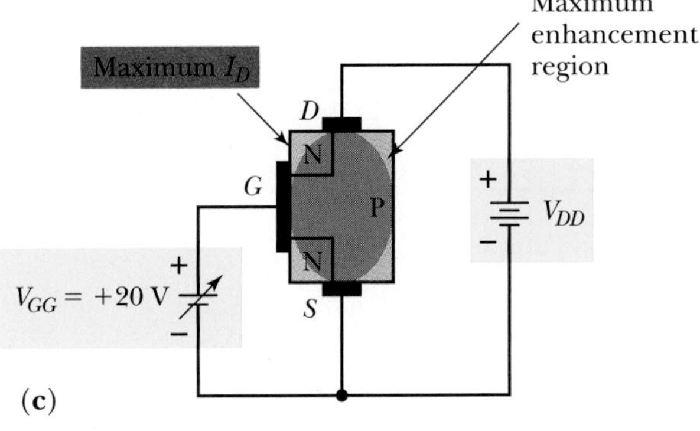

FIGURE 6–23 Action of an N-channel E-MOSFET: (a) V_{GG} is zero; (b) V_{GG} increased; (c) V_{GG} increased further

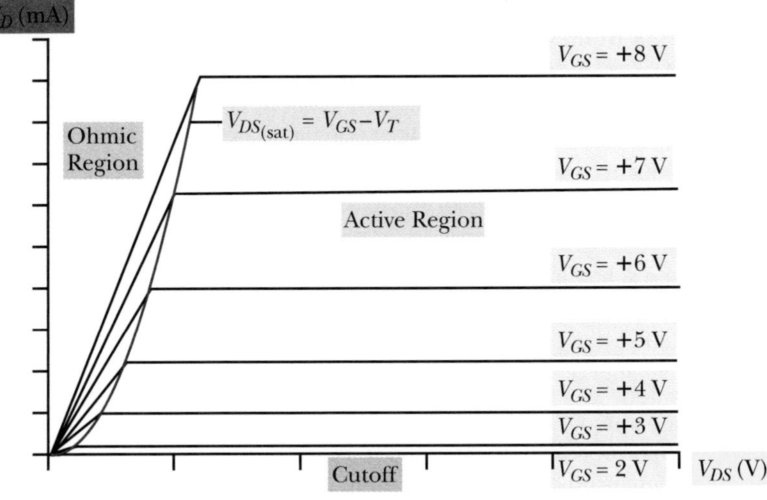

FIGURE 6–24 Family of drain characteristic curves for an N-channel E-MOSFET

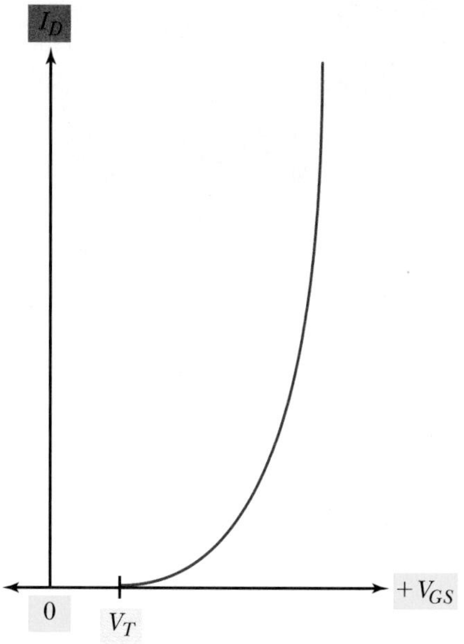

I_D

0 V_T $+V_{GS}$

FIGURE 6–25 A typical N-channel E-MOSFET transconductance curve (active region)

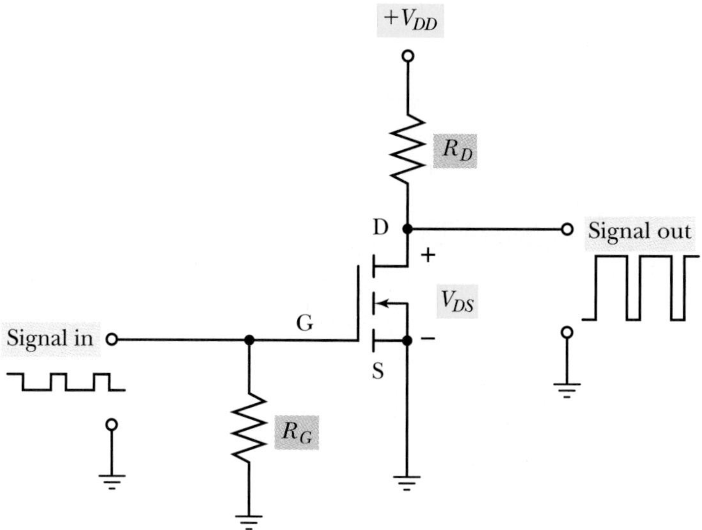

FIGURE 6–26 A simple switch circuit using an N-channel E-MOSFET

The transconductance curve for an N-channel E-MOSFET is shown in Figure 6–25. Consider the following features of E-MOSFETs that are common with D-MOSFETs:

1. Charge-carrier flow is from source to drain. For an N-channel E-MOSFET, this means electron flow is from source to drain—negative voltage is connected to the source and positive to the drain. For a P-channel version, hole flow is from source to drain—positive voltage is connected to the source and negative to the drain.
2. The type of semiconductor material used for the channel is opposite the type of material used for the substrate. An N-channel E-MOSFET uses a P-type semiconductor for its substrate. A P-channel version uses an N-type semiconductor for its substrate.
3. The arrow part of the schematic symbol indicates the type of material that is used for the substrate. An inward-pointing arrow signifies a P-type substrate material (therefore an N-channel device). An outward-pointing arrow represents an N-type substrate (and a P-channel device).

It is impossible to operate an E-MOSFET in a depletion mode. The structure of a D-MOSFET allows it to be operated in both modes, but the structure of an E-MOSFET prevents anything but enhancement-mode operation. The main reason an E-MOSFET

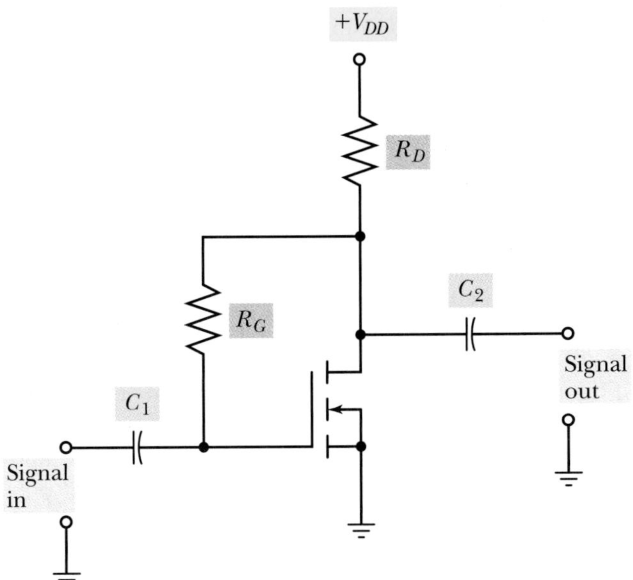

FIGURE 6–27 Drain-feedback bias

cannot be used in a depletion mode is there is never a continuous semiconductor material where carriers can be depleted from.

The circuit in Figure 6–26 is an example of how an E-MOSFET can be used as a switch. Since there is no dc bias to the gate, this amplifier is operating Class B. When the input signal is zero, the transistor is turned off and the V_{DS} level is at + V_{DD}. Applying a positive voltage level as an input signal turns the transistor on. As a consequence, the V_{DS} level at the output drops to some minimum level.

E-MOSFETS are similar to BJTs in the sense that they are switched off when zero bias is applied. Therefore, the biasing methods are much the same. With no bias, for example, the circuit behaves like a Class B amplifier where there is no current through the device until the signal forward biases it. To operate as a Class A amplifier, the gate is biased with the same polarity that is applied to the drain. Figure 6–27 shows an example of forward biasing a common-source E-MOSFET by a procedure called *drain-feedback biasing.*

■ IN-PROCESS LEARNING CHECK 3

Fill in the blanks as appropriate.

1. The term *MOS* stands for _______.
2. The gate terminal in a D-MOSFET is separated from the channel material by a thin layer of _______.
3. In a MOSFET, the direction of flow of charge carriers is always from the _______ terminal to the _______ terminal.
4. An outward-pointing arrow on a D-MOSFET symbol indicates a(n) _______-type substrate and a(n) _______ channel. An inward-pointing arrow on a D-MOSFET symbol indicates a(n) _______-type substrate and a(n) _______ channel.
5. The proper polarity for V_{DD} of an N-channel D-MOSFET is _______ to the source and _______ to the drain. For a P-channel D-MOSFET, the proper polarity for V_{DD} is _______ to the source and _______ to the drain.
6. When operating an N-channel D-MOSFET in the depletion mode, the polarity of V_{GG} is _______ to the gate terminal. And when V_{GS} is at its 0 V level, I_D is at its _______ level.
7. The _______ is the only FET that is non-conducting when the gate voltage is zero.
8. The proper polarity for V_{DD} of an N-channel E-MOSFET is _______ to the source and _______ to the drain. The proper polarity of V_{GG} is _______ to the gate terminal. ■

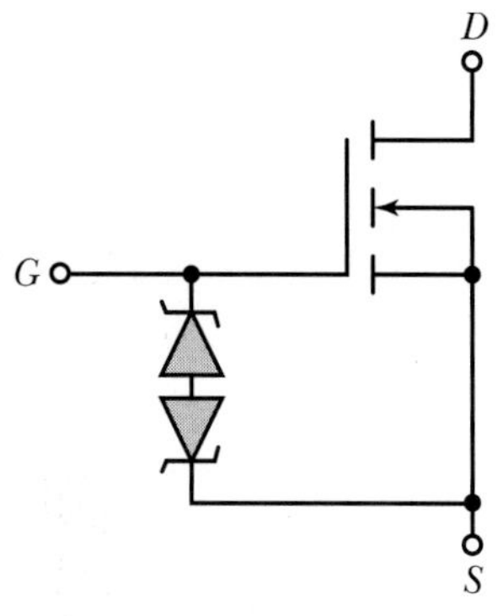

FIGURE 6–28 Diode clippers are sometimes used to prevent excessive static discharges in MOSFET devices

6–8 FET Handling Precautions

It was stated earlier that the silicon dioxide insulation between the gate and channel in MOSFETs is very thin. Although silicon dioxide is a good insulator, its thinness makes it prone to breakdown with voltages that are accidentally applied. In fact, the oxide layer can be destroyed by the simple electrical discharges we often feel as a result of brushing against certain materials in a dry atmosphere. P-N junctions (as in BJTs and JFETs) are not affected by electrical static discharges because they have relatively low impedances that harmlessly short out the static potential. The impedance of a MOSFET gate-channel connection is virtually infinite, so static potentials are not shorted out—at least not until the voltage punctures the insulation and creates a permanent short-circuit path!

Many MOSFET devices today (especially those found in integrated circuits) are protected to some extent by means of a bipolar zener clipper across the input terminals, Figure 6–28. Whenever the voltage at the input terminal attempts to exceed the breakdown voltage rating, the corresponding zener conducts and prevents the input voltage from rising above the clipping level.

Be sure your body and your soldering tools are not carrying a static electrical charge when you handle MOSFETs. ■

Practical Notes

MOSFETs that are packed and shipped in a conductive foam or conductive wrapping should be kept in that protective environment until you are ready to use them. When handling MOSFETs, use the proper anti-static precautions.

Voltage-sensitive MOSFETs are packaged in some sort of conductive wrapping, usually a graphite-impregnated plastic foam. The graphite content of this foam makes it conductive, thereby keeping all the terminals of the MOSFET electrically shorted together. This prevents any difference in potential from building up between the terminals.

Care should be exercised when handling MOSFETs. This applies to the time when you remove them from their conductive foam or wrapping to the time they are securely fastened into the circuit. It is important that you avoid picking up a static potential from your surroundings when you are about to handle a MOSFET. One way to make sure you are "discharged" is by touching the metal case of a piece of plugged-in electronic test equipment (such as an oscilloscope) before you handle a MOSFET. The most reliable procedure is to use a conductive wrist band. This wrist band has a wire and connector that is fastened to ground, thus preventing any static buildup on your body. Soldering tools used for installing MOSFETs should have similar ground-wire connections on their tips.

6–9 Power FETs

The NPN and PNP BJTs are current gain devices. For loads requiring high collector currents, the base current for the BJT must be high. The drain current for the N-channel and P-channel E-MOSFETs is controlled by the gate to source voltage. This MOSFET characteristic led the semiconductor industry to develop high power N-channel and P-channel E-MOSFETs called power FETs. (Siliconix manufactured the V-MOSFET, Motorola developed the TMOS power MOSFET, and International Rectifier developed the HEXFET. All of these names came from the internal architectural layout of the product.) Figure 6–29 provides a short list of some power FETs with some of their parameters. The power dissipation within the power FET is calculated using Formula 6–10.

FORMULA 6–10 $P_{FET} = |V_{DS} \times I_D|$

Device	Polarity	$R_{DS(on)}$ Ω	@V_{GS} V	I_D@ 25°C A	$V_{BR(DSS)}$ V	Power Dissipation P_D (W)	Package	Maufacturer
IRFZ44V	N	0.0165	10	55	60	115	TO-220AB	International Rectifier
IRLZ24	N	0.1	5	17	60	60	TO-220AB	International Rectifier
2N6764	N	0.055	10	38	100	150	TO-204AA	Motorola
IRL540NS	N	0.044	4.5	36	100	3.8	D2-Pak	International Rectifier
MTP20N10	N	0.15	10	20	100	100	TO-204AA	National Semiconductor
2N6758	N	0.4	10	9	200	75	TO-204AA	Motorola
2N6766	N	0.085	10	30	200	150	TO-204AA	International Rectifier
IRF640N	N	0.15	10	18	200	150	TO-220AB	International Rectifier
IRFPG50	N	2	10	6.1	1000	190	TO-247AC	International Rectifier
IRF9Z14	P	0.5	–10	6.7	–60	43	TO-220AB	International Rectifier
IRF5210	P	0.06	–10	40	–100	200	TO-220AB	International Rectifier
MTP12P10	P	0.3	–10	12	–100	75	TO-220AB	On Semiconductor
IRF9640	P	0.5	–10	11	–200	125	TO-220AB	International Rectifier
MTP8P20	P	0.7	–10	8	–200	125	TO-220AB	Motorola

FIGURE 6–29 Power FET parametric product table

The primary application of power FETs is switching circuits. The desire is to apply the total voltage to the load (on condition) or have no voltage applied to the load (off condition). Two circuit configurations are available to accomplish this task.

One circuit using the N-channel power FET provides the ground for the circuit. The *off* condition occurs when the gate to source voltage is less than the gate to source threshold voltage (low). While off, the power FET is in the cutoff region, the drain current is 0 A, and the voltage across the load is 0 V. The *on* condition occurs when the gate to source voltage is greater than the gate to source threshold voltage (high). While on, the power FET should be in the ohmic region, the drain current should be controlled by the load characteristics, and the voltage across the load should be close to the supply voltage.

EXAMPLE A 12 V, 10 Ω lamp is connected as shown in Figure 6–30. The power FET part number is IRLZ24, the supply voltage (V_{DD}) is 12 V, and the gate resistor (R_G) is 1 kΩ.

When the input voltage (V_{in}) is 0 V, find:

1. The gate to source voltage (V_{GS})
2. The drain current (I_D)
3. The drain to source voltage (V_{DS})
4. The load voltage (V_L)

Answers:

1. For the N-channel power FET, $V_{GS} = V_{in}$.

   Therefore, $V_{GS} = 0$ V.
2. Since $V_{GS} = 0$ V, the power FET is in cutoff. Therefore, $I_D = 0$ A.
3. Since the power FET is in cutoff,

   $V_{DS} = V_{DO} = 12$ V
4. Using the following equation, the load voltage can be calculated.

$$V_L = I_D \times R_L$$
$$V_L = 0\text{ A} \times 10\ \Omega$$
$$V_L = 0\text{ V}$$

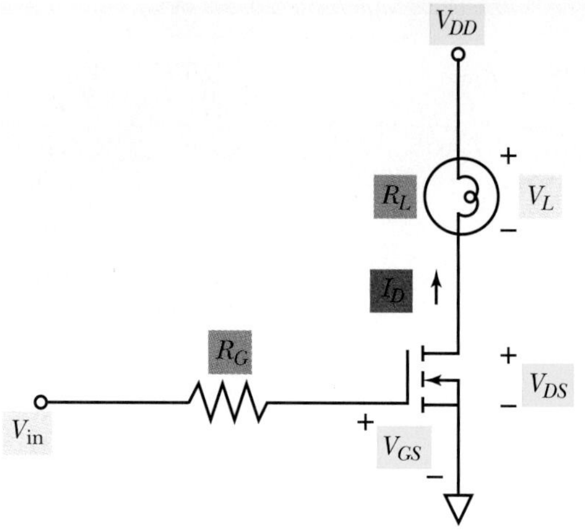

FIGURE 6–30 Lamp control circuit

When the input voltage is 5 V, find:

1. The gate to source voltage (V_{GS})
2. The drain current (I_D)
3. The drain to source voltage (V_{DS})
4. The load voltage (V_L)
5. The load power (P_L)
6. The power FET power (P_{FET})

Answers:

1. For the N-channel power FET, $V_{GS} = V_{in}$.
   Therefore, $V_{GS} = 5$ V.
2. Since $V_{GS} = 5$ V, the power FET is in the ohmic region and $R_{DS_{(on)}} = 0.1\ \Omega$.

$$I_D = \frac{V_{DD}}{(R_L + R_{DS_{(on)}})}$$
$$I_D = \frac{12\text{ V}}{(10\ \Omega + 0.1\ \Omega)}$$
$$I_D = 1.188\text{ A}$$

3.
$$V_{DS} = I_D \times R_{DS_{(on)}}$$
$$V_{DS} = 1.188\text{ A} \times 0.1\ \Omega$$
$$V_{DS} \approx 0.12\text{ V}$$

4. Using the following equation, the load voltage can be calculated.

$$V_L = I_D \times R_L$$
$$V_L = 1.188\text{ A} \times 10\ \Omega$$
$$V_L = 11.88\text{ V}$$

5.
$$P_L = V_L \times I_D$$
$$P_L = 11.88\text{ V} \times 1.188\text{ A}$$
$$P_L = 14.1\text{ W}$$

6. Using Formula 27–10 (restated below),

$$P_{FET} = |V_{DS} \times I_D|$$
$$P_{FET} = 0.12\text{ V} \times 1.188\text{ A}$$
$$P_{FET} = 143\text{ mW}$$ ◆

The other circuit using the P-channel power FET provides the positive voltage for the circuit. The *off* condition occurs when the gate to source voltage is less than the gate to source threshold voltage (low). While off, the power FET is in the cut-off region, the drain current is 0 A, and the voltage across the load is 0 V. The *on* condition occurs when the

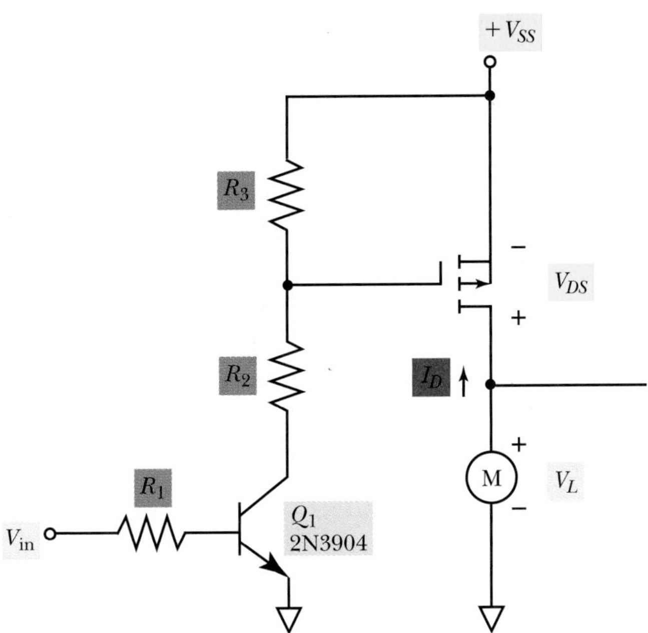

FIGURE 6–31 Motor control circuit

gate to source voltage is greater than the gate to source threshold voltage (high). While on, the power FET should be in the ohmic region, the drain current should be controlled by the load characteristics, and the voltage across the load should be close to the supply voltage.

EXAMPLE A 1 A motor is connected as shown in Figure 6–31. The power FET part number is IRF9Z14, the supply voltage (V_{SS}) is + 12 V, $V_1 = 100\ \text{k}\Omega$ $R_2 = 2\ \text{k}\Omega$, $R_3 = 10\ \text{k}\Omega$ and the gate resistor (R_C) is 1 kΩ.

When the input voltage (V_{in}) is 0 V, find:

1. The gate to source voltage (V_{GS})
2. The drain current (I_D)
3. The drain to source voltage (V_{DS})
4. The load voltage (V_L)

Answers:

1. Since $V_{in} = 0$ V, the NPN transistor is off and its collector current is 0 A.

   Therefore, $V_{GS} = V_G - V_S$

   where: $V_G = V_{SS} = 12\ \text{V}$

   and $V_S = V_{SS} = 12\ \text{V}$

   $V_{GS} = 12\ \text{V} - 12\ \text{V}$

   $V_{GS} = 0\ \text{V}$
2. Since $V_{GS} = 0$ V, the power FET is in cutoff.

   Therefore, $I_D = 0$ A.
3. Since the power FET is in cutoff.

   $V_{DS} = -V_{SS} = -12\ \text{V}$
4. Using the following equation, the load voltage can be calculated

$$V_L = I_D \times R_L$$
$$V_L = 0\ \text{A} \times 10\ \Omega$$
$$V_L = 0\ \text{V}$$

When the input voltage is 5 V, find:

1. The gate to source voltage (V_{GS})
2. The drain current (I_D)

3. The drain to source voltage (V_{DS})
4. The load voltage (V_L)
5. The load power (P_L)
6. The power FET power (P_{FET})

Answers:

1. Since $V_{in} = 5$ V, the NPN transistor is on and $V_{CE} \approx 0$ V. For the P-channel power FET,

$$V_G = V_{SS} \times \frac{R_2}{R_2 + R_3}$$
$$V_G = 12\text{ V} \times \frac{2\text{ k}\Omega}{2\text{ k}\Omega + 10\text{ k}\Omega}$$
$$V_G = 2\text{ V}$$
$$V_S = V_{SS} = 12\text{ V}$$
$$V_{GS} = V_G - V_S$$
$$V_{GS} = 2\text{ V} - 12\text{ V}$$
$$V_{GS} = -10\text{ V}$$

2. When enough voltage is applied to the motor, the current flowing through the motor will be 1 A.

   Therefore, $I_D = 1$ A.

3. Since $V_{GS} = -10$ V, the power FET is in the ohmic region and $R_{DS_{(on)}} = 0.5\ \Omega$.

$$V_{DS} = (I_D \times R_{DS(on)})$$
$$V_{DS} = -(1\text{ A} \times 0.5\ \Omega)$$
$$V_{DS} \approx -0.5\text{ V}$$

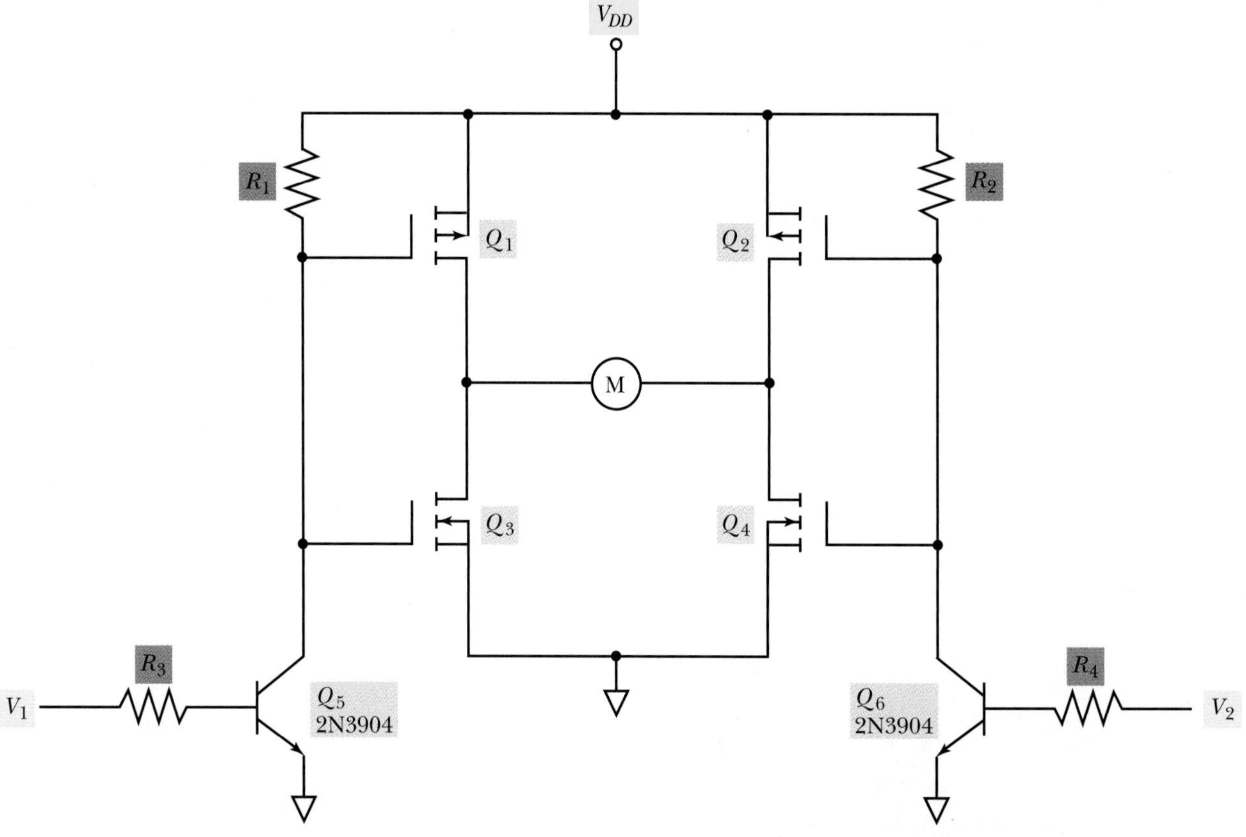

FIGURE 6–32 Power FET H-bridge circuit

4. Using the following equation, the load voltage can be calculated.

$$V_L = V_{SS} + V_{DS}$$
$$V_L = 12\text{ V} + (-0.5\text{ V})$$
$$V_L = 11.5\text{ V}$$

5. The load power is:

$$P_L = V_L \times I_D$$
$$P_L = 11.5\text{ V} \times 1\text{ A}$$
$$P_L = 11.5\text{ W}$$

6. Using Formula 6–10 (restated below), the power dissipated by the power FET is:

$$P_{\text{FET}} = |V_{DS} \times I_D|$$
$$P_{\text{FET}} = 0.5\text{ V} \times 1\text{ A}$$
$$P_{\text{FET}} = 500\text{ mW}$$ ◆

An application that combines the sinking circuit and the sourcing circuit is the H-bridge motor driver circuit shown in Figure 6–32. The P-channel power FET's are connected to the positive voltage source and the N-channel power FETs are connected to ground. The motor rotates in clockwise direction when V_1 equals 5 V and V_2 equals 0 V. Q_1 and Q_4 are turned on providing the path for current to flow through the motor. When V_1 equals 0 V and V_2 equals 5 V the motor turns counterclockwise Q_2 and Q_3 are turned on providing the path for current to flow through the motor.

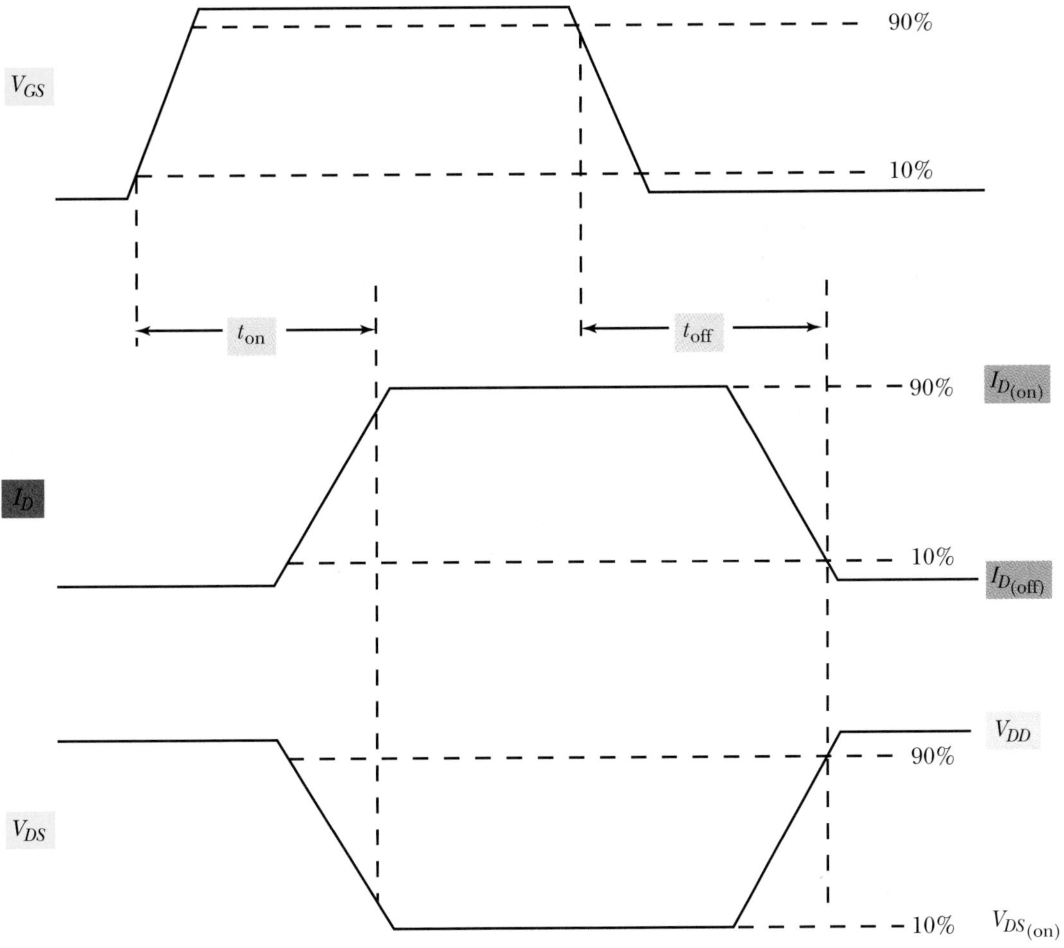

FIGURE 6–33 Resistive load switching waveforms

Application Problem

Communications

SATELLITE CONTROL

A satellite receiver located on top of a mountain utilizes several satellites. The realignment to each satellite is controlled by a microprocessor. The microprocessor is interfaced through its Port D to a precision stepper motor. A concept of the interface circuit is shown in the following figure. The motor drivers (ULN2003) are rated at 500 mA. The microprocessor outputs 5 mA to each driver. A 1-Amp-hour battery supplies power to the system and it is assumed that the system operates 10 minutes each hour. The battery voltage is 12 V.

(a) Calculate the power needed for the satellite dish realignment and compare that with a microcontroller power output.

(b) Calculate the battery life.

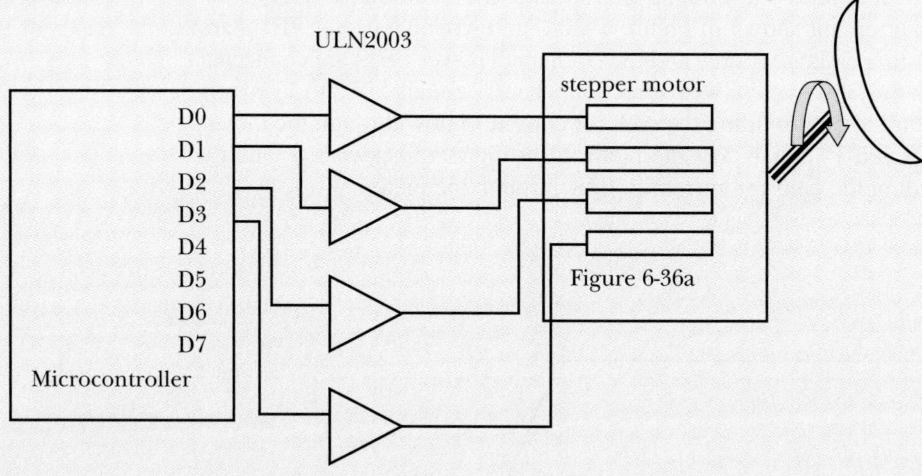

A satellite microprocessor control

Solution

(a) The stepper motor current is $I_{driver} = 500$ mA. A power delivered to the stepper motor by one driver.

$$P_{driver} = V \times I_{driver} = 12 \times 0.5 = 6 \text{ W}$$

The stepper motor interface has two drivers ON and two OFF at a time (Figure 6–38b). The total power delivered to the stepper motor.

$$P_{driver} = P_{driver} \times 2 = 6 \times 2 = 12 \text{ W}$$

The microcontroller supplies 5 mA to ULN2003. Two drivers are ON at a time. The power is

$$P_{CPU} = V \times I_{CPU} \times 2 = 5 \times 0.05 \times 2 = 0.5 \text{ W}$$

The FETs inside the microcontroller supply 0.5 W to the external circuit. The driver supplies 12 W to the antenna alignment.

(b) Next we calculate the battery life. Energy equation for the battery is

$$A = VI \times t = 12 \text{ V} \times 1\text{A} \times 3600 = 43200 \text{ J}$$

Using

$$A = P \times t$$

Then time is

$$t = \frac{A}{P} \times \frac{60}{10} = \frac{43200 \text{ J}}{12 \text{ W}} = 21600 \text{ sec}$$

A 1-Amp-hour battery would last only 6 hours. (There are 3600 seconds in an hour). We would need a solar cell to maintain the charge.

Practical Notes The gate resistor and the characteristics of the power FET input capacitance and output capacitance determine the turn-on and turn-off times of the power FETs (shown in Figure 6–33). The turn on time is the time between when the input voltage moves from a low to a high and the drain to source voltage moves from V_{DD} to $V_{DD_{(on)}}$. The turn-off time is the time between when the input voltage moves from a high to a low and the drain to source voltage moves from $V_{DD_{(on)}}$ to V_{DD}. Short turn-on and turn-off times are critical to prevent possible damage to the power FETs. During these transitions the drain to source voltage is high and the current is increasing or decreasing resulting in "momentary" high power dissipation within the part.

6–10 FET Summary

Figure 6–34 provides a summary of the FETs. For N-channel FETs, the drain current flows from source to drain, whereas for P-channel FETs, the drain current flows from drain to source. All of the depletion mode FETs conduct when $V_{GS} = 0$ V, whereas the E-MOSFETs do not conduct when $V_{GS} = 0$ V. $V_{GS_{(off)}}$ for N-channel depletion mode FETs is a negative voltage, whereas $V_{GS_{(off)}}$ for P-channel depletion mode FETs is a positive voltage. JFETs conduct between $V_{GS_{(off)}}$ and 0 V. The D-MOSFETs conduct between $V_{GS_{(off)}}$ and 0 V, and beyond 0 V. The gate to source threshold voltage for the N-channel E-MOSFETs is a positive voltage with a positive drain to source voltage. The threshold voltage for the P-channel E-MOSFETs is a negative voltage with a negative drain to source voltage.

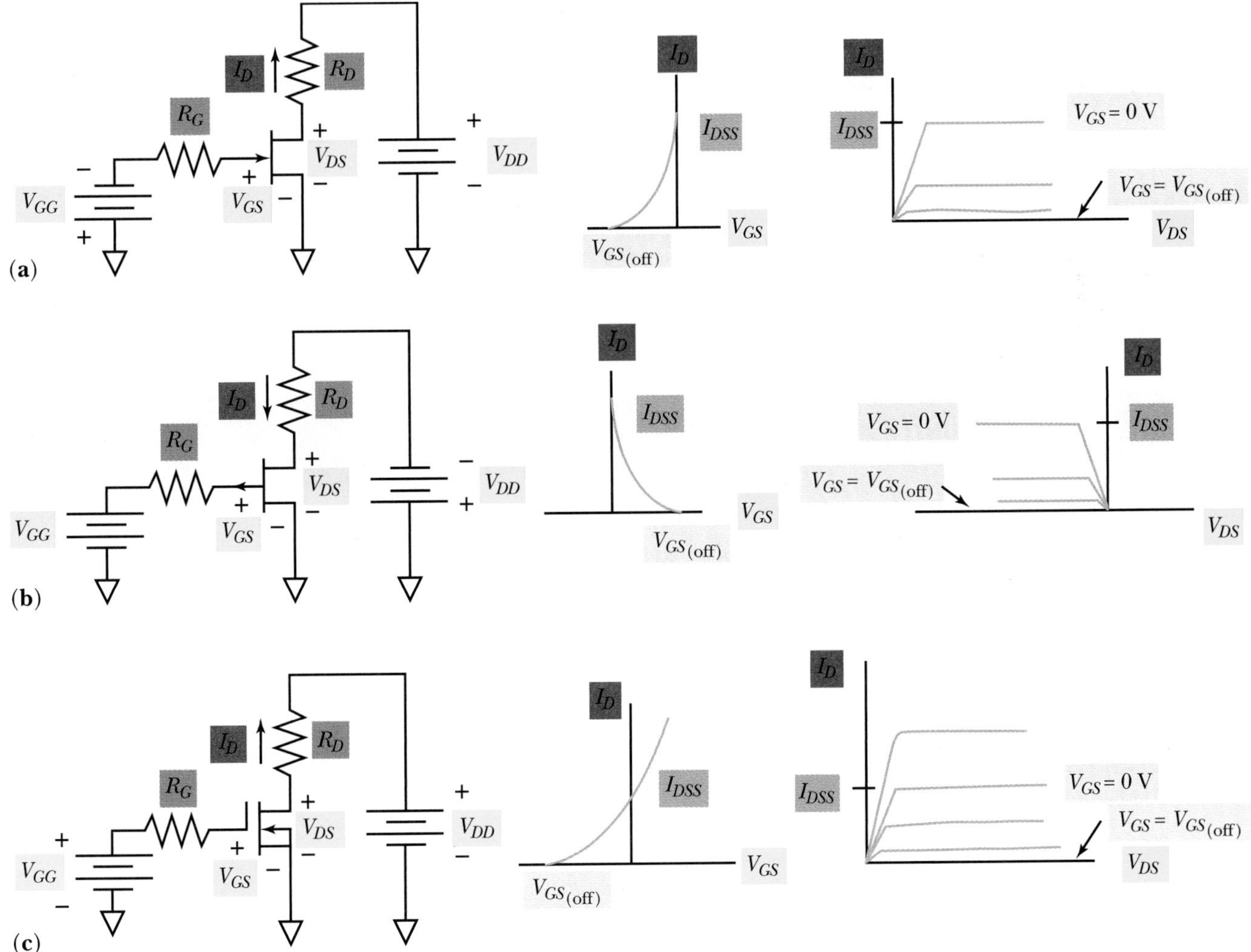

FIGURE 6–34 FET summary table: (a) N-channel JFET; (b) P-channel JFET (c) N-channel D-MOSFET;

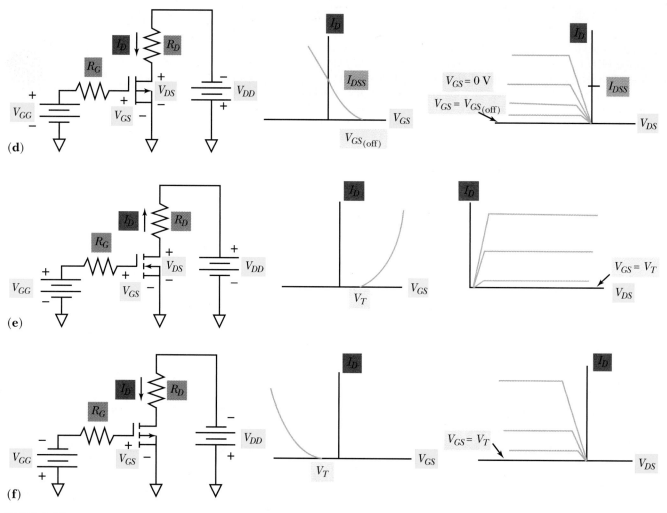

FIGURE 6–34 Continued. (d) P-channel D-MOSFET; (e) N-channel E-MOSFET; (f) P-channel E-MOSFET

Application Problem

LED INTERFACING

An indication LED (light emitting-diode) is connected to a computer port. The port driving capability is up to 10 mA. The figure 1 shows such an LED circuit. Calculate the current limiting resistor.

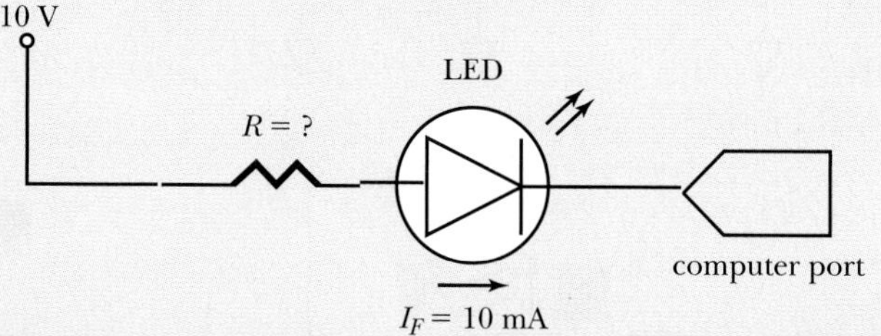

LED Circuit

Solution. The LED has a typical ON voltage of 2 V when conducting. A forward current is 10 mA. The voltage is about 0.15 to 0.2 V when the computer port is at its LOW state.

$$R = \frac{V_{CC} - 2.0\text{ V} - V_{\text{PIN}}}{I_F} = \frac{10\text{ V} - 2.0 - 0.2\text{ V}}{0.010} = 780\ \Omega$$

Choose the next resistor commercially available to increase a life expectancy of the LED.

$$R = 1\text{ k}\Omega$$

USING TECHNOLOGY: FETS USED IN A SYSTEM

FETs differ from BJTs studied earlier in two essential ways. FETs are unipolar devices—that is the current carriers are just one type of charge carrier, either holes or electrons. Also FETs are voltage-controlled not current-controlled devices.

FETs as a class only require a near zero gate current; instead FETs use the voltage applied between the gate and source to control the drain current. This voltage-controlled characteristic combined with their other ideal low signal handling properties are the reasons that designers have utilized the FET in their products. Semiconductor manufacturers have also integrated discrete FETs as a fundamental element for the design and construction of other components such as operation amplifiers, and more importantly, in the CMOS logic gates. The CMOS (complementary metal oxide semiconductor) family of digital logic devices is composed of both N- and P- channels, either depletion or enhancement mode MOSFETs, and is the principal design element in almost all computer circuits.

Reviewing the ideal characteristics of the N-channel, depletion-type MOSFET, they are specifically designed to operate in the depletion mode. That means with no gate bias and/or no gate input signal, the main charge carriers have a clear path through the channel and drain current flows freely. No turn-on bias is therefore required, the devices are operating in the linear portion of their operating curve, their input impedance is high and stable, and their noise is constant at the low level—in essence, a perfect Class A amplifier device.

One of the most common, almost universally used applications of MOSFETs, is in the tuner portion of every conceivable type broadcast receiver. These range from simple AM-FM radios, televisions, cordless telephones, cellular telephones, wireless receivers of every type, family radios (FRs), citizen band (CB) radios, to the sophisticated wireless network interface cards (NICs) used in laptop computers and now throughout the home as wireless computer networks. These tuner circuits, which are extremely sensitive, and working with only a few microvolts of input signal, are frequently completely shielded from stray RF signals. Figure 15 shows such RF enclosures, called "bath tub enclosures." The smaller enclosure shields the TV antenna/cable input components, while the larger enclosure shields the complete tuner circuits. Such enclosures are shown as dot-dash boxes around the circuit components

FIGURE 15 Photo of VHF/UHF Tuner

on schematic diagrams of tuner circuits. A photo of the circuitry inside the VHF-UHF tuner shielded enclosure is shown in Figure 16. You will see the shielded tuned circuits called "can" throughout the tuner circuit board. These are shown as dotted enclosures on the circuit schematic diagrams.

The system application block diagram for the VHF-UHF portion of the television from input to output is shown in Figure 17. The incoming RF signal from an antenna, a cable box, or a dish converter box is applied to the antenna/cable RF input inside a shielded enclosure to isolate the antenna or other components from the television chassis. The RF input is fed either to the UHF tuner stage for channels 14 through 69 or the VHF tuner stage for channels 2 through 13 and the cable channels. Tuned circuits select the desired channel and feed the selected band of frequencies to both the UHF and VHF RF amplifiers. These amplifiers significantly raise the level of the selected RF frequencies above the noise levels and pass them on to additional tuned circuits. These circuits further improve both the sensitivity and selectivity of the television's receiver circuits to the desired band of UHF and/or VHF frequencies. The UHF band of frequencies is now divided into the low band of channels (channels 14 through 51) and the high band of channels (channels 52 through 69). In a similar fashion, the VHF band of frequencies is now divided into the low band of channels (channels 2 through 6) and the high band of channels (channels 7 through 13) along with the cable channels (see Table 1.)

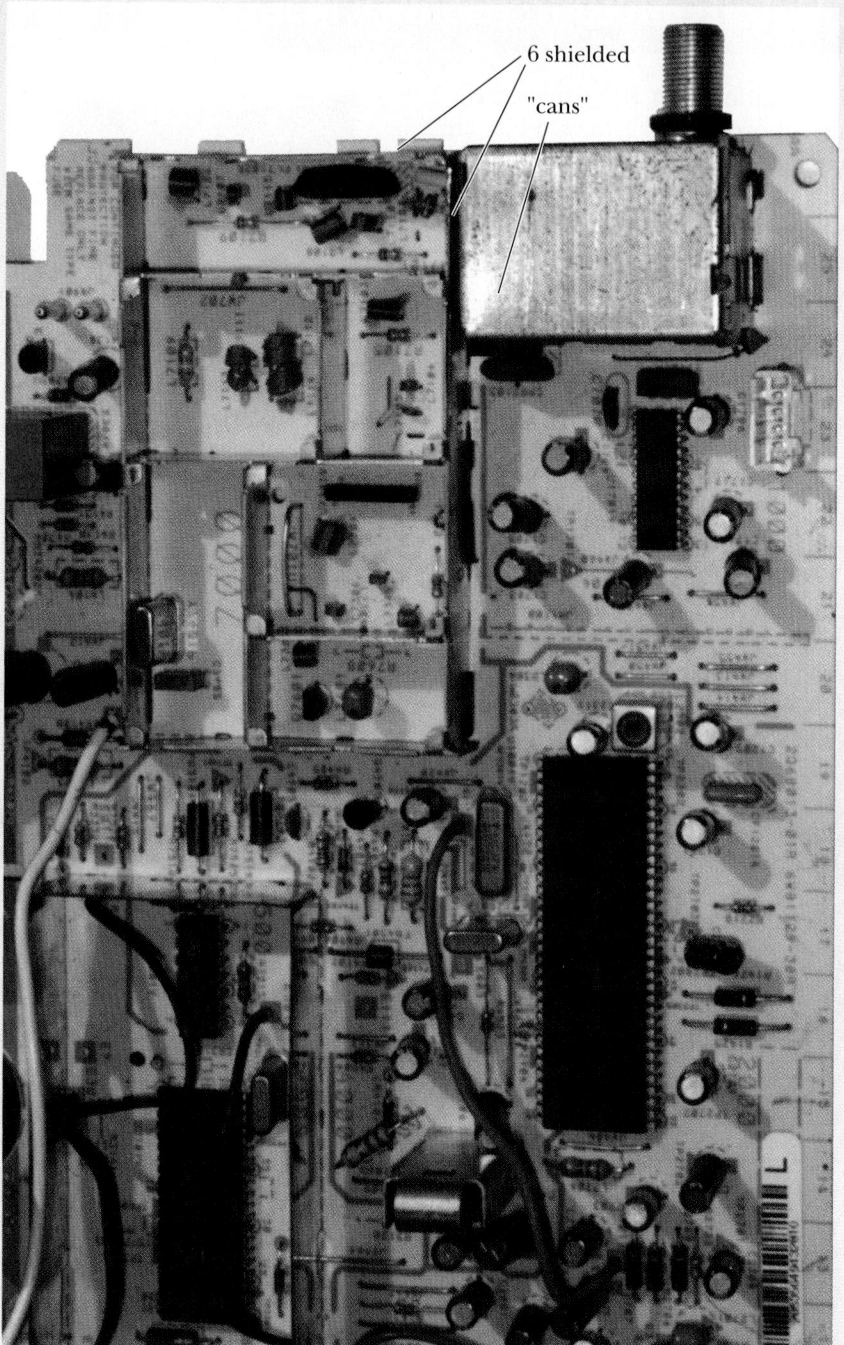

FIGURE 16 Interior view of VHF/UHF Television Tuner

The UHF and VHF mixer/oscillator adds these four bands of amplified RF input signals to the output of an independent oscillator circuit, one for UHF and one for VHF. The result is one fixed band of lower RF frequencies centered around a single intermediate frequency or IF frequency of 41.25 MHz for the sound carrier through 45.75 MHz for the picture carrier. The actual IF frequency ranges cover 6 MHz from 41 to 47 MHz in order to cover the complete 6 MHz band from the incoming broadcast signal. Table 1 shows these 6 MHz frequency bands for selected VHF, UHF, and cable channels. Also note the location of the unwanted FM radio stations right between the low and high VHF channels. Right above the FM stations in frequency are all the cable channels.

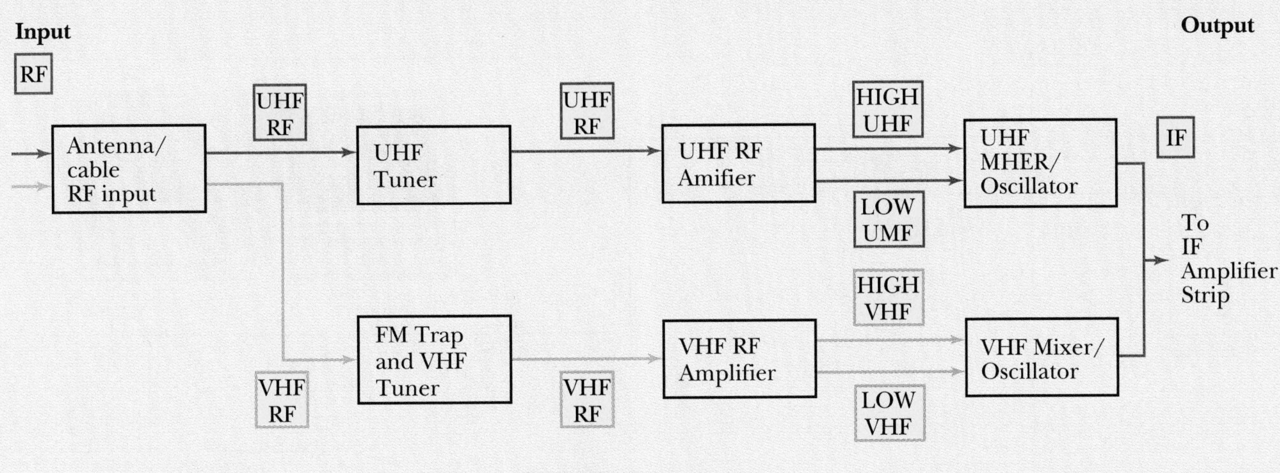

FIGURE 17

TABLE 1 Channels and Frequencies

Lower VHF Band		High VHF Band		UHF Band	
Channel	Frequency (MHz)	Channel	Frequency (MHz)	Channel	Frequency (MHz)
2	54–60	7	174–180	14	470–476
3	60–66	8	180–186	24	530–536
4	66–72	9	186–192	34	590–596
5	76–82	10	192–198	44	650–659
6	82–88	11	198–204	54	710–716
		12	204–210	64	770–776
FM Stations 88–108		13	210–216	69	800–806
Cable Channels 108–405 (uses letters not numbers)					

The circuit schematic in Figure 18 shows the details of the circuit. The full range of UHF, VHF, and cable channel frequencies is applied to the antenna/cable RF input. Five combination RC networks inside a shielded enclosure isolate the television chassis, provide impedance matching, and prevent strong stations or cable channels from overloading the tuner MOSFETs. The wide band signal is fed through an FM band pass filter trap to eliminate the undesirable FM frequencies from 88 to 108 MHz. The UHF band of frequencies are selected by the tuned circuit consisting of inductors L 7101 and L 7102 along with the varactor diodes CR 7101 and CR 7114 and applied to the gate G1 of MOSFET Q 7101, the UHF RF amplifier. In a similar fashion, the VHF and cable channels are selected by the tuned circuit consisting of inductors L 7106 and L7107 along with varactors CR 7106 and CR 7107 and applied to the gate G1 of MOSFET Q 7102, the VHF RF amplifier.

These N-channel depletion type MOSFETs, Q 7101 and Q 7102, are unique; they have two gates—G1 for the RF incoming signals and G2 for the automatic gain control bias. This separate G2 gate bias responds to changes in the received signal level, increasing the MOSFETs' gain for weak signals and decreasing the gain for strong local stations that might otherwise overload the RF amplifiers.

The choice of whether to amplify the UHF or the VHF signals is made by the user through their remote control and a sophisticated switch circuit which supplies the correct bias to turn on one or the other MOSFET at their gates G1.

Recall from Table 1, that the VHF TV channels are split into a low VHF band from 54 to 88 MHz and a high VHF band from 174 to 216 MHz. The output signal from MOSFET Q 7102 is further selected by inductors L 7111, L 7112, L 7113, and L 7114 along with varactors CR 7111 and CR 7113. The choice of either the low band or high band VHF channels is made by the user through their remote control and switching diodes CR 7109 and CR 7110 which select one or the other band to short circuit to ground.

Important

DC voltage measurements and AC oscilloscope patterns cannot be attempted in the RF amplifier section. Connecting a DVM or scope will affect the tuning, resulting in lost reception. Special RF voltmeters and RF probes working up to only 10 MHz and special low capacity scope leads and scopes working up to 500 MHz must be used when making measurements in the RF amplifier section of any receiver.

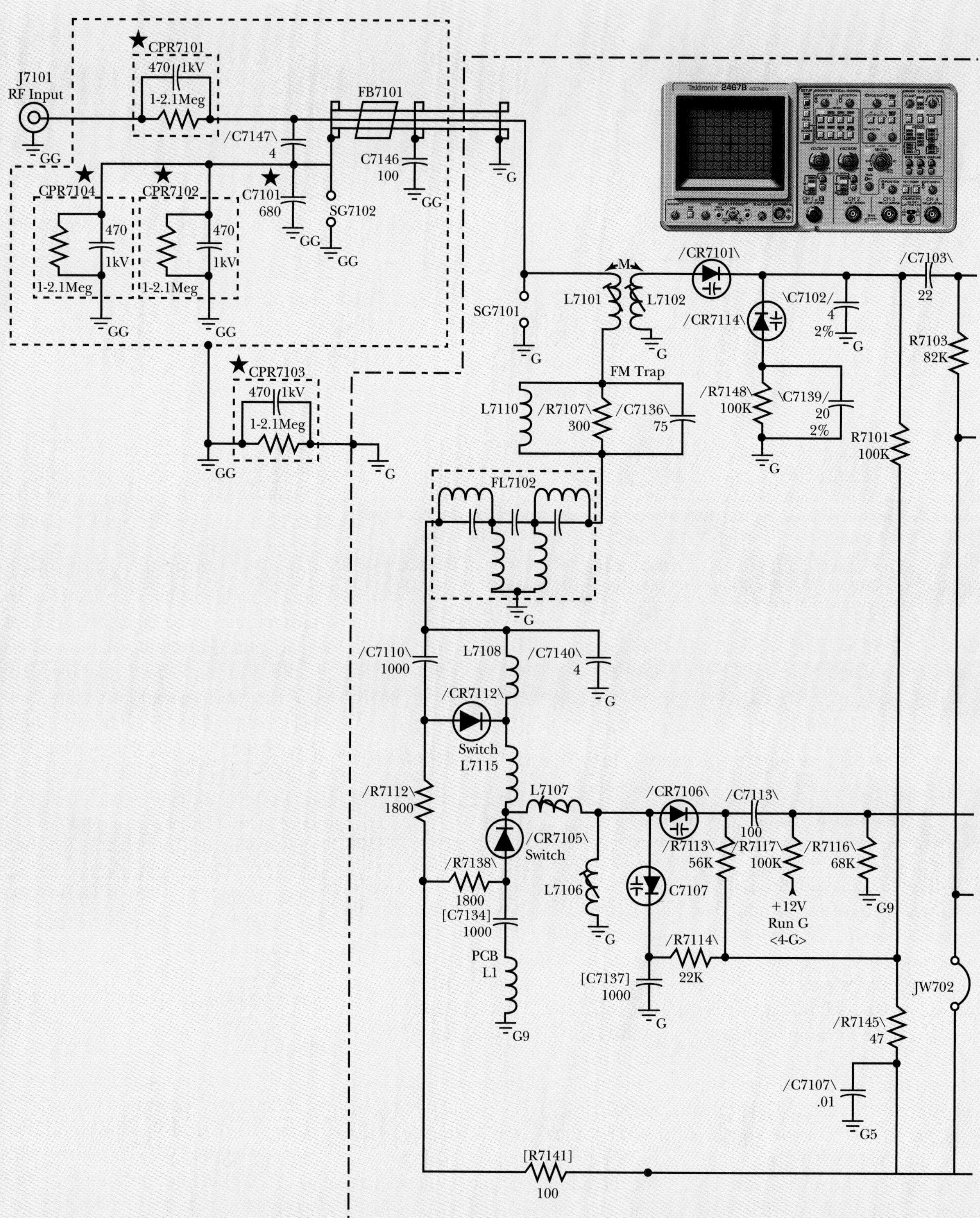

FIGURE 18 Circuit Schematic

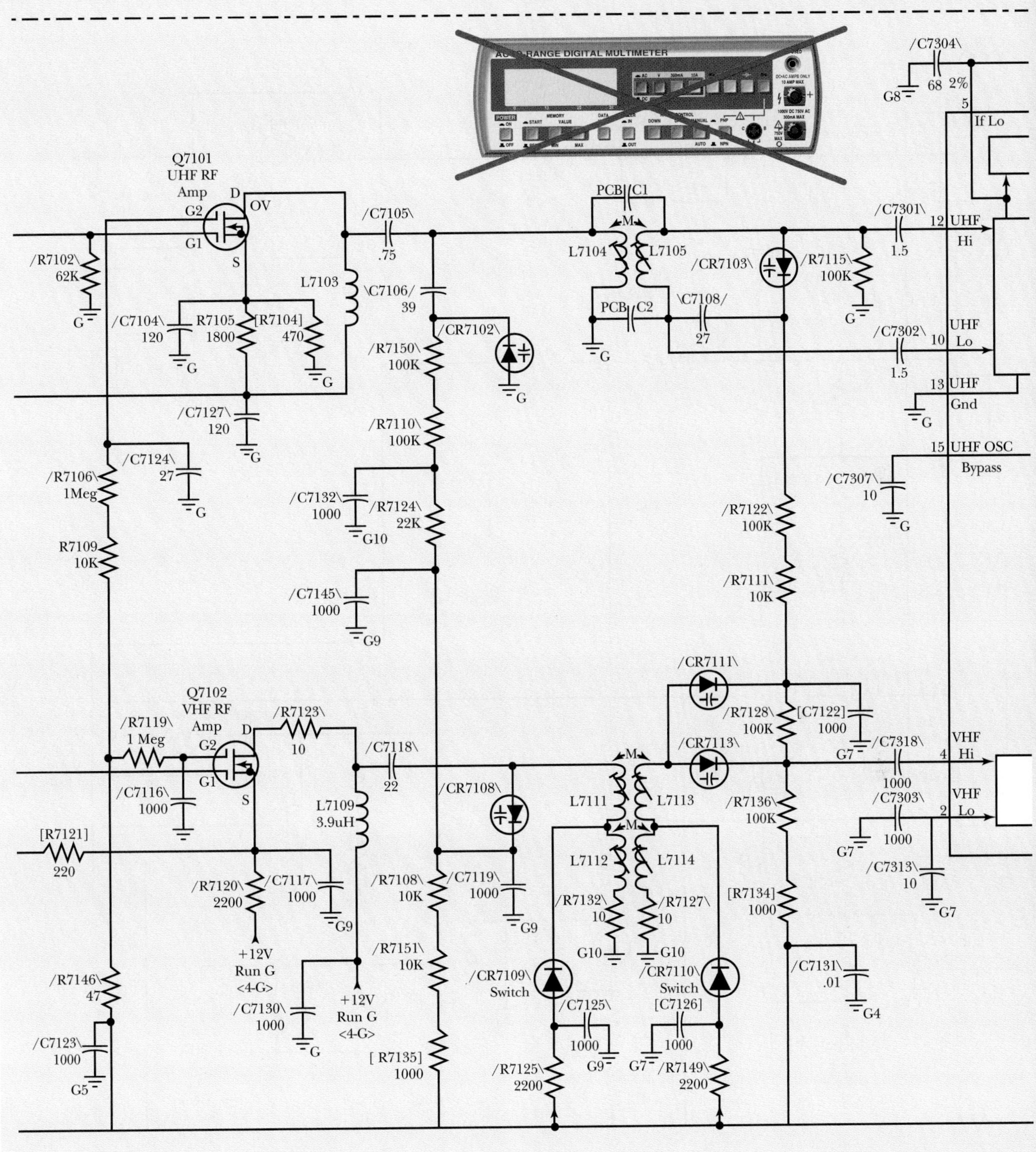
Q7101
UHF RF
Amp
D
OV
G2
G1
S
/R7102\
62K
G
/C7104\
120
G
R7105
1800
[R7104]
470
G
L7103
/C7105\
.75
\C7106/
39
/CR7102\
G
/R7150\
100K
/C7127\
120
G
/R7110\
100K
/C7124\
27
G
/R7106\
1Meg
R7109
10K
/C7132\
1000
G10
/R7124\
22K
/C7145\
1000
G9
PCB
C1
M
L7104
L7105
PCB
C2
G
\C7108/
27
/CR7103\
/R7115\
100K
G
/C7301\
1.5
12
UHF
Hi
/C7302\
1.5
10
UHF
Lo
13
UHF
Gnd
G
15
UHF OSC
Bypass
/C7307\
10
G
/C7304\
G8
68 2%
5
If Lo
/R7122\
100K
/R7111\
10K
/CR7111\
/R7128\
100K
[C7122]
1000
G7
/CR7113\
/C7318\
1000
4
VHF
Hi
/C7303\
1000
2
VHF
Lo
G7
/C7313\
10
G7
/R7136\
100K
[R7134]
1000
/C7131\
.01
G4
Q7102
VHF RF
Amp
D
G2
G1
S
/R7119\
1 Meg
/C7116\
1000
/R7123\
10
/C7118\
22
L7109
3.9uH
[R7121]
220
/R7120\
2200
/C7117\
1000
G9
+12V
Run G
<4-G>
/C7130\
1000
G
+12V
Run G
<4-G>
/R7146\
47
/C7123\
1000
G5
/CR7108\
/R7108\
10K
/C7119\
1000
G9
/R7151\
10K
[R7135]
1000
M
L7111
L7113
M
L7112
L7114
/R7132\
10
G10
/R7127\
10
G10
/CR7109\
Switch
/CR7110\
Switch
/C7125\
1000
G9
[C7126]
1000
G7
/R7125\
2200
/R7149\
2200

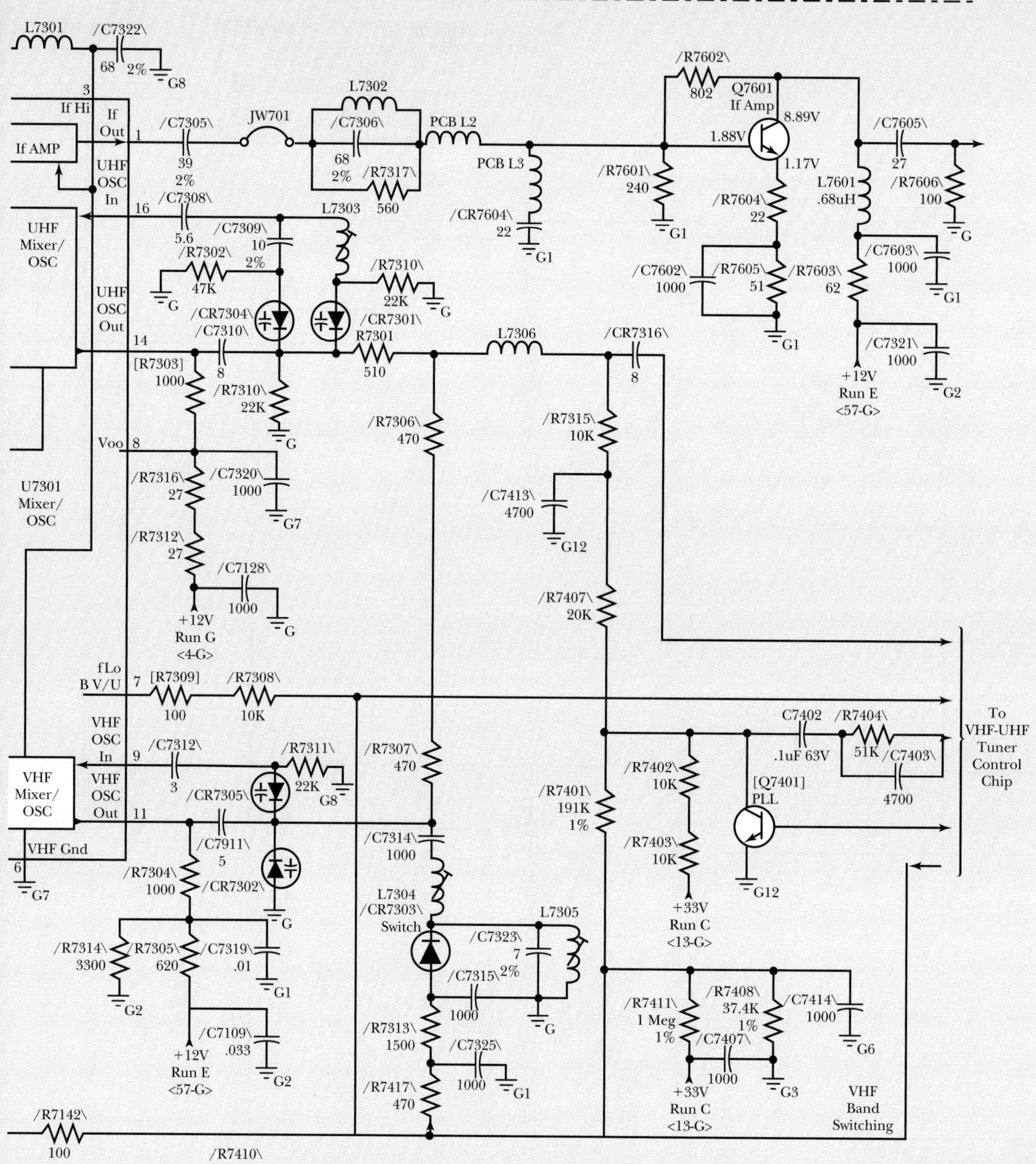

L7301
/C7322\
68
2%
G8
3
If Hi
If Out
1
If AMP
UHF OSC In
16
UHF Mixer/ OSC
/C7305\
39
2%
JW701
L7302
/C7306\
68
2%
/R7317\
560
PCB L2
PCB L3
/CR7604\
22
G1
/R7602\
802
Q7601
If Amp
8.89V
1.88V
1.17V
/R7601\
240
/R7604\
22
/C7602\
1000
/R7605\
51
/R7603\
62
/C7605\
27
L7601
.68uH
/R7606\
100
/C7603\
1000
/C7321\
1000
+12V Run E <57-G>
/C7308\
5.6
/C7309\
10
2%
/R7302\
47K
L7303
/R7310\
22K
/CR7304\
/C7310\
/CR7301\
R7301
510
UHF OSC Out
14
[R7303]
1000
8
/R7310\
22K
L7306
/CR7316\
8
/R7306\
470
/R7315\
10K
/C7413\
4700
G12
Voo
8
/R7316\
27
/C7320\
1000
G7
U7301 Mixer/ OSC
/R7312\
27
/C7128\
1000
+12V Run G <4-G>
/R7407\
20K
fLo B V/U
7
[R7309]
100
/R7308\
10K
To VHF-UHF Tuner Control Chip
C7402
.1uF 63V
/R7404\
51K
/C7403\
4700
VHF OSC In
9
/C7312\
3
/R7311\
22K
/R7307\
470
/R7402\
10K
[Q7401] PLL
VHF Mixer/ OSC
VHF OSC Out
11
/CR7305\
/C7911\
5
/R7401\
191K
1%
/R7403\
10K
VHF Gnd
6
G7
/R7304\
1000
/CR7302\
/C7314\
1000
L7304
/CR7303\
Switch
+33V Run C <13-G>
/R7314\
3300
/R7305\
620
/C7319\
.01
L7305
/C7323\
7
/C7315\
2%
1000
/R7411\
1 Meg
1%
/R7408\
37.4K
1%
/C7414\
1000
G6
/C7109\
.033
+12V Run E <57-G>
/R7313\
1500
/C7325\
1000
/C7407\
1000
G3
/R7417\
470
VHF Band Switching
/R7142\
100
/R7410\

Summary

- Field-effect transistors are unipolar devices. This means current through the devices is carried by just one type of charge carrier (either holes or electrons, depending on the type of semiconductor being used). This is in contrast to bipolar devices (BJTs), where holes carry the charge through one part and electrons carry the charge through a second part.
- The terminals on an FET are called gate, source, and drain. The flow of charge carriers is always from the source to the drain. The bias voltage that is used to control drain current is applied to the gate.
- FETs use a voltage (V_{GS}) to control the flow of current (I_D) through the device. This is in contrast to BJT, which uses a current (I_B) to control a current (I_C).
- The current flowing through the channel of an FET (I_D) rises sharply with increasing drain-source voltage (V_{DS}), but only to a certain point called the *pinch-off voltage.* Once the pinch-off voltage is reached, further increase in V_{DS} causes almost no increase in I_D.
- The family of drain characteristic curves and the transconductance curve for a JFET shows that it conducts fully (I_{DSS}) when no potential is applied to the gate-source junction ($V_{GS} = 0$ V). Increasing the amount of V_{GS} causes the amount of drain current (I_D) to decrease.
- MOSFETs differ from JFETs by the way the gate voltage is separated from the channel current. In JFETs, this separation is accomplished by means of a reverse-biased P-N junction. In MOSFETs, the gate voltage is separated from the channel current by means of an extremely good insulating material called silicon dioxide.
- The channel material in MOSFETs is opposite the type of material used as the substrate. So an N-channel MOSFET uses a P-type material as its substrate, and a P-channel MOSFET uses an N-type material as its substrate.
- The arrow portion of the symbols for MOSFETs indicates the type of substrate material used. An inward-pointing arrow indicates P-type substrate, and an outward-pointing arrow indicates an N-type substrate.
- The family of drain characteristic curves and the transconductance curve for a D-MOSFET show that it conducts fully when the maximum amount of enhancement-mode voltage is applied to the gate-source junction. For an N-channel D-MOSFET, the enhancement-mode polarity is positive; for a P-channel version, the polarity is negative. Reversing the polarity causes the D-MOSFET to enter its depletion-mode of operation, where increasing the amount of gate-source voltage causes drain current to decrease until cutoff occurs. A D-MOSFET can be operated in both the enhancement and depletion modes.
- A D-MOSFET is a natural choice for a Class A FET amplifier because the device operates about halfway between cutoff and saturation when there is no voltage applied to the gate.
- In an E-MOSFET, the source and drain are separated by two P-N junctions. Conduction is enabled only by creating a clear path for charge carriers (the inversion layer) between the source and drain segments of the structure.
- The family of drain characteristic curves and the transconductance curve for an E-MOSFET show that it is nonconducting when no voltage is applied to the gate-source junction. This means the device operates in the enhancement mode. For an N-channel E-MOSFET, increasing the positive polarity at the gate-source terminals increases drain current. For a P-channel E-MOSFET, increasing the negative polarity at the gate-source terminals increases drain current.
- Drain-feedback bias can be used with E-MOSFETs because the drain polarity is the same as the polarity required for controlling gate bias.
- The gate-source insulation on MOSFETs is especially sensitive to voltage breakdown because of an accidental buildup of static electricity. Therefore, MOSFETs are usually shipped and stored in conductive plastic foam. Also, it is recommended that you ground your body and soldering tools when installing MOSFET devices in a circuit.

Formulas and Sample Calculator Sequences

FORMULA 6–1
(Shockley's equation (JFETs and D-MOSFETs) to find drain current)

$$I_D = I_{DSS}\left(1-\left(\frac{V_{GS}}{V_{GS(\text{off})}}\right)\right)^2$$

1, [−], [(], V_{GS} value, [÷], $V_{GS_{(\text{off})}}$ value, [)], [=], x^2, , I_{DSS} value, [=]

FORMULA 6–2
(to find JFET Transconductance)

$$g_m = \frac{\Delta I_D}{\Delta V_{GS}}$$

ΔI_D value, [÷], ΔV_{GS} value, [=]

FORMULA 6–3
(to find JFET fixed bias gate to source voltage)

$V_{GS} = V_{GG}$

FORMULA 6–4
(to find JFET fixed bias drain to source voltage)

$V_{DS} = V_{DD} - I_D R_D$

V_{DD} value, [−], I_D value, , R_D value, [=]

FORMULA 6–5
(to find JFET self-bias gate to source voltage)

$V_{GS} = -I_D R_S$

I_D value, [×], R_S, [±], [=]

FORMULA 6–6
(to find JFET drain to source voltage (self and voltage divider bias))

$V_{DS} = V_{DD} - I_D R_D - I_D R_S$

V_{DD} value, [−], I_D value, [×], R_D values, [−], I_D value, [×], R_S value, [=]

FORMULA 6–7
(to find JFET voltage divider bias gate voltage)

$$V_G = V_{DD} \times \frac{R_2}{R_1 + R_2}$$

V_{DD} value, [×], R_2 value, [÷], [(], R_1 value, [+], R_2 value, [)], [=]

FORMULA 6–8
(to find JFET voltage divider bias gate to source voltage)

$V_{GS} = V_G - I_D R_S$

V_G value, [−], I_D value, [×], R_S value, [−]

FORMULA 6–9
(Shockley's equation (E-MOSFETs) to find drain current)

$I_D = k \times (V_{GS} - V_T)^2$

[(], V_{GS} value, [−], V_T value, [)], x^2, [×], k value, [=]

FORMULA 6–10
(to find FET power)

$P_{FET} = |V_{DS} \times I_D|$

V_{DS} value, [×], I_D value, [=]

Review Questions

1. Explain the meaning of the following terms for JFETs: V_{DS}, V_{GS}, I_D, I_{DSS}, and $V_{GS(\text{off})}$.
2. Draw and properly label the schematic symbols for P-channel and N-channel JFETs, including the names of the terminals.
3. State the polarity of V_{DD} and V_{GG} for Class A, gate-biased N-channel and P-channel JFET amplifiers.
4. Draw a common-source, gate-biased amplifier that uses a P-channel JFET. Be sure to indicate the polarity of the V_{DD} and V_{GG} connections.
5. Describe the main differences between the operations of a JFET and a BJT.
6. Draw an N-channel JFET amplifier that is using self-bias. What is the value of V_{GS} for this amplifier when $V_S = +4$ V?
7. State, in words, the formula for transconductance.
8. Describe the difference between drain characteristic curves and transconductance curves for FETs.
9. Explain why a D-MOSFET makes an ideal Class A amplifier.
10. Draw the schematic symbols for N-channel and P-channel D-MOSFETs, and label the terminals.
11. Provide the polarity of V_{DD} and V_{GS} for the following D-MOSFET configurations:
    a. N-channel depletion mode
    b. P-channel depletion mode
    c. N-channel enhancement mode
    d. P-channel enhancement mode
12. Draw the schematic symbols for N-channel and P-channel E-MOSFETs, and label the terminals.
13. Explain the difference between depletion modes and enhancement modes of operation for FETs.
14. Provide the polarity of V_{DD} and V_G for N-channel and P-channel, Class A E-MOSFET amplifiers.
15. Explain why an E-MOSFET makes an ideal electronic switching circuit.
16. Define the term *threshold gate-source voltage* ($V_{GS(\text{TH})}$) as it applies to E-MOSFET devices.
17. State the polarity of $V_{GS(\text{off})}$ for:
    a. N- and P-channel JFETs
    b. N- and P-channel D-MOSFETs
    c. N- and P-channel E-MOSFETs
18. List the types of FETs presented in this lesson that can be (or must be) operated in a depletion mode, and the types of FETs that can be (or must be) operated in an enhancement mode.
19. Explain the purpose of zener diodes that are internally connected between the gate and source terminals of some MOSFET devices.
20. Explain why MOSFETs are usually packaged and shipped in conductive foam.

Problems

1. For the circuit in Figure 6–35, let $-V_{GG} = 4$ V, $+V_{DD} = 12$ V, $R_G = 10$ mΩ, $R_D = 1$ kΩ, and $I_D = 8$ mA. Determine the following dc levels:
    a. Voltage measured between gate and common
    b. Voltage across resistor R_D
    c. Voltage measured between the source and drain
    d. Current through R_G
2. For the circuit in Figure 6–35, let $-V_{GG} = 5$ V, $+V_{DD} = 12$ V, $R_G = 4.7$ MΩ, and $R_D = 2.2$ kΩ. If the dc voltage measured between the source and drain is 6 V, determine the following:
    a. The amount of dc drain current
    b. The dc voltage across R_D
    c. The dc voltage measured from the gate to common
    d. The dc current through R_G
3. If a change in 2 V across R_G in Figure 6–35 causes the drain current to change 1 mA, determine the following:
    a. The value of transconductance
    b. The change in voltage between the source and drain

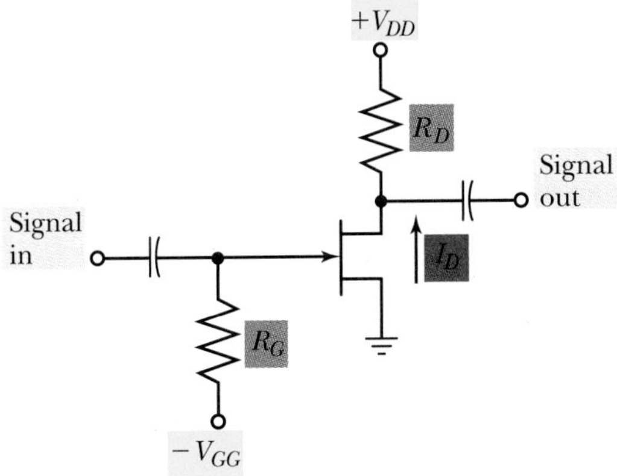

FIGURE 6–35

4. For the circuit in Figure 6–36, let $+V_{DD} = 12$ V, $R_G = 10$ MΩ, $R_S =$ kΩ, $R_D = 1.2$ kΩ, and $I_D = 5$ mA. Determine the following dc levels:
   a. V_{RS}
   b. V_{RD}
   c. V_{DS}
   d. V_{RG}
   e. V_{GS}
   f. I_{RG}
5. Referring to the circuit in Figure 6–36, let $+V_{DD} = 12$ V, $R_G = 4.7$ MΩ, $R_S = 1.2$ kΩ, $R_D = 2.2$ kΩ, and $V_{RS} = 2$ V. Determine the following dc levels:
   a. I_D
   b. V_{DS}
   c. Voltage measured between the gate and common
   d. V_{GS}
   e. V_{RG}
   f. I_{RG}
6. For the circuit in Figure 6–37, let $+V_{DD} = 18$ V, $R_1 =$ 100 kΩ, $R_2 = 10$ kΩ, $R_S = 330$ Ω, $R_D = 1$ kΩ, and $I_D =$ 10 mA. Determine the following dc values:
   a. I_{R_2}
   b. V_{R_2}
   c. V_{R_1}
   d. V_{RS}
   e. V_{GS}
   f. I_D
   g. V_{DS}
7. For the circuit in Figure 6–37, let $+V_{DD} = 12$ V, $R_1 =$ 100 kΩ, $R_2 = 33$ kΩ, $R_S = 470$ Ω, $R_D = 680$ Ω, and $V_{RS} = 4$ V. Determine the following dc values:
   a. I_{R_2}
   b. V_{R_2}
   c. V_{R_1}
   d. V_{RS}
   e. V_{GS}
   f. I_D
   g. V_{DS}

FIGURE 6–36

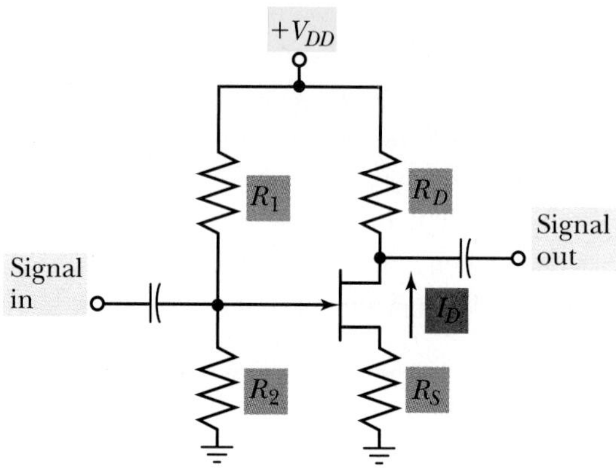

FIGURE 6–37

Analysis Questions

1. Explain why you should never attempt to operate a JFET in an enhancement mode.
2. Research and describe the construction of a high frequency version of a D-MOSFET that is called a *dual-gate* MOSFET. Explain where and why the dual-gate MOSFET is used.
3. Explain why a D-MOSFET can be operated in both the depletion and enhancement modes while an E-MOSFET can be operated only in the enhancement mode.
4. Research and describe the operation of a high power version of an E-MOSFET device that is called a *VMOSFET.*
5. Explain the operation of the unipolar stepper motor circuit shown in Figure 6–38.

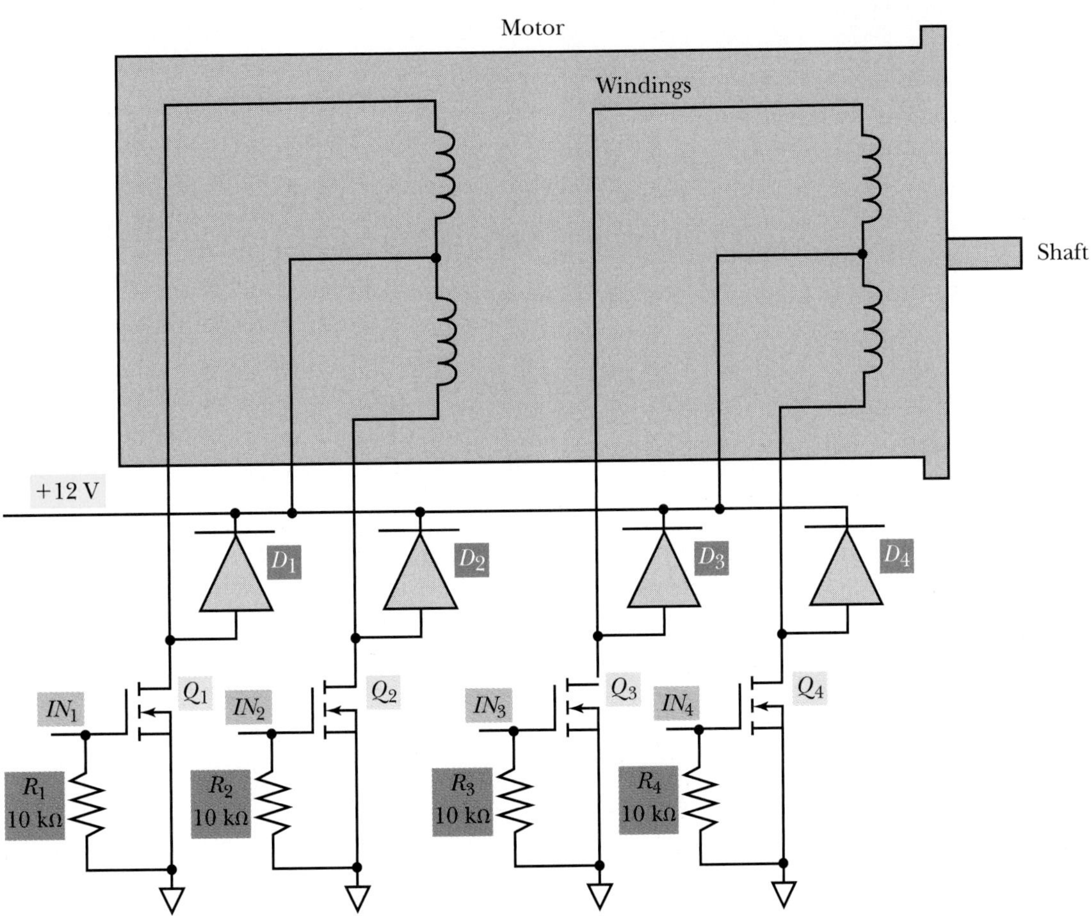

FIGURE 6–38 (a) A unipolar stepper motor circuit

FIGURE 6–38 (b) Clockwise Rotation

Step	IN1	IN2	IN3	IN4
1	5 V	0 V	5 V	0 V
2	0 V	5 V	5 V	0 V
3	0 V	5 V	0 V	5 V
4	5 V	0 V	0 V	5 V

FIGURE 6–38 (c) CounterClockwise Rotation

Step	IN1	IN2	IN3	IN4
1	5 V	0 V	5 V	0 V
2	5 V	0 V	0 V	5 V
3	0 V	5 V	0 V	5 V
4	0 V	5 V	5 V	0 V

Performance Projects Correlation Chart

Chapter Topic	Performance Project	Project Number
Biasing JFET Amplifiers and JFET Amplifier Circuits	Common-Source JFET Amplifier	14
Power FETs	The Power MOSFET	15
	Story Behind the Numbers: JFET Characteristics	

NOTE: It is suggested that after completing the above projects, the student should be required to answer the questions in the "Summary" at the end of this section of projects in the Laboratory Manual.

OUTLINE

OBJECTIVES

After studying this chapter, you should be able to:

1. Explain the derivation of the term operational amplifier **(op-amp)**
2. Draw op-amp symbol(s)
3. Define the term **differential amplifier**
4. Draw a block diagram of typical circuits used in op-amps
5. List the key characteristics of an ideal op-amp
6. Identify linear and nonlinear applications circuits for op-amps
7. Distinguish between **inverting** and **noninverting op-amp circuits**
8. Perform voltage gain and resistance calculations for standard inverting and noninverting op-amp circuits
9. Describe the operation of op-amps in voltage amplifiers, voltage followers, comparators, and Schmitt trigger amplifiers
10. Describe the function of op-amps in circuits originally designed for analog computers: summing amplifiers, differential amplifiers, differentiators, and integrators

CHAPTER 7

Operational Amplifiers

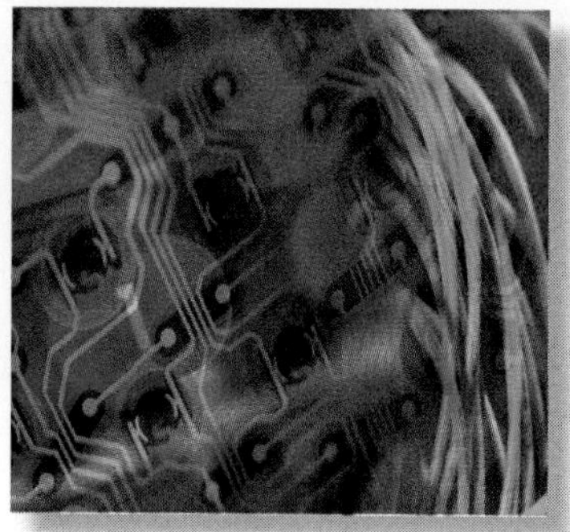

PREVIEW

The **operational amplifier (op amp)** is a very versatile device that is frequently used in modern electronic equipment of all types. For that reason, we feel it is important to provide an overview of this subject.

In this chapter you will learn about the basic features of op-amps. A variety of linear and nonlinear applications for these devices will be presented so you can understand their versatility. Additionally, you will learn that in many applications, a simple ideal op-amp approach can be used for amplifier analysis and design. Also, a brief discussion of some nonideal op-amp parameters will be presented. This chapter is an introduction to op-amps.

KEY TERMS

Common-made input
Differential amplifier
Differentiator circuit
Double-ended input
Hysteresis
Integrator circuit
Inverting amplifiers
Linear amplifier
Noninverting amplifiers
Nonlinear amplifier
Operational amplifier (op-amp)
Schmitt trigger
Single-ended input
Summing amplifier

7–1 Background Information

The name *operational amplifier* comes from the fact that this kind of circuit was once used for performing mathematical *operations* in analog computers. These early computer amplifier circuits were hand-wired circuits, using many discrete (individual) components, such as vacuum tubes, resistors, capacitors, and so on.

Today, the op-amp is a linear integrated-circuit (IC) device with all the components, including transistors, resistors, diodes, and their interconnections, fabricated on a single semiconductor substrate, Figure 7–1. It is easy to fabricate the IC transistors, resistors, and diodes, but more difficult to manufacture meaningful capacitor values on ICs. Therefore, we find that the op-amp stages are direct coupled (without coupling capacitors between stages). One stage's output is coupled directly into the next stage's input.

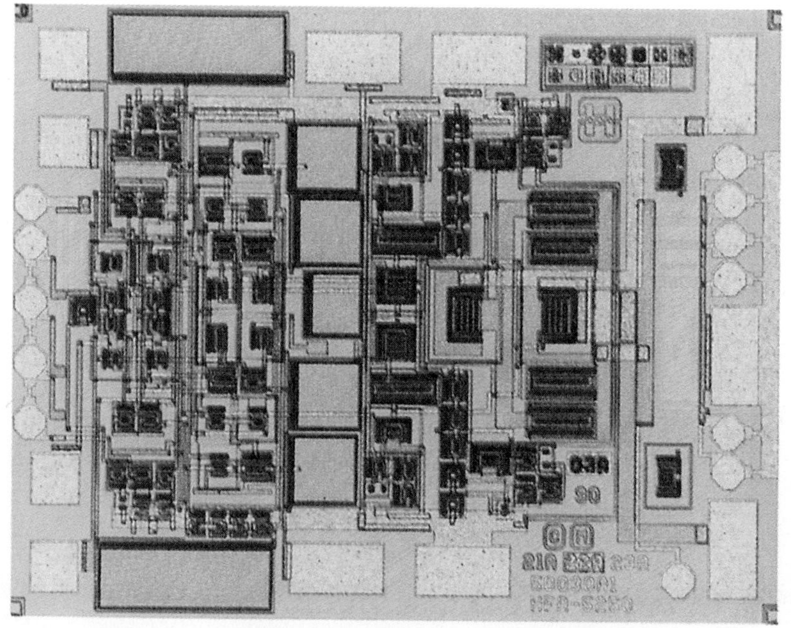

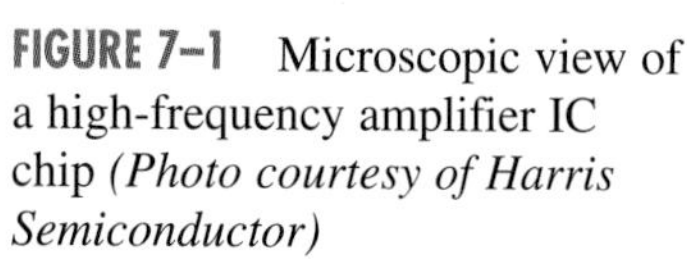

FIGURE 7–1 Microscopic view of a high-frequency amplifier IC chip *(Photo courtesy of Harris Semiconductor)*

Because dc is not blocked between stages, sometimes these amplifiers are called *dc amplifiers,* as well as direct-coupled amplifiers.

By definition, an op-amp is a very high gain, dc amplifier. By controlling and using appropriate feedback from output to input, important amplifier characteristics are controlled by design. Characteristics such as the amplifier gain and bandwidth features are controlled by the external components connected to the op-amp IC.

7–2 Op-amp Packaging Information

The op-amp symbol and sample IC packaging diagrams for a popular op-amp are shown in Figure 7–2. The packaging diagram identifies the pin locations and a picture of the physical appearance of the component (mechanical diagram). The packaging requires few external connections. Look at the schematic symbol and typical connection points in Figure 7–3. A simplified block diagram of the typical stages in an operational amplifier is shown in Figure 7–4.

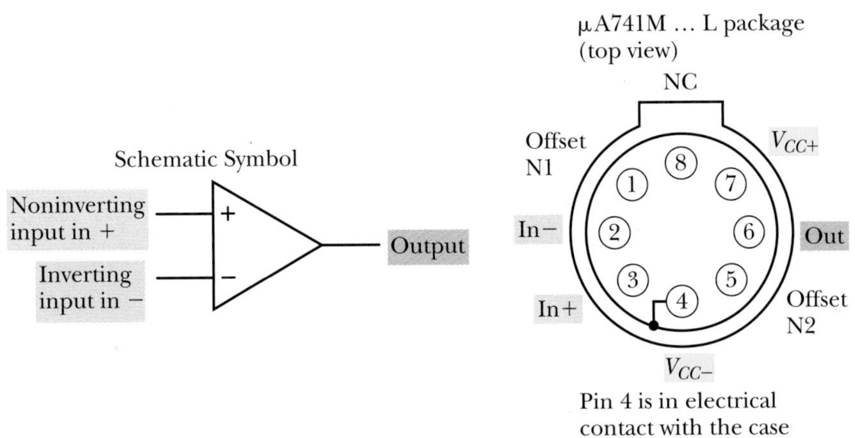

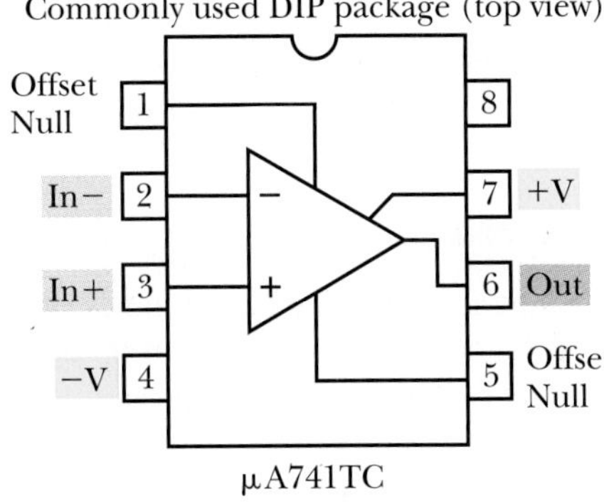

FIGURE 7–2 General-purpose operational amplifiers

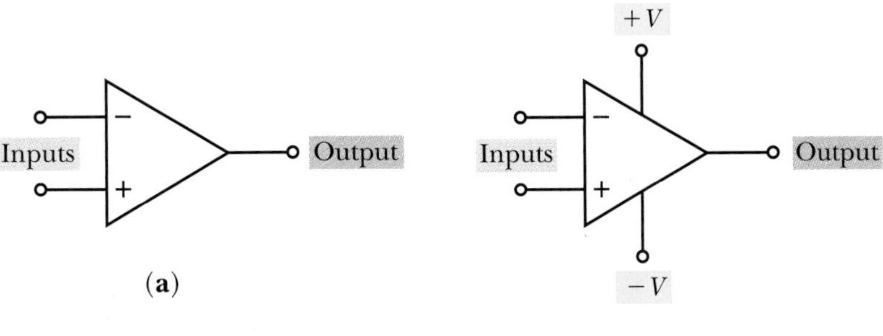

FIGURE 7–3 Op-amp symbols: (a) simplified symbol; (b) symbol showing *V* source connections

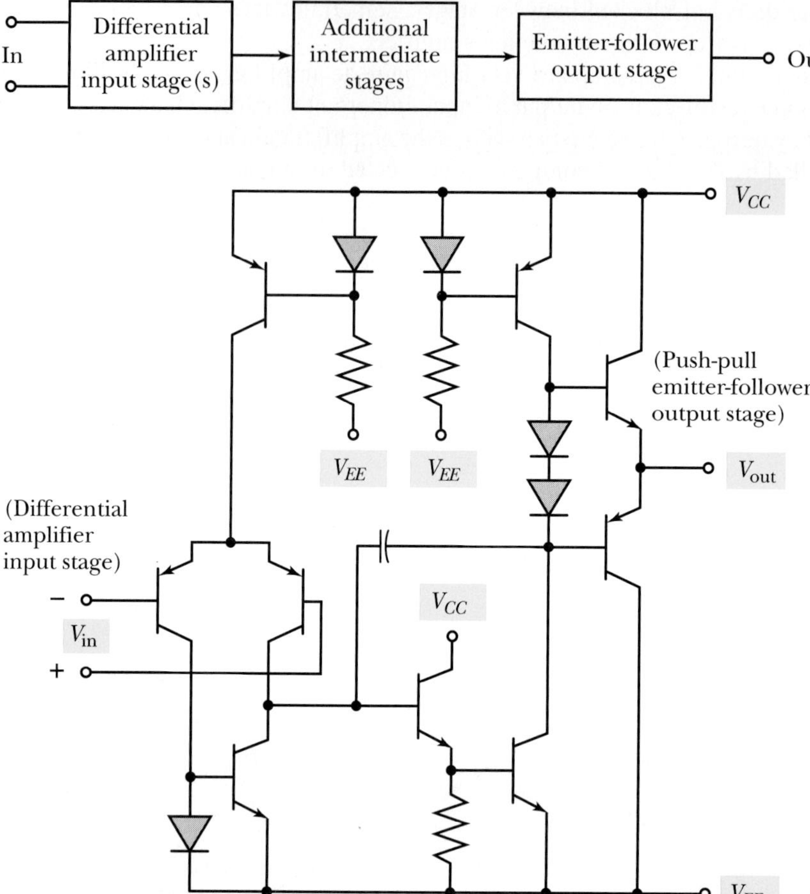

FIGURE 7–4 Simplified block diagram of an op-amp

FIGURE 7–5 Typical op-amp internal circuitry

As you can see from the preceding two figures, the schematic symbol, device connections, and block diagram for an op-amp are simple. However, the circuitry inside the device contains many circuits and connections shown in the typical IC op-amp diagram in Figure 7–5.

7–3 Op-amp Characteristics

With this background, let's now look at the characteristics of the op-amp. Figure 7–6 provides a circuit model for the op-amp. Some typical op-amp parameters are shown in Table 7–1 comparing the ideal amplifier with two common amplifiers, the LM741 and the LF347. From the table you can see that the ideal amplifier parameters are not achievable in practical devices. However, the ideal parameters are used to approximate operation when designing and evaluating circuits using op-amps.

The basic rules used with op-amps are:

1. No input bias current
2. No difference in voltage between the two inputs as long as the op-amp is not saturated
3. Infinite open-loop voltage gain for all frequencies

In practice, these devices have limitations similar to other semiconductor devices. That is, there are maximum current-carrying capabilities, maximum voltage ratings, and power limitations that must not be exceeded. Also, there is something less than a flat frequency response when attempting amplification from dc to radio frequencies.

As may be surmised from these discussions, real world op-amps cannot meet the absolute ideal parameters of infinite gain and so on. Table 7–1 shows that these

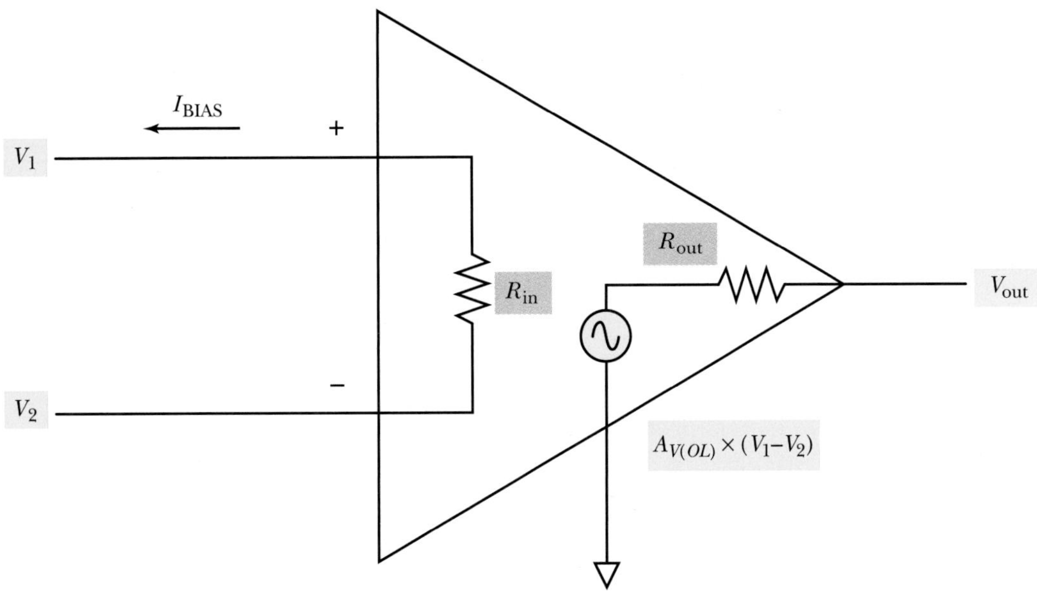

FIGURE 7–6 Op-amp circuit model

TABLE 7–1 Operational Amplifier Characteristics (Typical values at an ambient temperature of 25°C)

Parameter	Symbol	Ideal	LM741	LF347	Units
Technology			Bipolar	JFET	
Devices			Single	Quad	
Supply voltage	V_{CC}		5 to 15	3.5 to 18	V
	V_{EE}		0 to –15	–3.5 to –18	V
Supply current	I_{CC}	0	1.5	7.2	mA
Input bias current	I_B	0	80	0.05	nA
Output short curcuit current	I_{SC}	infinite	25	32	mA
Output voltage swing	$V_{O(P\text{-}P)}$	V_{EE} to + V_{CC}	(V_{EE} + 1.0) to (V_{CC} – 1.0)	(V_{EE} + 1.5) to (V_{CC} – 1.5)	V
Large signal voltage gain	$A_{V(OL)}$	infinite	200	100	V/mV
Bandwidth	BW	infinite	1	4	MHz
Slew rate	SR	infinite	0.5	13	V/us
Input resistance	R_{in}	infinite	2	1×10^6	M Ohms
Output resistance	R_{out}	0	100	30	Ohms
Input offset voltage	V_{os}	0	1	5	mV

devices exhibit high voltage gain, high input impedance, low output impedance, and wide bandwidth amplifying characteristics. Using this information, let's look at some basic op-amp circuits.

Elemental Op-amp Information

Op-amps can have open-loop voltage gains (the gains that occur when there is no feedback from the output to the input of the op-amp) from several thousand up to about a million. In Figure 7–7, the differential amplifier input stage of the op-amp has both an inverting input (at the negative input terminal) and a noninverting input (at the positive input terminal). If an input signal is applied to the inverting input with the noninverting input grounded, the output signal polarity is opposite the input signal polarity

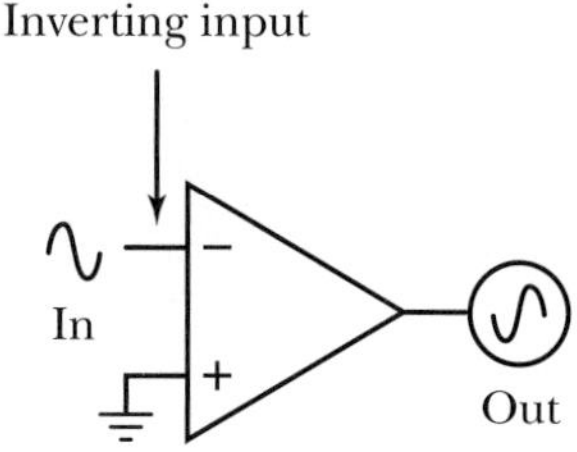

Signal fed to inverting input; output is inverted

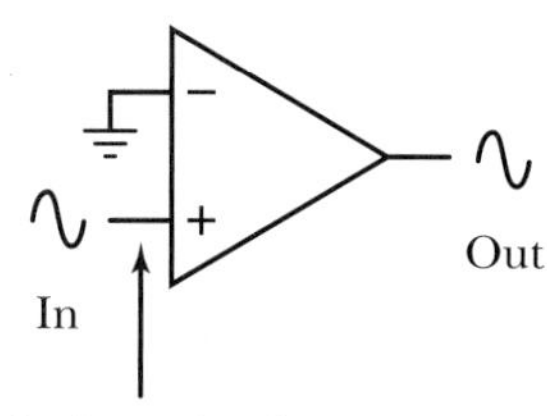

Signal fed to noninverting input; output is not inverted

FIGURE 7–7 Inverting and noninverting inputs

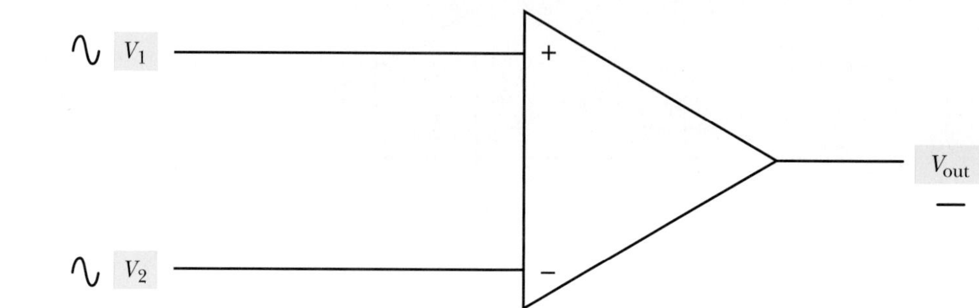

FIGURE 7–8 Difference amplifier

(Formula 7–1). Conversely, if the input signal is fed to the positive input terminal while the negative input terminal is grounded, the output signal polarity is the same polarity as the input one (Formula 7–2).

FORMULA 7–1 $V_{out} = -A_{VOL} \times V_{in}$

FORMULA 7–2 $V_{out} = -A_{VOL} \times V_{in}$

You have seen that a signal fed to the inverting input of an op-amp enables an inverted version at the output. Conversely, a signal fed to the noninverting input enables a noninverted version at the output. What will happen if you apply signals to both inputs at the same time? The result will be a difference waveform; that is, a waveform that represents the noninverted input minus the inverted input. This amplifier configuration is important in many applications today. It is called a **differential amplifier** because it amplifies the *difference* between the two input waveforms (Formula 7–3). See Figure 7–8. If you apply the exact same signal to both inputs of an op-amp, the differential amplifier effect produces no signal output.

FORMULA 7–3 $V_{out} = A_{VOL} \times (V_1 - V_2)$

This differential effect introduces another term called common-mode rejection ratio (CMRR). This is the degree the amplifier rejects (does not respond or produce output) for signals that are common to both inputs. This circuit feature provides useful rejection of undesired signals that can be present at both inputs. For example, undesired signals, such as 60-Hz pickup or other stray noise signals, may be present at the inputs. See Figure 7–9.

The enormous gain figures of the basic op-amp device (i.e., open-loop gain) make it necessary to control this gain by use of external components. In the next sections, resistors will be added to the circuit to provide negative feedback. This is the feedback that controls the amplifier gain.

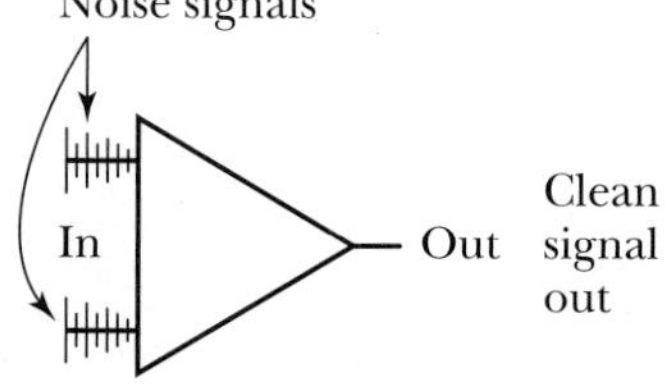

FIGURE 7–9 Noise signals common to both inputs cancel in the output

Linear and Nonlinear Op-amp Applications

Various **linear amplifier** applications (where the output waveform is similar to the input waveform) include **inverting amplifiers** (where the output is inverted from the input), **noninverting amplifiers,** voltage followers, summing amplifiers, differential amplifiers, amplifiers for instrumentation, stereo preamplifiers, logarithmic amplifiers, differentiators, integrators, sine wave oscillators, and active filters.

Several **nonlinear amplifier** applications (where the output waveform is different from the input waveform) include comparators, multivibrators, hysteresis oscillators, and other special-purpose circuits.

Practical Notes Manufacturers produce explicit data sheets on every device they manufacture. These data sheets accurately describe each device's capabilities and limitations. Thus, it is critical that a technician learns how to find, use, and properly interpret the data found on these data sheets. The LF347 data sheet (pp. 304–305) is provided as an example.

■ IN-PROCESS LEARNING CHECK 1

Fill in the blanks as appropriate.

1. An op-amp is a high gain ________-coupled amplifier.
2. The input stage of an op-amp uses a ________ amplifier configuration.
3. The name operational amplifier derives from their early use to perform mathematical ________.
4. The open-loop gain of an op-amp is much ________ than the closed-loop gain.
5. To achieve an output that is the inversion of the input, the input signal is fed to the ________ input of the op-amp.
6. Input signals to the op-amp can be fed to the ________ input, the ________ input, or to both inputs.

7–4 An Inverting Amplifier

In Figure 7–10 you see an amplifier with the input signal applied to inverting input through R_i. Part of the amplified and inverted output feeds back to the input through the feedback resistor, R_f. With the positive input grounded, the differential input signal equals the value of the input signal. The voltage gain of a circuit is defined in Formula 7–4. If ideal (infinite input impedance) conditions are assumed, the voltage gain of this amplifier circuit (commonly known as the closed-loop voltage gain) equals the ratio of the feedback resistor value to the input resistor value (Formula 7–5). The negative sign in the equation indicates that the output signal is inverted from the input signal. If the input signal is a positive dc voltage, then the output voltage is a negative dc voltage. If the input signal is a negative dc voltage, then the output signal is a positive dc voltage. If the input signal is a sine wave, then the output signal is an inverted sine wave (180° phase shift). This formula makes the design for a specific voltage gain easy.

FORMULA 7–4 $A_V = \dfrac{V_{out}}{V_{in}}$

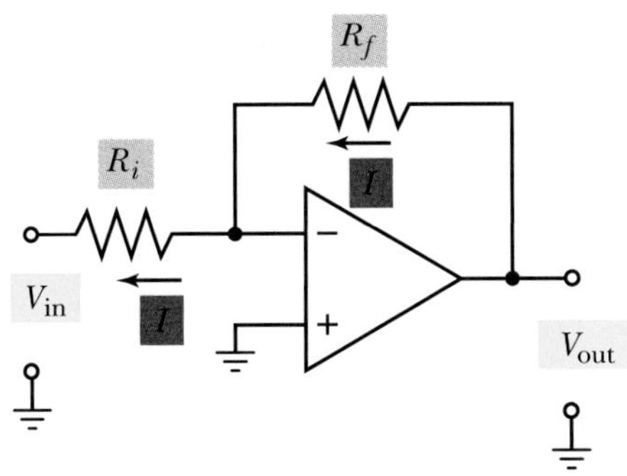

FIGURE 7–10 Inverting amplifier closed-loop gain

www.fairchildsemi.com

LF347

Quad Operational Amplifier (JFET)

Features

- Low input bias current
- High input impedance
- Wide gain bandwidth: 4 MHz Typ.
- High slew rate: 13 V/µs Typ.

Description

The LF347 is a high speed quad JFET input operational amplifier. This feature high input impedance, wide bandwidth, high slew rate, and low input offset voltage and bias current. LF347 may be used in circuits requiring high input impedance. High slew rate and wide bandwidth, low input bias current.

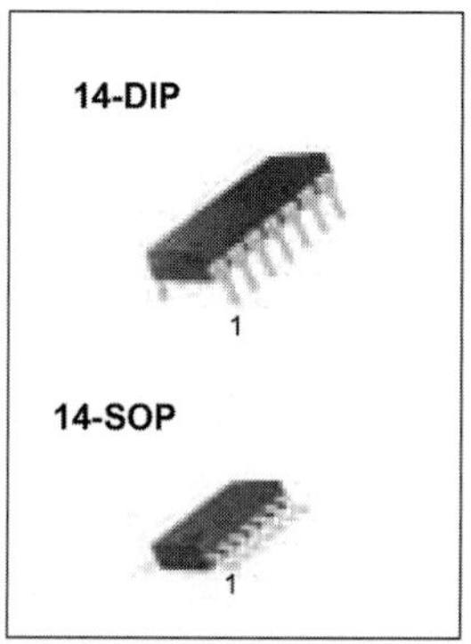

Internal Block Diagram

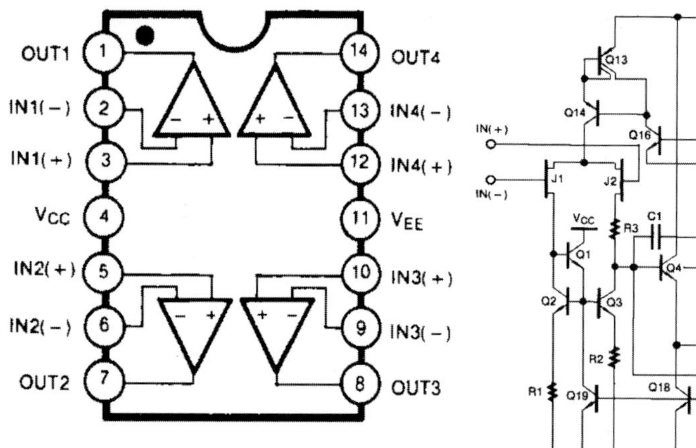

Schematic Diagram

(One Section Only)

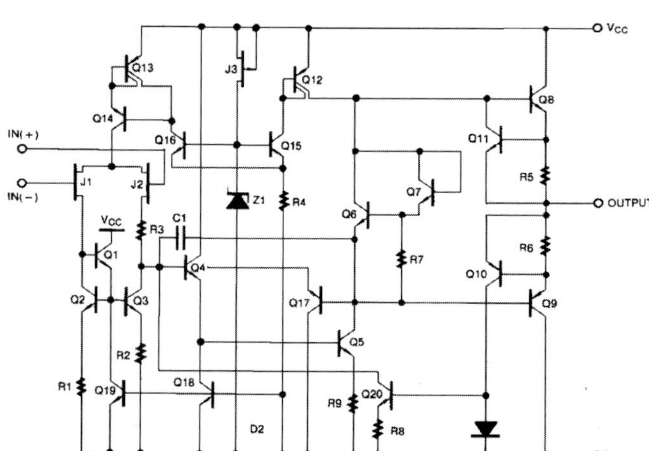

Rev. 1.0.0

LF347

Absolute Maximum Ratings

Parameter	Symbol	Value	Unit
Supply Voltage	V_{CC}	±18	V
Differential Input Voltage	$V_{I(DIFF)}$	30	V
Input Voltage Range	V_I	±15	V
Output Short Circuit Duration	-	Continuous	-
Power Dissipation	P_D	570	mW
Operating Temperature Range	T_{OPR}	0 ~ + 70	°C
Storage Temperature Range	T_{STG}	-65 ~ + 150	°C

Electrical Characteristics

(V_{CC}= +15V, V_{EE}= -15V, T_A=25 °C, unless otherwise specified)

Parameter	Symbol	Conditions		LF347 Min.	LF347 Typ.	LF347 Max.	Unit
Input Offset Voltage	V_{IO}	R_S = 10KΩ		-	5	10	mV
			Note 1	-	-	13	
Input Offset Voltage Drift(Note2)	$\Delta V_{IO}/\Delta T$	R_S = 10KΩ		-	10	-	µV/ °C
Input Offset Current	I_{IO}			-	25	100	pA
			Note 1	-	-	4	nA
Input Bias Current	I_{BIAS}			-	50	200	pA
			Note 1	-	-	8	nA
Large Signal Voltage Gain	G_V	R_L = 2KΩ		25	100	-	V/mV
		$V_{O(P-P)}$= ±10V	Note 1	15	-	-	
Output Voltage Swing	$V_{O(PP)}$	R_L = 10KΩ		±12	±13.5	-	V
Input Voltage Range	$V_{I(R)}$	-		±11	+15 -12	-	V
Common-Mode Rejection Ratio	CMRR	$R_S \leq$ 10KΩ		80	100	-	dB
Power Supply Rejection Ratio	PSRR	$R_S \leq$ 10KΩ		80	100	-	dB
Input Resistance	R_I	-		-	10^{12}	-	Ω
Supply Current	I_{CC}	-		-	7.2	11	mA
Slew Rate	SR	-		-	13	-	V/µS
Gain Bandwidth Product(Note2)	GBW	-		-	4	-	MHz
Channel Seperation	CS	f = 1Hz ~ 20Khz (input referenced)		-	120	-	dB
Equivalent Input Noise Voltage	e_N	R_S = 100Ω f = 1KHz		-	20	-	nV/ $\sqrt{Hz}$
Equivalent Input Noise Current	I_N	f = 1KHz		-	0.01	-	pA/$\sqrt{Hz}$

Note :

1. LF347 : 0≤T_A≤+70 °C
2. Guaranteed by design

Application Problem

General Electronics

PUBLIC ADDRESS SYSTEM

A public address system is connected to a microphone that has a maximum output voltage of 10 mV. The microphone is connected to a 10-watt audio amplifier system that is driving an 8 ohm speaker. The voltage amplifier is a noninverting op-amp circuit. Calculate the maximum voltage gain for the voltage amplifier stage and determine the resistor values to obtain the desired gain. Assume the power amplifier stage has a voltage gain of 1.

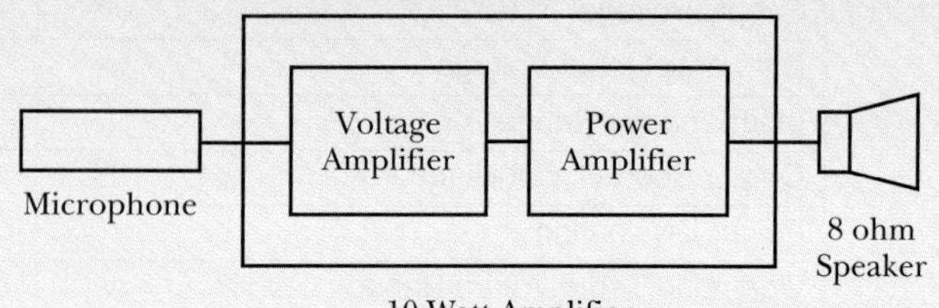

Solution First, find the formula for power, P.

$$P = \frac{V^2}{R}$$

solving for V we end up with:

$$V = \sqrt{R(\text{Power})} \quad V = \sqrt{8\,\Omega(10\text{ Watts})} = 8.94\text{ V}$$

The voltage gain for the amplifier is:

$$A_V = \frac{V_{\text{out}}}{V_{\text{in}}} = \frac{8.94\text{ V}}{10\text{ mV}} = 894$$

In selecting the resistor values for the noninverting amplifier, set the feedback resistor to 33 kΩ. The formula for the voltage gain of the noninverting amplifier is:

$$A_{VN1} = 1 + \frac{R_f}{R_i}$$

Solving for R_i we get the following result:

$$R_i = \frac{R_f}{A_V - 1} = \frac{33\text{ k}\Omega}{894 - 1} \cong 37\,\Omega$$

FORMULA 7–5 $A_V = -\frac{R_f}{R_i}$

The current flowing through R_i is $I = \frac{V_{\text{in}}}{R_1}$

The output voltage is $V_{\text{out}} = IR_i$

EXAMPLE For the circuit in Figure 7–10, if $R_f = 27$ kΩ, and $R_i = 2$ kΩ, what is the voltage gain of the amplifier? If the input voltage to this circuit is +0.2 V, what is the output voltage?

Answers:

From Formula 7–5:

$$A_V = \frac{R_e}{R_i}$$

$$A_V = -\frac{27\text{ k}\Omega}{2\text{ k}\Omega}$$

$$A_V = -13.5$$

From Formula 7–5:

$$V_{\text{out}} = A_V \times V_{\text{in}}$$

$$V_{\text{out}} = -13.5 \times (+0.2\text{ V})$$

$$V_{\text{out}} = -2.7\text{ V}$$

In this circuit, the negative terminal of the op-amp is referred to as a virtual ground since the voltage at this terminal is near 0 V. The input resistance for the inverting amplifier is equal to R_i.

Practical Notes

Be aware that the calculations generated through Formulas 7–4 and 7–5 may produce results that exceed the parameters for the op-amp that you are using. The op-amp's output voltage will not exceed the supply voltages applied to the op-amp.

PRACTICE PROBLEMS 1

1. What is the voltage gain of an inverting op-amp if $R_i = 100\ \text{k}\Omega$ and $R_f = 1\ \text{M}\Omega$
2. For the circuit in Figure 7–10 with $R_i = 2\ \text{k}\Omega$, what R_f value is needed to obtain a gain of –100?
3. What is the necessary value for R_i of an inverting op-amp if $R_f = 270\ \text{k}\Omega$ and the gain is to be –20?

7–5 A Noninverting Amplifier

The circuit for a noninverting op-amp is shown in Figure 7–11. As you can see, the junction of the feedback resistor (R_f) and input resistor (R_i) is still at the negative input terminal. However, the difference is that the input signal is fed to the positive input and R_i is grounded. R_f and R_i form a voltage divider for the feedback voltage. The differential voltage felt by the op-amp is the difference between the input voltage (V_{in}) fed to the positive input and the feedback voltage felt at the negative input terminal. This differential input is then amplified by the gain of the op-amp and produces the voltage output. The gain of the noninverting amplifier equals the feedback resistance divided by the input resistance, plus 1 (Formula 7–6).

FORMULA 7–6 $A_V = \dfrac{R_f}{R_i} \times 1$

The positive and negative op-amp inputs are the same as the input voltage. The current flowing through R_i and R_f is

$$I = \frac{V_{\text{in}}}{R_i}$$

The output voltage is

$$V_{\text{out}} = V_{\text{in}} + I \times R_f$$

EXAMPLE If the R_i and R_f values for a noninverting op-amp circuit are 1 kΩ and 12 kΩ, respectively, what is the voltage gain of the amplifier? If the input voltage to this circuit is 0.5 V, what is the output voltage?

Answers:

From Formula 7–6:

$$A_V = \frac{R_f}{R_i} + 1$$

$$A_V = \frac{12\ \text{k}\Omega}{1\ \text{k}\Omega} + 1$$

$$A_V = 13$$

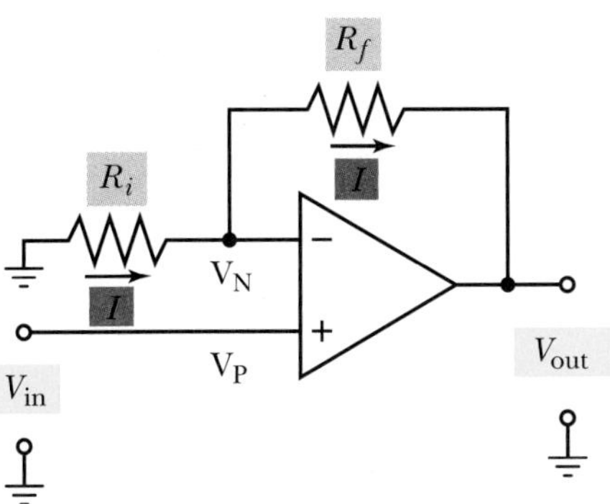

FIGURE 7–11 Noninverting amplifier closed-loop gain

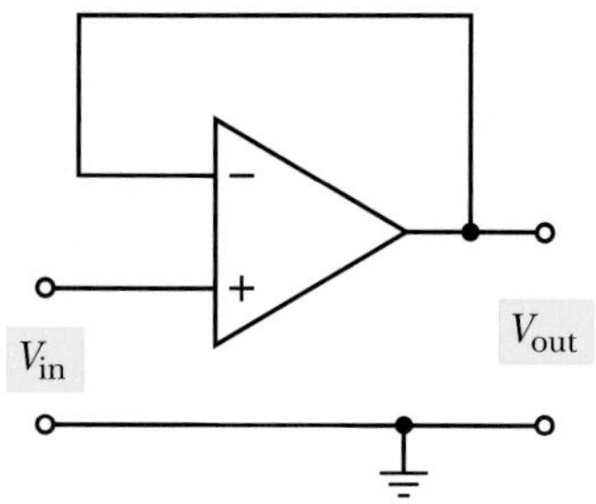

FIGURE 7–12 Voltage follower

From Formula 7–4:

$$V_{out} = A_V \times V_{in}$$
$$V_{out} = 13 \times (+0.5 \text{ V})$$
$$V_{out} = 6.5 \text{ V}$$ ◆

PRACTICE PROBLEMS 2

1. What is the voltage gain of a noninverting op-amp if $R_f = 20{,}000\ \Omega$ and $R_i = 2{,}000\ \Omega$?
2. For the circuit in Figure 7–11, what R_f value is needed to obtain a gain of 100 if $R_i = 10\ \text{k}\Omega$?
3. What is the value for R_i of a noninverting op-amp circuit if $R_f = 270\ \text{k}\Omega$ and the gain is to be 20?

A special application of a noninverting op-amp is one where the gain is 1, or unity. Any amplifier that has a gain of 1 and has no phase shift between the input signal and output signal is called a *voltage follower.* An op-amp voltage follower consists of a noninverting amplifier configuration where the R_i resistor has been removed and the feedback resistance is zero (the output is directly connected to the inverting input). See Figure 7–12.

One purpose of a voltage follower is to match a high impedance source to a low impedance load. The input impedance of an op-amp voltage follower is very close to the op-amp input resistance parameter, R_{in}. So if the op-amp has an input resistance of 2 MΩ, the input impedance is close to that value. This value can be changed by the addition of a resistor from the op-amp positive terminal to ground.

Figure 7–13 shows the addition of a voltage divider circuit to the input of the noninverting amplifier. The input resistance of this circuit is equal to the sum of resistors R_3 and R_4. The voltage gain of this circuit configuration is calculated by using Formula 7–7.

FORMULA 7–7 $A_V = \left(\frac{R_2}{R_1} + 1\right) \times \left[\frac{R_4}{(R_4 + R_3)}\right]$

◆ **EXAMPLE** Calculate the voltage divider noninverting amplifier voltage gain for the circuit in Figure 7–13 when $R_1 = 10\ \text{k}\Omega$, $R_2 = 50\ \text{k}\Omega$, $R_3 = 20\ \text{k}\Omega$, and $R_4 = 10\ \text{k}\Omega$. Find the output voltage when the input voltage is 4 V.

Answer:

From Formula 7–7:

$$A_V = \left(\frac{R_2}{R_1} + 1\right) \times \left[\frac{R_4}{(R_4 + R_3)}\right]$$
$$A_V = \left(\frac{50\ \text{k}\Omega}{10\ \text{k}\Omega}\right) \times \left[\frac{10\ \text{k}\Omega}{(10\ \text{k}\Omega + 20\ \text{k}\Omega)}\right]$$
$$A_V = 6 \times \frac{1}{3}$$
$$A_V = 2$$

Using Formula 7–4:

$$V_{out} = A_V \times V_{in}$$
$$V_{out} = 2 \times 4 \text{ V}$$
$$V_{out} = 8 \text{ V}$$ ◆

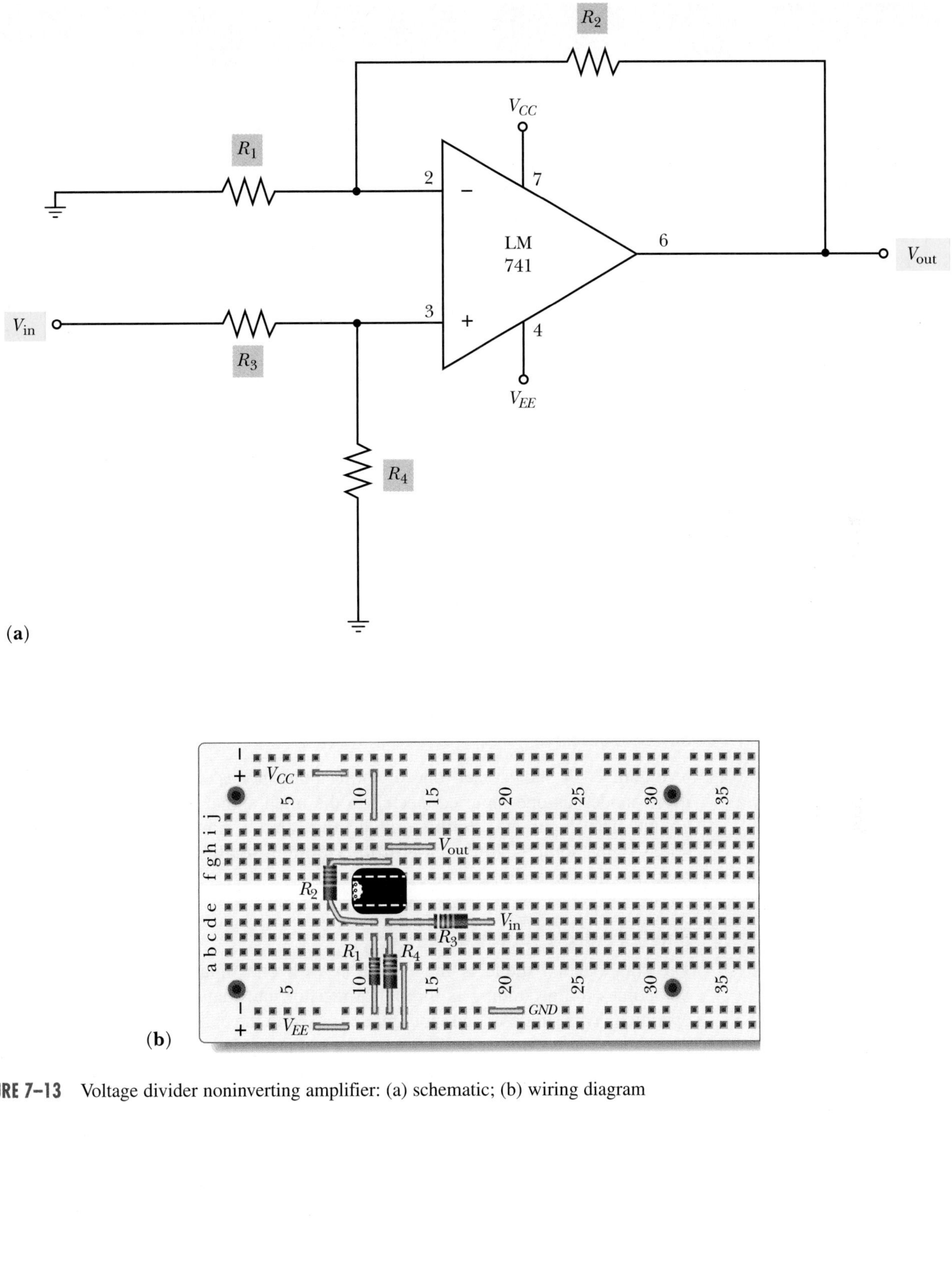

R_2

V_{CC}

R_1

2 − 7

LM 741

6 V_{out}

V_{in} 3 + 4

R_3

V_{EE}

R_4

(a)

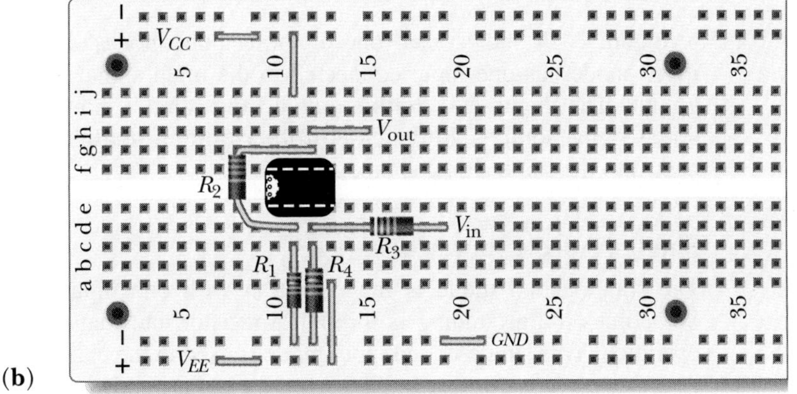

(b)

FIGURE 7–13 Voltage divider noninverting amplifier: (a) schematic; (b) wiring diagram

Application Problem

Industrial

PACKAGE COUNTING SYSTEM

The shipping department of a plastics manufacturing plant needs to monitor the number of packages traveling down a conveyer belt to keep track of the volume of packages the department ships daily. Each package has a silver reflective seal on it that will pass by a reflective optical sensor that will create a +1.5-V dc signal when it detects the seal. This voltage will need to be amplified and sent to a +24-V input module of a programmable logic controller (PLC) with an input resistance of 5 k Ω. Choose the correct amplifier to drive this low resistance load and select the resistor values to set the gain. The input voltage range for the PLC is +10 to +30 V dc.

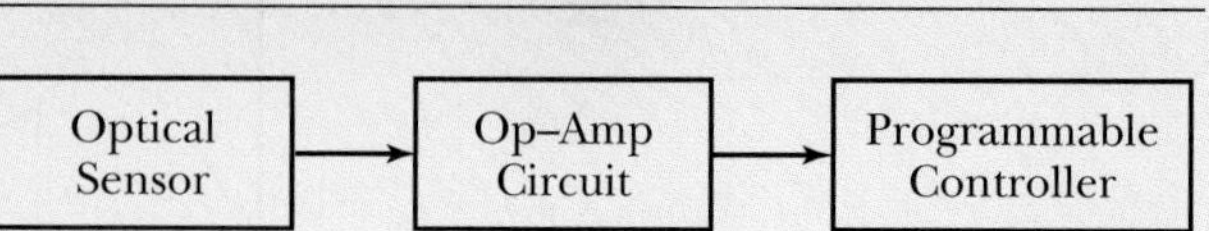

Solution The output of the sensor is +1.5 V and the required input voltage should range from +10 to +30 V. The noninverting amplifier would be a great match for this application since it won't invert the polarity of the input voltage. The amplifier has the ability to drive a low resistance load. A voltage gain of 11 would provide the PLC with 16.5 volts so it will recognize the presence of the object in front of the sensor. Let $R_f = 10\ \text{k}\Omega$ and $R_i = 1\ \text{k}\Omega$.

$$A_V = \frac{R_f}{R_i} + 1 = \frac{10\ \text{k}\Omega}{1\ \text{k}\Omega} + 1 = 11$$

7–6 Op-amp Input Modes

Op-amp circuits use one of three basic input modes of operation:

1. Single-ended
2. Double-ended (or difference)
3. Common-mode

Figure 7–14 shows each one of these input modes. Note that the **single-ended input** mode has one input connected to the input signal and the other connected to ground. The inputs of a **double-ended input** circuit, however, are connected to different signal sources. In the case of the **common-mode input,** both inputs are connected to the same signal source.

Single-Ended Input Mode

The single-ended mode is the operation with one input grounded, Figure 7–14a. Sometimes the grounding is through a resistor, but that doesn't affect the mode of operation. This single-ended input mode is used in the examples of inverting and noninverting op-amp circuits you have studied so far in this chapter. Most op-amp circuits in modern electronic equipment use the single-ended input mode. Formulas 7–1, 7–2, 7–5, and 7–6, relate to the single-ended input mode.

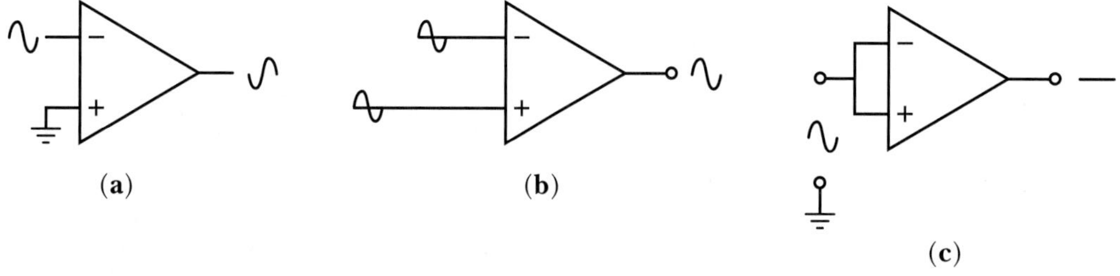

FIGURE 7–14 Modes of operation for op-amps: (a) single-ended; (b) double-ended; (c) common-mode

Op-amps are not as ideal as our approximations usually presume. When there is zero input to an inverting op-amp circuit, for instance, there should be zero output. But that is not precisely the case. The specification is called *input-offset voltage.* Likewise, there is always a little bit of current flowing into the junction point between R_f and R_i, even when there is no input voltage. This current is called the *input-bias current.*

The Double-Ended (Differential) Input Mode

The double-ended mode takes advantage of the fact that the op-amp is a differential amplifier. In this mode, signals are fed to both inputs at the same time. The result at the output is an amplified version of the difference between the two signals, Figure 7–14b.

Differential mode gain is the ratio of output voltage to the difference between the two signals fed to the two inputs of an op-amp that is functioning in the double-ended mode. Usually this gain figure is very high.

The Comparator Circuit

It often is useful to compare two voltages to see which is larger. An op-amp circuit that performs this function is shown in Figure 7–15. You can see that this is a basic double-ended input circuit. Because there is no feedback path, the gain of the circuit is extremely high. In fact, the output will normally be at one of the two extremes: close to +V supply voltage (V_{SAT+}) or close to –V supply voltage (V_{SAT-}). Just a few millivolts of difference between V_{REF} and V_{in} can drive the output to one of those extremes. It is the polarity of the output, rather than the actual voltage level, that is important to the function of this circuit. With this circuit, if V_{in} (noninverting input) is greater than V_{REF}(inverting input), the output goes to V_{SAT+}. On the other hand, if V_{in} less than V_{REF}, then the output goes to the V_{SAT-}.

Consider a specific application where V_{ref} is some dc voltage reference level and a sine-wave signal is fed to V_{in}. In this instance, the output is at its positive extreme only during the period when the sine-wave amplitude is greater than the dc reference value. During the remaining time, the output is at its low extreme. By changing the reference voltage level, the output can be varied from a pulse wave (duty cycle of 5%) to a pulse wave (duty cycle of 95%). One practical use of such a circuit is to convert a sine wave to a square wave (pulse wave with a 50% duty cycle).

A comparator circuit is an example of a nonlinear op-amp circuit. This means the waveform at the output does not necessarily resemble the waveforms at the input.

The Subtractor Circuit

A double-ended, or difference-mode, op-amp circuit with feedback is shown in Figure 7–16. This circuit performs the actual algebraic subtraction between two input voltage levels. It is an example of the mathematical computer applications of early operational amplifiers.

The output voltage (V_{out}) is related to the difference between the two input voltages (V_{in+} connected to the noninverting input and V_{in-} connected to the inverting input). The actual output voltage then depends on the amount of gain. Formula 7–8 provides the equation for this circuit. Formula 7–9 provides the equation to calculate the output

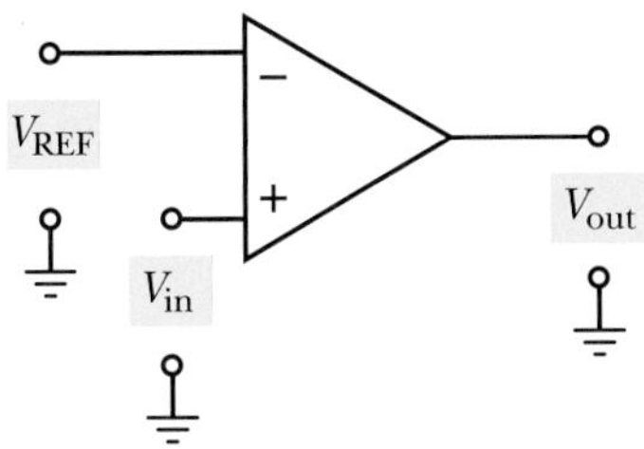

When noninverting V_{in} is greater than inverting V_{REF}, V_{out} is high.

When noninverting V_{in} is less than inverting V_{REF}, V_{out} is low.

FIGURE 7–15 Example of op-amp comparator

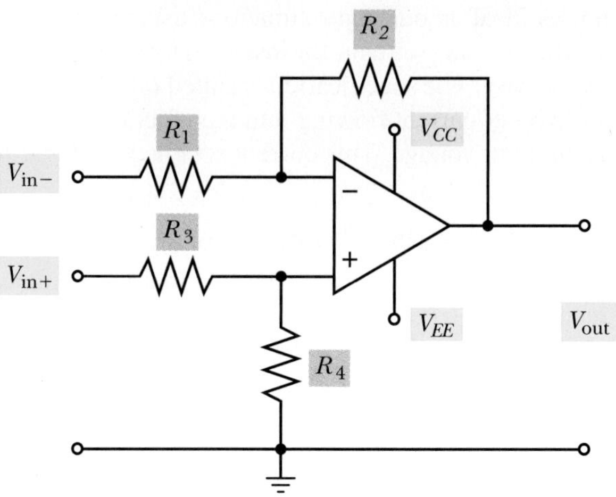

FIGURE 7–16 Op-amp as a difference amplifier

voltage for the case when the values of the resistors in Figure 7–16 are all the same. When all resistance values are the same for a difference amplifier, that gain is 1.

FORMULA 7–8 $V_{out} = V_{in+} \times \left(\frac{R_2}{R_1} + 1\right) \times \left[\frac{R_4}{(R_4 + R_3)}\right] + V_{in-} \times \left(-\frac{R_2}{R_1}\right)$

FORMULA 7–9 $V_{out} = V_{in+} - V_{in-}$

◆ **EXAMPLE**

1. Suppose the resistor values in Figure 7–16 all have a value of 10kΩ. What is the output voltage when $V_{in-} = +2$ V and $V_{in+} = +5$ V?

Answer:

From Formula 7–9:

$$V_{out} = V_{in+} - V_{in-}$$
$$V_{out} = 5 - 2 = 3 \text{ V}$$

2. What is V_{out} for question 1 when $V_{in-} = +2$ V and $V_{in+} = -5$ V?

Answer:

From Formula 7–9:

$$V_{out} = V_{in+} - V_{in-}$$
$$V_{out} = (-5) - (+2) = -7 \text{ V}$$ ◆

> *Practical Notes*
>
> You have seen that the output of a difference amplifier is the algebraic difference between the two input voltage levels. It is a mathematical substraction circuit. The output of a difference op-amp can go negative only if the device is supplied with both +V and –V power supplies as shown in the example offered here.
> Table 7–1 shows that some op-amps require dual-voltage power supplies while other op-amps can operate with a single-voltage power supply.

The Common-Mode Input

Although there is no practical application for an op-amp using the common-mode input, Figure 7–14c, the configuration is useful for defining the *common-mode gain* of the op-amp. If you apply the same signal to both the inverting and noninverting inputs, you do not see any output. Why not? Since the differential amplifier at the inputs causes the inverted and noninverted versions to cancel one another completely, the output will be 0 volts. That would be the ideal case. But no op-amp can be ideal, so there will always be a little bit of output signal. Recall that the common-mode rejection ratio (CMRR) is the ability of the op-amp to reject signals that are common to both inputs.

The common-mode gain specification for an op-amp is usually less than 1. The lower the value, the better the quality of the device.

Application Problem

RS232 LINE DRIVER

A computer will need to transmit serial digital data to an external modem. The TTL voltage (0 volts – Logic 0, +5 volts – Logic 1) will need to be converted to RS232 voltage levels (–3 V to –15 V for Logic 1, +3 V to +15 V for Logic 0) before being sent to the modem. A common voltage used is ±12 V. Determine the reference voltage needed to provide ±12 V at the output of a comparator circuit. Does the reference need to be on the inverting or noninverting input?

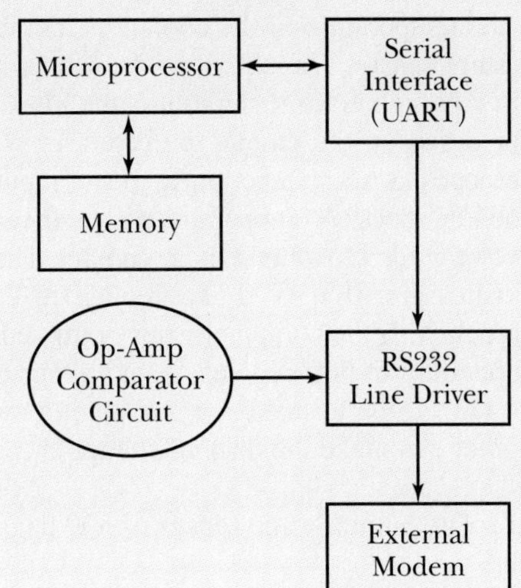

Solution The general formula for the output voltage for an op-amp is : $V_o = A(V_{ni} - V_i)$ V_o is the output voltage, V_{ni} is the voltage at the noninverting terminal, V_i is the voltage at the inverting terminal. We can use ±12 V supplies for the op-amp. Since the input voltage will be either 0 V or 5 V, the reference can be set at 2.5 V. Since we want the output voltage to be negative when the input is larger than the reference, the reference needs to be at the noninverting terminal.

Example: For a RS232 logic 1 using an op-amp with a gain of 10,000, the output voltage will be negative.

$$V_o = 10{,}000\ (2.5\text{ V} - 5\text{ V}) = -25{,}000\text{ V}$$

The maximum negative output voltage will be –12 V since we are using ±12 V supplies.

■ IN-PROCESS LEARNING CHECK 2

Fill in the blanks as appropriate.

1. For an op-amp inverting amplifier the input signal is applied through a resistor to the __________ input, and the feedback resistor is connected from the output to the ________________ input.
2. The voltage gain of an inverting amplifier is a negative value in order to indicate _______ of the signal.
3. In the schematic for an op-amp inverting amplifier, resistor R_f is located between the _______ of the circuit and the __________ input, Resistor R_i is connected to the __________ terminal.
4. For an op-amp noninverting amplifier the input signal is applied to the _______ input, and the feedback resistor is connected from the output to the ___________ input.
5. The special case of a noninverting op-amp circuit that has a voltage gain of 1 is called a(n) __________ circuit.
6. An op-amp circuit that has one signal input and one grounded input is called a(n) __________-ended circuit.
7. An op-amp circuit that has different signal sources connected to its inverting and noninverting inputs is called a(n) ___________-ended circuit.
8. A comparator circuit is an example of an op-amp operating in the __________-ended input mode.

Application Problem

Computers

IDLE SPEED MONITORING SYSTEM FOR AN AUTOMOBILE

In a car, the temperature of the coolant is monitored by a temperature sensor. The output of the sensor needs to be sent to a computer so it can adjust the idle speed of the motor based on the engine temperature. When the engine reaches a set temperature the computer will reduce the idle speed. A comparator will be the interface circuit between the sensor and the computer. The sensor characteristics are: 0 V @ 0° Fahrenheit to 2.5 V @ 800° Fahrenheit. If the computer is programmed to look for a temperature of 500 degrees, what voltage does the reference have to be to make the comparator switch so the computer can make the desired change?

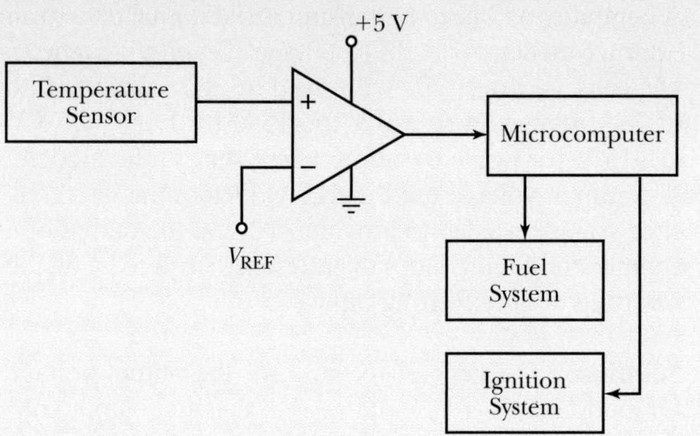

Solution The output of the sensor of 500 degrees would be:

$$\frac{V_0}{500} = \frac{2.5}{800}$$

Solving for V_0, the sensor output would be equal to 1.56 V. The reference of the comparator should be equal to 1.56 V.

7–7 Summing Amplifiers

You have already been introduced to the op-amp circuit that provides a subtraction operation. The **summing amplifier** is one that provides the algebraic sum of the input voltage levels. In Figure 7–17a, two input signals are applied through resistors into the inverting input of the amplifier. Formula 7–10 provides the equation to calculate the output voltage.

FORMULA 7–10 $V_{out} = V_{in1} \times \left(-\frac{R_f}{R_1}\right) + V_{in2} \times \left(-\frac{R_f}{R_2}\right)$

◆ **EXAMPLE** The resistor values in Figure 7–17a all have a value of 10 kΩ. What is the output voltage when $V_{in_1} = +2$ V and $V_{in_2} = +3$ V?

Answer:

From Formula 7–10:

$$V_{out} = V_{in1} \times \left(-\frac{R_f}{R_1}\right) + V_{in2} \times \left(-\frac{R_f}{R_2}\right)$$
$$V_{out} = (+2\text{ V}) \times \left(-\frac{10\text{ k}\Omega}{10\text{ k}\Omega}\right) + (+3\text{ V}) \times \left(-\frac{10\text{ k}\Omega}{10\text{ k}\Omega}\right)$$
$$V_{out} = (-2\text{ V}) + (-3\text{ V})$$
$$V_{out} = -5\text{ V}$$

◆

In Figure 7–17b, three input signals are applied through resistors into the inverting input of the amplifier. Formula 7–11 provides you with the equation to calculate the output voltage. Since the negative terminal is a virtual ground, all of the V_{in_1} voltage is dropped across R_1, all of the V_{in_2} voltage is dropped across R_2, and all of the V_{in_3} voltage is dropped across R_3. The currents flowing through these resistors combine to

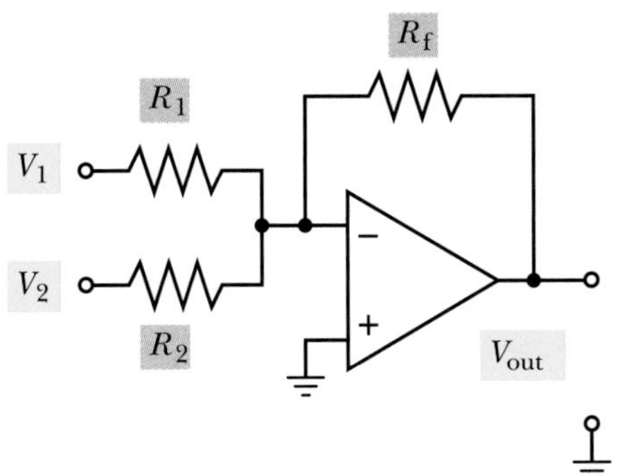

(a) 2-input summing amplifier (inverting input)

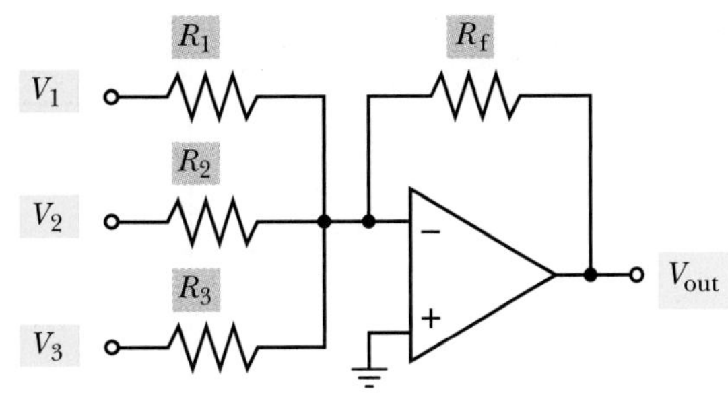

(b) 3-input summing amplifier (inverting input)

FIGURE 7–17 Examples of summing amplifiers

flow through the feedback resistor. Use Ohm's law to compute each input current. The sum of those currents passes through the feedback resistor (R_f), and the product of that current and the feedback resistor value provides the output voltage.

FORMULA 7–11 $V_{out} = V_{in_1} \times \left(-\frac{R_f}{R_1}\right) + V_{in_2} \times \left(-\frac{R_f}{R_2}\right) + V_{in_3} \times \left(-\frac{R_f}{R_3}\right)$

◆ **EXAMPLE** The resistor values in Figure 7–17b are: R_1 = 5 kΩ, R_2 = 10 kΩ, R_3 = 5 kΩ, and R_f = 1 kΩ. What is the output voltage when $V_{in_1} = V_{in_2} = V_{in_3} = +5$ V?

Answer:

From Formula 7–11:

$$V_{out} = V_{in_1} \times \left(-\frac{R_f}{R_1}\right) + V_{in_2} \times \left(-\frac{R_f}{R_2}\right) + V_{in_3} \times \left(-\frac{R_f}{R_3}\right)$$
$$V_{out} = (5\text{ V}) \times \left(-\frac{1\text{ k}\Omega}{5\text{ k}\Omega}\right) + (5\text{ V}) \times \left(-\frac{1\text{ k}\Omega}{10\text{ k}\Omega}\right) + (5\text{ V}) \times \left(-\frac{1\text{ k}\Omega}{5\text{ k}\Omega}\right)$$
$$V_{out} = (-1\text{ V}) + (-0.5\text{ V}) + (-1\text{ V})$$
$$V_{out} = -2.5\text{ V}$$

◆

The noninverting input of the amplifier can also be used for the summing amplifier as shown in Figure 7–18. Formula 7–12 provides the equation for this circuit. As can be observed from the equation, the value of R_3 impacts the gain for both inputs and the value of R_4 impacts the gain for both inputs.

FORMULA 7–12 $V_{out} = \left(1 + \frac{R_2}{R_1}\right) \times \left[V_{in_1} \times \frac{R_4}{(R_3 + R_4)} + V_{in_2} \times \frac{R_3}{(R_3 + R_4)}\right]$

◆ **EXAMPLE** In Figure 7–18, R_1 = 10 kΩ, R_2 = 20 kΩ, R_3 = 10 kΩ, and R_4 = 20 kΩ. What is the output voltage when V_{in_1} = +1.5 V and V_{in_2} = +2.5 V?

Answer:

From Formula 7–12:

$$V_{out} = \left(1 + \frac{R_2}{R_1}\right) \times \left[V_{in_1} \times \frac{R_4}{(R_3 + R_4)} + V_{in_2} \times \frac{R_3}{(R_3 + R_4)}\right]$$
$$V_{out} = \left(1 + \frac{20\text{ k}\Omega}{10\text{ k}\Omega}\right) \times \left[1.5\text{ V} \times \frac{20\text{ k}\Omega}{(10\text{ k}\Omega + 20\text{ k}\Omega)} + 2.5\text{ V} \times \frac{10\text{ k}\Omega}{(10\text{ k}\Omega + 20\text{ k}\Omega)}\right]$$
$$V_{out} = (3) \times \left[1.5 \times \frac{2}{3} + 2.5 \times \frac{1}{3}\right]$$
$$V_{out} = 5.5\text{ V}$$

◆

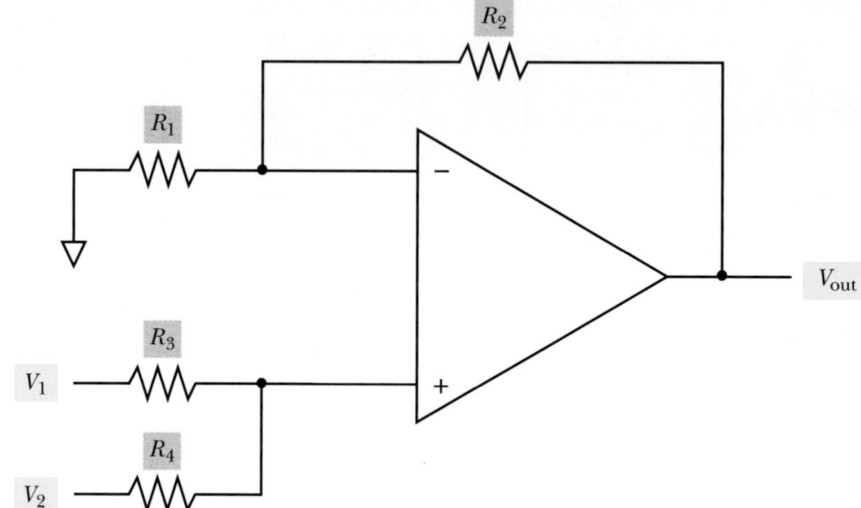

FIGURE 7–18 Noninverting input summing amplifier

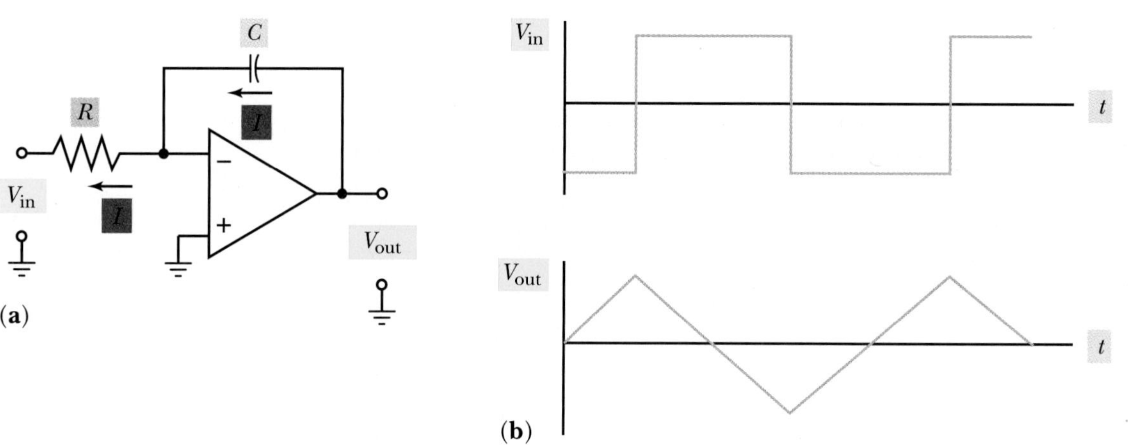

FIGURE 7–19 Example of (a) Op-amp integrator; (b) Integrator waveforms

7–8 Integrator Circuits

The **integrator circuit** uses a capacitor, instead of a resistor, as the feedback component, Figure 7–19. It performs the mathematical (calculus) function of integration. This means the output is related to the "area under the curve" for the input waveform. Formula 7–13 shows the relationship of the output voltage to the input voltage.

The current flowing through the resistor and capacitor is

$$I = \frac{V_{in}}{R}$$

The constant current flowing causes the capacitor to charge as evidenced by the output voltage.

One popular application of this circuit is to produce a triangular voltage output when its input is a rectangular wave.

FORMULA 7–13 $\Delta V_{out} = \left[-\frac{V_{in}}{(R \times C)} \right] \times \Delta t$

◆ **EXAMPLE** The resistor and capacitor values in Figure 7–19 are 1 kΩ and 0.01 μF respectively. With a 3 V input, what would be the change in the output voltage after 5 μsec?

Answer:

From Formula 7–13:

$$\Delta V_{out} = \left[-\frac{V_{in}}{(R \times C)}\right] \times \Delta t$$

$$\Delta V_{out} = \left[-\frac{3\text{ V}}{(1\text{ k}\Omega \times 0.01\ \mu\text{F})}\right] \times 5\ \mu\text{sec}$$

$$\Delta V_{out} = -1.5\text{ V}$$

◆

What would be the change in the output voltage after 20 μsec? The change in output voltage increases for longer time intervals. Is there a limitation to the output voltage? Remember that the output voltage of the op-amp will not be greater than V_{SAT+} will not be less than V_{SAT-}.

7–9 Differentiator Circuits

The **differentiator circuit,** Figure 7–20, is another example of a calculus function. In this case, the output is proportional to the derivative of the input voltage—the rate of change of voltage at the input. This operation is the opposite of integration, so you can see that a capacitor replaces the input resistor.

As the input voltage changes, current flows in the circuit charging the capacitor. The output voltage is equal to the current flow times the feedback resistor.

One application of this circuit is to detect pulsed edges (such as the leading or trailing edges of a rectangular wave). Another useful function of this circuit is to produce a rectangular output from a triangular input voltage.

FORMULA 7–14 $V_{out} = (-R \times C) \times \left(\frac{\Delta V_{in}}{\Delta t}\right)$

◆ **EXAMPLE** The resistor and capacitor values in Figure 7–20 are 1 kΩ and 0.01 μF respectively. With a ramp input voltage of 0.5 V/μsec, what would be the output voltage?

Answer:

From Formula 7–14:

$$V_{out} = (-R \times C) \times \left(\frac{\Delta V_{in}}{\Delta t}\right)$$

$$V_{out} = (-1\text{ k}\Omega \times 0.01\ \mu\text{F}) \times 0.5\text{ V}/\mu\text{sec}$$

$$V_{out} = -5\text{ V}$$

◆

As long as the input ramp voltage is applied, the output voltage will remain at –5 V. Once the input voltage goes to a constant dc voltage, the output voltage will go to 0 V. The output voltage will not be equal to 0 V as long as the input voltage is constantly changing.

7–10 Instrumentation Amplifier

The instrumentation amplifier (Figure 7–21) provides a similar result as the difference amplifier. The primary distinction of the instrumentation amplifier is the balanced high input impedance for both signal inputs. This provision increases common mode noise rejection and reduces voltage offsets observed with the difference amplifier. Formula 7–15 provides the equation to calculate the output voltage of the instrumentation amplifier where R_1, R_2, R_3, R_4, R_5, and R_6 are the same value *(R)*.

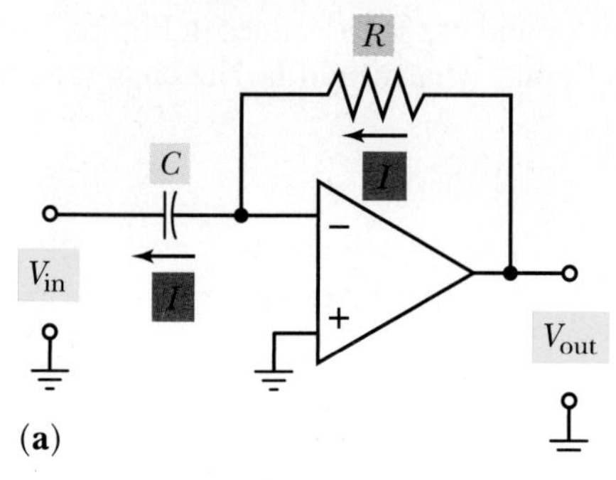

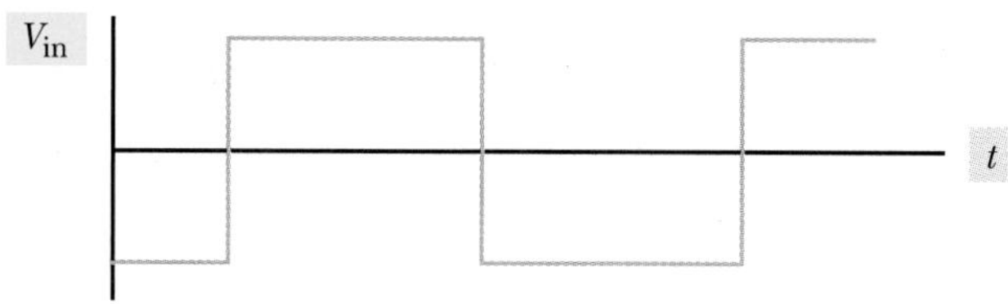

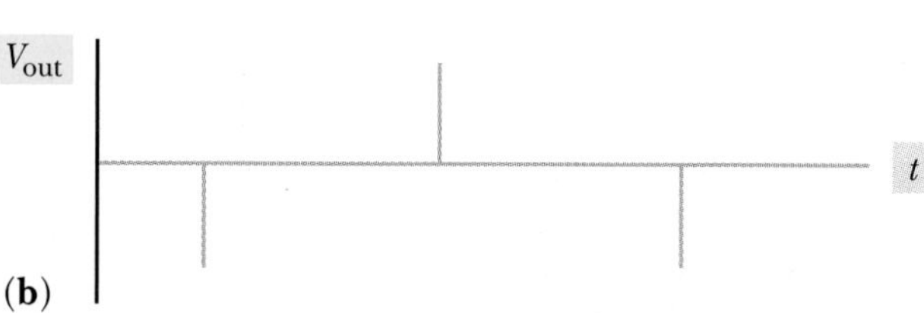

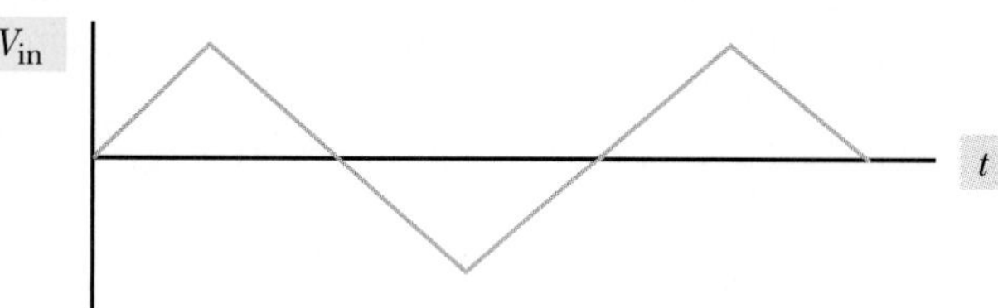

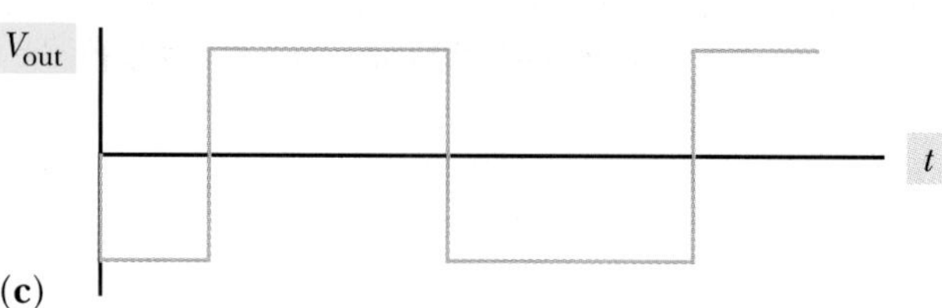

FIGURE 7–20 (a) Op-amp differentiator; (b) Differentiator waveforms (square wave input); (c) Differentiator waveforms (triangle wave input)

The voltage V_A is affected by both input voltages V_1 and V_2.

$$V_A = V_1 \times \left(1 + \frac{R_1}{R_G}\right) + V_2 \times \left(-\frac{R_1}{R_G}\right)$$

The voltage V_B is also affected by both input voltages.

$$V_B = V_2 \times \left(1 + \frac{R_2}{R_G}\right) + V_1 \times \left(-\frac{R_2}{R_G}\right)$$

The voltages V_A and V_B are applied to the difference amplifier. When resistors R_3, R_4, R_5 and R_6 are the same value of resistance, the difference amplifier output voltage will be affected by V_A and V_B equally.

$$V_{out} = V_B - V_A$$

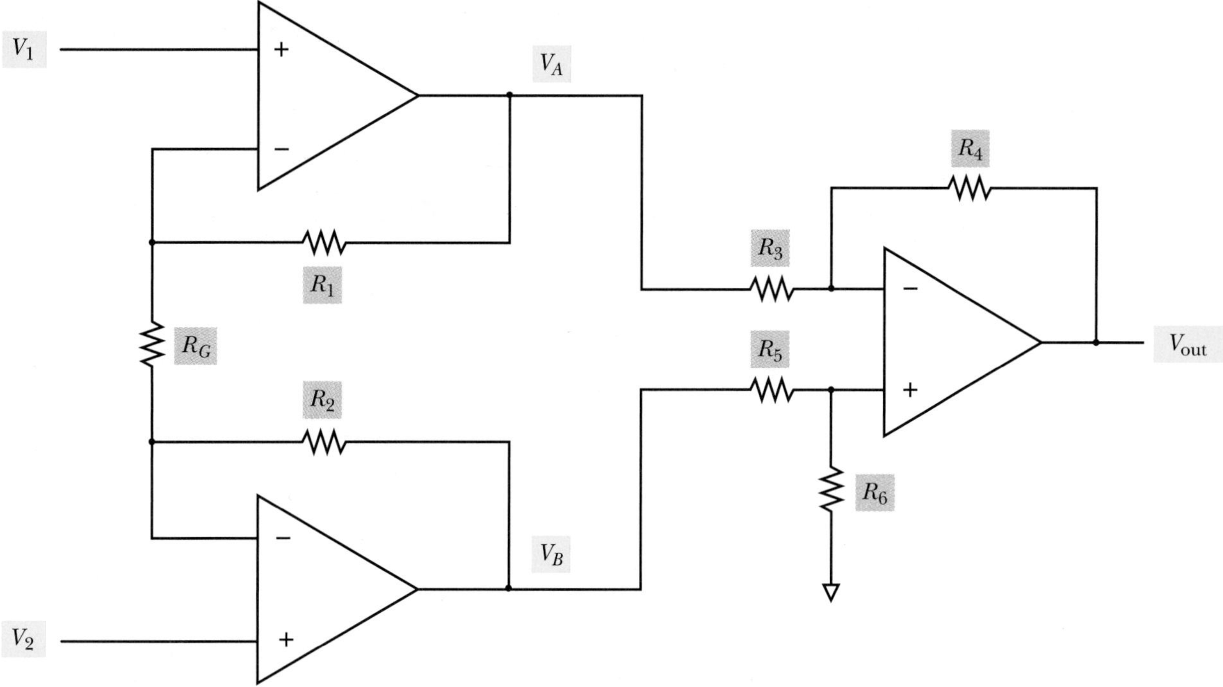

FIGURE 7–21 Instrumentation amplifier

FORMULA 7–15 $V_{out} = \left(1 + 2 \times \frac{R}{R_G}\right) \times (V_2 - V_1)$

EXAMPLE Calculate the output voltage for the instrumentation amplifier in Figure 7–21. The resistor values are: $R = 10\ k\Omega$ and $R_G = 5\ k\Omega$. The input voltages are: $V_1 = +0.5$ V and $V_2 = +2.0$ V.

Answer:

From Formula 7–15:

$$V_{out} = \left(1 + 2 \times \frac{R}{R_G}\right) \times (V_2 - V_1)$$
$$V_{out} = \left(1 + 2 \times \frac{10\ k\Omega}{5\ k\Omega}\right) \times (2.0\ V - 0.5\ V)$$
$$V_{out} = (5) \times (1.5\ V)$$
$$V_{out} = 7.5\ V$$

7–11 Comparators

The comparator circuit allows us to use an electrical circuit for the comparison of two voltages, a reference voltage and an input voltage, to determine which is larger. There are two output voltage levels. One voltage level indicates that the input voltage is greater than the reference voltage. The other voltage level indicates that the input voltage is less than the reference voltage. A variety of comparator circuits exist.

Noninverting Comparator

The schematic of a noninverting comparator using an operational amplifier is shown in Figure 7–22a. Figure 7–22b provides the waveforms for the circuit with $V_{CC} = +12$ V,

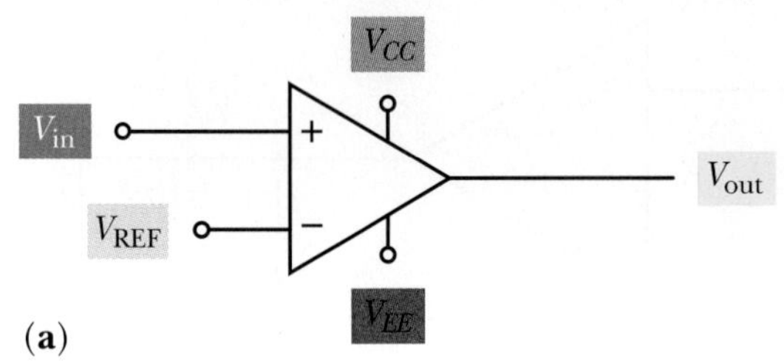

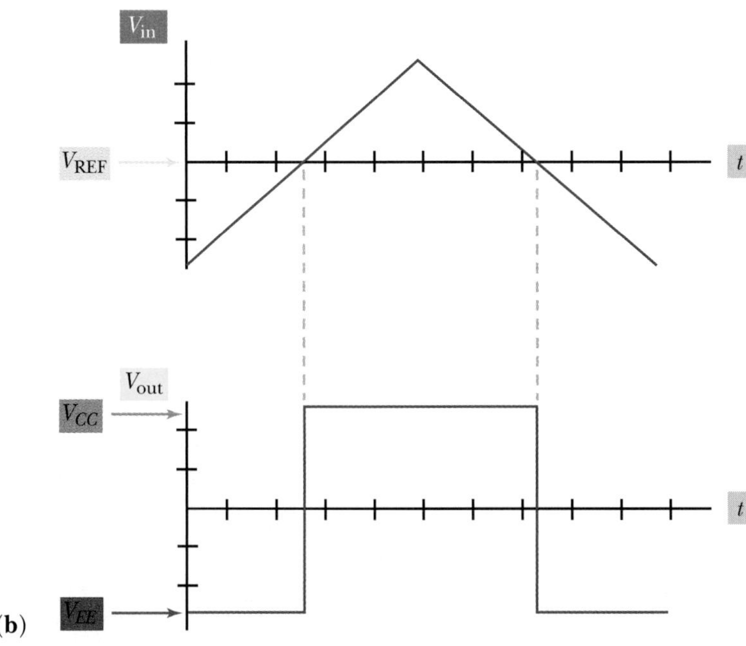

FIGURE 7–22 (a) Noninverted comparator input; (b) Noninverted comparator waveforms

$V_{EE} = -12$ V, and $V_{REF} = 0$ V. The reference voltage V_{REF} can be any voltage level between the positive supply voltage level V_{CC} and the negative supply voltage level V_{EE}.

Because there is no feedback path, the gain of the circuit is extremely high. In fact, the output will normally be at one of the two extremes: close to +V supply voltage (V_{SAT+}) or close to –V supply voltage (V_{SAT-}). Just a few millivolts of difference between V_{ref} and V_{in} can drive the output to one of those extremes. With this circuit, if V_{in} (noninverting input) is greater than V_{REF} (inverting input), the output goes to V_{SAT+}. On the other hand, if V_{in} is less than V_{REF}, then the output goes to V_{SAT-}.

◆ **EXAMPLE** The circuit in Figure 7–22a has $V_{CC} = +12$ V, $V_{EE} = -12$ V ($V_{SAT+} = 10.5$ V and $V_{SAT-} = -12$ V), and $V_{REF} = 4$ V.

a. If $V_{in} = 3$ V, what is the output voltage?

b. If $V_{in} = 5$ V, what is the output voltage?

Answers:

a. Since V_{in} (3 V) is less than V_{REF} (4 V), then V_{out} is equal to V_{SAT-} (–12 V).

b. Since V_{in} (5 V) is greater than V_{REF} (4 V), then V_{out} is equal to V_{SAT+} (+10.5 V). ◆

Inverting Comparator

The inverting comparator circuit schematic is provided in Figure 7–23a. The reference voltage is applied to the positive op-amp terminal while the input voltage is applied to the negative terminal of the op-amp. Figure 7–23b provides the waveforms for the circuit with $V_{CC} = +12$ V, $V_{EE} = -12$ V, and $V_{REF} = 0$ V. The reference voltage V_{REF} can be

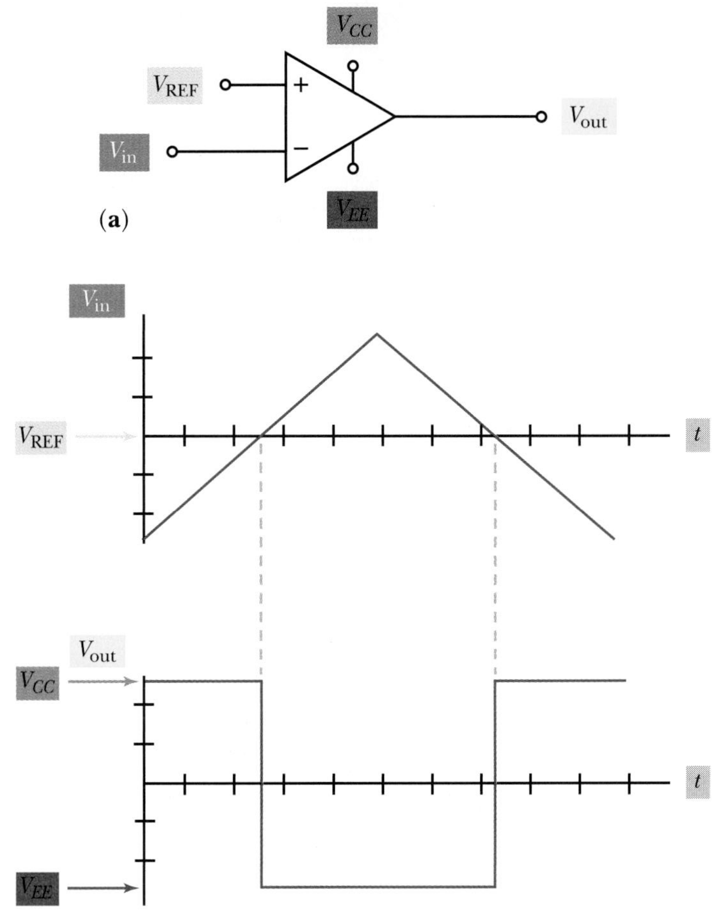

FIGURE 7–23 (a) Inverted comparator circuit; (b) Inverted comparator waveforms

any voltage level between the positive supply voltage level V_{CC} and the negative supply voltage level V_{EE}.

With this circuit, if V_{in} (inverting input) is less than V_{REF} (noninverting input), the output goes to $V_{\text{SAT+}}$. On the other hand, if V_{in} is greater than V_{REF}, then the output goes to the $V_{\text{SAT}-}$.

◆ **EXAMPLE** The circuit in Figure 7–23a has $V_{CC} = +12$ V, $V_{EE} = -12$ V ($V_{\text{SAT+}} = 10.5$ V and $V_{\text{SAT}-} = -12$ V), and $V_{\text{REF}} = -2$ V.

a. If $V_{\text{in}} = -3$ V, what is the output voltage?

b. If $V_{\text{in}} = 5$ V, what is the output voltage?

Answers:

a. Since V_{in} (–3 V) is less than V_{REF} (–2 V), then V_{out} is equal to $V_{\text{SAT}-}$ (+ 5 V).

b. Since V_{in} (5 V) is greater than V_{REF} (–2 V), then V_{out} is equal to $V_{\text{SAT+}}$ (–12 V). ◆

An Op-Amp Schmitt Trigger Circuit

The circuit in Figure 7–24a is an example of one of the few op-amp circuits that uses positive feedback. Note that the feedback resistor (R_2) is connected between the output of the op-amp and the noninverting input and resistor (R_1) is connected from the noninverting input to ground. Operated in this way, the circuit is known as a **Schmitt trigger** circuit, or **hysteresis** amplifier. This circuit uses a single-ended input mode, and it is further classified as a nonlinear application because the output waveform can be much different from the input waveform. There is some similarity in operation between the Schmitt trigger amplifier and the basic comparator circuit; see again Figure 7–15. Both types of circuits "square up" their input waveforms, but the use of

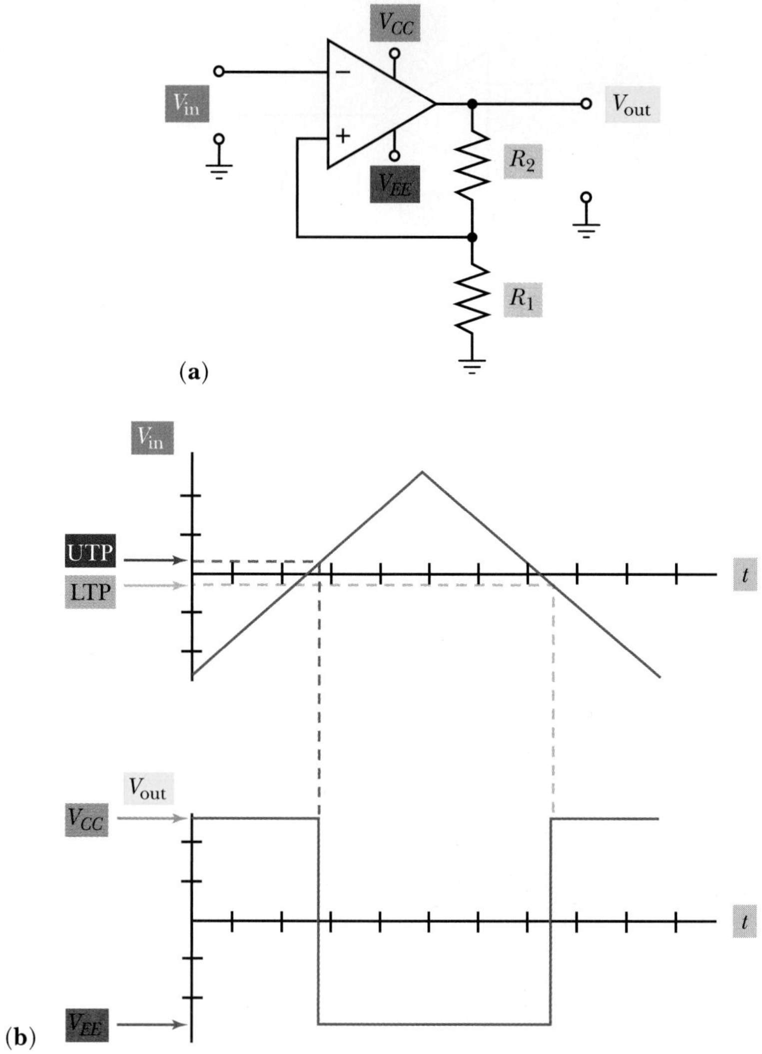

(a)

(b)

FIGURE 7–24 (a) Op-amp as Schmitt trigger amplifier; (b) Schmitt trigger amplifier waveforms

external resistors in a Schmitt trigger amplifier enables more control over the on/off triggering levels. When the input voltage is moving in a positive direction and reaches the upper threshold point (UTP), the output quickly switches (due to the positive feedback) to its negative output level. Then as the input waveform reaches it peak positive level and moves in the negative direction, it soon reaches the lower threshold point (LTP), where the output switches to its positive output level. With any Schmitt trigger circuit, there is always a difference between the UTP and LTP values. This difference is known as the hysteresis of the amplifier. You can calculate formulas for the threshold points and the hysteresis range (VH) by using Formulas 7–16, 7–17, and 7–18 while referring to the circuit in Figure 7–24a. V_{max} is the more positive of the dc supply voltages, and V_{min} is the more negative dc supply voltage.

FORMULA 7–16 $\text{UTP} = V_{max} \times \frac{R_1}{(R_1 + R_2)}$

FORMULA 7–17 $\text{LTP} = V_{min} \times \frac{R_1}{(R_1 + R_2)}$

FORMULA 7–18 $\text{VH} = \text{UTP} - \text{LTP}$

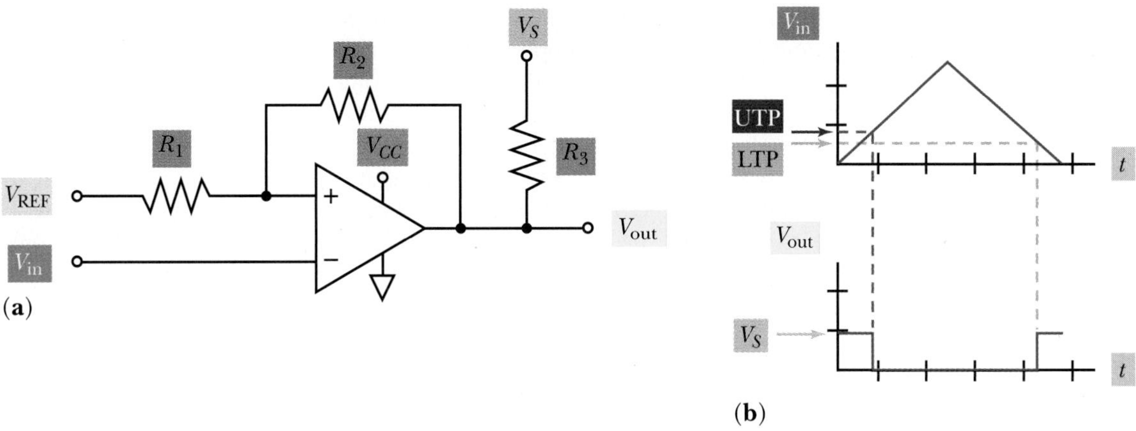

FIGURE 7–25 (a) Inverting comparator with hysteresis; (b) Inverting comparator with hysteresis waveforms

◆ **EXAMPLE** The circuit in Figure 7–24a has $V_{max} = +12$ V, $V_{min} = -12$ V, $R_1 = 10$ kΩ, and $R_2 = 100$ kΩ. Calculate the values of UTP, LTP, and the hysteresis range.

Answers:

From Formula 7–16:

$$\text{UTP} = V_{max} \times \frac{R_1}{(R_1 + R_2)}$$
$$\text{UTP} = 12\text{ V} \times \frac{10\text{ k}\Omega}{10\text{ k}\Omega + 100\text{ k}\Omega}$$
$$\text{UTP} = 1.09\text{ V}$$

From Formula 7–17:

$$\text{LTP} = V_{min} \times \frac{R_1}{(R_1 + R_2)}$$
$$\text{LTP} = (-12\text{ V}) \times \frac{10\text{ k}\Omega}{10\text{ k}\Omega + 100\text{ k}\Omega}$$
$$\text{LTP} = -1.09\text{ V}$$

The hysteresis range is calculated using Formula 7–18:

$$\text{VH} = \text{UTP} - \text{LTP}$$
$$\text{VH} = 1.09\text{ V} - (-1.09\text{ V})$$
$$\text{VH} = 2.18\text{ V}$$

Figure 7–24b shows the output voltage as the input voltage was varied from –12 V to +12 V and from +12 V back to –12 V. ◆

Inverting Comparator with Hysteresis

The inverting comparator with hysteresis circuit schematic is provided in Figure 7–25a. The reference voltage is applied through a resistor (R_1) to the positive op-amp terminal while the input voltage is applied to the negative terminal of the op-amp. The amplifier (LM311 or LM339) used within this circuit has been specially designed for comparator circuit applications. These devices can operate using a single or dual power supply. They have an open collector output stage what will require a pull-up resistor (R_3) to a supply voltage V_S. The reference voltage V_{REF} can be any voltage level between the positive supply voltage level V_{CC} and the negative supply voltage level V_{EE}. The positive feedback causes a change in the threshold voltage level when the output voltage changes from a logic low to a logic high or from a logic high to a logic low.

Figure 7–25b provides the waveforms for the circuit. When the output voltage is a logic high (close to the supply voltage V_S) and the input voltage (V_{in}) is less than the upper threshold voltage (UTP), the output voltage (V_{out}) will remain a logic high. Once the input voltage (V_{in}) exceeds the upper threshold voltage (UTP), the output voltage (V_{out}) will change to a logic low (close to the negative supply voltage V_{EE}). When the output voltage is a logic low and the input voltage (V_{in}) is greater than the upper threshold voltage (LTP), the output voltage (V_{out}) will remain a logic low. Once the input voltage (V_{in}) decreases below the lower threshold voltage (LTP), the output voltage (V_{out}) will change to a logic high. The threshold voltage is calculated using Formulas 7–19 and 7–20.

FORMULA 7–19 $$\text{UTP} = V_{REF} \times \frac{(R_2 + R_3)}{(R_1 + R_2 + R_3)} + V_S \times \frac{R_1}{(R_1 + R_2 + R_3)}$$

FORMULA 7–20 $$\text{LTP} = V_{REF} \times \frac{R_2}{(R_1 + R_2)} + V_{EE} \times \frac{R_1}{(R_1 + R_2)}$$

Noninverting Comparator with Hysteresis

The noninverting comparator circuit schematic is provided in Figure 7–26a. The LM339 comparator is used in this circuit. The reference voltage is applied to the negative comparator terminal while the input voltage is applied through a resistor (R_1) to the positive terminal of the comparator. The reference voltage is developed from a voltage divider circuit (R_4 and R_5) connected to the positive supply voltage (V_{CC}). The voltage applied to the positive terminal of the comparator (V_P) is affected by the state of the comparator output (V_{out}).

When the voltage that is applied to the positive terminal (V_P) is less than the reference voltage (V_{REF}), the comparator output (V_{out}) will be a logic low close to OV since the negative supply voltage (V_{EE}) is connected to ground. The voltage at the comparator positive terminal will be calculated using the following equation:

$$V_P = V_{in} \times \frac{R_2}{(R_1 + R_2)} + V_{EE} \times \frac{R_1}{(R_1 + R_2)}$$

The comparator transition from a logic low output to a logic high output will occur when the comparator positive terminal voltage (V_P) is equal to the reference voltage

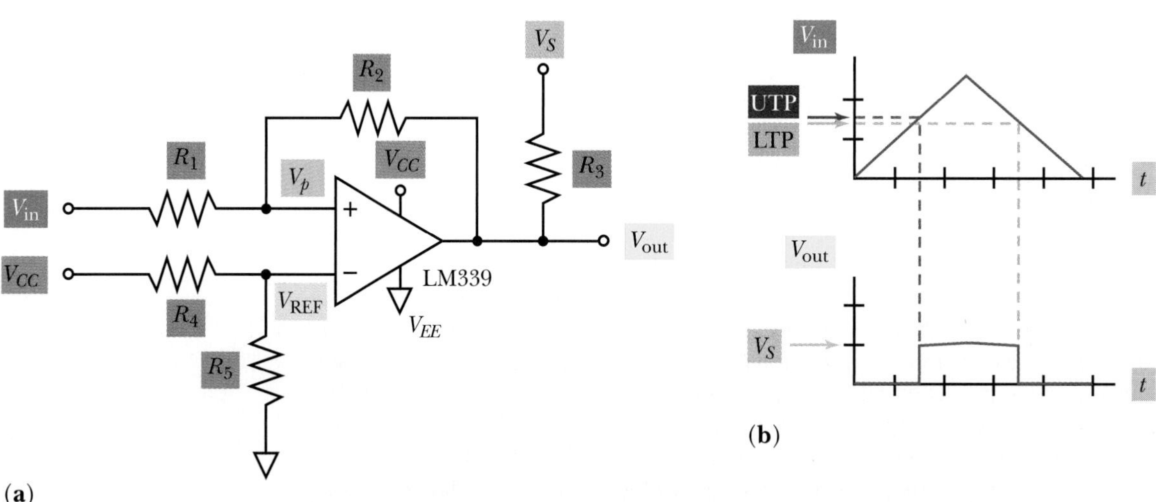

FIGURE 7–26 (a) Noninverting comparator with hysteresis; (b) Noninverting comparator with hysteresis waveforms

applied to the comparator negative terminal voltage (V_{REF}). Solving for the input voltage will calculate the upper threshold voltage as shown in Formula 7–21.

FORMULA 7–21 $UTP = V_{REF} \times \frac{(R_1 + R_2)}{R_2} - V_{EE} \times \frac{R_1}{R_2}$

When the voltage that is applied to the positive terminal (V_P) is greater than the reference voltage (V_{REF}), the comparator output (V_{out}) will be a logic high (close to the supply voltage (V_S). The voltage at the comparator positive terminal will be calculated using the following equation:

$$V_P = V_{in} \times \frac{(R_2 + R_3)}{(R_1 + R_2 + R_3)} + V_S \times \frac{R_1}{(R_1 + R_2 + R_3)}$$

The comparator transition from a logic high output to a logic low output will occur when the comparator positive terminal voltage is equal to the reference voltage applied to the comparator negative terminal voltage. Solving for the input voltage will calculate the lower threshold voltage as shown in Formula 7–22.

FORMULA 7–22 $\text{LTP} = V_{REF} \times \frac{(R_1 + R_2 + R_3)}{(R_2 + R_3)} - V_S \times \frac{R_1}{(R_2 + R_3)}$

Figure 7–26b provides the input and output waveforms for the noninverting comparator with hysteresis circuit.

Window Comparator

The window comparator circuit schematic is provided in Figure 7–27a. The inverting comparator and the noninverting comparator are combined together in this circuit. As the comparator outputs are connected, the open collector output stage used in the LM311 and LM339 is required. Two different reference voltages (V_{REF_1} and V_{REF_2}) are used. The input voltage (V_{in}) is applied to the positive terminal of one comparator and the negative terminal of the other comparator. When the reference voltage applied to the negative comparator terminal (V_{REF_1}) is greater than the reference voltage applied to the positive comparator terminal (V_{REF_2}), the circuit functions as an inverted window comparator. The output voltage will be determined using Formula 7–23 with the waveforms shown in Figure 7–27b.

FORMULA 7–23 V_{out} = "low" for $V_{REF_2} < V_{in} < V_{REF_1}$
V_{out} = "high" for $V_{in} < V_{REF_2}$
V_{out} = "high" for $V_{in} > V_{REF_1}$

When the reference voltage applied to the negative comparator terminal (V_{REF_1}) is less than the reference voltage applied to the positive comparator terminal (V_{REF_2}), the circuit functions as a noninverted window comparator. The output voltage will be determined using Formula 7–24 with the output waveform as shown in Figure 7–27c.

FORMULA 7–24 V_{out} = "high" for $V_{REF_1} < V_{in} < V_{REF_2}$
V_{out} = "low" for $V_{in} < V_{REF_1}$
V_{out} = "low" for $V_{in} > V_{REF_2}$

Bar Graph Display Driver

The LM3914 is a dot/bar graph display driver. One circuit application for this device is provided in Figure 7–28. The LM3914 is made of ten individual comparators referenced to a ten-level precision voltage divider connected to its own adjustable

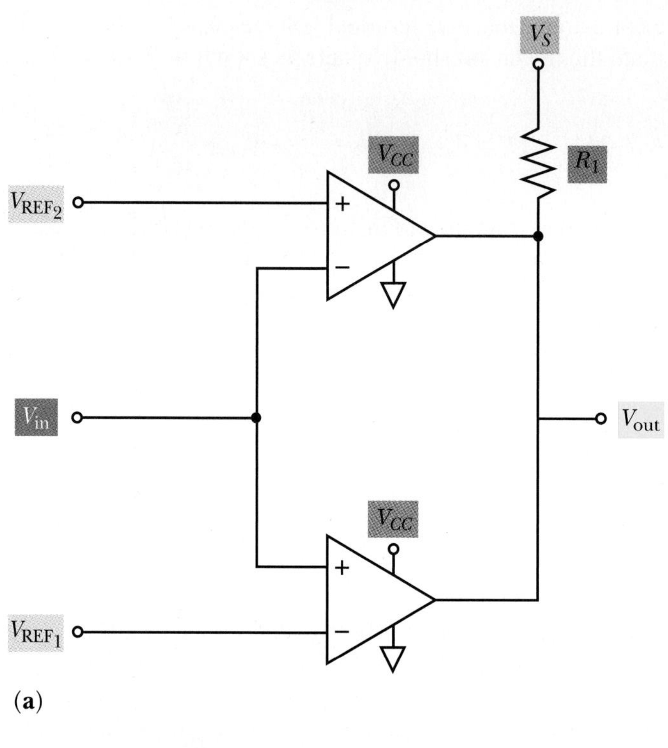

(a)

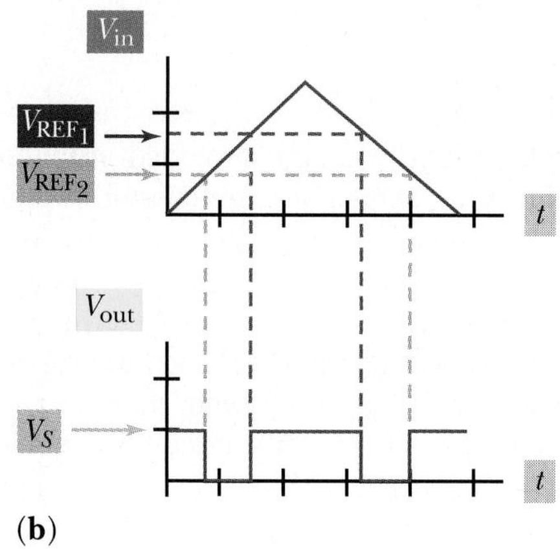

(b)

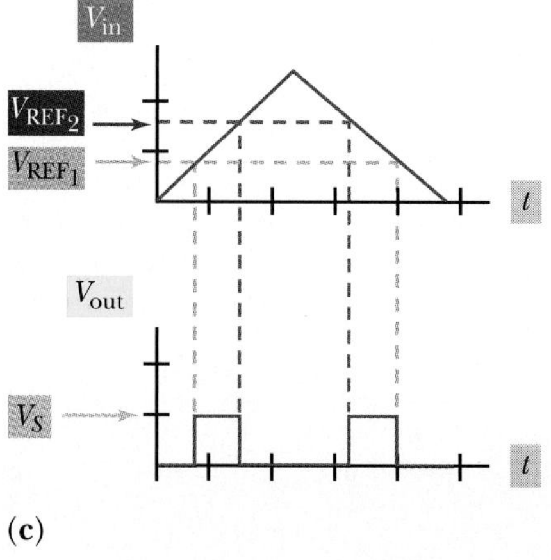

(c)

FIGURE 7–27 (a) Window comparator circuit; (b) Inverted window comparator waveform ($V_{REF_1} > V_{REF_2}$); (c) Noninverted window comparator waveform ($V_{REF_1} < V_{REF_2}$)

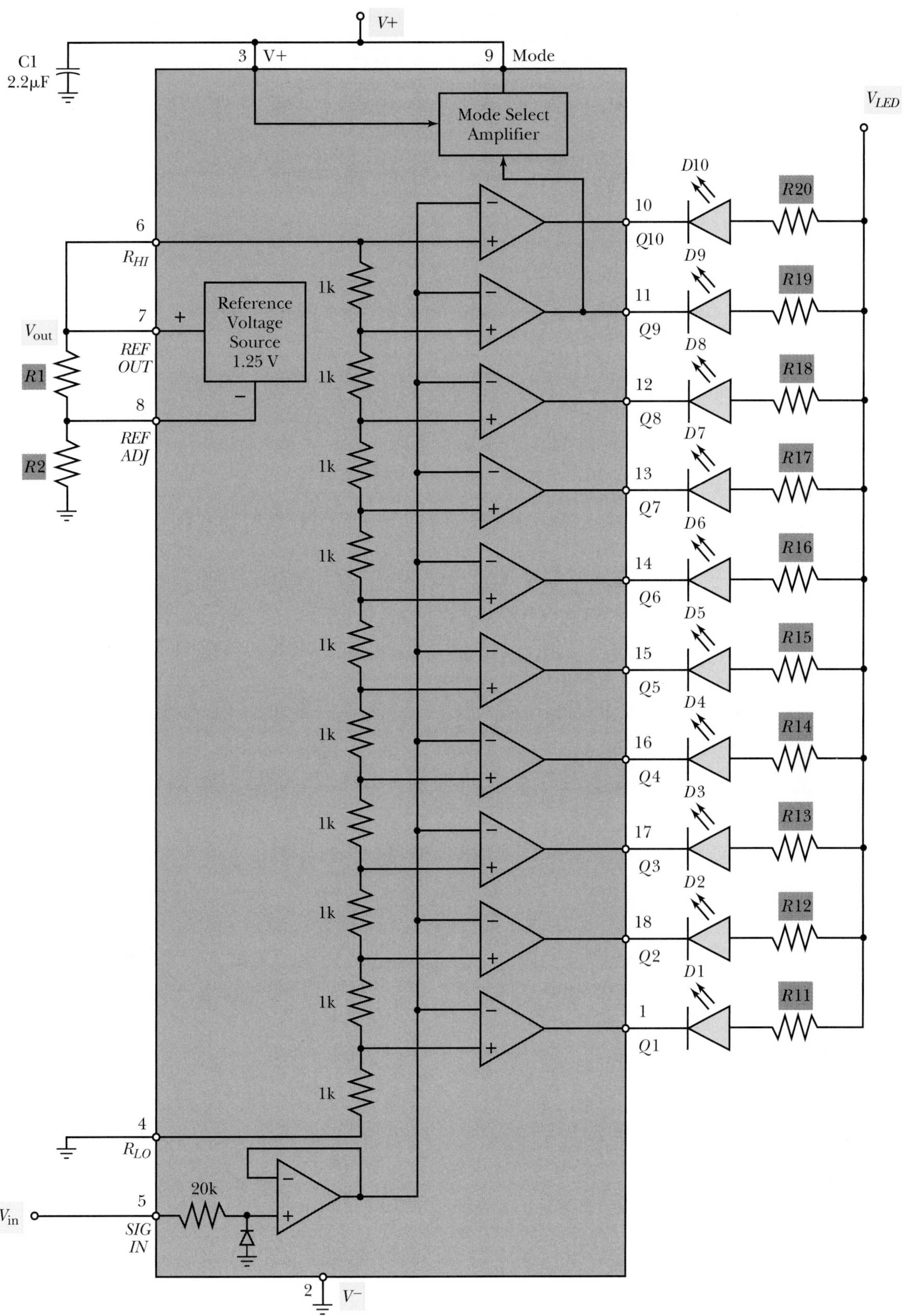

FIGURE 7–28 Dot/bar display driver (LM3914)

reference. The input voltage is applied to all of the comparator inputs. The circuit will function as a noninverting comparator (bar mode) or as a window comparator (dot mode). Connecting pin 9 to pin 3 (V^+) as shown in Figure 7–28 places the component into bar mode. Leaving pin 9 unconnected places the component into dot mode. The LED series resistors may not be needed as the R_1 resistor regulates the LED current and the LED supply voltage VLED may be a voltage between 3 V up to the circuit voltage V^+. Incandescent bulbs may be used instead of LEDs.

The comparator adjustable reference voltage is calculated using Formula 7–25.

FORMULA 7–25 $V_{out} = V_{REF} \times \left(1 + \frac{R_2}{R_1}\right) + I_{ADJ} \times R_2$

Where $V_{REF} = 1.25$ V (typical) and $I_{ADJ} = 75$ μA (typical)

The reference voltage applied to each comparator can be calculated using Formula 7–26.

FORMULA 7–26 $V_{REF(n)} = n \times \left(\frac{1\ \text{k}\Omega}{10\ \text{k}\Omega}\right) \times V_{out}$

Where $n = 1, 2, 3, 4, 5, 6, 7, 8, 9$, or 10

When operating in the bar mode and V_{in} is greater than $V_{REF(n)}$, then outputs Q_1 through $Q_{(n)}$ will be a logic "low" turning on LEDs D_1 through $D_{(n)}$. Outputs $Q_{(n+1)}$ through Q_{10} will be a logic "high" placing LEDs $D_{(n+1)}$ through D_{10} into an OFF condition.

When operating in the dot mode and V_{in} is greater than $V_{REF_{(n)}}$ and less than $V_{REF_{(n+1)}}$ then output $Q_{(n)}$ will be a logic "low" turning on LED $D_{(n)}$. Outputs Q_1 through $Q_{(n-1)}$ and $Q_{(n+1)}$ through Q_{10} will be a logic "high" placing LEDs D_1 through $D_{(n-1)}$ and $D_{(n+1)}$ through D_{10} into an OFF condition. The only condition that will cause all LEDs to be off for either mode is when V_{in} is less than V_{REF_1}.

EXAMPLE The inverted comparator with hysteresis circuit shown in Figure 7–28 operates with $V^+ = 12$ V, $V_{LED} = 12$ V, $R_1 = 1.2$ kΩ, $R_2 = 3.83$ kΩ, and R_{11} through $R_{20} = 1.2$ kΩ. Calculate V_{out} and the threshold voltage for each comparator. If $V_{in} = +2$ V, indicate the LEDs that will be turned ON.

Answers:

From Formula 7–25:

$$V_{out} = V_{REF} \times \left(1 + \frac{R_2}{R_1}\right) + I_{ADJ} \times R_2$$

where $V_{REF} = 1.25$ V (typical) and $I_{ADJ} = 75$ μA (typical)

$$V_{out} = 1.25\ \text{V} \times \left(1 + \frac{3.83\ \text{k}\Omega}{1.2\ \text{k}\Omega}\right) + 75\ \mu\text{A} \times 3.83\ \text{k}\Omega$$
$$V_{out} = 5.24\ \text{V} + 0.29\ \text{V}$$
$$V_{out} = 5.53\ \text{V}$$

The reference voltage applied to each comparator can be calculated using Formula 7–26. From Formula 7–26:

$$V_{REF(n)} = n \times \left(\frac{1\ \text{k}\Omega}{10\ \text{k}\Omega}\right) \times V_{out}$$

Where $n = 1, 2, 3, 4, 5, 6, 7, 8, 9$, or 10

$$V_{REF(1)} = 1 \times \left(\frac{1\ \text{k}\Omega}{10\ \text{k}\Omega}\right) \times 5.53\ \text{V}$$
$$V_{REF(1)} = 0.553\ \text{V}$$

$$V_{REF(2)} = 2\times\left(\frac{1\ \text{k}\Omega}{10\ \text{k}\Omega}\right)\times 5.53\ \text{V}$$
$$V_{REF(2)} = 1.106\ \text{V}$$
$$V_{REF(3)} = 3\times\left(\frac{1\ \text{k}\Omega}{10\ \text{k}\Omega}\right)\times 5.53\ \text{V}$$
$$V_{REF(3)} = 1.659\ \text{V}$$
$$V_{REF(4)} = 4\times\left(\frac{1\ \text{k}\Omega}{10\ \text{k}\Omega}\right)\times 5.53\ \text{V}$$
$$V_{REF(4)} = 2.212\ \text{V}$$
$$V_{REF(5)} = 5\times\left(\frac{1\ \text{k}\Omega}{10\ \text{k}\Omega}\right)\times 5.53\ \text{V}$$
$$V_{REF(5)} = 2.765\ \text{V}$$
$$V_{REF(6)} = 6\times\left(\frac{1\ \text{k}\Omega}{10\ \text{k}\Omega}\right)\times 5.53\ \text{V}$$
$$V_{REF(6)} = 3.318\ \text{V}$$
$$V_{REF(7)} = 7\times\left(\frac{1\ \text{k}\Omega}{10\ \text{k}\Omega}\right)\times 5.53\ \text{V}$$
$$V_{REF(7)} = 3.871\ \text{V}$$
$$V_{REF(8)} = 8\times\left(\frac{1\ \text{k}\Omega}{10\ \text{k}\Omega}\right)\times 5.53\ \text{V}$$
$$V_{REF(8)} = 4.424\ \text{V}$$
$$V_{REF(9)} = 9\times\left(\frac{1\ \text{k}\Omega}{10\ \text{k}\Omega}\right)\times 5.53\ \text{V}$$
$$V_{REF(9)} = 4.977\ \text{V}$$
$$V_{REF(10)} = 10\times\left(\frac{1\ \text{k}\Omega}{10\ \text{k}\Omega}\right)5.53\ \text{V}$$
$$V_{REF(10)} = 5.53\ \text{V}$$

$V_{REF(1)}$, $V_{REF(2)}$, and $V_{REF(3)}$ are less than +2 V, the voltage applied to the input (V_{in}). LEDs 1, 2, and 3 will be turned ON, while LEDs 4, 5, 6, 7, 8, 9, and 10 will be OFF. ◆

7–12 Special Purpose Op-amp Circuits

Series-Type Linear Voltage Regulator

Figure 7–29 shows a simple series-type voltage regulator circuit that uses an op-amp as one critical element. This circuit regulates its voltage output for any changes in input voltage (V_{in}) or for changes in output load current demand (I_L). It essentially keeps the output voltage constant for these changes.

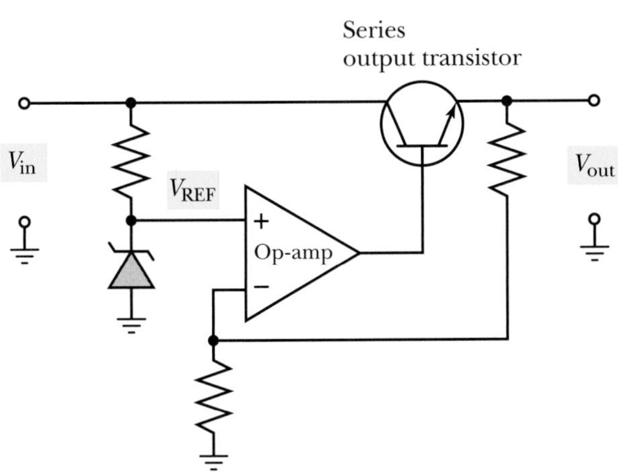

FIGURE 7–29 Example of op-amp series-type linear voltage regulator

The voltage-divider resistors from the V_{out} line to ground sense changes in output voltage. Since one op-amp differential input is held constant by the zener diode reference voltage circuit, any change fed from the output through the R-divider network to the other op-amp input causes a differential input between the op-amp inputs. This, in turn, is amplified and controls the series output transistor conduction and, consequently, its voltage drop from collector to emitter. If the output voltage attempts to decrease (from a decrease in V_{in} or an increased load at V_{out}), there is less drop across the transistor, bringing the output back up to the desired level. Conversely, if the output voltage attempts to increase (from V_{in} increasing or the output load decreasing), the transistor voltage drop increases, bringing the output voltage back down to the desired regulated level.

7–13 Operational Transconductance Amplifier (OTA) Circuits

Differential Amplifier

The OTA (CA 3080) is designed so that the input amplifier bias current controls the total output current and the circuit's transconductance. In Figure 7–30, two input signals are applied through resistors to the inverting and noninverting inputs of the amplifier. The bias voltage and bias resistor establish the bias current and the resulting output voltage. Formula 7–27 provides the equation to calculate the output voltage.

$$V_{out} = I_{out} \times R_L$$
$$V_{out} = (V_1 - V_2) \times g_m \times R_L$$

FORMULA 7–27 $V_{out} = (V_1 - V_2) \times k \times I_{BIAS} \times R_L$

where:

$$k \approx 20$$
$$I_{BIAS} = \frac{(V_{BIAS} - V^- - 0.7\text{ V})}{R_{BIAS}}$$

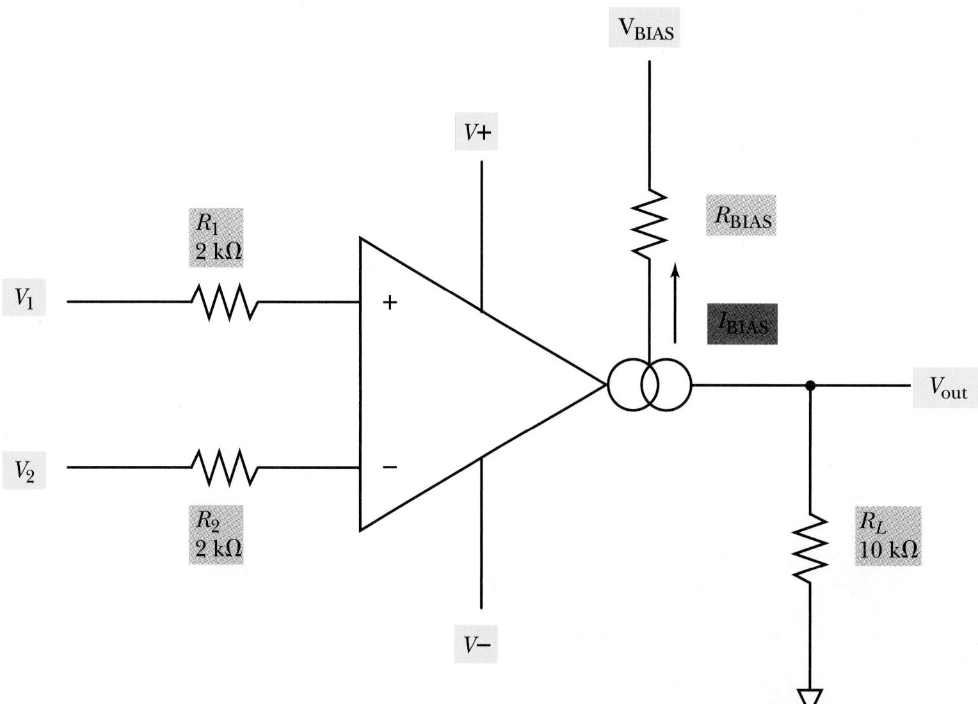

FIGURE 7–30 Operational transconductance amplifier (OTA) as a differential amplifier

EXAMPLE The circuit in Figure 7–30 has $V^+ = +15$ V, $V^- = -15$ V, $V_{BIAS} = +10$ V, $V_1 = 0.2$ V, $V_2 = 0.15$ V, $R_{BIAS} = 50$ kΩ, and $R_L = 10$ kΩ. Calculate V_{out}.

Answers:

From Formula 7–27:

$$I_{BIAS} = \frac{(V_{BIAS} - V^- - 0.7\text{ V})}{R_{BIAS}}$$

$$I_{BIAS} = \frac{[+10\text{ V} - (-15\text{ V}) - 0.7\text{ V}]}{50\text{ k}\Omega}$$

$$I_{BIAS} = 486\ \mu\Omega$$

$$V_{out} = (V_1 - V_2) \times k \times I_{BIAS} \times R_L$$
$$V_{out} = (0.2\text{ V} - 0.15\text{ V}) \times 20 \times 486\ \mu\text{A} \times 10\text{ k}\Omega$$
$$V_{out} = 4.86\text{ V}$$

Schmitt Trigger Comparator

You have already been introduced to the op-amp Schmitt trigger comparator circuit. By using positive feedback the OTA can be configured as a Schmitt trigger comparator. In Figure 7–31, the input signal is applied to the inverting input of the amplifier. Formula 7–28 provides the equation to calculate the upper trip point (UTP) voltage. Formula 7–29 is the equation to calculate the lower trip point (LTP) voltage. The output voltage of the device is equal to the trip point voltage. When the input voltage V_{in} is less than the UTP voltage, the output voltage V_{out} is equal to the UTP voltage. Once V_{in} goes greater than the UTP voltage, V_{out} changes to the LTP voltage. Now, while V_{in} is greater than the LTP voltage, V_{out} is equal to the UTP voltage. Once V_{in} goes less than the LTP voltage, V_{out} changes to the UTP voltage.

FORMULA 7–28 $\text{UTP} = I_{BIAS} \times R_1$

FORMULA 7–29 $\text{LTP} = -I_{BIAS} \times R_1$

EXAMPLE Calculate the upper trip point and lower trip point voltages for the circuit in Figure 7–31 where $V^+ = +15$ V, $V^- = -15$ V, $V_{BIAS} = +10$ V, $R_{BIAS} = 50$ kΩ, and $R_1 = 10$ kΩ.

What is the output voltage V_{out} if $V_{in} = +5$ V?

Answers:

From Formula 7–27:

$$I_{BIAS} = \frac{(V_{BIAS} - V^- - 0.7\text{ V})}{R_{BIAS}}$$

$$I_{BIAS} = \frac{[+10\text{ V} - (-15\text{ V}) - 0.7]}{50\text{ k}\Omega} = 486\ \mu\text{A}$$

From Formula 7–28:

$$\text{UTP} = I_{BIAS} \times R_1$$
$$\text{UTP} = 486\ \mu\text{A} \times 10\text{ k}\Omega = 4.86\text{ V}$$

From Formula 7–29:

$$\text{LTP} = -I_{BIAS} \times R_1$$
$$\text{LTP} = -486\ \mu\text{A} \times 10\text{ k}\Omega = -4.86\text{ V}$$

Since $V_{in} = +5$ V which is greater than the upper trip point voltage,

$$V_{out} = \text{LTP}$$
$$V_{out} = -4.86\text{ V}$$

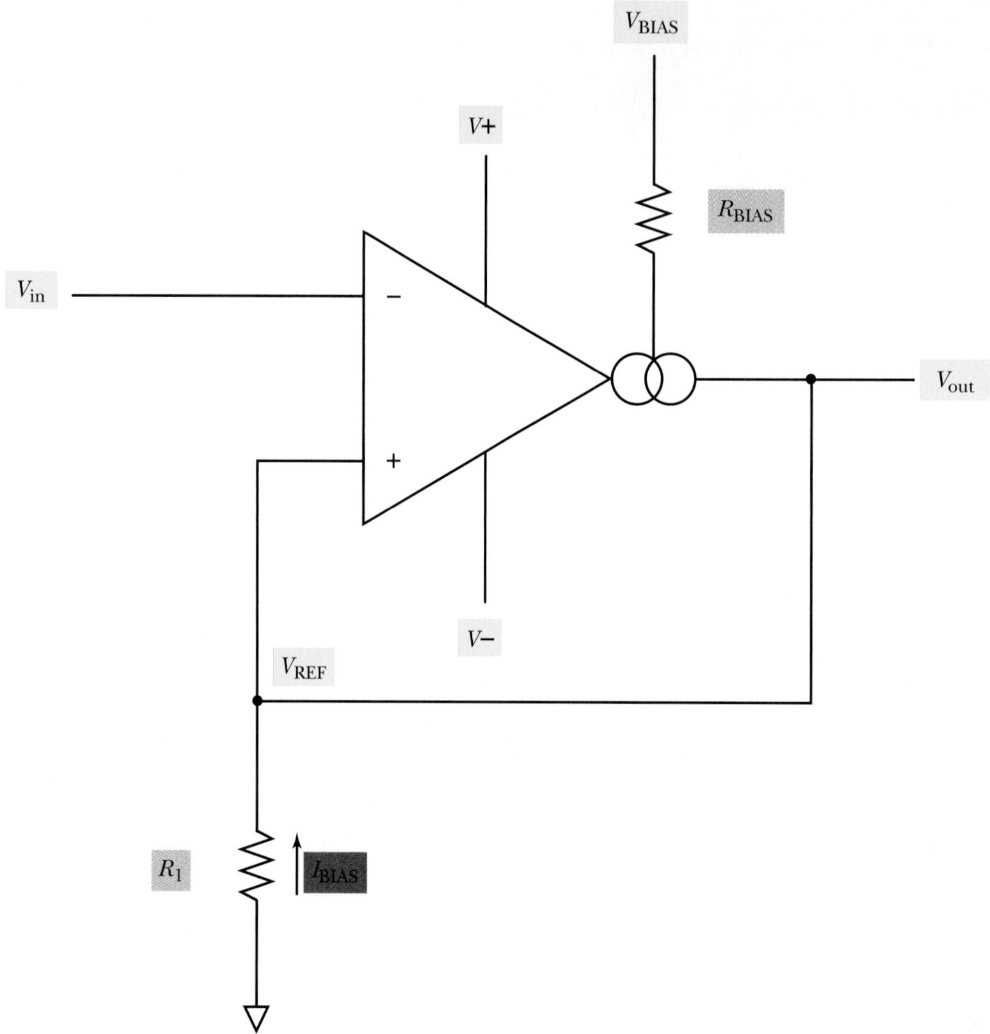

FIGURE 7–31 OTA as a Schmitt trigger

7–14 Digital Interface Circuits

Analog to Digital Converter (ADC)

The ability to convert analog signals such as voice and music into digital data is important. The telephone company and the recording studios in producing your CDs utilize this process. The analog signal is a continuous, uniformly changing signal. The digital signal is discrete steps defined as logic 0 and logic 1. A large variety of ADCs are manufactured with different features. The ADC0804 is the focus of our brief introduction to the ADC. Figure 7–32 provides the circuit schematic where the ADC is interfaced to a microcontroller. The ADC0804 is designed to operate with a single 5 V supply. Your input is an analog signal, a voltage between 0 V and 5 V. The output is an 8 bit digital value representing the sampled analog input. Knowing the sampled voltage, you can calculate the digital value using Formula 7–30.

To operate the ADC, the ADC must sample the input signal and convert it into a digital value. Pulsing the Chip Select (CS) and Write (WR) inputs to a logic low as shown in Figure 7–33 activates the ADC conversion process. The ADC indicates the completion of the conversion by a logic low on the Interrupt (INT) output. You can output the data from the ADC by pulsing the Chip Select and Read (RD) inputs to a logic low during which time the data (D_0 – D_7) will appear on the data bus.

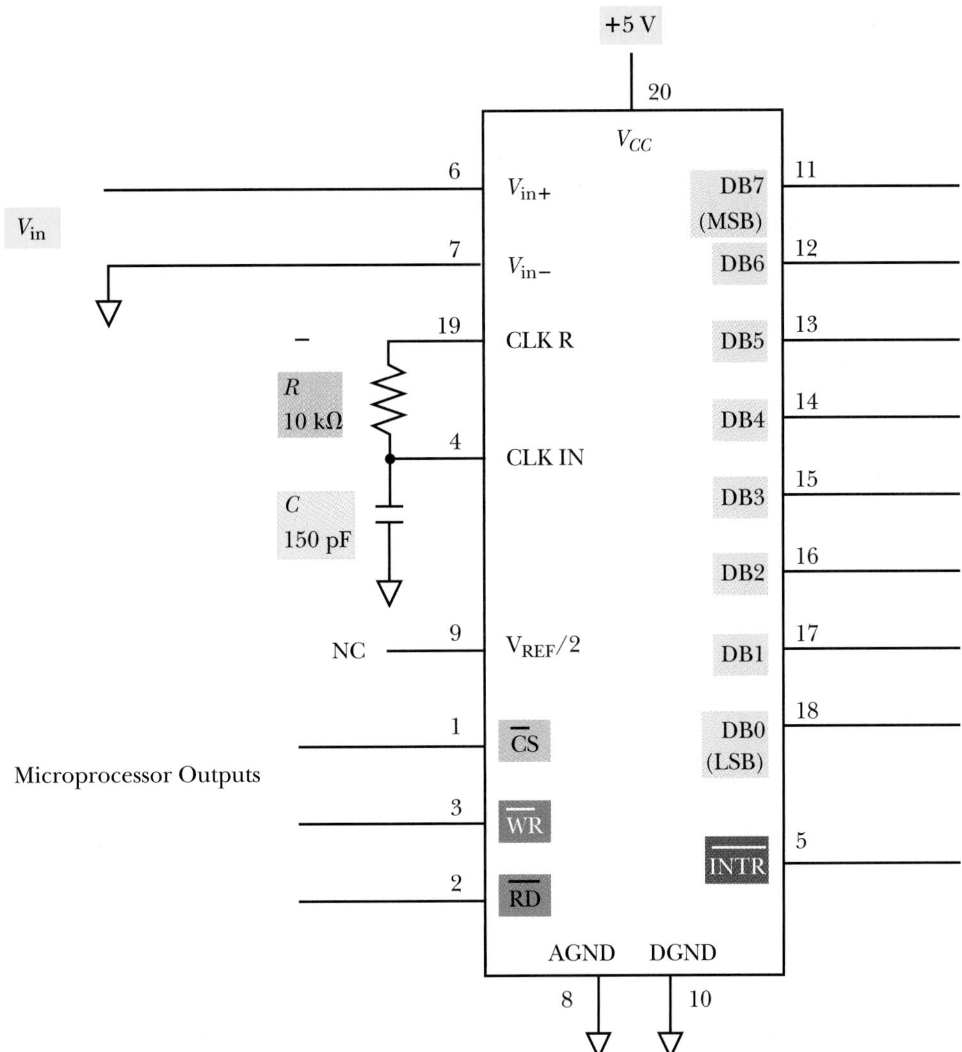

FIGURE 7–32 Analog to digital converter (ADC 0804)

FORMULA 7–30 $\text{DATA} = V_{\text{in}_1} \times \frac{255}{5\text{ V}}$

EXAMPLE Calculate the digital data for the Figure 7–32 ADC with an input voltage of 2 volts.

Answers:

From Formula 7–30:

$$\text{DATA} = V_{\text{in}} \times \frac{255}{5\text{ V}}$$

$$\text{DATA} = 2\text{ V} \times \frac{255}{5\text{ V}}$$

$$\text{DATA} = 102\ (\text{decimal})$$

$$102\ (\text{decimal}) = 0 \times 2^7 + 1 \times 2^6 + 1 \times 2^5 + 0 \times 2^4 + 0 \times 2^3 + 1 \times 2^2 + 1 \times 2^1 + 0 \times 2^0$$

$$102\ (\text{decimal}) = 01100110\ (\text{binary}) = 66\ (\text{hexadecimal})$$

D_7 output is a logic 0, D_6 output is a logic 1, D_5 output is a logic 1, D_4 output is a logic 0, D_3 output is a logic 0, D_2 output is a logic 1, D_1 output is a logic 1, and D_0 output is a

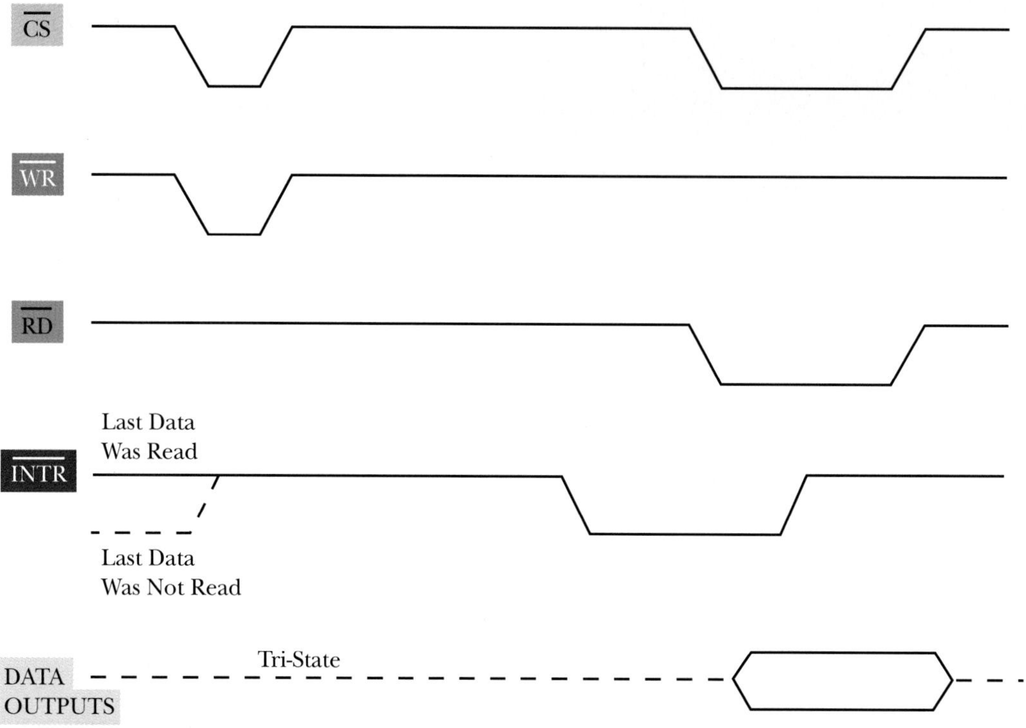

FIGURE 7–33 ADC timing diagram

logic 0. A logic 0 output (also known as a logic low) is a voltage value between 0 V and 0.4 V. A logic 1 output (also known as a logic high) is a voltage value between 2.4 V and 5.0 V. ◆

Practical Notes

The ADC sampling rate must be at least two times faster than the maximum frequency of the incoming analog signal. The time between samples must be longer than the component's conversion time.

Digital to Analog Converter (DAC)

The device to convert the digital data into an analog signal is the digital to analog converter (DAC). A large variety of DACs are available. Some main features in the comparison of DACs include serial or parallel data input, current or voltage output, single or multiple supply, number of data inputs, with or without handshaking, and linearity. Our brief overview of DACs will use the DAC 0808. The component can be connected as shown in Figure 7–34. Formula 7–31 provides the equation to calculate the output voltage for this circuit. The DAC 0808 output is a current output. Using the negative input of the operational amplifier converts the current into a voltage.

FORMULA 7–31

$$V_{\text{out}} = \left(-\frac{V_{\text{REF}}}{R_{14}}\right) \times \left(\frac{D_7}{2} + \frac{D_6}{4} + \frac{D_5}{8} + \frac{D_4}{16} + \frac{D_3}{32} + \frac{D_2}{64} + \frac{D_1}{128} + \frac{D_0}{256}\right) \times (-R_1)$$

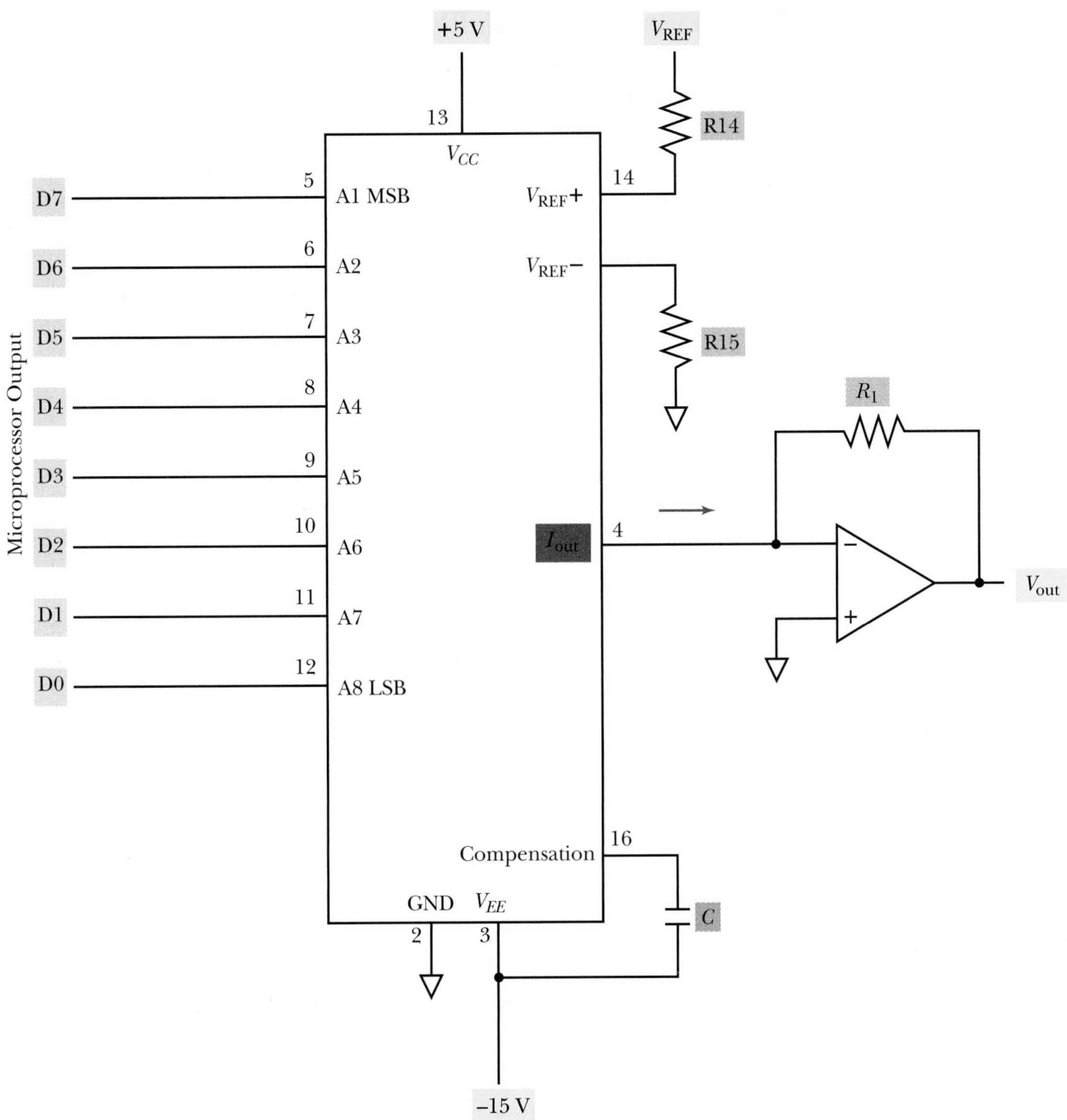

FIGURE 7–34 Digital to analog converter (DAC 0808)

As seen in the equation each digital input has a different weighted impact on the value for the current output. The data inputs (D_0 through D_7) are logic inputs either a 0 (a voltage between 0 and 0.8 V) or a 1 (a voltage between 2.0 and 5.0 V).

EXAMPLE Calculate the output voltage for the Figure 7–34 DAC with a digital input of 10001010 binary which is 8A hexadecimal. Circuit values are V_{REF} – 10 V, $R_1 = 10\ \text{k}\Omega$, $R_{14} = 10\ \text{k}\Omega$, $R_{15} = 10\ \text{k}\Omega$, and $C = 0.1\ \mu\text{F}$.

Answers:

From Formula 7–31:

$$V_{out} = \left(-\frac{V_{REF}}{R_{14}}\right)\times\left(\frac{D_7}{2}+\frac{D_6}{4}+\frac{D_5}{8}+\frac{D_4}{16}+\frac{D_3}{32}+\frac{D_2}{64}+\frac{D_1}{128}+\frac{D_0}{256}\right)\times(-R_1)$$

$$V_{out} = \left(-\frac{10\ \text{V}}{10\ \text{k}\Omega}\right)\times\left(\frac{1}{2}+\frac{0}{4}+\frac{0}{8}+\frac{0}{16}+\frac{1}{32}+\frac{0}{64}+\frac{1}{128}+\frac{0}{256}\right)\times(-10\ \text{k}\Omega)$$

$$V_{out} = (-1\ \text{mA})\times[0.53906]\times(-10\ \text{k}\Omega)$$

$$V_{out} = 5.3906\ \text{V}$$

7–15 Future Trends in Electronics

The development of the transistor in 1947 at Bell Laboratories launched the semiconductor industry. In 1958 the first integrated circuit was developed at Texas Instruments. The integrated circuit (IC) provides the ability of interconnecting several transistor circuits on one monolithic substrate. With the refinement of the semiconductor manufacturing process and enhancements in the semiconductor processing equipment, present day technology is able to produce submicron geometries. One micron is equal to 1×10^{-6} inches. Trace widths are below 0.01×10^{-6} inches. Traces are the interconnections between semiconductor devices on the substrate. Several million transistors can be placed on one monolithic substrate, which will be packaged as an integrated circuit.

The transistor initially was packaged as a 3 terminal device with different outlines such as TO-3 and TO-220. The integrated circuits were packaged in dual-in-line packages (DIPs). With the reduction in required substrate size, new topologies were introduced such as small outline integrated circuits (SOICs), plastic leadless chip carrier (PLCC), and ball grid arrays.

As can be seen through our discussion, the ability to package multiple transistors in one integrated circuit allowed for a reduction in the space required for a complete system. With the geometric density available over the past decade, the ability to place a complete radio receiver circuit on a single substrate has been accomplished. The radio receiver would be classified as an electrical system, and therefore you have a system on a chip.

The evolving concept of the system on a chip (SOC) is to provide several products where the user concept of the internal functional elements by programming the component. The in-circuit programmability of these devices allows for quick modifications of the system's operation. The ability to accomplish the user-defined configuration has existed for decades within the digital environment using the programmable read only memory (PROM), the programmable array logic (PAL), and complex programmable logic devices (CPLD).

Devices that allow a programmable analog architecture with the programmable digital architecture within one device (chip) have been under development over the past years and are now becoming available. The devices with this capability are called systems on a chip (SOC). Some versions of the SOCs include electromechanical devices. SOCs that will provide for optoelectronic inputs and outputs are technologically viable. Several manufacturers are involved with SOC architecture development, programming tools, and industry training components.

As a perspective of the impact that the electronics industry has made on society through component configuration and availability, compare the electronic systems that existed within a 1960 automobile to the electronics systems that exist in a new current model automobile. You can take any product or industry and observe the impact a reduction in component size, decrease in power requirements, and increase in component complexity has accomplished. The desired need to transfer more information and data between system elements is continuing to increase mandating the need for more compact complex circuit components to accomplish the task.

Summary

- Operational amplifiers were originally used in analog computers to perform mathematical operations.
- Integrated circuit op-amps are high gain, direct-coupled, de amplifiers. Op-amps perform various useful functions; therefore they are applied in many electronic systems and subsystems. They are used as inverting and noninverting amplifiers for numerous purposes. They are also used in nonlinear applications such as comparators and Schmitt triggers.
- The open-loop gain of op-amp devices can be in the hundreds of thousands. The closed-loop gain of an op-amp circuit is much less than the open-loop gain. The closed-loop gain is controlled by the values and types of external components connected to the input and output terminals. Using some ideal op-amp assumptions, the closed-loop gain is computed as the ratio of the feedback resistor value to the input resistor value.
- To produce an output that is inverted from the input, the input signal is connected to the inverting (negative) input of the op-amp. Conversely, to produce an output with the same phase or polarity as the input, the input signal is fed to the noninverting (positive) input.
- A circuit that produces a noninverted output (zero phase shift) and a gain of 1 (unity) is called a voltage follower. An op-amp voltage follower is created by using a feedback resistance of 0 Ω.
- The single-ended mode of operation for the input is achieved by grounding one input (sometimes through a resistor) and feeding the signal to the other input.
- The differential mode is used when different input signals are applied to the inverting and noninverting inputs.
- The common-mode operation of an op-amp is used when the same signal feeds both inputs. Although there is no practical application for this mode of operation, it provides the specification for common-mode rejection ratio.
- Operational amplifiers can perform mathematical operations on the values of input voltages. These operations are additional (summing amplifier), subtraction (subtractor circuit), differentiation (differentiator circuit), and integration (integrator circuit). Circuits such as signal mixers (summing amplifiers) and waveform converters, for example, can perform their mathematical functions in ways that do not seem to be mathematical.
- A Schmitt trigger circuit provides a rectangular output waveform that switches state (when the input voltage passes upward through a UTP (upper threshold point) and when the input voltage passes downward through a LTP (lower threshold point). These threshold voltage levels are determined by the values of the supply voltages and the values of external resistors. The difference between the UTP and LTP is the hysteresis range of the amplifier. The Schmitt trigger is one of a very few op-amp circuits that uses positive feedback.

CIRCUIT SUMMARY

Description	*Circuit*	*Formula*
Inverting Amplifier	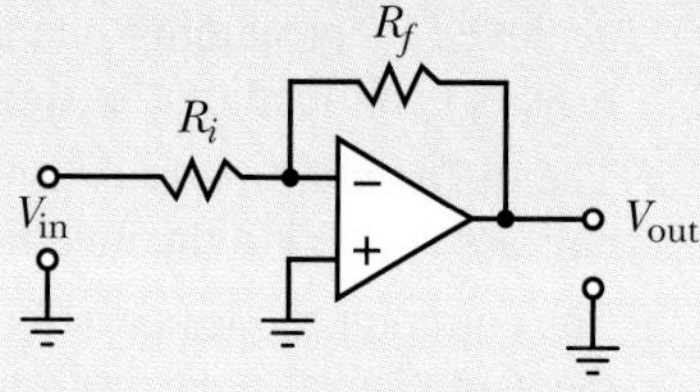	$A_V = -\frac{R_f}{R_i}$ $V_{out} = A_V V_{in}$
Non-inverting Amplifier	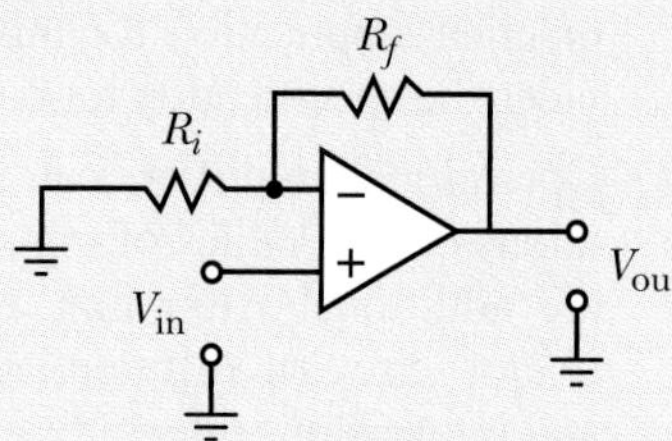	$A_V = 1 + \frac{R_f}{R_i}$ $V_{out} = A_V V_{in}$
Voltage Follower (Unity gain amplifier)	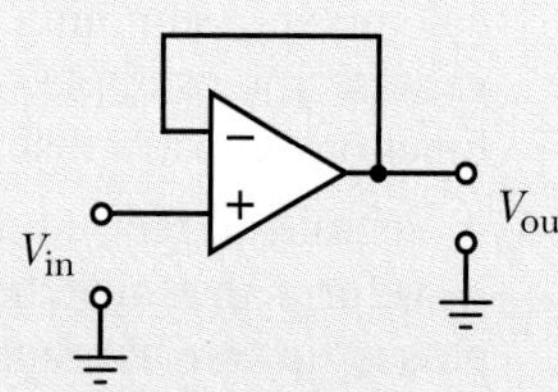	$A_V = 1$ $V_{out} = V_{in}$
Voltage Divider noninverting amplifier	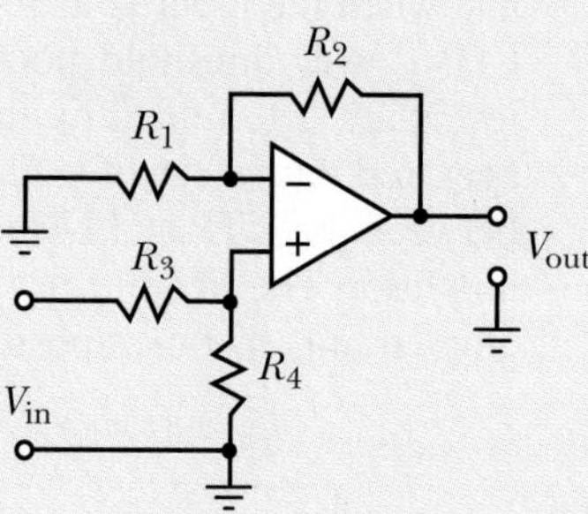	$A_V = \left(\frac{R_4}{R_3 + R_4}\right) \times \left(1 + \frac{R_2}{R_1}\right)$
Difference Amplifier	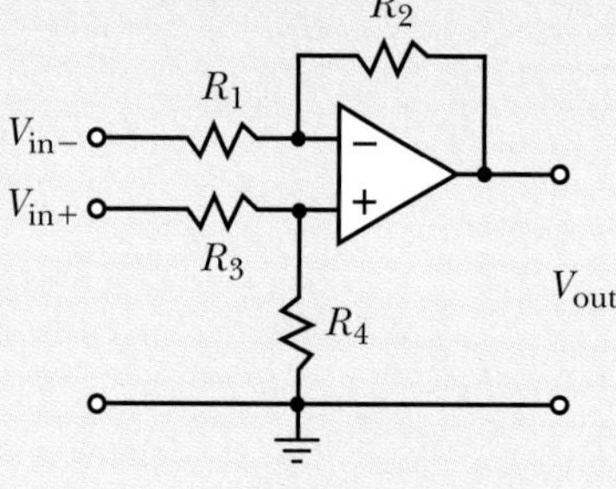	$V_{out} = V_{in+} \times \left(1 + \frac{R_2}{R_1}\right) \times \left(\frac{R_4}{R_3 + R_4}\right) + V_{in-} \times \left(-\frac{R_2}{R_1}\right)$ When $R_1 = R_2 = R_3 = R_4$ $V_{out} = V_{in+} - V_{in-}$
Summing Amplifier	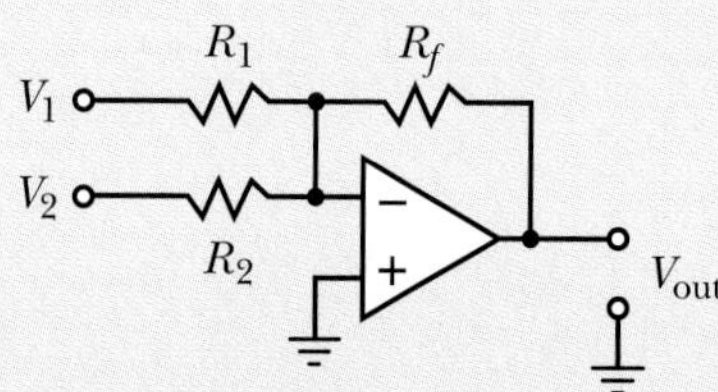	$V_{out} = V_1 \times \left(-\frac{R_f}{R_1}\right) + V_2 \times \left(-\frac{R_f}{R_2}\right)$

CIRCUIT SUMMARY

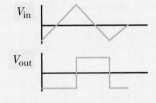

Description	*Circuit*	*Formula*
Integrator	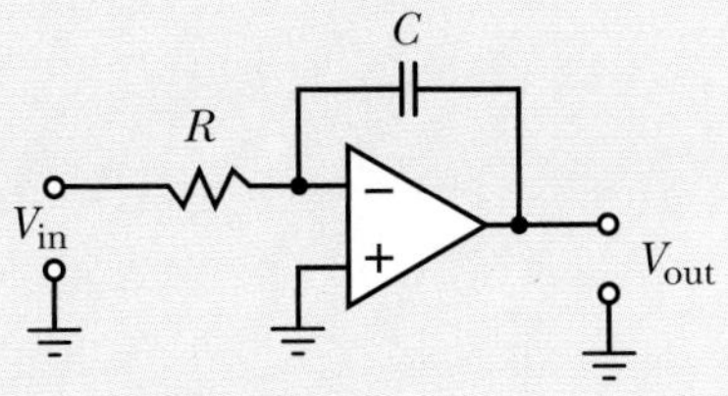	V_{in} V_{out}
Differentiator	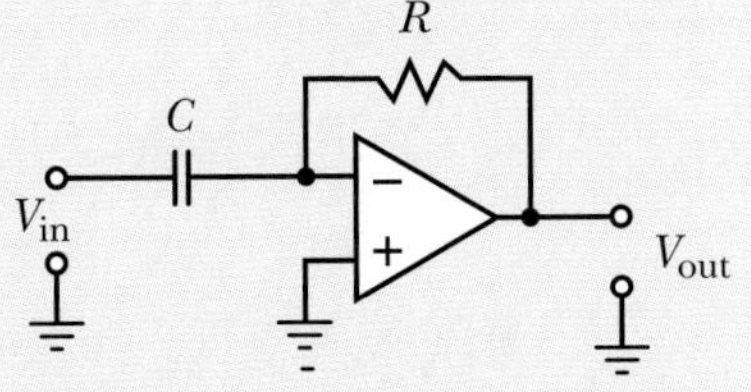	V_{in} V_{out}
Instrumentation Amplifier	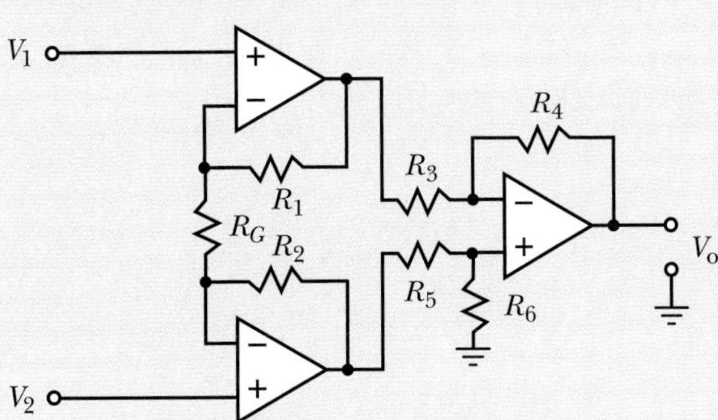	When $R_1 = R_2 = R_3 = R_4 = R_5 = R_6 = R$ $V_{out} = \left(1 + 2 \times \frac{R}{R_G}\right) \times \left(V_2 - V_1\right)$

Formulas and Sample Calculator Sequences

Operational amplifier without feedback.

FORMULA 7–1
(To find the output voltage (inverting input))

$V_{out} = -A_{VOL} \times V_{in}$

A_{VOL} value, [±], [×], V_{in} value, [=]

FORMULA 7–2
(To find the output voltage (noninverting input))

$V_{out} = A_{VOL} \times V_{in}$

A_{VOL} value, [×], V_{in} value, [=]

FORMULA 7–3
(To find the output voltage)

$V_{out} = A_{VOL} \times (V_1 V_2)$

A_{VOL} value, [×], [(], V_1 value, [−], V_2 value, [)], [=]

FORMULA 7–4
(To find the voltage gain)

$$A_V = \frac{V_{out}}{V_{in}}$$

V_{out} value, [÷], V_{in} value, [=]

Inverting amplifier

FORMULA 7–5
(To find the voltage gain)

$$A_V = \left(\frac{R_f}{R_i}\right)$$

R_f value, [±], [÷], R_i value, [=]

Noninverting amplifier

FORMULA 7–6
(To find the voltage gain)

$$A_V = \left(\frac{R_f}{R_i}\right) + 1$$

R_f value, [÷], R_i value, [=], [+], 1, [=]

Voltage divider noninverting amplifier

FORMULA 7–7
(To find the voltage gain)

$$A_V = \left[\left(\frac{R_2}{R_1}\right) + 1\right] \times \left[\frac{R_4}{(R_4 + R_3)}\right]$$

[(], [(], R_2 value, [÷], R_1 value, [)], [+], 1, [)], [×], R_4 value, [÷], [(], R_4 value, [+], R_3 value, [)], [=]

Subtractor circuit

FORMULA 7–8
(To find the output voltage)

$$V_{out} = V_{in+} \times \left(\left(\frac{R_2}{R_1}\right) + 1\right) \times \left(\frac{R_4}{(R_4 + R_3)}\right) + V_{in-} \times \left(-\frac{R_2}{R_1}\right)$$

V_{in+} value, [×], [(], [(], R_2 value, [÷], R_1 value, [)], [+], 1, [)], [×], R_4 value, [÷], [(], R_4 value, [+], R_3 value, [)], [+], V_{in-} value, [×], [(] R_2 value, [±], [÷], R_1 value, [)], [=]

Subtractor circuit where $R_1 = R_2 = R_3 = R_4$

FORMULA 7–9
(To find the output voltage)

$$V_{out} = V_{in+} - V_{in-}$$

V_{in+} value, [−], V_{in-} value, [=]

Summing amplifier (2-input inverting)

FORMULA 7–10
(To find the output voltage)

$$V_{out} = V_{in_1} \times \left(-\frac{R_f}{R_1}\right) + V_{in2} \times \left(-\frac{R_f}{R_2}\right)$$

V_{in_1} value, [×], [(], R_f value, [±], [÷], R_1 value, [)], [+], V_{in_2} value, [×], [(], R_f value, [±], [÷], R_2 value, [)], [=]

Summing amplifier (3-input inverting)

FORMULA 7–11
(To find the output voltage)

$$V_{out} = V_{in_1} \times \left(-\frac{R_f}{R_1}\right) + V_{in2} \times \left(-\frac{R_f}{R_2}\right) + V_{in3} \times \left(-\frac{R_f}{R_3}\right)$$

V_{in_1} value, [×], [(], R_f value, [±], [÷], R_1 value, [)], [+], V_{in_2} value, [×], [(], R_f value, [±], [÷], R_2 value, [)], [+], V_{in_3} value, [×], [(], R_f value, [±], [÷], R_3 value, [)], [=]

Summing amplifier (2-input noninverting)

FORMULA 7–12
(To find the output voltage)

$$V_{out} = \left(\left(\frac{R_2}{R_1}\right) + 1\right) \times \left[V_{in_1} \times \frac{R_4}{(R_3 + R_4)} + V_{in2} \times \frac{R_3}{(R_3 + R_4)}\right]$$

[(], [(], R_2 value, [÷], R_1 value, [)], [+], 1, [)], [×], [(], V_{in_1} value, [×], R_4 value, [÷], [(], R_3 value, [+], R_4 value, [)], [+], V_{in_2} value, [×], R_3 value, [÷], [(], R_3 value, [+], R_4 value, [)], [)], [=]

Integrator

FORMULA 7–13
(To find the change in the output voltage)

$$\Delta V_{out} = \left[\frac{-V_{in}}{(R \times C)}\right] \times \Delta t$$

[(], V_{in} value, [±], [÷], [(], R value, [×], C value, [)], [)], [×], Δ value, [=]

Differentiator

FORMULA 7–14
(To find the output voltage)

$$V_{out} = (-R \times C) \times \left(\frac{\Delta V_{in}}{\Delta t}\right)$$

[(], R value, [±], [×], C value, [)], [×], $\left(\frac{\Delta V_{in}}{\Delta t}\right)$ value, [=]

Instrumentation amplifier

FORMULA 7–15 *(To find the output voltage)*

$$V_{out} = \left(\frac{1+2\times R}{R_G}\right)\times(V_2 - V_1)$$

[(], 1, [+], 2, [×], R value, [÷], R_G value, [)], [×], [(], V_2 value, [−], V_1 value, [)], [=]

Schmitt trigger comparator

FORMULA 7–16 *(To find the upper trip point)*

$$\text{UTP} = V_{max}\left[\frac{R_1}{(R_1 + R_2)}\right]$$

V_{max} value, [×], [(], R_1 value, [÷], [(], R_1 value, [+], R_2 value, [)], [)], [=]

FORMULA 7–17 *(To find the lower trip point)*

$$\text{LTP} = V_{min}\left[\frac{R_1}{(R_1 + R_2)}\right]$$

V_{min} value, [×], [(], R_1 value, [÷], [(], R_1 value, [+], R_2 value, [)], [)], [=]

FORMULA 7–18 *(To find the hysteresis)*

$$V_H = \text{UTP} - \text{LTP}$$

UTP value, [−], LTP, [=]

Inverting comparator with hysteresis

FORMULA 7–19 *(To find the upper trip point)*

$$\text{UTP} = V_{REF}\left[\frac{(R_2 + R_3)}{(R_1 + R_2 + R_3)}\right] + V_S\left[\frac{R_1}{(R_1 + R_2 + R_3)}\right]$$

V_{REF} value, [×], [(], R_2 value, [+], R_3 value, [)], [÷], [(], R_1 value, [+], R_2 value, [+], R_3 value, [)], [+], V_S value, [×], R_1 value, [÷], [(], R_1 value, [+], R_2 value, [+], R_3 value, [)], [=]

FORMULA 7–20 *(To find the lower trip point)*

$$\text{LTP} = V_{REF}\left[\frac{R_2}{(R_1 + R_2)}\right] + V_{EE}\left[\frac{R_1}{(R_1 + R_2)}\right]$$

V_{REF} value, [×], R_2 value, [÷], [(], R_1 value, [+], R_2 value, [)], [+], V_{EE} value, [×], R_1 value, [÷], [(], R_1 value, [+], R_2 value, [)], [=]

Noninverting comparator with hysteresis

FORMULA 7–21 *(To find the upper trip point)*

$$\text{UTP} = V_{REF}\left[\frac{(R_1 + R_2)}{R_2}\right] + V_{EE}\left[\frac{R_1}{R_2}\right]$$

V_{REF} value, [×], [(], R_1 value, [+], R_2 value, [)], [÷], R_2 value, [−], V_{EE} value, [×], R_1 value, [÷], R_2 value, [=]

FORMULA 7–22 *(To find the lower trip point)*

$$\text{LTP} = V_{REF}\left[\frac{(R_1 + R_2 + R_3)}{(R_2 + R_3)}\right] - V_S\left[\frac{R_1}{(R_2 + R_3)}\right]$$

V_{REF} value, [×], [(], R_1 value, [+], R_2 value, [+], R_3 value, [)], [÷], [(], R_2 value, [+], R_3 value, [)], [−], V_S value, [×], R_1 value, [÷], [(], R_2 value, [+], R_3 value, [)], [=]

Window comparator

FORMULA 7–23 *(To find logic output)*

V_{out} > "low" for V_{REF_2}, < V_{in} < V_{REF_1}
V_{out} = "high" for $V_{in} < V_{REF_2}$
V_{out} = "high" for $V_{in} > V_{REF1}$

FORMULA 7–24 *(To find logic output)*

V_{out} = "high" for $V_{REF_1} < V_{in} < V_{REF_2}$
V_{out} = "low" for $V_{in} < V_{REF_1}$
V_{out} = "low" for $V_{in} > V_{REF_2}$

Bar graph display driver

FORMULA 7–25 *(To output reference voltage)*

$$V_{out} = V_{REF}\left(1 + \frac{R_2}{R_1}\right) + I_{ADJ}R_2$$

where V_{REF} = 1.25 V (typical) and I_{ADJ} = 75 μA (typical)

V_{REF} value, [×], [(], 1, [+], R_2 value, [÷], R_1 value, [)], [+], I_{ADJ} value, [×], R_2 value, [=]

FORMULA 7–26
(To comparator reference voltage)

$$V_{REF(n)} = n \times \left(\frac{1\ \text{k}\Omega}{10\ \text{k}\Omega}\right) \times V_{out}$$

where $n = 1, 2, 3, 4, 5, 6, 7, 8, 9,$ or 10

n value, [×], [(], 1000, [÷], 10000, [)], [×], V_{out} value, [=]

OTA differential amplifier

FORMULA 7–27
(To find the OTA output voltage)

$$V_{out} = (V_1 - V_2) \times k \times I_{BIAS} \times R_L$$

where:

$k \approx 20$

$$I_{BIAS} = \frac{(V_{BIAS} - V^{-} - 0.7)}{R_{BIAS}}$$

[(], V_1 value, [−], V_2 value, [)], [×], k value, [×], [(], V_{BIAS} value, [−], V^- value, [−], 0.7,[)], [÷], R_{BIAS} value, [=], [×], R_L value, [=]

OTA Schmitt trigger comparator

FORMULA 7–28
(To find the OTA upper trip point)

$$\text{UTP} = I_{BIAS} \times R_1$$

[(], V_{BIAS} value, [−], V^- value, [−], 0.7, [)], [÷], R_{BIAS} value, [=], [×], R_1 value, [=]

FORMULA 7–29
(To find the OTA lower trip point)

$$\text{LTP} = -I_{BIAS} \times R_1$$

[(], V_{BIAS} value, [−], V^- value, [−], 0.7,[)], [÷], R_{BIAS} value, [=], [×], R_1 value, [±], [=]

ADC

FORMULA 7–30
(To find the digital data (decimal representation))

$$\text{DATA} = V_{in_1} \times \left(\frac{255}{5\ \text{V}}\right)$$

V_{in_1} value, [×], 255, [÷], 5, [=]

DAC

FORMULA 7–31
(To find the output voltage)

$$V_{out} = \left(\frac{-V_{REF}}{R_{14}}\right) \times \left[\left(\frac{D_7}{2}\right) + \left(\frac{D_6}{4}\right) + \left(\frac{D_5}{8}\right) + \left(\frac{D_4}{16}\right) + \left(\frac{D_3}{32}\right) + \left(\frac{D_2}{64}\right) + \left(\frac{D_1}{128}\right) + \left(\frac{D_0}{256}\right)\right] \times (-R_1)$$

[(], V_{REF} value, [±], [÷], R_{14} value, [)], [×], [(], D_7, [÷], 2, [+], D_6, [÷], 4, [+], D_5, [÷], 8, [+], D_4, [÷], 16, [+], D_3, [÷], 32, [+], D_2, [÷], 64, [+], D_1, [÷], 128, [+], D_0, [÷], 256, [)], [×], R_1 value, [±], [=]

Using Excel

Op-amp Formulas

(Excel file reference: FOE7_01.xls)

DON'T FORGET! It is NOT necessary to retype formulas, once they are entered on the worksheet! Just input new parameters data for each new problem using that formula, as needed.

- Use the Formula 7–5 spreadsheet sample and the parameters given for Practice Problems 1 question 1. Solve for the voltage gain of the op-amp circuit. Check your answer against the answer for this question in the Appendix.
- Use the Formula 7–6 spreadsheet sample and the parameters given for Practice Problems 2 question 1. Solve for the voltage gain of the op-amp circuit. Check your answer against the answer for this question in the Appendix.

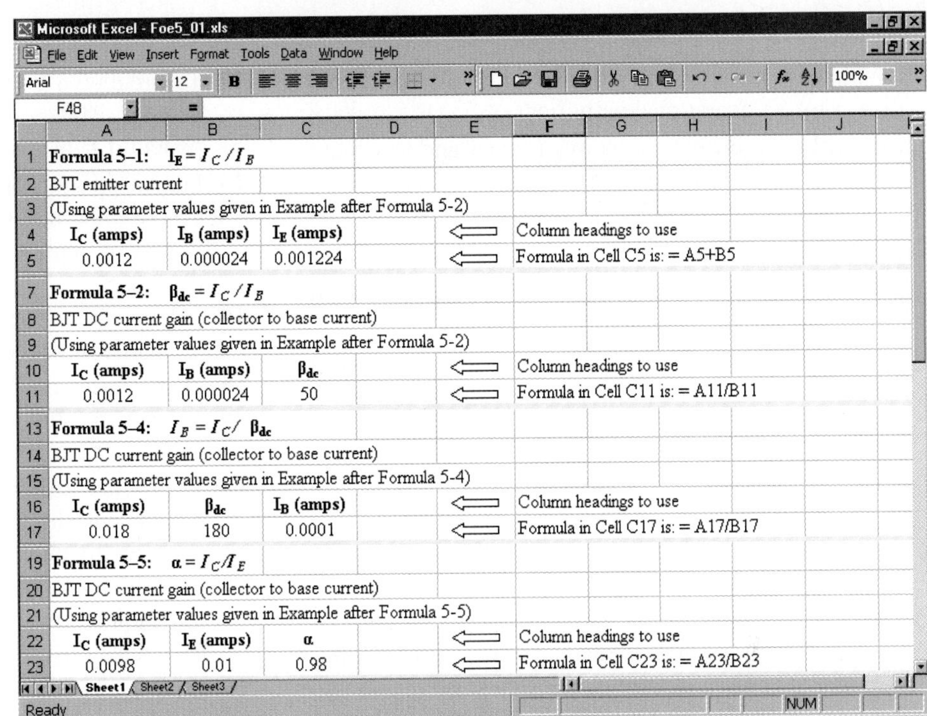

Microsoft Excel - Foe5_01.xls

	A	B	C	D	E	F
1	**Formula 5–1:** $I_E = I_C / I_B$					
2	BJT emitter current					
3	(Using parameter values given in Example after Formula 5-2)					
4	**I_C (amps)**	**I_B (amps)**	**I_E (amps)**		⟸	Column headings to use
5	0.0012	0.000024	0.001224		⟸	Formula in Cell C5 is: = A5+B5
7	**Formula 5–2:** $\beta_{dc} = I_C / I_B$					
8	BJT DC current gain (collector to base current)					
9	(Using parameter values given in Example after Formula 5-2)					
10	**I_C (amps)**	**I_B (amps)**	**β_{dc}**		⟸	Column headings to use
11	0.0012	0.000024	50		⟸	Formula in Cell C11 is: = A11/B11
13	**Formula 5–4:** $I_B = I_C / \beta_{dc}$					
14	BJT DC current gain (collector to base current)					
15	(Using parameter values given in Example after Formula 5-4)					
16	**I_C (amps)**	**β_{dc}**	**I_B (amps)**		⟸	Column headings to use
17	0.018	180	0.0001		⟸	Formula in Cell C17 is: = A17/B17
19	**Formula 5–5:** $\alpha = I_C / I_E$					
20	BJT DC current gain (collector to base current)					
21	(Using parameter values given in Example after Formula 5-5)					
22	**I_C (amps)**	**I_E (amps)**	**α**		⟸	Column headings to use
23	0.0098	0.01	0.98		⟸	Formula in Cell C23 is: = A23/B23

Using Excel

Op-amp Formulas

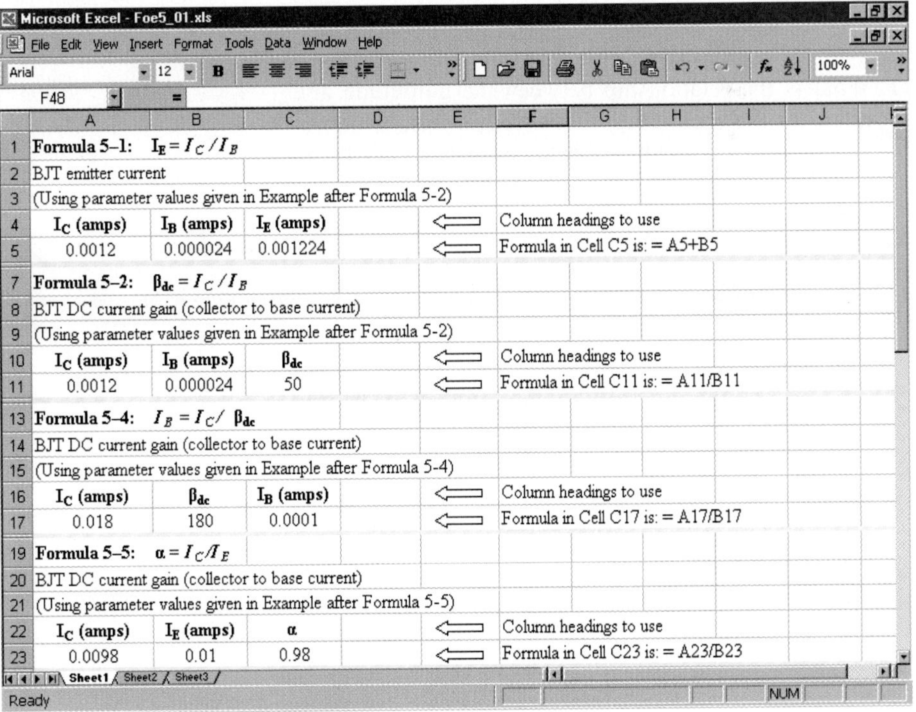

Microsoft Excel - Foe5_01.xls

	A	B	C	D	E	F
1	**Formula 5–1:** $I_E = I_C / I_B$					
2	BJT emitter current					
3	(Using parameter values given in Example after Formula 5-2)					
4	**I_C (amps)**	**I_B (amps)**	**I_E (amps)**		⟸	Column headings to use
5	0.0012	0.000024	0.001224		⟸	Formula in Cell C5 is: = A5+B5
7	**Formula 5–2:** $\beta_{dc} = I_C / I_B$					
8	BJT DC current gain (collector to base current)					
9	(Using parameter values given in Example after Formula 5-2)					
10	**I_C (amps)**	**I_B (amps)**	**β_{dc}**		⟸	Column headings to use
11	0.0012	0.000024	50		⟸	Formula in Cell C11 is: = A11/B11
13	**Formula 5–4:** $I_B = I_C / \beta_{dc}$					
14	BJT DC current gain (collector to base current)					
15	(Using parameter values given in Example after Formula 5-4)					
16	**I_C (amps)**	**β_{dc}**	**I_B (amps)**		⟸	Column headings to use
17	0.018	180	0.0001		⟸	Formula in Cell C17 is: = A17/B17
19	**Formula 5–5:** $\alpha = I_C / I_E$					
20	BJT DC current gain (collector to base current)					
21	(Using parameter values given in Example after Formula 5-5)					
22	**I_C (amps)**	**I_E (amps)**	**α**		⟸	Column headings to use
23	0.0098	0.01	0.98		⟸	Formula in Cell C23 is: = A23/B23

Review Questions

1. Which amplifier circuit is typically used as the input stage for an operational amplifier?
2. Describe the difference between linear and nonlinear op-amp circuits and give an example of each.
3. List the key parameters to describe the ideal op-amp.
4. Define *open-loop* gain.
5. Give an example of an op-amp circuit that is operated in the open-loop gain mode.
6. What is the closed-loop gain of an inverting op-amp circuit when the ratio of feedback to input resistance $\left(\frac{R_f}{R_i}\right)$ is 15? 10?
7. If R_i for an inverting op-amp is a variable resistor, increasing the value of R_i will (increase, decrease) _____ the gain of the amplifier.
8. What is the relationship between the amplitude and phase of the input and output signals from an inverting op-amp that has equal resistance values for R_i and R_f?
9. What is the closed-loop gain of a noninverting op-amp circuit when the ratio of feedback to input resistance $\left(\frac{R_f}{R_i}\right)$ is 15? 10?
10. If R_i for a noninverting op-amp is a variable resistor, increasing the value of R_i will (increase, decrease) the gain of the amplifier.
11. What is the relationship between the amplitude and phase of the input and output signals from a noninverting op-amp that has equal resistance values for R_i and R_f?
12. To produce a voltage follower with an op-amp what value resistance should be used as the feedback component if $R_i = 1\ \text{M}\Omega$?
13. For an op-amp operating in the single-ended mode, describe the output signal polarity when the positive input is grounded and the input signal feeds the negative input.
14. For an op-amp in common-mode operation, should the output level be large, small, or medium?
15. To produce integration in an op-amp, what type of component should be used as the feedback component? as the input component?
16. To produce differentiation in an op-amp, what type of component should be used as the feedback component? as the input component?
17. What type of op-amp circuit should be selected where it is required to change a rectangular waveform into a triangular waveform?
18. What type of op-amp circuit should be selected where it is required to change a rectangular waveform into brief pulses?
19. Assuming equal resistors throughout the circuit, write the formula for the output of a summing amplifier having three input voltages.
20. For the circuit in Figure 7–29, briefly explain how a condition where V_{out} trying to increase could be compensated for by the regulator circuit. Explain in terms of the coordination between the feedback circuit, the op-amp, and the series output transistor.

Problems

1. For the circuit in Figure 7–35, the feedback resistor is 250 kΩ and an input resistor is 1,200 Ω. What is the closed-loop gain?
2. What is the closed-loop gain of the circuit in Figure 7–35 when $R_i = 1\ \text{k}\Omega$ and $R_f = 1\ \text{k}\Omega$?
3. What value input resistor is required for an inverting op-amp if the feedback resistor is 1 MΩ and the closed loop gain is to be –120?

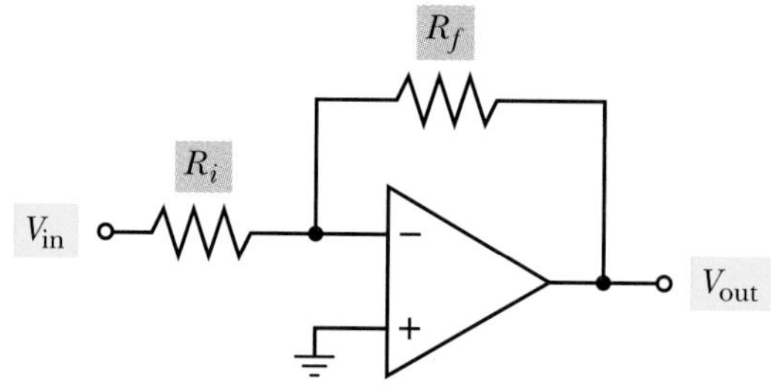

FIGURE 7–35

4. For the circuit in Figure 7–36, the feedback resistor is 250 kΩ and an input resistor is 1,200 Ω. What is the closed-loop gain?

5. What is the closed-loop gain of the circuit in Figure 7–36 when R_i = 1 kΩ and R_f = 1 kΩ?

6. What value input resistor is required for an inverting op-amp if the feedback resistor is 1 MΩ and the closed-loop gain is to be 120?

7. All resistors in a certain difference-mode op-amp (see Figure 7–16) have the same value. Determine the output of the amplifier when the following is true:

a. $V_{in+} = +2$ V and $V_{in-} = -3$ V

b. $V_{in+} = -2$ V and $V_{in} = -3$ V

c. $V_{in+} = +2$ V and $V_{in-} = +3$ V

d. $V_{in+} = -2$ V and $V_{in-} = +3$ V

8. All resistors in the summing op-amp of Figure 7–37 have the same value. Determine the output voltage, V_{out}, when the following is true:

a. $V_1 = +2$ V and $V_2 = +3$ V

b. $V_1 = -2$ V and $V_2 = +3$ V

c. $V_1 = +2$ V and $V_2 = -3$ V

d. $V_1 = -2$ V and $V_2 = -3$ V

9. For the summing op-amp of Figure 7–37, R_{i_1} = 10 kΩ, R_{i_2} = 15 kΩ, and R_f = 20 kΩ. Determine V_{out} when the following is true:

a. $V_1 = +2$ V and $V_2 = +3$ V

b. $V_1 = -2$ V and $V_2 = +3$ V

c. $V_1 = +2$ V and $V_2 = -3$ V

d. $V_1 = -2$ V and $V_2 = -3$ V

10. Calculate the UTP, LTP, and hysteresis for a Schmitt trigger circuit (Figure 7–24) where R_1 = 100 kΩ, R_2 = 27 kΩ, +V source = +12 V, and –V source = 0 V (grounded).

11. The input applied to a unity gain amplifier circuit is 4 V.

a. Draw the circuit schematic.

b. Calculate the output voltage.

12. All the resistors in a difference amplifier are 10 kΩ. $V_{in+} = 4$ V and $V_{in-} = 6$ V.

a. Draw the circuit schematic.

b. Calculate the output voltage.

13. A 10-kΩ resistor and a 2.2-nF capacitor are used in an integrator circuit. An input voltage of –5 V is applied for 1 μs. Calculate the change in the output voltage.

14. A 20-kΩ resistor and a 1.0-nF capacitor are used in a differentiator circuit. The ramp input voltage is 1.5 V/μs. Calculate the output voltage.

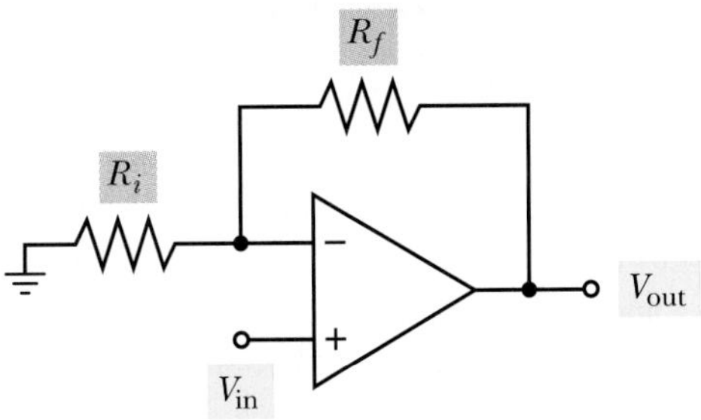

FIGURE 7–36

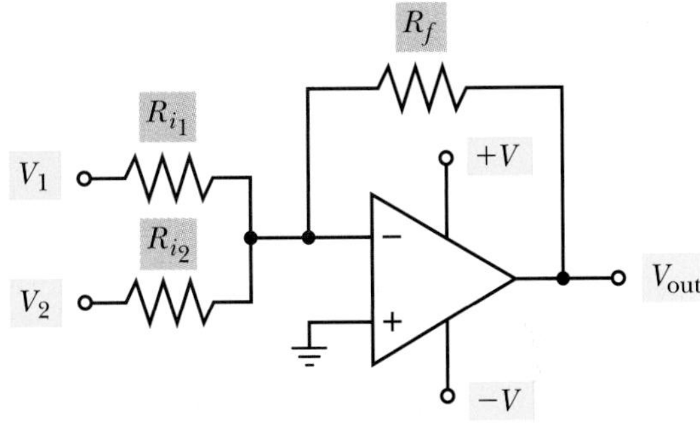

FIGURE 7–37

15. An instrumentation amplifier has all its resistors (R_1 through R_6 and R_G) equal to 10 kΩ. Calculate the output voltage when $V_1 = 2$ V and $V_2 = 4$ V.

16. The LM3914 is used in the circuit in Figure 7–28, where R_1 = 1.2 kΩ, R_2 = 5.6 kΩ, R_{11} through R_{20} = 1.2 kΩ, V^+ = 12 V, and V_{LED} = 12 V.

a. Calculate V_{out} and the threshold voltage for each comparator.

b. If $V_{in} = +4$ V, indicate the status of all ten LEDs.

17. An input voltage of 4 V is applied to the ADC in Figure 7–32. Calculate the digital data output. Represent the answer

a. in decimal

b. in binary

c. in hexadecimal

d. by the logic level of each output (DB0 through DB7)

18. A digital input of 63 (hexadecimal) is applied to the DAC in Figure 7–34. Circuit values are V_{REF} = 10 V, R_1 = 10 kΩ, R_{15} = 10 kΩ, R_{14} = 10 kΩ, and C = 0.1 μF. Calculate the output voltage.

Analysis Questions

1. Show how you can use the formula for the gain of a noninverting op-amp (formula 7–6) to prove that the gain of a voltage follower, Figure 7–12, is unity.
2. Find and list the data handbook parameters typically shown for a 741-type operational amplifier IC device.
3. A sensitive electronic thermometer produces a voltage that changes 10 mV for every degree of change in temperature. If the signal is connected to the input of an inverting op-amp circuit, what voltage gain is necessary so that the output of the amplifier produces a change of 2 V for every degree of change in temperature? If the input resistance to the op-amp is 2.2 kΩ, what is the value of the feedback resistor?
4. For the instrumentation amplifier circuit shown in Figure 7–21, show the algebra that confirms Formula 7–15 properly relates the output voltage with the input voltage. (Hint: Use the equations provided for V_R and V_B along with the equation for a difference amplifier.)
5. Define conversion time. Locate a datasheet for the ADC0804. What is the specified conversion time for this component?

MultiSIM Exercise for Op-amp Circuits

1. Use the MultiSIM program and utilize the circuit for the Example after Formula 7–6 on page 307.
2. Measure and record the output voltage.
3. Compare your MultiSIM results with the values given in the Example for this circuit. Was the output voltage that you made with MultiSIM reasonably close to the results of the Example?

Performance Projects Correlation Chart

All e-labs are available on the LabSource CD.

Chapter Topic	Performance Project	Project Number
An Inverting Amplifier	Inverting Op-Amp Circuit	16
A Noninverting Amplifier	Noninverting Op-Amp Circuit	17
An Op-Amp Schmitt Trigger Circuit	Op-Amp Schmitt Trigger Circuit	18
Digital-to-Analog Converter	Digital to Analog Converter (DAC0808)	21
Analog-to-Digital Converter	Analog to Digital Converter (ADC0804)	22
Digital Interface Circuits	Story Behind the Numbers: ADC and DAC Circuit	

NOTE: It is suggested that after completing the above projects, the student should be required to answer the questions in the "Summary" at the end of this section of projects in the Laboratory Manual.

Troubleshooting Challenge

CHALLENGE CIRCUIT 12

(Follow the SIMPLER sequence by referring to inside front cover.)

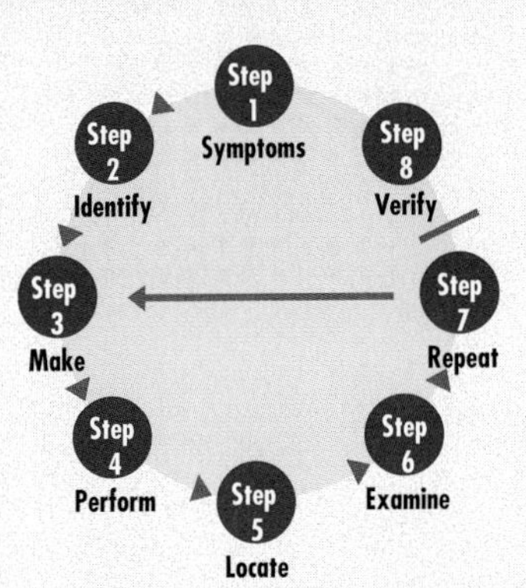

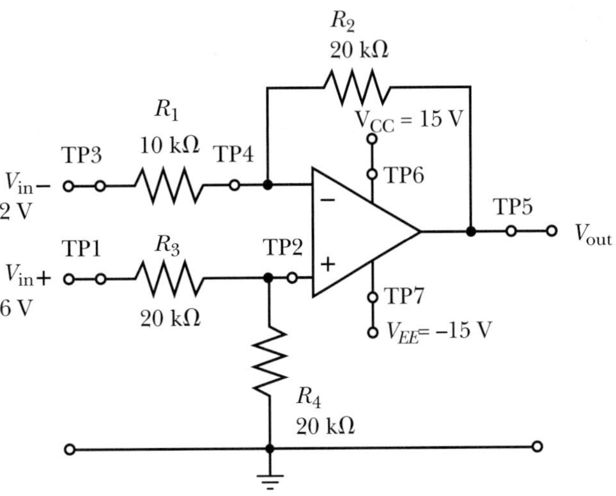

Challenge Circuit 12

STARTING POINT INFORMATION

1. Circuit diagram
2. The voltage at TP5 is at 13.5 V.

TEST	Results in Appendix C
DC voltage at TP1	(57)
DC voltage at TP2	(65)
DC voltage at TP3	(4)
DC voltage at TP4	(35)
DC voltage at TP5	(125)
DC voltage at TP6	(92)
DC voltage at TP7	(126)
Resistance of R_3	(117)
Resistance of R_4	(19)

CHALLENGE CIRCUIT 12

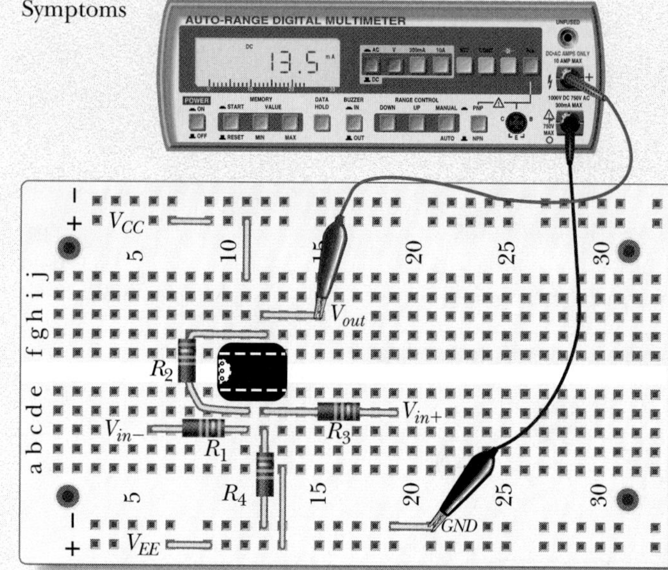

Symptoms

STEP [1]

SYMPTOMS The output voltage is at positive saturation.

STEP [2]

IDENTIFY initial suspect area. The high voltage can be caused by a failure of several components in the circuit.

STEP [3]

MAKE test decision. We will start by checking the voltage at the op-amp noninverting input (TP2).

STEP [4]

PERFORM **First Test:** Check the voltage at TP2. The result is a voltage that is higher than normal.

STEP [5]

LOCATE new suspect area. It looks like the voltage divider circuit connected to the noninverting op-amp input has a problem (either R_3 is shorted or R_4 is open).

STEP [6]

EXAMINE available data.

STEP [7]

REPEAT analysis and testing.

The next test is to measure the resistance of R_3.

Second Test: Separate the resistor R_3 from the circuit and measure its resistance. This measurement indicates that the resistance of R_3 is normal.

Third Test: Let us now separate the resistor R_4 from the circuit and measure its resistance. This measurement indicates that R_4 is open.

STEP [8]

VERIFY **Fourth Test:** Replace resistor R_4 and note the circuit parameters. When this is done, the circuit operates properly.

1st Test

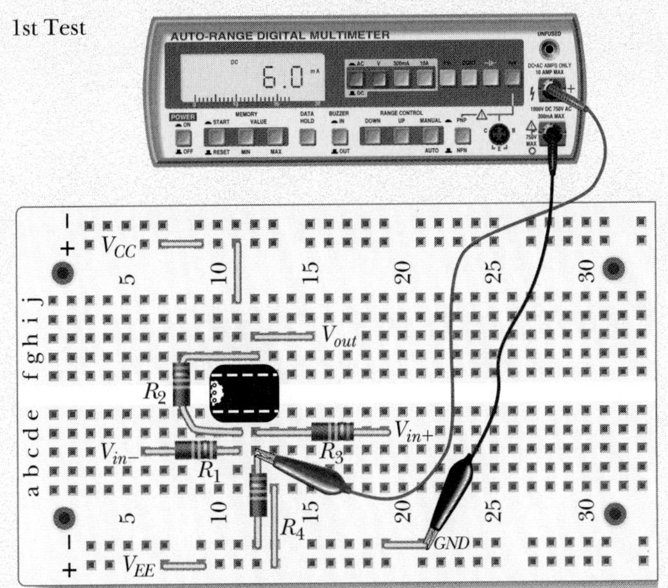

2nd Test

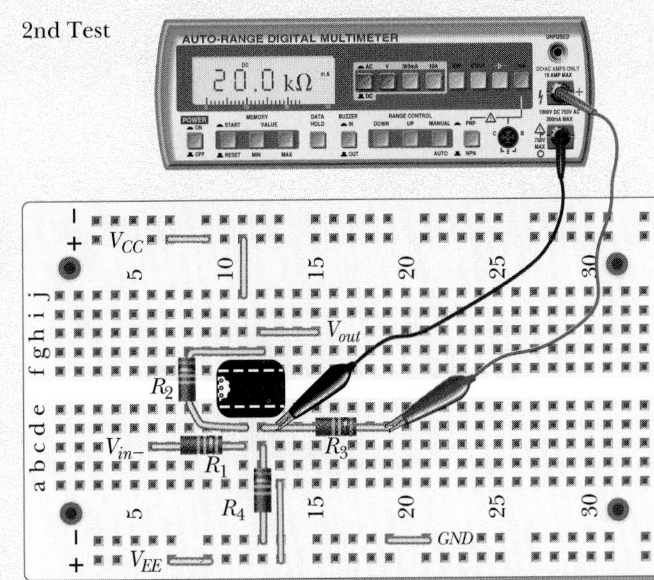

3rd Test

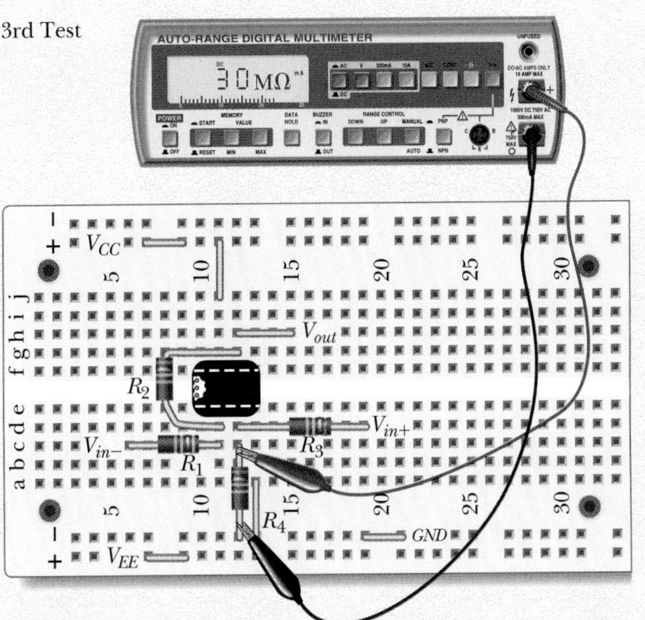

4th Test

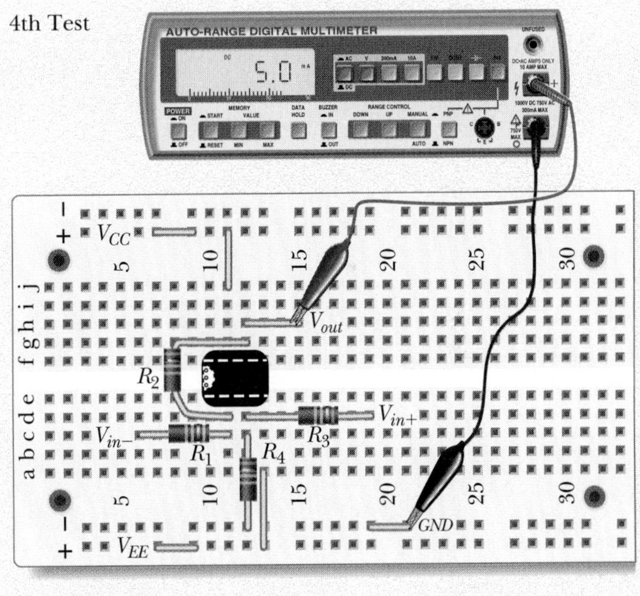

*Jump wire

Troubleshooting Challenge

CHALLENGE CIRCUIT 13 (Block Diagram)

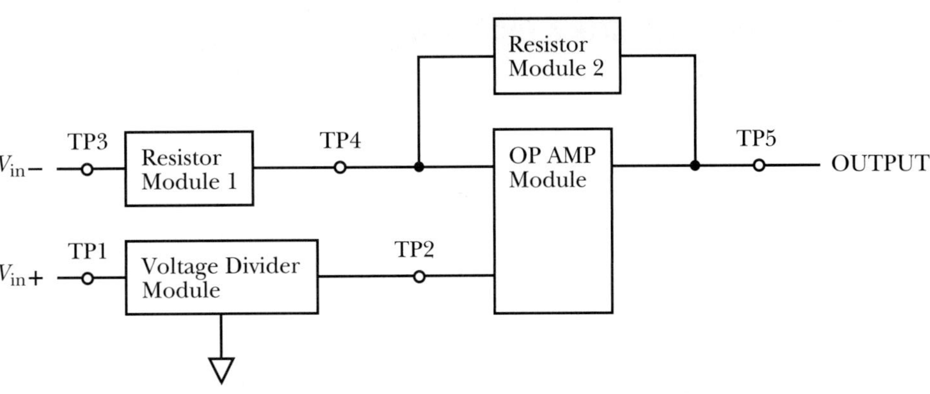

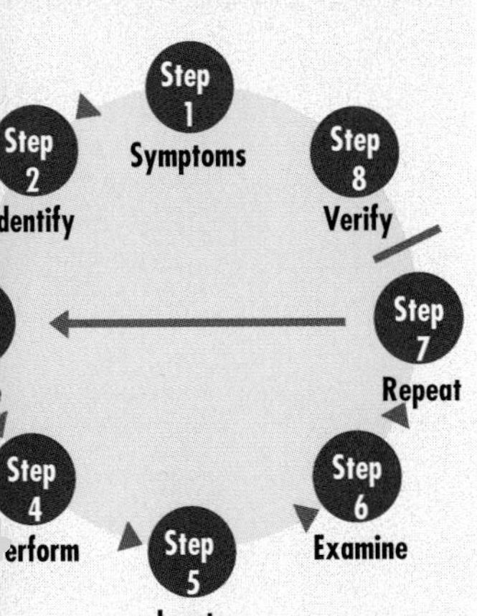

General Testing Instructions:

Measurement
Assumptions:
Signal = from **TP** to ground
I = at the **TP**
V = from **TP** to ground
R = from **TP** to ground (with the power source disconnected from the circuit)

Possible Tests & Results:
Signal (normal, abnormal, none)
Current (high, low, normal)
Voltage (high, low, normal)
Resistance (high, low, normal)

Starting Point Information	Test Points	Test Results in Appendix C			
		V	*I*	*R*	*Signal*
At TP_1	TP2	(276	277	278	NA)
	TP3	(279	280	NA	NA)
V = normal	TP4	(281	NA	NA	NA)
I = low	TP5	(282	NA	NA	NA)
R = high					
Signal = NA					

DIFFERENCE AMPLIFIER BLOCK DIAGRAM CHALLENGE PROBLEM

STEP [1]

SYMPTOMS At TP5, output voltage is at positive saturation.

STEP [2]

IDENTIFY initial suspect area. Since the supply voltages are normal, the problem could be caused by a failure of the operational amplifier, the feedback resistor, input resistor, or the voltage divider module.

STEP [3]

MAKE test decision based on the symptoms information. Let's start with the voltage measurements at strategic test points.

STEP [4]

PERFORM **First Test:** Voltage at TP2.
The voltage is higher than normal and is the same as V_{in+}. The voltage divider module appears to be faulty.

STEP [5]

LOCATE new suspect area. The voltage divider module is suspect.

STEP [6]

EXAMINE available data.

STEP [7]

REPEAT analysis and testing.
Second Test: Separate the voltage divider module from the circuit. Measure the resistance of the voltage divider module. The result is an open. The voltage divider module had failed.

STEP [8]

VERIFY **Third Test:** After replacing the voltage divider module with a new module that meets specifications, all circuit parameters come back to their norms.

Troubleshooting Challenge

CHALLENGE CIRCUIT 14
Find the Fault

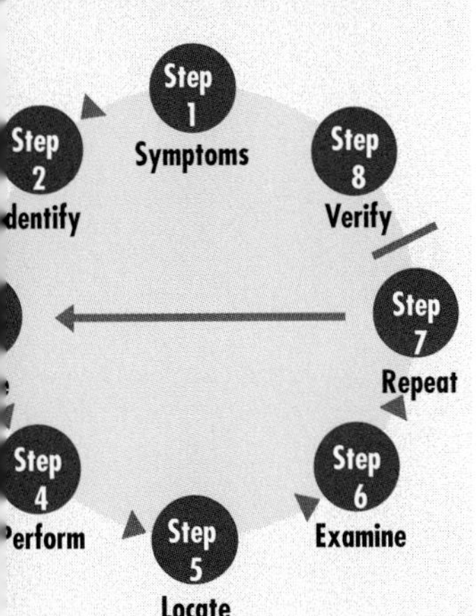

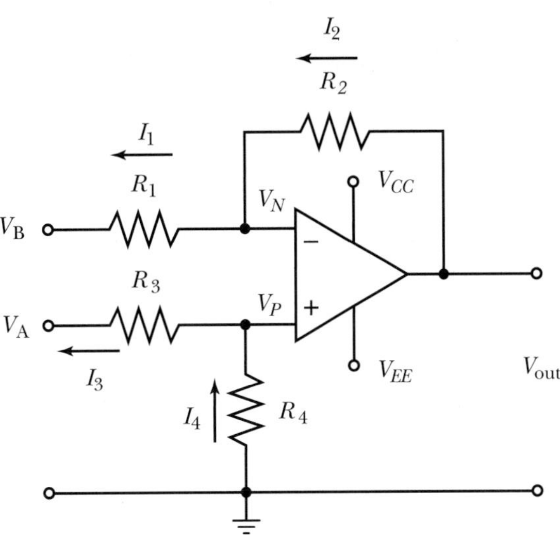

Challenge Circuit 14

STARTING POINT INFORMATION

1. Circuit diagram (R_1 = kΩ, R_2 = 10 kΩ, R_3 = 10 kΩ, R_4 = 10 kΩ)
2. V_{CC} = 12 V V_{EE} = 12 V V_A = 2 V V_B = – 2 V
3. LM324 op-amp
4. V_{out} is a voltage lower than expected

Test	Circuit Measurements	Theoretical Expected Value
DC Voltages:		
V_{CC}	12 V	______
V_{CC}	12 V	______
V_{EE}	–12 V	______
V_A	2 V	______
V_B	–2 V	______
V_{out}	1 V	______
V_P	1 V	______
V_N	1 V	______
I_1	0 mA	______
I_2	0 mA	______
I_3	0.1 mA	______
I_4	0.1 mA	______
R_1	....	______
R_2	....	______
R_3	....	______
R_4	....	______

STEP [1]

You will analyze the circuit to calculate the expected voltages, currents, and resistances for a properly functioning circuit. Record these values beside each test listed in the table provided. (Refer to Analysis Techniques)

STEP [2]

Starting with the symptom, document your steps for troubleshooting this circuit using the SIMPLER troubleshooting technique or instructions provided by your instructor.

STEP [3]

For each circuit measurement, you will indicate the test to be accomplished.

STEP [4]

Using the previous table, you will find the voltage and current measurements taken for the faulted circuit.

STEP [5]

You will now compare the faulted circuit measurement to the expected measurement for a properly functioning circuit. Are these values the same? (Refer to General Testing Instructions)

STEP [6]

You will repeat steps 2 through 5 until you have located the fault.

STEP [7]

When you have located the fault, you will identify the fault and the characteristics of the failure. (Refer to Types of Failures)

ANALYSIS TECHNIQUES (check with your instructor)

METHOD [1]

Using circuit theory and algebra, evaluate the circuit.

METHOD [2]

Using an Excel spreadsheet for this circuit, evaluate the circuit.

METHOD [3]

Using MultiSIM, build the circuit schematic and generate the results.

METHOD [4]

Assemble this circuit on a circuit board. Apply power and make the appropriate measurements using your DMM and oscilloscope.

General Testing Instructions:

Measurement Assumptions:
Signal = from **TP** to ground
I = at the **TP**
V = from **TP** to ground
R = from **TP** to ground (with the power source disconnected from the circuit)

Possible Tests & Results:
Signal (normal, abnormal, none)
Current (high, low, normal)
Voltage (high, low, normal)
Resistance (high, low, normal)

Types of Failures

Component failures
- wrong part value
- part shorted
- part open

Supply voltage—incorrect voltage
Ground—floating

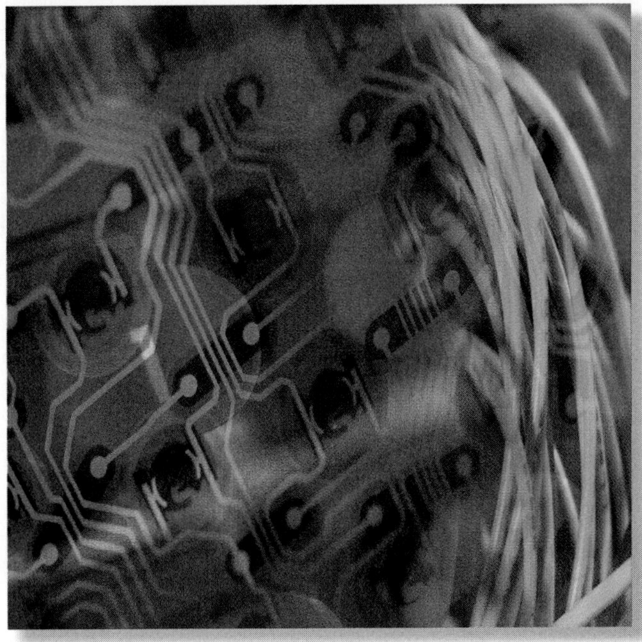

CHAPTER 8

Advanced Operational Amplifiers

OUTLINE

OBJECTIVES

After studying this chapter, you should be able to:

1. Explain the unity gain bandwidth product
2. Describe the op-amp slew rate
3. Determine the output signal for sine wave inputs
4. Draw the output signal for op-amp rectifier circuits
5. List the key characteristics of an ideal op-amp
6. Identify linear and nonlinear applications circuits for op-amps
7. Distinguish between **inverting** and **noninverting op-amp circuits**
8. Perform voltage gain and resistance calculations for standard inverting and noninverting op-amp circuits
9. Describe the operation of op-amps in voltage amplifiers, voltage followers, comparators, and Schmitt trigger amplifiers
10. Describe the function of op-amps in circuits originally designed for analog computers: summing amplifiers, differential amplifiers, differentiators, and integrators

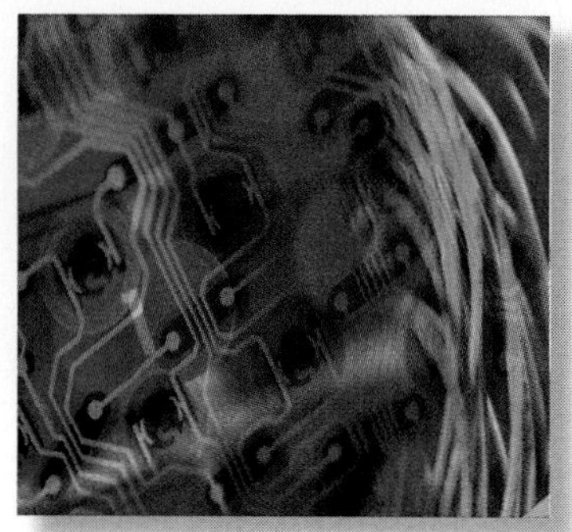

CHAPTER PREVIEW

The operational amplifier usage is not limited to dc applications. The emphasis in this chapter is the evaluation of sinusoidal signals applied to the op-amp. Audio signals (voice and music) are composed of sinusoidal waveforms containing frequencies from 20 Hz to 20 kHz. Electronic equipment use operational amplifier circuits to amplify the waveforms to acceptable levels to drive speakers. All periodic waveforms can be represented as the summation of sine and cosine signals and a dc offset according to Jean Baptiste Joseph Fourier (the Fourier series).

In this chapter you will learn about the basic features of op-amp frequency considerations. A variety of linear and nonlinear applications for these devices will be presented so you can understand their versatility. Additionally, you will learn that in many applications, a simple ideal op-amp approach can be used for amplifier analysis and design.

KEY TERMS

Active limiter
Actual closed-loop gain
Audio amplifier
Bypass capacitors
Peak detector
Positive clamper
Positive limiter
Precision full-wave rectifier positive output
Precision half-wave rectifier negative output
Precision half-wave rectifier positive output
Slew rate
Unity gain bandwidth product (GBW)

8–1 Background Information

Critical op-amp parameters as you consider using the op-amp for waveform applications are the unity gain bandwidth product and the slew rate.

Unity Gain Bandwidth Product

The **unity gain bandwidth product (GBW)** provides you with the frequency and open-loop gain limitations of the op-amp. For the ideal operational amplifier, the unity gain bandwidth product and open-loop gain were stated as infinite. There is a finite value for the unity gain bandwidth product and open-loop gain. These values for an LF347 op-amp will be different than these values for an LM324 op-amp. The GBW is a constant provided by the component manufacturer on the op-amp datasheet. Formula 8–1 provides the equation relating the unity gain bandwidth parameter to the signal frequency and the amplifier's open-loop gain. As the frequency of the signal applied to the amplifier circuit is increased, the open-loop voltage gain ($A_{V(OL)}$ or a) of the amplifier decreases. The open-loop voltage gain for a particular input frequency and op-amp can be calculated using Formula 8–2.

FORMULA 8–1 $GBW = a \times f$

FORMULA 8–2 $a = \dfrac{GBW}{f}$

◆ **EXAMPLE** The LM741 will be used in a circuit where the applied signal frequency is 100 Hz. Calculate the open-loop voltage gain for the op-amp at this frequency.

Answer:

The LM741 has a unity gain bandwidth product of 1.0 MHz. Figure 8–1 shows the gain versus frequency response for the LM741. Using Formula 8–2, you can calculate the open-loop voltage gain for a signal frequency of 100 Hz.

From Formula 8–2:

$$a = \frac{GBW}{f}$$

$$a = \frac{1.0\text{ MHz}}{100\text{ Hz}}$$

$$A_{V(OL)} = 10{,}000$$ ◆

The effect that this decrease in the op-amp's open-loop gain has on your usage of the op-amp will be considered when the inverting, noninverting, and differential amplifiers are presented within this chapter.

Power supply bypass capacitors are essential to maintain op-amp stability preventing unwanted high frequency oscillations at the output. The recommended component is a 0.1 μf ceramic capacitor connected between the positive op-amp power supply lead and the power supply ground. If you are operating the op-amp with a positive and negative power supply, you will connect another 0.1 μf ceramic capacitor from the negative op-amp power supply lead to the power supply ground as shown in Figure 8–2. The power supply bypass capacitors will help minimize power supply voltage variations.

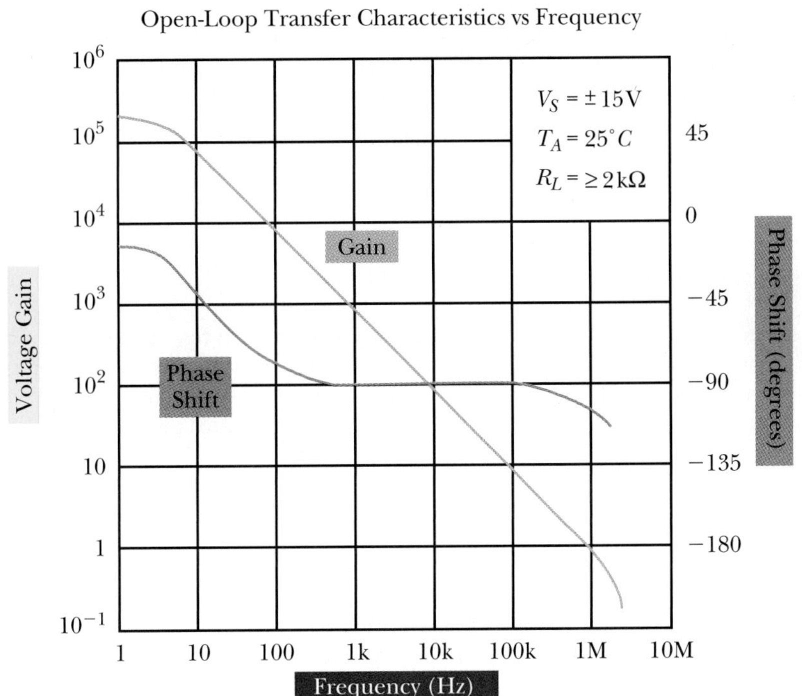

FIGURE 8–1 741 Open-loop gain

FIGURE 8–2 (a) Op-amp amplifier with power supply bypass capacitors (schematic); (b) Op-amp amplifier with power supply bypass capacitors (breadboard)

R_2 V_{CC} R_1 C_1 0.1 μF V_{out} V_{in} R_3 R_4 C_2 0.1 μF V_{EE}

(a)

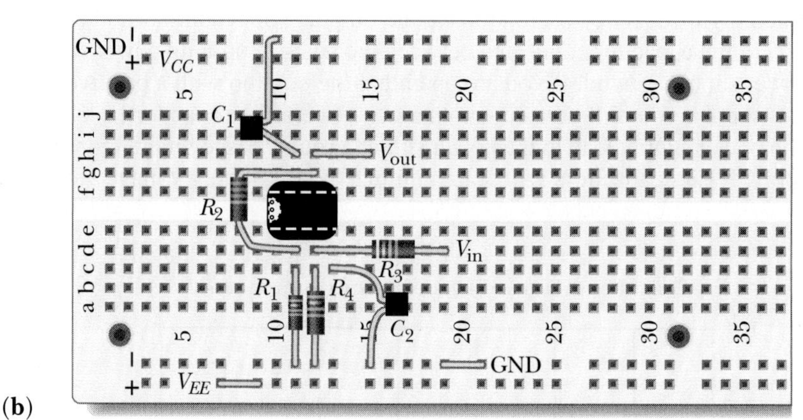

(b)

> ***Practical Notes***
>
> As the negative input to the op-amp is extremely sensitive and responsive to other circuit signals (electromechanical interference and electromechanical coupling), components connected to the negative op-amp terminal must be positioned as close as possible to this op-amp lead (Figure 8–2b) to prevent unwanted signal pickup.

Slew Rate

The **slew rate** (SR) is a measurement of the output voltage transition speed in volts per microsecond from one voltage level to another voltage level due to a change on the op-amp input (Figure 8–3b). Formula 8–3 provides the equation to calculate the time required for the voltage transition.

FORMULA 8–3 $\Delta t = \dfrac{\Delta V_{out}}{SR}$

FORMULA 8–4 $\Delta V_{out} = V_{out_2} - V_{out_1}$

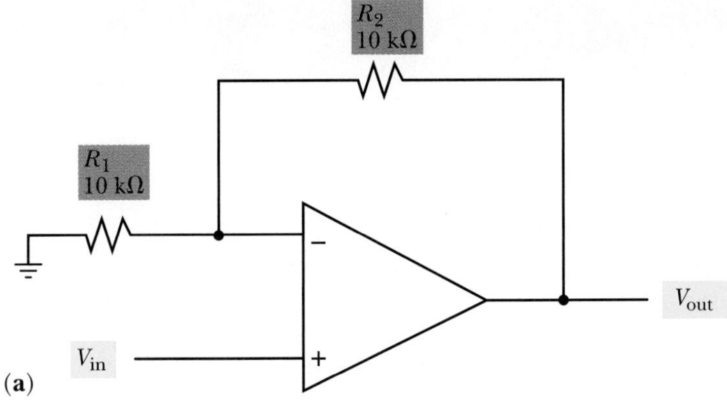

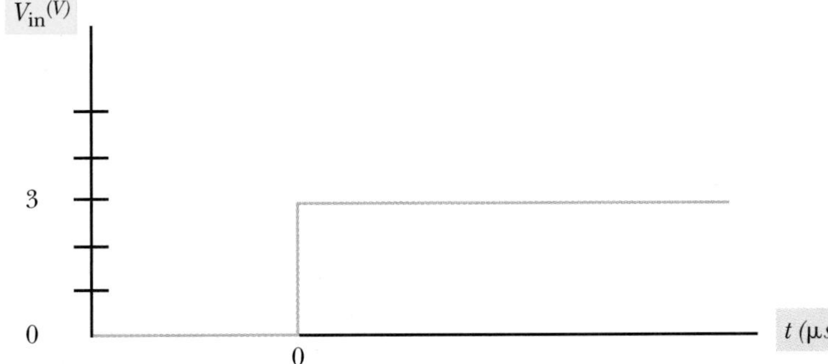

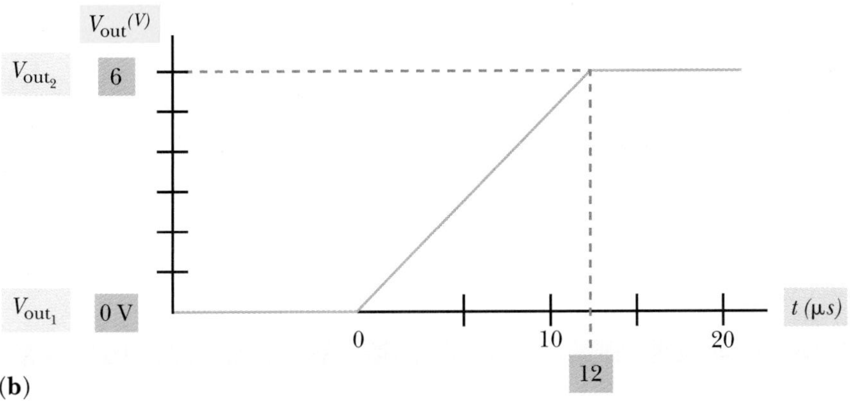

FIGURE 8–3 (a) Noninverting amplifier; (b) Slew rate

EXAMPLE The LM741 will be used in a circuit (Figure 8–3a) with a voltage gain of +2. The input voltage changes from 0 V to 3 V instantaneously. Calculate the change in output voltage and the time required for the output to reside at the new voltage level.

Answer:

The initial output voltage can be calculated using Formula 7–4:

$$V_{out_1} = A_V \times V_{in_1}$$
$$V_{out_1} = 2 \times 0 \text{ V}$$
$$V_{out_1} = 0 \text{ V}$$

The final output voltage can be calculated using Formula 7–4:

$$V_{out_2} = A_V \times V_{in_2}$$
$$V_{out_2} = 2 \times 3 \text{ V}$$
$$V_{out_2} = 6 \text{ V}$$

Application Problem

LEVEL SENSING SYSTEM

A pressure sensor is used to detect the fluid level in a mixing tank at a food processing plant. The sensor has the following characteristics: pressure range 0 to 50 pounds per square inch (psi), differential output voltage range 0 to .75 V. This voltage needs to be amplified by an instrumentation amplifier and sent to an analog input module of a programmable logic controller (PLC). The analog module has an input range of –10 V to +10 V. When the fluid has reached a maximum level, the pressure will be 45 psi. What will be the output voltage of the sensor and amplifier when the level is maximum?

Solution

$$\left(\frac{V_{\text{out sensor}}}{45}\right) = \left(\frac{.75}{50}\right)$$

$$V_{\text{out sensor}} = .675\text{ V}$$

$$V_{O_{\text{amp}}} = \left(1 + \frac{[(2)(1\text{k})]}{470}\right) \times 0.675\text{ V} = 3.55\text{ V}$$

The output voltage is 3.55 V.

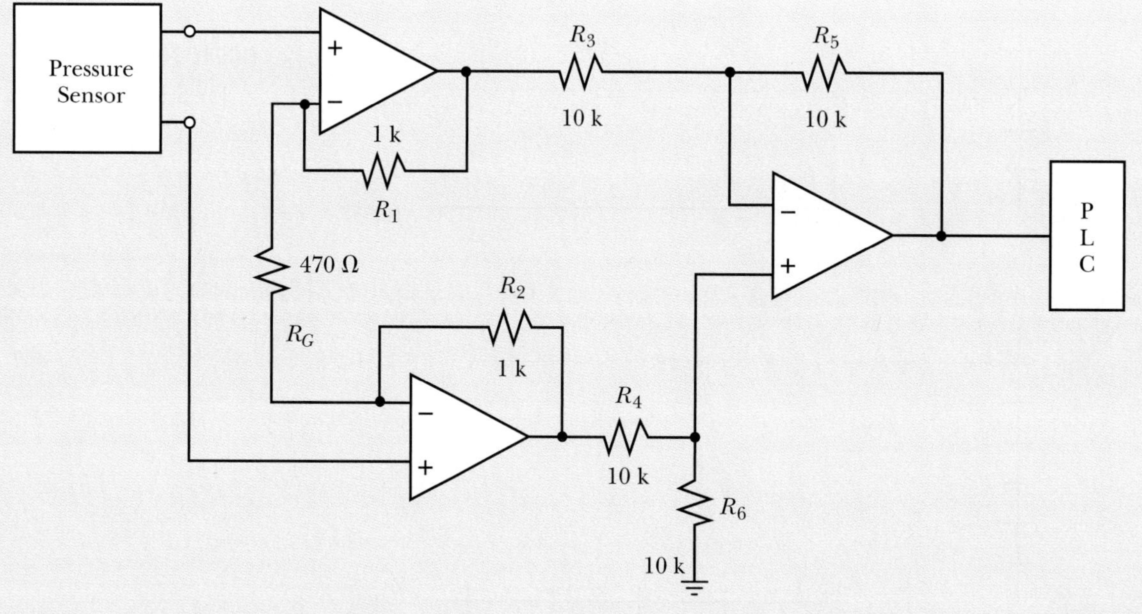

The change in output voltage can be determined by using Formula 8–4:

$$\Delta V_{\text{out}} = V_{\text{out}_2} - V_{\text{out}_1}$$
$$\Delta V_{\text{out}} = 6\text{ V} - 0\text{ V}$$
$$\Delta V_{\text{out}} = 6\text{ V}$$

The LM741 has a slew rate of 0.5 V/μs. Using Formula 8–3, you can calculate the time required for the output voltage to transition from 0 V to 6 V.

$$\Delta t = \frac{\Delta V_{\text{out}}}{SR}$$
$$\Delta t = \frac{6\text{ V}}{0.5\text{ V}/\mu s}$$
$$\Delta t = 12\ \mu s$$

Figure 8–3b shows the input and output voltage transitions with respect to time for the previous example.

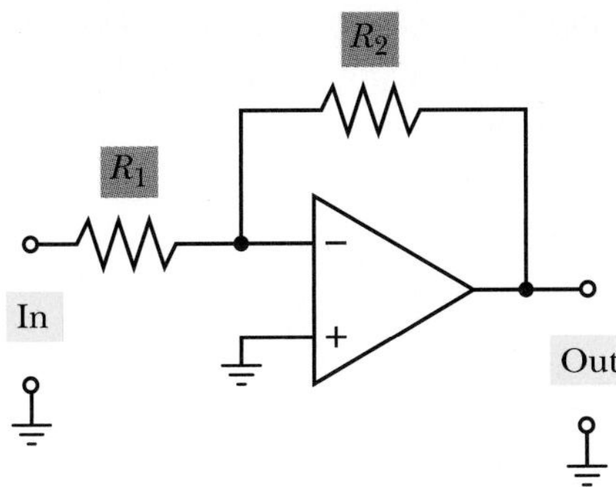

FIGURE 8–4 Inverting amplifier closed-loop gain multiSIM

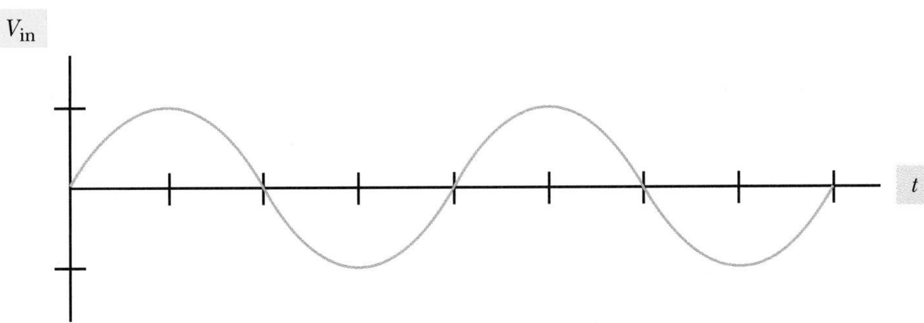

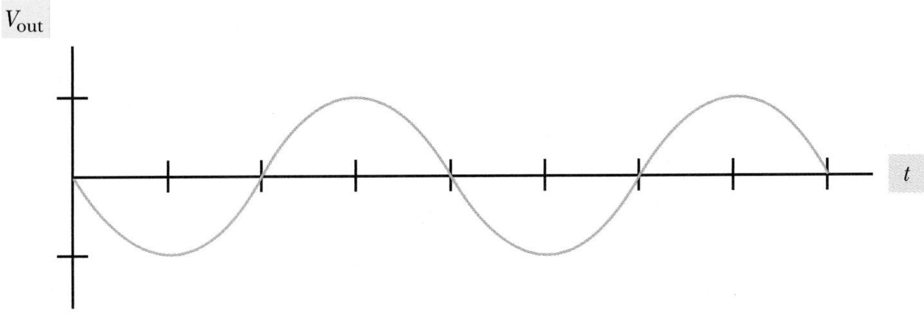

FIGURE 8–5 Inverting amplifier waveforms (ideal)

8–2 The Inverting Amplifier

In Figure 8–4 you see the inverting amplifier circuit. The voltage gain for the inverting amplifier circuit is calculated using Formula 8–5. Figure 8–5 shows the input and output waveforms for the situation when a sinusoidal waveform is applied to the ideal op-amp configured as an inverting amplifier. Notice that when the input signal voltage is positive that the output signal voltage is negative. When the input signal voltage is negative, the output signal voltage is positive.

The negative sign in the inverting amplifier gain equation indicates that the output signal is inverted from the input signal. When the input signal is a sine wave, the output signal is an inverted sine wave (180° phase shift).

FORMULA 8–5 $A_V = -\frac{R_2}{R_1}$

Practical Notes

The inverting amplifier voltage gain is also written

$$A_v = \frac{R_f}{R_i}$$

where R_f is the feedback resistor and R_i is the input resistor.

EXAMPLE For the circuit in Figure 8–4, $R_2 = 20\ \text{k}\Omega$ and $R_1 = 2\ \text{k}\Omega$; what is the voltage gain of the amplifier? If the input voltage to this circuit is a 0.25-V_{PP}, 10-kHz sine wave, what is the output voltage?

Answer:

From Formula 8–5:

$$A_V = -\frac{R_2}{R_1}$$

$$A_V = -\frac{20\ \text{k}\Omega}{2\ \text{k}\Omega}$$

$$A_V = -10$$

From Formula 7–4:

$$V_{out} = A_V \times V_{in}$$

$$V_{out} = -10 \times (0.25\ V_{PP})$$

$$V_{out} = 2.5\ V_{PP}$$

The negative sign indicates that the output signal is inverted from the input waveform. The output signal is a 2.5-V_{PP}, 10-kHz sine wave with a 180° phase shift.

Since the open-loop gain of the op-amp decreases as the input signal frequency applied to the op-amp increases, the closed-loop gain for the inverting amplifier circuit is also affected. The resulting close-loop gain for the inverting amplifier circuit shown in Figure 8–4 is calculated using Formula 8–6 where a is the open-loop gain of the amplifier corresponding to the input signal frequency.

FORMULA 8–6 $$A_V = \frac{\dfrac{-aR_2}{R_2 + R_1}}{1 + \dfrac{aR_1}{R_2 + R_1}}$$

Let's repeat the previous example evaluating the effect of the open-loop gain on the inverting amplifier's closed-loop voltage gain and resulting output signal.

EXAMPLE The circuit in Figure 8–4 uses the LM741 with $R_2 = 20\ \text{k}\Omega$ and $R_1 = 2\ \text{k}\Omega$; what is the voltage gain of the amplifier considering the open-loop gain at the input signal frequency of 10 kHz? If the input voltage to this circuit is a 0.25-V_{PP}, 10-kHz sine wave, what is the output voltage?

Answer:

From Figure 8–1, the open-loop voltage gain *(a)* at 10 kHz is about 80.

From Formula 8–6:

$$A_V = \frac{\dfrac{-aR_2}{R_2 + R_1}}{1 + \dfrac{aR_1}{R_2 + R_1}}$$

$$A_V = \frac{\dfrac{-80 \times 20\ \text{k}\Omega}{20\ \text{k}\Omega + 2\ \text{k}\Omega}}{1 + \dfrac{80 \times 20\ \text{k}\Omega}{20\ \text{k}\Omega + 2\ \text{k}\Omega}}$$

$$A_V = \frac{\dfrac{-1600}{22}}{1 + \dfrac{160}{22}}$$

$$A_V = \frac{-72.73}{8.273}$$

$$A_V = -8.79$$

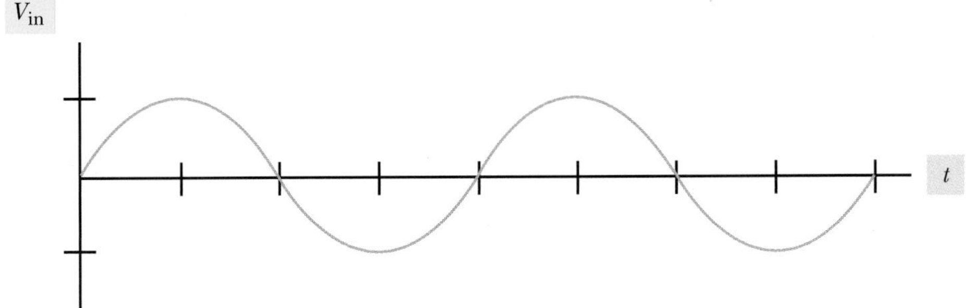

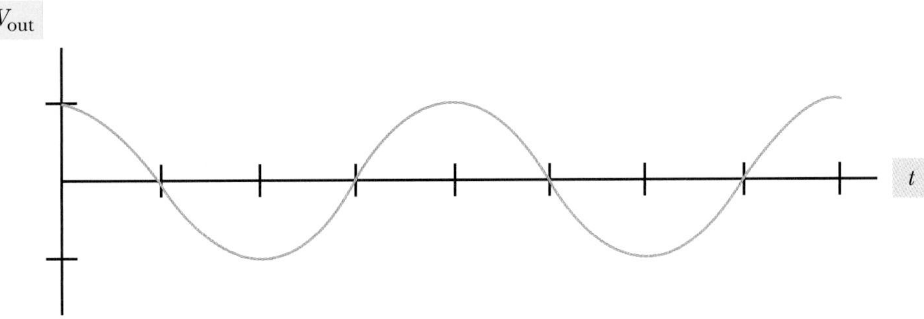

FIGURE 8–6 Inverting amplifier waveforms (LM741)

From Formula 7–4:

$$V_{out} = A_V \times V_{in}$$
$$V_{out} = -8.79 \times (0.25\ V_{PP})$$
$$V_{out} = -2.198\ V_{PP}$$

The output signal is a 2.198-V_{PP}, 10-kHz sine wave. The negative sign indicates that the output signal is inverted from the input waveform expressed as a 180° phase shift. However, the phase angle of the output signal is also affected due to the op-amp characteristics. From Figure 8–1, you can see that the phase shift at 10 kHz is –90°. Combining the –90° open-loop transfer phase-shift characteristics with the 180° phase-shift inverting amplifier characteristic results in a total phase shift of –270° for this circuit operating at 10 kHz. Figure 8–6 provides the observed input and output waveforms for this example. ◆

The open-loop gain characteristics of the op-amp resulted in reducing the closed-loop gain for the inverting amplifier. If the input signal frequency were changed, the open-loop voltage gain would be different resulting in a change in the closed-loop voltage gain of the circuit.

> *Practical Notes*
>
> Each operational amplifier has its own open-loop voltage gain characteristic. For most applications, you desire to maintain a constant closed-loop voltage gain over the input signal frequency range. The selection of op-amp and the circuit closed-loop voltage gain directly impact the ability to achieve your desire.

8–3 The Noninverting Amplifier

In Figure 8–7 you see the noninverting amplifier circuit. The voltage gain for the noninverting amplifier circuit is calculated using Formula 8–7. Figure 8–8 shows the input and output waveforms for the situation when a sinusoidal waveform is applied to the ideal op-amp configured as a noninverting amplifier. Notice that when the input signal voltage is positive that the output signal voltage is positive. When the input signal

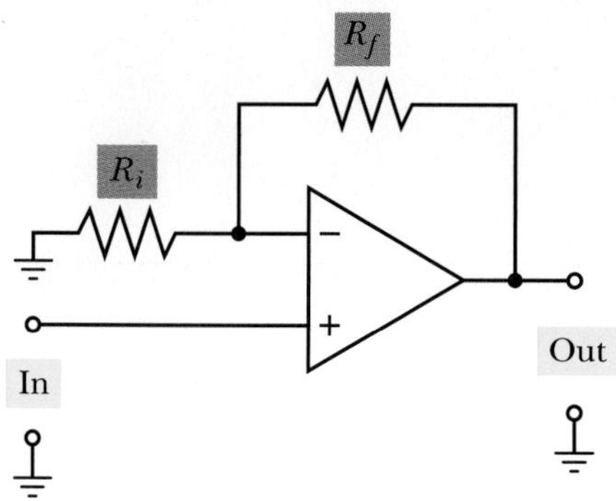

FIGURE 8–7 Noninverting amplifier closed-loop gain

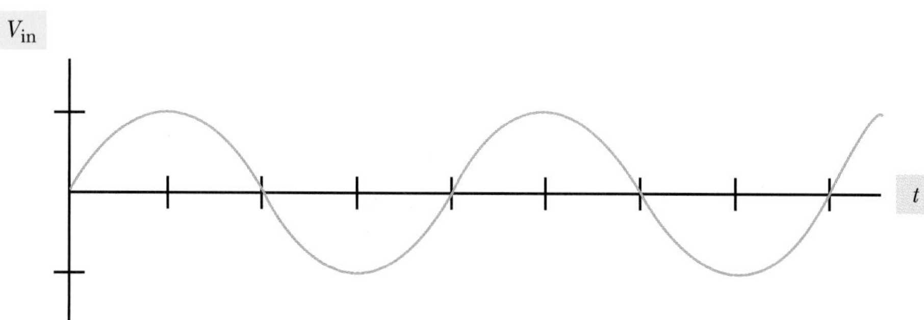

FIGURE 8–8 Noninverting amplifier waveforms (ideal)

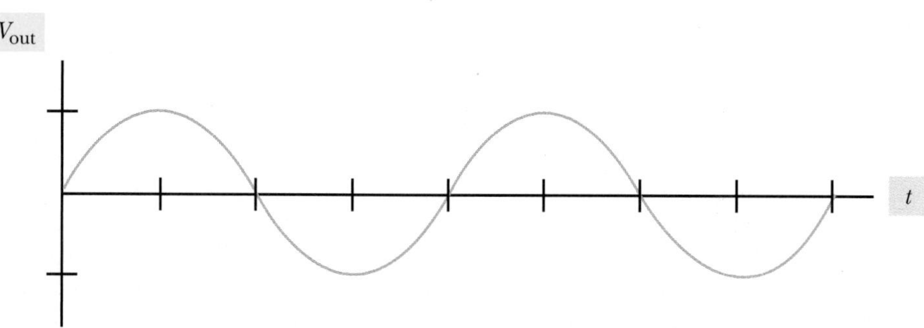

voltage is negative, the output signal voltage is negative. The positive sign in the noninverting amplifier gain equation indicates that the output signal is in-phase with the input signal. When the input signal is a sine wave, the output signal is a sine wave with 0° phase shift.

FORMULA 8–7 $A_V = 1 + \frac{R_2}{R_1}$

EXAMPLE For the noninverting amplifier circuit in Figure 8–7, $R_2 = 20\ \text{k}\Omega$ and $R_1 = 2\ \text{k}\Omega$; what is the voltage gain of the amplifier? If the input voltage to this circuit is a 0.25-V_{PP}, 10-kHz sine wave, what is the output voltage?

Answer:

From Formula 8–7:

$$A_V = 1 + \frac{R_2}{R_1}$$

$$A_V = 1 + \frac{20\ \text{k}\Omega}{2\ \text{k}\Omega}$$

$$A_V = +11$$

Practical Notes

The non-inverting voltage gain is also written

$$A_v = 1 + \frac{R_f}{R_i}$$

where R_f is the feedback resistor and R_i is the input resistor.

From Formula 7–4:

$$V_{out} = A_V \times V_{in}$$
$$V_{out} = +11 \times (0.25\ V_{PP})$$
$$V_{out} = 2.75\ V_{PP}$$

The positive sign indicates that the output signal is in phase with the input waveform. The output signal is a 2.75-V_{PP}, 10-kHz sine wave with a 0° phase shift. ◆

Since the open-loop gain of the op-amp decreases as the input signal frequency applied to the op-amp increases, the closed-loop gain for the noninverting amplifier circuit is also affected. The resulting closed-loop gain for the noninverting amplifier circuit shown in Figure 8–6 is calculated using Formula 8–8 where a is the open-loop gain of the amplifier corresponding to the input signal frequency.

FORMULA 8–8 $$A_V = \frac{a}{1 + \frac{a R_1}{R_2 + R_1}}$$

Let's repeat the previous example evaluating the effect of the open-loop gain on the noninverting amplifier's closed-loop voltage gain and resulting output signal.

◆ **EXAMPLE** The circuit in Figure 8–7 uses the LM741 with $R_2 = 20\ k\Omega$ and $R_1 = 2\ k\Omega$; what is the voltage gain of the amplifier considering the open-loop gain at the input signal frequency of 10 kHz? If the input voltage to this circuit is a 0.25-V_{PP}, 10-kHz sine wave, what is the output voltage?

Answer:

From Figure 8–1, the open-loop voltage gain (a) at 10 kHz is about 80. From Formula 8–8:

$$A_V = \frac{a}{1 + \frac{aR_1}{R_2 + R_1}}$$
$$A_V = \frac{80}{1 + \frac{80 \times 2\ k\Omega}{20\ k\Omega + 2\ k\Omega}}$$
$$A_V = \frac{80}{1 + \frac{160}{22}}$$
$$A_V = \frac{80}{8.273}$$
$$A_V = +9.67$$

From Formula 7–4:

$$V_{out} = A_V \times V_{in}$$
$$V_{out} = +9.67 \times (0.25\ V_{PP})$$
$$V_{out} = +2.418\ V_{PP}$$

The output signal is a 2.418-V_{PP}, 10-kHz sine wave. The positive sign indicates that the output signal is in phase with the input waveform expressed as a 0° phase shift. The phase angle of the output signal is also affected due to the op-amp characteristics. From Figure 8–1, you can see that the phase shift at 10 kHz is –90°. Combining the –90° open-loop transfer phase-shift characteristics with the 0° phase-shift noninverting amplifier characteristic results in a total phase shift of –90° for this circuit operating at 10 kHz. Figure 8–9 provides the observed input and output waveforms for this example. ◆

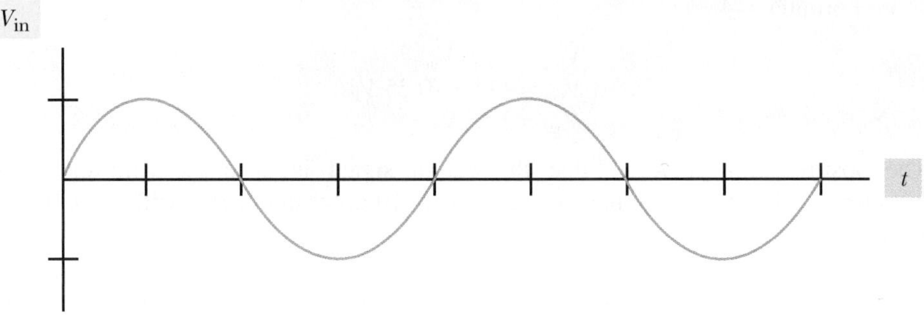

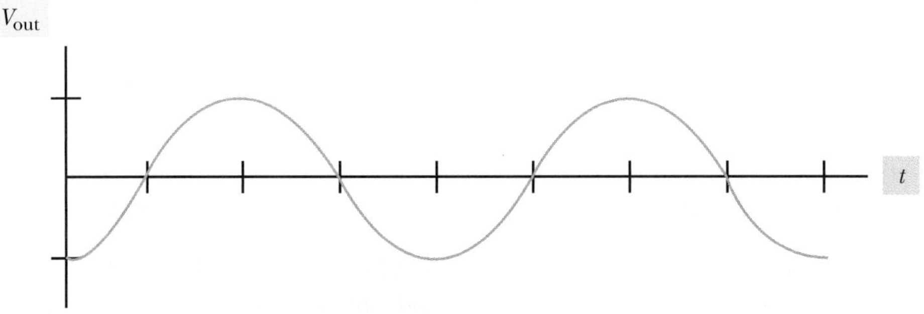

FIGURE 8–9 Noninverting amplifier waveforms (LM741)

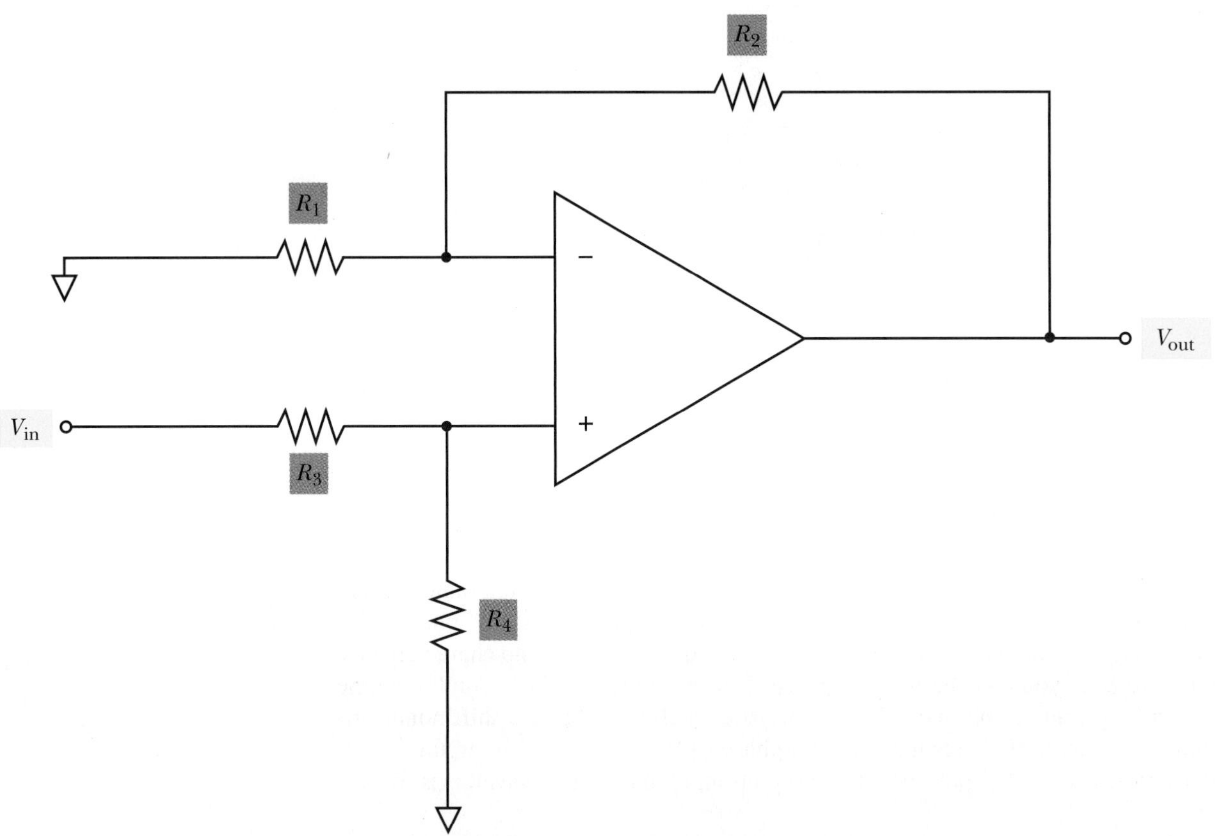

FIGURE 8–10 Voltage divider noninverting amplifier schematic

Application Problem

Communications

RELIABILITY TESTING SYSTEM

A temperature sensor is used to monitor the temperature inside a chamber that is used to cycle the temperature of electronic products that are being subjected to reliability testing over a two-week period. The sensor has the following characteristics: O V @ 0 degrees Centigrade to –.75 V @ 150 degrees Centigrade. The sensor output will be amplified by an inverting amplifier and sent to a modem for transmission to a data logger that is located in a building 20 miles away. What will be the output of the amplifier when the temperature is 125 degrees?

Solution

$$\frac{V_{\text{out sensor}}}{125} = \frac{-.75}{150}$$

$$V_{\text{out}_{\text{sensor}}} = -.625$$

$$\text{V}_{\text{O}_{\text{amp}}} = -(10\text{K}/1\text{K})\ (-.625) = 6.25\ \text{V}$$

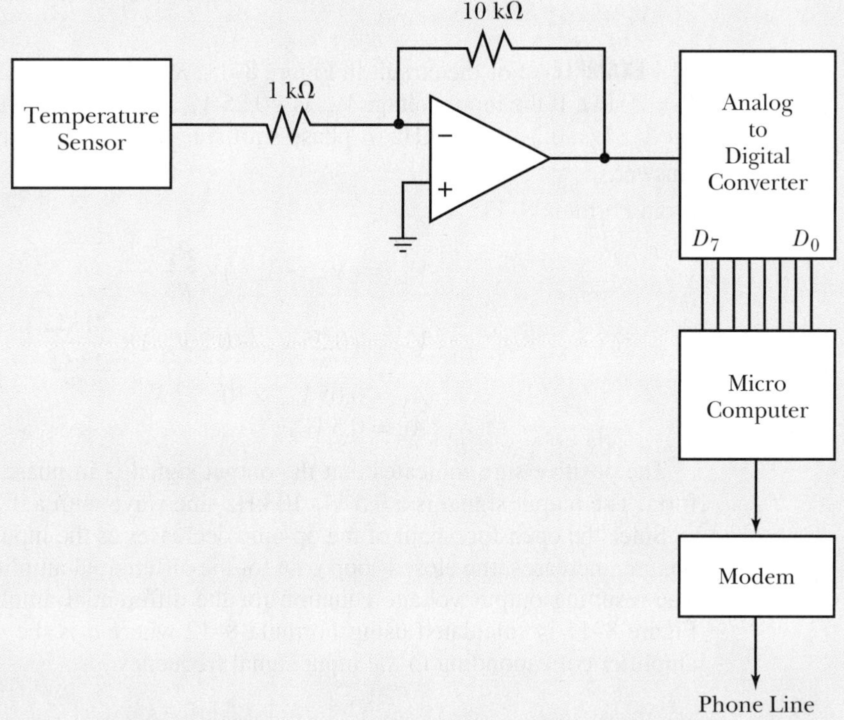

The open-loop gain characteristics of the op-amp resulted in reducing the closed-loop gain for the noninverting amplifier. If the input signal frequency were changed, the open-loop voltage gain would be different resulting in a change in the closed loop voltage gain of the circuit. Figure 8–10 shows the addition of a voltage divider circuit to the input of the noninverting amplifier. The input resistance of this circuit is equal to the sum of resistors R_3 and R_4. The voltage gain of this circuit configuration is calculated by using Formula 8–9.

FORMULA 8–9 $$A_V = \left(1 + \frac{R_2}{R_1}\right) \times \frac{R_4}{R_4 + R_3}$$

Even though the voltage divider circuit configured of resistors R_3 and R_4 directly influences the total circuit gain, the attenuation for the voltage divider is constant for all frequencies when metal film or carbon resistors are used in the circuit. The op-amp noninverting amplifier circuit closed-loop gain is affected by open-loop voltage gain and the input signal frequency.

8–4 The Differential Amplifier

The differential amplifier uses both the inverting and noninverting inputs to the op-amp as shown in Figure 8–11. The output voltage for this amplifier circuit is calculated by using Formula 8–10. In this mode, signals are fed to both inputs at the same time. The result at the output is an amplified version of the difference between the two signals, Figure 8–12.

FORMULA 8–10 $$V_{out} = V_{in+} \times \left(1 + \frac{R_2}{R_1}\right) \times \frac{R_4}{R_4 + R_3} - V_{in-} \times \left(\frac{R_2}{R_1}\right)$$

When $R_4 = R_2$ and $R_3 = R_1$, Formula 8–10 simplifies into Formula 8–11.

FORMULA 8–11 $$V_{out} = (V_{in+} - V_{in-}) \times \frac{R_2}{R_1}$$

EXAMPLE For the circuit in Figure 8–11, $R_1 = 2\ \text{k}\Omega$, $R_2 = 20\ \text{k}\Omega$, $R_3 = 2\ \text{k}\Omega$, and $R_4 = 20\ \text{k}\Omega$. If the input voltage V_{in+} is a 0.25-V_{PP}, 10-kHz sine wave and the input voltage V_{in-} is a 0.2-V_{PP}, 10-kHz 0° phase-shift sine wave, what is the output voltage?

Answer:

From Formula 8–11:

$$V_{out} = (V_{in+} - V_{in-}) \times \frac{R_2}{R_1}$$

$$V_{out} = (0.25\ V_{PP} - 0.2\ V_{PP}) \times \frac{20\ \text{k}\Omega}{2\ \text{k}\Omega}$$

$$A_V = 0.05\ V_{PP} \times 10$$

$$A_V = 0.5\ V_{PP}$$

The positive sign indicates that the output signal is in phase with the input waveform. The output signal is a 0.5 V_{PP} 10 kHz sine wave with a 0° phase shift.

Since the open-loop gain of the op-amp decreases as the input signal applied to the op-amp increases, the closed-loop gain for the differential amplifier circuit is affected. The resulting output voltage equation for the differential amplifier circuit shown in Figure 8–11 is calculated using Formula 8–12 where *a* is the open-loop gain of the amplifier corresponding to the input signal frequency.

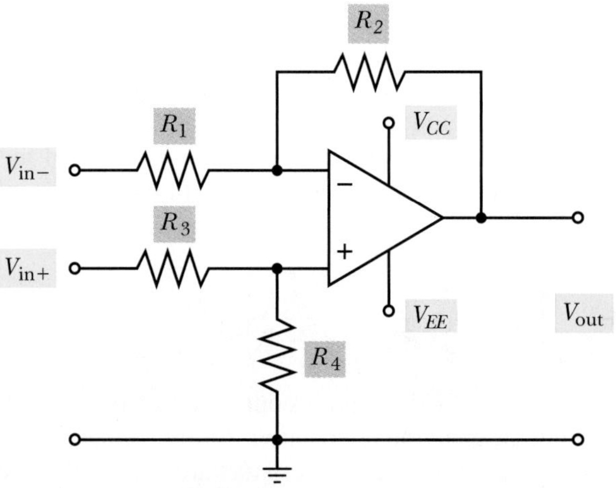

FIGURE 8–11 Op-amp as a difference amplifier

FORMULA 8–12 $$V_{\text{out}} = V_{\text{in+}}\left(\frac{R_4}{R_3+R_4}\right)\left(\frac{a}{1+\frac{aR_1}{R_2+R_1}}\right)+V_{\text{in-}}\left(\frac{\frac{-aR_2}{R_2+R_1}}{1+\frac{aR_1}{R_2+R_1}}\right)$$

When $R_4 = R_2$ and $R_3 = R_1$, Formula 8–12 simplifies into Formula 8–13.

FORMULA 8–13 $$V_{\text{out}} = (V_{\text{in+}} - V_{\text{in-}})\left(\frac{\frac{aR_2}{R_2+R_1}}{1+\frac{aR_1}{R_2+R_1}}\right)$$

Let's repeat the previous example evaluating the effect of the open-loop gain on the differential amplifier's output signal.

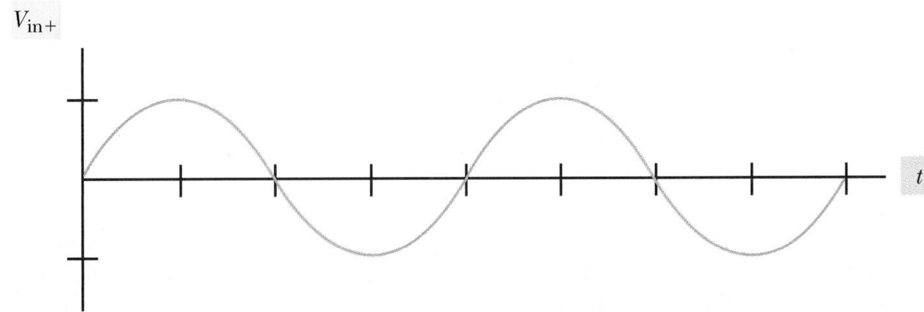

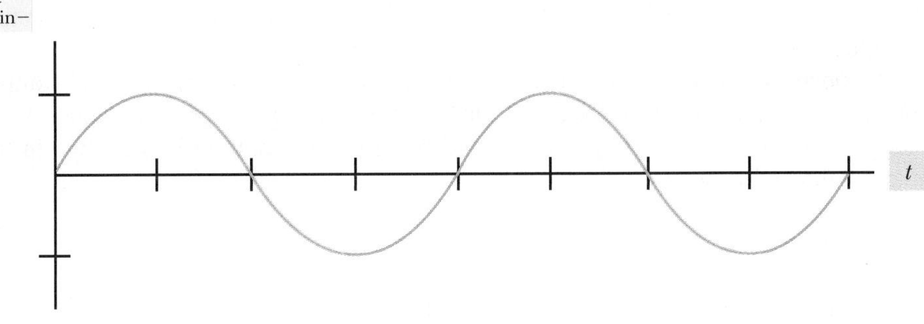

FIGURE 8–12 Differential amplifier waveforms (ideal)

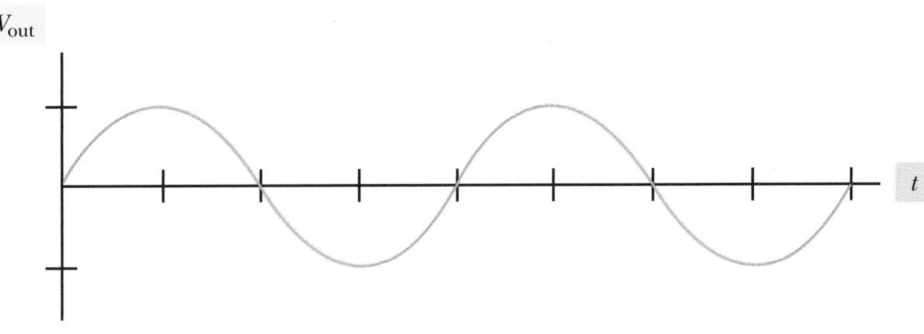

EXAMPLE The Figure 8–11 differential amplifier circuit uses the LM741 with $R_1 = 2\ \text{k}\Omega$, $R_2 = 20\ \text{k}\Omega$, $R_3 = 2\ \text{k}\Omega$, and $R_4 = 20\ \text{k}\Omega$. If the input voltage V_{in+} is a 0.25-V_{PP}, 10-kHz sine wave and the input voltage V_{in-} is a 0.2-V_{PP}, 10-kHz 0° phase-shift sine wave, what is the output voltage?

Answer:

From Figure 8–1, the open-loop voltage gain *(a)* at 10 kHz is about 80. From Formula 8–13:

$$V_{out} = (V_{in+} - V_{in-})\left(\frac{\dfrac{aR_2}{R_2 + R_1}}{1 + \dfrac{aR_1}{R_2 + R_1}}\right)$$

$$V_{out} = (0.25 - 0.2)\left(\frac{\dfrac{80 \times 20\ \text{k}\Omega}{20\ \text{k}\Omega + 2\ \text{k}\Omega}}{1 + \dfrac{80 \times 2\ \text{k}\Omega}{20\ \text{k}\Omega + 2\ \text{k}\Omega}}\right)$$

$$V_{out} = (0.25 - 0.2)\left(\frac{\dfrac{1600}{22}}{1 + \dfrac{160}{22}}\right)$$

$$V_{out} = 0.05 \times \left(\frac{72.73}{8.273}\right)$$

$$V_{out} = +0.05\ \text{V} \times 8.791$$

$$V_{out} = +0.4396\ V_{PP}$$

The output signal is a 0.4396-V_{PP}, 10-kHz sine wave. The positive sign indicates that the output signal is in phase with the input waveform expressed as a 0° phase shift. The phase angle of the output signal is also affected due to the op-amp characteristics. From Figure 8–1, you can see that the phase shift at 10 kHz is –90°. Combining the –90° open-loop transfer phase-shift characteristics with the 0° phase-shift differential amplifier characteristic results in a total phase shift of –90° for this circuit operating at 10 kHz. Figure 8–13 provides the observed input and output waveforms for this example.

The open-loop gain characteristics of the op-amp resulted in reducing the output voltage amplitude for the differential amplifier. If the input signal frequency were changed, the open-loop voltage gain would be different resulting in a change in the output voltage for the circuit.

Practical Notes

For the inverting and noninverting op-amp configurations the circuit voltage gain is calculated with the following formula where *A* is the forward gain of the amplifier circuit, while $A\beta$ is the loop gain for the circuit.

$$A_V = \frac{A}{1 + A\beta}$$

IN PROCESS LEARNING CHECK

1. As the input signal frequency increases, the op-amp's unity gain bandwidth product causes the open-loop voltage gain of the op-amp to __________ (decrease, increase).
2. At an input signal frequency of 100 kHz, the op-amp's unity gain bandwidth product causes the actual op-amp circuit gain to be __________ (the same as, greater than, less than) the ideal op-amp circuit gain.
3. The slew rate is a measure of the output signal __________ change with respect to __________.

Application Problem

AUTOMATED ORDER FILLING SYSTEM

The output of a programmable logic controller's analog interface module controls the speed of a motor that drives parts bins used in an automatic order filling system. The speed must remain constant. The speed is monitored by a tachometer that is using a Hall Effect sensor that detects magnetic pulses coming from a magnet mounted on the shaft on the motor. If the analog output is 500 mV, what is the output of the amplifier?

Solution

$$V_{O_{amp}} = \left(1 + \frac{250\text{ k}}{33\text{ k}}\right) 500\text{ mV} = 4.29\text{ V}$$

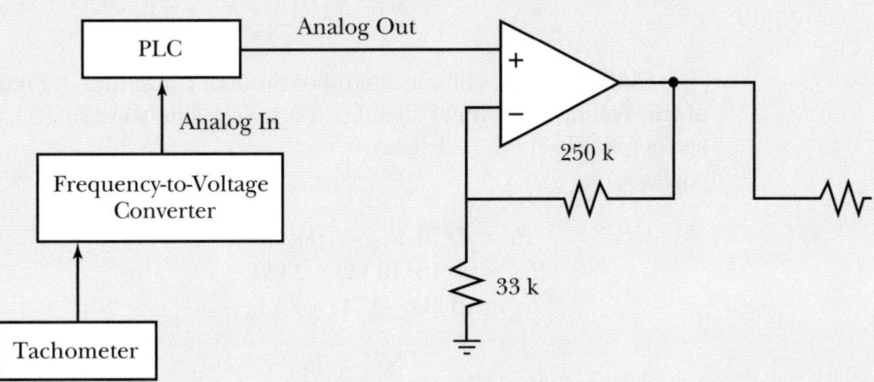

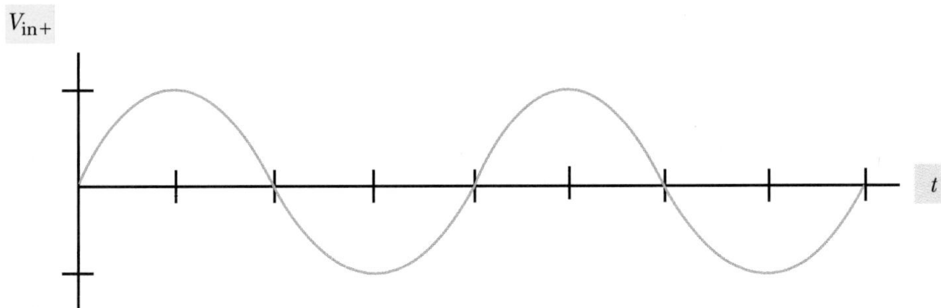

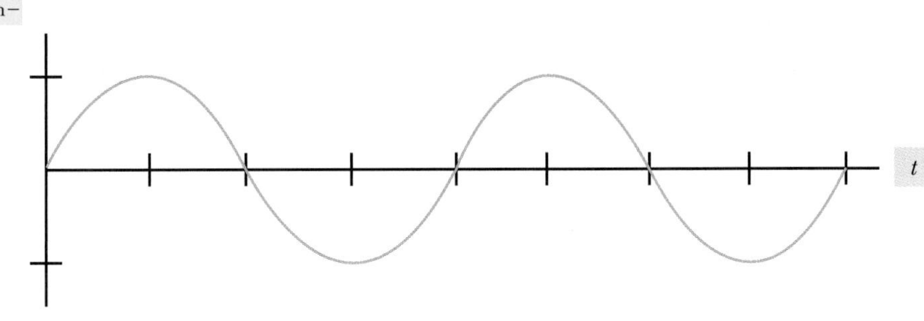

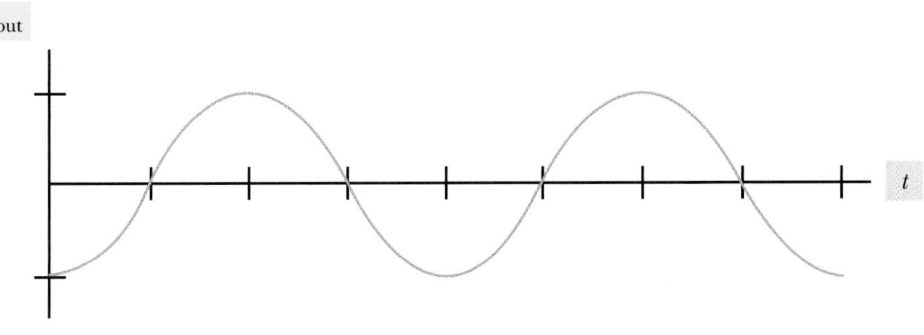

FIGURE 8–13 Differential amplifier waveforms (LM741)

8–5 Audio Amplifier

The LM386 audio amplifier is configured to amplify an incoming signal by 20. The LM386 output is designed to power a 500-m W 8-ohm speaker. Figure 8–14 provides the schematic with component values. The potentiometer provides volume control for this circuit. The coupling capacitors block the dc voltage of the input and output signals. This circuit is designed to operate from 20 Hz to 20 kHz. Formula 8–14 provides the equation to calculate the output voltage.

FORMULA 8–14 $V_{out} = V_{in} \times \left(\dfrac{R_{1B}}{R_{1A} + R_{1B} - jX_{C1}} \right) \times 20 \times \left(\dfrac{R_{SPK}}{R_{SPK} - jX_{C2}} \right)$

EXAMPLE The volume control of the audio amplifier in Figure 8–14 is set at 20% of full volume. The input signal is a 0.1 V_{RMS} sine wave at 100 Hz. What is the voltage amplitude at the 8 Ω speaker?
Answer:

$$R_1 = R_{1A} + R_{1B} = 10\ \text{k}\Omega$$
$$R_{1B} = 20\% \times 10\ \text{k}\Omega = 2\ \text{k}\Omega$$
$$R_{1A} = 10\ \text{k}\Omega - 2\ \text{k}\Omega = 8\ \text{k}\Omega$$

$$V_{out} = V_{in} \times \left(\frac{R_{1B}}{R_{1A} + R_{1B} - jX_{C1}} \right) \times 20 \times \left(\frac{R_{SPK}}{R_{SPK} - jX_{C2}} \right)$$

$$X_{C_1} = \frac{1}{(2 \times \pi \times f \times C_1)}$$
$$X_{C_1} = \frac{1}{(2 \times \pi \times 100\ \text{Hz} \times 2.2\ \mu\text{F})}$$
$$X_{C_1} = 723.4\ \Omega$$

$$X_{C_2} = \frac{1}{(2 \times \pi \times f \times C_2)}$$
$$X_{C2} = \frac{1}{(2 \times \pi \times 100\ \text{Hz} \times 2200\ \mu\text{F})}$$
$$X_{C_2} = 0.7234\ \Omega$$

From Formula 8–14:

$$V_{out} = V_{in} \times \left(\frac{R_{1B}}{R_{1A} + R_{1B} - jX_{C1}} \right) \times 20 \times \left(\frac{R_{SPK}}{R_{SPK} - jX_{C2}} \right)$$
$$V_{out} = 0.1 V_{rms} \times \frac{2\ \text{k}\Omega}{\sqrt{(2\ \text{k}\Omega + 8\ \text{k}\Omega)^2 + (723.4\ \Omega)^2}} \times 20 \times \frac{8\ \Omega}{\sqrt{(8\ \Omega)^2 + (0.7234\ \Omega)^2}}$$
$$V_{out} = 0.397\ V_{rms}$$

Practical Note

The low frequency component of the incoming waveform will be attenuated by the reactance of the capacitors. You will have a reduction of the base portion of your music. The higher frequency components of the waveform will have less attenuation since the capacitive reactance will be lower.

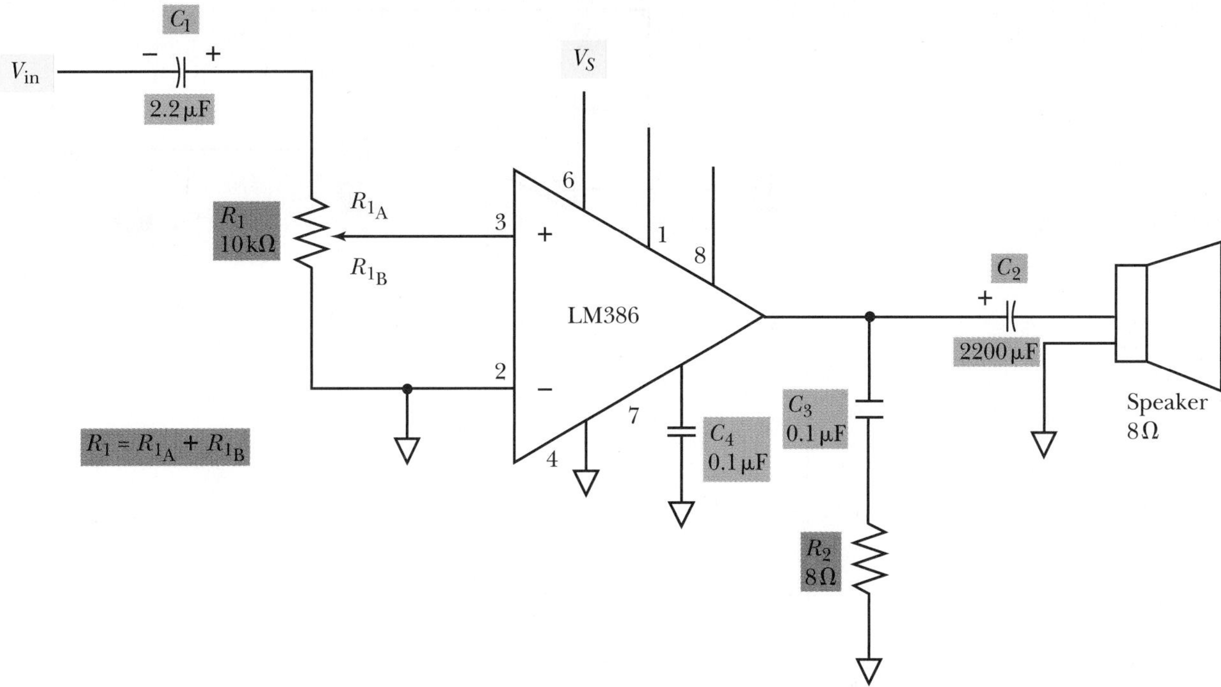

FIGURE 8-14 LM386 audio power amplifier

8–6 Op-amp Rectifier Circuits

The diode can be used with the op-amp to provide a variety of limiter, clipper, clamper, and precision rectifier circuits. You do recall that the diode controls the direction of current flow. When the diode is forward biased, the diode conducts current. When the diode is reverse biased, the current will not conduct through the diode.

The **precision half-wave rectifier positive output** circuit is shown in Figure 8–15a. Without the diode D_1, this circuit would be a unity gain noninverting amplifier. Due to the placement of the diode, the output voltage will be identical to the input voltage whenever the input voltage is greater than 0 V. Figure 8–15b provides the input and output waveforms for the circuit. When the input signal is less than 0 V, the diode is reverse biased and "no" current (only bias currents which are extremely low) will flow in the circuit. With no current flowing, the output voltage, observed across the load resistor, is 0 V. When the input signal is at a positive 350 nV, the op-amp output voltage would be at 0.7 V (assuming an op-amp open-loop gain of 200,000) at which point the diode would be forward biased (satisfying the 0.7 V diode forward voltage drop) resulting in a unity gain noninverting amplification. The output voltage will be identical to the input voltage of 350 nV. As the input voltage is increased, the output voltage will remain identical to the input voltage until the op-amp output positive saturation voltage is reached. Figure 8–16 provides a circuit where resistors R_1 and R_2 were added to provide a positive gain greater than 1.

Figure 8–17a provides a positive output half-wave rectifier circuit for high frequency applications. The circuit input and output signals are provided in Figure 8–17b. Diode D_2 will be forward biased when diode D_1 is reverse biased. This condition occurs when the input signal is greater than 0 V. As long as the input voltage is greater than 0 V, the current flowing through the resistor R_1 will be conducting through the diode D_2 and the op-amp output voltage will be 0 V. Since no current will be flowing through the resistor R_2, the circuit output voltage will be 0 V. Once the input signal is less than 0 V, diode D_2 will be reverse biased while diode D_1 is forward biased. As long as the input voltage is less than 0 V, the current flowing through the resistor R_1 will be conducting through the resistor R_2. In this condition the output signal will be equal to the op-amp inverting gain times the input voltage level as long as the op-amp output voltage is less than the positive saturation voltage.

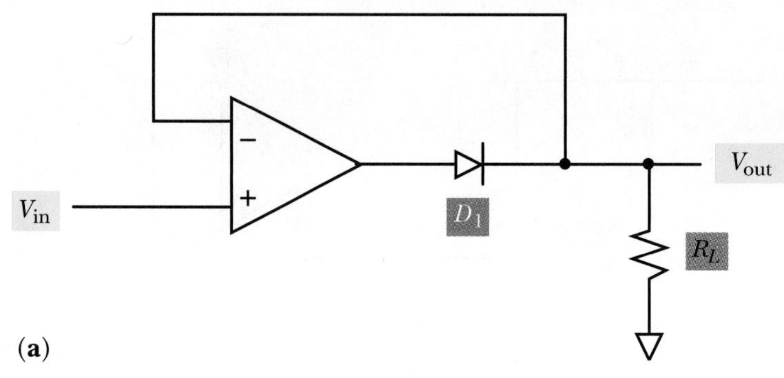

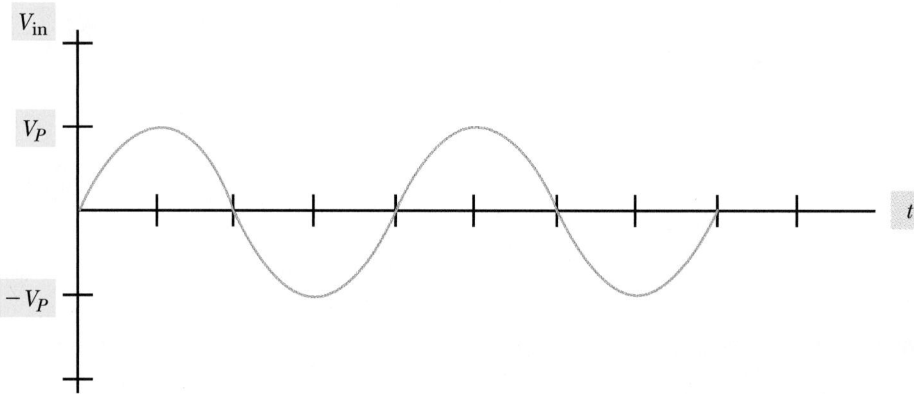

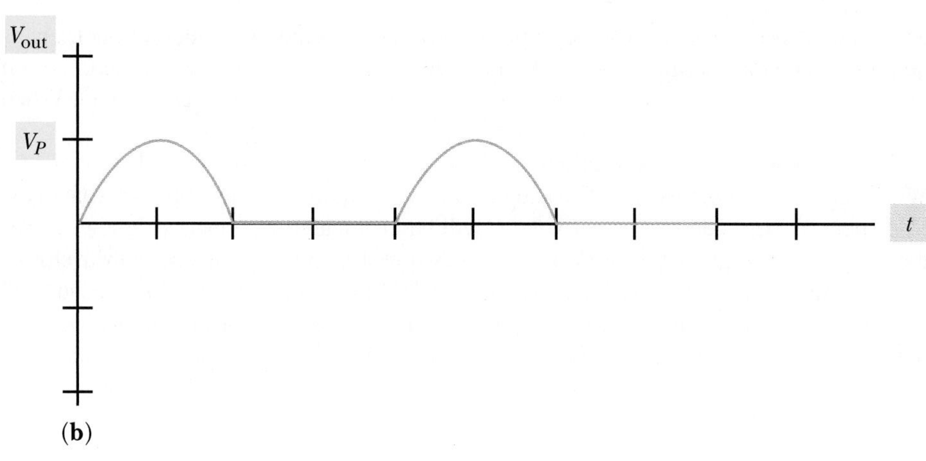

FIGURE 8–15 (a) Precision half-wave rectifier (positive); (b) Waveforms

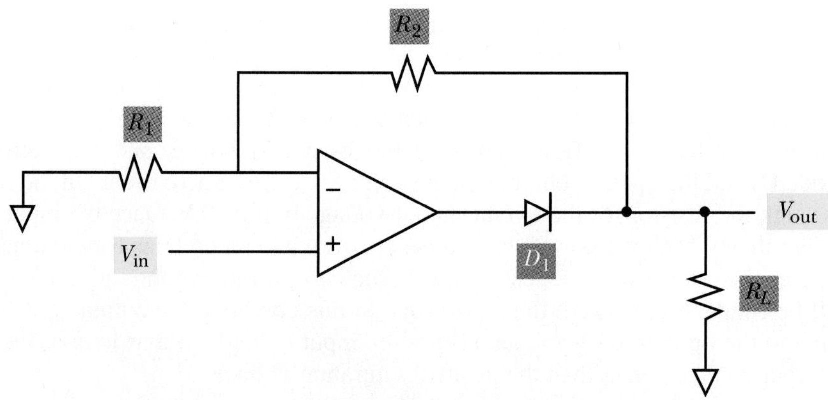

FIGURE 8–16 Precision half-wave rectifier (positive) with gain

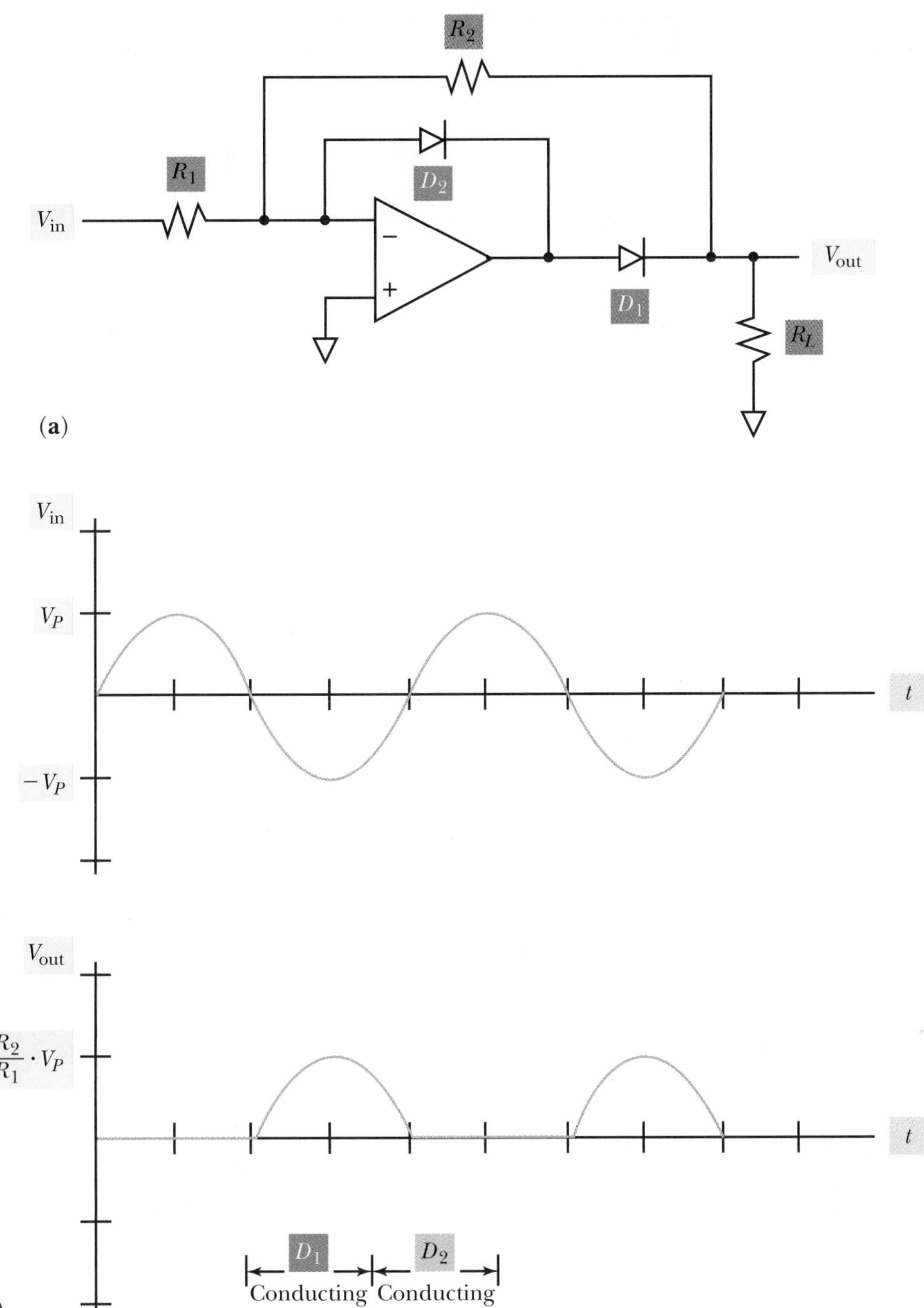

FIGURE 8–17 (a) Precision high frequency half-wave rectifier positive output; (b) Waveforms

The **precision half-wave rectifier negative output** circuit is shown in Figure 8–18a. Without the diode D_1, this circuit would be a unity gain noninverting amplifier. Due to the placement of the diode, the output voltage will be identical to the input voltage whenever the input voltage is less than 0 V. Figure 8–18b provides the input and output waveforms for the circuit. When the input signal is greater than 0 V, the diode is reverse biased and "no" current (only bias currents which are extremely low) will flow in the circuit. With no current flowing, the output voltage, observed across the load resistor, is 0 V. When the input signal is at a negative 350 nV, the op-amp output voltage would be at 0.7 V (assuming an op-amp open-loop gain of 200,000) at which point the diode would be forward biased (satisfying the 0.7 V diode forward voltage drop) resulting in a unity gain noninverting amplification. The output voltage will be identical to the input voltage of –350 nV. As the input voltage is decreased, the

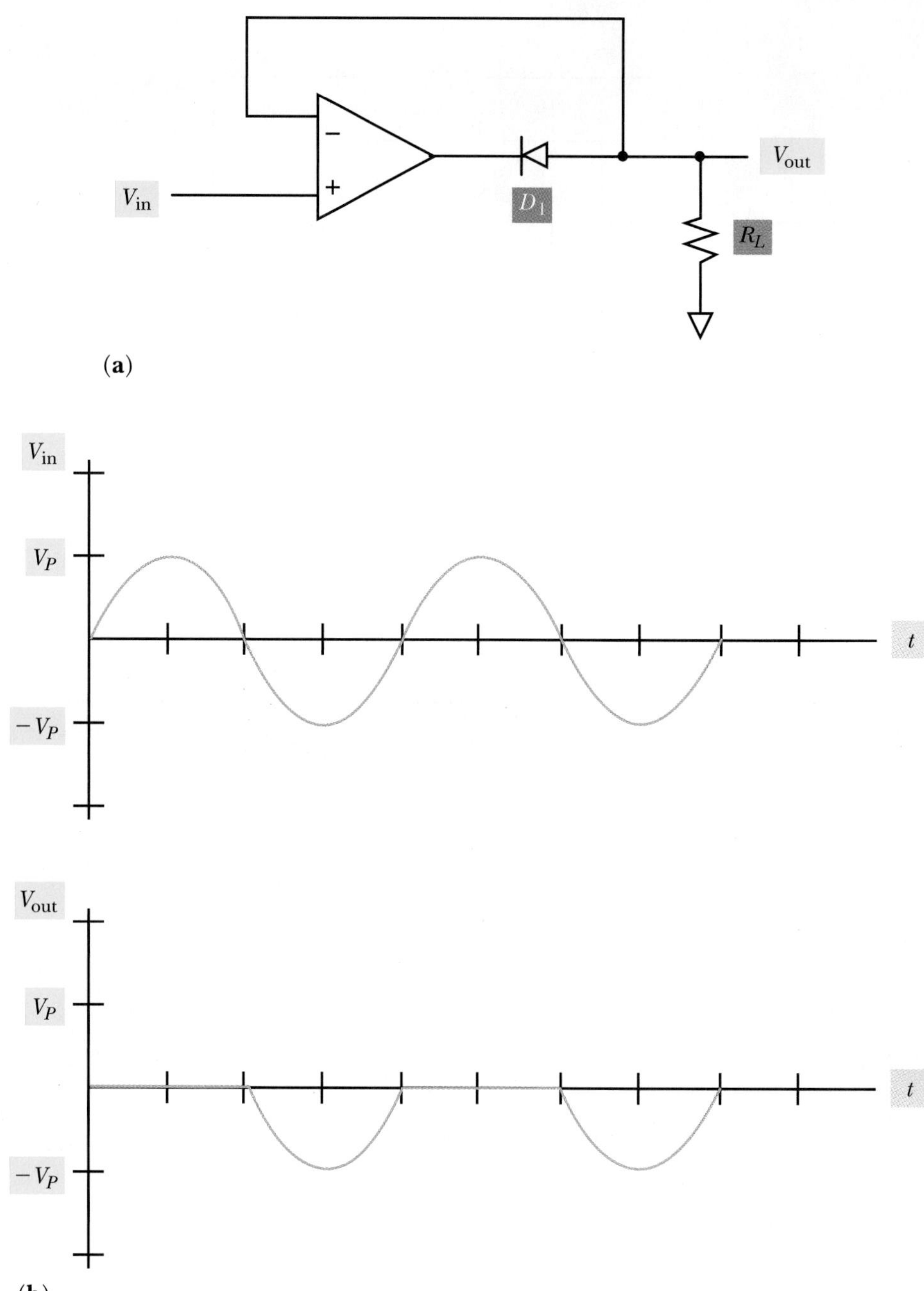

FIGURE 8–18 (a) Precision half-wave rectifier (negative); (b) Waveforms

output voltage will remain identical to the input voltage until the op-amp output negative saturation voltage is reached. Resistors R_1 and R_2 can be added to the Figure 8–18a circuit to provide a positive gain greater than 1.

Figure 8–19a provides a negative output half-wave rectifier circuit for high frequency applications. The circuit input and output signals are provided in Figure 8–19b. Diode D_2 will be forward biased when diode D_1 is reverse biased. This condition occurs when the input signal is less than 0 V. As long as the input voltage is less than 0 V, the current flowing through the resistor R_1 will be conducting through the diode D_2 and the op-amp output voltage will be 0 V. Since no current will be flowing through the resistor R_2, the circuit output voltage will be 0 V. Once the input signal is greater than 0 V, diode D_2 will be reverse biased while diode D_1 is forward biased. As long as the input voltage is greater than 0 V, the current flowing through the resistor R_1 will be conducting through the resistor R_2. in this condition the output signal will

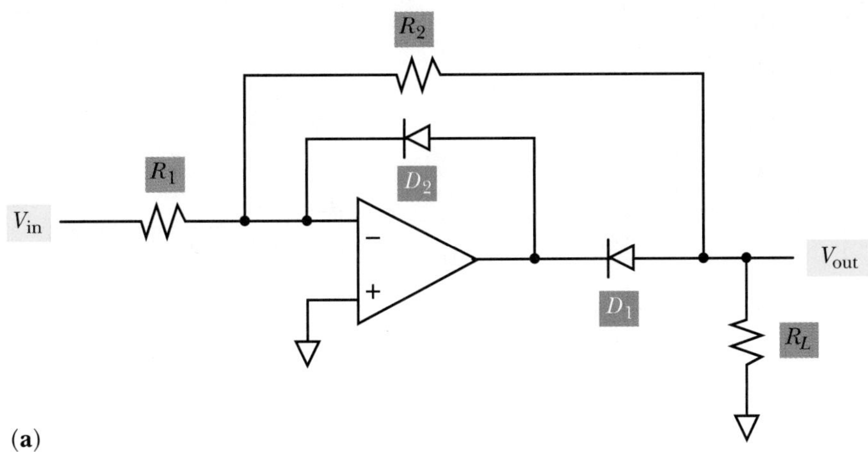

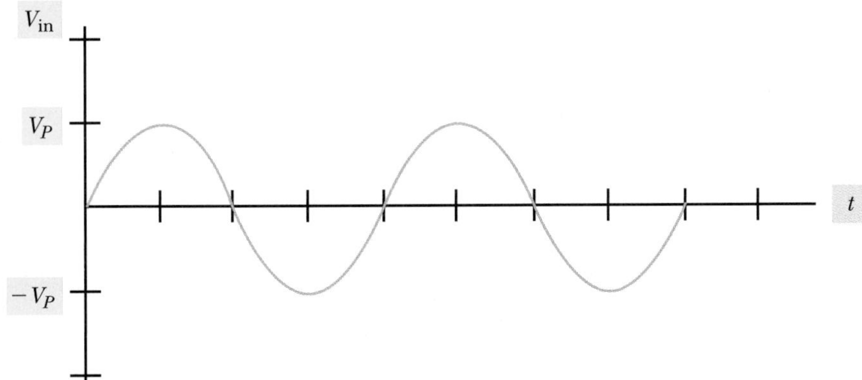

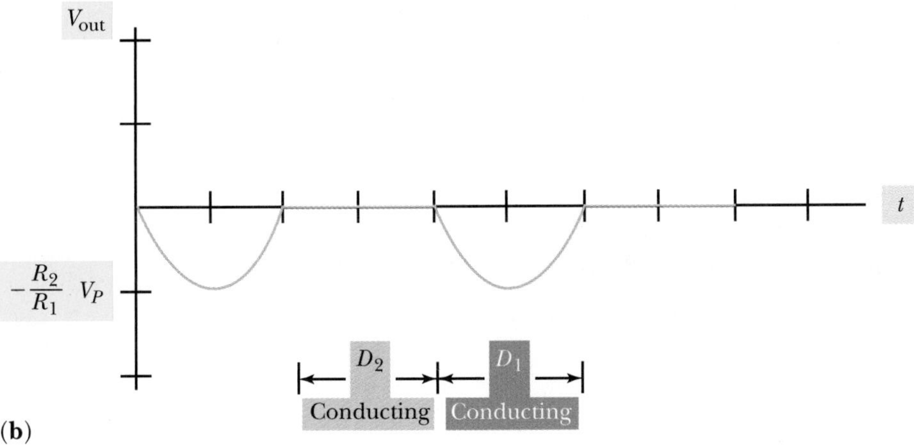

FIGURE 8–19 (a) Precision high frequency half-wave rectifier negative output; (b) Waveforms

be equal to the op-amp inverting gain times the input voltage level. The circuit output voltage will be limited by the negative saturation voltage of the op-amp.

By combining the output of a high frequency precision half-wave rectifier negative output circuit with the input signal using a summing amplifier, a **precision full-wave rectifier positive output** circuit has been configured as shown in Figure 8–20a where $R_1 = R_2 = R_4 = R_5 = 2 \times R_3$. The input and output waveforms for the circuit are shown in Figure 8–20b.

Additional circuits that combine the diode and op-amp include the **positive clamper** shown in Figure 8–21, the **positive limiter** shown in Figure 8–22, the **active limiter** shown in Figure 8–23, and the **peak detector** circuit shown in Figure 8–24. Changing the direction of the diode in most of the circuits will result in a different output signal.

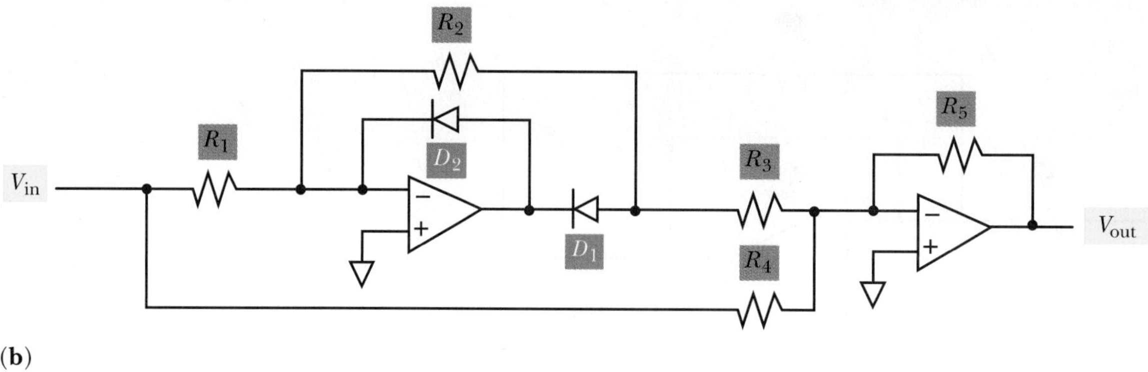

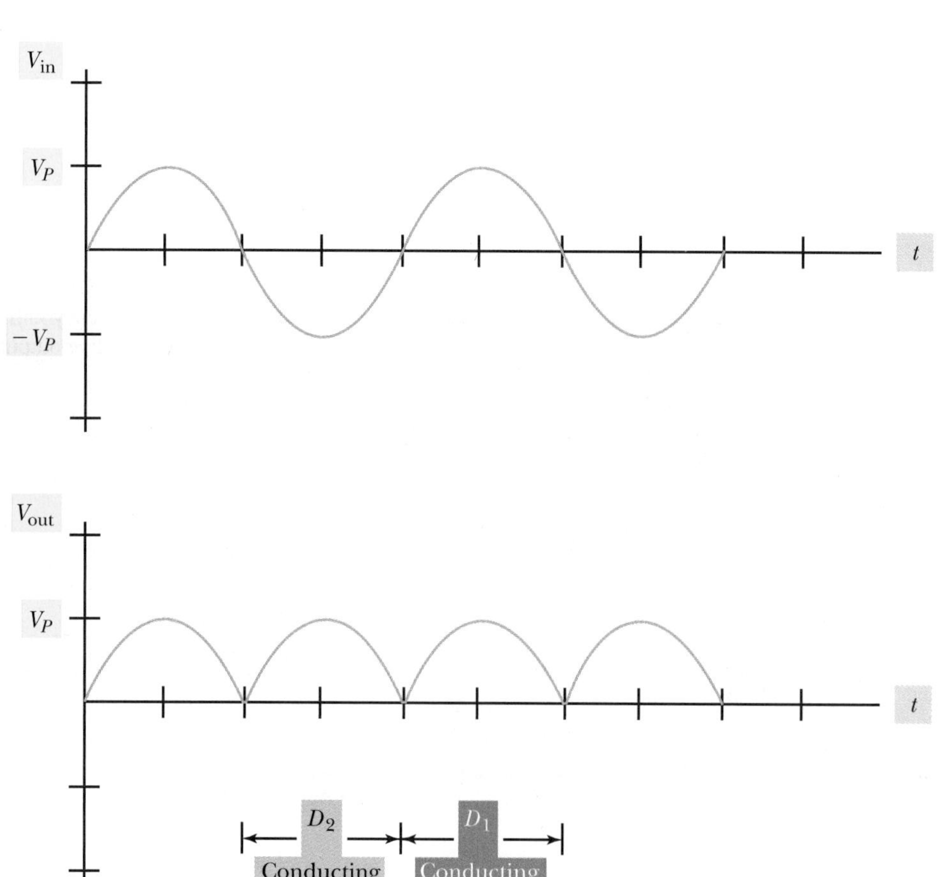

FIGURE 8–20 (a) Precision full-wave rectifier (positive output); (b) Waveforms

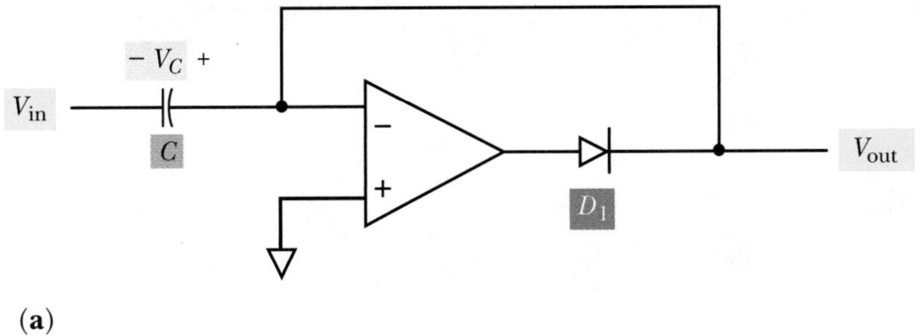

(a)

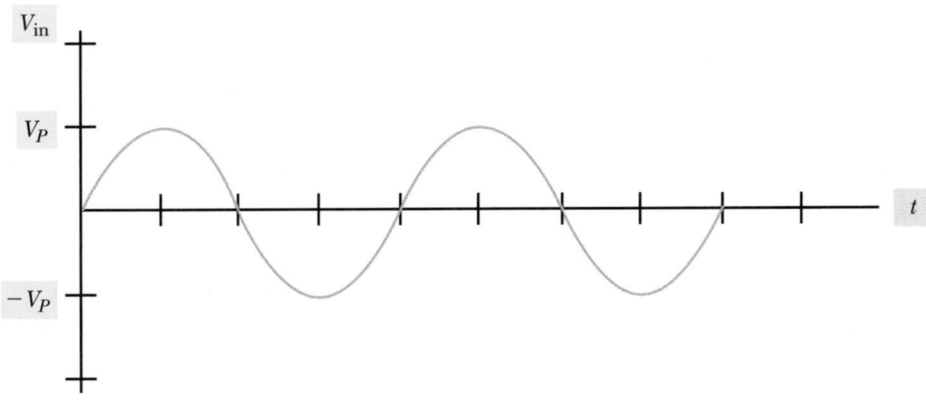

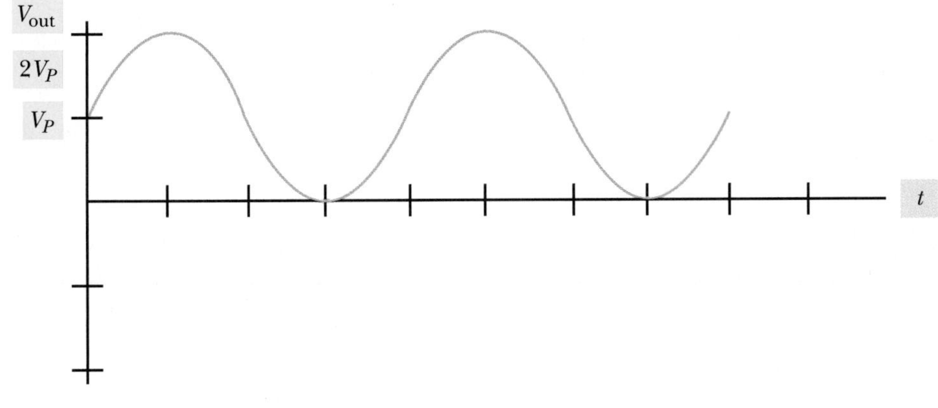

(b)

FIGURE 8–21 (a) Active positive clamper; (b) Waveforms

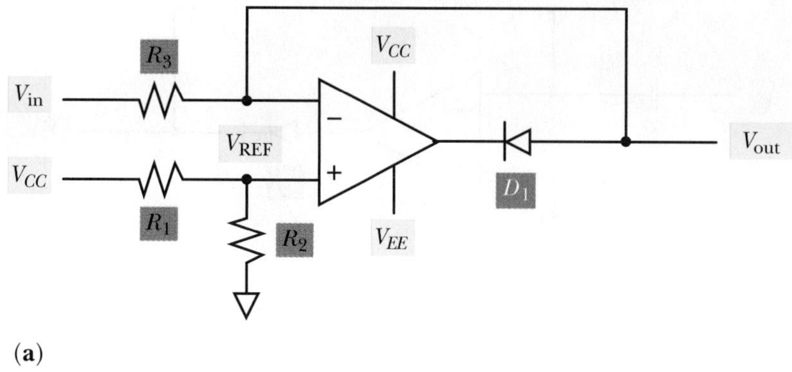

(a)

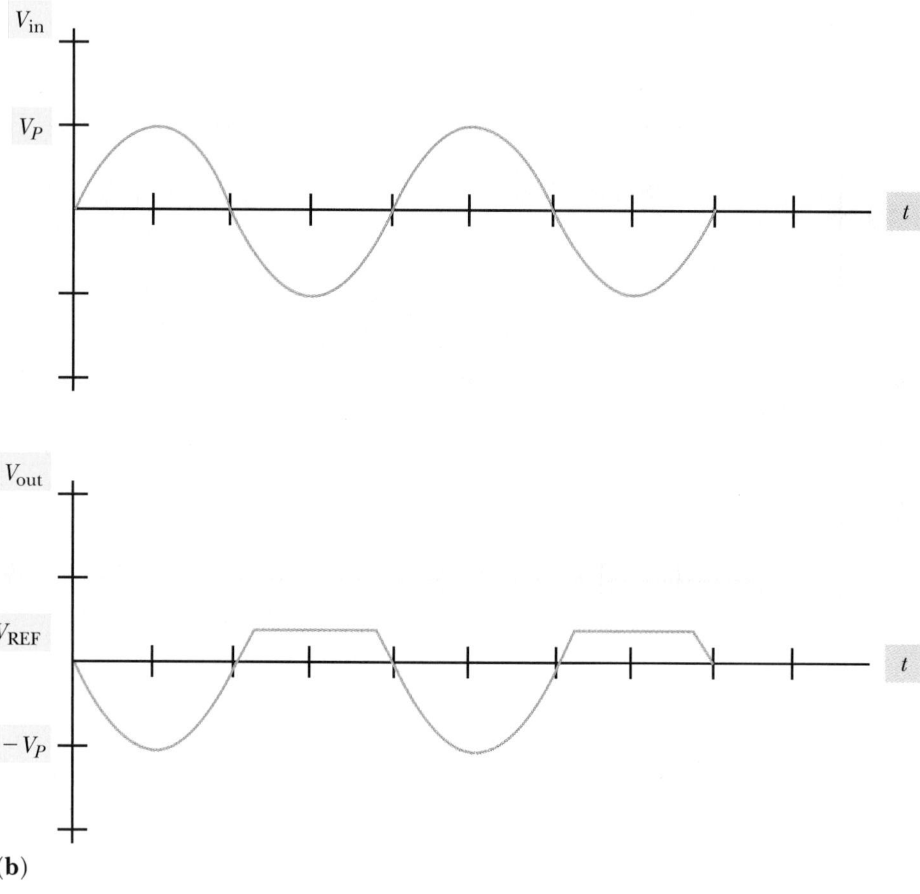

(b)

FIGURE 8–22 (a) Active positive limiter; (b) Waveforms

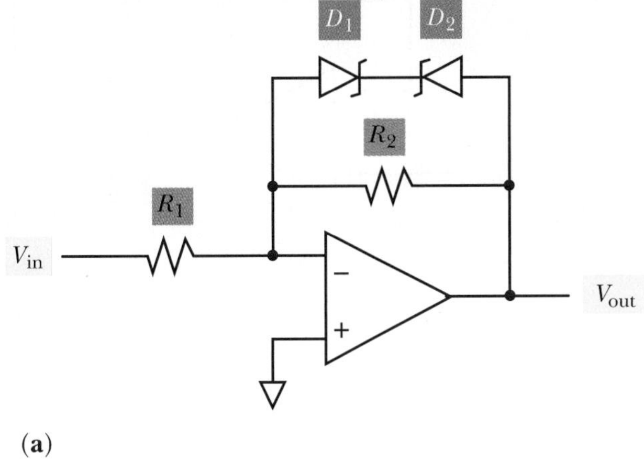

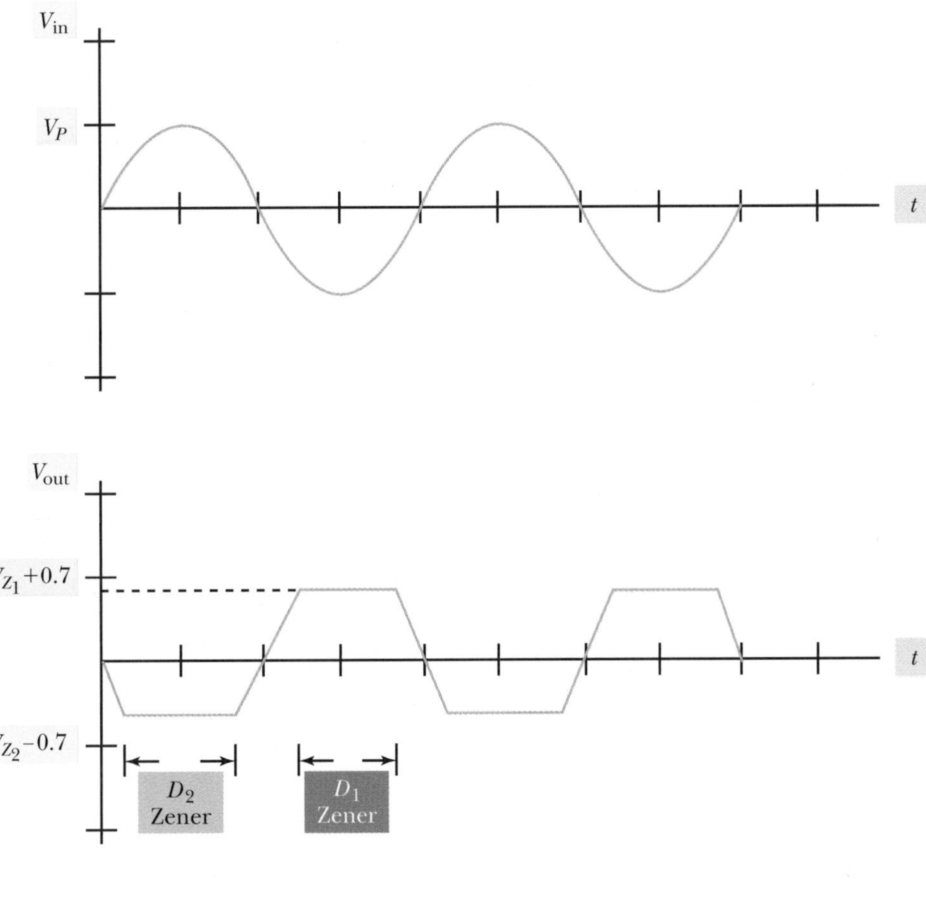

FIGURE 8–23 (a) Active limiter; (b) Waveforms

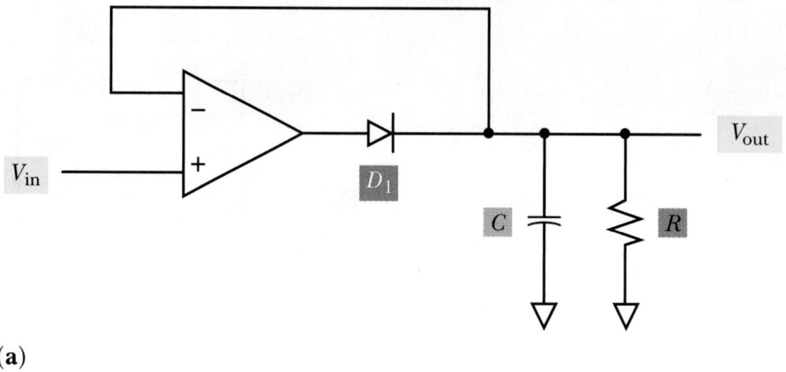

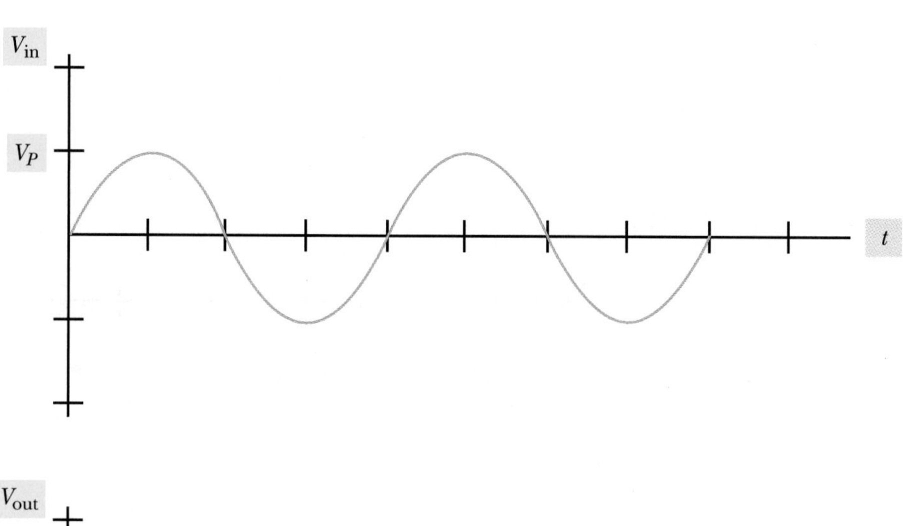

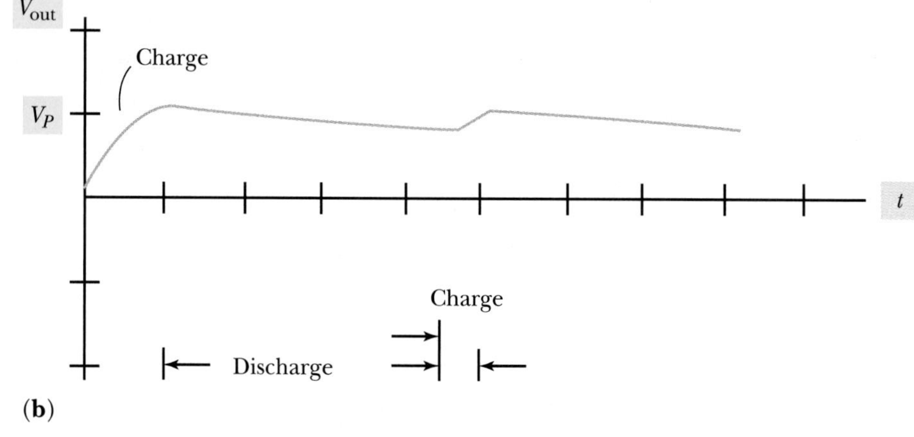

FIGURE 8–24 (a) Peak detector circuit; (b) Waveforms

Application Problem

General Electronics

OSCILLOSCOPE AMPLIFIER

An oscilloscope uses an instrumentation amplifier to increase the signal coming from the input jacks for channel 1 and 2. The output of the amplifier will drive the oscilloscope's video circuits. From the figure below, determine the voltage gain.

Solution

$$A_{V_{CL}} = 1 + \left(\frac{2 \times 12\text{ k}}{33} \right) = 728$$

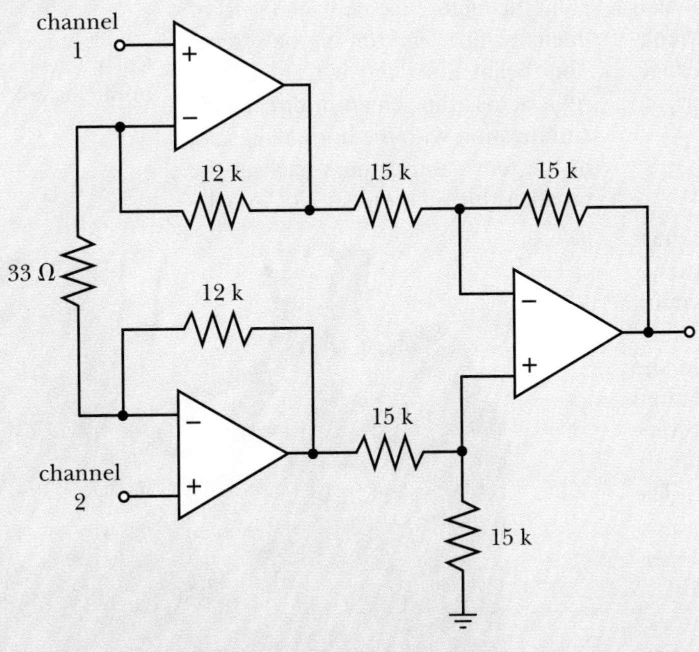

8–7 Complex Waveforms

To evaluate the circuit's output signal when a complex signal was applied to the circuit input, it is beneficial to separate the complex waveform into a combination of sine waves containing different frequencies with their appropriate amplitudes. Jean Baptiste Joseph Fourier stated that all periodic waveforms can be represented as the summation of sine and cosine signals and a dc offset. This relationship is known as the Fourier series. Due to the op-amp frequency response characteristic, higher frequency components of a signal might be amplified less than lower frequency components of a signal.

USING TECHNOLOGY: OP-AMPS USED IN A SYSTEM

In this as in preceding chapters you will study operational amplifiers in detail including the circuit operation, circuit waveforms, and troubleshooting tips. Up to this point, these devices, namely, the diode and the junction and the field effect transistor, are all separate devices with their elements interconnected and contained within an individual package. Such complete functional devices are called discrete components.

Circuit Operation

In this part, we will study both linear and nonlinear integrated circuits (ICs) in which the individual circuit components, namely, the resistors, the capacitors, the diodes, and the transistors along with the required wiring interconnections, are combined on a single chip of specially doped semiconductor material. The required internal connection points for input and output signals, dc operating voltages, and ground are brought out to external pins for easy connection into the product's circuitry. These integrated circuits (ICs) range from simple 8-pin devices up to complete microcomputers with over 192 pins. The T circuit chip discussed in this part is such a microprocessor commonly used in television receivers.

Reviewing the ideal characteristics of the operational amplifier, it can be seen that this device possesses high voltage, current, and power gain abilities, stable operation under varying conditions of temperature, and operating voltages and uniform response over a wide range of frequency.

The op-amp integrated circuit combines the three basic circuit elements of an op-amp:

a. the differential amplifier input stage;

b. intermediate stabilized amplifiers stages; and

c. an emitter-follower output stage.

One of the most common practical applications of the op-amp is in the vertical drive portion of every television. To display information on the face of a television picture tube, streams of highly focused electrons, in essence a beam of electrons, must paint the picture elements in terms of intensity, hue, and tint on the screen. This beam must move from left to right across the screen in direct synchronization with the horizontal scan of the television camera back at the broadcasting station studio. At the

Color Television Application: High resolution color monitor

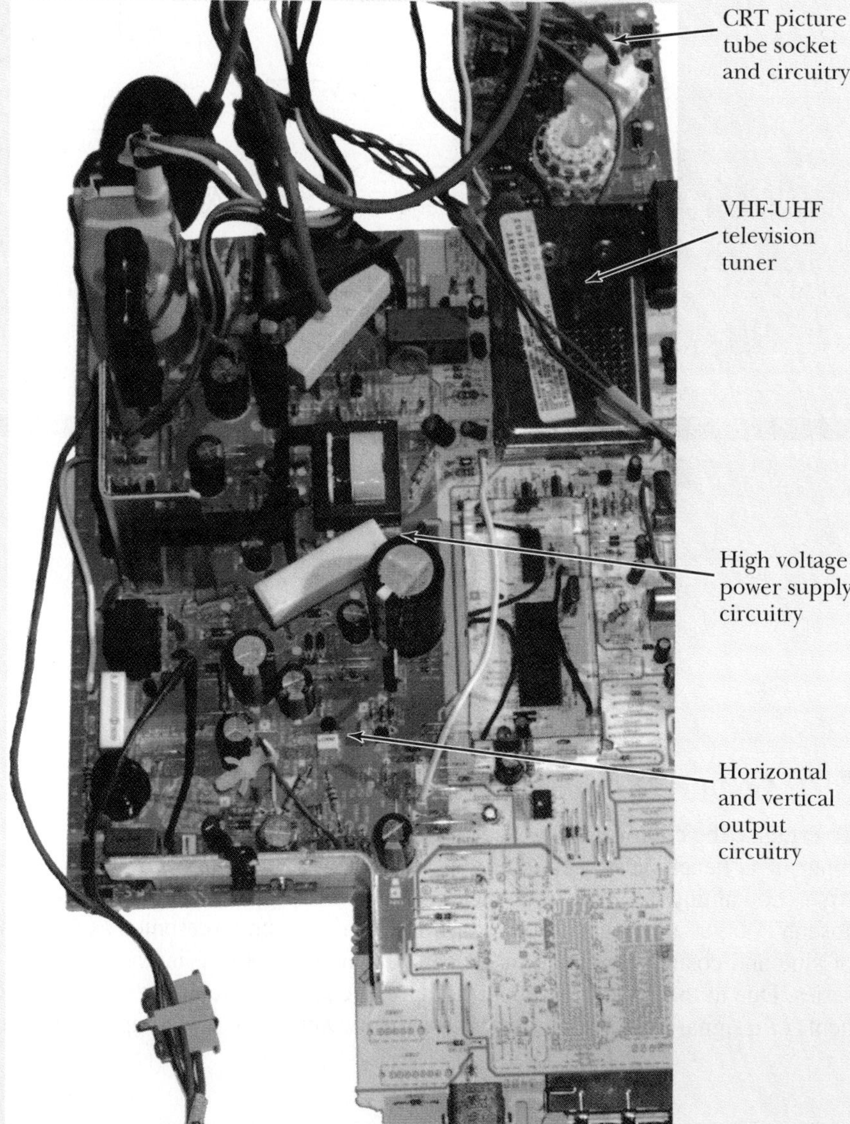

FIGURE 19 Color Television Open Chassis with CRT Socket

same time, the beam must move from top to bottom of the screen, again indirect synchronization with the vertical scan. This signal is called the vertical drive signal, since it drives the internal circuitry of the television receiver to form the vertical part of this television picture on the screen.

Figure 22 shows the most common circuitry of a representative 14″ to 21″ color television receiver. The op-amp is used in the summing amplifier mode in the inverting configuration. The op-amp essentially is operating as a power amplifier to supply a large sawtooth drive current to the vertical deflection coils. These coils are mounted in a deflection yoke on the neck of the picture tube just after the focusing elements. The vertical drive circuit working in conjunction with the coils in the yoke set up an electromagnetic field to deflect the scanning beam of electrons. The sawtooth drive current increases at a fixed 60 Hz rate to deflect the scanning beam of electrons down the face of the picture tube from the extreme left top corner, to the bottom right corner.

Referring to the central part of IC U1001, the T-chip, both horizontal and vertical sync pulses from the incoming television signal are applied each to their respective constant diode circuit (C/D), which passes only the actual H or V sync

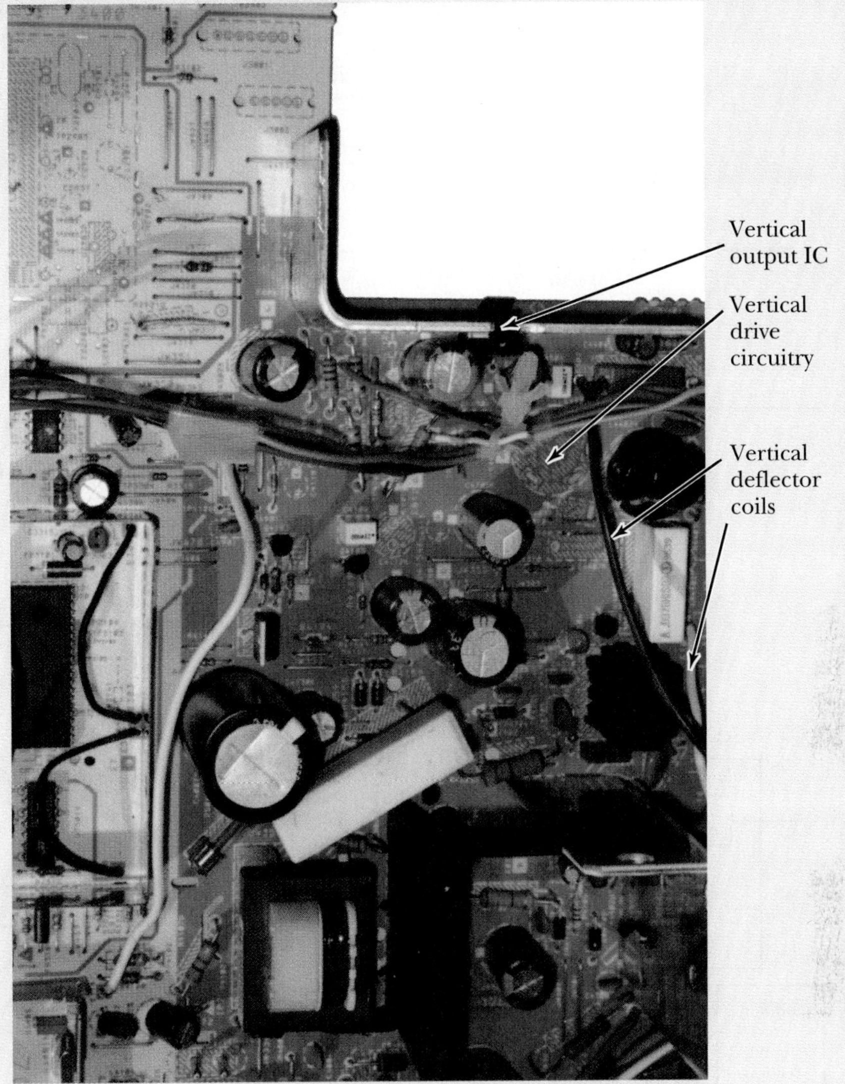

FIGURE 20 TV printed circuit board vertical drive circuitry

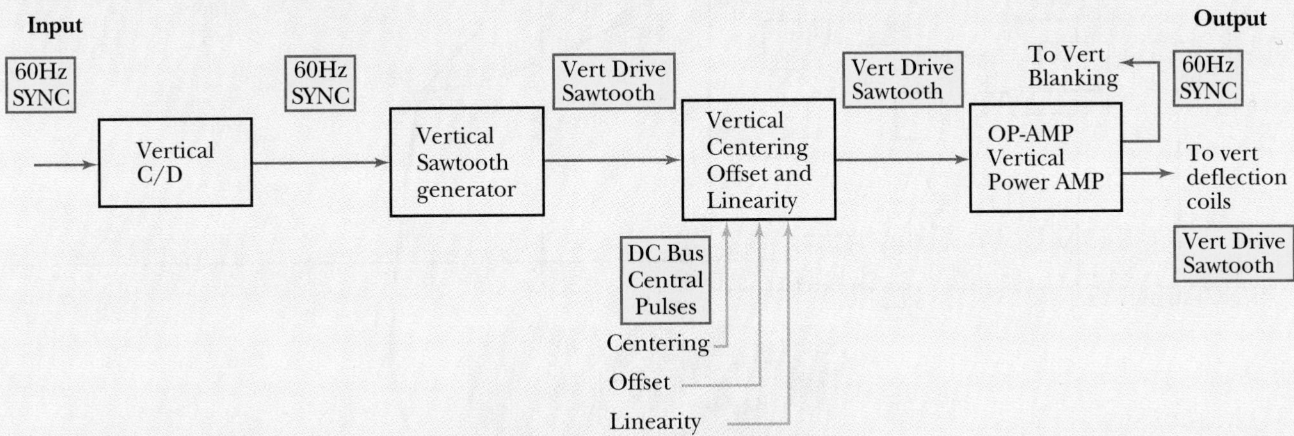

FIGURE 21 Block Diagram

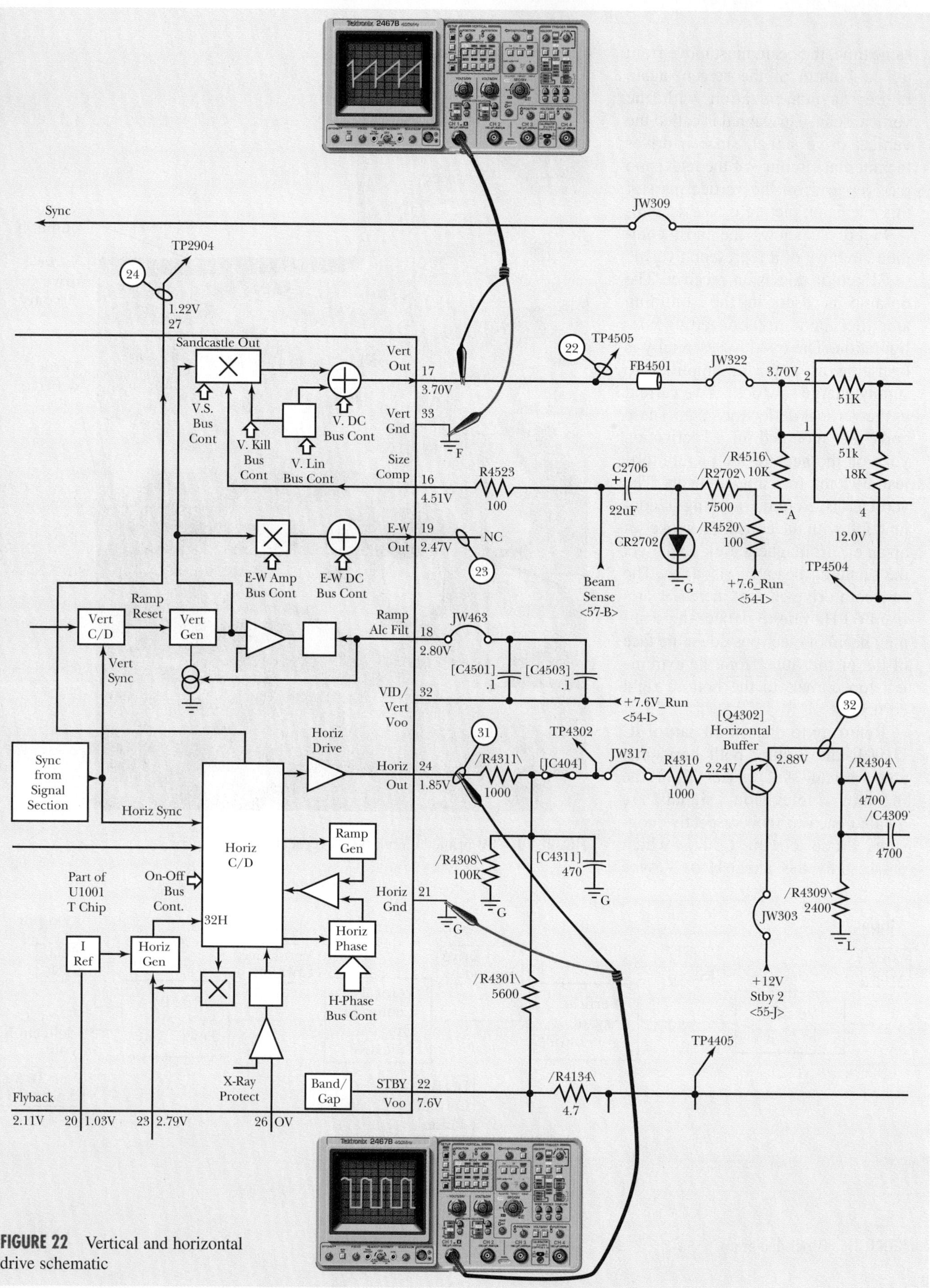

FIGURE 22 Vertical and horizontal drive schematic

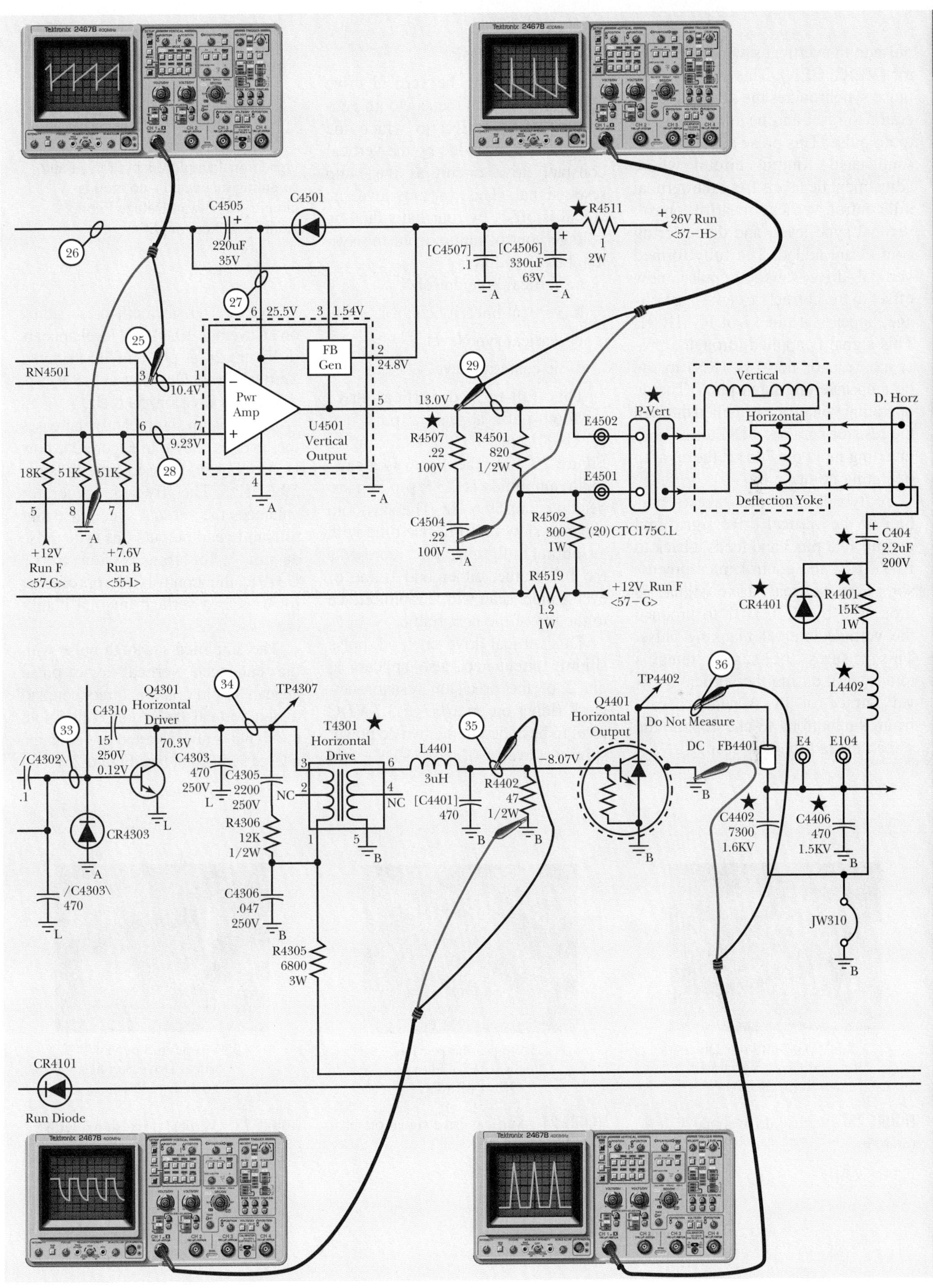

pulse to the vertical sawtooth generator (VERT GEN). This vertical sync pulse synchronizes the sawtooth generator to reset and begin a new sawtooth pulse. This pulse appears at the sand castle output circuit where adjustment dc levels from the vertical kill control, vertical linearity control, vertical sync level, and dc centering control are added. The fully formed vertical drive sawtooth pulse, now offset to the correct dc centering voltage, appears at pin 17 of IC U1001. This signal is applied through a precision resistor network (RN) to the op-amp inverting input pin 1. The dc operating voltages are applied through the resistor network (RN) to the noninverting input pin 7 to set the op-amp operating point.

A flyback generator takes a portion of the vertical drive signal and outputs it at pin 3 and feeds it back to IC U3101, the control microprocessor. This vertical drive signal is processed by IC U3101 to produce the vertical interval blanking pulse. This ensures the screen remains totally dark during the vertical interval retrace of the scanning beam from the bottom to the top of the screen to start the new scan.

Waveform Analysis

The formation of the vertical drive signal begins with the sawtooth generator (VERT GEN) to which the vertical sync is added by the vertical constant drive circuit as the ramp reset signal. This varying level ac signal is offset by four individual dc voltage control inputs from the system bus as follows:

a. vertical kill control

b. vertical linearity control

c. vertical sync level

d. dc centering level

This full-formed vertical drive sawtooth pulse appears at pin 17 of IC U1001, the T chip. Referring to Figure 23, note the signal is a sawtooth, amplitude is 2.5 × p-p, 5 msec per division, 59.6 Hz. The sawtooth shows a slow uniform rise time with an abrupt fall time; this represents the full frame, either odd frame or even frame, scan with a rapid retrace to the top of the next frame.

This vertical drive sawtooth pulse slightly attenuated, also appears at pin 2 of the precision resistor network riding on an average 3.7 VDC level to be added to the two dc power supply levels of 12 VDC and 7.6 VDC. The sawtooth pulse riding on an average 10.4 VDC level appears in Figure 24 at pin 1 of IC #U4501, Vertical Drive Op-Amp at 1.5 V p-p, 5 msec per division, 59.6 Hz.

The op-amp significantly increases the level as shown in Figure 25, now at 55 V p-p, 5 msec per division, 59.6 Hz. The flyback generator enhances the vertical sync pulse significantly and outputs this sync pulse at pin 3 for transmission to IC U3101, the control microprocessor for uses as a vertical internal blanking pulse.

The amplified sawtooth pulse with the enhanced vertical sync pulse appears at pin 5 and is direct coupled to the vertical deflection coils. Note the signal is now inverted and contains a large sync burst (see Figure 25).

Operational amplifiers used in an auto summing mode and auto muting circuitry: 120-watt Deluxe Public Address Mixer/Amplifier

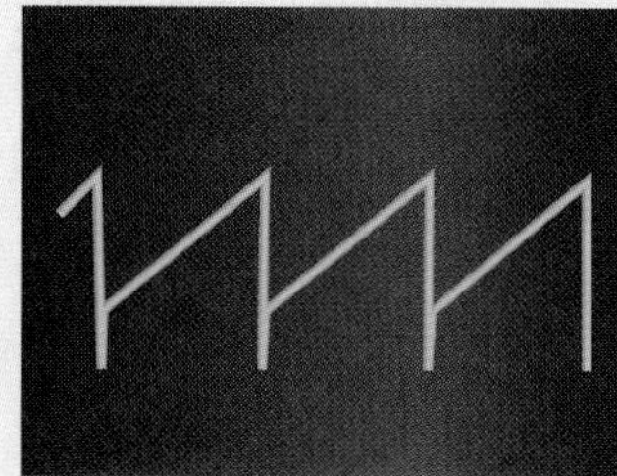

2.5 V p-p 5 mS per Div.
Vertical Drive Signal with
Ramp Reset

FIGURE 23 Vertical drive signal with ramp reset

1.5 V p-p 5 mS per Div.
Vertical Drive Signal
Offset with DC

FIGURE 24 Vertical drive signal offset with dc

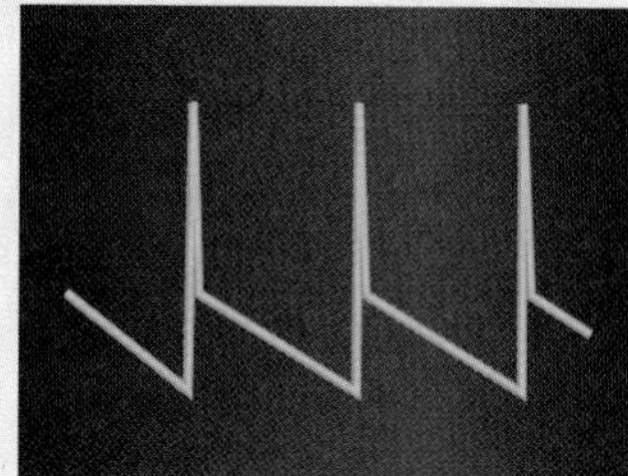

55 V p-p 5 mS per Div.
Vertical Drive Signal
with Sync Pulse

FIGURE 25 Vertical drive signal with sync pulse

Troubleshooting Hints

The most common trouble found in the use of op-amps as vertical drive circuitry is the loss of the incoming sync signal. This signal is essential in setting the repetition rate and the sawtooth ramp reset time. Both the horizontal H sync and vertical V sync pulse should be checked with a scope at the output of the sync separator circuit. The observed troubles are vertical tearing, missing, or compressed top and bottom picture elements.

Another common trouble is the loss of the dc levels and control pulses from IC U3101, the control microprocessor. These control bus signals should be checked with an oscilloscope set to dc to check the dc operating levels and then set to ac to observe the control pulses riding on the dc levels. The observed troubles are an off-centered picture and missing or compressed top or bottom picture elements.

Finally, either an open or short may develop in the vertical deflection coils or a short may occur between the vertical and the horizontal deflection coils. The vertical horizontal coil drive signal can be checked with an oscilloscope. The observed troubles are loss of raster.

CAUTION: Extremely dangerous high voltages are present at the horizontal output circuits, certain pins on the picture tube, and on the outside surface of the picture tube. Be cautious.

Summary

- The unity gain bandwidth product (GBW) is a device specific parameter relating the op-amp's open-loop voltage gain with the applied signal frequency. The GBW affects the actual closed-loop gain for all amplifier circuits (inverting, noninverting, and differential amplifiers). High input signal frequencies and/or high closed-loop gains result in the greatest affects on the actual op-amp output signal amplitudes.
- The slew rate (SR) is another op-amp parameter indicating the time required for the op-amp output voltage to change from the initial output voltage to the final output voltage due to an instantaneous change of the input voltage.
- Precision rectifier circuits are developed by placing the diode in the feedback loop of the op-amp circuits. The traditional diode drop to the output signal amplitude experienced by resistor-diode rectifier circuits does not exist with the precision rectifier circuits.
- The audio amplifier circuit using the LM386 allows the user to amplify audio frequency signals. The LM386 output is configured to deliver up to 0.5 W of power to an 8 Ω speaker.

CIRCUIT SUMMARY

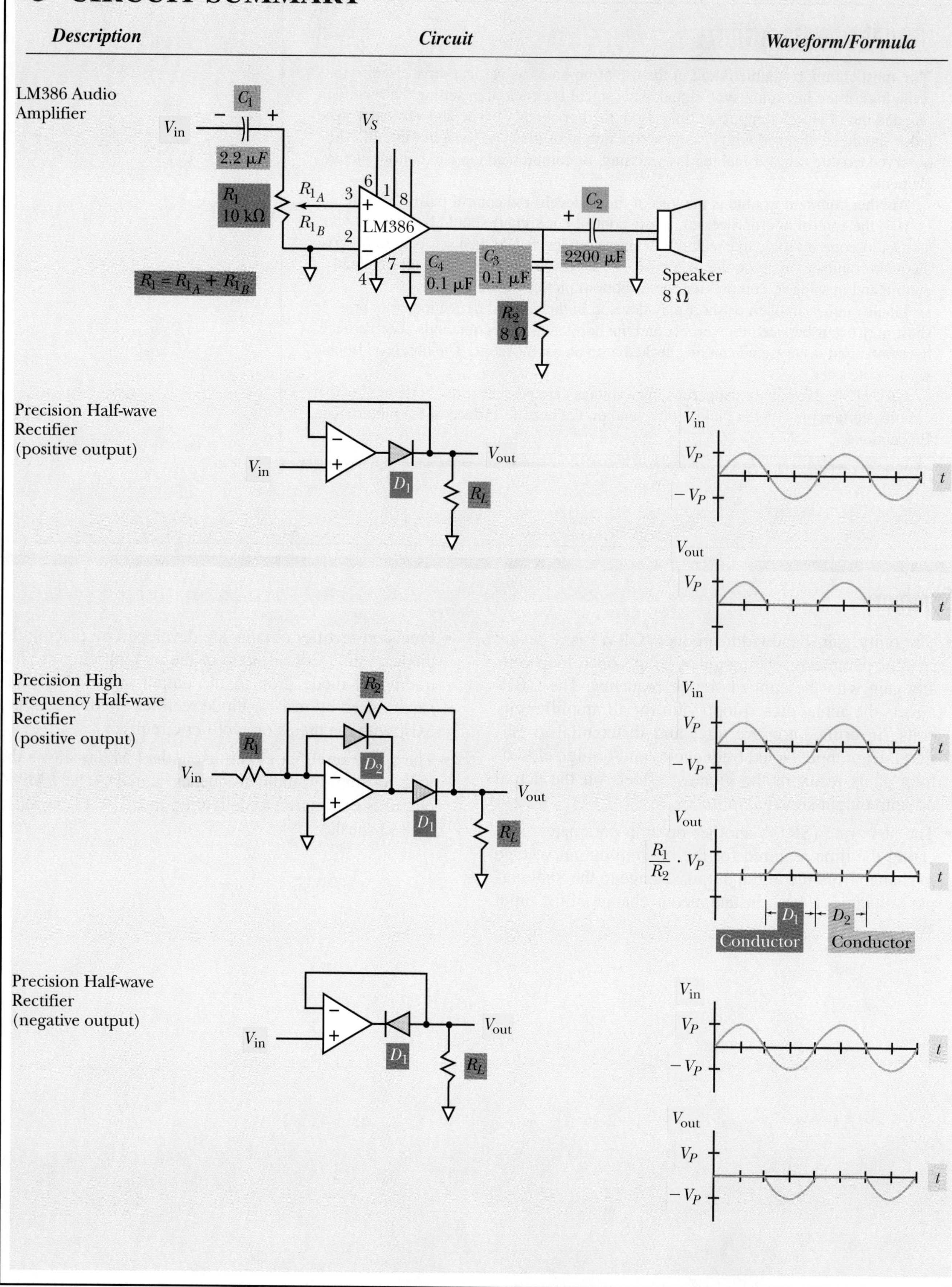

CIRCUIT SUMMARY

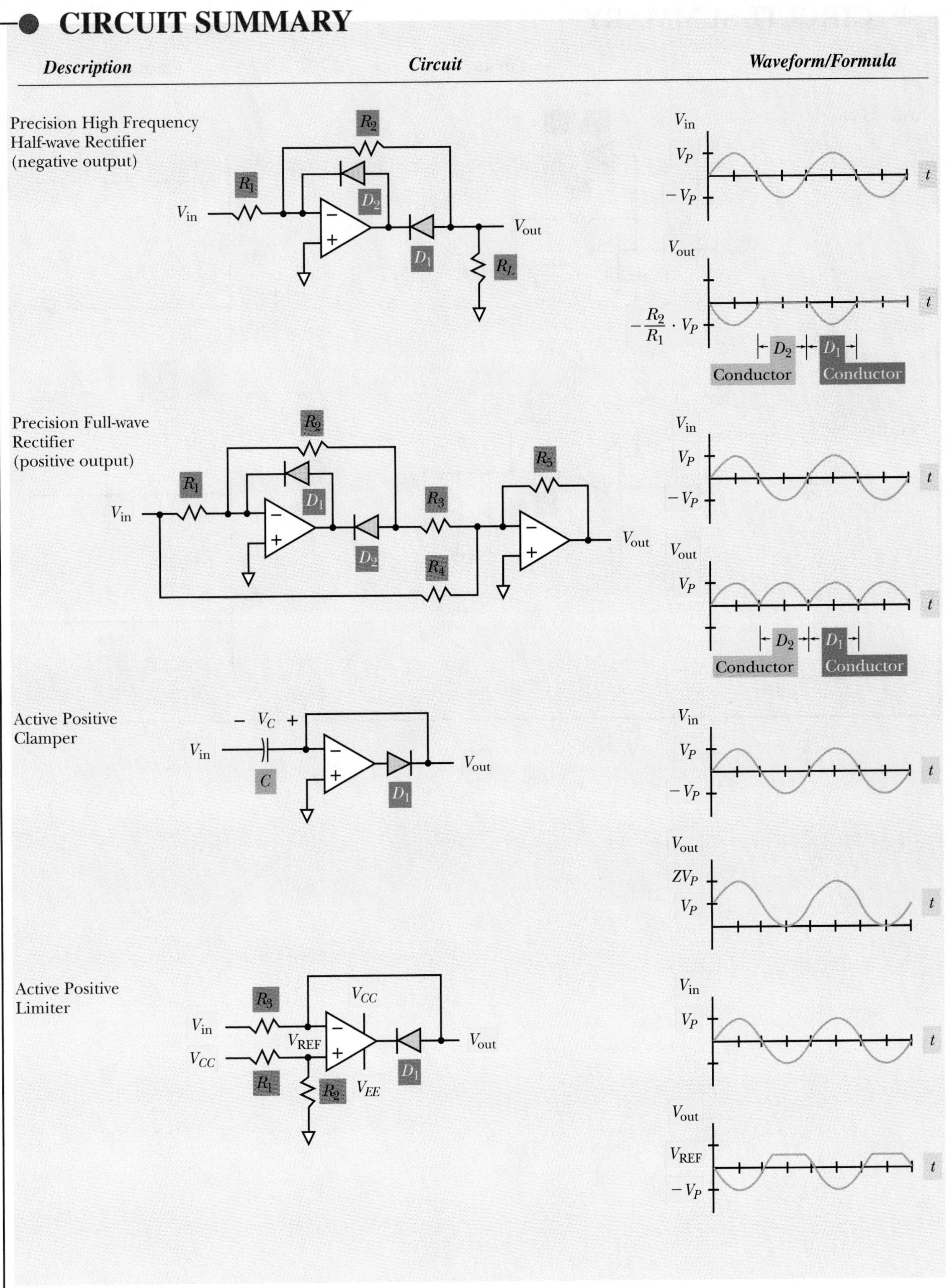
Description
Circuit
Waveform/Formula
Precision High Frequency Half-wave Rectifier (negative output)
R2
R1
D2
D1
RL
Vin
Vout
VP
− VP
t
−R2/R1 · VP
D2 Conductor
D1 Conductor
Precision Full-wave Rectifier (positive output)
R1
R2
R3
R4
R5
D1
D2
Active Positive Clamper
− VC +
C
D1
ZVP
Active Positive Limiter
VCC
VEE
VREF
R1
R2
R3
D1

CIRCUIT SUMMARY

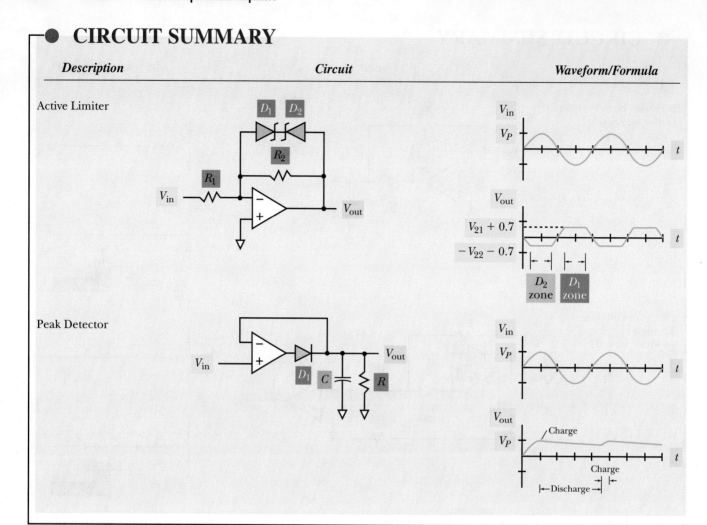

Description	*Circuit*	*Waveform/Formula*
Active Limiter	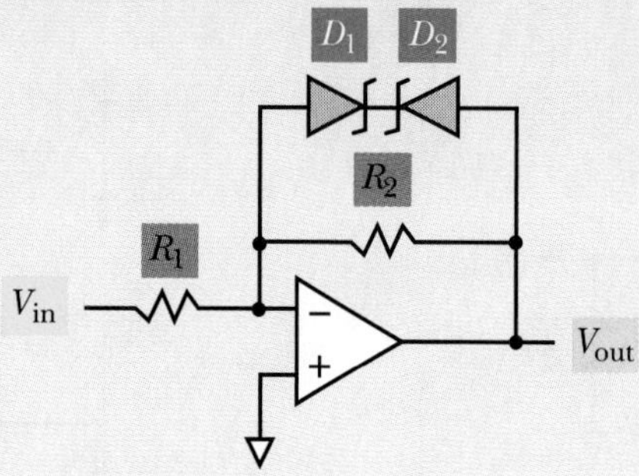 D_1 D_2 R_2 R_1 V_{in} V_{out}	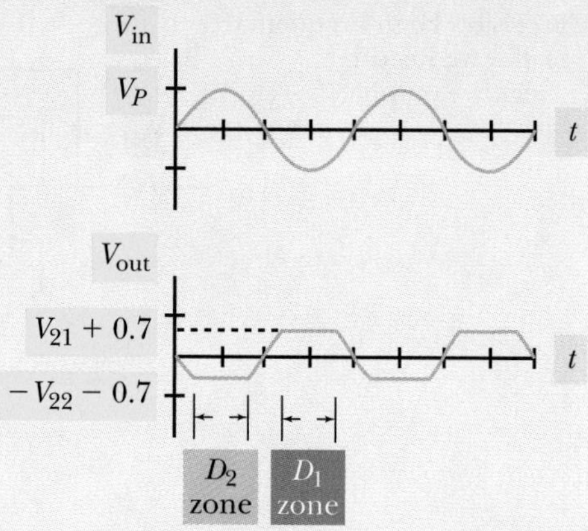 V_{in} V_P t V_{out} $V_{21} + 0.7$ $-V_{22} - 0.7$ t D_2 zone D_1 zone
Peak Detector	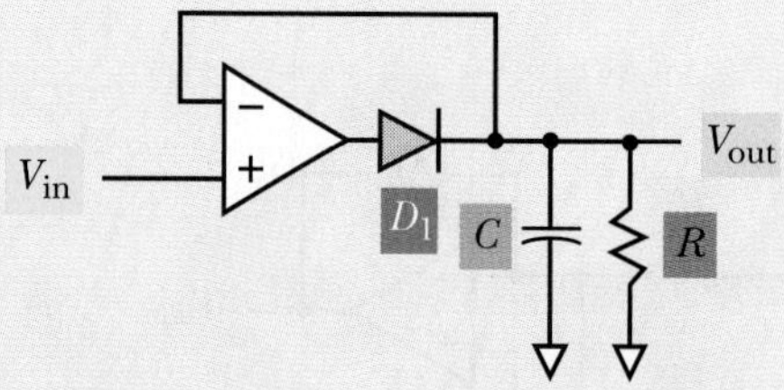 V_{in} D_1 C R V_{out}	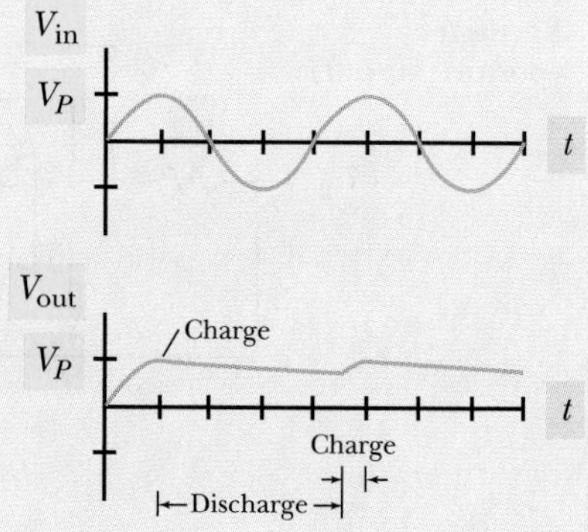 V_{in} V_P t V_{out} V_P Charge t Charge Discharge

Formulas and Sample Calculator Sequences

FORMULA 8–1
(To find the gain bandwidth product)

$GBW = a \times f$

a value, ⊠, f value, ⊟

FORMULA 8–2
(To find the open loop voltage gain)

$$a = \frac{GBW}{f}$$

GBW value, ÷, f value, =

FORMULA 8–3
(To find the time change for the voltage change)

$$\Delta t = \frac{\Delta V_{out}}{SR}$$

ΔV_{out} value, ÷, SR value, =

FORMULA 8–4
(To find the open loop voltage gain)

$\Delta V_{out} = V_{out_2} - V_{out_1}$

V_{out_2} value, −, V_{out_1} value, =

Inverting Amplifier

FORMULA 8–5
(To find the ideal voltage gain)

$$A_V = -\frac{R_2}{R_1}$$

R_2 value, ±, ÷, R_1 value, =

FORMULA 8–6
(To find the actual voltage gain)

$$A_V = \frac{\dfrac{-aR_2}{R_2 + R_1}}{1 + \dfrac{aR_1}{R_2 + R_1}}$$

a value, ±, ×, R_2 value, ÷, (, R_2 value, +, R_1 value,), ÷, (, 1, +, (, a value, ×, R_1 value, ÷, (, R_2 value, +, R_1 value,),), =

Noninverting Amplifier

FORMULA 8–7
(To find the ideal voltage gain)

$$A_V = 1 + \frac{R_2}{R_1}$$

R_2 value, ÷, R_1 value, =, +, 1, =

FORMULA 8–8
(To find the actual voltage gain)

$$A_V = \frac{a}{1 + \dfrac{aR_1}{R_2 + R_1}}$$

a value, ÷, (, 1, +, (, a value, ×, R_1 value, ÷, (, R_2 value, +, R_1 value,),), =

Voltage Divider Noninverting Amplifier

FORMULA 8–9
(To find the ideal voltage gain)

$$A_V = \left(1 + \frac{R_2}{R_1}\right) \times \frac{R_4}{R_4 + R_3}$$

(, (, R_2 value, ÷, R_1 value,), +, 1,), ×, R_4 value, ÷, (, R_4 value, +, R_3 value,), =

FORMULA 8–10
(To find the ideal voltage gain)

$$V_{out} = V_{in+} \times \left(1 + \frac{R_2}{R_1}\right) \times \frac{R_4}{R_4 + R_3} - V_{in-} \times \left(\frac{R_2}{R_1}\right)$$

V_{in+} value, ×, (, 1, +, (, R_2 value, ÷, R_1 value,),), ×, R_4 value, ÷, (, R_4 value, +, R_3 value,), −, (, V_{in-} value, ×, R_2 value, ÷, R_1 value,), =

FORMULA 8–11
(To find the ideal voltage gain)

$$V_{out} = (V_{in+} - V_{in-}) \times \frac{R_2}{R_1}$$

(, V_{in+} value, −, V_{in-} value,), ×, R_2 value, ÷, R_1 value, =

FORMULA 8–12
(To find the actual voltage gain)

$$V_{\text{out}} = V_{\text{in+}}\left(\frac{R_4}{R_3 + R_4}\right)\left(\frac{a}{1+\dfrac{aR_1}{R_2 + R_1}}\right) + V_{\text{in-}}\left(\frac{\dfrac{-aR_2}{R_2 + R_1}}{1+\dfrac{aR_1}{R_2 + R_1}}\right)$$

When $R_4 = R_2$ and $R_3 = R_1$, Formula 8–12 simplifies into Formula 8–13.

FORMULA 8–13
(To find the actual voltage gain)

$$V_{\text{out}} = (V_{\text{in+}} - V_{\text{in-}})\left(\frac{\dfrac{aR_2}{R_2 + R_1}}{1+\dfrac{aR_1}{R_2 + R_1}}\right)$$

[(], $V_{\text{in+}}$ value, [−], $V_{\text{in-}}$ value, [)], ↔, [(], a value, ↔, R_2 value, ⌉, [(], R_2 value, [+], R_1 value, [)], [)], ⌉, [(], 1, [+], [(], a value, ↔, R_1 value, ⌉, [(], R_2 value, [+], R_1 value, [)], [)], [)], [=]

Audio Amplifier

FORMULA 8–14
(To find the output voltage)

$$V_{\text{out}} = V_{\text{in}} \times \left(\frac{R_{1_B}}{R_{1_A} + R_{1_B} - jX_{C1}}\right) \times 20 \times \left(\frac{R_{\text{SPK}}}{R_{\text{SPK}} - jX_{C2}}\right)$$

V_{in} value, [×])[(], R_{1_B} value, [÷], [(], [(], R_{1_A} value, [+], R_{1_B} value, [)], x^2, [+], X_{C_1} value, x^2, [)], $\sqrt{x}$, [)], [×], 20, [×], [(], R_{SPK} value, [÷], [(], R_{SPK} value, x^2, [+], X_{C_1} value, x^2, [)], $\sqrt{x}$, [)], [=]

Review Questions

1. The ___________ provides the op-amp frequency and open-loop gain limitations.
2. The op-amp open-loop gain increases as the frequency of the applied signal is increased (True or False).
3. The power supply bypass capacitors help to maintain op-amp stability (True or False).
4. The ___________ is a measurement of the output voltage change with respect to time due to a change in the input voltage.
5. The actual voltage gain of an inverting amplifier is larger than the ideal voltage gain (True or False).
6. The actual voltage of a noninverting amplifier is smaller than the ideal voltage gain (True or False).
7. The ideal voltage phase shift (output compared to input) for a noninverting amplifier with a 1-kHz sine wave input is ___________.
8. The ideal phase shift (output compared to input) for an inverting amplifier with a 1-kHz sine wave input is ___________.
9. The differential amplifier output is the ___________ of the two input signals.
10. The ___________ is used to amplify audio signals providing power to an 8 Ω speaker.
11. The peak output of precision rectifier circuits are reduced by 0.7 V (True or False).

Problems

1. The LM741 will be used in a circuit where the applied signal frequency is 1.0 kHz. Calculate the open-loop voltage gain for the op-amp at this frequency.
2. The LM741 will be used in a circuit (Figure 8–3a) with a voltage gain of +10. The input voltage changes from 0 V to 1.0 V instantaneously. Calculate the change in output voltage and the time required for the output to reside at the new voltage level. Sketch the output signal.
3. The Figure 8–4 inverting amplifier circuit uses the LM741 with R_2 = 10 kΩ and R_1 = 2 kΩ. A 0.20-V_{PP}, 100-kHz since wave is applied as the input signal to the circuit.
   a. Calculate the voltage gain using the ideal op-amp.
   b. Calculate the voltage gain of the amplifier considering the op-amp open-loop gain.
   c. Calculate the output voltage using the ideal op-amp.
   d. Calculate the output voltage gain considering the op-amp open-loop gain.

4. The Figure 8–7 noninverting amplifier circuit uses the LM741 with R_2 = 10 kΩ and R_1 = 2 kΩ. A 0.20-V_{PP}, 100-kHz sine wave is applied as the input signal to the circuit.
   a. Compare the actual circuit voltage gain to the ideal op-amp circuit voltage gain.
   b. Compare the actual circuit output signal to the ideal op-amp circuit output signal.
5. The Figure 8–11 differential amplifier circuit uses the LM741 with R_1 = 2 kΩ, R_2 = 10 kΩ, R_3 = 2 kΩ, and R_4 = 10 kΩ. The input voltage V_{in+} is a 0.4-V_{PP}, 10-kHz sine wave and the input voltage V_{in-} is a 0.5-V_{PP}, 10-kHz 0° phase shift sine wave.
   a. Compare the actual circuit voltage gain to the ideal op-amp circuit voltage gain.
   b. Compare the actual circuit output signal to the ideal op-amp circuit output signal.

Analysis Questions

1. For the unity gain noninverting amplifier using an LM741, determine the maximum frequency at which the actual voltage gain is within 10% of the ideal voltage gain.
2. The LM741 op-amp is used in an inverting amplifier circuit with an ideal closed-loop gain of –1. Determine the maximum frequency at which the actual voltage gain is within 10% of the ideal voltage gain.
3. The LM741 op-amp is used in a differential amplifier circuit. If the ideal closed-loop gain for the inverting and noninverting signal inputs, evaluate the affects of the gain bandwidth product on the same frequency signal applied to both V_{in+} and V_{in-}.
4. The LM741 op-amp is used in a differential amplifier circuit. The ideal closed-loop gain for the inverting signal input is –1 and the ideal closed-loop gain for the noninverting signal input is +1. A 4-V_{PP}, 10-kHz sine wave is applied to both inputs. Calculate the ideal differential amplifier output signal. Calculate the actual differential amplifier output signal. Are they the same? Explain your answer.
5.
   a. Using Formula 8–2 with the LM741 1.0 MHz unity gain bandwidth product, create an Excel spreadsheet graph of open-loop voltage gain on the vertical *(y)* axis with respect to the frequency on the horizontal *(x)* axis. The frequency will be configured into a logarithmic scale. The calculated frequencies in Hz should include: 10, 20, 50, 100, 200, 500, 1 k, 5 k, 10 k, 20 k, 50 k, 100 k, 200 k, 500 k, and 1 M.
   b. Add to the LM741 unity gain bandwidth product bode plot, a graph of the actual closed-loop gain for an inverting amplifier with a gain of –10.
   c. Add to the LM741 unity gain bandwidth product bode plot, a graph of the actual closed-loop gain for a noninverting amplifier with a gain of +1.

Performance Projects Correlation Chart

Alle-labs are available on the LabSource CD.

Chapter Topic	Performance Project	Project Number
Inverting and Noninverting Amplifiers	Inverting and Noninverting Amplifiers with Sine Wave Input	19
Audio Amplifier	Audio Amplifier (LM386)	20

OUTLINE

OBJECTIVES

After studying this chapter, you should be able to:

1. Explain the function of a filter
2. Define the filter passband
3. Define the filter stopband
4. Calculate the filter cutoff frequency
5. Determine the order of a filter
6. Identify a low-pass filter
7. Identify a high-pass filter
8. Identify a bandpass filter
9. Identify a notch filter
10. Describe how the order of a filter affects the output signal response

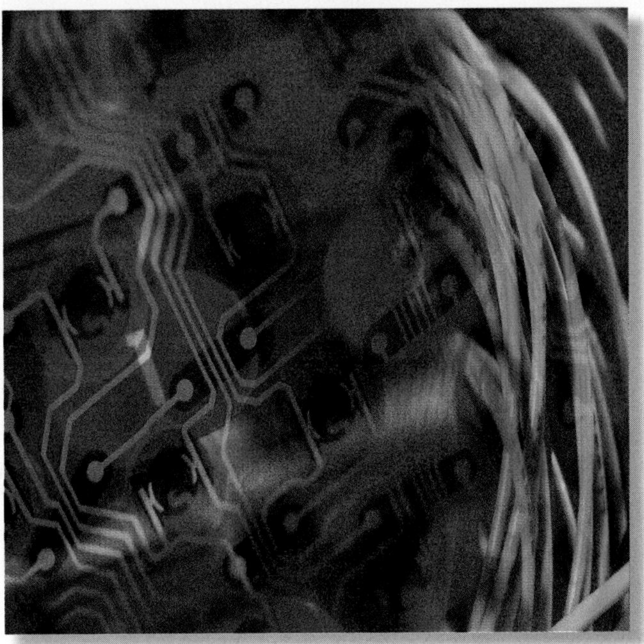

CHAPTER 9

Filters

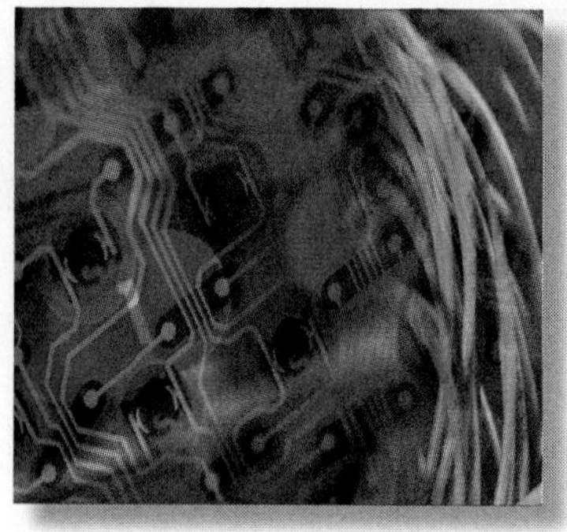

CHAPTER PREVIEW

The **filter** performs an important task in the separation of desired signal frequencies from undesired signal frequencies. Complex waveforms such as music are comprised of a variety of sine waves combined together producing the sound that you enjoy. Within the electrical circuitry, a filter can be used to separate the different frequency components of the music. The stereo band equalizer allows you the ability to adjust the gain for each frequency range. The circuitry to accomplish this task consists of several bandpass filters having different upper and lower cutoff frequencies.

Working with signals a need often arises to block a range of frequencies allowing only a fixed range of frequencies to pass (be delivered) to the output. You select the TV channel of your preference. Within a geographical area, each station broadcasts its video and audio information at a different carrier frequency. By selecting the channel you desire to watch, you have actually selected the frequencies for the bandpass filter to obtain the desired video and audio signal associated with that channel.

KEY TERMS

Band reject filter	Cutoff frequency	Notch filter
Bandpass filter	Filter	Order
Bandstop filter	High-pass filter	Pass-band
Bandwidth	Low-pass filter	Stopband
Bode plot		

9–1 Background Information

The filter is used to pass signals within a range of frequencies and block signals outside that frequency range. To evaluate the operation of a filter, you will apply different frequency sine waves, each having the same amplitude to the filter circuit observing the filter output amplitude. The signal will be passed to the output if the signal frequency is within the filter's passband frequencies. The output signal will have the same amplitude as the input signal amplitude (unity gain). (Note: a filter can be designed to have other gains within the passband.) The signal will be blocked from passing to the output if the signal frequency is outside the filter's passband frequency. Ideally, the output voltage at this frequency will be 0 V.

The five primary classifications for filters are low-pass, high-pass, bandpass, bandstop (notch), and all-pass (phase-shift filter). Our discussion in the text will concentrate on the low-pass, high-pass, bandpass, and bandstop. The ideal filter will have a phase shift of 0°, a fixed gain within the passband, and an output voltage of 0 V within the **stopband.**

The ideal low-pass filter will allow signals having frequencies below the cutoff frequency to pass to the output while blocking the signals having frequencies above the cutoff frequency as shown in Figure 9–1. The ideal high-pass filter will block signals having frequencies below the cutoff frequency while passing the signals having frequencies above the cutoff frequency as shown in Figure 9–2. The ideal bandpass filter will block signals having frequencies below the lower cutoff frequency, pass the signals with frequencies between the lower cutoff frequency and the upper cutoff frequency, and block the signals with frequencies above the upper cutoff frequency as shown in Figure 9–3. The ideal bandstop filter will pass signals having frequencies below the lower cutoff frequency, block the signals with frequencies between the lower cutoff frequency and the upper cutoff frequency, and pass the signals with frequencies above the upper cutoff frequency as shown in Figure 9–4. The ideal all-pass

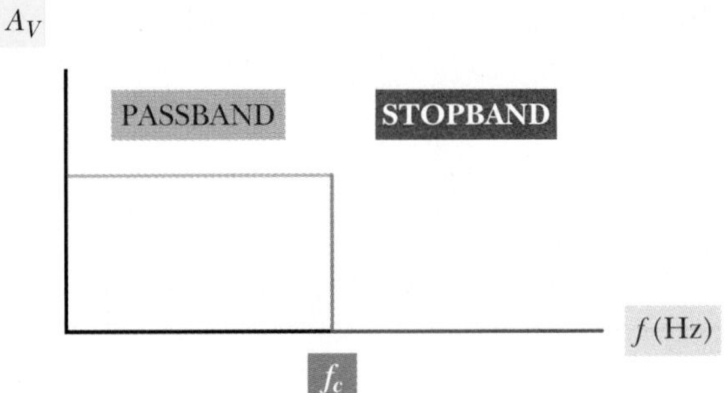

FIGURE 9–1 Ideal low-pass filter response

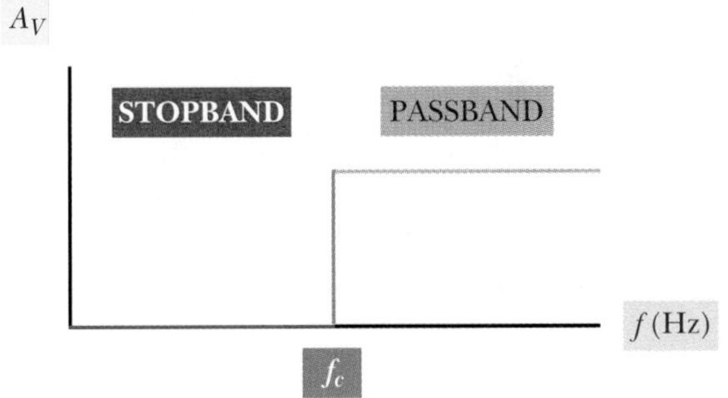

FIGURE 9–2 Ideal high-pass filter response

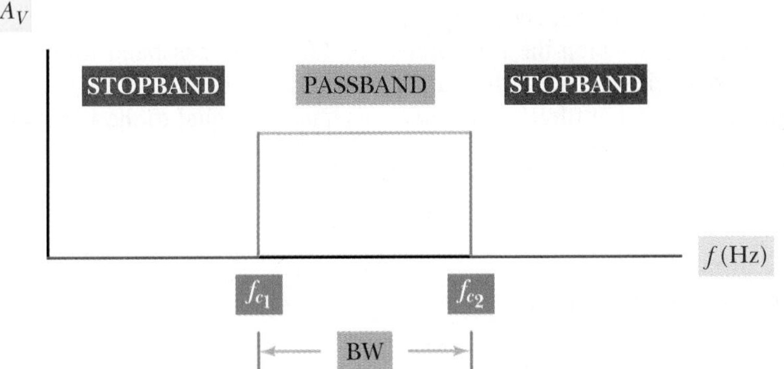

FIGURE 9–3 Ideal bandpass filter response

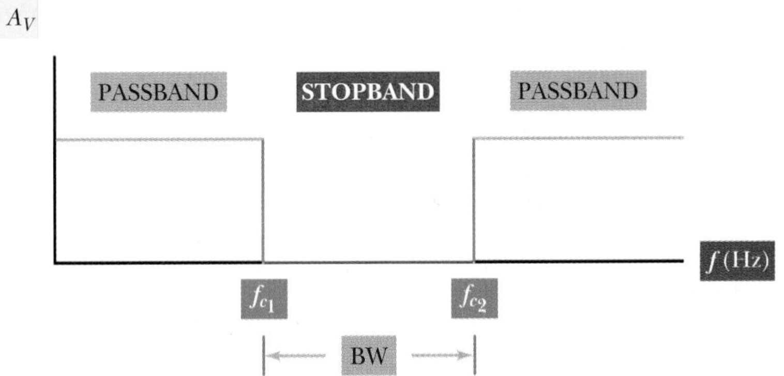

FIGURE 9–4 Ideal bandstop (notch) filter response

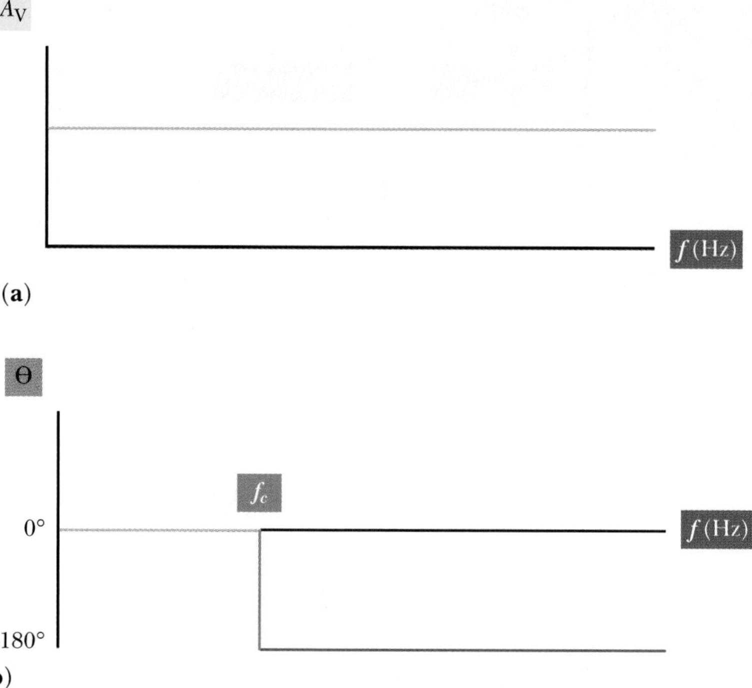

FIGURE 9–5 Ideal all-pass filter response: (a) voltage gain; (b) phase shift

filter will have the same amplitude gain for all signal frequencies (Figure 9–5a) with the output signal phase shift at one value below the cutoff frequency and a different value above the cutoff frequency (Figure 9–5b).

The frequency at which the filter transitions from the passband frequencies to the stopband frequencies is called the **cutoff frequency.** The cutoff frequency occurs at the frequency where the filter's capacitive reactance is equal to the filter's resistance.

$$X_C = R$$
$$\frac{1}{2\pi fC} = R$$

Solving for the frequency:

FORMULA 9–1 $f_c = \dfrac{1}{2\pi RC}$

When the applied signal frequency is the same as the cutoff frequency, the output voltage amplitude is 0.707 times the input voltage amplitude (assuming unity gain passband). Expressed as the filter's voltage gain, the voltage gain at the cutoff frequency is 0.707.

FORMULA 9–2 $V_{out} = A_v \times V_{in}$
$V_{out} = 0.707 \times V_{in}$

FORMULA 9–3 $A_V = \dfrac{V_{out}}{V_{in}}$
$A_V = 0.707$

Representing the voltage gain at the cutoff frequency in units of decibels (dB),

FORMULA 9–4 $A(\text{dB}) = 20 \log (A_v)$
$A(\text{dB}) = 20 \log (0.707)$
$A(\text{dB}) = 3 \text{ dB}$

The filter power gain is the filter output power divided by the filter input power. At the cutoff frequency, the power gain for the filter circuit is 0.5. Since half of the power applied to the filter circuit input will be at the filter circuit output, the cutoff frequency is also known as the half power point.

Filters are also characterized by the output response. The Butterworth filter approximation has a flat gain within the passband, a slow roll-off rate in comparison with other filter types, a monotonic stopband (gain changes with frequency), and a nonlinear phase response. The Chebyshev filter approximation has a variation of gain (ripples) within the passband frequency (as you increase the frequency within the passband frequencies the filter gain will increase and decrease), a faster roll-off rate, monotonic stopband, and a nonlinear phase response. The Inverse Chebyshev filter approximation has a flat gain within the passband, a fast roll-off rate, a rippled stopband response, and a nonlinear phase response. The Elliptic filter approximation (also known as Cauer filter approximation) has a rippled passband, a very fast roll-off rate, a rippled stopband response, and a nonlinear phase response. The Bessel filter approximation has a flat passband, slow roll-off rate, a monotonic stopband, and a linear phase shift requiring complex circuitry.

The emphasis within the text will be on the Butterworth filter circuits.

The number of reactive components (capacitors and inductors) located within the filter circuit determines the **order** of the filter. The filter's order (n) establishes the slope (m) of the filter voltage gain with respect to the input signal frequency outside of the passband frequency range.

FORMULA 9–5 $\text{m} = \text{n} \times \left(\frac{-20 \text{ dB}}{\text{decade}} \right)$

The filter voltage gain at a frequency outside of the passband is calculated by

FORMULA 9–6 $A_{\text{SIGNAL}}\,(\text{dB}) = A_{\text{PASSBAND}}\,(\text{dB}) + \text{m} \times \left| \text{LOG}\left(\frac{f_{\text{SIGNAL}}}{f_{\text{CUTOFF}}} \right) \right|$

EXAMPLE Calculate the slope of a second order low-pass filter having a cutoff frequency of 1 kHz. If the voltage gain for this filter circuit is 0 dB at 100 Hz, what is the filter voltage gain at 100 kHz? With an input voltage of 10 V_{RMS}, what would be the output voltage?

Answer:

Since the filter is a second order filter, n is 2. Using Formula 9–5, the slope m can be calculated.

$$\text{m} = \text{n} \times (-20 \text{ dB/decade})$$
$$\text{m} = 2 \times (-20 \text{ dB/decade})$$
$$\text{m} = -40 \text{ dB/decade}$$

The filter voltage gain at 100 kHz is calculated by using Formula 9–6

$$A_{\text{SIGNAL}}\ (\text{dB}) = A_{\text{PASSBAND}}\ (\text{dB}) + \text{m} \times \left|\text{LOG}\left(\frac{f_{\text{SIGNAL}}}{f_{\text{CUTOFF}}}\right)\right|$$

$$A_{\text{SIGNAL}}\ (\text{dB}) = 0\ \text{dB} + (-40) \times \left|\text{LOG}\left(\frac{100\ \text{kHz}}{1\ \text{kHz}}\right)\right|$$

$$A_{\text{SIGNAL}}\ (\text{dB}) = 0\ \text{dB} - 40 \times |\text{LOG}(100)|$$

$$A_{\text{SIGNAL}}\ (\text{dB}) = 0\ \text{dB} - 40 \times |2|$$

$$A_{\text{SIGNAL}}\ (\text{dB}) = -80\ \text{dB}$$

From Formula 9–4, you can determine the unitless voltage gain of the low-pass filter at 100 kHz.

$$A_V = 10^{(A_{\text{SIGNAL}}(\text{dB}))20}$$

$$A_V = 10^{(-80\text{dB})/20}$$

$$A_V = 10^{-4}$$

With Formula 9–3, the output voltage is calculated for this low-pass filter circuit having a 10 V_{rms} 100 kHz sine wave applied to the input.

$$V_{\text{out}} = A_V \times V_{\text{in}}$$

$$V_{\text{out}} = 10^{-4} \times 10\ \text{V}_{\text{RMS}}$$

$$V_{\text{out}} = 1.00\ \text{mV}_{\text{RMS}}$$ ◆

9–2 Low-Pass Filter Circuits

The **low-pass filter** shown in Figure 9–6 is a first order filter with unity gain. The equation to calculate the cutoff frequency is provided in Formula 9–7. The output voltage is frequency dependent and can be calculated using Formula 9–8. The **Bode plot** is a graph of the circuit voltage gain as a function of the frequency. The Bode plot displays the frequency on a logarithmic scale. The Bode plot for the unity gain low-pass filter is shown in Figure 9–7.

FORMULA 9–7 $f_c = \dfrac{1}{2\pi RC}$

FORMULA 9–8 $V_{\text{out}} = V_{\text{in}} \times \dfrac{-\text{j}X_C}{R - \text{j}X_C}$

$$A_V = \frac{V_{\text{out}}}{V_{\text{in}}} = \frac{X_C}{\sqrt{R^2 + X_C^2}}$$

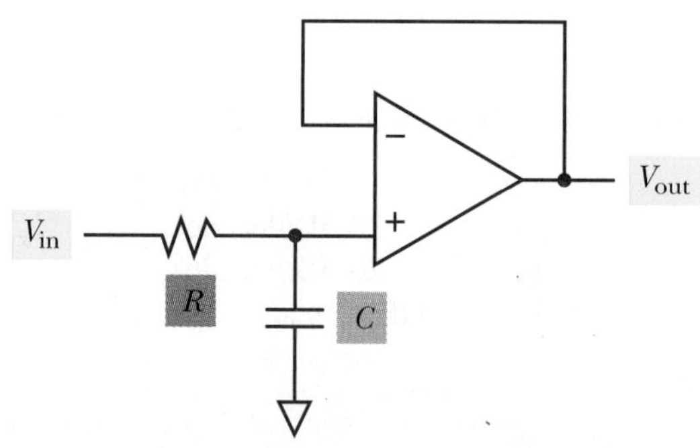

FIGURE 9–6 First order low-pass filter with unity gain

At very low frequencies, the output voltage will be at its maximum voltage based on the closed-loop gain of the op-amp circuit (+1). At these low frequencies, the capacitor's reactance is much larger than the resistor *(R)* value. The capacitor appears like an open circuit with all of the input signal voltage being applied to the op-amp's positive terminal. As the signal frequency increases, the capacitor's reactance decreases resulting in a decrease of the voltage applied to the op-amp's positive terminal. The increasing of the signal frequency caused the filter circuit voltage gain to decrease. At the cutoff frequency, the filter voltage gain is –3 dB ($V_{out} = 0.707 \times V_{in}$). When the signal frequency is at 10 times the cutoff frequency, the filter voltage gain is –20 dB ($V_{out} = 0.1 \times V_{in}$). At very high frequencies, the capacitor's reactance appears like a short circuit with 0 V being applied to the op-amp's positive terminal. The low-pass filter's pass range is defined as all signal frequencies that are below the filter's cutoff frequency. The low-pass filter's block range exists for all signal frequencies that are above the filter's cutoff frequency.

EXAMPLE The low-pass filter in Figure 9–6 has resistor value $R = 2.0\ \text{k}\Omega$, and capacitor value $C = 0.022\ \mu\text{F}$. Calculate the cutoff frequency and the output voltage amplitude when the input waveform is a 1 V_{rms} sine wave having a frequency equal to the cutoff frequency.

Answer:

From Formula 9–7

$$f_c = \frac{1}{2\pi RC}$$

$$f_c = \frac{1}{2\pi \times 2000\ \Omega \times 0.022\ \mu\text{F}}$$

$$f_c = 3.617\ \text{kHz}$$

To calculate the capacitive reactance, use

$$X_c = \frac{1}{2\pi fC}$$

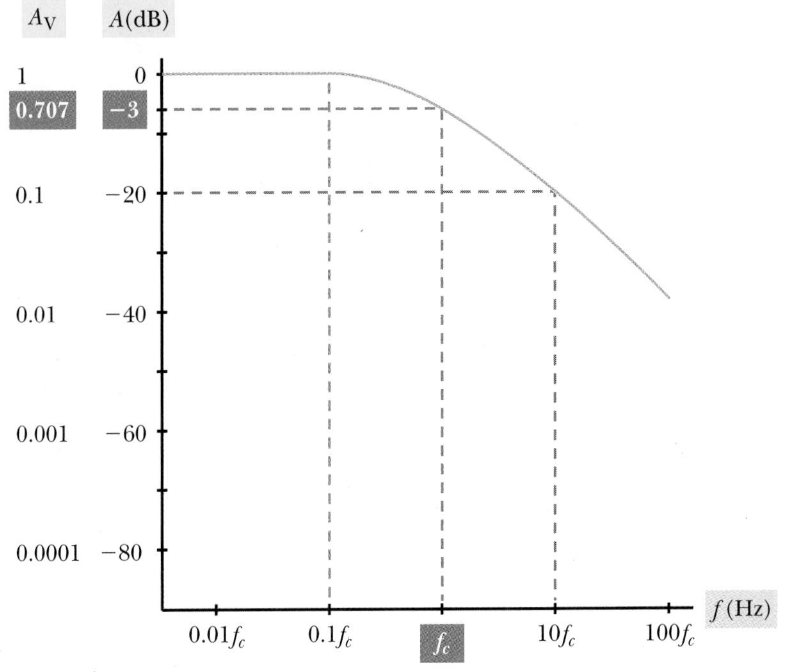

FIGURE 9–7 First order low-pass filter with unity gain response

When $f_{in} = f_c = 3.617$ kHz,

$$X_c = \frac{1}{2\pi \times 3.617\ \text{kHz} \times 0.022\ \mu\text{F}}$$
$$X_c = 2000\ \Omega$$

FORMULA 9–9 $$V_{out} = V_{in} \times \frac{-jX_C}{R - jX_C}$$

$$V_{out} = 1\ \text{V}_{rms} \times \left(\frac{2000}{\sqrt{(2000)^2 + (2000)^2}}\right)$$
$$V_{out} = 0.707\ \text{V}_{rms}$$ ◆

The low-pass filter shown in Figure 9–8 is a first order filter with gain. The equation to calculate the cutoff frequency is provided in Formula 9–7. The output voltage is frequency dependent and can be calculated using Formula 9–10. For very low frequencies, the output voltage will be at its maximum based on the closed-loop gain of the op-amp circuit $\left(1 + \frac{R_2}{R_1}\right)$. The output voltage of the low-pass filter with gain circuit will decrease as the applied signal frequency is increased. The filter's pass range is all frequencies that are less than the filter's cutoff frequency. The filter's block range is all frequencies that are greater than the filter's cutoff frequency.

FORMULA 9–10 $$V_{out} = V_{in} \times \left(\frac{-jX_C}{R - jX_C}\right) \times \left(1 + \frac{R_2}{R_1}\right)$$

$$A_V = \frac{V_{out}}{V_{in}} = \frac{X_C}{\sqrt{R^2 + X_C^2}} \times \left(1 + \frac{R_2}{R_1}\right)$$

◆ **EXAMPLE** The low-pass filter in Figure 9–8 has resistor values $R = 1.0$ kΩ, $R_1 = 10$ kΩ, $R_2 = 10$ kΩ, and capacitor value $C = 0.001$ μF. Calculate the cutoff frequency and the output voltage amplitude when the input waveform is a 1 V_{rms} sine wave having a frequency equal to the cutoff frequency.

Answer:

From Formula 9–7

$$f_c = \frac{1}{2\pi RC}$$
$$f_c = \frac{1}{2\pi \times 1000\ \Omega \times 0.001\ \mu\text{F}}$$
$$f_c = 159.15\ \text{kHz}$$

To calculate the capacitive reactance, use

$$X_c = \frac{1}{2\pi fC}$$

When $f_{in} = f_c = 159.15$ kHz,

$$X_c = \frac{1}{2\pi \times 159.15\ \text{kHz} \times 0.001\ \mu\text{F}}$$
$$X_c = 1000\ \Omega$$

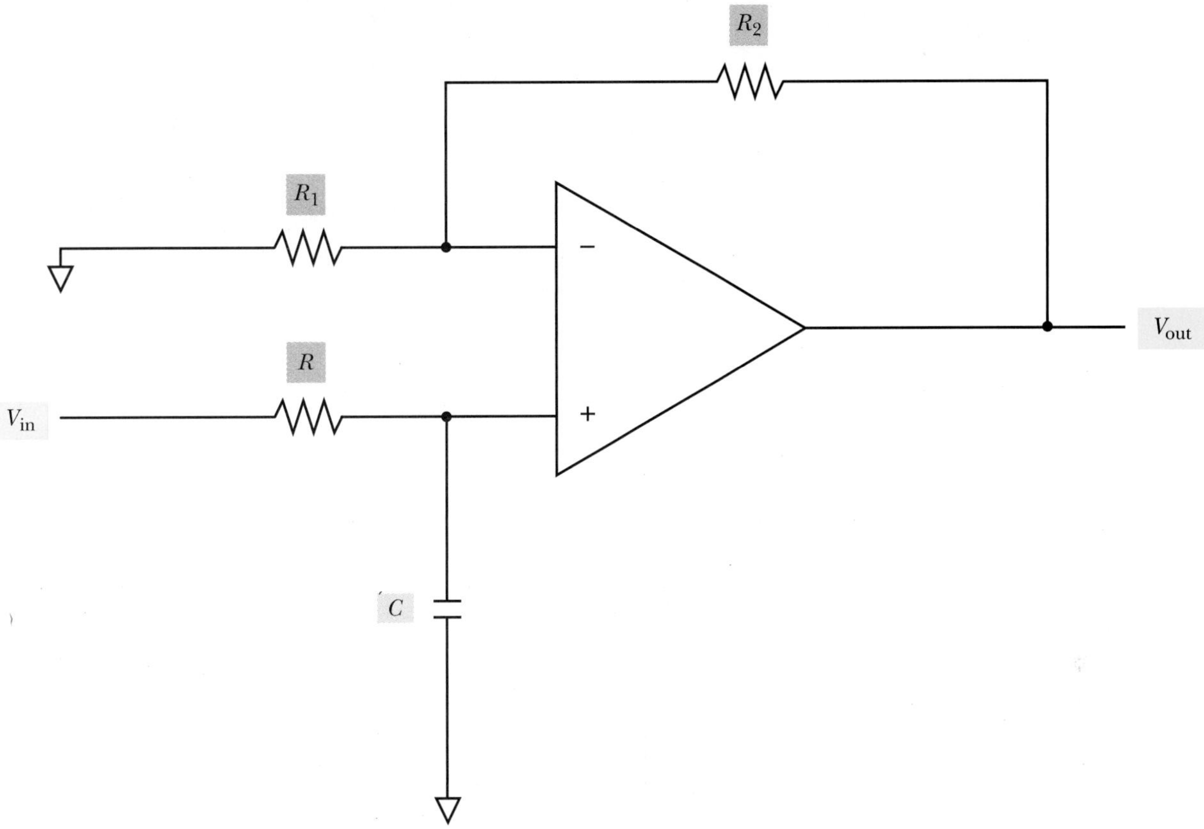

FIGURE 9–8 First order active low-pass filter

Using Formula 9–10, the output voltage can be calculated.

$$V_{out} = V_{in} \times \left(\frac{-jX_C}{R - jX_C}\right) \times \left(1 + \frac{R_2}{R_1}\right)$$

$$V_{out} = 1\ V_{rms} \times \left(\frac{1000}{\sqrt{(1000)^2 + (1000)^2}}\right) \times \left(1 + \frac{10000}{10000}\right)$$

$$V_{out} = 1.414\ V_{rms}$$

◆

The **low-pass filter** shown in Figure 9–9 is a second order filter with unity gain. The filter type is known as a Sallen-Key low-pass filter as well as a Voltage Controlled Voltage Source (VCVS) filter. For proper operation with a Butterworth filter characteristic, the following component selections must be maintained: $R_2 = R_1$ and $C_2 = 2 \times C_1$ while $A_V = 1.0$. The equation to calculate the cutoff frequency is provided in Formula 9–11.

FORMULA 9–11 $f_c = \dfrac{1}{2\pi\sqrt{R_1 R_2 C_1 C_2}}$

Two capacitors (C_1 and C_2) exist within the second order low pass filter. At very low frequencies, the output voltage will be at its maximum voltage based on the closed loop gain of the op-amp circuit (+1). At these low frequencies, the capacitors' reactance is very large. The capacitor appears like an open circuit with all of the input signal voltage being applied to the op amp's positive terminal.

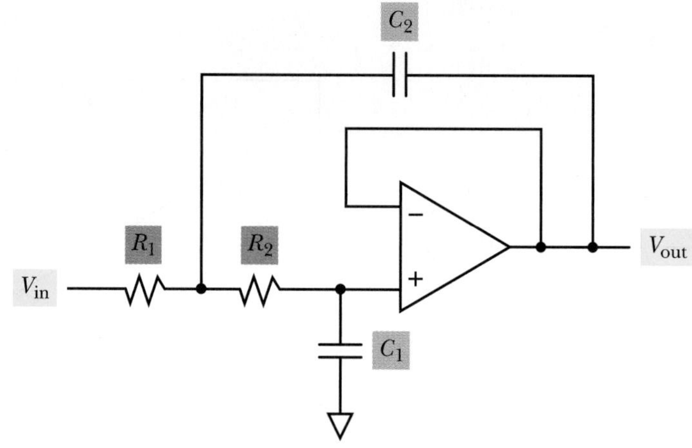

FIGURE 9–9 Second order low-pass filter with unity gain

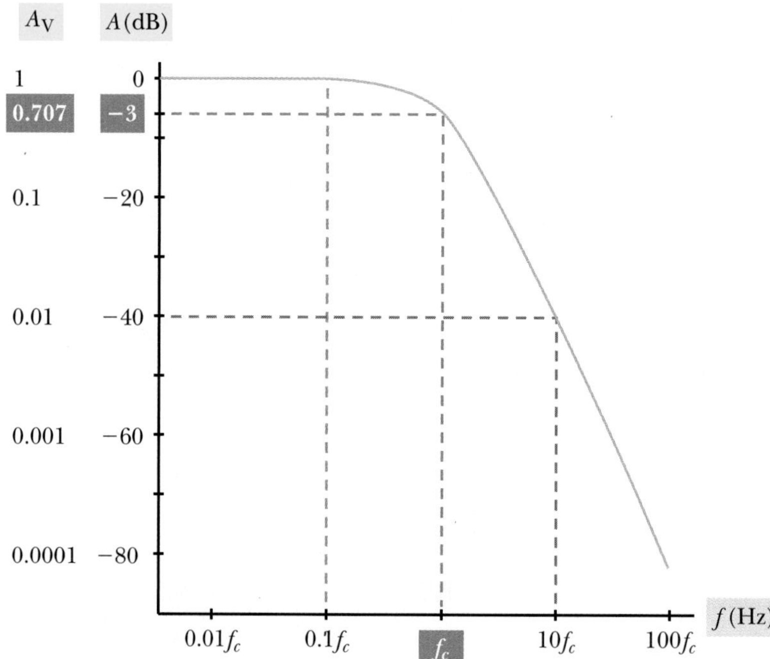

FIGURE 9–10 Second order low-pass filter with unity gain response

When the signal frequency increases, the capacitor reactances of C_1 and C_2 decrease resulting in a decrease of the voltage applied to the op amp's positive terminal. At the cutoff frequency, the voltage at the positive op amp terminal would be –6 dB as the second order low pass filter is comprised of two filter elements. R_1 and C_2 are the first filter with R_2 and C_1 being the second filter. The connection of the capacitor C_2 to provide positive feedback from the op amp output to the op amp positive input terminal results in an increase of the voltage applied to the positive op amp terminal. At the cutoff frequency, the portion of the output signal that combines with the input signal will result in a –3 dB signal ($V_{out} = 0.707 \times V_{in}$) applied to the op amp's positive terminal. The output voltage will be identical to the positive op amp terminal voltage since the op amp has been configured with unity gain amplification. The increasing of the signal frequency caused the filter circuit voltage gain to decrease. When the signal frequency is at 10 times the cutoff frequency, the filter voltage gain is –40 dB ($V_{out} = 0.01 \times V_{in}$). At very high frequencies, the capacitor's reactance appears like a short circuit with 0 V being applied to the op amp's positive terminal. The low-pass filter's pass range is defined as all signal frequencies that are below the filter's cutoff frequency. The low-pass filter's block range exists for all signal frequencies that are above the filter's cutoff frequency. The Bode plot showing the filter's voltage gain as a function of frequency is provided in Figure 9–10.

Application Problem

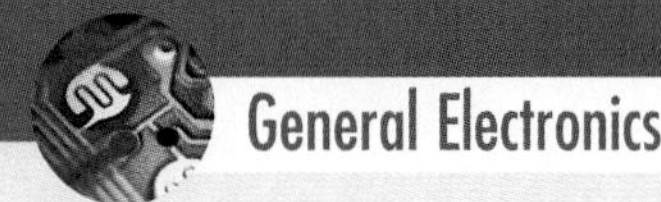

HIGH FREQUENCY REJECTION CIRCUIT IN AN OSCILLOSCOPE

An oscilloscope uses a low-pass active filter to reject high frequencies when a user selects that function. What is the cutoff frequency of this filter shown in the schematic?

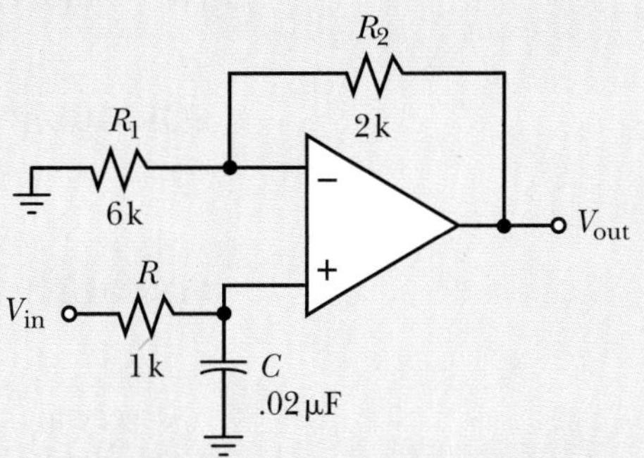

Solution

$$f_c = \frac{1}{[2\pi \times (1\text{ k})(.02\ \mu\text{F})]} = 7.96\text{ kHz}$$

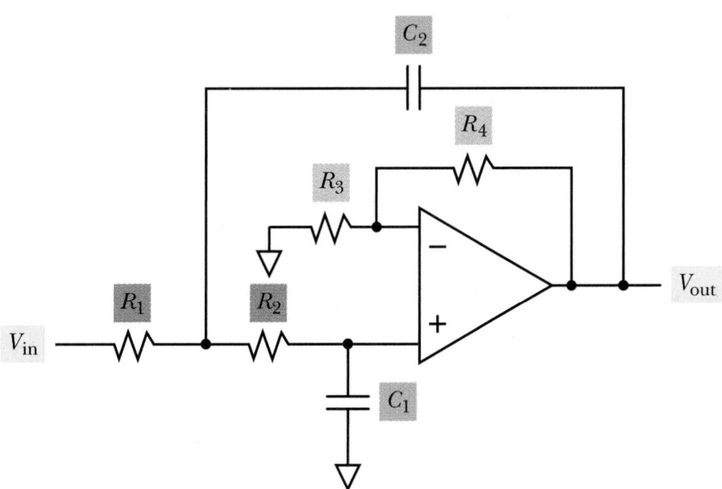

FIGURE 9–11 Second order low-pass filter

The low-pass filter shown in Figure 9–11 is a second order filter with gain. For proper operation with a Butterworth filter characteristic, the following component selections must be maintained: $R_2 = R_1$ and $C_2 = C_1$ while $A_v = 1.586$. The equation to calculate the cutoff frequency is provided in Formula 9–11. Resistors R_4 and R_3 establish the voltage gain within the passband using Formula 9–12. This gain also affects the circuit's gain beyond the cutoff frequency.

FORMULA 9–12 $A_V = \left(1 + \frac{R_4}{R_3}\right)$

IN-PROCESS LEARNING CHECK 1

1. For a first order low-pass filter, the _________ band is less than the cutoff frequency and the _________ band is greater than the cutoff frequency.
2. At the cutoff frequency, the voltage gain for the unity gain first order low-pass filter is _________.
3. The order of a filter is determined by the number of _________ in the circuit.
4. For a second order low-pass filter, the gain-frequency slope above the cutoff frequency is _________.

9–3 High-Pass Filter Circuits

The **high-pass filter** shown in Figure 9–12 is a first order filter with unity gain. The equation to calculate the cutoff frequency is provided in Formula 9–13, which is the same equation to calculate the cutoff frequency for the first order low-pass filter. The output voltage is frequency dependent and can be calculated using Formula 9–14. The Bode plot for the unity gain high-pass filter is shown in Figure 9–13.

FORMULA 9–13 $f_c = \dfrac{1}{2\pi RC}$

FORMULA 9–14 $V_{out} = V_{in} \times \dfrac{R}{R - jX_C}$

At very high frequencies, the output voltage will be at its maximum voltage based on the closed-loop gain of the op-amp circuit (+1). At these high frequencies, the capacitor's reactance is much smaller than the resistor *(R)* value. The capacitor appears like a short circuit with all of the input signal voltage being applied to the op-amp's positive terminal. As the signal frequency decreases, the capacitor's reactance increases resulting in a decrease of the voltage applied to the op-amp's positive terminal. The decreasing of the signal frequency caused the filter circuit voltage gain to decrease. At the cutoff frequency, the filter voltage gain is 3 dB ($V_{out} = 0.707 \times V_{in}$). When the signal frequency is at 0.1 times the cutoff frequency, the filter voltage gain is –20 dB ($V_{out} = 0.1 \times V_{in}$). At very low frequencies, the capacitor's reactance appears like an open circuit with 0 V being applied to the op-amp's positive terminal. The high-pass filter's pass range is defined as all signal frequencies that are greater than the filter's cutoff frequency. The high-pass filter's block range exists for all signal frequencies that are less than the filter's cutoff frequency.

EXAMPLE The high-pass filter in Figure 9–12 has resistor value $R = 2.0\ \text{k}\Omega$, and capacitor value $C = 0.022\ \mu\text{F}$. Calculate the cutoff frequency and the output voltage amplitude when the input waveform is a 1 V_{RMS} sine wave having a frequency equal to the cutoff frequency.

Answer:

From Formula 9–13

$$f_c = \frac{1}{2\pi RC}$$

$$f_c = \frac{1}{2\pi \times 2000\ \Omega \times 0.022\ \mu\text{F}}$$

$$f_c = 3.617\ \text{kHz}$$

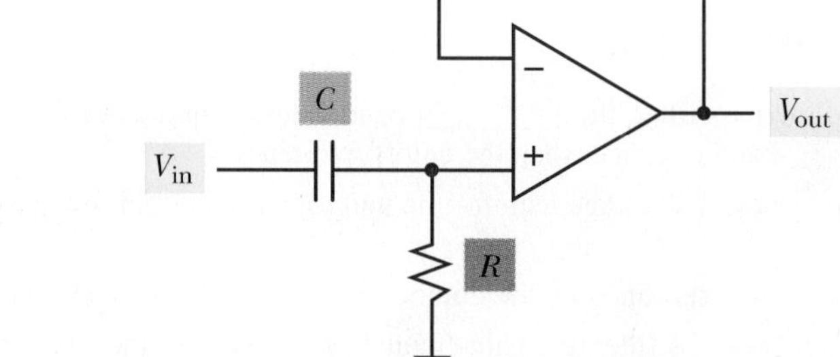

FIGURE 9–12 First order high-pass filter with unity gain

To calculate the capacitive reactance, use

$$X_c = \frac{1}{2\pi f C}$$

When $f_{in} = f_c = 3.617$ kHz,

$$X_c = \frac{1}{2\pi \times 3.617 \text{ kHz} \times 0.022\ \mu\text{F}}$$
$$X_c = 2000\ \Omega$$

FORMULA 9–15 $$V_{out} = V_{in} \times \frac{R}{R - jX_C}$$

$$V_{out} = 1\ V_{rms} \times \left(\frac{2000}{\sqrt{(2000)^2 + (2000)^2}} \right)$$

$$V_{out} = 0.707\ V_{rms}$$

The high-pass filter shown in Figure 9–14 is a first order filter with gain. The equation to calculate the cutoff frequency is provided in Formula 9–13. The output voltage is frequency dependent and can be calculated using Formula 9–16. For very high frequencies, the output voltage will be at its maximum based on the closed-loop gain of the op-amp circuit ($1 + R_2/R_1$). The output voltage of the high-pass filter with gain circuit will decrease as the applied signal frequency is decreased. The filter's pass range is all frequencies that are greater than the filter's cutoff frequency. The filter's block range is all frequencies that are less than the filter's cutoff frequency.

FORMULA 9–16 $$V_{out} = V_{in} \times \left(\frac{R}{R - jX_C} \right) \times \left(1 + \frac{R_2}{R_1} \right)$$

$$A_V = \frac{V_{out}}{V_{in}} = \frac{R}{\sqrt{R^2 + X_C^2}} \times \left(1 + \frac{R_2}{R_1} \right)$$

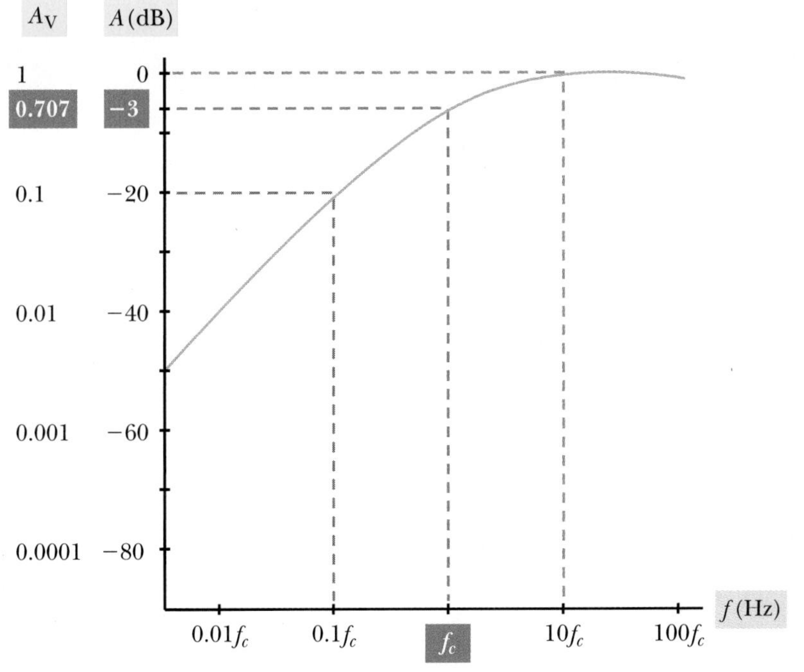

FIGURE 9–13 First order high-pass filter with unity gain response

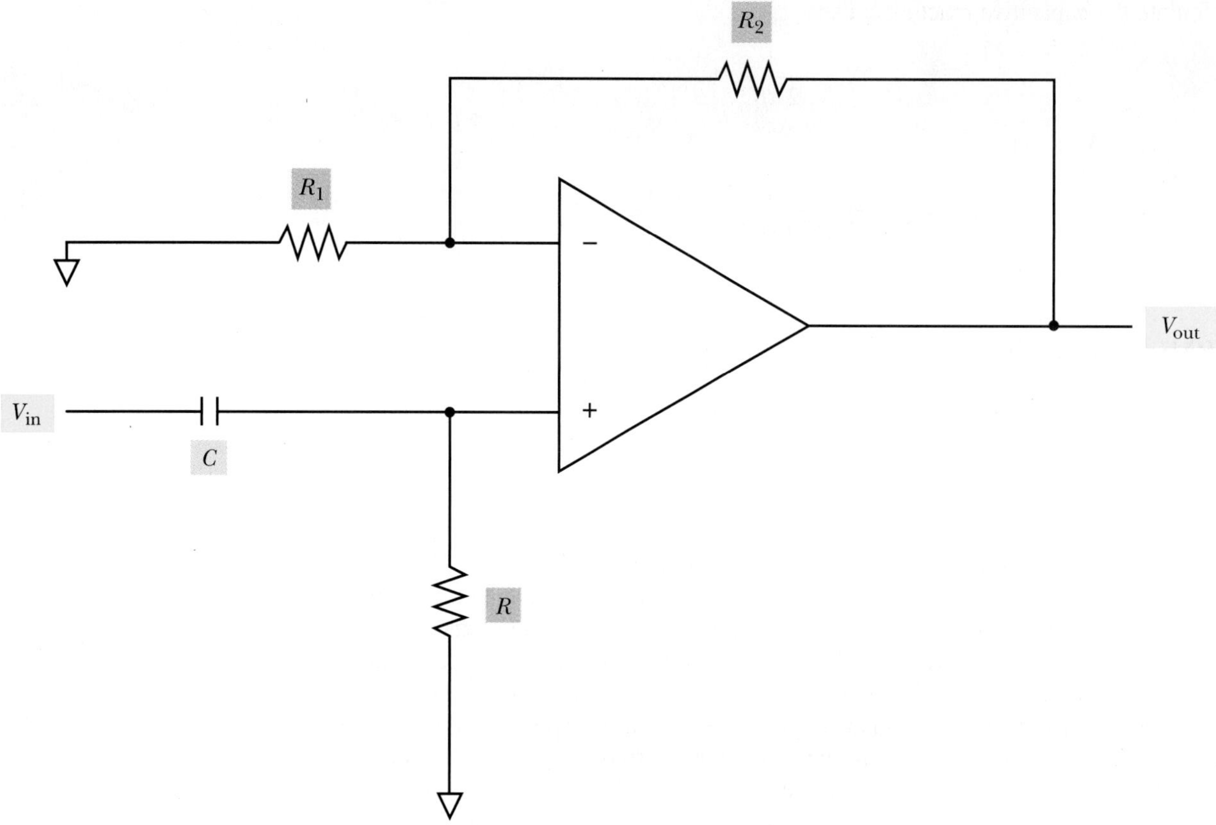

FIGURE 9–14 First order active high-pass filter

◆ **EXAMPLE** The high-pass filter in Figure 9–14 has resistor values R = 1.0 kΩ, R_1 = 10 kΩ, R_2 = 10 kΩ, and capacitor value C = 0.001 μF. Calculate the cutoff frequency and the output voltage amplitude when the input waveform is a 1 V_{rms} sine wave having a frequency equal to the cutoff frequency.

Answer:

From Formula 9–13

$$f_c = \frac{1}{2\pi RC}$$

$$f_c = \frac{1}{2\pi \times 1000\ \Omega \times 0.001\ \mu\text{F}}$$

$$f_c = 159.15\ \text{kHz}$$

To calculate the capacitive reactance, use

$$X_C = \frac{1}{2\pi fC}$$

When $f_{in} = f_c$ = 159.15 kHz,

$$X_C = \frac{1}{2\pi \times 159.15\ \text{kHz} \times 0.001 \mu\text{F}}$$

$$X_C = 1000\ \Omega$$

Using Formula 9–16, the output voltage can be calculated.

$$V_{out} = V_{in} \times \left(\frac{R}{R - jX_C} \right) \times \left(1 + \frac{R_2}{R_1} \right)$$

$$V_{out} = 1\ V_{rms} \times \left(\frac{1000}{\sqrt{(1000)^2 + (1000)^2}} \right) \times \left(1 + \frac{10000}{10000} \right)$$

$$V_{out} = 1.414\ V_{rms}$$

◆

◆ **EXAMPLE** The high-pass filter in Figure 9–14 has resistor values $R = 10\ \text{k}\Omega$, $R_1 = 10\ \text{k}\Omega$, $R_2 = 10\ \text{k}\Omega$, and capacitor value $C = 0.0022\ \mu\text{F}$. Calculate the cutoff frequency and the output voltage amplitude when the input waveform is a 1 V_{rms} sine wave having a frequency equal to the cutoff frequency.

Answer:

From Formula 9–13

$$f_c = \frac{1}{2\pi RC}$$

$$f_c = \frac{1}{2\pi \times 10\ \text{k}\Omega \times 0.0022\ \mu\text{F}}$$

$$f_c = 7.234\ \text{kHz}$$

When $f_{in} = f_c = 7.234$ kHz,

$$X_c = \frac{1}{2\pi fC}$$

$$X_c = \frac{1}{2\pi \times 7.234\ \text{kHz} \times 0.0022\ \mu\text{F}}$$

$$X_c = 10000\ \Omega$$

From Formula 28–25

$$V_{out} = V_{in} \times \left(\frac{R}{R - jX_C} \right) \times \left(1 + \frac{R_2}{R_1} \right)$$

$$V_{out} = 1\ V_{rms} \times \left(\frac{10000}{\sqrt{(10000)^2 + (10000)^2}} \right) \times \left(1 + \frac{10000}{10000} \right)$$

$$V_{out} = 1.414\ V_{rms}$$

◆

The high-pass filter shown in Figure 9–15 is a second order filter with unity gain. The filter type is known as a Sallen-Key high-pass filter as well as a Voltage Controlled Voltage Source (VCVS) filter. For proper operation with a Butterworth filter characteristic, the following component selections must be maintained: $R_2 = 2 \times R_1$ and $C_2 = C_1$ while $A_V = 1.0$. The equation to calculate the cutoff frequency is provided in Formula 9–17.

FORMULA 9–17 $$f_c = \frac{1}{2\pi\sqrt{R_1 R_2 C_1 C_2}}$$

Two capacitors (C_1 and C_2) exist within the second order high-pass filter. At very high frequencies, the output voltage will be at its maximum voltage based on the closed-loop gain of the op-amp circuit (+ 1). At these high frequencies, the capacitors' reactance is very small. The capacitor appears like a short circuit with all of the input signal voltage being applied to the op-amp's positive terminal.

Application Problem

TONE CONTROL IN A STEREO RECEIVER

An audio amplifier in a stereo receiver is using two active filters to adjust the bass and treble. Determine the filter type and the critical frequency of each filter from the figure.

Solution For the high-pass filter,

$$f_c = \frac{1}{[2\pi(1.59\ \text{k})(0.01\ \mu\text{F})]} = 10\ \text{kHz}$$

For the low-pass filter,

$$f_c = \frac{1}{[2\pi(1\ \text{k})(.01\ \mu\text{F})]} = 15.9\ \text{kHz}$$

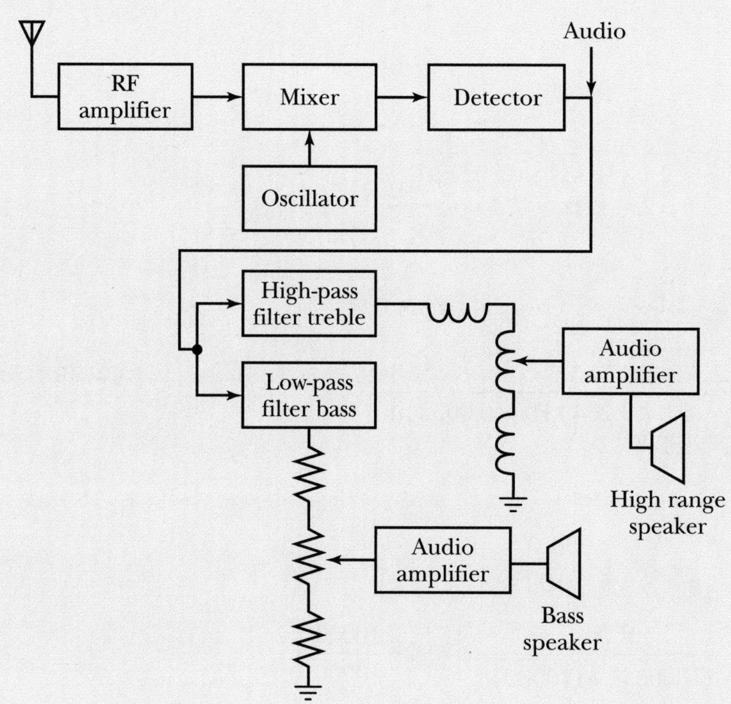

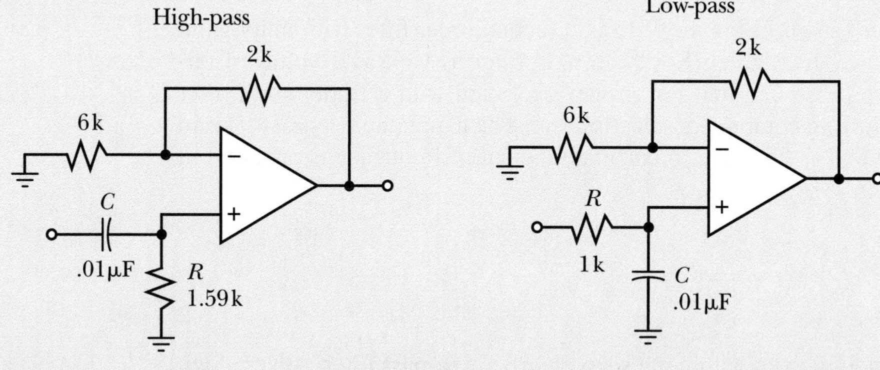

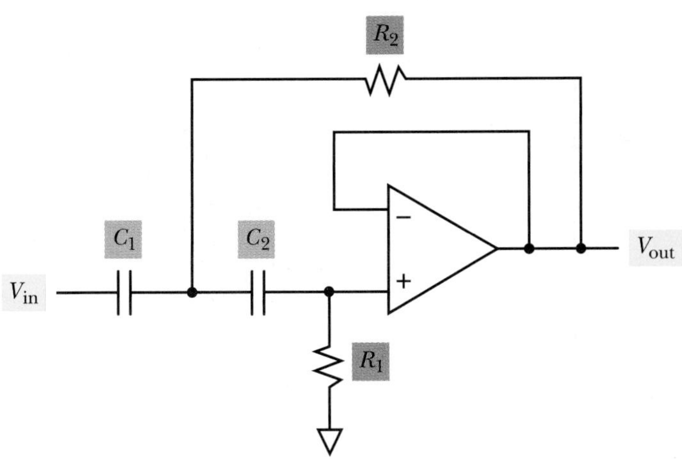

FIGURE 9–15 Second order high-pass filter with unity gain

When the signal frequency decreases, the capacitor reactances C_1 and C_2 increase resulting in a decrease of the voltage applied to the op-amp's positive terminal. At the cutoff frequency, the voltage at the positive op-amp terminal would be –6 dB as the second order high pass filter is comprised of two filter elements. C_1 and R_2 are the first filter with C_2 and R_1 being the second filter. The connection of the resistor R_2 to provide positive feedback from the op-amp output to the op-amp positive input terminal results in an increase of the voltage applied to the positive op-amp terminal. At the cutoff frequency, the portion of the output signal that combines with the input signal will result in a –3 dB signal ($V_{out} = 0.707 \times V_{in}$) applied to the op amp's positive terminal. The output voltage will be identical to the positive op-amp terminal voltage since the op-amp has been configured with unity gain amplification. The decreasing of the signal frequency caused the filter circuit voltage gain to decrease. When the signal frequency is at 0.1 times the cutoff frequency, the filter voltage gain is –40 dB ($V_{out} = 0.01 \times V_{in}$). At very low frequencies, the capacitor's reactance appears like an open circuit with 0 V being applied to the op amp's positive terminal. The high pass filter's pass range is defined as all signal frequencies that are above the filter's cutoff frequency. The low pass filter's block range exists for all signal frequencies that are below the filter's cutoff frequency.

The Bode plot showing the filter's voltage gain as a function of frequency is provided in Figure 9–16. The high-pass filter shown in Figure 9–17 is a second order filter with gain. Resistors R_4 and R_3 establish the voltage gain within the passband using Formula 9–12. This gain also affects the circuit's gain below the cutoff frequency.

Practical Notes

The unity gain bandwidth product for an ideal operational amplifier is infinite. However, the operational amplifiers have a finite unity gain bandwidth product. As such the unity gain bandwidth product causes the operational amplifier to operate as a low-pass filter. With a low-pass filter circuit, you control the cutoff frequency. For a high-pass filter circuit using the operational amplifier, you actually have a bandpass filter due to the operational amplifier characteristics. The passband is defined by your designed cutoff frequency (which must be less than the op-amp's unity gain bandwidth product) and the unity gain bandwidth product characteristic of the op-amp.

■ IN-PROCESS LEARNING CHECK 2

1. For a first order high-pass filter, the _________ band is all frequencies less than the cutoff frequency and the _________ band is all frequencies greater than the cutoff frequency.
2. At the cutoff frequency, the voltage gain for the unity gain first order high-pass filter is _________.
3. The second order filter is also known as _________________ filter.
4. For a second order high-pass filter, the gain-frequency slope above the cutoff frequency is _________.

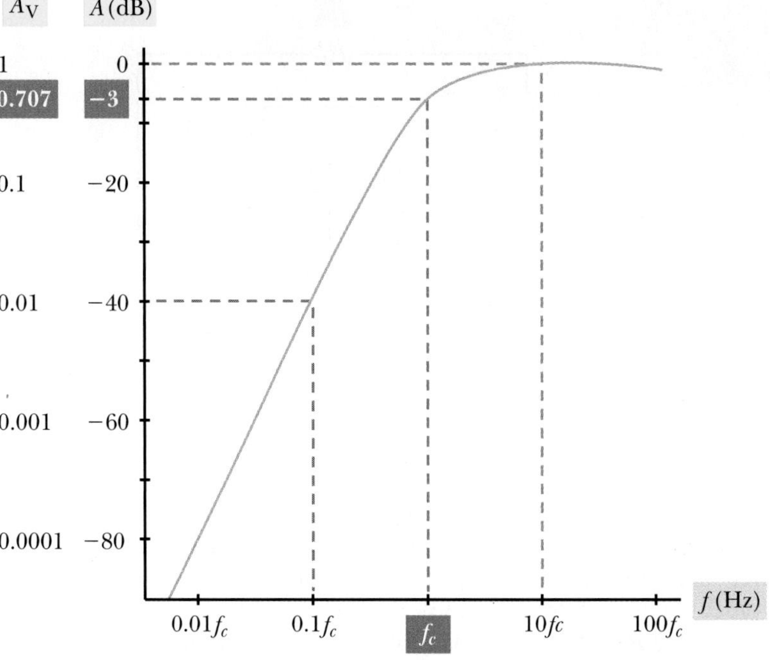

FIGURE 9–16 Second order high-pass filter response

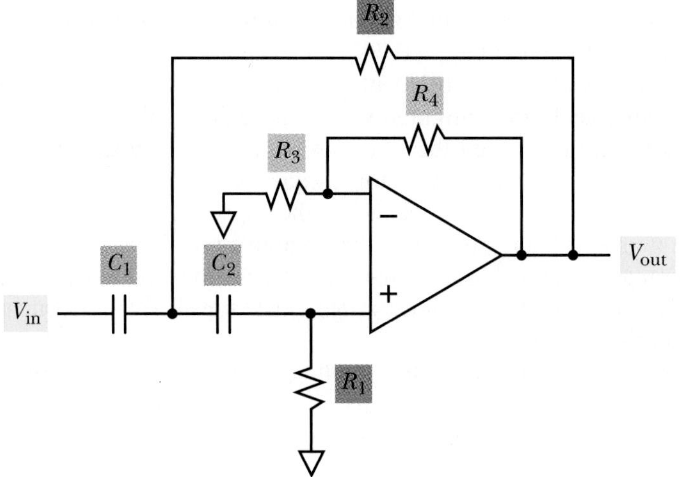

FIGURE 9–17 Second order high-pass filter

9–4 Bandpass Filter Circuits

The **bandpass filter** is shown in Figure 9–18. The maximum output voltage occurs within the **pass-band** of the bandpass filter circuit. The filter response is provided in Figure 9–19. The center frequency is calculated using Formula 9–18. The range of frequencies passed is the **bandwidth,** which is calculated using Formula 9–19. The quality factor Q of the circuit is the ratio of the energy storage and the energy dissipated. After calculating the center frequency and the bandwidth, the low cutoff frequency and the upper cutoff frequency can be calculated using Formulas 9–20 and 9–21 respectively. The voltage inputs having frequency components below the lower cutoff frequency are blocked from reaching the output and voltage inputs having frequency components above the upper cutoff frequency are blocked from reaching the output.

FORMULA 9–18

$$f_o = \frac{1}{2\pi\sqrt{\left(\frac{R_1\,R_2}{R_1 + R_2}\right)R_3\,C_1\,C_2}}$$

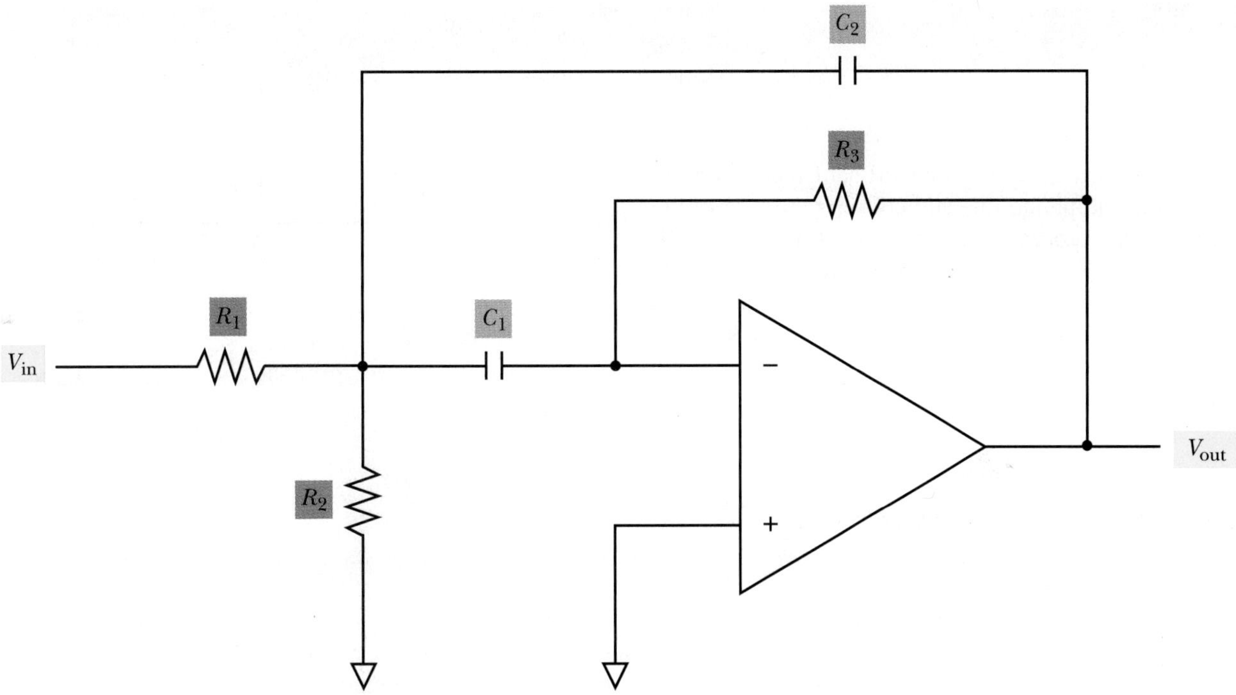

FIGURE 9–18 Bandpass filter

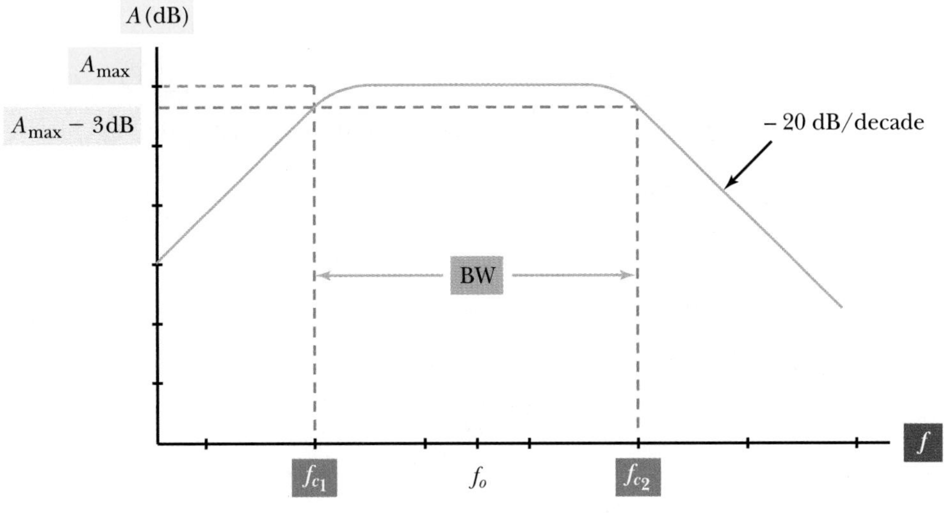

FIGURE 9–19 Bandpass filter frequency response

FORMULA 9–19 $BW = \frac{f_o}{Q}$ where $Q = \pi f_o R_3 \sqrt{C_1 C_2}$

FORMULA 9–20 $f_{C_1} = f_o - \frac{BW}{2}$

FORMULA 9–21 $f_{C_2} = f_o + \frac{BW}{2}$

Application Problem

MODEM

A modem on a computer motherboard receives audio tones from the phone line and converts them to digital codes that the processor can understand. We need to filter out any unwanted frequencies that could cause errors in transmission. Observe the filter below and determine the Q of the filter if the center frequency is 10 kHz.

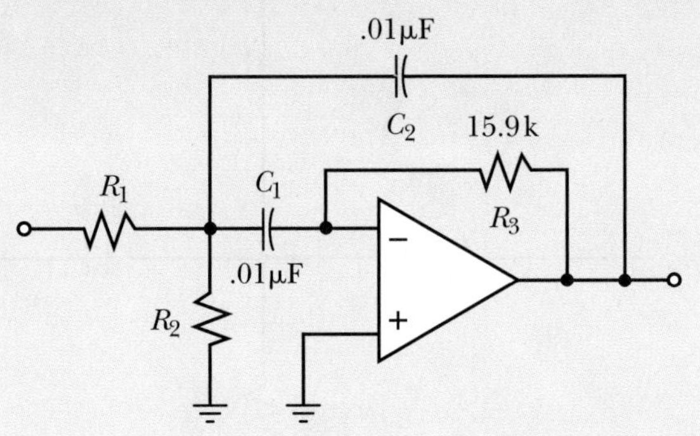

Solution

$$Q = \pi f_o R_3 \sqrt{C_1 C_2}$$

Substituting the values from the schematic, the $Q = 5$.

◆ **EXAMPLE** The bandpass filter in Figure 9–18 has resistor values $R_1 = 1\ \text{k}\Omega$, $R_2 = 1\ \text{k}\Omega$, $R_3 = 20\ \text{k}\Omega$, and capacitor values $C_1 = 0.01\ \mu\text{F}$, $C_2 = 0.0047\ \mu\text{F}$. Calculate the center frequency, bandwidth, low cutoff frequency, and the high cutoff frequency.

Answer:

From Formula 9–18

$$f_o = \frac{1}{2\pi\sqrt{\left(\dfrac{R_1 R_2}{R_1 + R_2}\right) R_3 C_1 C_2}}$$

$$f_o = \frac{1}{2\pi\sqrt{\left(\dfrac{1000 \times 1000}{1000 + 1000}\right) \times 20000 \times 0.01\ \mu\text{F} \times 0.0047\ \mu\text{F}}}$$

$$f_o = 7.341\ \text{kHz}$$

$$Q = \pi\ f_o\ R_3 \sqrt{C_1 C_2}$$

$$Q = \pi\ 7.341\ \text{kHz} \times 20\ \text{k}\Omega \times \sqrt{0.01\ \mu\text{F} \times 0.0047\ \mu\text{F}}$$

$$Q = 3.16$$

From Formula 9–19

$$\text{BW} = \frac{f_o}{Q}$$

$$\text{BW} = \frac{7.341\ \text{kHz}}{3.16}$$

$$\text{BW} = 2.322\ \text{kHz}$$

From Formula 9–20

$$f_{C_1} = f_o - \frac{\textbf{BW}}{2}$$

$$f_{C_1} = 7.341\ \textbf{kHz} - \frac{2.322\ \textbf{kHz}}{2}$$

$$f_{C_1} = 6.180\ \textbf{kHz}$$

From Formula 9–21

$$f_{C_2} = f_o + \frac{\text{BW}}{2}$$

$$f_{C_2} = 7.341\ \text{kHz} + \frac{2.322\ \text{kHz}}{2}$$

$$f_{C_2} = 8.502\ \text{kHz}$$

◆

9–5 Bandstop Filter Circuits

A **bandstop filter** circuit is shown in Figure 9–20. The bandstop filter is also known as a **notch filter,** a **band reject filter,** and a band block filter. The bandstop filter circuit of Figure 9–20 consists of the bandpass filter circuit combined with a noninverting input of the original signal. For the range of frequencies where the "bandpass" inverted signal is at its maximum, the output voltage will be blocked. The bandstop filter response is shown in Figure 9–21. For all other frequencies the output voltage comes from the input signal through the noninverting input. The center frequency of the bandstop filter is calculated using Formula 9–22.

FORMULA 9–22 $f_o = \dfrac{1}{2\pi\sqrt{R_1\ R_2\ C_1\ C_2}}$

EXAMPLE The bandstop filter in Figure 9–20 has resistor values $R_1 = 1\ k\Omega$, $R_2 = 20\ k\Omega$, $R_3 = 1\ k\Omega$, and $R_4 = 20\ k\Omega$, and capacitor values $C_1 = 0.01\ \mu F$, $C_2 = 0.01\ \mu F$. Find the center frequency.

Answer:

From Formula 9–22

$$f_o = \frac{1}{2\pi\sqrt{R_1 R_2 C_1 C_2}}$$

$$f_o = \frac{1}{2\pi\sqrt{1000\times 20000\times 0.01\ \mu F\times 0.01\ \mu F}}$$

$$f_o = 3.559\ kHz$$

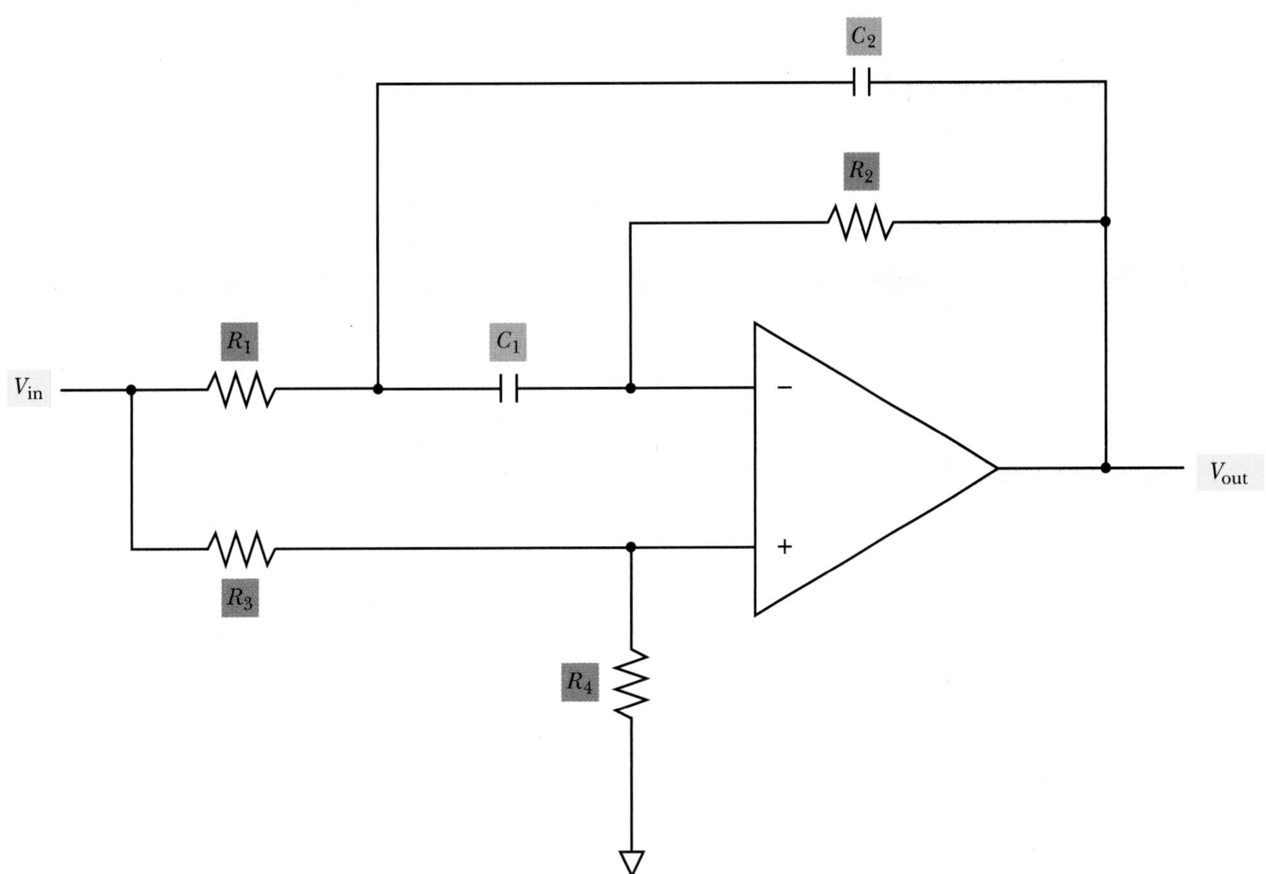

FIGURE 9–20 Bandstop filter

Application Problem

Industrial

ROBOTIC ARM CONTROL

A radio transmitter controls a robotic arm. An audio tone generated in the transmitter represents a specific movement. There are 5 separate tones. The receiver in the robot's control system must be able to select the tone to perform the correct function. The receiver has 5 active bandpass filters. The filter for the upward movement of the arm is shown in the schematic. What is the center frequency of the filter?

Solution:

$$f_o = \frac{1}{2\pi\sqrt{\dfrac{R_1 R_2 R_3 C_1 C_2}{(R_1 + R_2)}}}$$

Substituting the values from the schematic, $f_o = 358$ Hz

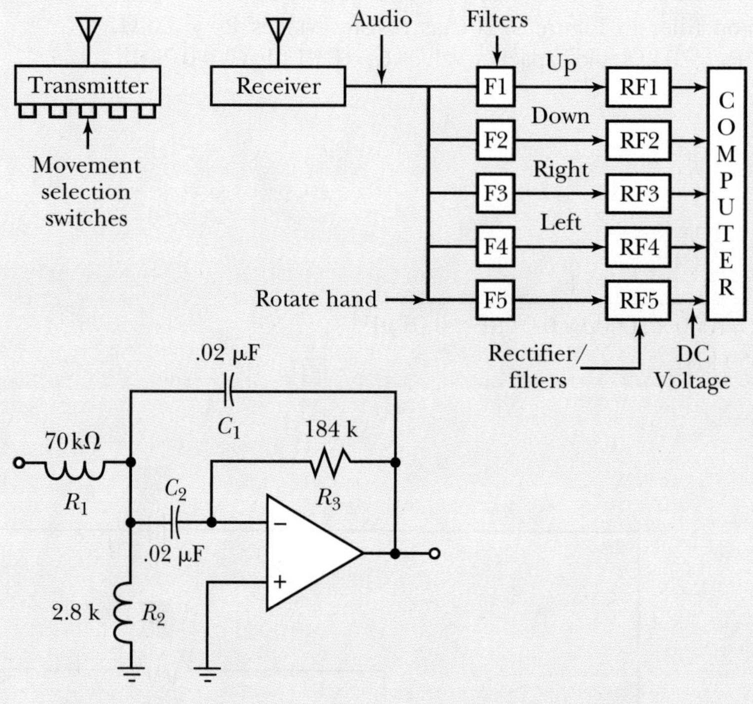

FIGURE 9–21 Notch filter frequency response

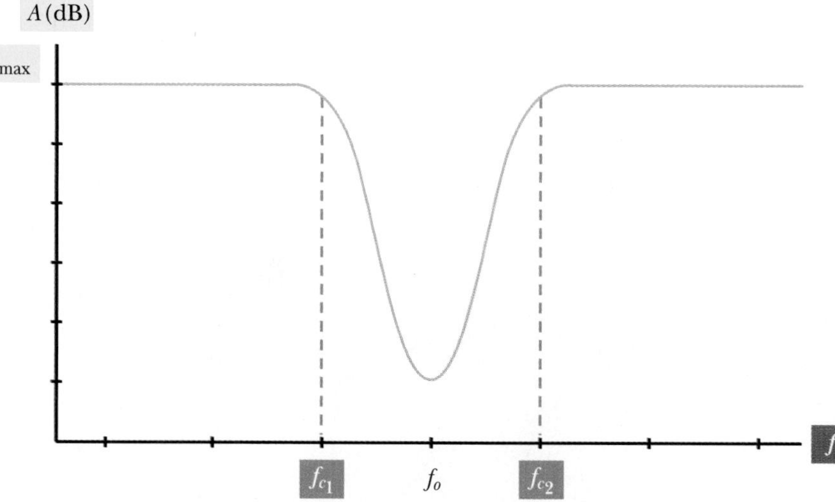

Summary

- Filters are an important element of electronic systems. An active filter is a circuit using capacitors and/or inductors combined with an active component (transistor or op-amp) and resistors. The purpose of the filter is to pass a specified range of frequencies to the output and block (prevent the passing) of all other frequencies from reaching the output. A large selection of filter circuits exists. For our discussion, the filters will be classified into four categories: low-pass, high-pass, bandpass, and bandstop.
- All filters have a cutoff frequency. The bandpass and bandstop filters have a lower cutoff frequency and an upper cutoff frequency. The cutoff frequency is the frequency at which the reactance of the circuit is equal to the circuit resistance. For first order systems (one reactive component within the circuit), the cutoff frequency occurs at the half-power point, also known as the 3 dB point, or when the output voltage at this frequency equals 0.707 times the maximum output voltage.

CIRCUIT SUMMARY

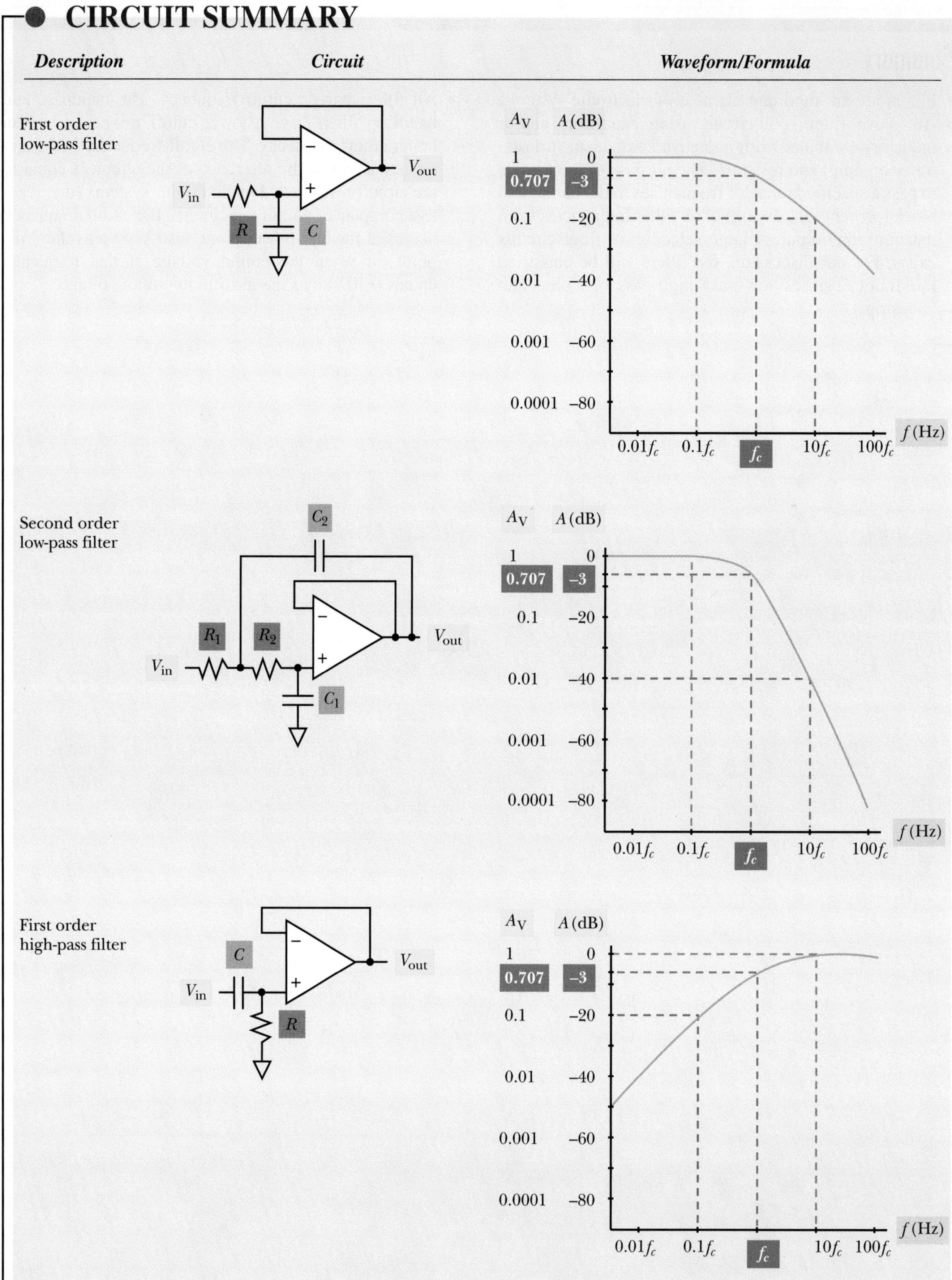

Filter Circuit Summary

CIRCUIT SUMMARY

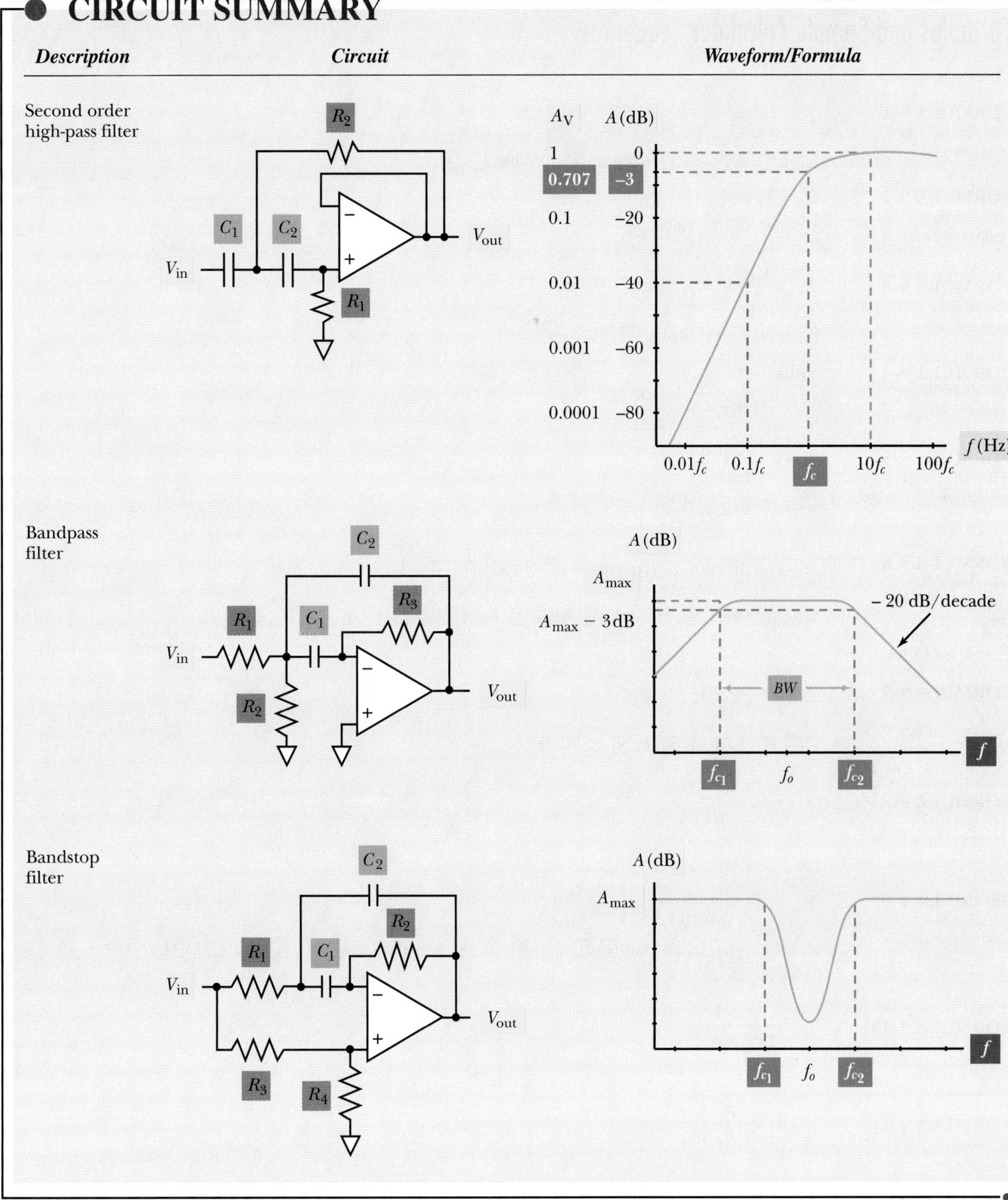

Formulas and Sample Calculator Sequences

FORMULA 9–1
(To find the cutoff frequency)

$$f_c = \frac{1}{2\pi RC}$$

1, ÷, (, 2, ×, π, ×, R value, ×, C value,), =

FORMULA 9–2
(To find the output voltage)

$$V_{out} = A_V \times V_{in}$$

A_V value, ×, V_{in} value, =

FORMULA 9–3
(To find the voltage gain)

$$A_V = \frac{V_{out}}{V_{in}}$$

V_{out} value, ÷, V_{in} value, =

FORMULA 9–4
(To find voltage gain in dB)

$$A\ (\text{dB}) = 20 \log (A_V)$$

A_V value, log, ×, 20, =

FORMULA 9–5
(To find voltage gain slope)

$$\text{m} = \text{n} \times \frac{-20\,\text{dB}}{\text{decade}}$$

n value, ±, ×, 20, =

FORMULA 9–6
(To find voltage gain in dB at the frequency f_{SIGNAL})

$$A_{SIGNAL}\,(\text{dB}) = A_{PASSBAND}\,(\text{dB}) + \text{m} \times \left|\text{LOG}\left(\frac{f_{SIGNAL}}{f_{CUTOFF}}\right)\right|$$

$A_{PASSBAND}$ value, +, m value, ×, ABS, (, log, (, f_{SIGNAL}, ÷, f_{CUTOFF},),), =

Low-Pass Filter

FORMULA 9–7
(To find the low-pass filter cutoff frequency)

$$f_c = \frac{1}{2\pi RC}$$

1, ÷, (, 2, ×, π, ×, R, ×, C,), =

FORMULA 9–8
(To find the output voltage)

$$V_{out} = V_{in} \times \frac{-\text{j}X_C}{R - \text{j}X_C}$$

FORMULA 9–9
(To find the output voltage)

$$V_{out} = V_{in} \times \left(\frac{-\text{j}X_C}{R - \text{j}X_C}\right) \times \left(1 + \frac{R_2}{R_1}\right)$$

V_{in} value, ×, (, X_C value, ÷, (R value, x^2, +, X_C value, x^2,), $\sqrt{x}$,), ×, (, 1, +, R_2 value, ÷, R_1 value,), =

FORMULA 9–10
(To find the cutoff frequency)

$$f_c = \frac{1}{2\pi\sqrt{R_1 R_2 C_1 C_2}}$$

FORMULA 9–11
(To find the voltage gain)

$$A_V = \left(1 + \frac{R_4}{R_3}\right)$$

High-Pass Filter

FORMULA 9–12
(To find the high-pass filter cutoff frequency)

$$f_c = \frac{1}{2\pi RC}$$

1, ÷, (, 2, ×, π, ×, R, ×, C,), =

FORMULA 9–13
(To find the output voltage)

$$V_{out} = V_{in} \times \frac{R}{R - jX_C}$$

V_{in} value, ⊠, ⊏(⊐, R value, ⊟÷, ⊏(⊐, R value, x^2, ⊞, X_c value, x^2, ⊏)⊐, $\sqrt{x}$, ⊏)⊐, ⊟=

FORMULA 9–14
(To find the output voltage)

$$V_{out} = V_{in} \times \left(\frac{R}{R - jX_C}\right) \times \left(1 + \frac{R_2}{R_1}\right)$$

V_{in} value, ⊠, ⊏(⊐, R value, ⊟÷, ⊏(⊐, R value, x^2, ⊞, X_C value, x^2, ⊏)⊐, $\sqrt{x}$, ⊏)⊐, ⊠, ⊏(⊐, 1 , ⊞, R_2 value, ⊟÷, R_1 value, ⊏)⊐, ⊟=

FORMULA 9–15
(To find the cutoff frequency)

$$f_c = \frac{1}{2\pi\sqrt{R_1 R_2 C_1 C_2}}$$

Bandpass Filter

FORMULA 9–16
(To find the filter center frequency)

$$f_o = \frac{1}{2\pi\sqrt{\left(\frac{R_1\ R_2}{R_1 + R_2}\right)R_3 C_1 C_2}}$$

1, ⊟÷, ⊏(⊐, 2, ⊠, π, ⊠, ⊏(⊐, ⊏(⊐, R_1 value, ⊠, R_2 value, ⊟÷, ⊏(⊐, R_1 value, ⊞, R_2 value, ⊏)⊐, ⊏)⊐, ⊠, R_3 value, ⊠, C_1 value, ⊠, C_2 value, ⊏)⊐, $\sqrt{x}$, ⊏)⊐, ⊟=

FORMULA 9–17
(To find the bandpass filter bandwidth)

$$\text{BW} = \frac{f_o}{Q}$$

where $Q = \pi\ f_o\ R_3\sqrt{C_1 C_2}$

f_o value, ⊟÷, ⊏(⊐, π, ⊠, f_o value, ⊠, R_3 value, ⊠, ⊏(⊐, ⊏(⊐, C_1 value, ⊠, C_2 value, ⊏)⊐, $\sqrt{x}$, ⊏)⊐, ⊟=

FORMULA 9–18
(To find the filter lower cutoff frequency)

$$f_{C_1} = f_o - \frac{\text{BW}}{2}$$

f_o value, ⊟, BW value, ⊟÷, 2, ⊟=

FORMULA 9–19
(To find the filter upper cutoff frequency)

$$f_{C_2} = f_o + \frac{\text{BW}}{2}$$

f_o value, ⊞, BW value, ⊟÷, 2, ⊟=

Bandstop Filter

FORMULA 9–20
(To find the center frequency)

$$f_o = \frac{1}{2\pi\sqrt{R_1 R_2 C_1 C_2}}$$

1, ⊟÷, ⊏(⊐, 2, ⊠, π, ⊠, ⊏(⊐, ⊏(⊐, R_1 value, ⊠, R_2 value, ⊠, C_1 value, ⊠, C_2 value, ⊏)⊐, $\sqrt{x}$, ⊏)⊐, ⊟=

Review Questions

1. The bode plot is a graph of voltage amplitude versus frequency (True or False).
2. A low-pass filter has a pass-band greater than the cutoff frequency and a stop band less than the cutoff frequency (True or False).
3. A ______________ filter has a pass-band greater than the cutoff frequency and a stopband less than the cutoff frequency.
4. A ______________ filter has a pass-band less than the cutoff frequency and a stopband greater than the cutoff frequency.
5. The voltage gain of a unity gain low-pass filter at the cutoff frequency is 0.707 (True or False).
6. The voltage gain of a unity gain high-pass filter at the cutoff frequency is –3 dB (True or False).
7. A second order filter in the stopband has a voltage gain decrease of –40 dB/decade (True or False).
8. A first order filter in the stopband has a voltage gain decrease of –40 dB/decade (True or False).
9. The bandpass filter and notch filter are the same type of filter (True or False).

Problems

1. Calculate the slope of a second order low-pass filter having a cutoff frequency of 10 kHz. If the voltage gain for this filter circuit is 0 dB at 100 Hz, what is the filter voltage gain at 100 kHz? With an input voltage of 10 V_{rms}, what would be the output voltage?
2. The low-pass filter in Figure 9–6 has resistor value $R = 10.0\ \text{k}\Omega$, and capacitor value $C = 0.0047\ \mu\text{F}$. Calculate the cutoff frequency and the output voltage amplitude when the input waveform is a 1 V_{rms} sine wave having a frequency equal to the cutoff frequency.
3. The low-pass filter in Figure 9–8 has resistor values $R = 1.0\ \text{k}\Omega$, $R_1 = 10\ \text{k}\Omega$, $R_2 = 10\ \text{k}\Omega$, and capacitor value $C = 0.22\ \mu\text{F}$. Calculate the cutoff frequency and the output voltage amplitude when the input waveform is a 1 V_{rms} sine wave having a frequency equal to the cutoff frequency.
4. The high-pass filter in Figure 9–12 has resistor value $R = 4.7\ \text{k}\Omega$, and capacitor value $C = 0.022\ \mu\text{F}$. Calculate the cutoff frequency and the output voltage amplitude when the input waveform is a 1 V_{rms} sine wave having a frequency equal to the cutoff frequency.
5. The high-pass filter in Figure 9–14 has resistor values $R = 2\ \text{k}\Omega$, $R_1 = 10\ \text{k}\Omega$, $R_2 = 10\ \text{k}\Omega$, and capacitor value $C = 0.0047\ \mu\text{F}$. Calculate the cutoff frequency and the output voltage amplitude when the input waveform is a 1 V_{rms} sine wave having a frequency equal to the cutoff frequency.
6. The bandpass filter in Figure 9–18 has resistor values $R_1 = 1\ \text{k}\Omega$, $R_2 = 1\ \text{k}\Omega$, $R_3 = 20\ \text{k}\Omega$, and capacitor values $C_1 = 0.022\ \mu\text{F}$, $C_2 = 0.01\ \mu\text{F}$. Calculate the center frequency, bandwidth, low cutoff frequency, and the high cutoff frequency.
7. The bandstop filter in Figure 9–20 has resistor values $R_1 = 1\ \text{k}\Omega$, $R_2 = 20\ \text{k}\Omega$, $R_3 = 1\ \text{k}\Omega$, and $R_4 = 20\ \text{k}\Omega$, and capacitor values $C_1 = 0.022\ \mu\text{F}$, $C_2 = 0.022\ \mu\text{F}$. Find the center frequency.

Analysis Questions

1. For the first order low-pass filter, create an Excel spreadsheet that will graph the voltage gain (vertical axis) as a function of the frequency (horizontal axis—logarithmic scale). Allow the user to enter the resistor and capacitor component values. The spreadsheet will display the bode plot and the cutoff frequency. Provide a sample result.

2. For the first order low-pass filter, create an Excel spreadsheet that will graph the voltage gain (vertical axis) as a function of the frequency (horizontal axis—logarithmic scale). Allow the user to enter the resistor and capacitor component values. The spreadsheet will display the bode plot and the cutoff frequency. Provide a sample result.

Performance Projects Correlation Chart

Chapter Topic	Performance Project	Project Number
First Order Low-Pass Filter	First Order Low-Pass Filter	23
Second Order Low-Pass Filter	Second Order Low-Pass Filter	24
First Order High-Pass Filter	First Order High-Pass Filter	25
Second Order High-Pass Filter	Second Order High-Pass Filter	26
Bandpass Filter	Bandpass Filter	27
Bandstop (Notch) Filter	Bandstop (Notch) Filter	28

OUTLINE

OBJECTIVES

After studying this chapter, you should be able to:

1. Identify from schematic diagrams the BJT, FET, and op-amp versions of the **Hartley, Colpitts,** and **Clapp oscillators**
2. Identify the tuning components and describe the procedure for determining the oscillating frequency of the Hartley, Colpitts, and Clapp oscillators
3. Explain the operation of a **crystal oscillator**
4. Identify from schematic diagrams the **phase-shift** and **Wien-bridge oscillators**
5. Identify the tuning components and describe the procedure for determining the oscillating frequency of the phase-shift and Wien-bridge oscillators
6. Define the operation of a **monostable multivibrator** and calculate the duration of the output pulse
7. Define the operation of an **astable multivibrator** and determine the operating frequency for both symmetrical and nonsymmetrical output waveforms

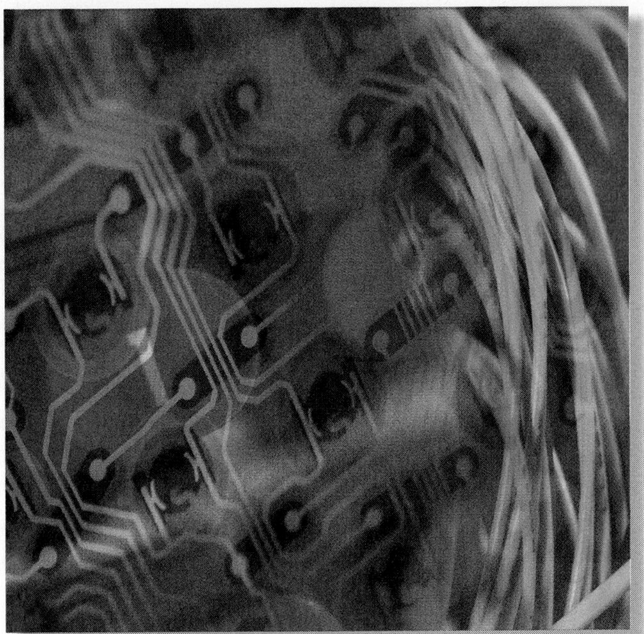

CHAPTER 10

Oscillators and Multivibrators

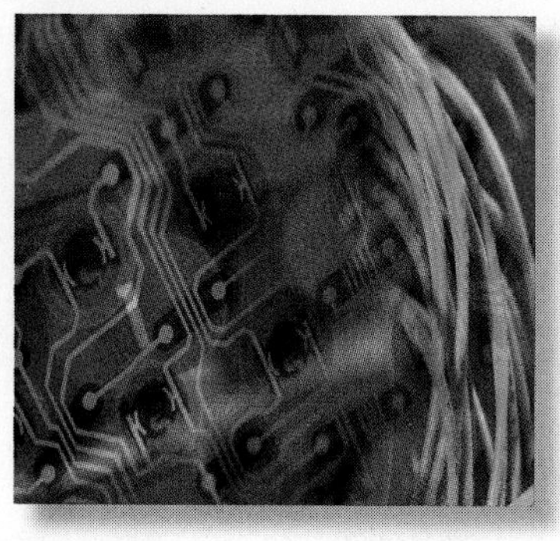

PREVIEW

Someone once said that an oscillator is a "hot rod" or "turbo" amplifier. Let's examine that comment. An oscillator is a circuit that generates and sustains an output signal without an input signal supplied by another circuit or source. This capability enables oscillators to be used as signal sources for many circuit applications.

There are basically three kinds of electronic circuits: those that amplify signals, those that control the flow of signals, and those that generate signals. You have already studied amplifiers in previous chapters, and you will study control circuits in a later course of study. This chapter is about the oscillator circuits that generate signals.

You will find that the function of oscillators depends upon the qualities of *LC* tuned circuits or *RC* phase-shift circuits. A special type of oscillator, an astable (or free-running) multlvibrator, generates its signal according to the charge and discharge time of *RC* circuits. You will also be introduced to a monostable (or one-shot) multivibrator.

KEY TERMS

Astable multivibrator
Clapp oscillator
Colpitts oscillator
Crystal oscillator
Free-running multivibrator
Hartley oscillator
Monostable multivibrator
Multivibrators
One-shot multivibrator
Oscillator
Phase-shift oscillator
Piezoelectric effect
Positive feedback
Rectangular waveforms
Wien-bridge oscillator

10–1 Background Information

As stated earlier, an **oscillator** is a circuit that generates and sustains an output signal without an input signal supplied by another circuit or source. For example, an oscillator generates the waveform that produces the tones we hear from an electronic organ or keyboard. An oscillator generates the high frequency waveforms that carry radio and television signals around the earth and into space. Also, oscillators synchronize the operation of industrial controls and personal computers. Figure 10–1 shows the two most common kinds of waveforms—sinusoidal and rectangular—generated by oscillator circuits. These are the waveforms generated by the oscillator circuits discussed in this lesson.

The Barkhaven criteria for oscillation states that:

1. The loop gain must equal one.
2. The loop phase shift must be $n \times 360°$, where $n = 1, 2, 3, \ldots$

Four conditions are absolutely required to begin and sustain the operation of an oscillator:

1. A power source
2. A device or components that determine the frequency of oscillation
3. Amplification
4. **Positive feedback**

The power source for oscillator circuits is taken from batteries or a dc power supply. Oscillator circuits sometimes create signals that find their way into the wiring for the power supply. This is highly undesirable because the oscillations can thus show up in circuits where they do not belong, At least one *decoupling capacitor* is used with oscillator circuits to bypass (short-circuit) any of these stray oscillations from the

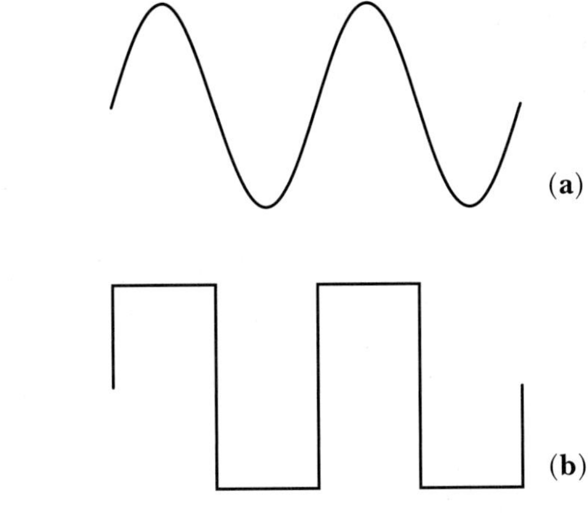

FIGURE 10–1 Standard waveforms from oscillators and multivibrators: (a) sinusoidal waveform; (b) rectangular waveform

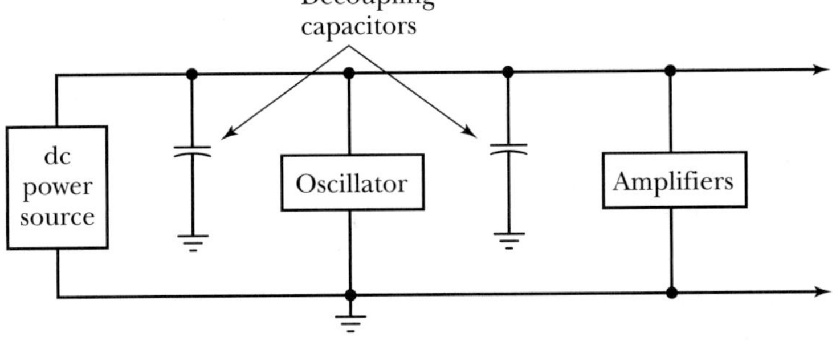

FIGURE 10–2 Power supply decoupling capacitors

power supply lines to ground. See the example in Figure 10–2. Decoupling capacitors generally have values on the order of 10 pF to 100 pF, and they are physically located very close to the oscillator circuit.

The devices or components that control the frequency of oscillation include combinations of resistors, capacitors, inductors, and quartz crystals. You will soon find that the frequency determining components for some oscillators are *RC* circuits—combinations of resistors and capacitors. A number of different oscillators use *LC* tank circuits (a parallel circuit composed of inductors and capacitors) as the frequency determining components. One type of oscillator uses a quartz crystal to fix the rate of oscillation.

The amplification for oscillators can be provided by BJTs, FETs, and op-amps. Every oscillator must contain a device that provides signal amplification greater than 1.

Positive feedback contributes to the operation of an oscillator by feeding a portion of the amplifier's output back to its own input. The feedback signal must be phased in such a way that it enhances the amplifier's input/output signal (positive feedback). The amplifier part of the oscillator usually inverts the signal by 180°, then a phase-shifting component (capacitor or inductor) in the feedback path provides the necessary additional phase shift that is required for positive feedback. Positive feedback is also called *regenerative feedback.*

10–2 *LC* Sine-Wave Oscillators

An important and commonly used class of oscillators are those that use a parallel *LC* (tank) circuit to determine the operating frequency. As long as these circuits are operating properly, they produce sine-wave output waveforms. The frequency of these waveforms is the same as the resonant frequency of the tank circuit; and as you learned in an earlier chapter, you can compute the resonant frequency of a tank circuit by means of the formula shown here as Formula 10–1. In this formula, f_r is the resonant frequency (frequency of the oscillator), L is the value of the inductance, and C is the value of the capacitance.

FORMULA 10–1 $$f_r = \frac{1}{2\pi\sqrt{LC}}$$

LC sine-wave oscillators contain various combinations of two or more capacitor and inductors. You will find that the values of the inductors can be combined into a single L value and, likewise, the values of two or more capacitors can be combined into a single C value.

In all of these resonant circuits, increasing the value of the capacitance decreases the operating frequency, while decreasing the capacitance increases operating frequency. This is also true for the inductances: increasing the inductance decreases frequency, while decreasing inductance increases frequency. We say that the operating frequency of an LC oscillator is inversely proportional to the square root of the product of the capacitance and inductance values in the oscillator's tank circuit.

Hartley Oscillators

The circuit in Figure 10–3 is known as a **Hartley oscillator.**

1. The power source is shown as + V_{CC}.
2. The frequency determining part of the oscillator is the LC tank circuit composed of L_{1_A}, L_{1_B}, and C_1.
3. Amplification is provided by the NPN common-emitter transistor (Q_1) and its associated components.
4. Positive feedback is provided from the collector of the amplifier through capacitor C_3 to the tapped coil in the tank circuit.

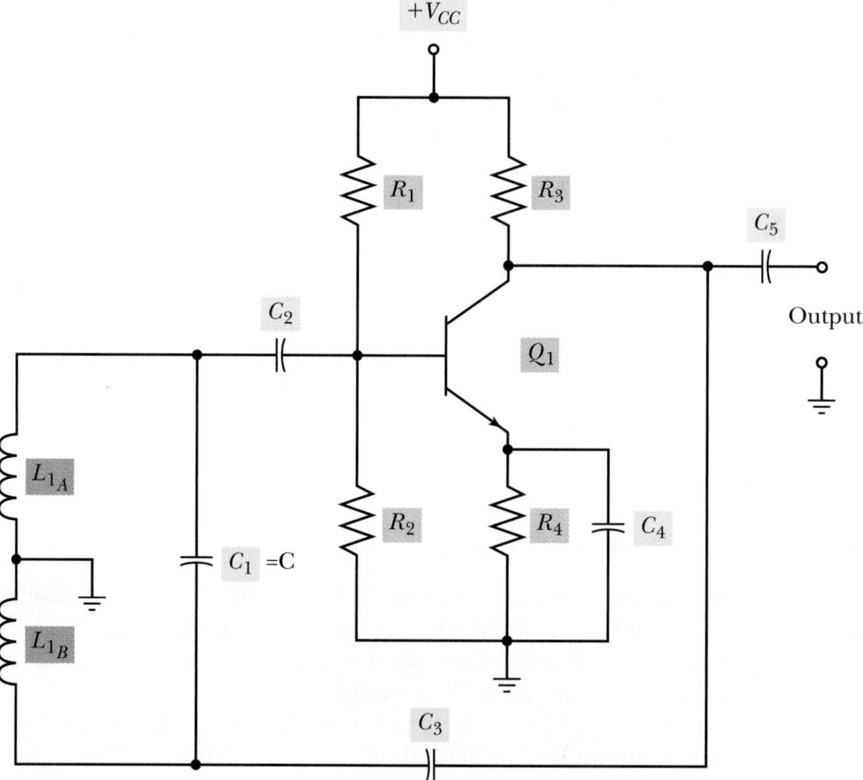

FIGURE 10–3 BJT Hartley oscillator

The Hartley oscillator is characterized by having a rapped inductor in the *LC* resonant circuit. The tapped point for the inductor is grounded. You will sometimes see two separate inductors instead of a single, tapped inductor.

You can calculate the operating frequency of a Hartley oscillator by applying the basic parallel *LC* circuit formula for resonance:

FORMULA 10–2 $f_r = \dfrac{1}{2\pi\sqrt{L_t C}}$

In Formula 10–2, L_1 is the total inductance and is found by:

FORMULA 10–3 $L_1 = L_{1_A} + L_{1_B}$

EXAMPLE What is the operating frequency of the Hartley oscillator shown in Figure 10–3 if $C = 100$ pF, $L_{1_A} = 20\ \mu\text{H}$, and $L_{1_B} = 20\ \mu\text{H}$?

Answer:

From Formula 10–3:

$$L_1 = L_{1_A} + L_{1_B} = 20\ \mu\text{H} + 20\ \mu\text{H} = 40\ \mu\text{H}$$

From Formula 10–2:

$$f_r = \frac{1}{2\pi\sqrt{L_t C}}$$
$$f_r = \frac{1}{2\pi\sqrt{(40\ \mu\text{H})(100\ \text{pF})}}$$
$$f_r = 2.5\ \text{MHz}$$

Figure 10–4 shows a version of the Hartley oscillator that uses an op-amp as the amplifier element, Resistor R_2 in this circuit provides the dc negative feedback that is necessary for the operation of an inverting op-amp. The regenerative feedback that is required for oscillation is from the output of IC_1, through coupling capacitor C_2, to the tank circuit, and through resistor R_1 to the inverting input of the op-amp.

A JFET Hartley oscillator is shown in Figure 10–5.

The formulas for finding the oscillating frequency of op-amp and FET versions of the Hartley oscillator are exactly the same as for the BJT versions. It is important you understand that, in *LC* oscillators, the type of amplifying device being used has little effect on the operating frequency. The operating frequency is determined by the values of the inductances and capacitors in the tank circuit. The amplifying device merely ensures continuous oscillation of the tank circuit.

PRACTICE PROBLEMS 1

Refer to the JFET Hartley oscillator circuit in Figure 10–5 to answer the following questions.

1. Which component is the active amplifying device in this oscillator?
2. Which components make up the frequency determining part of the circuit?
3. Capacitor C_3 is part of the __________ portion of this oscillator circuit.
4. What is the total inductance of the tank circuit if $L_{1_A} = 0.1$ mH and $L_{1_B} = 0.1$ mH?
5. Suppose the inductances have the values indicated in question 4. Further suppose the value of the capacitor is 1,000 pF. What is the frequency of oscillation of this circuit?
6. What happens to the frequency of oscillation of this circuit if you decrease the value of *C*? Increase the value of L_{1_A}?

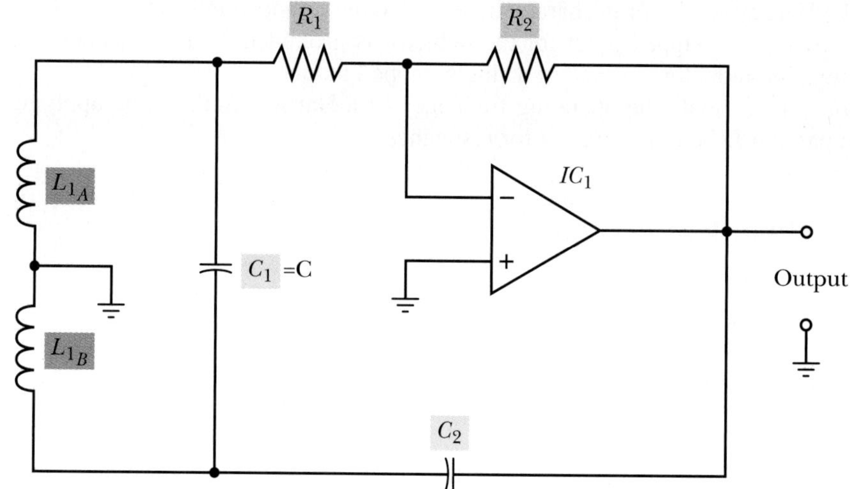

FIGURE 10–4 Op-amp Hartley oscillator

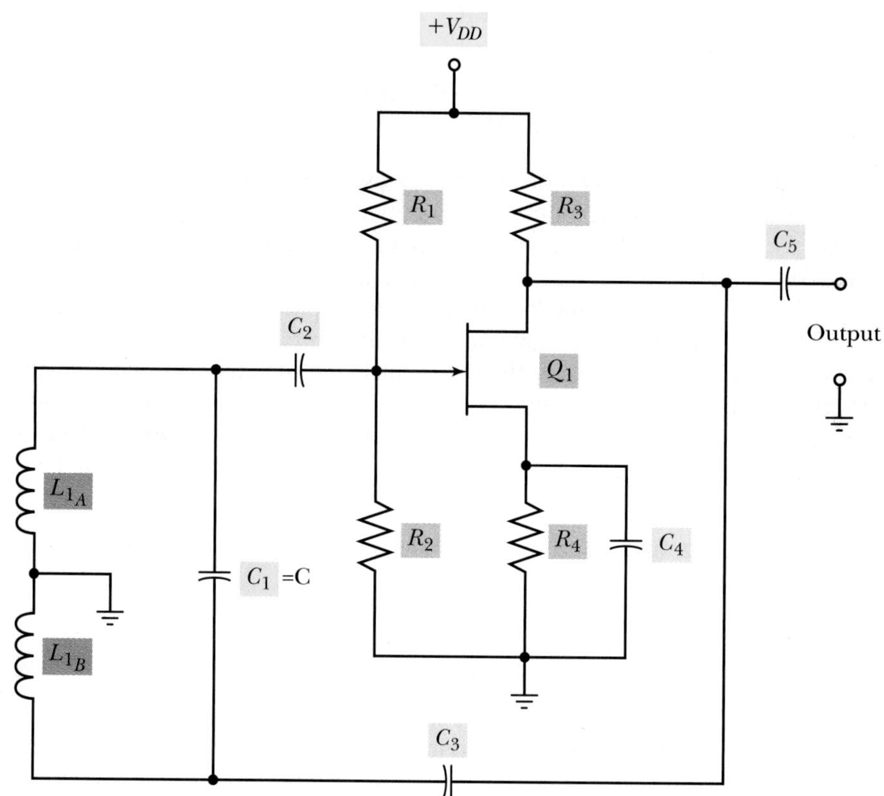

FIGURE 10–5 JFET Hartley oscillator

Application Problem

Communications

RADIO RECEIVER

The basic block diagram for a FM radio receiver is shown in the figure. When the consumer changes the channel, the oscillator frequency changes. The frequency of the radio signal coming out of the mixer will always be 10.7 MHz regardless of the channel selected. If the oscillator frequency is 99.2 MHz, what is the frequency of the station selected? If a Hartley tuned oscillator is being used with a variable capacitor in the *LC* circuit, what is the value of capacitance if the total inductance is 2.5 μH?

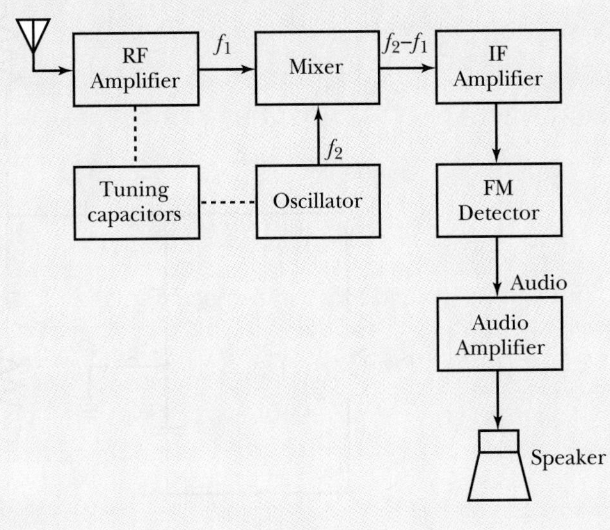

Solution The output of the mixer stage is the difference of the two input frequencies.

$$f_{station} = f_{oscillator} - f_{mixer} = 99.2\text{ MHz} - 10.7\text{ MHz} = 88.5\text{ MHz}$$

The formula for the frequency for the oscillator is:

$$f = \frac{1}{2\pi\sqrt{LC}}$$

Solving for *C* we have:

$$C = \left(\frac{1}{2\pi f}\right)^2\left(\frac{1}{L}\right)$$

If $f = 99.2$ MHz and $L = 2.5$ μH, $C = 1.03$ pF

Colpitts Oscillators

The circuit in Figure 10–6 is the BJT version of a **Colpitts oscillator.** It closely resembles a Hartley oscillator, Figure 10–3. The main difference is the location of the grounded tap in the resonant circuit. Recall that a Hartley oscillator has a grounded tap between two series inductors. The Colpitts oscillator has a grounded tap between a pair of series capacitors.

1. The dc power source is the $+V_{CC}$ connection.
2. The frequency determining element of the oscillator is the tank circuit composed of L_1, C_1, and C_2.
3. Amplification is provided by the NPN common-emitter amplifier transistor circuit.
4. Positive feedback is accomplished by the connection from the collector of the amplifier transistor, through capacitor C_4, to the tank circuit.

You can calculate the operating frequency of a Colpitts oscillator by applying a slightly modified version of the resonance formula:

FORMULA 10–4 $$f_r = \frac{1}{2\pi\sqrt{LC_T}}$$

Total capacitance (C_T) is determined by:

FORMULA 10–5 $$C_T = \frac{C_1 C_2}{(C_1 + C_2)}$$

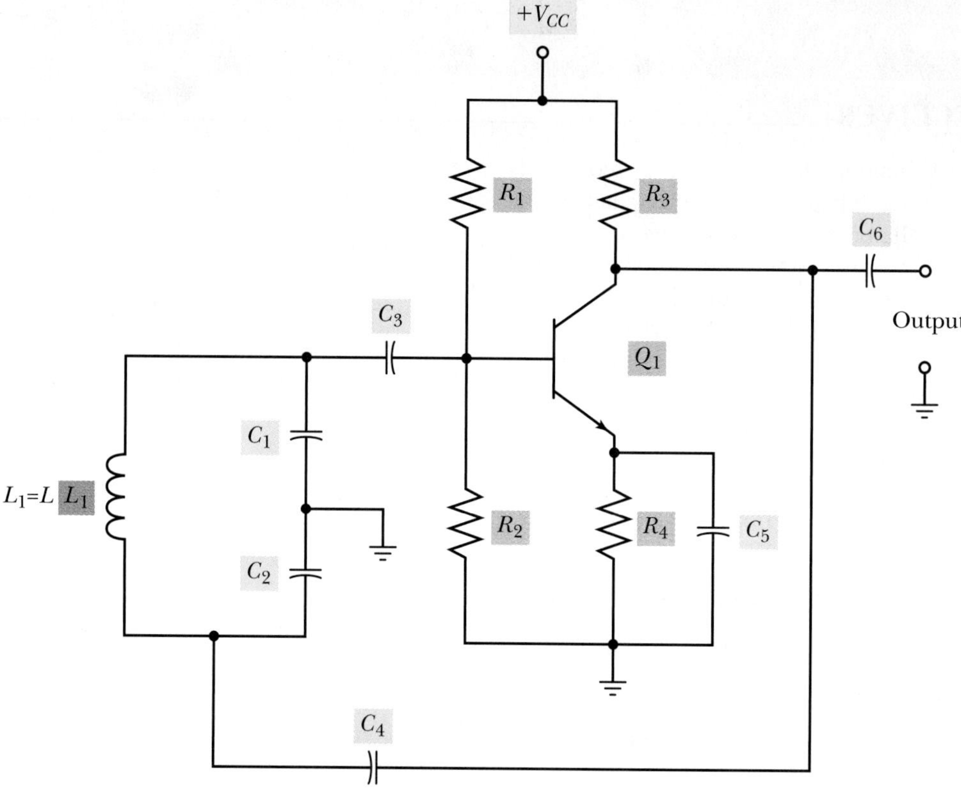

FIGURE 10–6 BJT Colpitts oscillator

EXAMPLE Referring to the Colpitts oscillator circuit in Figure 10–6, determine the operating frequency if $L = 100\ \mu\text{H}$, $C_1 = 1{,}000\ \text{pF}$, and $C_2 = 500\ \text{pF}$.

Answer:

From Formula 10–5:

$$C_T = \frac{C_1 C_2}{(C_1 + C_2)}$$

$$C_T = \frac{1000\ \text{pF} \times 500\ \text{pF}}{(1000\ \text{pF} + 500\ \text{pF})}$$

$$C_T = 333\ \text{pF}$$

From Formula 10–4:

$$f_r = \frac{1}{2\pi\sqrt{LC_T}}$$

$$f_r = \frac{1}{2\pi\sqrt{(100\ \mu\text{H}) \times (333\ \text{pF})}}$$

$$f_r = 872\ \text{kHz}$$

Figure 10–7 shows a version of the Colpitts oscillator that uses an op-amp as the amplifier element. This is basically an inverting amplifier where R_1 is the input resistor and R_2 is the feedback resistor. Proper phase inversion is automatic by means of this type of amplifier.

The Colpitts oscillator in Figure 10–8 is built around a JFET. Note transformer T_1 at the output. The stability of operation of Colpitts oscillators is often affected by changes in the amount of loading (impedance) across its output connections. Direct coupling or capacitive coupling to the next stage of the circuit is sometimes inade-

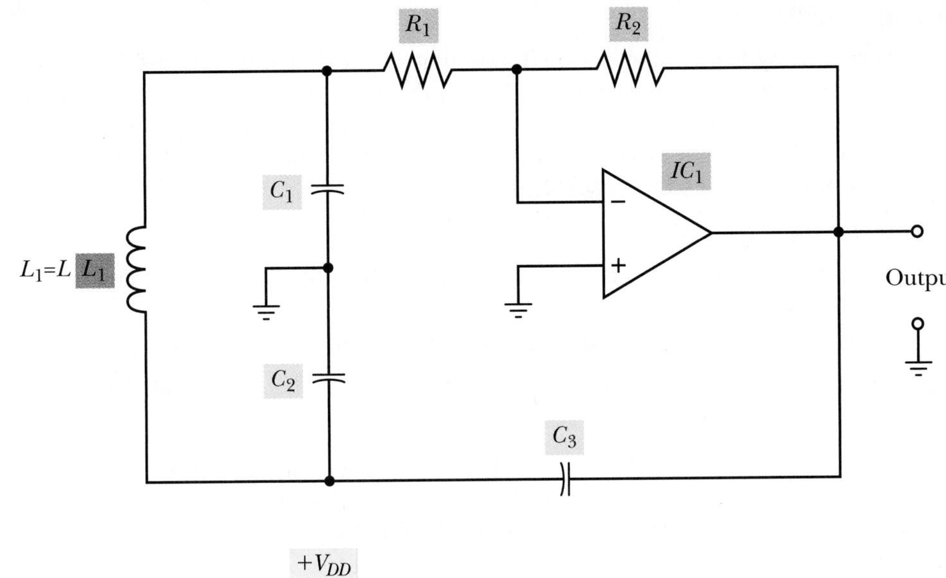

FIGURE 10–7 Op-amp Colpitts oscillator

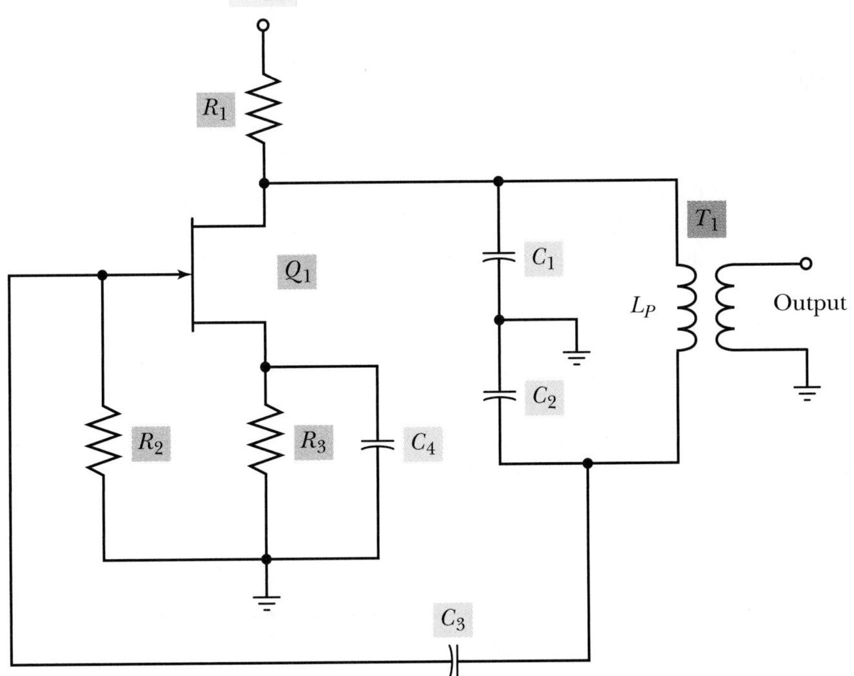

FIGURE 10–8 JFET Colpitts oscillator

quate, so you will often find transformer coupling at the output of a Colpitts oscillator stage. Notice that the inductance value L_P of the primary winding of coupling transformer T_1 is the inductance value of the oscillator's tank circuit.

PRACTICE PROBLEMS 2

Refer to the JFET Colpitts oscillator circuit in Figure 10–8 to answer the following questions.

1. Which component is the active amplifying device in this oscillator?
2. Which components make up the frequency determining part of the circuit?
3. What is the total capacitance of the tank circuit if $C_1 = 100$ pF and $C_2 = 100$ pF?
4. What is the frequency of oscillation of this circuit if the total capacitance is the value found in question 3 and the inductor LP is rated at 750 μH?
5. What happens to the frequency of oscillation of a Colpitts oscillator when you decrease the total capacitance? increase the value of inductance?

Practical Notes

You can remember that a Hartley oscillator has a divided inductance by recalling that H is the unit of measure for inductance (and *Hartley* begins with the letter *H*). Colpitts and Clapp oscillators, on the other hand, have split capacitances and begin with the letter *C* (also the first letter in *capacitance*).

Clapp Oscillators

You have already learned that a Colpitts oscillator stage often uses transformer coupling at the output to stabilize operation when the input impedance of the next stage changes. Another way to help stabilize operation is to insert a third capacitor into the tank circuit, Figure 10–9. When this is done, the circuit is no longer called a Colpitts oscillator. It is called a **Clapp oscillator.**

1. The dc power source is the $+V_{CC}$ connection.
2. The frequency determining element of the oscillator is the tank circuit composed of L_1, C_1, C_2, and C_3.
3. Amplification is provided by the NPN common-emitter amplifier transistor circuit.
4. Positive feedback is accomplished by the connection from the collector of the amplifier transistor to the tank circuit through C_5.

The main formula for finding the oscillating frequency for a Clapp oscillator is the same one used for a Colpitts oscillator (Formula 10–4). Because there are three capacitors in the tank circuit for a Clapp oscillator, the formula for total capacitance (C_T) is different:

FORMULA 10–6

$$C_T = \frac{1}{\left(\frac{1}{C_1} + \frac{1}{C_2} + \frac{1}{C_3}\right)}$$

The values of C_2 and C_3 are selected to reduce the effects of the transistor's capacitance on the oscillating frequency. The values of the capacitor C_1 and the inductor C_1 determine the frequency of the Clapp oscillator. (NOTE: $C_T \approx C_1$).

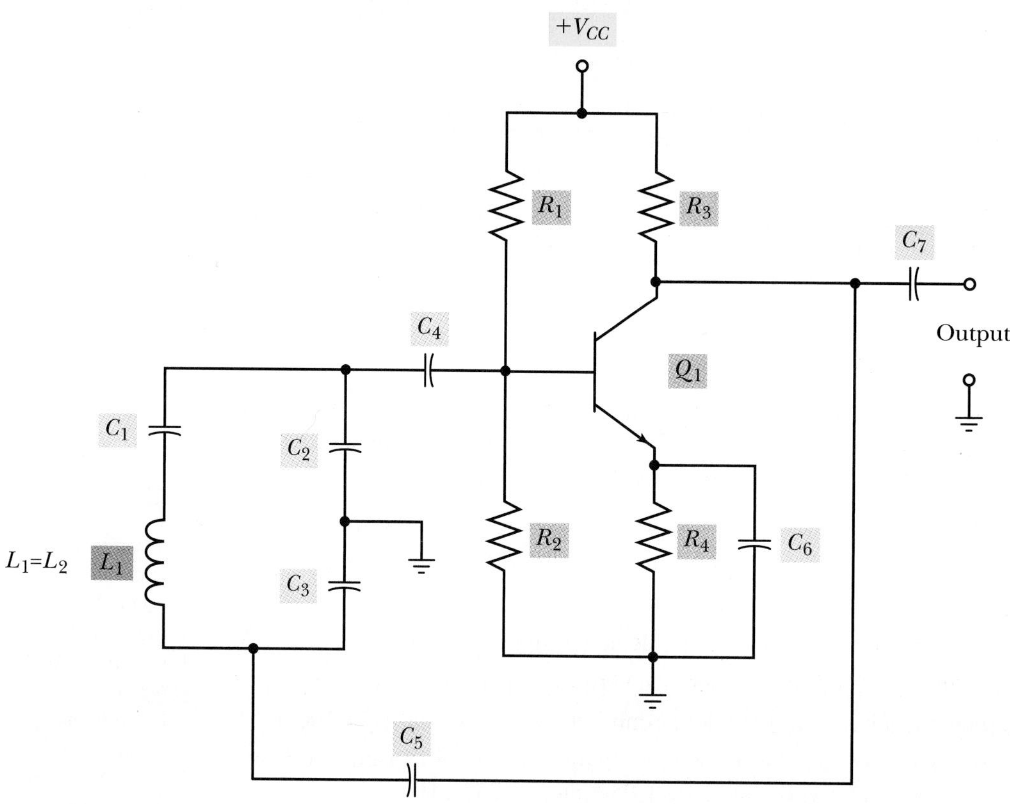

FIGURE 10–9 BJT Clapp oscillator

◆ **EXAMPLE** Suppose you find a Clapp oscillator, such as the one in Figure 10–9, that has $L = 100\ \mu H$, $C_1 = 0.001\ \mu F$, $C_2 = 0.1\ \mu F$, and $C_3 = 0.1\ \mu F$. Calculate the operating frequency of the circuit.

Answer:

From Formula 10–6:

$$C_T = \frac{1}{\left(\frac{1}{C_1} + \frac{1}{C_2} + \frac{1}{C_3}\right)}$$

$$C_T = \frac{1}{\left(\frac{1}{0.001\ \mu F} + \frac{1}{0.1\ \mu F} + \frac{1}{0.1\ \mu F}\right)}$$

$$C_T = 980\ pF$$

From Formula 10–4:

$$f_r = \frac{1}{2\pi\sqrt{LC_T}}$$

$$f_r = \frac{1}{2\pi\sqrt{(100\ \mu H)\times(980\ pF)}}$$

$$f_r = 508\ kHz$$

Repeating the calculation with the approximation of $C_T = C_1$, $C_T = 1000$ pF
From Formula 10–4:

$$f_r = \frac{1}{2\pi\sqrt{LC_T}}$$

$$f_r = \frac{1}{2\pi\sqrt{(100\ \mu H)\times(1000\ pF)}}$$

$$f_r = 503\ kHz$$

The approximate calculation for the oscillator's frequency is very close to the exact calculation of the oscillator's frequency. ◆

10–3 Crystal Oscillators

Quartz crystals possess a piezoelectric quality that permits their use in very precise sine-wave oscillators. The **piezoelectric effect** is one where the application of a bending force to crystal material causes it to generate a small amount of voltage; then applying a voltage to the same material causes it to bend or warp as long as the voltage is applied. When a quartz crystal is connected to an amplifying device as shown in Figure 10–10, the amplifier can provide a burst of voltage that will cause the crystal to bend. Then when the voltage is removed and the crystal is allowed to return to its normal shape, the crystal generates a small voltage that is detected and amplified by the transistors. This impulse is then amplified and returned to the crystal to cause the crystal to bend again. This positive feedback cycle continues, thus producing oscillations determined by the rate that the crystal vibrates. The rate that the crystal vibrates is determined by its physical shape and size.

The crystal in a **crystal oscillator** replaces the tank circuit found in *LC*-type oscillators. The schematic symbol and equivalent *RCL* circuit are shown in Figure 10–11. In this equivalent circuit, *R* is the natural resistance of the crystal, *L* is the equivalent inductance, C_N is the "natural" capacitance, and C_M is the "mounting" capacitance (which includes the stray capacitance of the metal case, mounting terminals, etc.).

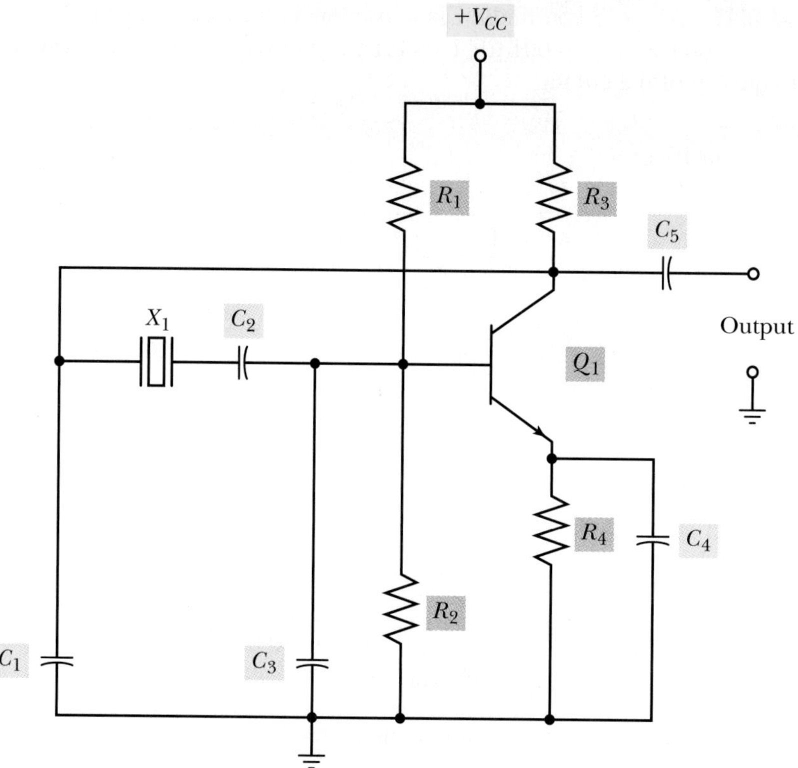

FIGURE 10–10 BJT crystal oscillator

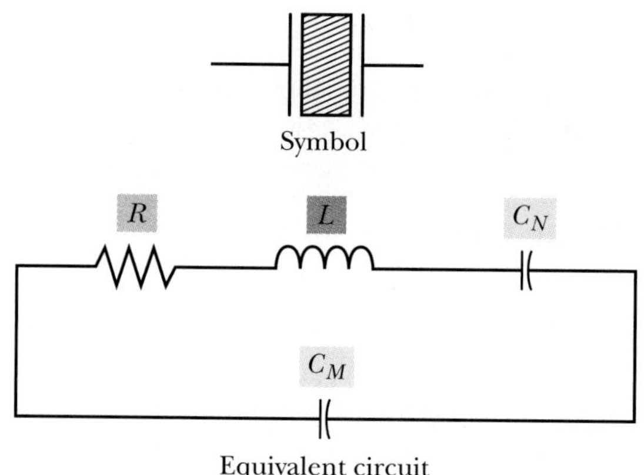

FIGURE 10–11 Symbol and equivalent circuit for a quartz crystal

Figure 10–12 shows an op-amp crystal oscillator with the following features:

1. The power source is the $+V$ and $-V$ connections.
2. The frequency determining element of the oscillator is crystal X_1.
3. Amplification is provided by the op-amp.
4. Positive feedback is accomplished by the connection from the output of the amplifier through the crystal to the noninverting input.

Application Problem

Computers

MICROPROCESSOR CLOCK CIRCUIT

A crystal oscillator drives a microprocessor chip. The frequency of the oscillator exceeds the maximum frequency specification for the processor. A clock divider is used to reduce the frequency. If the microprocessor's current instruction time is 80nS and the clock divider is a divide by four circuit, determine the resonant frequency of the crystal used in the oscillator.

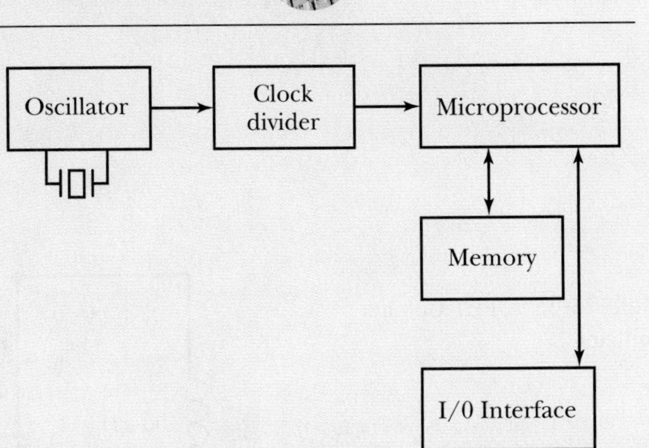

Solution Microprocessor Frequency = 1/Instruction Time = 1/80nS = 12.5 MHz

The crystal frequency = 4 × 12.5 MHz = 50 MHz crystal

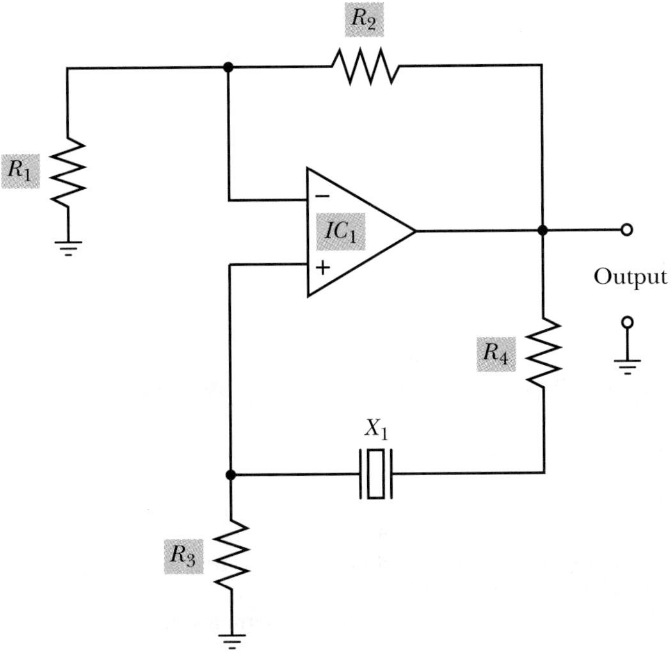

FIGURE 10–12 Op-amp crystal oscillator

IN-PROCESS LEARNING CHECK 1

Fill in the blanks as appropriate.

1. The *LC* oscillator that has a tapped coil or a set of two coils in the tank circuit is called the ________ oscillator.
2. The *LC* oscillator that has a tank circuit of two capacitors and a single inductor is called the ________ oscillator.
3. The *LC* oscillator that has three capacitors in the tank circuit is called the ________ oscillator.
4. The output of a ________ *LC* oscillator is often coupled to the next stage by means of a transformer whose primary winding is part of the tank circuit.
5. Referring to the oscillator circuit in Figure 10–13, determine the upper and lower tunable limits of output frequency when (a) C_1 is adjusted for 100 pF and (b) C_1 is adjusted for 850 pF. Upper f ________, lower f ________

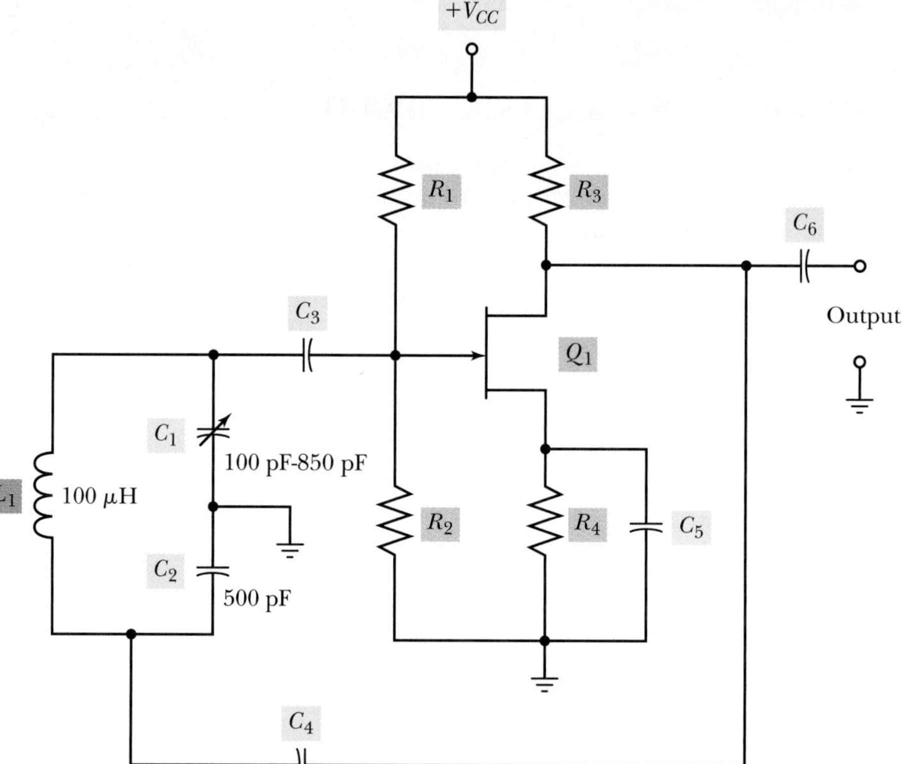

FIGURE 10–13 JFET Colpitts oscillator

Troubleshooting Hints

In actual practice, *LC* oscillators tend to work properly or not at all. The best laboratory tool for checking the operation of an oscillator is an oscilloscope. An electronic frequency measuring instrument (sometimes called a "freq meter") is valuable for checking the operating frequency of an oscillator very accurately, but this meter can give you misleading information if you haven't determined the correct voltage levels by oscilloscope measurement.

The two most common reasons an *LC* oscillator stops operating are a defective amplifier element and excessive loading at the output. The best way to check for a defective amplifier element is to replace it with one that is known to be in good working order. As far as output loading is concerned, a trouble in the next stage in the circuit might place a low impedance across the output of the oscillator and stop it from running. You can check for this trouble by temporarily disconnecting the output of the oscillator from the next circuit along the line. If the oscillator begins functioning normally under these circumstances, the trouble is more likely in the next stage rather than in the oscillator circuit itself.

Avoid taking frequency and signal amplitude measurements while touching the components in a high frequency oscillator. The stray capacitance of your fingers and hand can shift the frequency and amplitude levels.

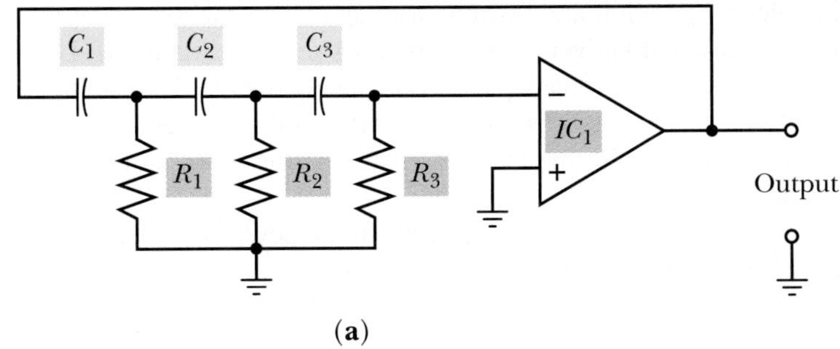

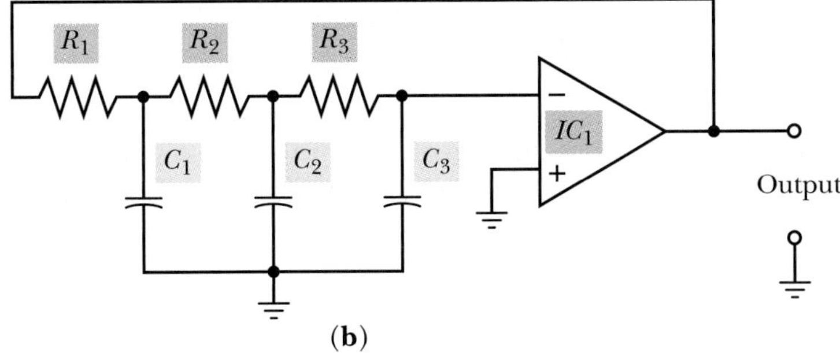

FIGURE 10–14 Phase-shift oscillators: (a) lead network; (b) lag network

10–4 *RC* Sine-Wave Oscillators

A second major group of sine-wave oscillators is built around *RC* timing components rather than *LC* resonant circuits and quartz crystals. Whereas the operation of *LC* oscillators is based on resonant frequency conditions, the operation of *RC* oscillators is based on phase-shift conditions. The four basic elements and conditions for oscillation, however, are the same; that is, a power source, frequency determining elements, amplifier, and positive feedback.

RC Phase-Shift Oscillator

The circuit in Figure 10–14 shows two versions of the **phase-shift oscillator.** You can always identify a phase-shift oscillator by means of its three-stage *RC* network. The amplifying device shown in these examples is an op-amp, but the three-stage *RC* network is readily identified on a schematic where the amplifying element is a BJT or FET.

1. The dc power source is the $+V_{CC}$ connection.
2. The frequency determining element of the oscillator is the *RC* phase-shift network composed of R_1, C_1, R_2, C_2, R_3, and C_3.
3. Amplification is provided by the op-amp circuit.
4. Positive feedback is accomplished by the connection from the output of the op-amp, through the phase-shift network, to the inverting (–) input of the op-amp.

As the name of this oscillator suggests, its operation is based on phase-shift principles. Recall that an *RC* circuit will shift the phase of an ac waveform by a certain number of degrees, depending on the values of *R* and *C* and on the frequency. The *RC* network in a phase-shift oscillator is capable of producing a 180° phase shift at one particular frequency. When this 180° phase shift (which occurs at only one particular frequency) is combined with the normal 180° phase shift of the inverting amplifier, we have the positive feedback condition required for starting and maintaining oscillation. Oscillation occurs only at the frequency that causes a 180° phase shift through the *RC* network. At any other frequency, the feedback is some value other than the required 180° + 180° = 360° needed for positive feedback.

The principle of operation and formula for calculating the operating frequency are the same for the lead and lag versions. In the lead version, Figure 10–14a, the capacitors are in series with the feedback path, thus causing the feedback voltage developed across the *R*s to lead the signal from the amplifier. In the lag version, Figure 10–14b, the resistors are in series with the feedback path and the *C*s in parallel, thus enabling a lagging feedback voltage.

The formula for determining the operating frequency of a phase-shift oscillator is complicated unless you assume all of the capacitors in the *RC* network have the same value and all the resistors in the network have the same value. Assuming all three capacitors and all three resistors have the same value:

FORMULA 10–7 $f_r = \dfrac{1}{2\pi RC\sqrt{6}}$

◆ **EXAMPLE** What is the operating frequency of the phase-shift oscillator in Figure 10–14a if $R_1 = R_2 = R_3 = 3.3\ \text{k}\Omega$ and $C_1 = C_2 = C_3 = 0.01\ \mu\text{F}$?

Answer:

From Formula 10–7:

$$f_r = \frac{1}{2\pi RC\sqrt{6}}$$

$$f_r = \frac{1}{2\pi(3.3\ \text{k}\Omega)(0.01\ \mu\text{F})\sqrt{6}}$$

$$f_r = 1.97\ \text{kHz}$$

◆

Phase-shift oscillators are known for their relative simplicity (no inductors), low cost, and reliability. Their main disadvantage is that it is difficult to fine-tune them. This means they are used only in circuits where a very exact frequency is not required. Another disadvantage is the loss in feedback signal level as it passes through the series of three *RC* combinations. To meet the voltage gain requirements for oscillation, the amplifier in a phase-shift oscillator must have a gain greater than 30.

Wien-Bridge Oscillator

The **Wien-bridge oscillator** in Figure 10–15 is a simple, stable, and popular *RC* oscillator configuration. Its identifying feature is a set of two *RC* combinations called a lead-lag network. A series *RC* combination is connected in series with the feedback path, and a parallel *RC* combination is connected in parallel with the feedback path. Another unusual feature is that it uses a noninverting amplifier instead of the usual inverting configuration.

1. The dc power source is the $+V_{CC}$ connection.
2. The frequency determining element of the oscillator is the lead-lag *RC* network composed of C_1, R_1, C_2, and R_2.
3. Amplification is provided by the op-amp circuit.
4. Positive feedback is accomplished by the connection from the output of the op-amp, through the *RC* network, to the noninverting input (+) of the op-amp.

The frequency of oscillation is determined by the values of C_1, R_1, C_2, and R_2 in the lead-lag network. The important property of this kind of *RC* network is that the signal fed back from the output of the amplifier undergoes a phase shift of 0° at the oscillator frequency. Because the lead-lag network causes zero phase shift at the operating frequency, the amplifier must be a noninverting type to enable a 0° phase shift as well.

It is instructive to compare the operation of a Wien-bridge oscillator with the phase-shift oscillator you studied in the previous section of this chapter.

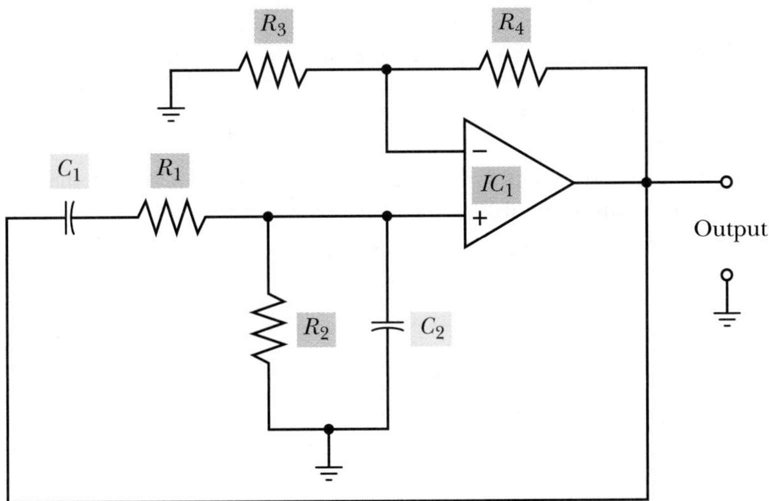

FIGURE 10–15 Wien-bridge oscillator

1. In the phase-shift oscillator, the *RC* section produces a 180° phase shift at the oscillation frequency. In the Wien-bridge oscillator, the *RC* section produces a 0° phase shift at the oscillation frequency.
2. The amplifier in a phase-shift oscillator is of the inverting type, thereby producing a 180° phase shift. The amplifier in a Wien-bridge oscillator is of the noninverting type, thereby producing zero phase shift between input signals and output signals.
3. Considering the combined *RC* network and amplifier phase shifts for a phase-shift oscillator:

$$\text{Total phase shift} = 180° + 180° = 360° = 0°$$

Considering the combined *RC* lead-lag and amplifier phase shifts for a Wien-bridge oscillator:

$$\text{Total phase shift} = 0° + 0° = 0°$$

You can see from this comparison that the phase-shift and Wien-bridge oscillators both provide a positive-feedback phase shift of 0° at the oscillating frequency.

As with the phase-shift oscillator, determining the oscillating frequency of the Wien-bridge oscillator is made simpler by making both capacitor values and both resistor values in the lead-lag network the same. In other words, let $C = C_1 = C_2$ and $R = R_1 = R_2$. Formula 10–8 provides the oscillating frequency of the Wien-bridge oscillator.

FORMULA 10–8 $f_r = 1/2\pi RC$

EXAMPLE Referring to the Wien-bridge circuit in Figure 10–15, determine the output frequency if $C_1 = C_2 = 0.1\ \mu\text{F}$, and $R_1 = R_2 = 100\ \text{k}\Omega$.

Answer:

From Formula 10–8:

$$f_r = \frac{1}{2\pi RC}$$

$$f_r = \frac{1}{2\pi(100\ \Omega)(0.1\ \mu\text{F})}$$

$$f_r = 15.9\ \text{Hz}$$

PRACTICE PROBLEMS 3

Refer to the Wien-bridge oscillator circuit in Figure 10–15 for the following questions.

1. Determine the output frequency of the Wien-bridge oscillator when $C_1 = C_2 = 0.002\ \mu F$, and $R_1 = R_2 = 4.7\ k\Omega$.
2. What happens to the output frequency if you double the value of the capacitors? Cut the resistor values in half?
3. Suppose $C_1 = C_2 = 0.1\ \mu F$ and you want to obtain a frequency of 1,000 Hz. What value should be used for the resistors?

IN-PROCESS LEARNING CHECK 2

Fill in the blanks as appropriate.

1. Whereas _____ oscillators operate according to resonance, *RC* oscillators operate according to _____.
2. The four basic elements and conditions for starting and sustaining oscillation in *RC* sine-wave oscillators are _____, _____, _____, and _____.
3. The sine-wave oscillator that uses a three-stage *RC* network for achieving a 180° phase shift at the frequency of oscillation is the _____ oscillator.
4. The sine-wave oscillator that uses a lead-lag *RC* network to produce 0° phase shift at the frequency of oscillation is the _____ oscillator.
5. The amplifier element of a phase-shift oscillator must produce a phase shift of _____°, while the amplifier element of a Wien-bridge oscillator must produce a phase shift of _____°.

10–5 Multivibrators

Multivibrators are special types of circuits that produce **rectangular waveforms** that switch approximately between ground and the supply voltage level. If the supply voltage level happens to be +5.2 V, for example, a multivibrator circuit will produce a rectangular waveform that switches between +0.3 V (close to ground potential) and about +5.0 V (close to the supply voltage). These voltage levels are those of a BJT or FET that is operating as a switch—either fully on (saturation) or switched off (cutoff).

Another characteristic of multivibrator circuits is that their timing and frequency determining components are always *RC* circuits. Do not expect to find inductors doing anything significant in a multivibrator circuit.

There are three types of multivibrator circuits: monostable multivibrators (or one-shot multivibrators), bistable multivibrators (or flip-flops), and astable multivibrators. We will discuss monostable and astable multivibrators in this chapter. You will learn about bistable multivibrators (flip-flops) in your later studies of digital electronics.

Monostable Multivibrators

A **monostable multivibrator** is a circuit that produces a rectangular pulse of a fixed duration in response to a brief input trigger pulse. A monostable multivibrator does not run continuously as do oscillators (described earlier) and the astable multivibrator (described next). This circuit is also called a **one-shot** because one trigger pulse causes one output pulse to occur.

Figure 10–16 shows the schematic diagram for a BJT monostable multivibrator. The trigger pulse that starts the output timing interval is applied to the base of transistor Q_1 through resistor R_3. The output pulse is taken from the collector of transistor Q_2. The waveforms indicate the action of the circuit in response to a trigger pulse being applied at the *Trig in* point.

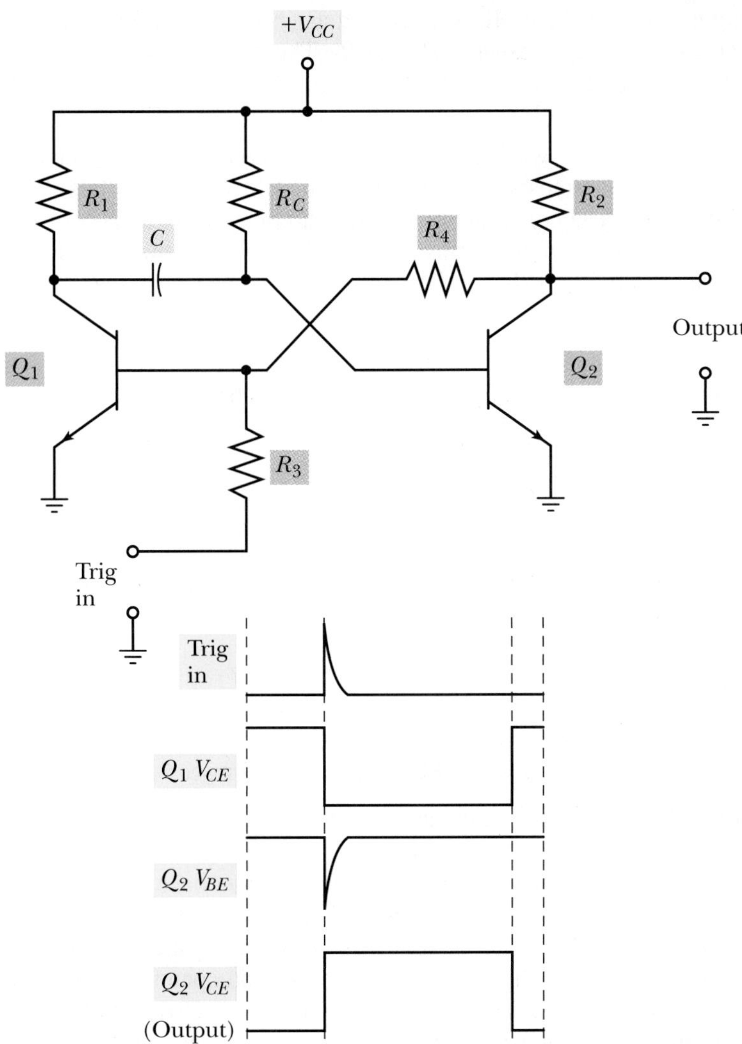

FIGURE 10–16 Discrete-component monostable multivibrator circuit and waveforms

Monostable multivibrators have four states of operation: the resting state, the trigger-on state, the active state, and the return state. The following is a summary of the four operating states for the monostable multivibrator shown in Figure 10–16.

1. The resting (quiescent) state is the standby state. It is where transistor Q_1 is turned off and Q_2 is turned on. You can see the effect in the waveforms where the V_{CE} for Q_1 is at the source voltage level (+V_{CC}) and the V_{CE} for Q_2 is close to ground potential. The circuit remains in this state until some external condition causes the positive-going spike to occur at the *Trig in* connection.
2. The trigger-on state begins when the *Trig in* terminal sees the positive-going edge of a trigger pulse waveform. This positive transition immediately biases transistor Q_1 to its ON state. This effectively places the Q_1 side of capacitor C at ground potential, which effectively grounds the base of transistor Q_2. This action turns off transistor Q_2. Note how the waveforms reflect these changes when the *Trig in* signal goes positive:
   - **a.** The V_{CE} for Q_1 drops to 0 V. (Q_1 is switched on.)
   - **b.** The base voltage (V_{BE}) for Q_2 immediately drops to 0 V. (This is what turns off Q_2.)
   - **c.** The V_{CE} for Q_2 rises to + V_{CC}. (Q_2 is switched off.)
3. The active state begins as capacitor C starts charging with the collector current of switched-on transistor Q_1. As long as capacitor C is charging toward +V_{CC} voltage

level, transistor Q_2 is held off. And as long as Q_2 is held off, base current for transistor Q_1 flows through resistors R_4 and R_2. The waveforms show that the *Trig in* signal has returned to 0 V, Q_1 is still turned on, capacitor C is charging (through Q_1 and R_C), and transistor Q_2 being held off.

4. The return state begins when capacitor C has charged very close to the $+V_{CC}$ level. When this happens, transistor Q_2 is biased on. The V_{CC} for Q_2 thus drops close to ground potential. The collector of Q_2 is direct-coupled to the base of Q_1, so when Q_2 is switched on, it switches transistor Q_1 off. Now everything has returned to the resting state where it remains until the next *Trig in* pulse is delivered to the base of Q_1 to turn it on again.

It is possible to estimate very closely the active timing interval for this monostable multivibrator. By studying this formula, you will find that the active timing interval is proportional to the values of the timing resistor (R_C) and timing capacitor (C).

FORMULA 10–9 $t_{on} = 0.69R_CC$

◆ **EXAMPLE** What is the time of the active state of a monostable multivibrator (Figure 10–16) when $R_C = 100\ \text{k}\Omega$ and $C = 0.1\ \mu\text{F}$?

Answer:

From Formula 10–9:

$$t_{on} = 0.69R_CC$$
$$t_{on} = 0.69 \times 100\ \text{k}\Omega \times 0.1\ \mu\text{F} = 6.9\ \text{msec}$$ ◆

Monostable multivibrator circuits are used for performing electronic timing operations. The timing in such applications takes place from the moment the trigger waveform occurs to the end of the active output interval. You have already seen how you can closely estimate that time with the aid of Formula 10–9. A monostable multivibrator is used in the circuits for controlling the overhead light in modern automobiles. In this system, the overhead light goes on whenever you open the car door, but then it remains on for a fixed period from the moment the door is closed. In this example, the monostable multivibrator is triggered to its active state when the car door is closed. Barring any interruption (such as turning on the ignition key), the monostable multivibrator causes the overhead light to remain on for a preset period that is on the order of 1 minute.

Today's monostable multivibrators are rarely constructed from discrete components. Most are constructed from integrated circuits that replace both transistors and most of the resistors. Only the timing resistors and capacitors are external to the circuit. This makes sense when you consider that the timing will be different for different kinds of applications, and that we need to have direct access to the timing components. Figure 10–17 shows a monostable multivibrator circuit that uses a 555 timer integrated circuit.

Let's point out the sequence of events that transpires during a complete cycle of monostable operation.

1. The quiescent state is the standby state. In this state, IC pin 7 is internally connected to ground so that capacitor C is held fully discharged in spite of the path to $+V_{CC}$ through R_C. Also, the output is pulled down very close to 0 V. The circuit remains in this state until the positive-going edge of a pulse occurs at the *Trig in* connection.

2. The active phase begins when the *Trig in* terminal at IC pin 2 sees the positive-going edge of a trigger pulse waveform. This positive transition immediately opens the internal ground connection at IC pin 7 and allows capacitor C to begin charging through R_C. Also, the triggering action switches the output to a voltage that is very close to $+V_{CC}$, where it remains until capacitor C is discharged again.

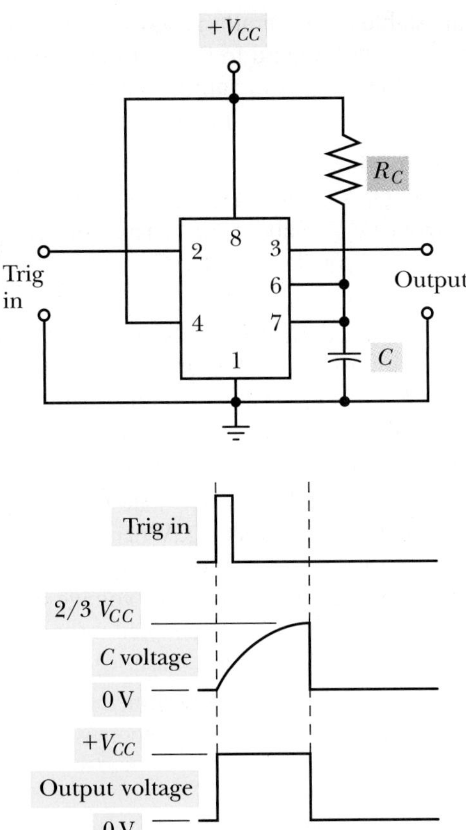

FIGURE 10–17 555 version of a monostable multivibrator circuit and waveforms

3. The timing interval ends when IC pin 6 senses that the voltage charge across capacitor C has reached 2/3 the amount of $+V_{CC}$. At that instant, IC pin 7 is internally shorted to ground. This shorting action immediately returns the output to 0 V and discharges capacitor C to ground. The circuit is now in its quiescent state until another trigger pulse occurs.

The timing interval of a 555-type monostable multivibrator is easily determined by Formula 10–10. In this formula, the output timing interval is in seconds, R_C is the value of the timing resistor, and C is the value of the timing capacitor.

FORMULA 10–10 $t_w = 1.1R_CC$

EXAMPLE What is the time of the active state of a 555-type monostable multivibrator, Figure 10–17, when $R_C = 1\ \text{M}\Omega$ and $C = 10\ \mu\text{F}$?

Answer:

From Formula 10–10:

$$t_w = 1.1R_CC$$
$$t_w = 1.1 \times 1\ \text{M}\Omega\ 10\ \mu\text{F} = 11\ \text{sec}$$

Astable Multivibrators

An **astable multivibrator** is a type of RC oscillator that generates square or rectangular waveforms. The term *astable* means not stable. This type of a circuit has a no stable state, and it produces its output as a result of continually seeking a stable state. The frequency of oscillation is determined by the charge time of the RC networks. Astable multivibrators are also known as **free-running multivibrators** and/or relaxation oscillators.

A BJT version of an astable multivibrator is shown in Figure 10–18. Let's see how the circuit works, especially with regard to the fact that one of the two transistors is always in saturation while the other is in cutoff. The operation of the circuit is based on the saturation and cutoff states alternating between the two transistors.

1. The active state for Q_1 begins when it is switched on. (How this happens will be shown in a moment.) As Q_1 is switched from cutoff (off) to saturation (on), its collector voltage drops from $+V_{CC}$ to about 0.7 V. This negative-going transition at the collector of Q_1 is coupled through capacitor C_{C_1} as a negative polarity to the base of Q_2. This reverse-bias voltage at the base of Q_2 turns it off.
2. During the active state of Q_1, capacitor C_{C_1} is charging through a path that begins with the ground connection to the emitter of Q_1, through the emitter and collector of Q_1, through C_{C_1} and R_{C_1} to $+V_{CC}$. The rate of charge of capacitor C_{C_1} depends on its own value and the value of resistor R_{C_1}. The *RC* time constant of R_{C_1} and C_{C_1} determines how long the active state of Q_1 persists.
3. The active state for Q_2 begins (and the active state for Q_1 ends) when C_{C_1} becomes charged through the path described in Step 2. When this happens, there is no longer a negative bias applied to the base of Q_2, so Q_2 begins to conduct. Q_2 thus switches from cutoff to a state of saturation. Its collector voltage drops from $+V_{CC}$ to about 0.7 V, and this negative-going transition is coupled through capacitor C_{C_2} as a negative polarity to the base of Q_1. This action biases Q_1 off.
4. Q_2 remains in its active state for a period determined by the *RC* time constant of the circuit composed of C_{C_2} and R_{C_2}. When C_{C_2} becomes fully charged, the negative

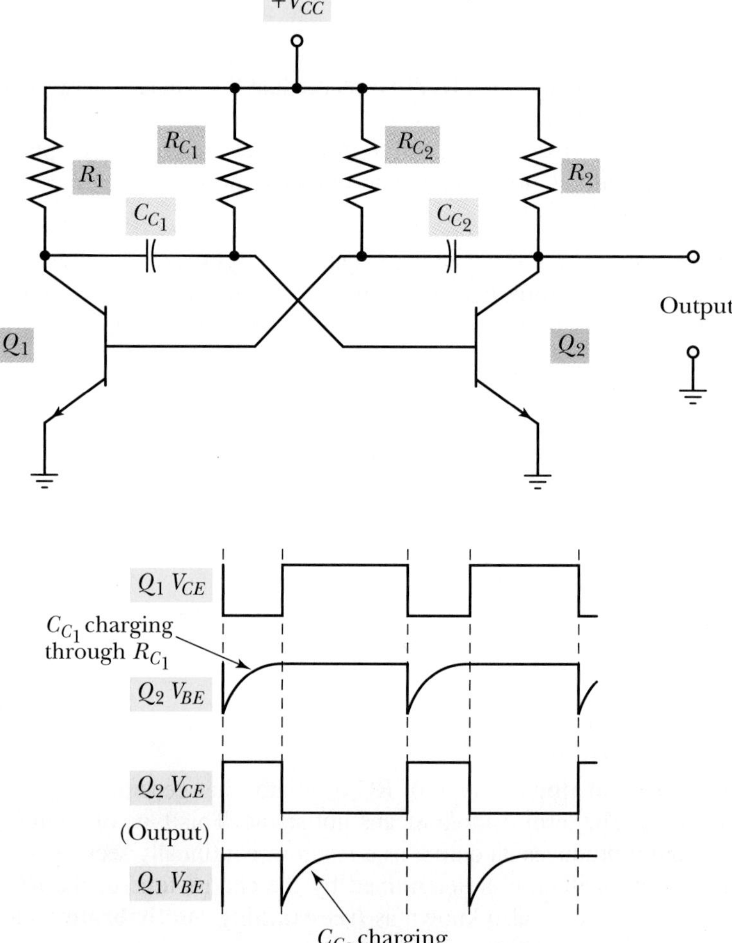

FIGURE 10–18 Discrete-component astable multivibrator circuit and waveforms

bias (due to the charging of C_{C_2}) disappears from the base of Q_1, and Q_1 is allowed to switch on. As Q_1 is switched on, this cycle returns to Step 1, where the switching-on action of Q_1 switches off Q_2.

The action described in these four states represents the operation of an astable multivibrator. The frequency of operation (or oscillating frequency) of this circuit depends on the *RC* time constant of R_{C_1} and C_{C_1} and the *RC* time constant of R_{C_2} and C_{C_2}. The general formula for calculating either one of these time intervals is shown in Formula 10–11, where *R* is the value of the resistor and *C* is the value of the capacitor.

FORMULA 10–11 $t = 0.7RC$

As part of the procedure for determining the operating frequency of an astable multivibrator, you must solve Formula 10–11 for *both RC* paths in the circuit. If desired, you can plot the rectangular waveforms found at the output of the multivibrator.

◆ **EXAMPLE** Referring to the circuit in Figure 10–18, determine the active time intervals for Q_1 and Q_2, and plot the waveform at the collector of Q_2 when:

$$R_{C_1} = 10\ \text{k}\Omega,\ C_{C_1} = 0.01\ \mu\text{F}$$
$$R_{C_2} = 22\ \text{k}\Omega,\ C_{C_2} = 0.001\ \mu\text{F}$$

Answer:

The active time for Q_1 is determined by the values of R_{C_1} and C_{C_1}. From Formula 10–11:

$$t_1 = 0.7\ R_{C_1}\ C_{C_1} = 0.7(10\ \text{k}\Omega)(0.01\ \mu\text{F})$$
$$t_1 = 70\ \mu\text{sec}$$

The active time for Q_2 is determined by the values of R_{C_2} and C_{C_2}. From Formula 10–11:

$$t_2 = 0.7\ R_{C_2}\ C_{C_2} = 0.7(22\ \text{k}\Omega)(0.001\ \mu\text{F})$$
$$t_2 = 15\ \mu\text{sec}$$ ◆

The waveform shown in the example represents the output taken from the collector of Q_2. The output of an astable multivibrator can also be taken from the collector of the opposite transistor, Q_1. The waveforms from either Q_1 or Q_2 are the same except they are inverted. That is, when one output is near 0 V, the other is near $+V_{CC}$.

The waveform in the example also shows that the output of an astable multivibrator is not necessarily symmetrical—the time interval of one of the states is not necessarily equal to the time of the opposite state. In this example, one interval is 70 µsec and the other is only 15 µsec.

Once you know the time interval of the two states, you can sum the intervals to arrive at the total period of the waveform: $T = t_1 + t_2$. And once you know the total period of the waveform, you can calculate frequency by the familiar period-to-frequency formula: $f = 1/T$.

◆ **EXAMPLE** What is the frequency of oscillation of an astable multivibrator that has an output waveform with states of 100 µsec and 250 µsec?

Answer:

The total period is given by:

$$T = t_1 + t_2 = 100\ \mu\text{sec} + 250\ \mu\text{sec} = 350\ \mu\text{sec}$$

The frequency is given by:

$$f = \frac{1}{T}$$
$$f = \frac{1}{350\ \mu\text{sec}}$$
$$f = 2.86\ \text{kHz}$$

When the *RC* circuits in an astable multivibrator contain identical values of resistance and capacitance, the output waveform is symmetrical. This means that the two states have the same interval. This is often the case in actual practice, and the procedure for calculating the operating frequency is much simpler as indicated by Formula 10–12. In this formula, it is assumed that $R_{C_1} = R_{C_2} = R_C$, and $C_{C_1} = C_{C_2} = C$. ◆

FORMULA 10–12 $f = \dfrac{1}{1.38 R_C C}$

◆ **EXAMPLE** What is the frequency of oscillation of the astable multivibrator in Figure 10–18 if $C_{C_1} = C_{C_2} = 0.01\ \mu\text{F}$ and $R_{C_1} = R_{C_2} = 47\ \text{k}\Omega$?

Answer:

From Formula 10–12:

$$f = \frac{1}{1.38 R_C C}$$

$$f = \frac{1}{(1.38)\times(47\ \text{k}\Omega)\times(0.01\ \mu\text{F})}$$

$$f = 1.54\ \text{kHz}$$

◆

Just as there is a simpler, integrated-circuit (IC) version of the monostable multivibrator, Figure 10–17, there is also a simpler, IC version of the astable multivibrator. In fact both versions can be designed around the same IC device—the 555 timer IC. See the 555-type astable multivibrator in Figure 10–19.

1. The active state begins when the output is switched to its $+V_{CC}$ level. The instant this happens, capacitor *C* begins charging toward the $+V_{CC}$ level through resistors R_A and R_B.
2. During the active state of the output, capacitor *C* continues to charge until pin 6 senses a capacitor voltage that is 2/3 times $+V_{CC}$. This marks the end of the active state.
3. At the end of the active state, the output voltage drops to zero. Also pin 7 is internally connected to ground, enabling the capacitor to begin discharging through resistor R_B.
4. The circuit remains in this zero-output state until pin 6 senses the capacitor voltage has discharged to 1/3 times $+V_{CC}$. At that time, the output switches to its $+V_{CC}$ level, the internal ground connection to pin 7 is switched open, and capacitor *C* is again allowed to charge towed $+V_{CC}$ through resistors R_A and R_B.

The frequency of operation of this circuit depends on the *RC* time constant of $R_A + R_B$ and *C* during the charge (active) time, and the *RC* time constant of R_B and *C* during the discharge (zero-output) time. The formula for calculating the frequency of oscillation is given in Formula 10–13.

FORMULA 10–13 $f = 1/0.69C(R_A + 2R_B)$

◆ **EXAMPLE** What is the frequency of operation of the 555-type astable multivibrator in Figure 10–19 if $R_A = 10\ \text{k}\Omega$, $R_B = 47\ \text{k}\Omega$, and $C = 0.001\ \mu\text{F}$?

Answer:

From Formula 10–13:

$$f = \frac{1}{0.69C(R_A + 2R_B)}$$

$$f = \frac{1}{0.69\times(0.001\ \mu\text{F})\times(10\ \text{k}\Omega + 2\times 47\ \text{k}\Omega)}$$

$$f = 13.9\ \text{kHz}$$

◆

Application Problem

MOTOR SPEED CONTROL

A motor is connected to a conveyor belt that is transporting cartons of Pepsi-Cola. The speed of the motor is adjusted by changing the average voltage across the motor. A 555 timer is configured as an astable multivibrator driving an op-amp comparator circuit. Adjusting the reference voltage of the comparator will change the duty cycle of the comparator output. What voltage would V_{REF} have to be for the motor to run at maximum speed?

Solution The voltage at the noninverting input will cycle between 1/3 and 2/3 of V_{CC}. The motor will run at maximum speed when there is maximum voltage across the motor. This will occur when the voltage coming out of the comparator is consistently positive. In order for the MOSFET transistor to remain in an ON state, the reference must be below (1/3)(5 V) = 1.67 V.

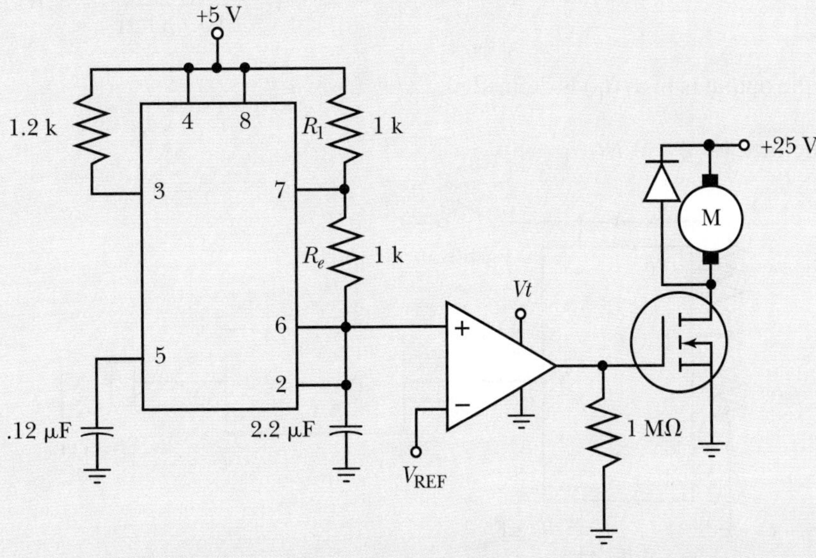

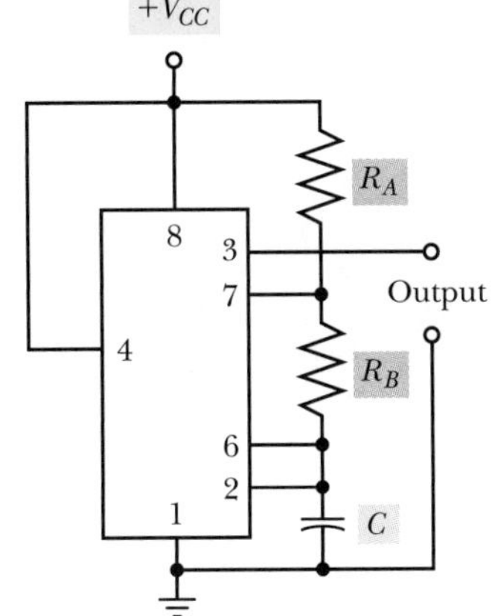

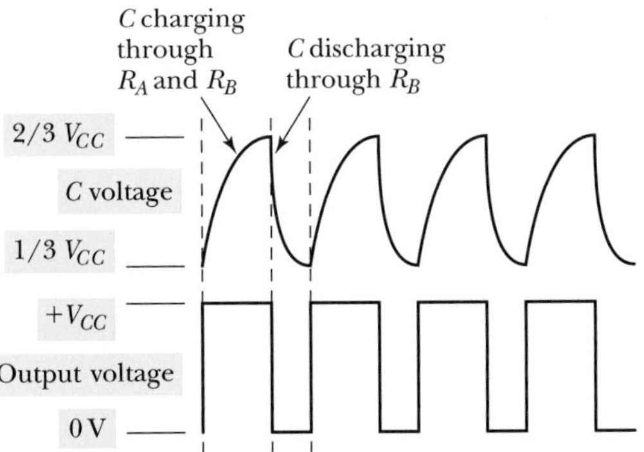

FIGURE 10–19 555 version of an astable multivibrator circuit and waveforms

Application Problem

General Electronics

AUDIO ALARM

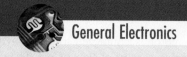

Commercial customers coming to pick up materials at a local hardware store must come to a parts counter and push a button to sound a buzzer to get the employees' attention. A 555 timer is configured as an astable multivibrator that is used to drive an amplifier. The amplifier output is going to a speaker. If the frequency is 700 Hz and the duty cycle is 60%, determine the R_1 and R_2 values if C_1 is 1 μF.

Solution The time the output is high (t_H) is defined as:
$t_H = 0.7(R_1 + R_2)C_1$
The low time (t_L) is defined as: $t_L = 0.7R_2C_1$

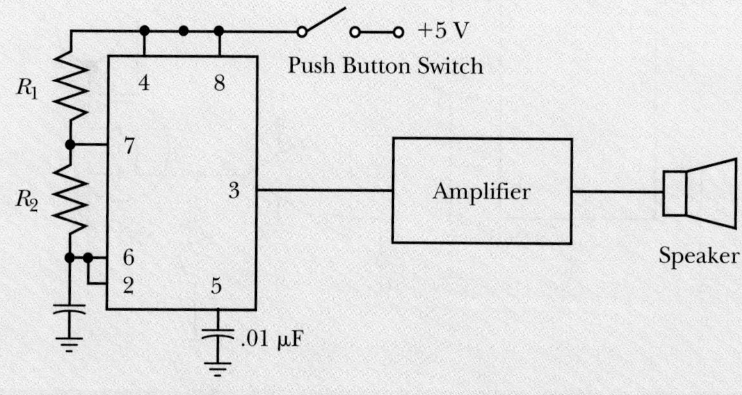

The period is $t_H + t_L$
Period = 1/frequency = 1/700 Hz = 1.43 ms
The low time is 40% of 1.43 ms, which equals 571.4 μs
R_2 = 571.4 ms/0.7 (1μF) = 816 ohms
t_H = 1.43 ms – 571.4 μs = 857.2 μs

$$R_1 = \frac{857.2\ \mu\text{s}}{(0.7)(1\ \mu\text{F})} - 816\ \Omega = 408.5\ \Omega$$

USING TECHNOLOGY: OSCILLATORS USED IN A SYSTEM

There are three basic types of electronic circuits. The simplest are those that control the flow of signals; Chapter 3 Diodes and Chapter 4 Power Supply Circuits covered this type. The second and by far the most common are those that amplify signals. Chapter 5 BJTs, Chapter 6 FETs, and Chapters 7, 8, and 9 on operational amplifiers described the multitude of types. This chapter covers the third type—oscillators or a circuit that generates and sustains an output signal without an input signal from another circuit or source.

Reviewing the requirements for an oscillator circuit, four basic conditions or circuit functions are needed for an oscillator to begin and then to sustain itself. First, a power source is needed. Second, components are required that determine the frequency of the oscillation. Third, amplification of the generated oscillation is necessary. Finally, positive feedback must occur from outback to the input to sustain the oscillation. This positive feedback is also called regenerative feedback; the name illustrates its purpose.

The tuning fork is a great example of a mechanical oscillator. Power is supplied by striking the tongs, while the length, width, and cross-section of the tongs determine the audible

CCTV Video-Recorder Application: Digital video recorder with built-in video server

sound waves and frequency generated. The design and materials allow the oscillation to be initially amplified and sustained through positive feedback of the sound waves as they move back and forth between the two tongs. Finally inertia and mechanical losses cause the sound level and the resultant positive feedback to decrease and then finally cease together.

Radio receivers of almost every type utilize the oscillator in a universal circuit configuration called the superheterodyne. This most popular receiver circuit operates on the principle of converting all incoming frequencies, which vary all over the band, to a single intermediate frequency. In this way the band of frequencies so converted can be amplified by tuned circuits fixed in frequency and optimized for the best sensitivity and the selectivity. Most of the gain, the sensitivity, and selectivity of a superheterodyne receiver is obtained by the use of these fixed tuned intermediate fre-

Voice Recorder Application: Digital player with flash memory

quency (IF) amplifiers. The oscillator is paired with another circuit called the mixer, which acts like a simple modulator to produce the sum (incoming RF plus the oscillator frequency) and the difference (incoming RF minus the oscillator frequency) frequencies needed to produce the IF band of frequencies.

The system application block diagram for the VHF-UHF mixer/oscillator portion of the television from input to output is shown in Figure 25. The incoming RF signal from

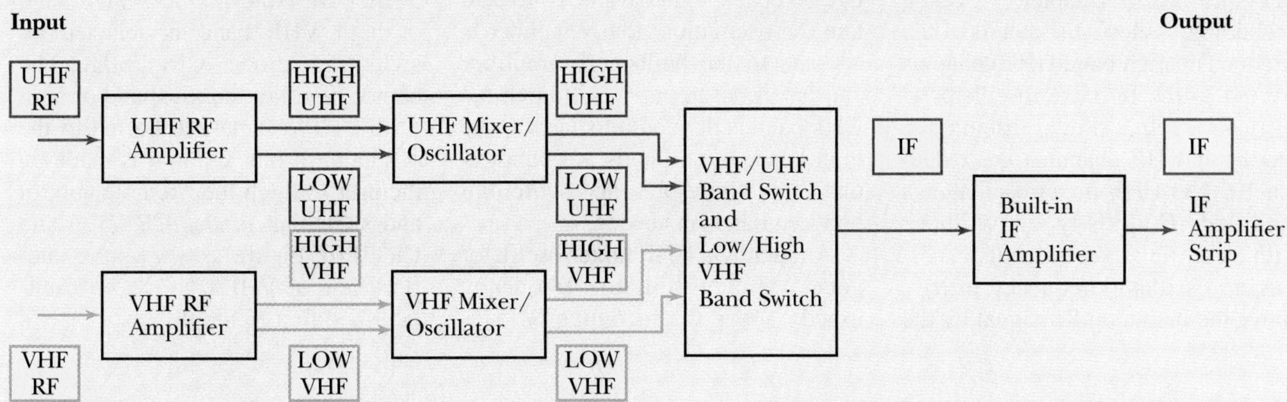

FIGURE 25 Block Diagram

an antenna, a cable box, or a dish converter box is applied to the antenna/cable RF input. The RF input is fed either to the UHF RF amplifier stage for channels 14 through 69 or to the VHF RF amplifier stage for channels 2 through 13 and the cable channels. The UHF band of frequencies is now divided into the low band of channels (channels 14 through 51) and the high band of channels (channels 52 through 69). In a similar fashion, the VHF band of frequencies is now divided into the low band of channels (channels 2 through 6) and the high band of channels (channels 7 through 13) along with the cable channels.

The mixer/oscillator is a combination Colpitts-type oscillator with a self-contained mixer. Such combination circuits are also called autodyne converters or simply converters. The UHF and the separate VHF mixer/oscillator each operate at four separate frequencies as follows: the RF input frequency, the oscillator frequency, the sum, and the difference of these two frequencies. The VHF/UHF band switch along with the low/high VHF band switch selects only one of these four as the IF frequency and feeds this signal to the built-in IF amplifier. This amplified IF signal is passed into the first IF amplifier causing a further increase in gain and then to the remaining IF amplifiers for additional amplification.

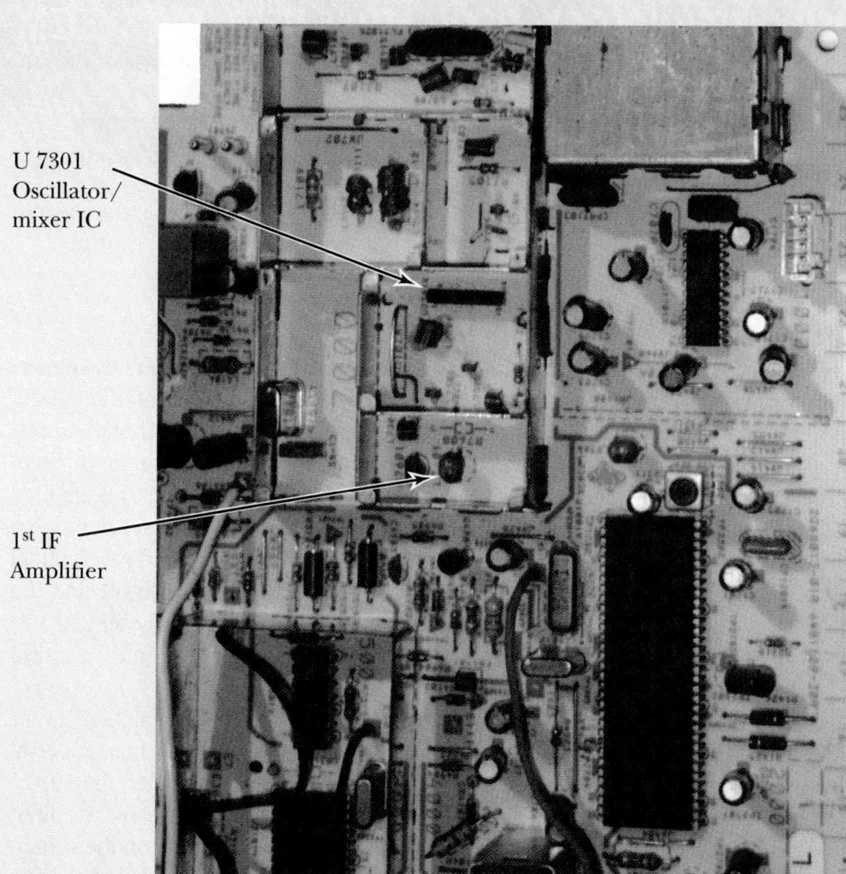

FIGURE 26 Interior view of VHF/UHF mixer/oscillator and first IF amplifier

Figure 17 in Chapter 6 "Using Technology" shows the details of the circuit. The high band UHF channels are fed to the IC mixer/oscillator U 7301 pin 12. In a similar fashion, the low band UHF channels are fed to pin 10. The UHF mixer/oscillator, a modified Colpitts-type oscillator with a built-in mixer function, generates an oscillator frequency exactly above the incoming RF signal by the IF frequency. The UHF oscillator frequency is set by varactor diodes CR 7304 and CR 7301 along with tunable inductor L 7303. A portion of this frequency is fed back as a positive feedback signal at pin 16 to sustain the oscillation; the remainder is coupled to the built-in IF amplifier and is fed out at pin 1. Although the UHF band is divided into a low and a high band, these bands are contiguous and therefore no switching between bands is needed.

An identical VHF mixer/oscillator generates an oscillator frequency exactly above the incoming RF signal by the IF frequency. The VHF oscillator frequency is set by varactor diodes CR 7305 and CR 7302 along with tunable inductors L 7304 and L 7305. The VHF oscillator frequency for either the low VHF band or high VHF band is selected by switching diode CR 7303. The choice of either the low band or high band VHF channels to be fed to the oscillator at pins 2 and 4 is made by the user through their remote control and switching diodes CR 7109 and CR 7110. Figure 26 shows the interior view of VHF/UHF Mixer/oscillator and first IF amplifier.

Summary

- Four elements or conditions are absolutely required to begin and sustain the operation of an oscillator; a power source, frequency determining device or components, amplification greater than 1, and positive feedback.
- The Hartley oscillator is an *LC* sine-wave oscillator that has a tank circuit composed of a capacitor and a tapped inductor (or two inductors in series) that is grounded at the tap (or common connection between the two inductors). The total inductance of the Hartley tank circuit is equal to the sum of the two inductance values.
- The Colpitts oscillator is an *LC* sine-wave oscillator that has a tank circuit composed of a single inductor and two series-connected capacitors that are grounded at their common connection. The total capacitance of the Colpitts tank circuit is found by the product-over-sum rule as applied to the values of the two capacitors.
- Because of undesirable loading effects, the output of a Colpitts *LC* oscillator is often transformer coupled to the next stage. The primary winding of the coupling transformer serves as the inductance for the tank circuit.
- The Clapp oscillator is an *LC* sine-wave oscillator that is identical with the Colpitts oscillator except for the existence of three capacitors in the tank circuit. The extra capacitor provides greater stability against changes in the output loading. The total capacitance in the Clapp tank circuit is found by the formula used for the total capacitance of three capacitors in series.
- One way to remember the difference between Hartley and Colpitts oscillators is to remember: Hartley = H = Henrys (two inductance values), and Colpitts = C = Capacitance (two capacitance values).
- The frequency of a crystal-controlled oscillator is fixed by the mechanical shape and size of the crystal. The crystal is the equivalent of a parallel tuned (*LC* tank) circuit.
- The phase-shift oscillator is an *RC* sine-wave oscillator that uses a phase-shift network composed of three *RC* elements to produce a 180° phase shift at one particular frequency. When capacitors are connected in series with the feedback path, the feedback voltage leads the output of the amplifier. When the resistors are connected in series with the feedback path, the feedback voltage lags the output of the amplifier.
- The Wien-bridge oscillator is an *RC* sine-wave oscillator that uses a lead-lag network comprised of two *RC* elements (one in series and one in parallel) to produce a 0° phase shift at the circuit's oscillating frequency. The amplifier is designed for noninverting operation. Thus the overall feedback at the oscillating frequency remains at 0°—as required for starting and sustaining oscillation.
- Multivibrators are characterized by generating rectangular waveforms and using only *RC* circuits for setting the timing or frequency of oscillation.
- A monostable multivibrator (also called a one-shot multivibrator) generates a single, *RC*-timed, output pulse in response to a single input trigger pulse.
- An astable multivibrator (also called a free-running multivibrator) generates a continuous rectangular waveform whose active and inactive times are determined by the charge times of capacitors in the circuit.

CIRCUIT SUMMARY

Description	*Circuit*	*WaveForm/Formula*
Op-amp Hartley oscillator	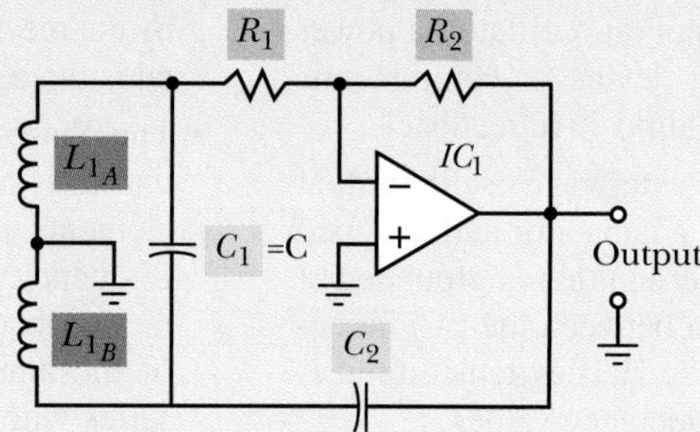	$f_r = \frac{1}{2\pi\sqrt{L_t C}}$ $L_t = L_{1_A} + L_{1_B}$
Op-amp Colpitts oscillator	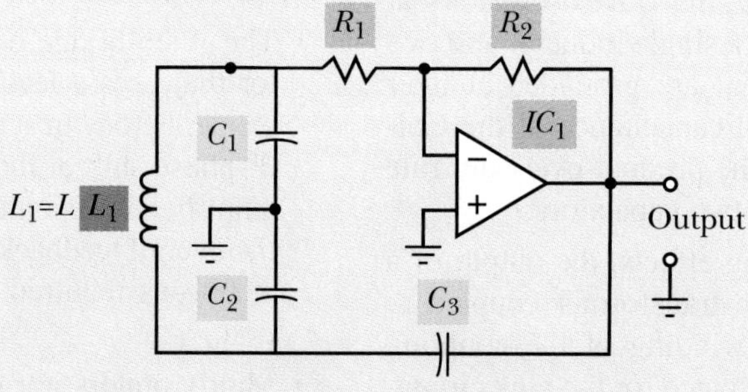	$f_r = \frac{1}{2\pi\sqrt{LC_T}}$ $C_T = \frac{C_1 C_2}{C_1 + C_2}$
BJT Clapp oscillator	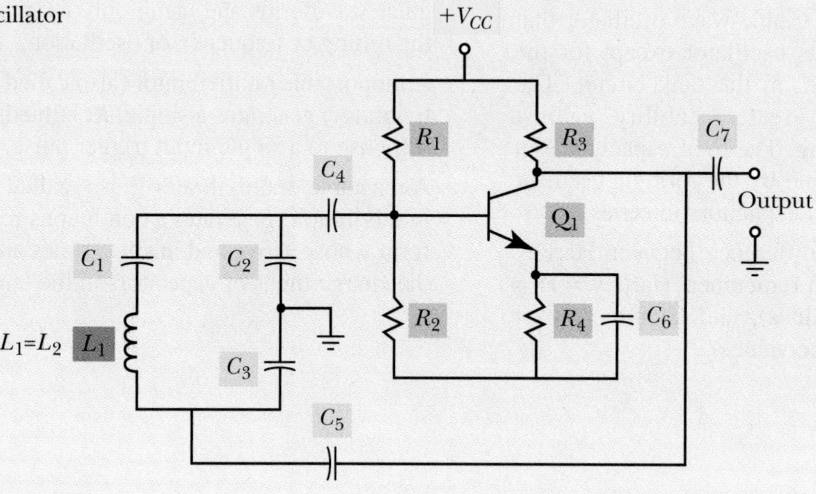	$f_r = \frac{1}{2\pi\sqrt{LC_T}}$ $C_T = \frac{1}{\frac{1}{C_1} + \frac{1}{C_2} + \frac{1}{C_3}}$
Op-amp Crystal oscillator	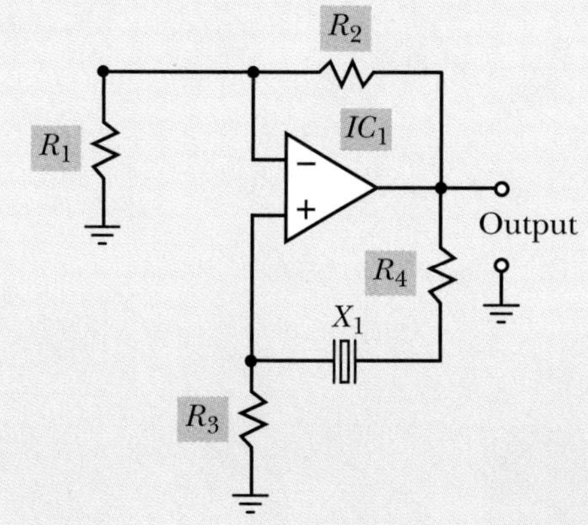	

Formulas and Sample Calculator Sequences

FORMULA 10–1
(To find the resonant frequency)

$$f_r = \frac{1}{2\pi\sqrt{LC}}$$

L value, [×], C value, [=], √‾, [×], 6.28, [=], [1/x]

FORMULA 10–2
(To find the Hartley oscillator frequency)

$$f_r = \frac{1}{2\pi\sqrt{L_t C}}$$

(Use the keystrokes for Formula 29–1)

FORMULA 10–3
(To find the Hartley total inductance)

$L_t = L_{1_A} + L_{1_B}$

L_{1_A} value, [+], L_{1_B} value, [=]

FORMULA 10–4
(To find the Colpitts and Clapp oscillator frequency)

$$f_r = \frac{1}{2\pi\sqrt{LC_T}}$$

(Use the keystrokes for Formula 29–1)

FORMULA 10–5
(To find the Colpitts total capacitance)

$$C_T = \frac{C_1 C_2}{(C_1 + C_2)}$$

C_1 value, [×], C_2 value, [÷], [(], C_1 value, [+], C_2 value, [)], [=]

FORMULA 10–6
(To find the Clapp total capacitance)

$$C_T = \frac{1}{\left(\frac{1}{C_1} + \frac{1}{C_2} + \frac{1}{C_3}\right)}$$

C_1 value, [1/x], [+], C_2 value, [1/x], [+], C_3 value, [1/x], [=], [1/x]

FORMULA 10–7
(To find the RC phase-shift oscillator)

$$f_r = \frac{1}{2\pi RC\sqrt{6}}$$

6, √‾, [×], 6.28, [×], R value, [×], C value, [=], [1/x]

FORMULA 10–8
(To find the Wien-bridge oscillator frequency)

$$f_r = \frac{1}{2\pi RC}$$

R value, [×], C value, [×], 6.28, [=], [1/x]

FORMULA 10–9
(To find active timing interval for monostable multivibrator)

$t_{on} = 0.69\ R_C C$

0.69, [×], R_C value, [×], C value, [=]

FORMULA 10–10
(To find timing interval for monostable multivibrator)

$t_w = 1.1 R_C C$

1.1, [×], R_C value, [×], C value, [=]

FORMULA 10–11
(To find active time interval for astable multivibrator)

$t = 0.7RC$

0.7, [×], R value, [×], C value, [=]

FORMULA 10–12 *(To find frequency for discrete astable multivibrator)*

$$f = \frac{1}{1.38 R_C}$$

1.38, [×], R_C value, [×], C value, [=], [1/x]

FORMULA 10–13 *(To find frequency for 555 astable multivibrator)*

$$f = \frac{1}{0.69C(R_A + 2R_B)}$$

[(], 2, [×], R_B value, [+], R_A value, [)], [×], 0.69, [×], C value, [=], [1/x]

Using Excel

Oscillator Formulas

(Excel file reference: FOE10_01.xls)

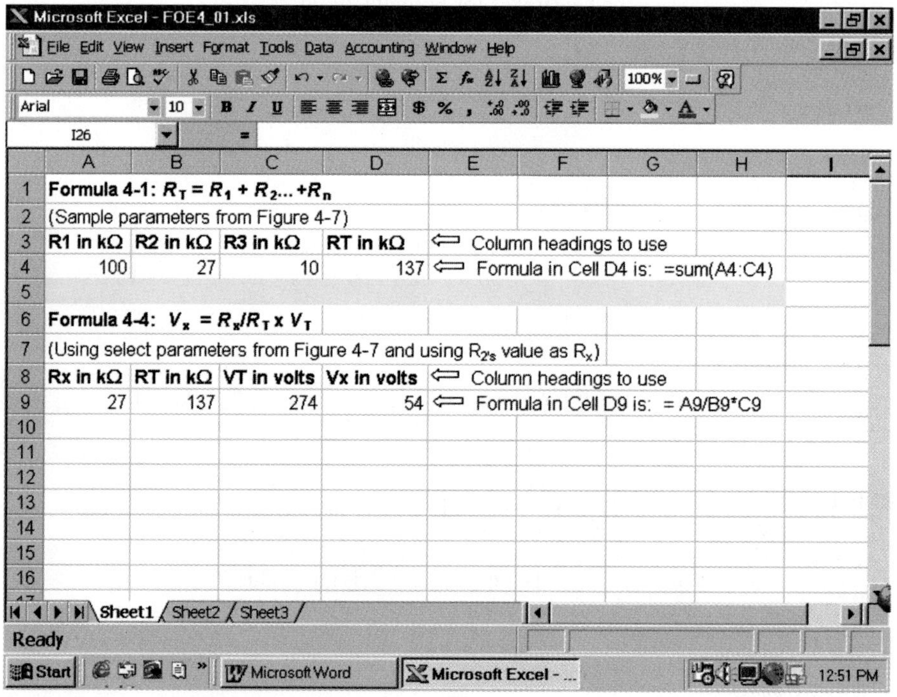

DON'T FORGET! It is NOT necessary to retype formulas, once they are entered on the worksheet! Just input new parameters data for each new problem using that formula, as needed.

1. Use the Formula 10–2 and Formula 10–3 spreadsheet samples and the parameters given for Chapter Problems #3. Solve for the total inductance and the output frequency of the oscillator. Check your answer against the answer for this question in the Appendix.

2. Use the Formula 10–4 and Formula 10–5 spreadsheet samples and the parameters given for Chapter Problems #5. Solve for the total capacitance and the output frequency of the oscillator. Check your answer against the answer for this question in the Appendix.

Microsoft Excel - FOE4_01.xls

	A	B	C	D	E	F	G	H	I
1	Formula 4-1: $R_T = R_1 + R_2 ... + R_n$								
2	(Sample parameters from Figure 4-7)								
3	R1 in kΩ	R2 in kΩ	R3 in kΩ	RT in kΩ	⇐ Column headings to use				
4	100	27	10	137	⇐ Formula in Cell D4 is: =sum(A4:C4)				
5									
6	Formula 4-4: $V_x = R_x/R_T \times V_T$								
7	(Using select parameters from Figure 4-7 and using R_2's value as R_x)								
8	Rx in kΩ	RT in kΩ	VT in volts	Vx in volts	⇐ Column headings to use				
9	27	137	274	54	⇐ Formula in Cell D9 is: = A9/B9*C9				
10									
11									
12									
13									
14									
15									
16									

Sheet1 / Sheet2 / Sheet3

Ready

Review Questions

1. List the four elements and conditions that are absolutely required to begin and sustain the operation of an oscillator.
2. Write the main formula that applies for finding the frequency of oscillation of all *LC*-type oscillators. Specify the meaning of each term.
3. Which type of *LC* oscillator uses a tapped coil? Two capacitors with a ground connection between them? Three capacitors and a single inductor in the tank circuit?
4. Draw the tank circuit portion of the following:
   - **a.** A Hartley oscillator
   - **b.** A Clapp oscillator
   - **c.** A Colpitts oscillator
5. Suppose the components for the tank circuit of the Hartley oscillator you drew in question 4 have the following values: L_1 = 10 mH, L_2 = 10 mH, and C = 0.1 μF. What is the frequency of oscillation?
6. Assign values of your own choosing to the components for the tank circuit of the Clapp oscillator you drew in question 4 Calculate the frequency of oscillation. (Be sure to show all your work.)

7. Suppose the components for the tank circuit of the Colpitts oscillator you drew in question 4 has the following values: $L = 10$ mH, $C_1 = 0.1$ μF, and $C_2 = 0.1$ μF. What is the frequency of oscillation?
8. In any of the *LC* sine-wave oscillators studied in this chapter, you can increase the frequency of operation by _____ the value(s) of inductance or by _____ the value(s) of capacitance.
9. Which type of *RC* sine-wave oscillator studied in this chapter uses a network of three identical *RC* elements? Which uses a lead-lag network?
10. What element(s) of a crystal-controlled oscillator determine(s) its frequency of oscillation?
11. At the frequency of oscillation, the *RC* network in a phase-shift oscillator produces a phase shift of _____°, and the amplifier produces a phase shift of _____°.
12. Sketch the frequency determining *RC* network for (a) a phase-shift oscillator and (b) a Wien-bridge oscillator.
13. The capacitors in the *RC* network of a Wien-bridge oscillator have a value of 1 μF and the resistors have a value of 1 kΩ. What is the frequency of oscillation?
14. What is another common name for a monostable multivibrator?
15. In order to decrease the active time of a monostable multivibrator, you must _____ the value of the timing capacitor and/or _____ the value the timing resistors.
16. What is another common name for an astable multivibrator?
17. What is the frequency of oscillation of an astable multivibrator that produces a waveform near $+V_{CC}$ for 10 μsec and near 0 V for 8 μsec?
18. Explain why the $+V_{CC}$ time of the output waveform of a 555-type astable multivibrator is always longer than the 0-V time.
19. In a 555-type astable multivibrator, $R_A = 10$ kΩ, $R_B = 10$ kΩ, and $C = 0.1$ μF. What is the frequency of oscillation?
20. In order to increase the frequency of operation of the 555-type astable multivibrator, you must _____ the value of the timing capacitor, or you can the value of either or both of the timing resistors.

Problems

1. Determine the resonant frequency of a parallel *LC* circuit where $L = 1$ mH and $C = 1$ nF.
2. A certain tank circuit is to resonate at 10.2 MHz. If you already know that $L = 20$ μH, determine the required value of *C*.
3. A BJT Harley oscillator (see Figure 10–3) uses inductances of 90 μH and 40 μH. If the capacitor in the tank circuit is 500 μF, what is the operating frequency of the circuit?
4. Determine the frequency of a Hartley oscillator that uses a tank circuit composed of a 1-mH center-tapped coil and a 100-nF capacitor.
5. An op-amp Colpitts oscillator (see Figure 10–7) uses capacitances of 90 pF and 40 pF. If the inductor in the tank circuit is 5 μH, what is the operating frequency of the circuit?
6. Referring to the FET Colpitts oscillator to Figure 10–8, replace C_2 with a variable capacitor that can be adjusted between 10 pF and 90 pF. If C_1 is fixed at 400 pF and the primary inductance of T_1 is 10 μH, calculate the lowest and highest operating frequencies for the circuit.
7. All three capacitors in the tank circuit for a Clapp oscillator (see Figure 10–9) are listed as 22 nF. What is the value for C_T?
8. If the operating frequency of a Clapp oscillator is to be 1.1 MHz when the inductor in the tank circuit is rated at 120 μH, what is the necessary value for C_T?
9. Calculate the frequency of a phase-shift oscillator (see Figure 10–14) where all the capacitors are rated as 1 μF and all the resistors at 1 MΩ.
10. A certain Wien-bridge oscillator (see Figure 10–15) is operating a 100 kHz when the timing resistor R_2 is equal to 1 MΩ. What is the value of the timing capacitor C_2?

Analysis Questions

1. Draw a JFET and an op-amp version of the Clapp oscillator.
2. Research and briefly explain the operation of the Armstrong oscillator circuit.
3. Explain why the amplifier element of a phase-shift oscillator must be an inverting amplifier, while the amplifier element in a Wien-bridge oscillator must be a noninverting amplifier.
4. Research and draw the schematic diagram for a twin-T *RC* oscillator.
5. Explain why the amplifier element in a phase-shift oscillator must have a voltage gain that is greater than those for the other types of oscillators described in this lesson.
6. Research and explain the meanings of *interruptible* and *noninterruptible* monostable multivibrator.
7. Research and describe the function of a bistable multivibrator, or flip-flop.
8. Explain why the charge time for the timing capacitor of a 555 astable multivibrator (see Figure 10–19) is always longer than the discharge time. Further explain how this accounts for the fact that the time the output voltage is near $+V_{CC}$ is always longer than the time it is near 0 V.
9. Describe what happens to the operating frequency of a 555 astable multivibrator when the value of the timing capacitor is increased and when it is decreased.
10. Explain why the timing for 555 monostable and astable multivibrators is independent of the power supply voltage.

Performance Projects Correlation Chart

All e-labs are available on the LabSource CD.

Chapter Topic	Performance Project	Project Number
Wien-Bridge Oscillator	Wien-Bridge Oscillator Circuit	29
Astable Multivibrators	555 Astable Multivibrator	30

NOTE: It is suggested that after completing the above projects, the student should be required to answer the questions in the "Summary" at the end of this section of projects in the Laboratory Manual.

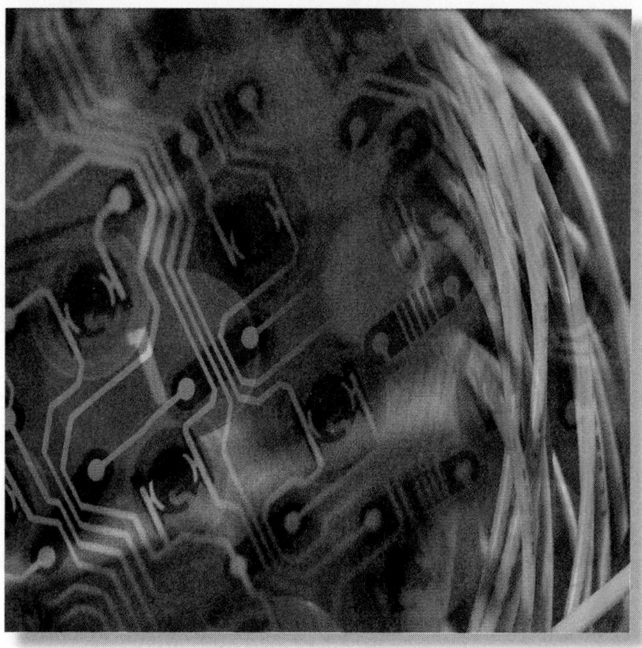

CHAPTER 11

Thyristors

OUTLINE

OBJECTIVES

After studying this chapter, you should be able to:

1. Describe what a **thyristor** is
2. Describe in detail the way an **SCR** can be switched on and off.
3. Explain the operation of simple SCR circuits including a power "on/off" push-button control circuit and an electronic "crowbar"
4. Identify the symbols for and describe the operation of the **gate-controlled switched, silicon-controlled switch,** and **light-activated SCR**
5. Identify and explain the purpose of thyristors connected in inverse parallel, or back-to-back
6. Identify the schematic symbol and explain the operation of a **diac**
7. Explain the details for starting and ending the conduction of a **triac**
8. Identify phase-control power circuits that use thyristors
9. Describe basic troubleshooting procedures for thyristors

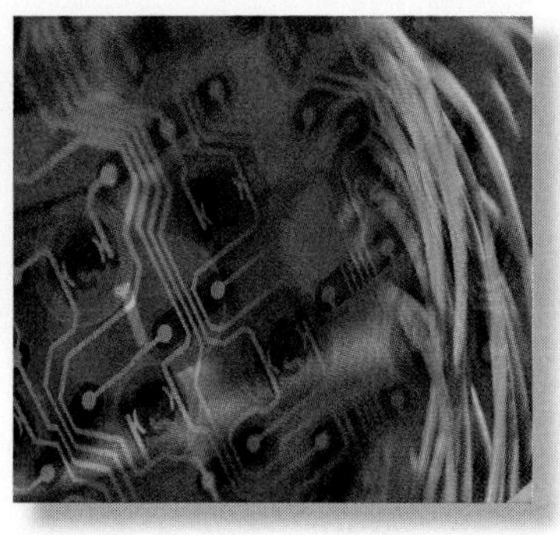

PREVIEW

A **thyristor** is a four-layer semiconductor device that has only two states of operation: fully conducting or fully nonconducting. Recall that a junction diode is composed of two layers, P-N. Thyristors, with their four layers, are like two diodes connected in series: P-N-P-N. (It is important to realize, however, that two junction diodes connected in series cannot be made to perform like a thyristor.)

There are different types of thyristors that respond differently with regard to the ways they are turned on and off. The thyristors discussed in this chapter are the silicon-controlled rectifier (SCR), diac, and triac. Compared with the other semiconductor devices you have studied in previous chapters, thyristors are built for high current and voltage ratings. Whereas a typical BJT could have a maximum collector current of 100 mA at 30 V, a typical SCR could handle 5 or 10 A at 220 V.

Thyristors are found in consumer electronic devices, most notably in wall switch light dimmers and automobile electronic ignition systems. By far the greatest quantity and variety of thyristors, however, are used in industrial power control equipment.

KEY TERMS

Bidirectional
Diac
Gate-controlled switch (GCS)
Holding current
Inverse parallel
Light-activated SCR (LASCR)
Silicon-controlled rectifier (SCR)
Silicon-controlled switch (SCS)
Thyristor
Triac

11–1 Silicon-Controlled Rectifiers (SCRs)

The **silicon-controlled rectifier (SCR)** is a simple four-layer thyristor device. The semiconductor structure shown in Figure 11–1 indicates the four layers and shows the standard schematic symbol. The terminals are called cathode *(K),* anode *(A),* and gate *(G).*

If the gate terminal is left open or shorted directly to the cathode, the device cannot conduct current between the cathode and anode. The reason is that at least one of the three P-N junctions will be reverse biased, no matter which polarity of voltage you apply between the cathode and anode. The SCR can be made to conduct, however, by meeting two conditions:

1. A voltage is applied between the cathode and anode such that the anode is positive compared to the cathode.
2. A voltage is applied to the gate such that the P-N junction between the gate and cathode is forward biased (at least momentarily).

Once an SCR begins conducting under the conditions just described, the only way to turn it off is by removing the voltage applied between the anode and cathode, or by reverse biasing the anode-cathode connection. You cannot turn off a conducting SCR by removing the gate voltage, and you cannot turn off an SCR by reversing the voltage applied to the gate. The gate of an SCR is intended to initiate conduction between the cathode and anode. Once this condition begins, the gate plays no further role in the operation. The SCR will turn off when the anode current is less than the **holding current** (the current required to sustain SCR conduction).

Theory of SCR Operation

The diagrams in Figure 11–2 should help you understand and remember the conditions for turning an SCR on and off. The four layers have been drawn to clarify the fact that they can be regarded as a set of two BJTs. The equivalent BJT that includes the anode and gain connections is a PNP transistor shown as Q_1 on the schematic. The equivalent BJT that includes the cathode is an NPN transistor shown as Q_2 on the schematic.

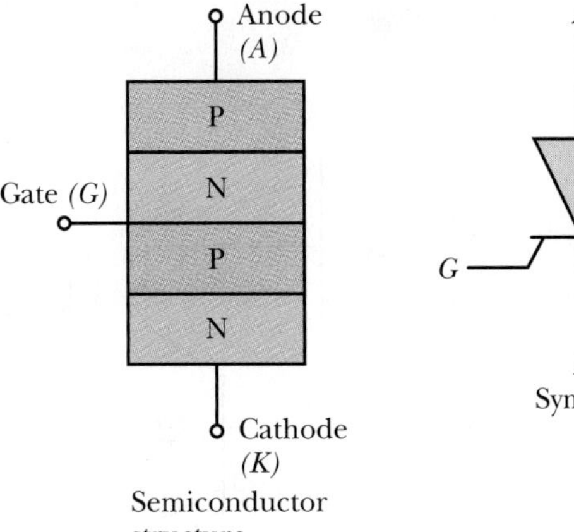

FIGURE 11–1 SCR semiconductor structure and schematic symbol

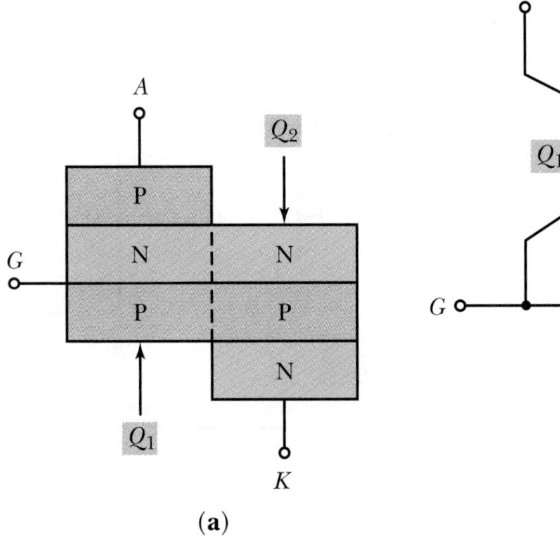

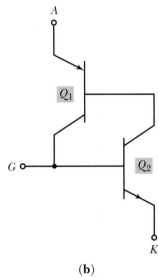

FIGURE 11–2 SCR as a set of two BJTs: (a) semiconductor structure; (b) equivalent BJT circuit

With the equivalent BJT circuit in Figure 11–2, it is impossible for just one BJT to conduct. Either they both conduct or neither one conducts. This is because forward-biasing base current for Q_1 cannot flow (to turn it on) unless Q_2 is turned on. The collector current for Q_2 provides the base current for Q_1. If Q_2 is *not* conducting, there is no base current for Q_1, so it cannot be conducting either.

Now suppose a positive pulse from an outside voltage source is applied to the gate terminal. The equivalent circuit in Figure 11–3 indicates the sequence of events that takes place. Observe the numbers (indicating the sequence of events) and paths for current flow:

Step 1. The positive pulse at the gate terminal causes forward-biasing base current (I_{B_2}) to flow at BJT Q_2.

Step 2. The forward-biasing base current for Q_2 turns on that transistor and causes collector current (I_{C_2}) to flow.

Step 3. The collector current for Q_2 provides forward-biasing base current (I_{B_1}) for BJT Q_1.

Step 4. The forward-biasing base current for Q_1 turns on that transistor and causes collector current (I_{C_1}) to flow.

By the time the process reaches Step 4, both transistors are in saturation. Even after the trigger potential at the gate terminal returns to zero, the transistors remain latched on. This is ensured by collector current I_{C_1} maintaining base current for Q_2. For all

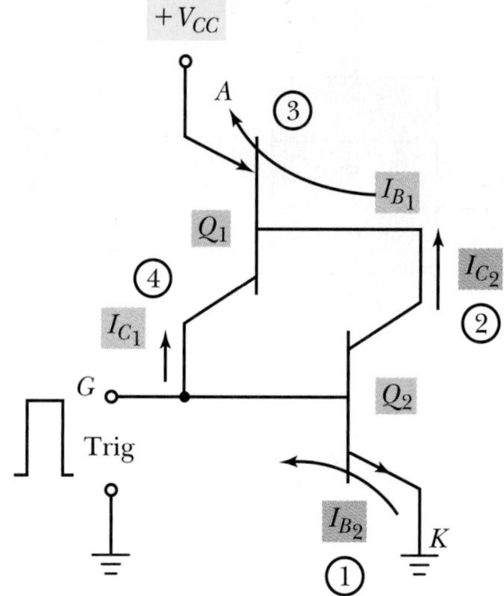

FIGURE 11–3 Turning on an SCR

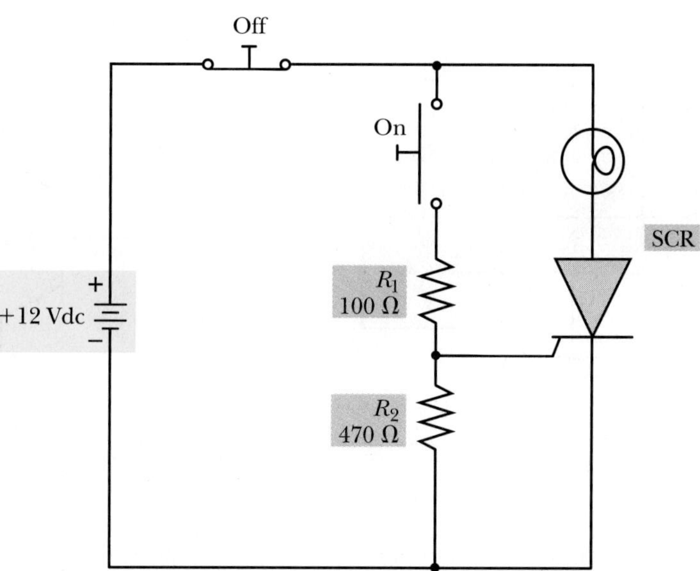

FIGURE 11–4 Simple SCR power "on/off" circuit

practical purposes, the collector current for Q_1 acts as a steady gating pulse that keeps both transistors latched on. The only way to interrupt this latching effect is by removing the supply voltage or reversing its polarity.

Figure 11–4 shows a simple SCR demonstration circuit. It indicates how an SCR can be latched into conduction and turned off by interrupting the source of power to it. The power source is a 12-V battery, the load to be controlled is a 12-V automobile lamp. The normally closed "off" push-button switch is connected in series between the positive side of the dc power source and anode of the SCR. When this button is depressed, power is totally disconnected from the circuit. The normally open "on" push-button switch is connected in series with the positive side of the power source and the gate of the SCR. Whenever the "on" push button is pressed, a positive bias is applied to the gate of the SCR. This circuit is thus turned on by momentarily pressing the "on" button; and it is turned off by depressing the "off" button.

The most useful ratings for an SCR are listed here:

- *Peak forward blocking voltage:* the largest amount of forward-bias voltage the SCR can hold off without breaking down and conducting. The range of available values is between 30 V and 1,200 V.

FIGURE 11–5 80-ampere stud-mounted SCRs

- *Maximum forward current:* the largest amount of current, dc or rms, that the SCR can handle continuously without special heat sinking. The range of available values is between 0.8 A and 100 A, Figure 11–5.
- *Peak reverse voltage:* the largest amount of reverse voltage that can be applied between the cathode and anode without breaking into conduction. The value of the peak reverse voltage is often the same as the peak forward blocking voltage.
- ***Holding current:*** the smallest amount of cathode-anode forward current that can sustain conduction of the SCR. This value is on the order of 5 mA to 20 mA.
- *Forward "on" voltage:* the forward voltage drop between cathode and anode while the SCR is conducting. Under normal operating conditions, this voltage is between 1 to 2 V.

There are also current and voltage ratings for the gate circuit, including forward cathode-gate voltage, maximum forward gate current, reverse breakdown voltage for the cathode-gate junction, and reverse gate leakage current.

Now that you have seen the wide range of SCR ratings, let's look at some further examples of SCR circuits.

SCR Circuits

The circuit in Figure 11–6 provides an effective means for protecting sensitive electronic components (such as computer memory circuits) from sudden, unexpected surges of high voltage from a dc power supply. The power supply in this example is designed to provide +12 Vdc to the circuit being protected. Note that the zener diode is rated at 14 V, so you can realize that the zener does not normally conduct. In fact, the zener will not conduct at all unless the dc power supply voltage rises above 14 V (which occurs from a surge of high voltage). As long as the zener diode does not conduct, the SCR cannot conduct because the zener current is necessary for gating the SCR. But the moment the zener diode conducts due to a surge of high voltage, the SCR is gated on.

Once the SCR is gated on, it acts as a short circuit across the dc power supply. This causes a surge of relatively high current that blows the fuse to remove the dc power supply from the circuit being protected. This circuit is known as an electronic *crowbar.*

The circuit in Figure 11–7 represents the most common application of SCRs. Such a circuit is used for smoothly adjusting the amount of ac power that is applied to a wide range of devices. For lamps, the circuit can be used to control their intensity from off to full brightness; for ac motors, the circuit can be used to control their speed; and for heating elements, the circuit can be used to control their amount of heating.

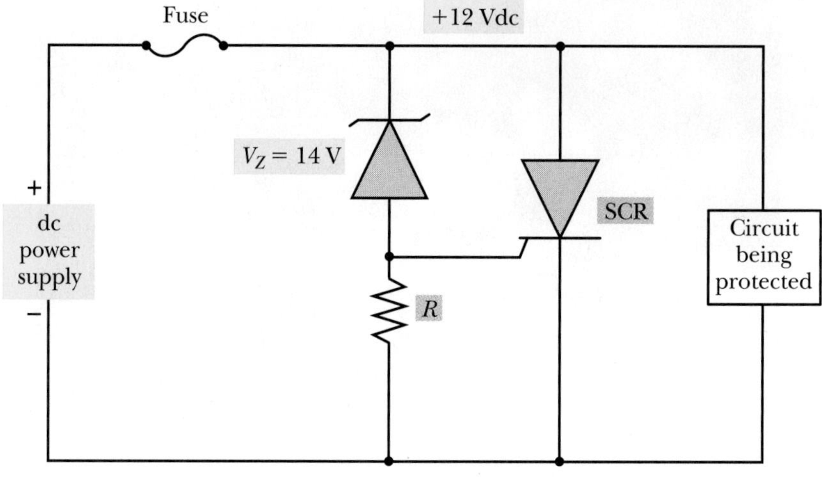

FIGURE 11–6 SCR used in a "crowbar" protection circuit

FIGURE 11–7 AC power control circuit using two SCRs connected in inverse parallel

$T_{1\text{-S1}}$
SCR$_1$
SCR$_2$
$T_{1\text{-S2}}$
ac
source
ac
load
Phase-shift
control
$T_{1\text{-P}}$

Before getting into the details of operation of this power control circuit, let's look at a few of its general features. First notice that the circuit is operated from an ac source. The ac power source is usually 120 Vac or 240 Vac as supplied by the local power utility. The ac load is the device that is to be operated from the ac source. Normally, the ac load runs at full power (light, speed, heat, etc.) at the rated ac voltage level; so an SCR control circuit of this type is expected to reduce the power level from the full-power level down to some other desired level.

Next, notice how the two SCRs (SCR$_1$ and SCR$_2$) are connected with respect to one another. Note that the cathode of one is connected to the anode of the other. This circuit arrangement is known as **inverse parallel** or back-to-back. Using two SCRs in inverse parallel is necessary in an ac power control circuit because it enables ac power to reach the load. If only one SCR is used, only half-wave pulsating dc could reach the load. By using the inverse parallel connection, SCR$_1$ can pass current to the load on one half-cycle, and SCR$_2$ can pass current to the load on the opposite half-cycle. The SCR that is switched on will be turned off at the end of its forward-biasing half-cycle because its anode voltage goes through zero and to the reverse-biasing polarity.

Finally, notice the arrangement of windings for transformer T_1. The primary of this transformer ($T_{1\text{–P}}$) is shown connected to the output of a phase-shift control circuit. The transformer has two separate secondaries, which are $T_{1\text{–S1}}$ at the gate of SCR$_1$, and $T_{1\text{–S2}}$ at the gate of SCR$_2$. The phasing dots on the transformer windings indicate the same phase of signal is applied to both SCRs.

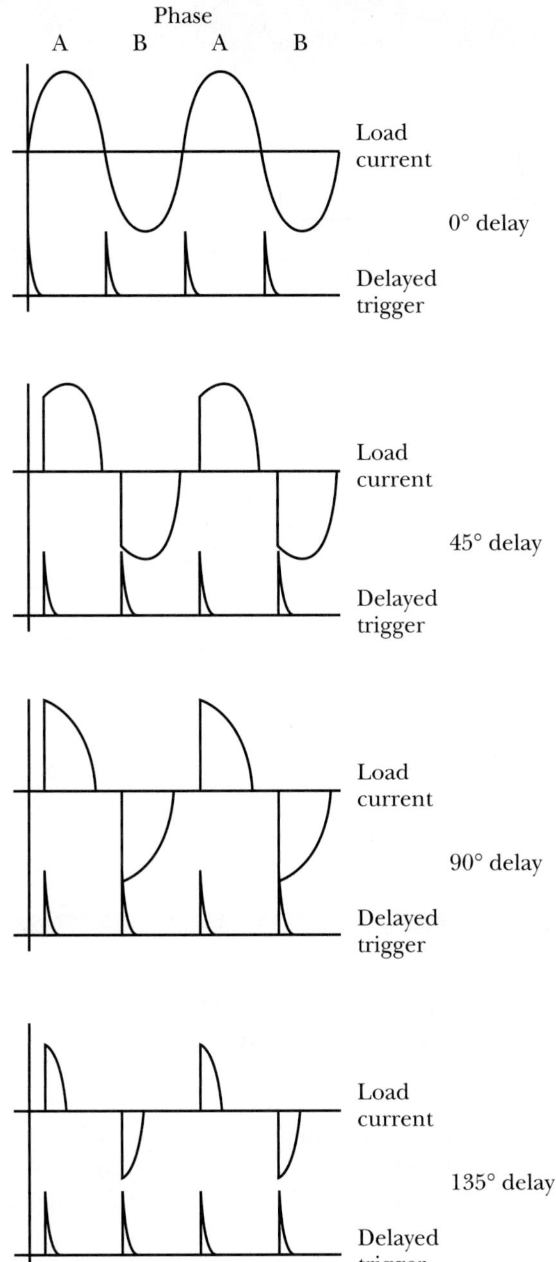

FIGURE 11–8 AC power phase-control waveforms

Details of the phase-shift control circuit are not shown on the schematic diagram. It is sufficient for our purposes to describe the signal that is applied to $T_{1\text{-P}}$. The signal from the phase-shift control circuit consists of positive pulses that occur at twice the frequency of the ac source. If the ac source is 60 Hz, for example, the positive pulses from the phase-shift control circuit occur at 120 Hz. These pulses have a fixed amplitude (about 3 V at the secondary windings) and are applied to the gate of both SCRs. The occurrence of a gate pulse turns on the SCR that happens to be forward biased between its cathode and anode. The next gate pulse will occur on the next ac half-cycle and thus switch on the opposite SCR. The effect is that the SCRs conduct alternately.

Now you are ready to consider the phase-control waveforms shown in Figure 11–8. The trigger waveforms from the phase-control circuit are delayed by a variable amount compared with the applied ac waveform. The longer the delay, the later in each half-cycle the SCRs are turned on and the lower the amount of power applied to the ac load.

Application Problem

General Electronics

GARAGE DOOR OPENER

An SCR is used in a garage door opening system illustrated below. The specifications for the SCR are as follows: trigger ON current and voltage are 30 mA @ 1 V, cathode to anode voltage when the SCR is conducting is 0.7 V, and the holding current is 30 mA. The motor's resistance is 200 Ω. Calculate the voltage at which the SCR will turn OFF.

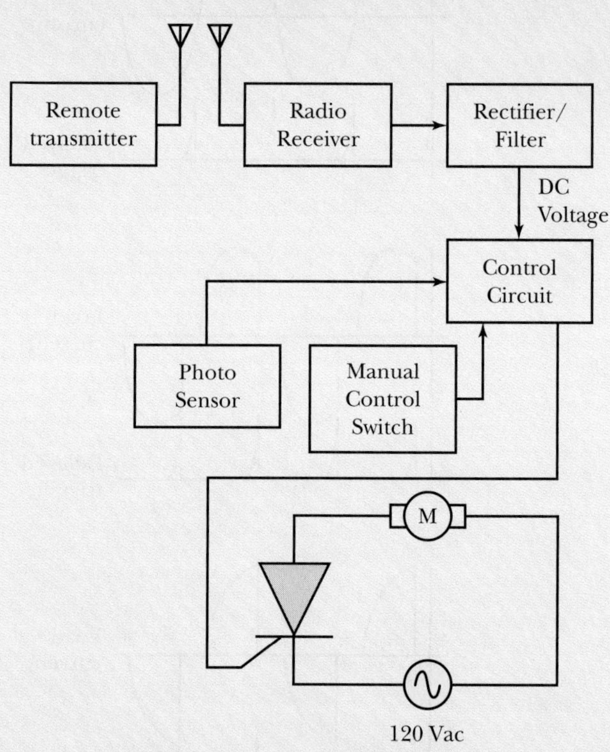

Solution

At 30 mA the voltage across the motor is:

$$V_{\text{motor}} = I_{\text{motor}} \times R_{\text{motor}} = 30 \text{ mA} \times 200\ \Omega = 6 \text{ V}$$

The supply voltage will be

$$V_{\text{scr}} + V_{\text{motor}} = 6 \text{ V} + 0.7 \text{ V} = 6.7 \text{ V}$$

So if the supply goes below 6.7 V, the SCR will turn OFF.

The amount of delay is measured in terms of degrees of the sinusoidal waveform. When the trigger pulses occur at 0°, for instance, there is no delay and the SCRs are triggered at the beginning of each ac half-cycle (maximum ac power to the load). When the trigger pulses are delayed 45°, however, the SCRs are not gated on until 45° into each half-cycle (about one quarter power applied to the load). And at a 90° delay, the SCRs conduct through the last half of each half-cycle. If you delayed to 180°, the SCRs would be gated at the end of each half-cycle, and no power would be applied to the ac load.

Other Types of SCRs

The SCRS you have just studied are the most commonly used types. There are several other types of SCRs that are not quite as popular but that possess some features that are advantageous in certain applications.

Gate-Controlled Switch (GCS)

The **gate-controlled switch** is very similar to an SCR except that a GCS can be gated off as well as gated on. Recall the only way to stop the conduction of an SCR is by removing or reversing the voltage applied between the cathode and anode. A GCS can also be turned off by applying a reverse-bias voltage across the gate-cathode P-N junction. See Figure 11–9.

The two conditions that must be met for switching on a GCS are the same as for an SCR:

1. A voltage is applied between the cathode and anode such that the anode is positive compared with the cathode.
2. A voltage is applied to the gate such that the P-N junction between the gate and cathode is forward biased (at least momentarily).

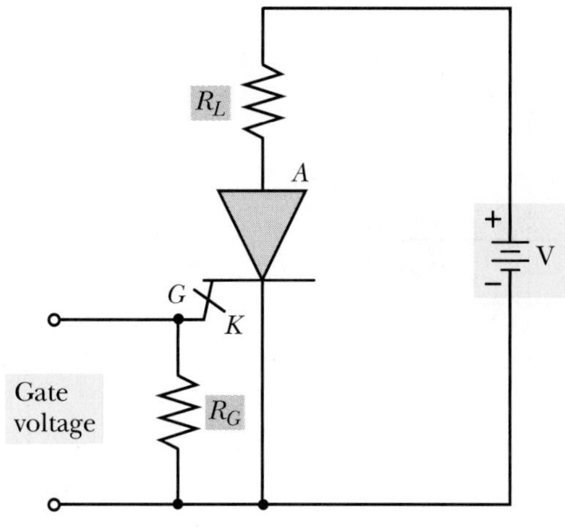

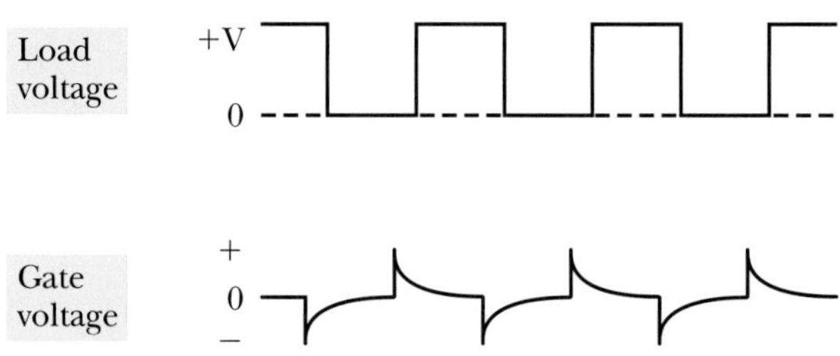

FIGURE 11–9 Gate-controlled switch (GCS) basic operation

A GCS, however, can be turned off in two different ways:

1. The voltage applied between the cathode and anode is removed or reversed so that the cathode is positive with respect to the anode.
2. A voltage is applied to the gate such that the P-N junction between the gate and cathode is reverse biased (at least momentarily).

The example circuit in Figure 11–9 is a chopper circuit that changes a dc current into a rectangular, or pulsating dc, waveform.

Silicon-Controlled Switch (SCS)

The **silicon-controlled switch** can be compared to an SCR that happens to have two gate terminals. Compare the equivalent BJT circuit for an SCS in Figure 11–10 to the SCR version in Figure 11–2. You can see the difference is that only one of the transistors for an SCR can be used for gating the device on, while either (or both) of the transistors for an SCS can be used for gating the device on.

Light-Activated SCR (LASCR)

The **light-activated SCR** can be gated by applying a positive voltage to the gate terminal or by directing light through a lens. Once the LASCR is conducting, it is turned off only by removing the anode voltage or reversing it. See the symbol and demonstration circuit in Figure 11-11.

In the demonstration circuit, light falling onto the LASCR causes it to switch to its conducting state. Dc power is thus applied to the alarm circuit. The circuit is reset (switched off) by momentarily depressing the normally closed reset push button.

The greatest amount of sensitivity to light occurs when there is no connection to the gate terminal. The variable resistor in the diagram serves as a sensitivity adjustment. The greater the amount of resistance, the greater the sensitivity to light. Setting the variable resistor for minimum resistance provides minimum sensitivity.

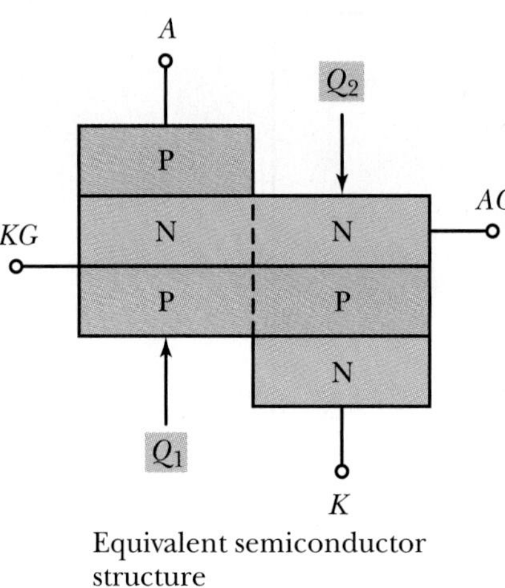

Equivalent semiconductor structure

A
AG
KG
K
Symbol

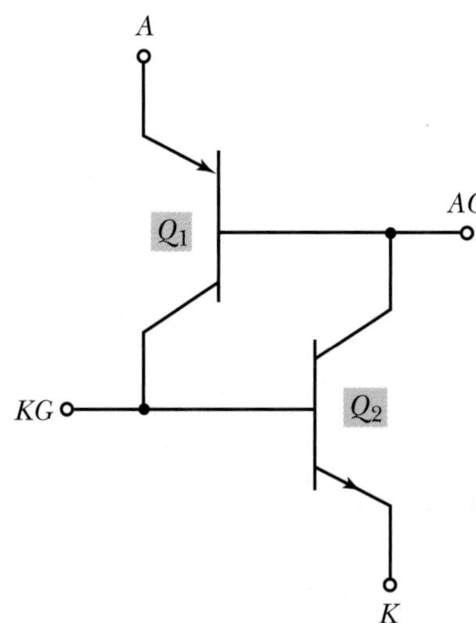

Equivalent BJT structure

FIGURE 11–10 Silicon-controlled switch (SCS) semiconductor structure, symbol, and equivalent BJT circuit

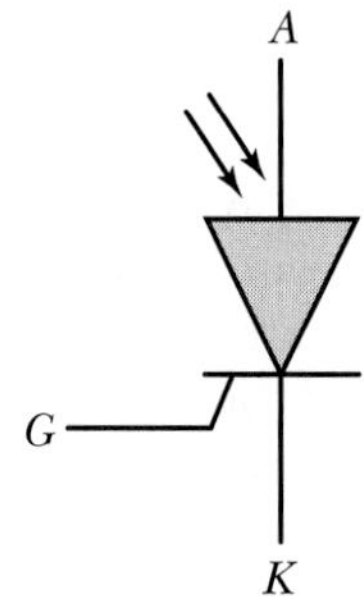

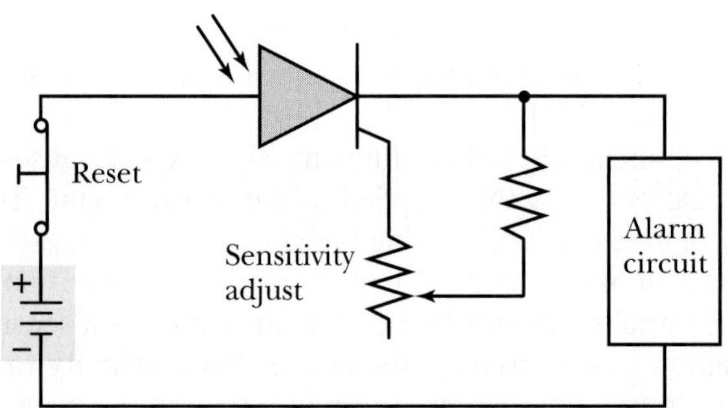

FIGURE 11–11 Light-activated SCR (LASCR) and alarm circuit application

Application Problem

Industrial

HEATER CONTROL SYSTEM

A heater is used in a temperature cycling system to treat glass products coming out of production to make the item stronger. A diagram is shown in the figure below. The specifications for four different SCRs are indicated in the table below. Which SCR is a better fit for this application?

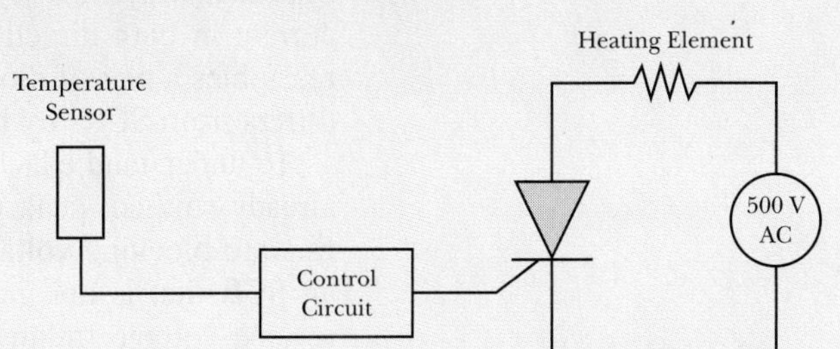

Peak Forward Voltage	Model
400V	2N6397
600V	2N6398
700V	MCR220-9
800V	2N6399

Solution Determine the peak value of AC $V_p = 500 \times 1.414 = 707\ V_p$. We don't want the AC supply to trigger the SCR ON. The control circuit is supposed to trigger the SCR. The 2N6399 would be the best fit for this application.

■ IN-PROCESS LEARNING CHECK 1

Fill in the blanks as appropriate.

1. A thyristor is a _______–layer device that has just _______ operating states.
2. An SCR is gated on by applying a _______-bias voltage to the _______ P-N junction. A _______ can also be gated on by shining a light onto its lens.
3. An SCR will conduct when it is gated and a voltage is applied between the cathode and anode such that the anode polarity is _______ and the cathode is _______.
4. The SCR rating that specifies the maximum forward voltage an SCR can handle is called the _______.
5. The SCR rating that specifies the maximum reverse voltage an SCR can handle is called the _______.
6. The smallest amount of cathode-anode forward current that can sustain conduction of an SCR is called the _______.
7. The _______ is a type of SCR that can be gated off as well as gated on.
8. The _______ is a type of SCR that has two gate terminals.
9. In order to control both cycles of ac power, two SCRs must be connected in reverse with respect to one another in an arrangement that is known as _______ or _______.
10. When a pair of SCRs are gated halfway through their respective half-cycles of the ac power waveform, the circuit is said to be firing at _______°.

11–2 Diacs

The **diac** (Figure 11–12) is another thyristor, or four-layer device. The semiconductor structure is identical with that of an SCR. The diac is different from an SCR in two important ways. First, the diac is a **bidirectional** device. This means that it conducts current in both directions. You can see this bidirectional nature by the symbol that resembles a pair of diodes connected in inverse parallel (back-to-back). The diac also differs from SCRs by having no gate terminal.

To understand exactly how a diac works, first consider two SCR ratings you have already studied: peak forward blocking voltage and peak reverse voltage. The peak forward blocking voltage is the largest amount of forward bias that can be applied to an SCR that is not gated on without the SCR breaking into conduction. The peak reverse voltage rating is the largest amount of reverse voltage that can be applied

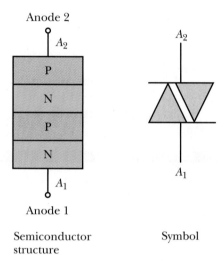

FIGURE 11–12 Diac semiconductor structure and symbol

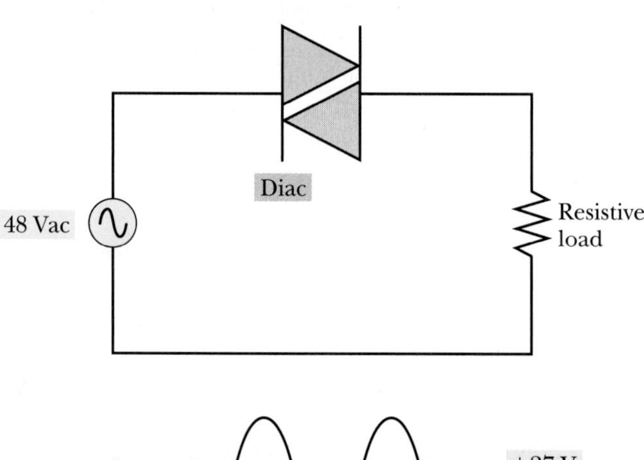

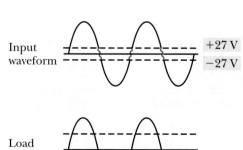

FIGURE 11–13 A simple diac circuit

Application Problem

Communications

REMOTE CONTROL ENGINE STARTING SYSTEM

A remote controlled starting system for a car is indicated in the figure below. The data sheet for the SCR indicates that the required gate current and voltage to turn the SCR on is 5 mA and 0.7 V respectively. If the starter motor is running, what is the voltage coming out of the control circuit?

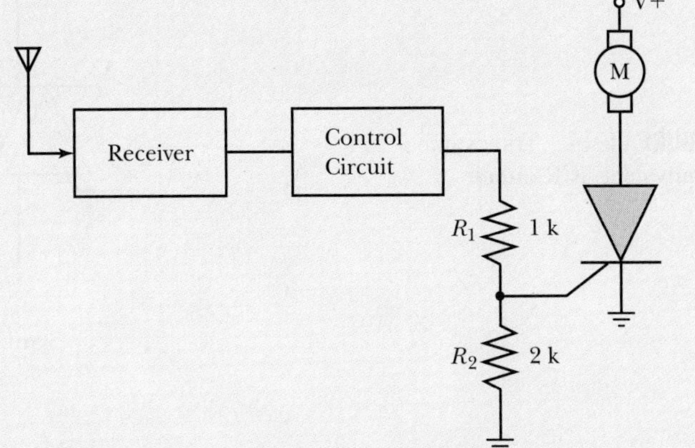

Solution $V_{R_2} = 0.7\text{ V}$ $\quad I_{R_2} = 0.7\text{ V}/2\text{ k} = 350\ \mu\text{A}$

$I_{R_1} = 5\text{ mA} + 350\ \mu\text{A} = 5.35\text{ mA}$

$V_{R_1} = 5.35\text{ mA }(1\text{ k}) = 5.35\text{ V}$

$V_{\text{control}} = 0.7\text{ V} + 5.35\text{ V} = 6.05\text{ V}$

The voltage coming out of the control circuit is 6.05 V.

between the cathode and anode without breaking down the junctions. It is also important to recall that the forward blocking voltage and reverse breakdown voltage for an SCR are usually the same voltage levels. A diac has the same semiconductor structure as an SCR, so it follows that a diac also has forward breakover and reverse breakover voltage ratings. In fact, the only way to get a diac to conduct is by exceeding its forward breakover and reverse breakover ratings.

Figure 11–13 shows a diac connected in series with an ac power source and resistive load. The voltage ratings for this version are 27 V. The waveforms show that no current flows through the circuit on either half-cycle until the applied voltage exceeds 27 V. The diac then remains conducting until the applied voltage waveform goes through zero. On the positive half-cycle, the diac does not conduct until the diac's forward breakover voltage is exceeded. Now, the diac conducts through the remainder of the half-cycle. For the negative half-cycle, the diac does not conduct until the diac's reverse breakover voltage is exceeded. Then, the diac conducts through the remainder of the half-cycle.

The voltage ratings for diacs are typically between 25 V and 50 V. Current ratings are on the order of several amperes. Since diacs require only two terminals, they look much like rectifier diodes.

The most common commercial and industrial applications of diacs are as triggering devices for triacs as described in the next section.

11–3 Triacs

The **triac** bidirectional thyristor works as a pair of SCRs connected in inverse parallel, Figure 11–14. A single gate terminal can switch on the triac in either direction. The triac was originally devised to replace sets of two inverse-parallel SCRs with a single device. It is said to be an "ac SCR," and the two conditions required for switching on a triac are as follows:

1. A voltage is applied between the cathode and anode (polarity is not important).
2. A forward-bias voltage is applied to the gate-cathode junction (at least momentarily).

A conducting triac is turned off by removing the anode voltage or reversing the present polarity applied to it.

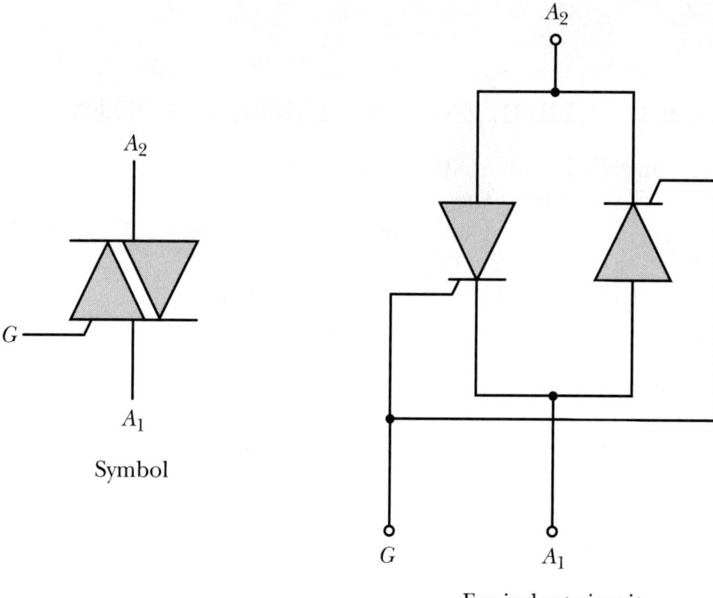

FIGURE 11–14 Triac symbol and equivalent SCR circuit

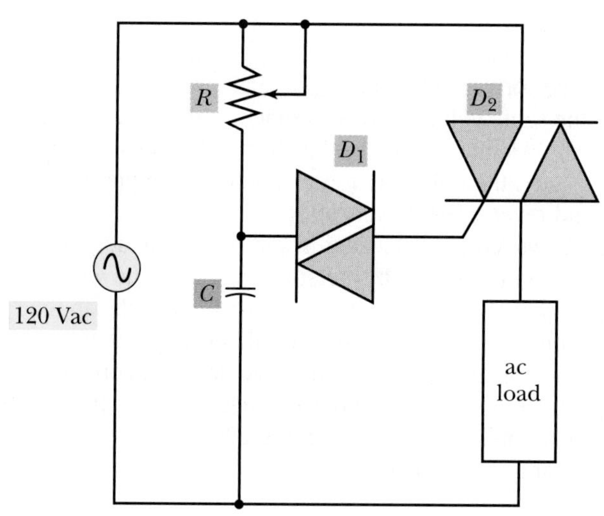

FIGURE 11–15 A light dimmer circuit using a diac and triac

> **Practical Notes**
>
> SCRs and TRIACs are often used in place of electromechanical relays and switches especially in potentially explosive environments. Since the electromechanical relay and mechanical switch will open at any time during the signal sine wave (from when there is no current flowing to maximum current flowing), arcing can occur. The thyristors will only turn off when the anode current is less than the holding current, and no arcing occurs.

Figure 11–15 shows one of the most popular thyristor circuits in the consumer marketplace. This circuit is typical of the circuits that are used for adjusting the brightness of the lights in a room. In this drawing, the lights would be the ac load. The ac source is the utility power supplied to the home. The diac (D_1) is connected to the gate of the triac (D_2)—the triac is gated each time the diac breaks into conduction. The adjustable resistor *(R)* is the power control adjustment. Let's look at how this circuit works.

Resistor *R* and capacitor *C* form an ac phase-shift circuit. The ac waveform that is applied to D_1 is thus a phase-shifted, level-adjusted version of the ac source voltage. This means the diac will start conducting at a point on each half-cycle that can be varied from 0° to 180°. In turn, this means the triac is triggered into conduction at various points on each of its half-cycles, thereby enabling the control of power to the ac load, from minimum power to maximum power.

Application Problem

CAR WASH WATER PRESSURE MONITORING SYSTEM

A pressure sensor is monitoring the water pressure in an automated car wash sprayer system. The output of the sensor is amplified and sent to a microcomputer. If the pressure goes below 50 pounds per square inch (psi), a pump will be switched on to increase the pressure and switched off when the desired pressure is reached. A unijunction transistor will be used to trigger the triac. If the voltage spikes on the gate are 2 volts and the transformer's secondary resistance is 250 ohms, what is the maximum gate current?

Solution:

$$I_G = \frac{V_G}{R_{sec}} = \frac{2\text{ V}}{250\ \Omega} = 8\text{ mA}$$

The maximum gate current is 8 mA.

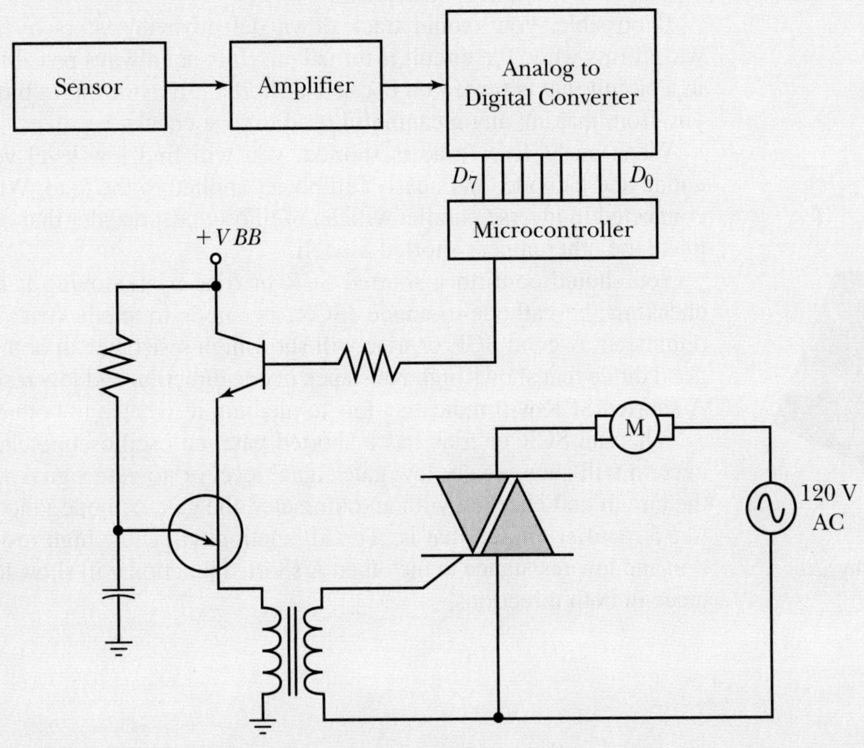

■ IN-PROCESS LEARNING CHECK 2

Fill in the blanks as appropriate.

1. Thyristors that can conduct in two directions (such as a diac and triac) are said to be _______ devices.
2. A diac has _______ gate terminal(s), while a triac has _______ gate terminal(s).
3. The only way to get a diac to conduct is by exceeding its _______ breakover and reverse _______ ratings.
4. A triac operates like a pair of _______ connected back-to-back.
5. The most common commercial and industrial applications of _______ are as triggering devices for triacs.
6. In order to apply full ac power to a load, a triac should be fired at the _______° point on each half-cycle.

Safety Hints

Thyristors (especially SCRs and triacs) are often used in electrical power control circuits. This means you can expect them to be handling lethal levels of current and voltage. Also, thyristors can be operated at temperatures high enough to cause painful burns when touched. So take special care when placing oscilloscope leads, test clips, and so on, in an operating thyristor circuit.

11–4 Troubleshooting Thyristors

When a thyristor fails, the trouble is usually due to one or more shorted P-N junctions. Most often, the short is between the anode and cathode, thereby allowing the device to conduct all the time. Also, gated thyristors (SCRs and triacs) can fail due to a short circuit in one of the gate junctions. The way the device responds to a shorted gate depends on the nature of the short; sometimes the device is not gated on at all, sometimes it is gated on all the time, and sometimes it is gated on at the wrong time. Internal open circuits in a thyristor are fairly rare.

Thyristors are normally mounted on heat sinks to prevent overheating, Figure 11–16.

One thing to bear in mind about thyristors is that they are by nature all-or-nothing devices. They are either switched completely on or switched completely off. So when you find a thyristor that is operating somewhere between cutoff and saturation, you may have found a shorted thyristor.

If possible, you should track down defective thyristors by noting their voltage waveforms while the circuit is turned on. (It is not always possible to check thyristors in a circuit that is turned on because a shorted thyristor might blow fuses and prevent you from making any meaningful oscilloscope checks.)

When an SCR or triac is shorted, you will find low-level voltages between the anode and cathode, and nearly full power applied to the load. When observing SCRs connected in inverse parallel with an oscilloscope, consider that one shorted SCR will make the other appear shorted as well.

FIGURE 11–16 Power thyristor mounted on a heat sink

You should confirm a shorted SCR or triac by removing it from the circuit and checking the cathode-to-anode (SCR) or anode-to-anode (triac) resistance with an ohmmeter. A good SCR or triac will show high resistance in both directions (unlike a good diode that shows high resistance in one direction and low resistance in the other). A shorted SCR will indicate a low to medium resistance in both directions.

When an SCR or triac has a shorted gate, an oscilloscope check at the gate connection will show a very low gate signal level or no gate signal at all. Removed from the circuit and checked with an ohmmeter, the gate-cathode junction will test exactly like a rectifier diode. That is, a good junction will show high resistance in one direction and low resistance in the other. A shorted junction will show low to medium resistance in both directions.

USING TECHNOLOGY: THYRISTORS USED IN A SYSTEM

Thyristors are one of the most common devices used in both ac and dc power supply system applications. Thyristors' ideal characteristics of high maximum current carrying capacity, low holding current to maintain the turn on or conduction, along with high peak forward blocking voltage and almost identical peak reverse voltage, make it one of the most popular and almost universally used applications in ac powered products. Common applications include light and lamp dimmers, ac power tools, especially variable speed drills, fan speed controls, and air conditioner compressors. Any electronic circuit that requires precise continuous control as opposed to step controls is a suitable application for a thyristor. A thyristor is combined with some control or reference voltage source such as a zener diode or a phase-shift or phase variable voltage to supply the precise and continuous control required.

Thyristors are fired (start conducting) when the turn-on voltage and current requirements are met. After being placed into conduction, the thyristor will turn off only after the current drops below the thyristors' rated holding current. The Thyristors' turn-off characteristic eliminates the back EMF experienced by other switches controlling inductive loads. The extremely low current existing when the thyristor turns off will not produce a high voltage across the thyristor. A dc-to-dc power converter using thyristors is shown in Figure 30. Such converters change a constant 12.6 Vdc voltage from a car battery to a pulsating dc voltage, which is converted back to a constant 16 Vdc at 20 mA to supply a laptop computer and recharge its battery.

The circuit schematic in Figure 27 shows a triac used to control the

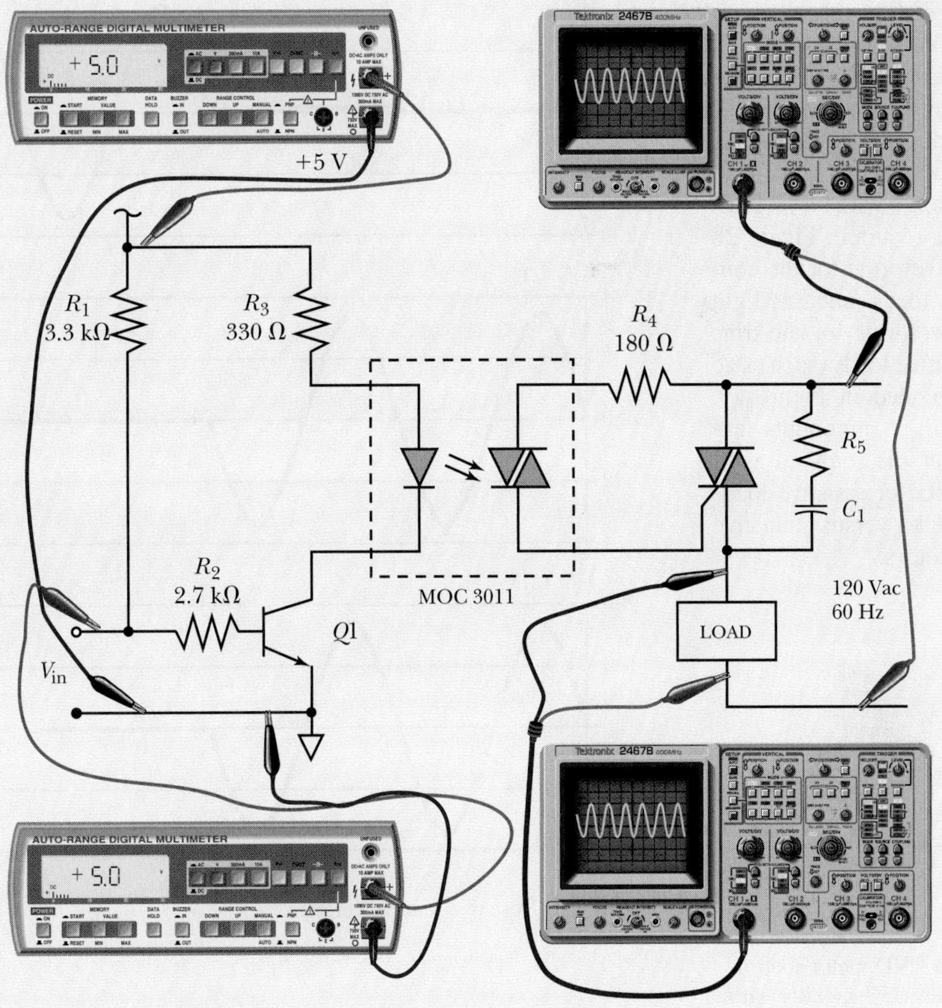

FIGURE 27 Thyristors used in a triac load power regulator

application of 120 Vac 60 Hz to the load from the power cord. When the input voltage (V_{in}) equals 5 V, the optoisolator will conduct. The triac will then fire every half-cycle applying power to the load. If the input voltage (V_{in}) equals 0 V, the optoisolator will not conduct. The triac will then turn off resulting in no power applied to the load. The series resistor (R_5) and capacitor (C_1) in parallel with the triac are the snubber circuit. The snubber circuit prevents incorrect triac firing by limiting the voltage transition (dV/dt) to a few volts per microsecond. The back and forth dV/dt and the 120 V system power turn on are conditions when the triac could experience high transition voltages if the snubber circuit was not connected in parallel with the triac. The recommended snubber circuit consists of a 620-ohm, 1/2-watt resistor R_5 along with a 0.047-μF capacitor C_1 for currents at or below 12 amps rms. For currents from 12 amps to 40 amps rms, 330 ohm, a 1/2-watt resistor R_5 along with a 0.1-μF capacitor C_1 are required. Figure 28 provides the waveforms for the control of resistive loads (such as light bulbs). The waveforms for the triac controlled inductive loads (such as ac motors) are provided in Figure 29. Different circuit components are required to meet the required voltages and currents for resistive loads such as light bulbs versus inductive loads such as motors.

Thyristors used in DVD motor control side loading system: TFT monitor with built-in DVD player

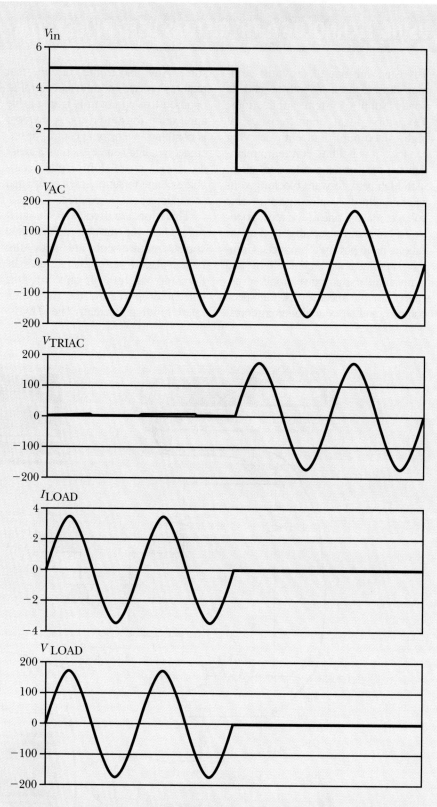

FIGURE 28 Resistive load waveforms

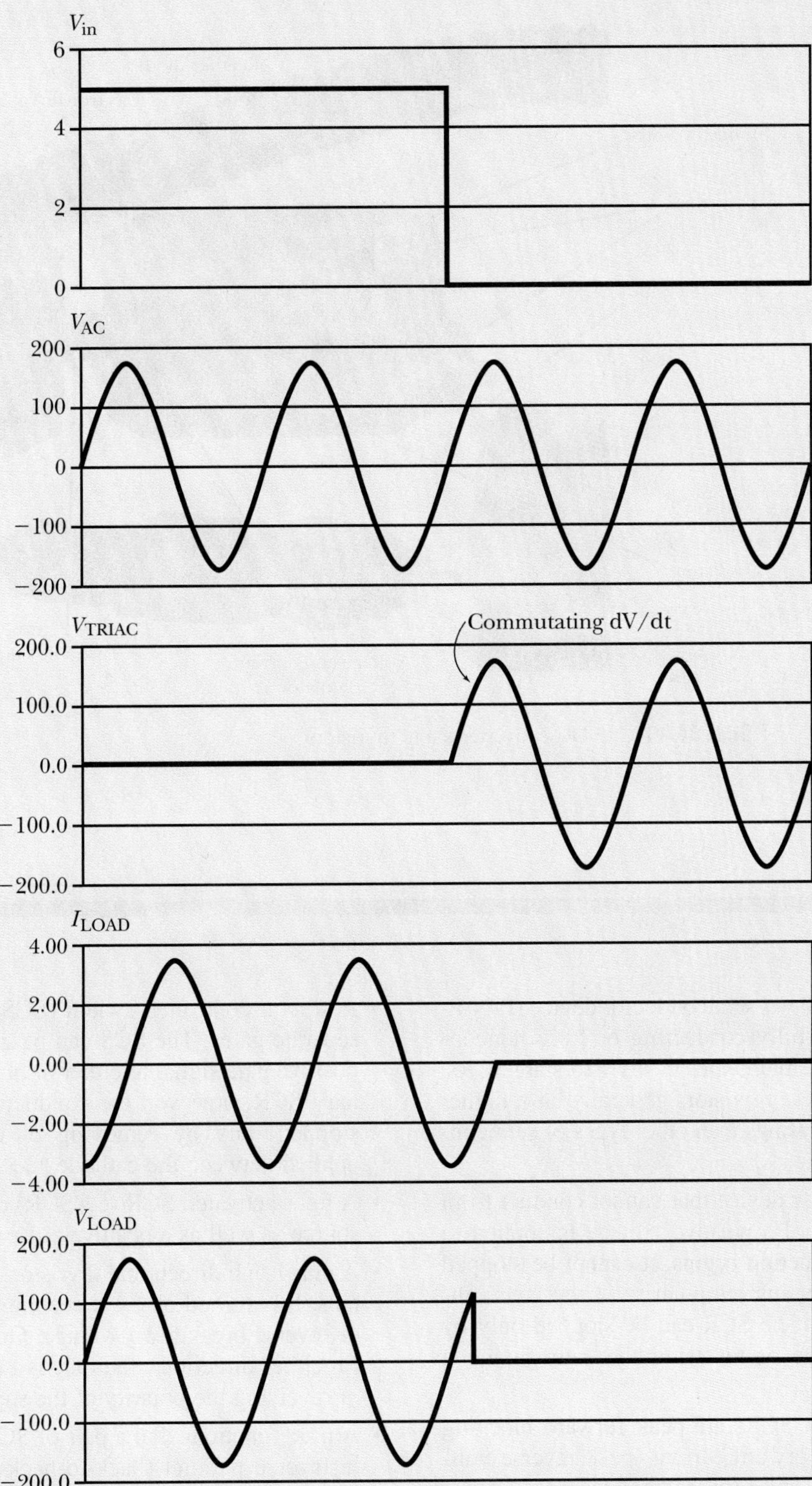

FIGURE 29 Inductive load waveforms

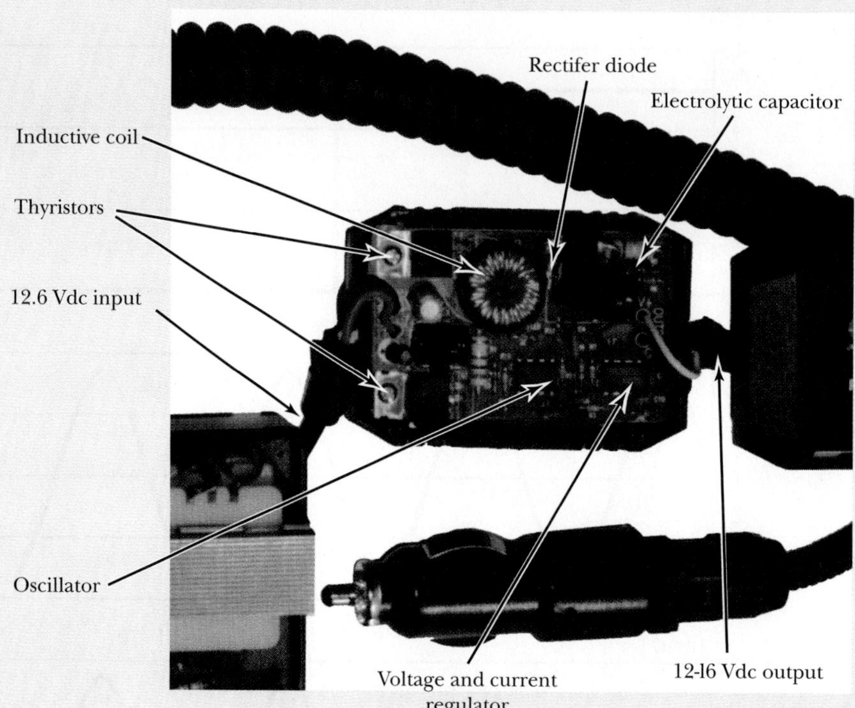

FIGURE 30 DC to DC converter using thyristors

Summary

- A thyristor is a four-layer semiconductor device (P-N-P-N), which is either fully conducting or fully nonconducting. Three of the main types of thyristors are SCRs, diacs, and triacs. These thyristors generally have higher current and voltage ratings than other types of semiconductor devices.
- An SCR is a thyristor device that cannot conduct from cathode to anode until a positive voltage is applied to the gate. Once conduction begins, it cannot be stopped by removing or reversing the polarity of the gate voltage. The condition of an SCR can be stopped only by removing the voltage or reversing the cathode-anode voltage.
- The main ratings for SCRs are peak forward blocking voltage, maximum forward current, peak reverse voltage, holding current, and forward "on" voltage.
- A gate-controlled switch (GCS) is similar to an SCR except that a GCS can be gated off as well as gated on. The GCS is gated on by forward biasing the gate-cathode junction, and it is gated off by reverse biasing the gate-cathode junction. A GCS can also be turned off by reversing or eliminating the voltage applied between the cathode and anode.
- A silicon-controlled switch (SCS) is an SCR with two separate gates. The SCS can be gated on by applying a positive gate signal to either or both gates. Like an ordinary SCR, however, the conduction of an SCS can be stopped only by removing or reversing the voltage applied between the cathode and anode.
- A light-activated SCR (LASCR) can be gated by a light source as well as a positive pulse to the gate.
- A diac is a bidirectional thyristor that has no gate terminals. It is turned on by exceeding the forward blocking or reverse breakdown voltage. Once conduction begins (in either direction), the diac is turned off by removing or reversing the polarity of the supply voltage.
- A triac functions like a pair of SCRs that are connected in inverse parallel (back-to-back). Triacs are designed mainly for phase-shift control of ac power.
- In an ac phase-shift power control circuit, the later the triac is fired in each half-cycle the lower the amount of power applied to the load.

CIRCUIT SUMMARY

Device name	*Schematic symbol*
Silicon Controlled Rectifier (SCR)	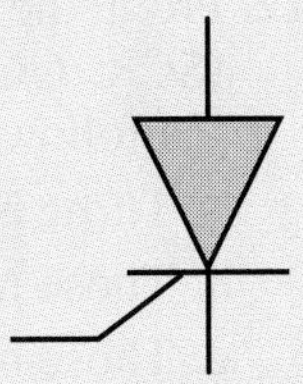
Silicon Controlled Switch (SCS)	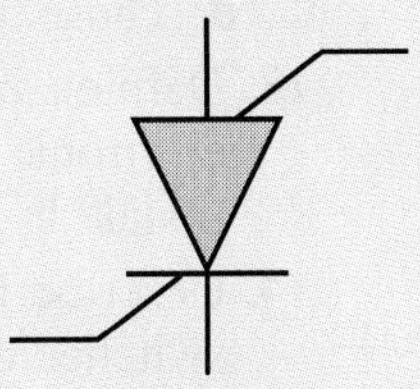
Light-Activated SCR (LA SCR)	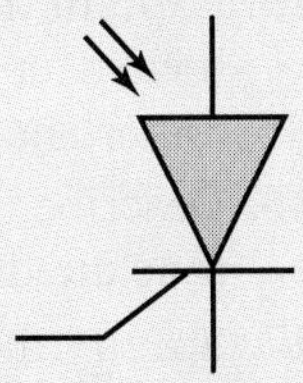
Diac	
Triac	

Thyristor summary chart

Review Questions

1. Specify the two possible states of operation of a thyristor.
2. Complete the following series of statements:
   a. A junction diode is a _____ -layer device. An example is _____.
   b. A BJT is a _____ -layer device. An example is _____.
   c. A thyristor is a_____ -layer device. An example is _____.
3. Specify the conditions required for turning on an SCR, and describe how to turn it off.
4. Sketch the BJT equivalent circuit for an SCR and describe the process of gating the circuit on and show the paths for current that keep the circuit latched on.
5. Define *holding current* as it applies to thyristor devices, especially SCRs.
6. Define and compare the thyristor ratings of *peak forward blocking voltage* and *peak reverse voltage.*
7. Specify completely the meaning of the following abbreviations: GCS, LASCR, SCR, and SCS.
8. Demonstrate your understanding of inverse-parallel connections by drawing two junction diodes connected in inverse parallel. State another name for this type of connection.
9. Describe the most important difference between an SCR and a GCS.
10. Describe the most important differences between an SCR and an SCS.
11. Cite two different ways a LASCR can be switched on.
12. From the list of thyristors studied in this chapter (SCR, GCS, SCS, LASCR, diac, and triac), specify the following:
    a. Those having no gate terminal
    b. Those having one gate terminal
    c. Those having two gate terminals
    d. Those that are bidirectional
13. Define *bidirectional* in terms of current flow through electronic devices. Is a common resistor bidirectional? Is a junction diode bidirectional?
14. Describe how a diac is gated on and how it can be switched off.
15. Explain how the thyristor ratings (i.e., forward breakover voltage and reverse voltage) are extremely important to the theory of operation of a diac.
16. Describe how the schematic symbols for a diac and a triac are different.
17. Specify the conditions required for turning on a triac, and describe how to turn it off.
18. Explain the basic difference between the application of an SCR and a triac.
19. Describe the main purpose of an ac phase-control circuit.
20. Sketch the output voltage waveform for an ac phase control circuit that is being fired at 90° on each half-cycle.

Analysis Questions

1. Explain how an SCR can be considered a nonlinear power amplifier.
2. Research a parts catalog or thyristor data book, identify the part number, and list the ratings for the SCR having the lowest forward current rating and the SCR having the highest current rating.
3. Explain the similarities and differences between the action of a diac and a zener diode.
4. Research and explain the operation of a unijunction transistor (UJT) and show how it can be used to trigger SCRs or a triac in an ac phase-control circuit.

Performance Projects Correlation Chart

Chapter Topic	Performance Project	Project Number
Theory of SCR Operation		
SCR Circuits	SCRs in DC Circuits	31
SCR Circuits	SCRs in AC Circuits	32

NOTE: It is suggested that after completing the above projects, the student should be required to answer the questions in the "Summary" at the end of this section of projects in the Laboratory Manual.

OUTLINE

12–1 Optical sources (photoemitters)
12–2 Photo-reactive devices (photodetectors)
12–3 Opto-isolators
12–4 Fiber-optic cable systems
12–5 Optoelectronic application circuits
12–6 The liquid crystal display

OBJECTIVES

After studying this chapter, you should be able to:

1. Describe the operation of **LEDs** and **photodiodes**
2. Determine the value of resistor to be placed in series with an LED for proper operation
3. Describe the purpose of **laser diodes**
4. Understand the operation seven segment displays
5. Understand the operation of optocouplers
6. Understand fiber-optic cables
7. Understand photoemitters and photodetectors

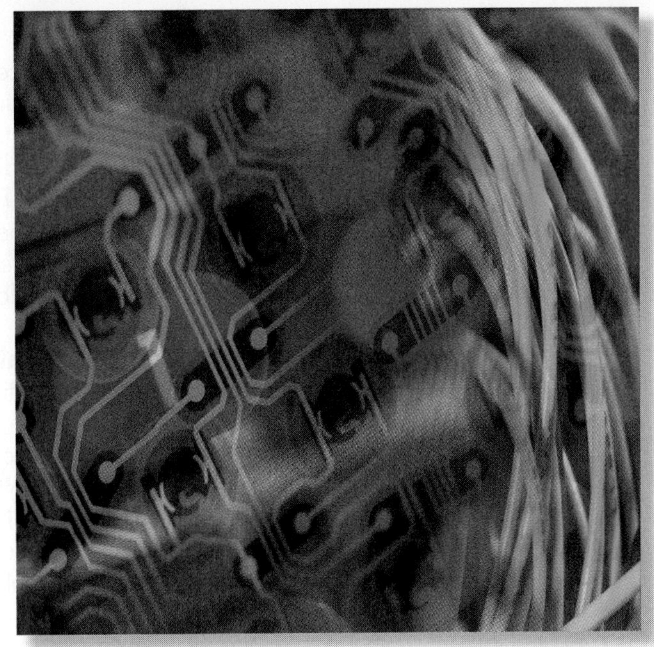

CHAPTER 12

Optoelectronics

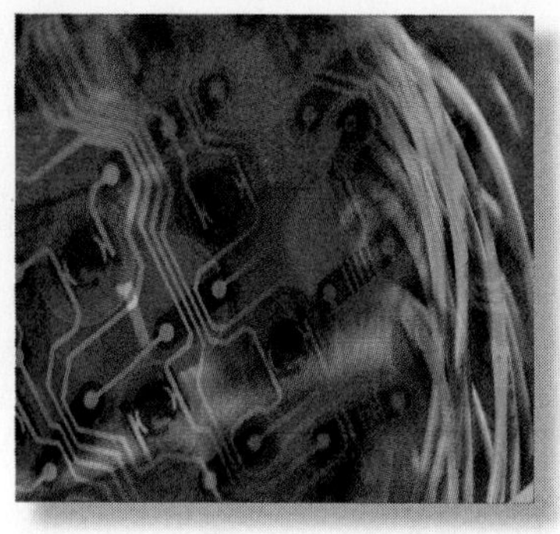

PREVIEW

Optoelectronic devices combine the effects of light and electricity. The **electromagnetic frequency spectrum** is provided in Figure 12–1. The electromagnetic spectrum lists the frequency (Hz) and the wavelength (meters) associated with signals from audio signals to gamma rays. The frequencies associated with light are highlighted. For each color of light, the photons have a unique frequency and wavelength. Some optical devices (optical sources) have been developed with the ability to emit a fixed color of light. Some optical devices **(photo-reactive devices)** were developed to respond to the reception of photons within a range of wavelengths.

An **optoelectronic device** is a transducer that converts electrical energy to optical energy or optical energy to electrical energy. The optoelectronic devices discussed in this chapter are light-emitting diodes, photodiodes, laser diodes, optocouplers, and photodetectors. A transducer is any device that converts energy from one form to another form.

Certain kinds of P-N junction diodes are used in optoelectronics. Light-emitting diodes (LEDs), for example, emit light energy when current flows through them. Photodiodes, on the other hand, vary in their P-N junction conduction as the intensity of light falling onto them varies. Another important optoelectronic diode is the **laser diode.** The **fiber-optic cable** is a prominent component of optoelectronic communication systems.

KEY TERMS

Electromagnetic frequency spectrum
Fiber-optic cable
Laser diode
Light-emitting diode
Optoelectronic device
Opto-isolator
Photodiode
Photo-reactive device

12–1 Optical Sources (Photoemitters)

Light-Emitting Diodes

Whenever electrical energy is applied to a semiconductor material, electrons in their valence bands absorb some of that energy and jump up to the higher energy conduction-band levels. As a result, the number of free electrons and holes increases. Conduction-band electrons cannot hold onto their extra energy levels for very long, however. Within a few nanoseconds after jumping up into a conduction band, an electron gives up its extra energy and falls back into a more stable valence band. This extra energy is usually given up in the form of heat, but certain kind of semiconductor materials favor giving up this extra energy in the form of light. So it is possible to create semiconductor devices that actually emit light that can be used for practical purposes. These devices are **light-emitting diodes** (abbreviated **LEDs**).

The semiconductor materials used for making LEDs include gallium, arsenic, phosphorus, and carbon. The composition of the semiconductor materials used in the LED determines the wavelength and the color of the light (infrared, red, orange, amber, green, blue, and violet) that is emitted. The light emitted from an LED is considered monochromatic (having only one color). The most common type of LED is made of a compound, gallium arsenide. These LEDs emit a bright red light that can be used for all kinds of indicators and displays for electronic equipment. Silicon carbide LEDs emit blue light. Figure 12–2 summarizes the specifications for several common LEDs.

Figure 12–3 shows the electronics symbol for an LED. This looks very much like the symbol for a common junction diode, but also includes arrows that point away from the diode. The idea of the arrows is to suggest that the diode emits light energy. LEDs are easy to identify because a prominent feature is the transparent lens assembly.

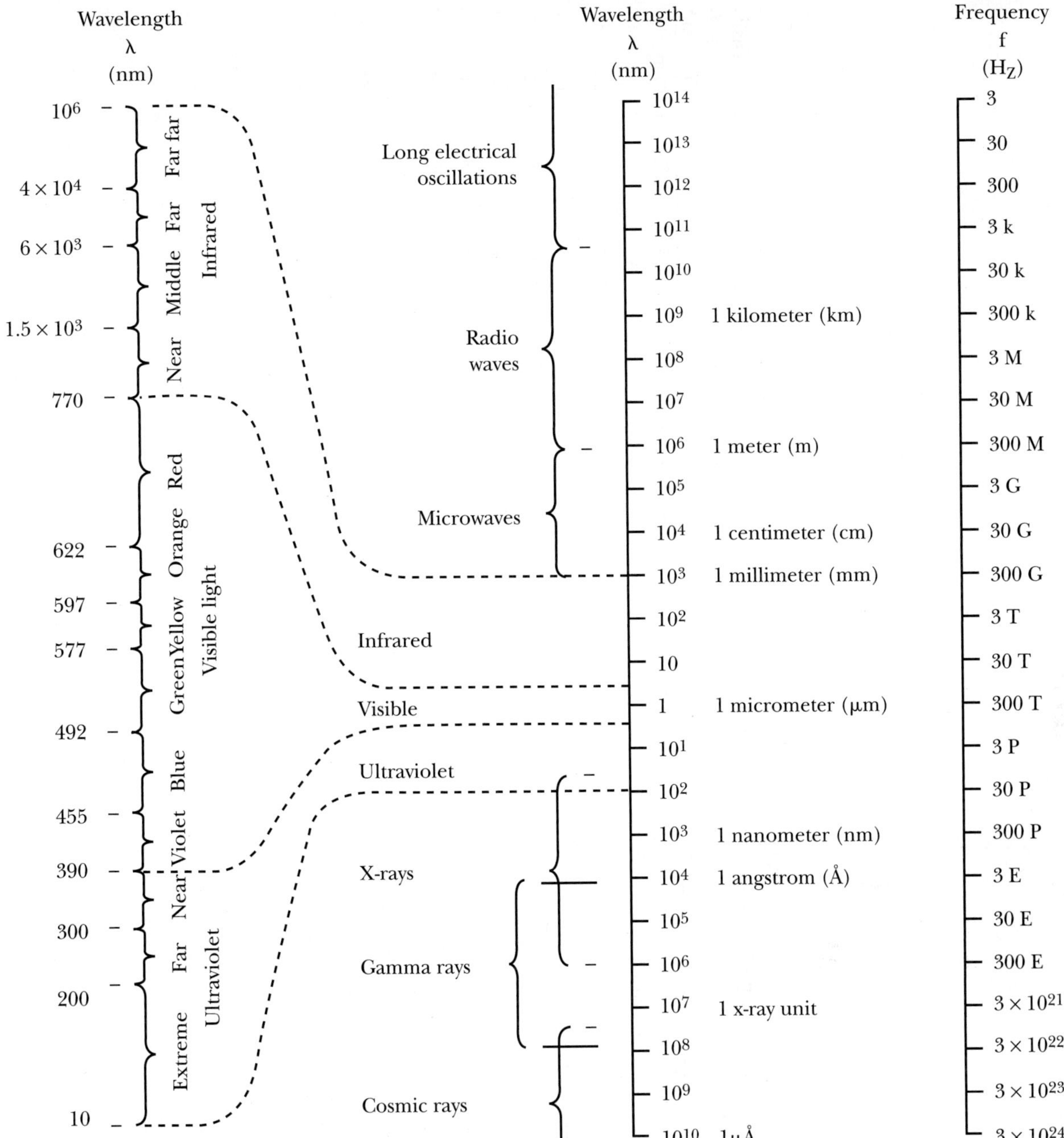

FIGURE 12–1 Electromagnetic spectrum showing the approximate boundaries between spectral regions

An LED emits light whenever it is forward biased and conducting current. Refer to the simple circuit in Figure 12–4. If you are using a gallium arsenide LED, the barrier potential (forward voltage drop) is 1.5 V, and you need about 15 mA of forward current to generate a useful amount of light. The value of the current limiting resistor connected in series with the LED depends on the amount of supply voltage. Formula 12–1 lets you determine the amount of voltage that has to be dropped across the resistor (V_R) in terms of the source voltage (V_S) and the forward voltage drop across the diode (V_D).

FORMULA 12–1 $V_R = V_S - V_D$

Part Number	Composition	Emitted Color	Peak Wavelength λ_{PEAK} nm	Power Dissipation P_D mW	Forward Voltage V_F V	Forward Current I_F mA	Reverse Voltage V_R V	Viewing Angle $2 \times \theta$ degrees	Luminous Intensity I_V	units	at I_F mA
MCD-CLE130W	GaAs	InfraRed	940	200	1.9	100	5	35	2.4	mW/sr	100
MCDL-593HD	GaP	Bright Red	700	45	2.1	15	5	60	5	mcd	10
MCDL-5C3SRC	GaA1As	Super Red	660	115	1.8	20	5	120	150	mcd	20
MCDL5V3UEC-610-8D	A1InGaP	Orange	610	120	2.0	40	5	8	8500	mcd	20
MCDL-5C3UOC	A1InGaP	Amber	605	115	2.0	40	5	120	500	mcd	20
MCDL-5C3UYC-9D	A1InGaP	Yellow	592	115	2.0	40	5	120	500	mcd	20
MCDL-593GC	GaP	Green	565	100	2.0	30	5	45	30	mcd	10
MCDL-513SBC4	InGaN	Blue	470	100	3.6	20	5	25	3000	mcd	20
MCDL-5013VC	InGaN	Super Violet	405	200	3.8	20	5	20	800	mcd	20
MCDL-5013BGC**	InGaN	Green	505	200	3.8	20	5	25	2850	mcd	21
MCDL-5013UAC**	GaA1InP	Super Amber	605	200	2.0	20	5	25	2300	mcd	20
MCDL-5013UEC**	GaA1InP	Super Orange Red	625	200	2.0	20	5	25	2300	mcd	23

*Information acquired from www.mcdelectronics.com

**5 mm (T 1-3/4) Traffic signal LEDs

FIGURE 12–2 Typical LED ratings

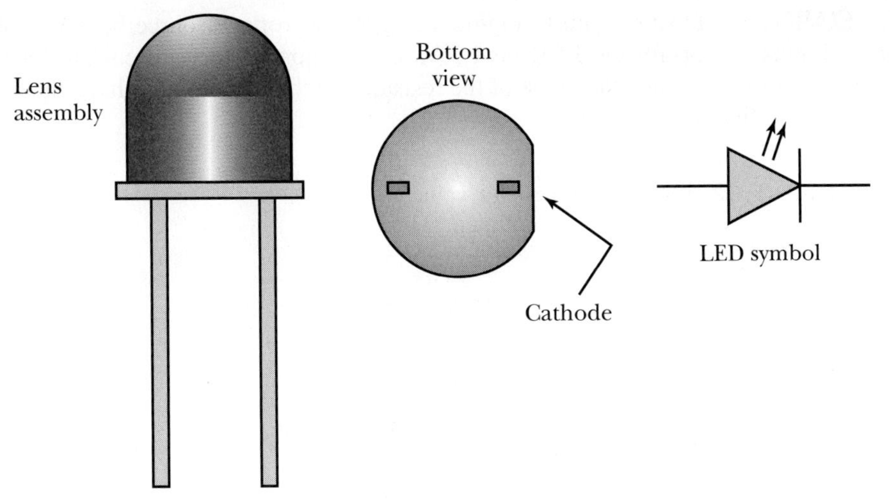

FIGURE 12–3 The light emitting diode (LED)

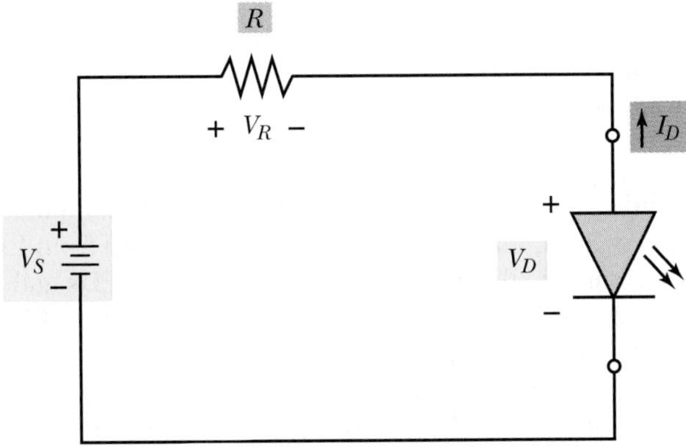

FIGURE 12–4 A practical LED circuit multiSIM

EXAMPLE Referring to the circuit in Figure 12–4, let $V_S = 12$ V, and assume that the forward voltage drop for the LED (V_D) is 1.5 V. Calculate the voltage across the current limiting resistor (V_R).

Answer:

From Formula 12–1:

$$V_R = V_S - V_D$$
$$V_R = 12\text{ V} - 1.5\text{ V} = 10.5\text{ V}$$
$$V_R = 10.5\text{ V}$$

So the voltage across the current limiting resistor in this example is 10.5 V.

When you also know the amount of forward current (I_D) that is necessary for operating the LED, you can calculate the value of the series resistor *(R)*. Formula 12–2 allows you to calculate the value of *R* on the basis of the voltage across the resistor and the desired amount of LED current. Formula 12–3 lets you calculate the value of the resistor in terms of source voltage, diode voltage, and forward current.

FORMULA 12–2 $R = \dfrac{V_R}{I_D}$

FORMULA 12–3 $R = \dfrac{(V_S - V_D)}{I_D}$

◆ **EXAMPLE** Referring again to Figure 12–4, let the voltage source be 9 Vdc, the forward voltage drop for the LED be 1.5 V, and the normal forward current for the LED be 15 mA. Calculate the value of the resistor and the amount of voltage that you can expect to find dropped across it.

Answer:

Using Formula 12–3 to find the resistor value:

$$R = \frac{(V_S - V_D)}{I_D}$$

$$R = \frac{(9\text{ V} - 1.5\text{ V})}{15\text{ mA}}$$

$$R = 500\ \Omega$$

Using Formula 12–1 to find the voltage across the resistor:

$$V_R = V_S - V_D$$

$$V_R = 9\text{ V} - 1.5\text{ V}$$

$$V_R = 7.5\text{ V}$$ ◆

PRACTICE PROBLEMS 1

Use Formula 12–3 to calculate the value of the resistor in Figure 12–4 if $V_S = 9$ V, the LED's rated forward voltage drop is 1,2 V, and the required forward current is 24 mA.

Aside from emitting light energy, LEDs operate like common junction diodes. You have already seen, however, that the forward-bias voltage drop from LEDs tends to be larger than that of common silicon diodes (about 1.5 V for LEDs and about 0.7 V for silicon junction diodes). Also, the reverse-breakdown voltage for LEDs is much lower than for practical switching and rectifier diodes. Whereas the reverse-breakdown voltages for ordinary diodes are commonly on the order of 100, 200, and 300 V, the reverse-breakdown voltage for LEDs may be typically in the range of 4 V or so. Because LEDs have such low reverse-breakdown voltage ratings, they cannot be used in practical rectifier circuits.

Figures 12–5, 12–6, and 12–7 provide three different circuit configurations to turn the LED on and off. The NPN and PNP transistors in these circuits can be replaced with N-channel and P-channel power FETs. The resistor in series with the LED limits the current flow through the LED. The current limiting resistor is critical and prevents damage to the LED.

The circuit in Figure 12–5, 12–6, and 12–7 will be evaluated to determine the input condition that will cause the photoemitter to emit light as well as the input condition that will turn the photoemitter off.

◆ **EXAMPLE** Evaluate the Figure 12–5 circuit when the input voltage is 0 V. Repeat the circuit evaluation when the input voltage is 5 V. The following components are used in the circuit:

$$V_{CC} = 5\text{ V}$$

$$R_1 = 1\text{ k}\Omega$$

$$R_2 = 100\ \Omega$$

$$Q_1 = \text{2N3904 } (100 < \beta < 400)$$

$$D_1 = \text{SFH450 – photoemitter}$$

Launched optical power $P_L = \left(\dfrac{9\ \mu\text{W}}{10\text{ mA}}\right) \times I_D$

Diode forward voltage $V_F = 1.3$ V

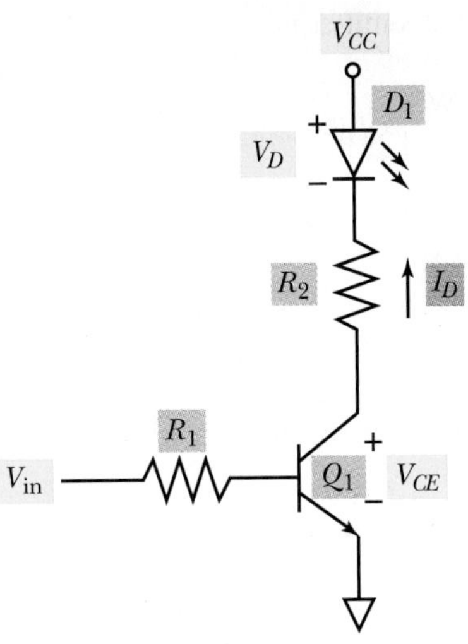

FIGURE 12–5 Noninverting photoemitter circuit with NPN transistor

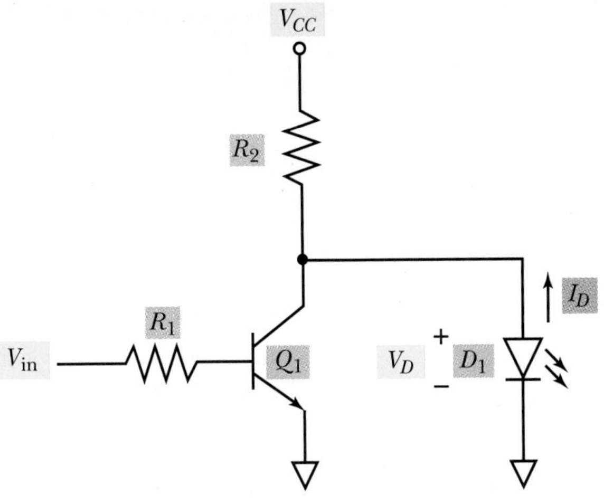

FIGURE 12–6 Inverting photoemitter circuit using an NPN transistor

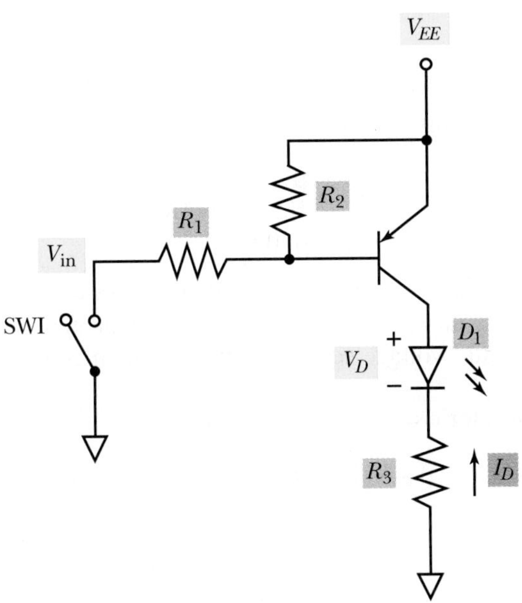

FIGURE 12–7 Inverting photoemitter circuit with PNP transistor

Answer:

For $V_{in} = 0$ V:

To evaluate the circuit operation, you will first consider the operational state of the transistor. The equation for the base-emitter loop is:

$$V_{in} - I_B \times R_1 - V_{BE} = 0$$

Since V_{in} is less than 0.7 V, the transistor is in cutoff ($I_B = 0$ A and $I_C = 0$ A). In the ideal condition with the transistor in cutoff, no current will be flowing through the photoemitter. As a result the photoemitter will not be emitting any light. (Actual condition, the transistor will have a very small leakage collector current. This current is inadequate to cause the photoemitter to emit light.)

The optical power launched from the photoemitter is

$$P_L = \left(\frac{90\ \mu\text{W}}{10\ \text{mA}}\right) \times I_D$$
$$P_L = \left(\frac{90\ \mu\text{W}}{10\ \text{mA}}\right) \times 0\ \text{mA}$$
$$P_L = 0\ \mu\text{W}$$

For $V_{\text{in}} = 5$ V:

Again, you will first consider the operational state of the transistor. The equation for the base-emitter loop is:

$$V_{\text{in}} - I_B \times R_1 - V_{BE} = 0$$

Since V_{in} is greater than 0.7 V, the transistor will be operating in either the linear region ($I_C = \beta \times I_B$) or the saturation region ($V_{CE(\text{sat})} \leq 0.3$ V). Using the base-emitter loop equation, the base current will be calculated.

$$5\ \text{V} - I_B \times 1\ \text{k}\Omega - 0.7\ \text{V} = 0$$
$$I_B = \frac{4.3\ \text{V}}{1\ \text{k}\Omega}$$
$$I_B = 4.3\ \text{mA}$$

If the transistor were operating in the linear region, the collector current would be:

$$I_C = \beta \times I_B$$
$$I_C = 100 \times 4.3\ \text{mA}$$
$$I_C = 430\ \text{mA}$$

Will the transistor collector-emitter circuit allow a collector current of 430 mA? To answer this question, you will need to evaluate the collector-emitter loop equation with the collector-emitter voltage equal to 0 V (saturation condition) solving for the collector current.

The transistor collector-emitter loop equation is:

$$V_{CC} - V_D - I_C \times R_2 - V_{CE} = 0$$
$$5\ \text{V} - 1.3\ \text{V} - I_C \times 100\ \Omega - 0\ \text{V} = 0$$
$$I_C = \frac{3.7\ \text{V}}{100\ \Omega}$$
$$I_C = 37\ \text{mA}$$

37 mA is the maximum current that the circuit will be able to conduct. Therefore, the transistor will be operating in the saturation region with 37 mA flowing through the photoemitter.

The optical power launched from the photoemitter is:

$$P_L = \left(\frac{90\ \mu\text{W}}{10\ \text{mA}}\right) \times I_D$$
$$P_L = \left(\frac{90\ \mu\text{W}}{10\ \text{mA}}\right) \times 37\ \text{mA}$$
$$P_L = 333\ \mu\text{W}$$

◆

Through this example, you observed the analysis of the on and off conditions for the photoemitter used in the Figure 12–5 circuit. Can you determine if the photoemitter is on? Since the SFH450 component is an infrared photoemitter, you are unable to observe any light being emitted from the device without an optical detector with a spectral response that includes the frequency being emitted by the photoemitter.

◆ **EXAMPLE** Evaluate the Figure 12–6 circuit when the input voltage is 0 V. Repeat the circuit evaluation when the input voltage is 5 V. The following components are used in the circuit:

$$\begin{aligned} V_{CC} &= 5\text{ V} \\ R_1 &= 1\text{ k}\Omega \\ R_2 &= 100\text{ k}\Omega \\ Q_1 &= 2\text{N}3904\,(100 < \beta < 400) \\ D_1 &= \text{MFOE71} - \text{photoemitter} \end{aligned}$$

$$\text{Launched optical power} \quad P_L = \left(\frac{165\ \mu\text{W}}{100\text{ mA}}\right) \times I_D$$

$$\text{Diode forward voltage} \quad V_F = 1.5\text{ V}$$

Answer:

For $V_{in} = 0$ V:

To evaluate the circuit operation, you will first consider the operational state of the transistor. The equation for the base-emitter loop is:

$$V_{in} - I_B \times R_1 - V_{BE} = 0$$

Since V_{in} is less than 0.7 V, the transistor is in cutoff ($I_B = 0$ A and $I_C = 0$ A). In the ideal condition with the transistor in cutoff, current will be allowed to conduct through the photoemitter circuit. The photoemitter current is calculated using the following equation:

$$\begin{aligned} V_{CC} - I_2 \times R_2 - V_D &= 0 \\ 5\text{ V} - I_2 \times 100\ \Omega - 1.5\text{ V} &= 0 \\ I_2 &= 35\text{ mA} \end{aligned}$$

The optical power launched from the photoemitter is:

$$\begin{aligned} P_L &= \left(\frac{165\ \mu\text{W}}{100\text{ mA}}\right) \times I_D \\ P_L &= \left(\frac{165\ \mu\text{W}}{100\text{ mA}}\right) \times 35\text{ mA} \\ P_L &= 57.75\ \mu\text{W} \end{aligned}$$

For $V_{in} = 5$ V:

Again, you will first consider the operational state of the transistor. The equation for the base-emitter loop is:

$$V_{in} - I_B \times R_1 - V_{BE} = 0$$

Since V_{in} is greater than 0.7 V, the transistor will be operating in either the linear region ($I_C = \beta \times I_B$) or the saturation region ($V_{CE(sat)} \leq 0.3$ V). Using the base-emitter loop equation, fine base current will be calculated.

$$\begin{aligned} 5\text{ V} - I_B \times 1\text{ k}\Omega - 0.7\text{ V} &= 0 \\ I_B &= \frac{4.3\text{ V}}{1\text{ k}\Omega} \\ I_B &= 4.3\text{ mA} \end{aligned}$$

The transistor collector-emitter loop equation is:

$$V_{CC} - I_2 \times R_2 - V_{CE} = 0$$

If the transistor were operating in the linear region, the collector current would be:

$$\begin{aligned} I_C &= \beta \times I_B \\ I_C &= 100 \times 4.3\text{ mA} \\ I_C &= 430\text{ mA} \end{aligned}$$

Will the transistor collector-emitter circuit allow a collector current of 430 mA? To answer this question, you will need to evaluate the collector-emitter loop equation with the collector-emitter voltage equal to 0 V (saturation condition) solving for the collector current.

$$5\text{ V} - I_C \times 100\ \Omega - 0\text{ V} = 0$$
$$I_C = \frac{5\text{ V}}{100\ \Omega}$$
$$I_C = 50\text{ mA}$$

50 mA is the maximum current that the circuit will be able to conduct. Therefore, the transistor will be operating in the saturation region ($V_{CE(\text{sat})} \leq 0.3$ V). Since the transistor collector-emitter circuit is connected in parallel with the photoemitter, no current will be flowing through the photoemitter. As a result the photoemitter will not be emitting any light.

The optical power launched from the photoemitter is:

$$P_L = \left(\frac{165\ \mu\text{W}}{100\text{ mA}}\right) \times I_D$$
$$P_L = \left(\frac{165\ \mu\text{W}}{100\text{ mA}}\right) \times 0\text{ mA}$$
$$P_L = 0\text{ mA}$$

◆

Through the analysis of the Figure 12–6 circuit, you noticed that an input of 0 V caused the photoemitter to emit infrared light. With an input of 5 V, the photoemitter is off.

◆ **EXAMPLE** Evaluate the Figure 12–7 circuit when the input voltage is 0 V (switch closed). Repeat the circuit evaluation when no voltage is applied to the input (switch open). The following components are used in the circuit:

$$V_{EE} = 12\text{ V}$$
$$R_1 = 1\text{ k}\Omega$$
$$R_2 = 1\text{ k}\Omega$$
$$R_3 = 200\ \Omega$$
$$Q_1 = 2\text{N}3907\ (100 < \beta < 400)$$
$$D_1 = \text{SFH}450 - \text{photoemitter}$$

Launched optical power $P_L = \left(\frac{90\ \mu\text{W}}{10\text{ mA}}\right) \times I_D$

Diode forward voltage $V_F = 1.3\text{ V}$

Answer:

For the switch closed ($V_{\text{in}} = 0$ V):

To evaluate the circuit operation, you will first consider the operational state of the transistor. The equation for the emitter-base loop is:

$$V_{EE} - V_{EB} - I_B \times R_t - V_{\text{in}} = 0$$

Since $V_{EE} - V_{\text{in}}$ is greater than 0.7 V, the transistor will be operating in either the linear region ($I_C = \beta \times I_B$) or the saturation region ($V_{CE(\text{sat})} \leq 0.3$ V). Using the base-emitter loop equation, the base current will be calculated.

$$12\text{ V} - 0.7\text{ V} - I_B \times 1\text{ k}\Omega - 0\text{ V} = 0$$
$$I_B = \frac{11.3\text{ V}}{1\text{ k}\Omega}$$
$$I_B = 11.3\text{ mA}$$

The transistor emitter-collector loop equation is:

$$V_{EE} - V_{EC} - V_D - I_C \times R_3 = 0$$

If the transistor were operating in the linear region, the collector current would be:

$$I_C = \beta \times I_B$$
$$I_C = 100 \times 11.3 \text{ mA}$$
$$I_C = 1.130 \text{ A}$$

Will the transistor collector-emitter circuit allow a collector current of 1.13 A? To answer the question, you will evaluate the collector-emitter loop equation with the collector-emitter voltage equal to 0 V (saturation condition) solving for the collector current.

$$12 \text{ V} - 0 \text{ V} - 1.3 \text{ V} - I_C \times 200\ \Omega = 0$$
$$I_C = \frac{10.7 \text{ V}}{200\ \Omega}$$
$$I_C = 53.5 \text{ mA}$$

53.5 mA is the maximum current that the circuit will be able to conduct. Therefore, the transistor will be operating in the saturation region ($V_{CE(\text{sat})} \leq 0.3$ V). As a result, the 53.5 mA current will be flowing through the photoemitter.

The optical power launched from the photoemitter is:

$$P_L = \left(\frac{90\ \mu\text{W}}{10 \text{ mA}}\right) \times I_D$$
$$P_L = \left(\frac{90\ \mu\text{W}}{10 \text{ mA}}\right) \times 53.5 \text{ mA}$$
$$P_L = 481.5\ \mu\text{W}$$

For the switch open:

Again, you will first consider the operational state of the transistor. The equation for the base-emitter loop is:

$$V_{EE} - V_{EB} - I_B \times R_1 - V_{\text{in}} = 0$$

Since the switch is open and the resistor R_2 is connected in parallel with the transistor emitter-base circuit, the transistor is in cutoff ($I_B = 0$ A and $I_C = 0$ A). In the ideal condition with the transistor in cutoff, the emitter current, which is the current that will conduct through the photoemitter, is also 0 A. With no current conducting through the photoemitter, no light will be emitted from the photoemitter.

The optical power launched from the photoemitter is:

$$P_L = \left(\frac{90\ \mu\text{W}}{10 \text{ mA}}\right) \times I_D$$
$$P_L = \left(\frac{90\ \mu\text{W}}{10 \text{ mA}}\right) \times 0 \text{ mA}$$
$$P_L = 0\ \mu\text{W}$$

◆

Through the analysis of the Figure 12–7 circuit, you noticed that a closed switch caused the photoemitter to emit infrared light and an open switch caused the photoemitter to be off.

Other components and circuit configurations can be utilized to accomplish the task of controlling the current being provided to the photoemitter.

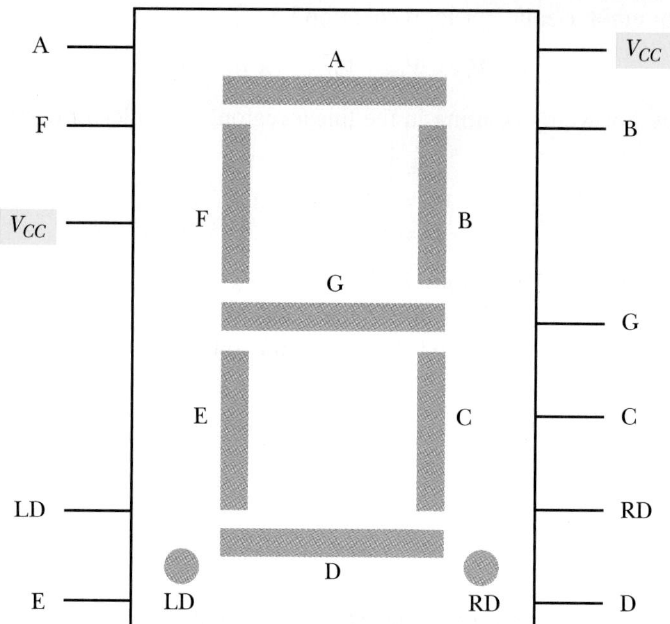

FIGURE 12–8 Seven segment display

Seven Segment Display

Another common configuration for LEDs is represented by seven segment displays (shown in Figure 12–8). A seven segment display is made up of seven individual LEDs that have a long, narrow bar shape. These narrow bars are known as *segments.* Lighting certain combinations of these segments allows you to indicate decimal numbers 0 through 9. As shown in Figure 12–8, some seven segment displays contain right-hand and/or left-hand decimal points. The seven segment displays are configured as common anode (Figure 12–8) or common cathode. The common anode displays require you to connect the positive terminal of the power supply to the common terminal and provide a path to ground through series resistors for each segment connection. Figure 12–9 shows a circuit, which allows the user to select the segment to be illuminated by closing the appropriate switch. The common cathode displays require you to connect the ground of the power supply to the common terminal and provide a path to the positive voltage source through series resistors for each segment connection.

PRACTICE PROBLEMS 2

Using Figure 12–9 indicate which switches must be closed to illuminate the number **5** (no decimal points lit). The display forward voltage drop is 2.0 V. The supply voltage (V_4) is 5 V. The resistors are all the same value of 330 Ω. Calculate how much current the supply must provide to the seven segment display.

Laser Diodes

The **laser diode** is growing in popularity these days. Like the LED, a laser diode emits light energy. But the light energy from a laser diode is far more intense and is at a single frequency *(monochromatic).* The effects of laser diodes are clearly seen as the criss-crossing pattern of ruby-colored light in the price scanners of modern supermarket checkout counters. Laser diodes are also used in compact disk (CD) players and optical disk drives for computers.

Laser diodes must be pulsed with short bursts of high current. The pulses occur so rapidly the light seems to be steady, however.

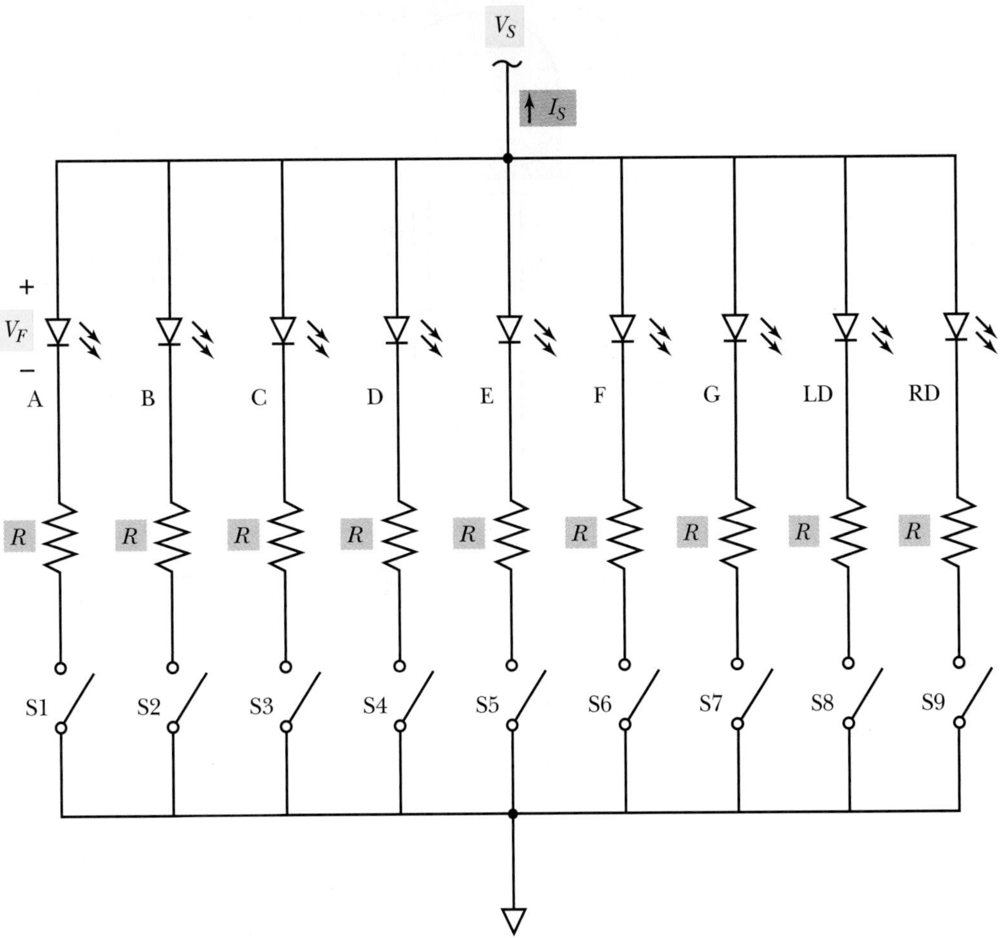

FIGURE 12–9 Common anode seven segment display control circuit multiSIM

12–2 Photo-Reactive Devices (Photodetectors)

Photodiodes

Earlier discussions of semiconductor materials and P-N junctions described these junctions as sensitive to light energy. Semiconductor devices are normally encased in light-tight enclosures, so they are unable to respond to changes in surrounding light level. In a **photodiode,** there is a lens that focuses light onto the P-N junction. This light creates available charge carriers–increasing the amount of light falling onto the junction increases the current flow. Figure 12–10 shows the typical enclosure and schematic symbol for the photodiode.

A photodiode is operated in its reverse-biased configuration, Figure 12–11. This means the amount of light falling onto the depletion region changes the amount of reverse leakage current. As the light intensity increases, the photodiode allows more current to flow through the circuit; this means the voltage across the output resistor increases with increasing light level. Likewise, the voltage across the output resistor decreases with a decreasing light level.

The measurement for the amount of current flowing in relationship to the received power is called responsivity *(R)* also known as spectral response. The MFOD71 photodiode has a responsivity of 0.2 μA/μW. The amount of current that will conduct in the circuit due to the received optical power can be calculated using Formula 12–4

FORMULA 12–4 $I = R \times P_{RCVD}$

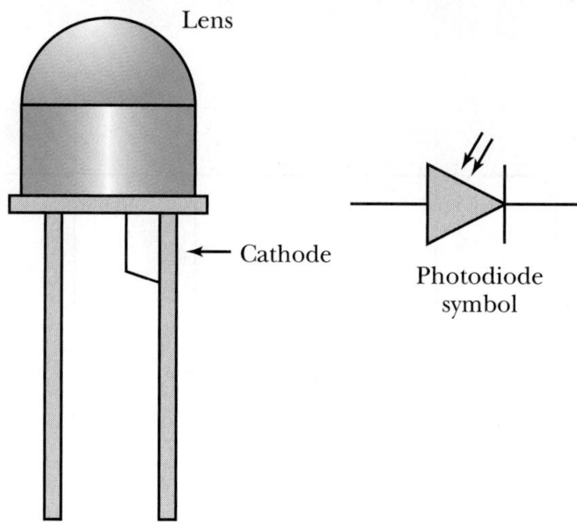

FIGURE 12–10 The photodiode

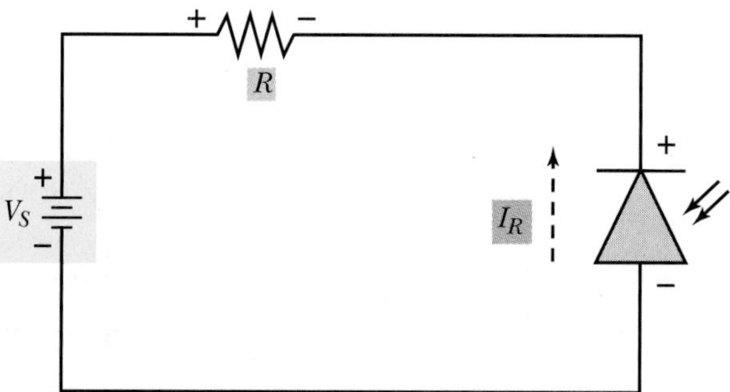

FIGURE 12–11 Elements of a photodiode circuit

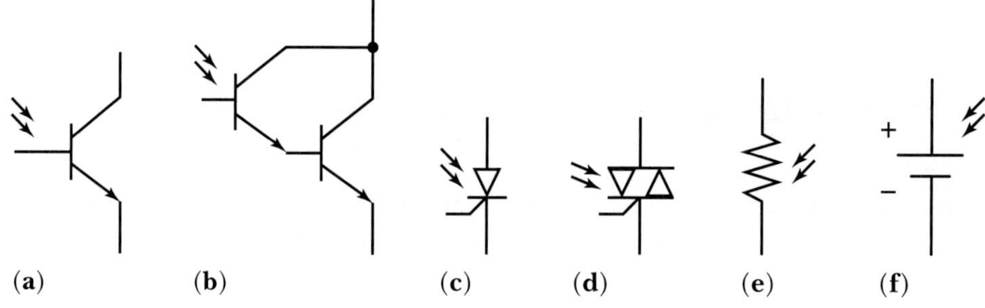

FIGURE 12–12 Photo-reactive devices: (a) phototransistor; (b) photodarlington; (c) photoSCR; (d) photoTRIAC; (e) photoresistor; (f) photovoltaic cell

Phototransistors and Photodarlingtons

Phototransistors (Figure 12–12a) and photodarlingtons (Figure 12–12b) are available as both two-terminal (collector and emitter) and three-terminal devices (base, collector, and emitter). The received light produces a base current to flow. Due to the gain of the transistor, the collector current is high. The MFOD72 phototransistor has a responsivity of 125 μA/μW, whereas the MFOD73 photodarlington has a responsivity of 1500 μA/μW.

PhotoSCRs and PhotoTRIACs

PhotoSCRs (Figure 12–12c) is also known as light-activated SCR (LASCR). The light activates the gate current, which results in the SCR turning on. Once on, the, SCR will stay on until the anode current drops below the hold current. Photo TRIACs (Figure 12–12d) operate in the same manner as the PhotoSCRs.

Photoresistors (Photoconductive Cells)

Photoresistors are two-terminal devices (Figure 12–12e). The resistance of the device varies inversely proportional to the light intensity. When no light is received within the response wavelengths, the photoresistor has a high resistance (R_{OFF} between 500 kΩ to 200 MΩ). When enough light in the response wavelength is detected, the photoresistor has a low resistance (R_{ON} between 1 kΩ to 100 kΩ).

Photovoltaic Cells (Solar Cells)

Photovoltaic cells are two-terminal devices (Figure 12–12f). The voltage across the terminals is proportional to the received light intensity. The photovoltaic cell is configured to provide power to external loads. On a clear day with the sun directly overhead, a silicon photovoltaic cell with a 0.5 cm² area can deliver 25 mA at 0.4 V. Under the same conditions, a silicon photovoltaic cell with a 75 cm² area can deliver 2 A at 0.4 V. To provide usable power levels, solar cells are connected in parallel to increase the current flow, or solar cells are connected in series to increase the output voltage.

EXAMPLE Using the circuit in Figure 12–13, calculate the output voltage.

$V_{CC} = 5$ V
$R_1 = 1$ kΩ
Received optical power (P_{RCVD}) 20 μW
Phototransistor detector (SFH350)

Responsivity $R = \left(\frac{800\ \mu A}{10\ \mu W}\right)$

Photocurrent $I_C = R \times P_{RCVD}$ (Formula 12–4)
Collector dark current $I_{CEO} = 2$ nA (typical) 50 nA (max)

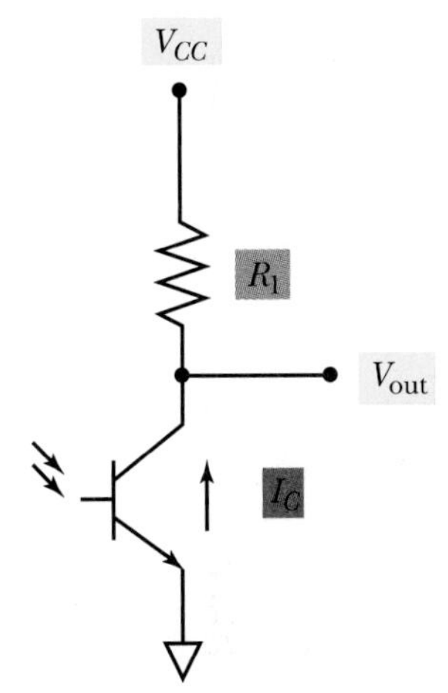

FIGURE 12–13 Phototransistor detector circuit (collector output)

Answer:

Using Formula 12–4 find the phototransistor collector current:

$$I_C = R \times P_{RCVD}$$
$$I_C = \left(\frac{800\ \mu A}{10\ \mu W}\right) \times 20\ \mu W$$
$$I_C = 1.6\ mA$$

The output voltage will be calculated next.

$$V_{CC} - I_C \times R_1 - V_{out} = 0$$
$$5\ V - 1.6\ mA \times 1\ k\Omega - V_{out} = 0$$
$$V_{out} = 3.4\ V$$

EXAMPLE Using the circuit in Figure 12–14, calculate the output voltage.

$V_{CC} = 5$ V
$R_1 = 1$ kΩ
Received optical power (P_{RCVD}) 20 μW
Phototransistor detector (MFOD72)

Responsivity $R = \left(\frac{125\ \mu A}{1\ \mu W}\right)$

Photocurrent $I_C = R \times P_{RCVD}$ (Formula 12–4)
Collector dark current $I_{CEO} = 100$ nA (max)

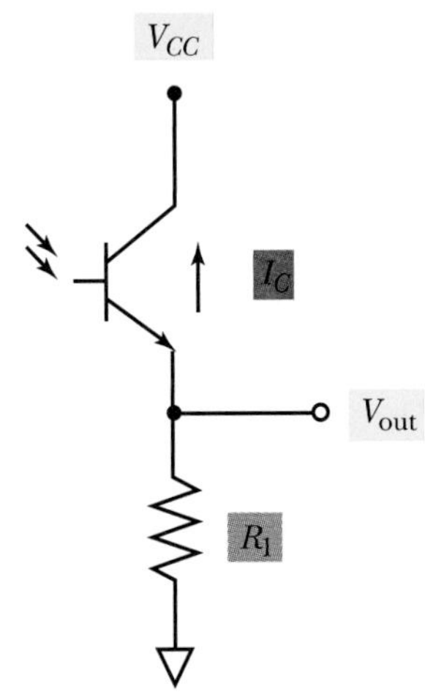

FIGURE 12–14 Phototransistor detector circuit (emitter output)

Answer:

Using Formula 12–4 find the phototransistor collector current:

$$I_C = R \times P_{\text{RCVD}}$$
$$I_C = \left(\frac{125\ \mu\text{A}}{1\ \mu\text{W}}\right) \times 20\ \mu\text{W}$$
$$I_C = 2.5\ \text{mA}$$

The output voltage is calculated as the voltage dropped across R_1.

$$V_{\text{out}} = I_C \times R_1$$
$$V_{\text{out}} = 2.5\ \text{mA} \times 1\ \text{k}\Omega$$
$$V_{\text{out}} = 2.5\text{V}$$ ◆

In these examples if the calculation for $I_C \times R_1$ is greater than V_{CC}, the phototransistor will be in saturation. In the last example, the replacement of the R_1 1 kΩ with a 10 kΩ would result in a calculated output voltage (V_{out}) of 25 V. Since the calculated value exceeds the 5 V supply connected to the circuit, the phototransistor is driven into saturation ($V_{\text{CE}} \leq 0.3$ V), the output voltage (V_{out}) will be close to 5 V, and the photodetector current will be 0.5 mA (V_{out}/R_1).

12–3 Opto-Isolators

Sometimes an LED (used as a source of light) and a photodiode (used as a detector of light) are used to provide isolation between circuits or processes. The information is transmitted from the source to the detector by light waves. Such a device is called an **opto-isolator** (also known as optocouplers) and is illustrated in Figure 12–15.

Common applications for opto-isolators include preventing a low impedance load from loading down a high impedance source, enabling a low-current source to control a high-current detector, and interfacing two systems with different ground references. The distance between the source and detector diodes is usually small (a tenth of an inch or less), but the electrical isolation (above 2000 V) is better than those of a transformer (about 700 V). The efficiency of the optocoupler is higher than the transformer since the power dissipated by the optocoupler is lower than the power dissipated by the transformer.

With all optical systems, it is important that the source emits light frequencies to which the detector is sensitive. Therefore, sources and detectors are matched according to the frequencies emitted by the source and the most sensitive frequency range of the detector. The emitter and source diodes are usually enclosed in a light-proof package. However, some applications require that the package is open and exposed to out-

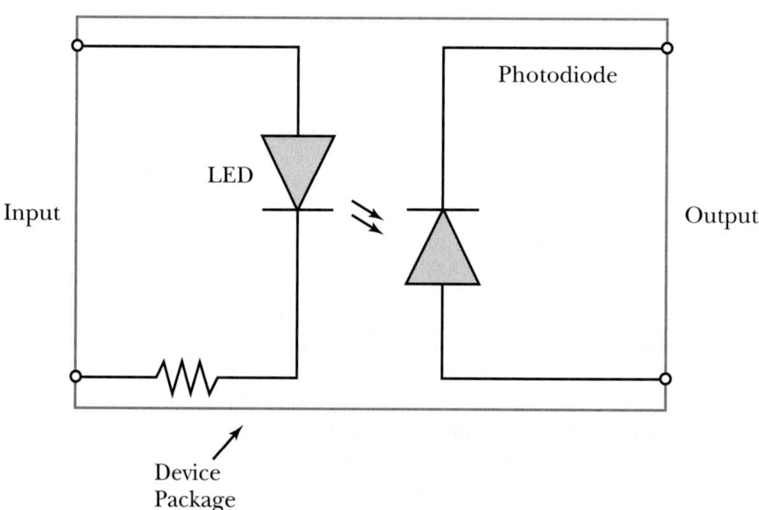

FIGURE 12–15 Opto-isolator symbol

side light sources. In this case, the light from the source diode is sharply focused onto the photodiode, thereby drowning out the effects of outside light sources.

The opto-isolator has been packaged with a variety of configurations. All opto-isolators traditionally have a photoemitter (LED). Different options for the detector portion of the opto-isolators are available to comply with your circuit applications.

The photovoltaic MOSFET drivers convert the light from the photoemitter into a dc voltage with a small current available to power other circuitry. More information is available about these devices at *www.dionics-usa.com.* Figure 12–16a shows the schematic for one of Dionics photovoltaic MOSFET drivers.

The photovoltaic relay is configured for a dc input (3 mA at 1.5 VDC) and will control the switching of ac or dc loads (2.5 A at 20 V_{peak}). For this device, power MOSFETs are used in the construction of the output circuit. The schematic for a photovoltaic relay manufactured by International Rectifier is shown in Figure 12–16b.

The 4N25 opto-isolator shown in Figure 12–16c is constructed with a gallium arsenide infrared emitting diode coupled with a phototransistor. With 10 mA of current applied to the diode, the voltage drop across the diode will be about 1.15 V. The maximum output collector current is 100 mA, and is directly related to the current flowing through the diode. The current transfer ratio (CTR) is defined as the collector

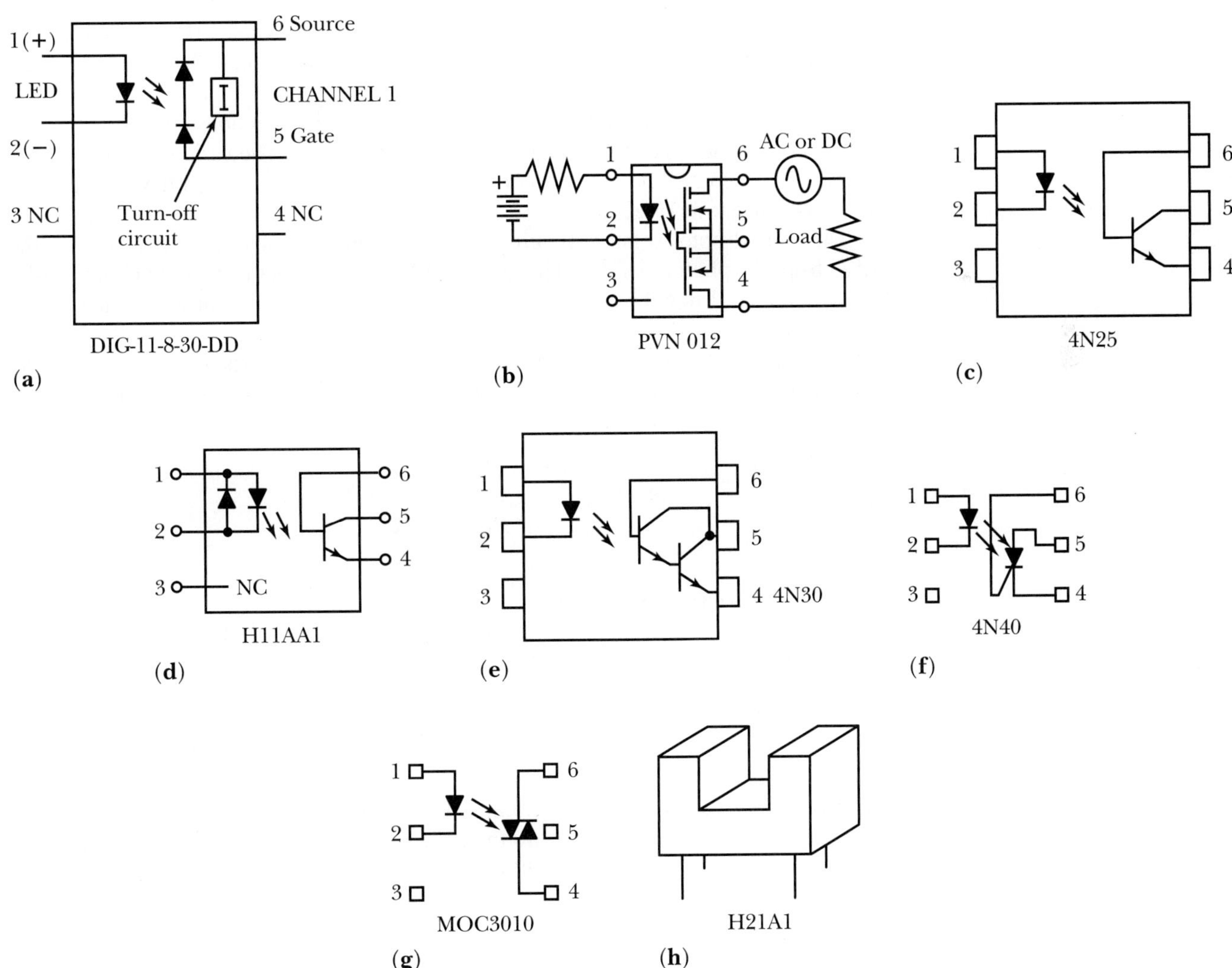

FIGURE 12–16 (a) photovoltaic MOSFET driver; (b) power MOSFET photovoltaic relay; (c) opto-isolator/transistor output; (d) opto-isolator ac input/transistor output; (e) opto-isolator Darlington transistor output; (f) opto-isolator SCR output; (g) opto-isolator triac driver output; (h) slotted opto-isolator

current divided by the diode current. The typical CTR for this device is 100%. If you know the diode current, the collector current can be calculated by using Formula 12–5.

FORMULA 12–5 $I_C = \text{CTR} \times I_D$

◆ **EXAMPLE** Using the 4N25 opto-isolator with a CTR of 100% and a diode current of 10 mA, calculate the available collector current.

Answer:

From Formula 12–5:

$$I_C = \text{CTR} \times I_D$$
$$I_C = 100\% \times 10 \text{ mA}$$
$$I_C = 10 \text{ mA}$$ ◆

The H11AA1 opto-isolator shown in Figure 12–16d is manufactured with two infrared emitting diodes connected in an inverse parallel configuration. This allows you the ability to apply ac signals to the input. The output of the device is an NPN transistor. The 4N30 opto-isolator is configured with a photo Darlington transistor output as shown in Figure 12–16e. The device is designed for use in applications where you need output currents up to 150 mA.

The 4N39 and MOC3010 are opto-isolators with thyristor driver outputs. The thyristor in the 4N39 is a SCR, and the thyristor in the MOC3010 is a Triac. The schematics for these devices are provided in Figures 12–16f and 12–16g respectively. Both of these devices are designed for switching ac signals.

The devices mentioned provide you an overview of available opto-isolators. As stated, each device is configured for different applications. The variety of components allows you to control dc or ac motors, solenoids, and lights to list a few applications with low current dc or ac inputs.

The slotted opto-isolator (Figure 12–16h) is designed with a slot separating the LED from the phototransistor. The infrared LED light output causes the phototransistor to conduct. When the light beam is blocked, the phototransistor will be in cutoff.

◆ **EXAMPLE** The circuit in Figure 12–17 is used to count rotations of a motor by the connection of a slotted wheel to the motor shaft. Calculate the output voltage when the slot is open and when the slot is blocked.

$$V_{CC} = 5 \text{ V}$$
$$V_S = 5 \text{ V}$$
$$R_1 = 120\ \Omega$$
$$R_2 = 2.5\ \text{k}\Omega$$

Slotted opto-isolator (H21A1)

Parameter	Test Condition	Value
Input forward voltage	$I_F = 60$ mA	$V_F \leq 1.7$ V
Input forward voltage	$I_F = 30$ mA	$V_F = 1.4$ V (typical)
Off-state collector current	$V_{CE} = 25$ V	$I_{\text{CEO}} \leq 100$ nA
On-state collector current	$I_F = 5$ mA, $V_{CE} = 5$ V	$I_{C(\text{on})} \geq 0.15$ mA
On-state collector current	$I_F = 20$ mA, $V_{CE} = 5$ V	$I_{C(\text{on})} \geq 1.0$ mA
On-state collector current	$I_F = 30$ mA, $V_{CE} = 5$ V	$I_{C(\text{on})} \geq 1.9$ mA
Turn-on time	$I_F = 30$ mA, $V_{CC} = 5$ V, $R_L = 2.5$ kΩ	$t_{\text{on}} = 8$ μs (typical)
Turn-off time	$I_F = 30$ mA, $V_{CC} = 5$ V, $R_L = 2.5$ kΩ	$t_{\text{off}} = 50$ μs (typical)

Answer:

Writing the KVL equation for the photoemitter circuit, you can calculate the photoemitter current:

$$V_S - I_D \times R_1 - V_D = 0$$
$$5 \text{ V} - I_D \times 120\ \Omega - 1.4 \text{ V} = 0$$
$$I_D = 30 \text{ mA}$$

Application Problem

Industrial

CONVEYER BELT SPEED MONITOR

The speed of a conveyer belt in an automotive assembly plant is monitored by a computer. The shaft on the motor driving the belt has a disk with a hole in it and is mounted to a slotted opto-isolator. When the light passes from the emitter to the receiver, the switch conducts. When it is blocked, it doesn't. If the frequency at the output of the opto-isolator is 2 Hz, what is the RPM of the shaft.

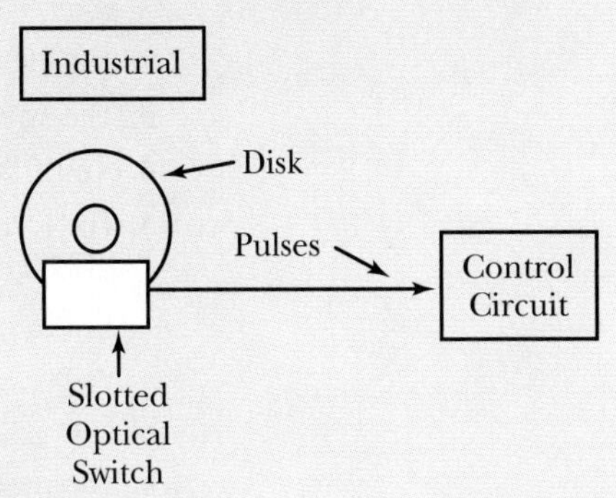

Solution A complete 360 degree rotation is one cycle, so 2 Hz = 2 cycles per second. We need to convert this value to cycles per minute.

2 cycles/sec = Xrevolutions/60 sec
Xrevolutions = 2 (60) = 120 revolutions per minute

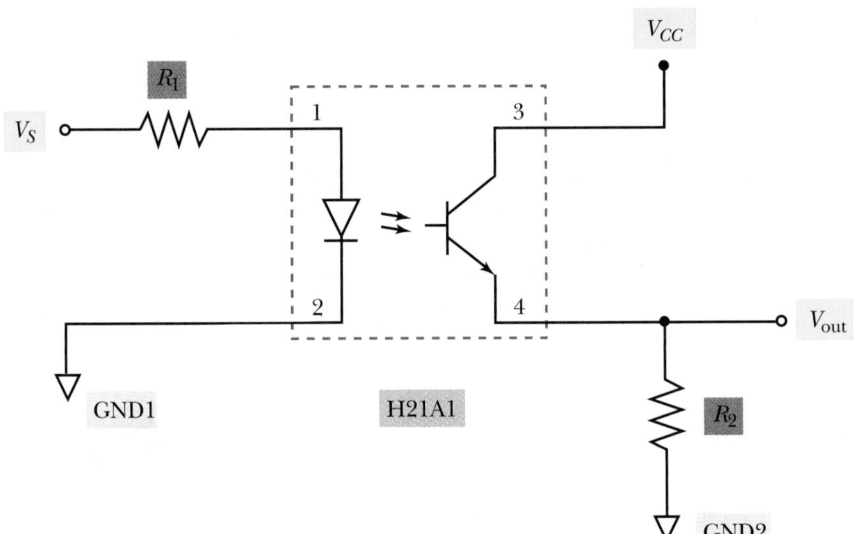

FIGURE 12–17 Slotted opto-isolator circuit

With the *open slot,* the on-state collector current will be greater than or equal to 1.9 mA for the 30 mA photoemitter current (obtained from the component data sheet). Using the 1.9 mA collector current the output voltage can be calculated.

$$V_{out} = I_C \times R_2$$
$$V_{out} \geq 1.9 \text{ mA} \times 2.5 \text{ k}\Omega$$
$$V_{out} \geq 4.75 \text{ V}$$

With the *slot blocked,* the off-state collector current will be less than or equal to 100 nA (obtained from the component data sheet). Using the 100 nA collector current the output voltage can be calculated.

$$V_{out} = I_C \times R_2$$
$$V_{out} \leq 100 \text{ nA} \times 2.5 \text{ k}\Omega$$
$$V_{out} \leq 0.25 \text{ mV}$$ ◆

■ **IN-PROCESS LEARNING CHECK 1**

Fill in the blanks as appropriate.

1. An LED emits light energy when an electron in the _____ band falls to the _____ band.
2. The choice of _____ determines the color of the light from an LED.
3. The arrow on an LED symbol points _____ the diode.
4. An LED emits light when it is _____ biased.
5. The formula for calculating the value of the resistor to be used in series with an LED and its dc voltage source is $R = (V_S - V_D)/I_D$, where:

   V_S is _____

   V_D is _____

   I_D is _____
6. The reverse-breakdown voltage of an LED is generally _____ than that of a typical rectifier diode, and the forward voltage drop of an LED is _____ than that of a rectifier diode.
7. Light falling onto the depletion region of a reverse-biased diode _____ the amount of reverse leakage current.
8. The arrow on a photodiode symbol points _____ the diode.
9. A photodiode is connected into the circuit so that it is _____ biased.
10. In an opto-isolator, a(n) _____ is the source and a(n) _____ is the detector of light energy.

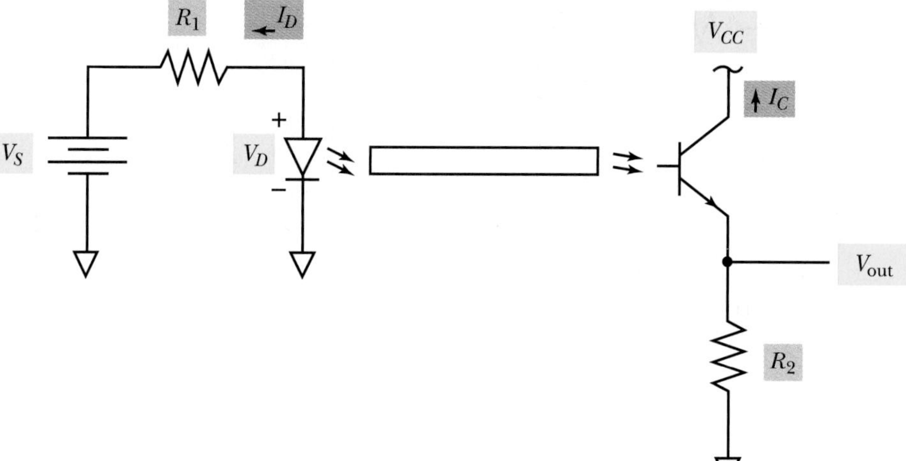

FIGURE 12–18 Fiber-optic cable system

12–4 Fiber-Optic Cable Systems

Another type of optoelectronic system uses an LED or laser diode as a source of light and a photodiode (or phototransistor) as the detector. The system works much like an opto-isolator except that light from the source is transmitted through a length of thin fiber-optic cable as shown in Figure 12–18.

Fiber-Optic Cable

Fiber-optic cable is constructed using very pure glass or plastic. Fiber-optic cables (Figure 12–19) consist of a core (the center of the optical fiber), the cladding (surrounding the core), and the jacket. Fiber-optic cable is designed to guide light from an

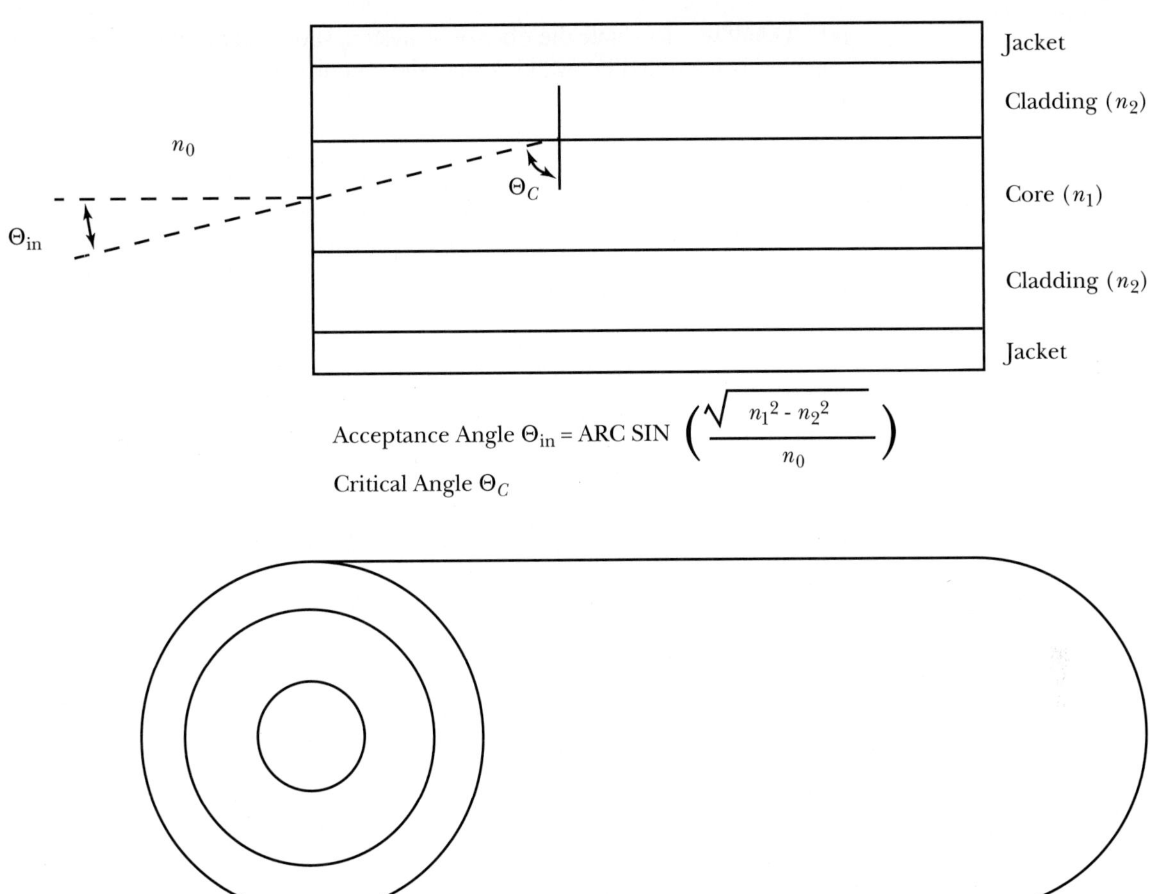

FIGURE 12–19 Fiber-optic cable

optical source to an optical detector. Due to the construction of the fiber-optic cable, light energy can be transmitted over long distances with a very small loss in light intensity without interference from outside electrical and optical sources or poor atmospheric conditions. The optical power loss (attenuation) in fiber-optic cables is measured in dB/km (decibels per kilometer). Power gain is defined as the output power divided by the input power. Formula 12–6 is used to calculate the power gain in decibels. A negative gain indicates that the output power is lower than the input power. If the input power and the gain are known, you can calculate the output power using Formula 12–7.

FORMULA 12–6 $A = 10 \log_{10}\left(\frac{P_{out}}{P_{in}}\right)$

FORMULA 12–7 $P_{out} = P_{in} \times (10^{A/10})$

Optoelectronic/fiber-optic systems are already replacing ordinary copper wire for telephone communications. They are quickly becoming the preferred systems for communications between computers in modern office buildings.

EXAMPLE Evaluate the fiber-optic system shown in Figure 12–18 with the device characteristics stated below. Calculate the output voltage.

Circuit

$$V_S = 5\text{ V}$$
$$R_1 = 100\ \Omega$$
$$V_{CC} = 15\text{ V}$$
$$R_2 = 1.00\text{ k}\Omega$$

Photoemitter (SFH450)

$$\text{Launched optical power} \quad P_L = \left(\frac{90\ \mu\text{W}}{10\text{ mA}}\right) \times I_D$$
$$\text{Diode forward voltage} \quad V_F = 1.3\text{ V}$$

Phototransistor detector (SFH350)

$$\text{Responsivity} \quad R = \left(\frac{800\ \mu\text{A}}{10\ \mu\text{W}}\right)$$
$$\text{Photocurrent} \quad \text{I}_\text{C} = R \times P_{\text{RCVD}} \text{ (Formula 12–4)}$$

Fiber-optic cable (1000 micron)

Length	4 km
Attenuation	2.5 dB/km

Connector losses

Photoemitter to cable	8 dB
Cable to photodetector	2 dB

Answer:

Manipulating Formula 12–3 where $R = (V_S - V_D)/I_D$, you can solve for I_D.

$$I_D = \frac{(V_S - V_D)}{R_1}$$
$$I_D = \frac{(5\text{ V} - 1.3\text{ V})}{100\ \Omega}$$
$$I_D = 37\text{ mA}$$

The optical power launched from the photoemitter can now be calculated.

$$P_L = \left(\frac{90\ \mu\text{A}}{10\text{ mA}}\right) \times I_D$$
$$P_L = \left(\frac{90\ \mu\text{A}}{10\text{ mA}}\right) \times 37\text{ mA}$$
$$P_L = 333\ \mu\text{W}$$

Total optical power gain (negative sign indicates a loss)

$$\text{A} = (-8\text{ dB}) + (-2.5\text{ dB/km}) + 4\text{ km} + (-2\text{ dB})$$
$$\text{A} = (-20\text{dB})$$

The optical power received at the photodetector (Formula 12–7)

$$P_{\text{RCVD}} = P_L + 10^{A/10}$$
$$P_{\text{RCVD}} = 333\ \mu\text{W} \times 10^{-20/10}$$
$$P_{\text{RCVD}} = 3.33\ \mu\text{W}$$

Application Problem

Communications

LOCAL AREA NETWORK COMMUNICATION

Company A recently had another building constructed to provide more room to add additional departments. The new employees had computers in each office. They needed to be able to access files on a server in the main building. In the process of development, a decision needed to be made as to how the computers between the 2 buildings were to be connected. A choice needs to be made between using copper wire underground or a fiber-optic connection. Which system would be more reliable and less susceptible to unwanted noise signals being injected into the system?

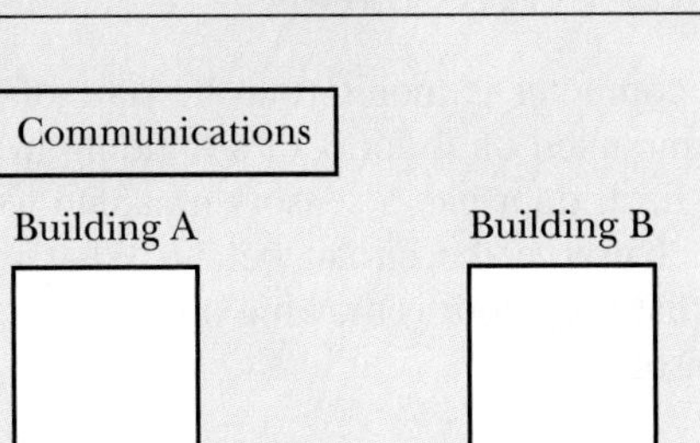

Solution Copper wire is more susceptible to interference. Fiber-optic systems have the following advantages over copper wire: higher bandwidth (can transmit data at several gigabits per second), cables are light in weight, immune to inductive coupling, impossible to tap into without being detected, low loss, and very reliable.

The photodetector current can be calculated using Formula 12–4.

$$I_C = R \times P_{\text{RCVD}}$$

$$I_C = \left(\frac{800\ \mu\text{A}}{10\ \mu\text{W}}\right) \times 3.33\ \mu\text{W}$$

$$I_C = 266.4\ \mu\text{A}$$

The output voltage is calculated using Ohm's law.

$$V_{\text{out}} = I_C \times R_2$$

$$V_{\text{out}} = 266.4\ \mu\text{A} \times 1.00\ \text{k}\Omega$$

$$V_{\text{out}} = 266.4\ \text{mV}$$ ◆

Digital Communication System

A block diagram of a digital communications system utilized by the telephone company is shown in Figure 12–20. The input to the communications system is your voice (a sound wave with frequency components between 300 to 3400 Hz). The telephone microphone converts the sound wave into an electrical signal (an analog waveform). The signal is carried on twisted pair wires to the telephone company's central office. At the central office, the analog signal is converted to a digital signal within the coder portion of the codec. The coder is an eight-bit analog-to-digital converter having a sample rate of 8 kHz. The sample rate is greater than two times the maximum frequency component of the analog signal. The phone company groups the sampled data into 64 kbps (kilobits per second) bit streams. The digital signal is applied to the light source. The light is sent through fiber-optic cables and repeaters to a photodetector. The photodetector converts the optical energy back into electrical voltages (digital format). The signal then enters the decoder section of the codec at another central office, converting the digital data into an analog signal. The analog signal is sent over twisted pair wires to the speaker of the destination telephone. The telephone speaker converts the electrical signal back to a sound wave, which is heard by the listening individual.

Application Problem

Computers

SECURITY SYSTEM

A microcomputer is monitoring the state of 20 optical sensors mounted on 4 doors of a building in a security system. Each door has 5 sensors mounted to detect an intruder. Observe the circuit below. What is the total current that each door is drawing on the receiver side of each sensor?

Solution

$$I_C = \frac{V_{CC}}{R} = \frac{12\text{ V}}{470\ \Omega} = 25.5\text{ mA}$$

$$I_{\text{total}} = 5I_C = 5 \times 25.5\text{ mA} = 127.7\text{ mA}$$

The total current is 127.7 mA.

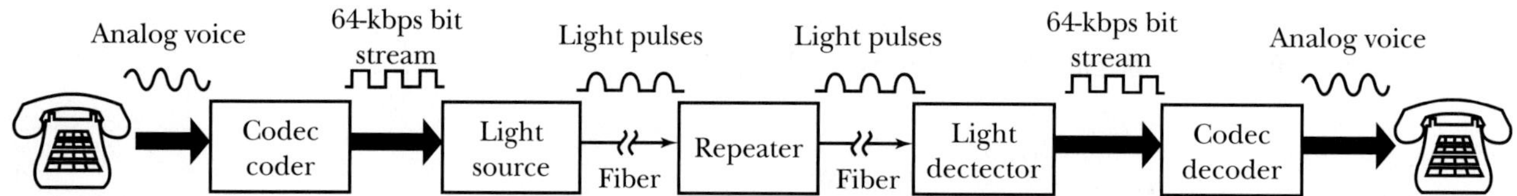

FIGURE 12–20 Fiber-optic communication system

Advancements in technology have resulted in the ability for 200 windows of light (different optical frequencies from infrared to blue) to be emitted into a fiber-optic cable and received by a photodetector. With only 32 windows of light, the telephone company is able to send over 516,000 phone calls "simultaneously" down one fiber-optic cable pair (one cable for transmit, one cable for receive). The greatest volume of telephone traffic is now non-voice communications (such as Internet and fax machine data).

12–5 Optoelectronic Application Circuits

The usage of optoelectronic devices is continually increasing. Telephone companies have been utilizing optoelectronic components for years. Computer products and computer networks are using optoelectronic devices. Optical devices (laser) are used for speed and distance measurement devices. The medical profession has been revolutionized with the usage of optics (laser, small fiber-optic cables, and imaging devices). The transportation industry is replacing traditional incandescent bulbs in traffic lights and railroad crossings with LED grid arrays. Buildings are replacing the traditional light bulb signs with LED signs. The TV remote control uses an IR LED as the optical transmitter (Figure 12–21). The infrared light is received by an IR photodetector. Consumer products are prominently implementing optoelectronic devices.

Can the fiber optic cable be used for sine wave applications? The following example evaluates the application of a sine wave to the fiber-optic cable.

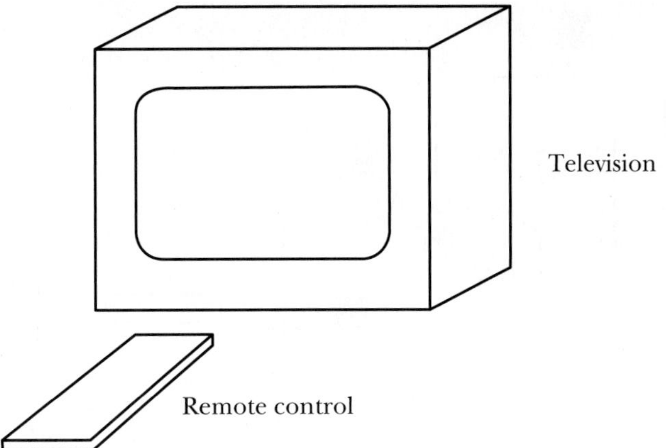

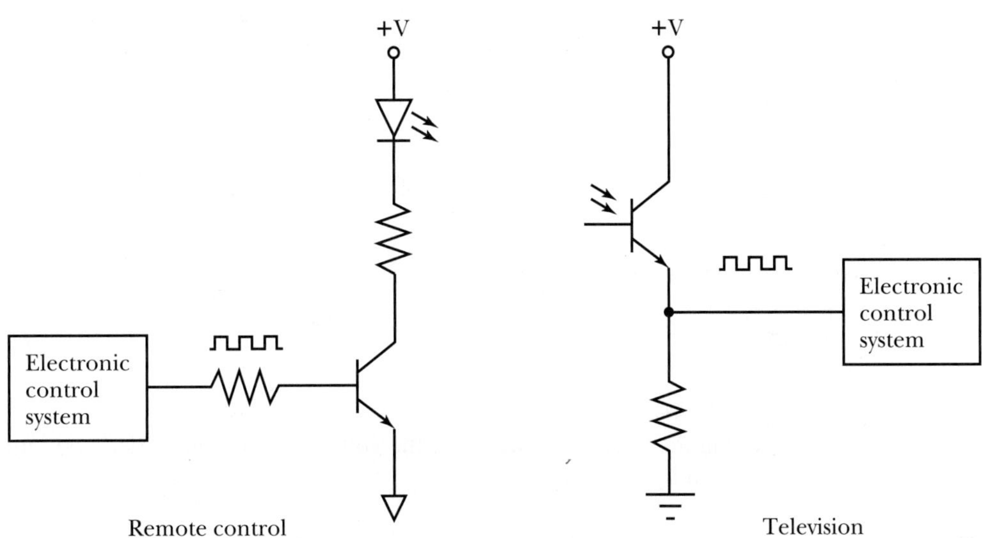

FIGURE 12–21 Television remote control optical interface

EXAMPLE Evaluate the analog fiber optic system shown in Figure 12–22 with the device characteristics stated below. Calculate the output voltage.

Circuit

$V_{CC_1} = 15$ V, $V_{EE_1} = -15$ V, $V_{CC_2} = 15$ V, $V_{EE_2} = -15$ V
$R_1 = 10$ kΩ, $R_2 = 10$ kΩ, $R_3 = 30$ kΩ, $R_4 = 10$ kΩ, $R_5 = 1$ kΩ, $R_6 = 10$ kΩ
$V_{in} = 6\ V_{pp}$ 1 kHz sine wave

Op-amps (LM741)

Photoemitter (MFOE71)

Launched optical power $P_L = \left(\dfrac{165\ \mu\text{W}}{100\ \text{mA}}\right) \times I_D$

Diode forward voltage $V_F = 1.5$ V

Phototransistor detector (MFOD72)

Responsivity $R = \left(\dfrac{125\ \mu\text{A}}{1\ \mu\text{W}}\right)$

Photocurrent $\text{I}_\text{C} = R \times P_{\text{RCVD}}$ (Formula 12–4)

Collector dark current $I_{\text{CEO}} = 100$ nA (max)

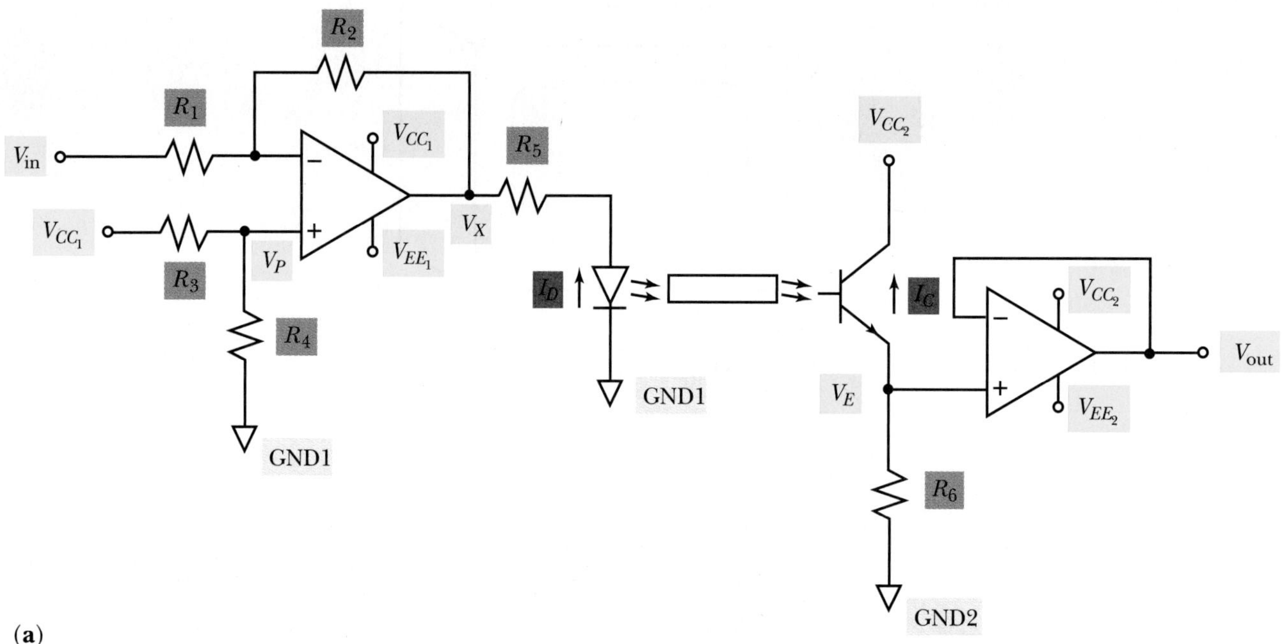

FIGURE 12–22 (a) Fiber-optic analog circuit schematic; (b) Fiber-optic analog circuit waveforms

Fiber-optic cable system current transfer ratio

$$\text{CTR} = 0.1$$

Answer:

When $V_{in} = 0$ V:

Using the voltage divider rule, the voltage at the transmitter's op-amp positive terminal can be calculated:

$$V_P = V_{CC_1} \times \frac{R_4}{(R_3 + R_4)}$$
$$V_P = 15\text{ V} \times \frac{10\text{ k}\Omega}{(10\text{ k}\Omega + 30\text{ k}\Omega)}$$
$$V_P = 3.75\text{ V}$$

Using the positive input op-amp gain equation, the op-amp's output voltage can be found:

$$V_X = V_P \times \left(1 + \frac{R_2}{R_1}\right)$$
$$V_X = 3.75\text{ V} \times \left(1 + \frac{10\text{ k}\Omega}{10\text{ k}\Omega}\right)$$
$$V_X = 7.5\text{ V}$$

Knowing the op-amp output voltage, you can determine the current flowing through the photoemitter:

$$V_X - I_D \times R_5 - V_D = 0$$
$$7.5\text{ V} - I_D \times 1\text{ k}\Omega - 1.5\text{ V} = 0$$
$$I_D = 6.0\text{ mA}$$

With the known fiber-optic system current transfer ratio (CTR), the photodetector current can be calculated using Formula 12–5.

$$I_C = \text{CTR} \times I_D$$
$$I_C = 0.1 \times 6.0\text{mA}$$
$$I_C = 0.6\text{ mA}$$

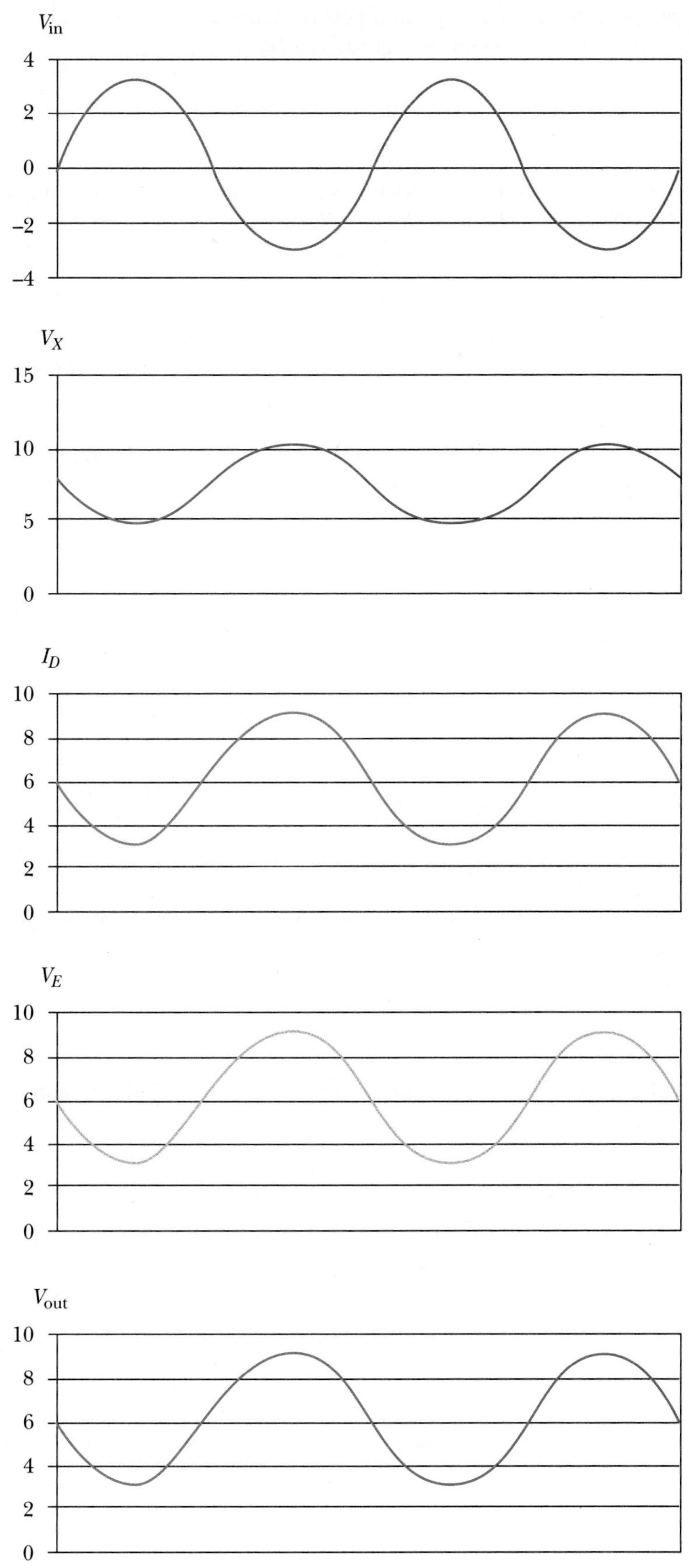

FIGURE 12–22 Continued

The voltage at the receiver's op-amp positive terminal is found by calculating the voltage across the photodetector emitter resistor (R_6):

$$V_E = I_C \times R_6$$
$$V_E = 0.6 \text{ mA} \times 10\text{k}\Omega$$
$$V_E = 6 \text{ V}$$

Since the receiver op-amp is connected as a unity gain amplifier, the output voltage is the same as the positive terminal input voltage.

$$V_{\text{out}} = V_E$$
$$V_{\text{out}} = 6 \text{ V}$$

Now that you have established the operating voltages throughout the circuit when the input voltage (V_{in}) was at 0 V, we will now consider the circuit voltage when the input signal is at +3 V_{peak}:

As V_{in} is applied through a resistor to the inverting input of the transmitter op-amp in addition to the dc voltage applied through the voltage divider to the nonverting op-amp input, the transmitter op-amp's output voltage can be calculated using the following equation:

$$V_X = V_{\text{in}} \times \left(-\frac{R_2}{R_1}\right) + V_P \times \left(1 + \frac{R_2}{R_1}\right)$$

$$\text{where } V_P = V_{CC} \times \frac{R_4}{(R_3 + R_4)}$$

$$V_P = 15 \text{ V} \times \frac{10 \text{ k}\Omega}{(10 \text{ k}\Omega + 30 \text{ k}\Omega)}$$

$$V_P = 3.75 \text{ V}$$

$$V_X = 3 \text{ V} \times \left(-\frac{10\text{k}\Omega}{10\text{k}\Omega}\right) + 3.75 \text{ V} \times \left(1 + \frac{10 \text{ k}\Omega}{10 \text{ k}\Omega}\right)$$

$$V_X = 4.5 \text{ V}$$

Knowing the op-amp output voltage, the current flowing through the photoemitter can be determined.

$$V_X - I_D \times R_5 - V_D = 0$$
$$4.5 \text{ V} - I_D \times 1 \text{ k}\Omega - 1.5 \text{ V} = 0$$
$$I_D = 3.0 \text{ mA}$$

With the known fiber-optic system current transfer ratio (CTR), the photodetector current can be calculated using Formula 12–5.

$$I_C = \text{CTR} \times I_D$$
$$I_C = 0.1 \times 3.0 \text{ mA}$$
$$I_C = 0.3 \text{ mA}$$

The voltage at the receiver's op-amp positive terminal is found by calculating the voltage across the photodetector emitter resistor (R_6):

$$V_E = I_C \times R_6$$
$$V_E = 0.3 \text{ mA} \times 10 \text{ k}\Omega$$
$$V_E = 3 \text{ V}$$

Since the receiver op-amp is connected as a unity gain amplifier, the output voltage is the same as the positive terminal input voltage.

$$V_{\text{out}} = V_E$$
$$V_{\text{out}} = 3 \text{ V}$$

You now know the output voltage when the input sine wave reaches its positive peak amplitude, we will now consider the system output voltage when the input sine wave reaches its negative peak amplitude of –3 V_{peak}:

As V_{in} is applied through a resistor to the inverting input of the transmitter op-amp in addition to the dc voltage applied through the voltage divider to the noninverting op-amp input, the transmitter op-amp's output voltage can be calculated using the following equation:

$$V_X = V_{in} \times \left(-\frac{R_2}{R_1}\right) + V_P \times \left(1+\frac{R_2}{R_1}\right)$$

$$\text{where } V_P = V_{CC} \times \frac{R_4}{(R_3 + R_4)}$$

$$V_P = 15\text{ V} \times \frac{10\text{ k}\Omega}{(10\text{ k}\Omega + 30\text{ k}\Omega)}$$

$$V_P = 3.75\text{ V}$$

$$V_X = (-3\text{ V}) \times \left(-\frac{10\text{ k}\Omega}{10\text{ k}\Omega}\right) + 3.75\text{ V} \times \left(1+\frac{10\text{ k}\Omega}{10\text{ k}\Omega}\right)$$

$$V_X = 10.5\text{ V}$$

Knowing the op-amp output voltage, the current flowing through the photoemitter can be determined.

$$V_X - I_D \times R_5 - V_D = 0$$

$$10.5\text{ V} - I_D \times 1\text{ k}\Omega - 1.5\text{ V} = 0$$

$$I_D = 9.0\text{ mA}$$

With the known fiber-optic system current transfer ratio (CTR), the photodetector current can be calculated using Formula 12–5.

$$I_C = \text{CTR} \times I_D$$

$$I_C = 0.1 \times 9.0\text{ mA}$$

$$I_C = 0.9\text{ mA}$$

The voltage at the receiver's op-amp positive terminal is found by calculating the voltage across the photodetector emitter resistor (R_6):

$$V_E = I_C \times R_6$$

$$V_E = 0.9\text{ mA} \times 10\text{ k}\Omega$$

$$V_E = 9\text{ V}$$

Since the receiver op-amp is connected as a unity gain amplifier, the output voltage is the same as the positive terminal input voltage.

$$V_{out} = V_e$$

$$V_{out} = 9\text{ V}$$ ◆

Figure 12–22b provides the waveform at several locations throughout the circuit allowing you to observe the effects of the sine wave input. Through this example you can see that a fiber-optic cable can be used in analog signal applications. Limitations in the usage of fiber-optic cable for the transfer of complex analog signals such as voice, music, and video exist due to fiber-optic cable characteristics (such as modal dispersion). In addition, the photoemitter, photodetector, and amplifier characteristics need to be evaluated to establish the input signal requirements (amplitude and frequency, as well as acceptable output waveform distortion).

PRACTICE PROBLEMS 3

Using the parameters for the previous example, evaluate the analog fiber-optic system shown in Figure 12–22 calculating the maximum amplitude for the input sine wave that produces a nondistorted output signal. Calculate the maximum output waveform for this input signal. What device(s) limit(s) the amplitude of the input waveform?

Practical Notes

As the optoelectronic devices provide electrical isolation between the transmitter circuit and the receiver circuit, the power sources and ground used for the transmitter circuit can be totally different than the power sources and ground used for the receiver circuit.

Application Problem

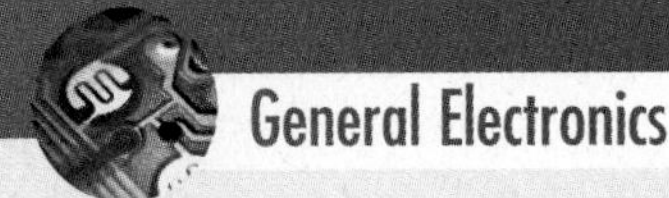

SAFETY SYSTEM FOR A GARAGE DOOR OPENER SYSTEM

The safety system of a garage door opener consists of an opposed sensing method (a light transmitter module and a receiver module). If the light beam is blocked when the door is going down, it will stop and go back up. Observe the circuit below. Calculate the current flow in the receiver circuit when the beam is not blocked.

Solution

$$I_C = \frac{10\text{ V}}{330}$$

$$I_C = 30\text{ mA}$$

V+

+10 V

330 Ω

Control Circuit

12–6 The Liquid Crystal Display

The liquid crystal display (LCD) is another optoelectronic component. LCDs can be obtained as monochrome and color. The LCD is available in text format (5 × 7 or 5 × 8 characters) and in graphical format. Common configurations for the text format are structured by the number of characters per line and the number of text lines. 16 × 1 LCD modules will have sixteen characters per line with only one line of text, while 20 × 4 LCD modules will have twenty characters per line with four text lines. The graphical format LCD module is specified by the number of horizontal dots, the colors (if a color LCD), and the number of vertical dots. A 160 × 128 dot LCD graphical module is a monochrome display with 160 dots in the horizontal direction with 128 dots in the vertical direction. A 240 × 3 (R.G.B) × 160-dot LCD graphical module is a color display with red, green, and blue (the component colors of white light allowing full color—the same as your color television) with 240 dots in the horizontal direction and 160 dots in the vertical direction. A color graphic LCD module is shown in Figure 12–23.

Reflective, transmissive, and transflective technology LCDs are manufactured. The reflective LCD are units where the ambient (room) light is selectively reflected back through the liquid crystal cells to the viewer. The power requirements are greatly reduced with the reflective LCD modules since they do not use a backlight. In dim to dark ambient light conditions, the information contained on the reflective LCD cannot be seen. Transmissive LCD modules use a backlight to selectively transmit the light through the liquid crystal cell to the viewer. The best usage for the transmissive LCD module is low ambient light conditions (indoor light levels or nighttime). Transflective LCD modules combine the features of the reflective LCD and the reflective LCD. Since each subpixel is both reflective and transmissive, the transflective LCD module is readable in any ambient light condition. A backlighting is used to pass light through the module in addition to the reflection of light from the front of the module by the ambient light. As backlighting is used in both the transmissive and transflective LCD modules, more power is required to operate the module.

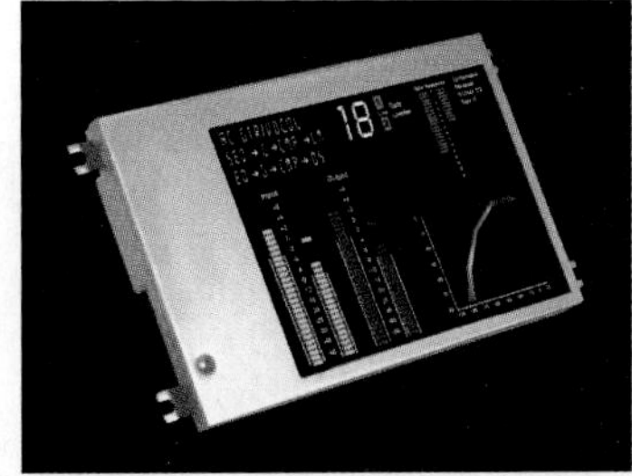

FIGURE 12–23 Color graphic LCD module (Courtesy of Optrex)

LCD modules are manufactured with the LCD and the display driver circuitry. Typical interface pins for the LCD module include the power $-V_{CC}$, ground, brightness control and backlighting (transmissive and transflective modules only); read input; write input; and eight data lines (D0 through D7).

LCD technology is used in a multitude of applications including cell phone displays, automobile dashboards, televisions, and computer monitors.

USING TECHNOLOGY: OPTOELECTRONICS USED IN A SYSTEM

Almost all electronic systems of every type require some type of output display. Consider for a moment all the consumer electronics products that use an output display, either a numeral or an alphabetical letter. These devices range in size from a digital watch and handheld calculator, up to the latest liquid crystal and light-emitting diodes displays and television screens.

One of the most common uses for an optoelectronic light-emitting diode or LED is the POWER ON-OFF indicator on most everyday electronic products. Another common use is the seven segment LED display used in digital counters, frequency generators, and industrial displays. The following description also applies to the liquid crystal display or LCD. The essential difference between an LED display is that it emits light energy while an LCD display blocks or allows either reflected light from the surroundings or from a low-intensity fluorescent tube.

The system application block diagram for the seven segment common anode LED display is shown in Figure 31. The seven segment LED display gets its name from the seven separate LEDs which are integrated into a single device. Referring to Figure 32, observe that a minimum of seven segments are needed to form any digit from 0 to 9, with the numeral 8 requiring all seven segments or LEDs to be illuminated. Also note from the drawing that at least one decimal-point LED is required. Most such LED displays include two decimal-point LEDs, one on the left side and another on the right side of the numeral LEDs to allow the decimal point to be positioned, anywhere in the displayed numbers. At the input side of the block diagram, a 4-bit Binary Coded Decimal number is fed to a decoder which translates the binary code of zeros (0's) and ones (1's) into a seven-segment signal. The negative-going dc output is fed through current

DVD Laser Application: Portable laptop style DVD player with rotating screen

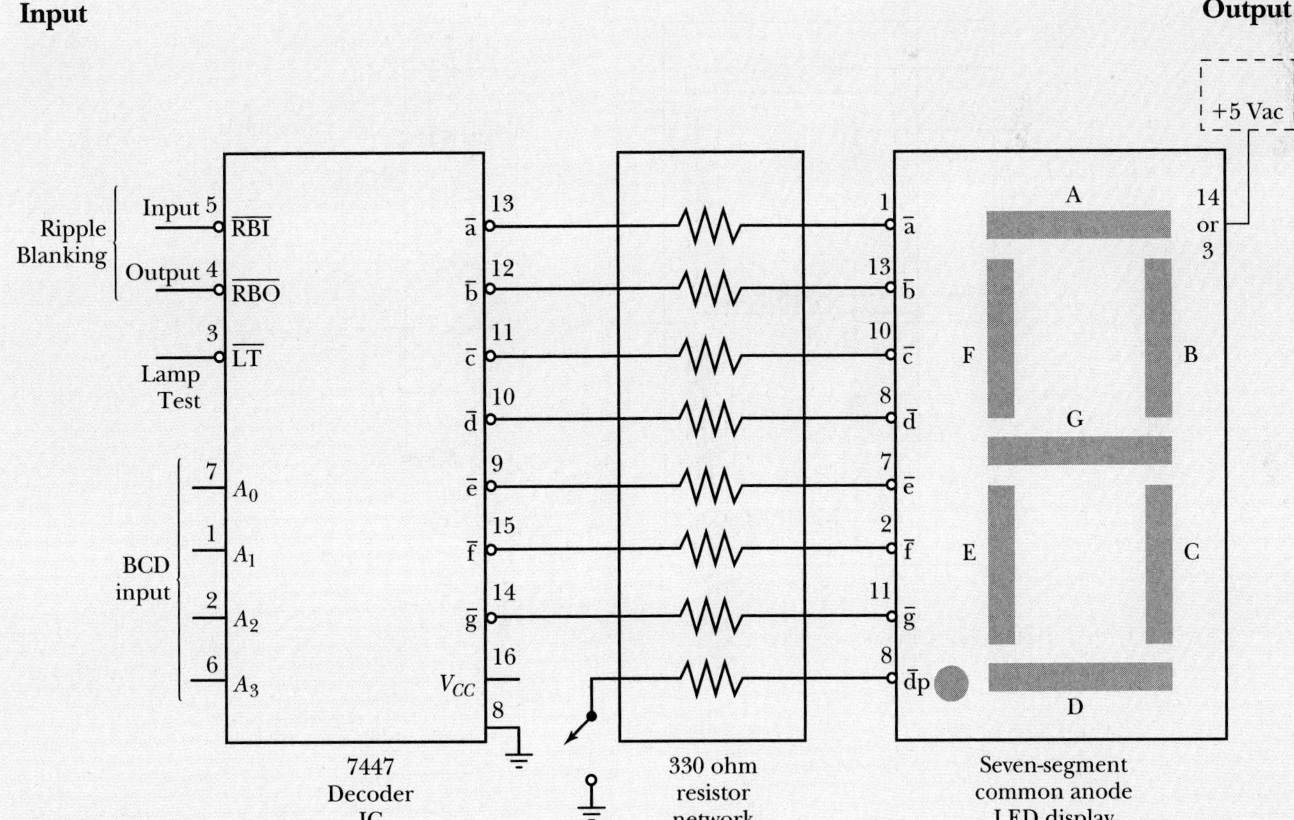

FIGURE 31 Block diagram

limiting or dropping resistors to the inputs of the seven-segment common anode LED display. These negative-going dc voltages will turn on the appropriate LED seven segments to form and display the correct decimal number.

Two additional negative-going dc voltages are applied at the $\overline{\text{RBI}}$ and $\overline{\text{RBO}}$ inputs to the decoder to maintain the LEDs in the ON or lighted condition for any condition of an ac ripple signal riding on the dc supply or input code signals. One additional negative-going dc voltage is also applied at the $\overline{\text{LT}}$ to allow for testing or lighting all the seven segments. The bar atop the signals along with the zero or bubble indicates a negative (–) signal is required to activate that input and turn on the associated LED.

The circuit schematic in Figure 33 shows the details of the seven segment common anode LED display. Observe that each of the anodes of the individual LEDs are connected together in common mode, hence the term *common anode.* The LED anodes are connected together to a +5 vdc supply source. The individual cathodes of each segment LED are connected to the associated input: a, b, decimal point dp, etc. A negative-going dc voltage from the decoder is applied through a 330-ohm current limiting or dropping resistor to the associated LED cathode. This turns on or illuminates the LED connected to this input pin forming the correct numeral. A separate negative-going dc voltage, usually a circuit ground, is connected through the same 330 ohm resistor network from a separate counter circuit to turn on and illuminate the decimal point LED.

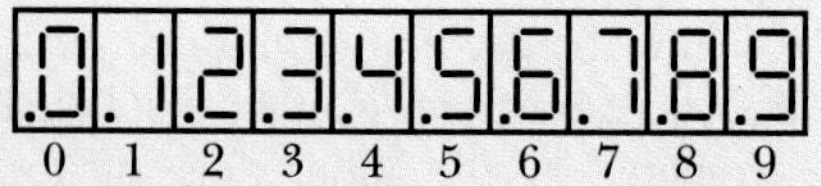

FIGURE 32 Seven segment LED display

DVD Recorder Application: Home DVD player/recorder

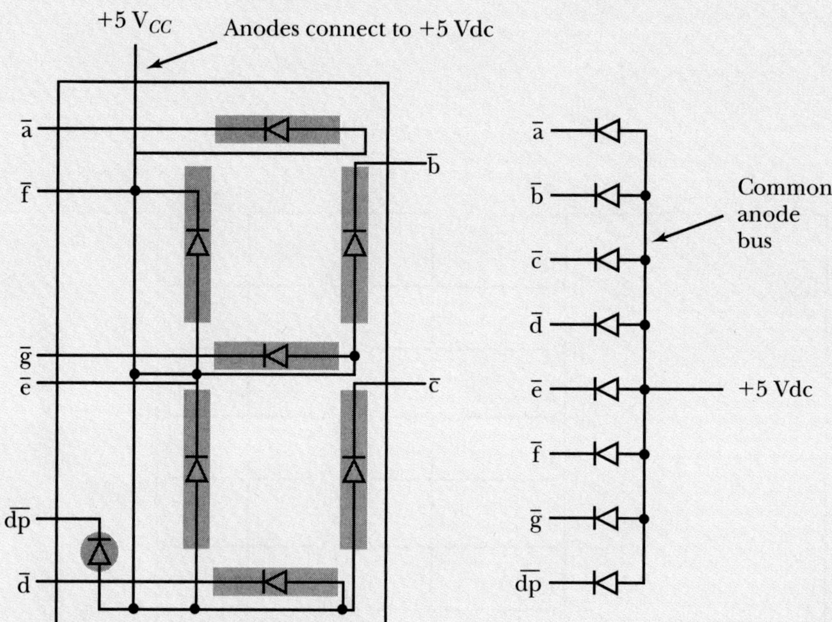

FIGURE 33 Circuit schematic

Summary

- An optoelectronic device is one that interacts with light energy. Common optoelectronic diodes are light-emitting diodes (LEDs), photodiodes, and laser diodes. LEDs and laser diodes both emit light when they are forward biased and carrying current. The color of light from an LED depends on the semiconductor materials selected for its construction. The light from LEDs and lasers is monochromatic (at a single frequency). A photodiode, when reverse biased, conducts in proportion to the amount of light focused onto its P-N junction.
- An opto-isolator, also known as an optocoupler, combines the LED with a photodetector. Opto-isolators come with a variety of different output circuitry (photodetectors). The photodetector portion includes photovoltaic, photodiodes, and phototransistors, photodarlingtons, photoMOSFETs, photoSCRs, and photoTRIACS.
- The fiber-optic cable has altered the interconnection of electronic communications systems. Its immunity to electrical and atmospheric conditions has made the fiber-optic cable system the preferred medium for communications systems.

Formulas and Sample Calculator Sequences

Resistor—LED circuit (Figure 12–4)

FORMULA 12–1
(To find voltage drop across the series resistor)

$V_R = V_S - V_D$

V_S value, [−], V_D value, [=]

FORMULA 12–2
(To find series resistance knowing resistor voltage and current)

$$R = \frac{V_R}{I_D}$$

V_R value, [÷], I_D value, [=]

FORMULA 12–3
(To find series resistance knowing diode voltage and current and supply voltage)

$$R = \frac{(V_S - V_D)}{I_D}$$

[(], V_S value, [−], V_D value, [)], [÷], I_D value, [=]

FORMULA 12–4
(To find photodetector current)

$I = R \times P_{\text{RCVD}}$

R value, [×], P_{RCVD} value, [=]

FORMULA 12–5
(To find photodetector current knowing photoemitter current)

$I_C = \text{CTR} \times I_D$

CTR value, [×], I_D value, [=]

FORMULA 12–6
(To find power gain)

$$A = 10 \log_{10}\left(\frac{P_{\text{out}}}{P_{\text{in}}}\right)$$

[(], P_{out} value, [÷], P_{in} value, [)], log, [×], 10, [=]

FORMULA 12–7
(To find output power)

$P_{\text{out}} = P_{\text{in}} \times (10^{A/10})$

[(], A value, [÷], 10, [)], [×], P_{in} value, [=]

Using Excel

Optoelectronics Formulas

(Excel file reference: FOE12_01.xls)

DON'T FORGET! It is NOT necessary to retype formulas, once they are entered on the worksheet! Just input new parameters data for each new problem using that formula, as needed.

- Use the Formula 12–3 spreadsheet sample and the parameters given for Practice Problem 1. Solve for the value of the resistor. Check your answer against the answer for this question in the Appendix.

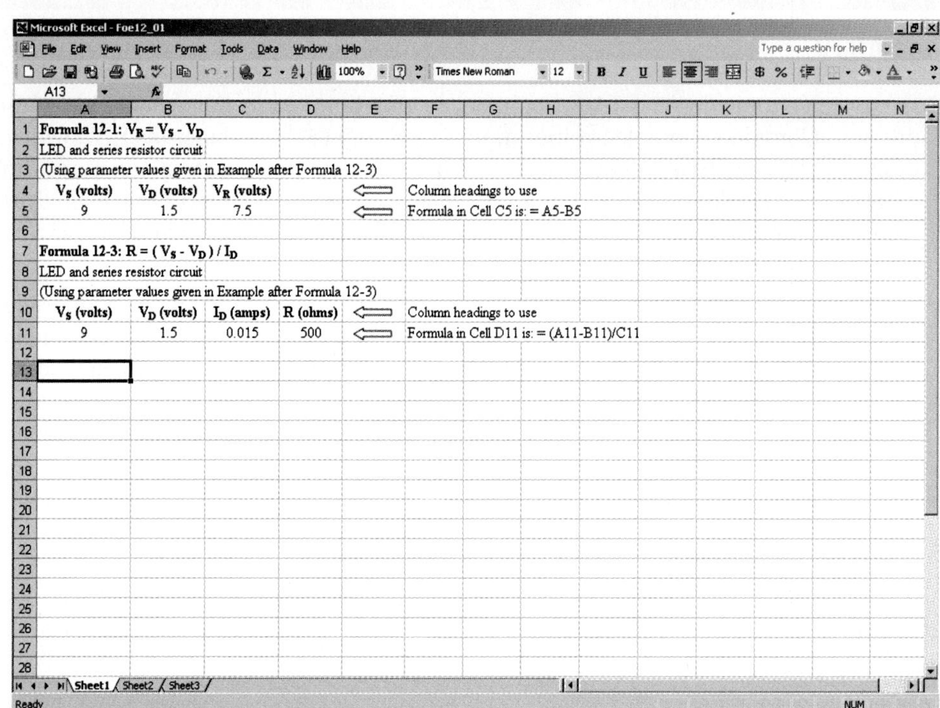

	A	B	C	D	E	F
1	Formula 12-1: $V_R = V_S - V_D$					
2	LED and series resistor circuit					
3	(Using parameter values given in Example after Formula 12-3)					
4	V_S (volts)	V_D (volts)	V_R (volts)		⇐	Column headings to use
5	9	1.5	7.5		⇐	Formula in Cell C5 is: = A5-B5
6						
7	Formula 12-3: $R = (V_S - V_D) / I_D$					
8	LED and series resistor circuit					
9	(Using parameter values given in Example after Formula 12-3)					
10	V_S (volts)	V_D (volts)	I_D (amps)	R (ohms)	⇐	Column headings to use
11	9	1.5	0.015	500	⇐	Formula in Cell D11 is: = (A11-B11)/C11

Review Questions

1. Describe the conditions required for getting an LED to emit light energy.
2. Sketch a simple LED circuit that consists of a dc power source, resistor, and LED. Arrange the polarities of the power source and LED so that the LED will light.
3. Explain how light falling onto the junction of a photodiode affects its conductance.
4. Draw a circuit showing a dc power source, resistor, and photodiode connected in series. Show the polarities and direction of the photodiode for normal operation.
5. Describe in words the difference between the schematic symbol for an LED and a photodiode.
6. Name the two types of diodes that can be found in an opto-isolator.
7. Describe how the light energy from an LED is different from that of a laser diode.

Problems

1. For the LED circuit in Figure 12–4, let $V_S = 6$ V, $R_S = 180\ \Omega$, and $V_D = 1.5$ V. Determine:
   - **a.** the amount of current through the circuit.
   - **b.** the current through the LED.
   - **c.** the power dissipation of R_S.
2. You want to operate an LED from a 9-V battery. The specifications for the LED list I_F at 150 mA and V_D at 1.5 V. Calculate the value of resistance you should connect between the battery and LED.
3. Refer to the information on the seven segment display shown in Figures 12–6 and 12–7. Indicate the switches that must be closed to illuminate 7 (include the right decimal point—RD). The display forward voltage drop is 2.0 V. The supply voltage (V_s) is 5 V. The resistors are all the same value of 330 Ω. Calculate how much current the supply must provide to the seven segment display.

4. Evaluate the circuit in Figure 12–5. The following components are used in the circuit:
   $V_{CC} = 12$ V
   $R_1 = 10$ kΩ
   $R_2 = 470$ Ω
   Q_1 = 2N3904 ($100 < \beta < 400$)
   D_1 = MCDL-5C3SRC0

   Indicate the LED status (ON/OFF) and the current flowing through the LED when the input voltage is
   a. $V_{in} = 0$ V
   b. $V_{in} = 5$ V

Analysis Questions

1. Explain how laser diodes are used in bar-code readers.
2. Compare the current transfer ratio (CTR) of an opto-isolator transistor output to the CTR of an opto-isolator Darlington transistor output. Indicate the component part numbers that you selected to compare. (Hint: Research the data sheets of the devices.)
3. Search for information on the DMC-16249 LCD module at *www.optrex.com.* Provide a brief description of this component.
4. Visit *www.digikey.com.* Type LCD in "Paris Search." Under the optoelectronic category, look at the available LCD module options. List the filter categories for LCD Character Displays/Modules.

Performance Projects Correlation Chart

All e-labs are available on the LabSource CD.

Chapter Topic	Performance Project	Project Number
	Story Behind the Numbers: Fiber-optic systems characteristics	

NOTE: It is suggested that after completing the above projects, the student should be required to answer the questions in the "Summary" at the end of this section of projects in the Laboratory Manual.

Troubleshooting Challenge

CHALLENGE CIRCUIT 15 (Block Diagram)

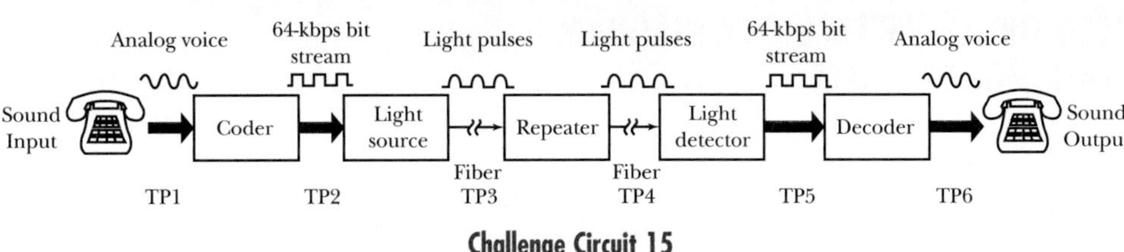

Challenge Circuit 15

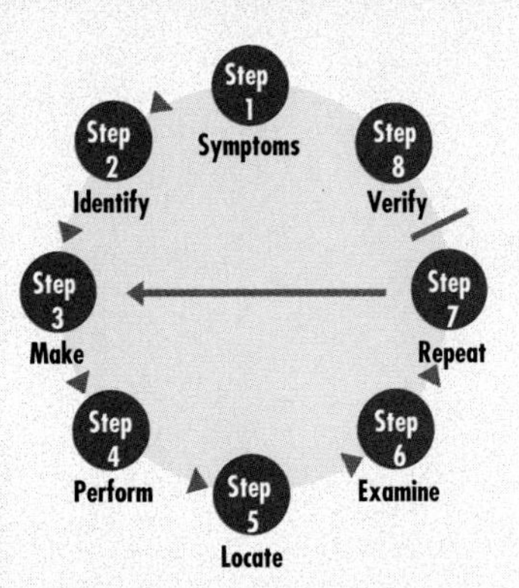

General Testing Instructions:

Measurement
Assumptions:
Signal = from **TP** to ground
I = at the **TP**
V = from **TP** to ground
R = from **TP** to ground (with the power source disconnected from the circuit)

Possible Tests & Results:
Signal (normal, abnormal, none)
Current (high, low, normal)
Voltage (high, low, normal)
Resistance (high, low, normal)

Starting Point Information	Test Points	Test Results in Appendix C			
		V	*I*	*R*	*Signal*
At TP_1	TP2	(NA	277	NA	283)
	TP3	(NA	280	NA	284)
V = normal	TP4	(NA	NA	NA	285)
I = NA	TP5	(NA	NA	NA	286)
R = NA	TP5	(287	288	289	290)
Signal = normal					

FIBER-OPTIC SYSTEM BLOCK DIAGRAM CHALLENGE PROBLEM

STEP [1]

SYMPTOMS No signal at the sound output.

STEP [2]

IDENTIFY initial suspect area. Any of the modules within the fiber-optic communication system could have failed.

STEP [3]

MAKE test decision is based on the symptoms information. Let's start with the voltage measurements at strategic test points.

STEP [4]

PERFORM **First Test:** Waveform at TP6.
The signal is a fixed dc voltage of 6 V. The signal is not normal for a voice input.

STEP [5]

LOCATE new suspect area. The decoder module is suspect.

STEP [6]

EXAMINE available data.

STEP [7]

REPEAT analysis and testing.
Second Test: The input to the decoder module is a digital bit stream. The decoder appears to have failed.

STEP [8]

VERIFY **Third Test:** After replacing the decoder module with a new module that meets specifications, all circuit parameters come back to their norms. Sound is now heard at the telephone speaker.

Troubleshooting Challenge

CHALLENGE CIRCUIT 16
Find the Fault

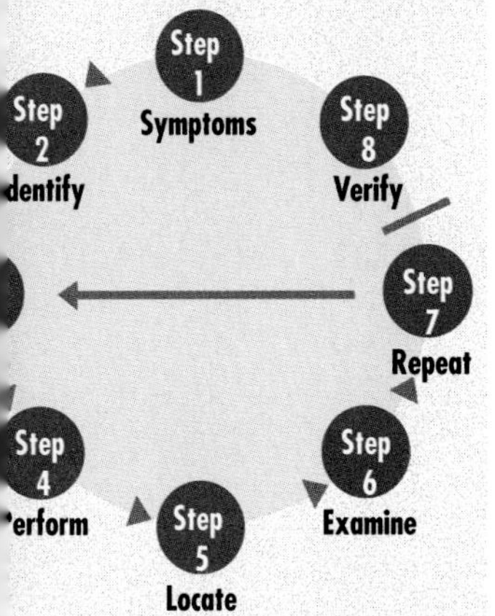

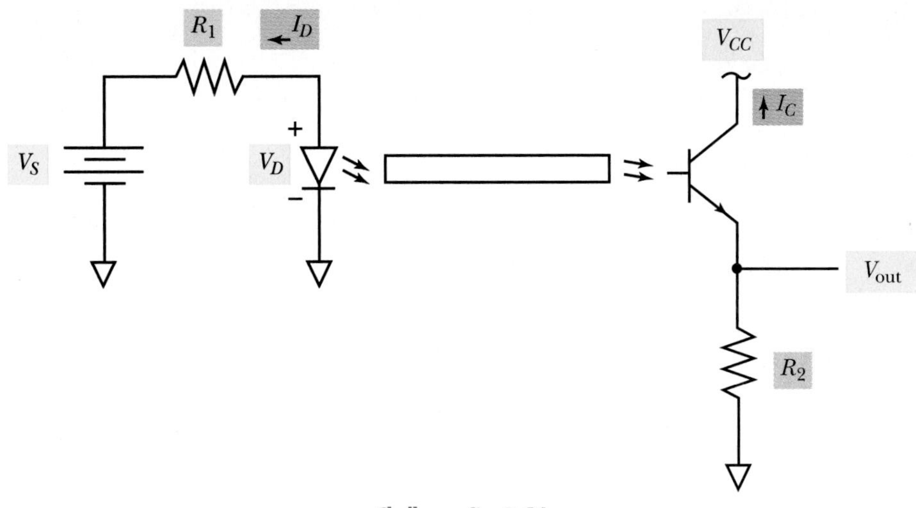

Challenge Circuit 16

STARTING POINT INFORMATION

1. Circuit diagram ($R_1 = 1\ \text{k}\Omega$, $R_2 = 20\ \text{k}\Omega$)
2. $V_{CC} = 12$ V $V_S = 25$ V
3. Photoemitter (MFOE71)
   Launched optical power $P_L = (165\ \mu\text{W} / 100\ \text{mA}) \times I_D$
   Diode forward voltage $V_F = 1.5$ V
4. Phototransistor detector (MFOD72)
   Responsivity $R = (125\ \mu\text{A} / 1\ \mu\text{W})$
   Photocurrent $I_C = R \times P_{RCVD}$
   Collector dark current $I_{CEO} = 100$ nA (max)
5. Fiber-optic cable (1000 micron)
   Length 2 km
   Attenuation 2.5 dB/km
6. Connector losses
   Photoemitter to cable 10 dB
   Cable to photodetector 3 dB
7. V_{out} is a voltage lower than expected

Test	Circuit Measurements	Theoretical Expected Value in Appendix C
V_S	25 V	______
V_{CC}	12 V	______
V_D	1.5 V	______
V_{out}	0.5 V	______
I_D	23.5 mA	______
I_C	25 μA	______
P_L	38.8 μW	______
P_{RCVD}	200 nW	______
R_1	10 kΩ	______
R_2	20 kΩ	______

STEP [1]

You will analyze the circuit to calculate the expected voltages, currents, and resistances for a properly functioning circuit. Record these values beside each test listed in the above table. (Refer to Analysis Techniques)

STEP [2]

Starting with the symptom, document your steps for troubleshooting this circuit using the SIMPLER troubleshooting technique or instructions provided by your instructor.

STEP [3]

For each circuit measurement, you will indicate the test to be accomplished.

STEP [4]

Using the table above, you will locate the test that you performed and the corresponding Appendix C reference number. In Appendix C, you will find the measurement for the faulted circuit.

STEP [5]

You will now compare the faulted circuit measurement to the expected measurement for a properly functioning circuit. Are these values the same? (Refer to General Testing Instructions)

STEP [6]

You will repeat steps 2 through 5 until you have located the fault.

STEP [7]

When you have located the fault, you will identify the fault and the characteristics of the failure. (Refer to Types of Failures)

ANALYSIS TECHNIQUES (check with your instructor)

METHOD [1]

Using circuit theory and algebra, evaluate the circuit.

METHOD [2]

Using an Excel spreadsheet for this circuit, evaluate the circuit.

METHOD [3]

Using MultiSIM, build the circuit schematic and generate the results.

METHOD [4]

Assemble this circuit on a circuit board. Apply power and make the appropriate measurements using your DMM and oscilloscope.

General Testing Instructions:

Measurement
Assumptions:
Signal = from **TP** to ground
I = at the **TP**
V = from **TP** to ground
R = from **TP** to ground (with the power source disconnected from the circuit)

Possible Tests & Results:
Signal (normal, abnormal, none)
Current (high, low, normal)
Voltage (high, low, normal)
Resistance (high, low, normal)

Types of Failures

Component failures
- wrong part value
- part shorted
- part open
- defective fiber-optic cable

Supply voltage—incorrect voltage

Ground—floating

APPENDIX A

Color Codes

A. RESISTOR COLOR CODE

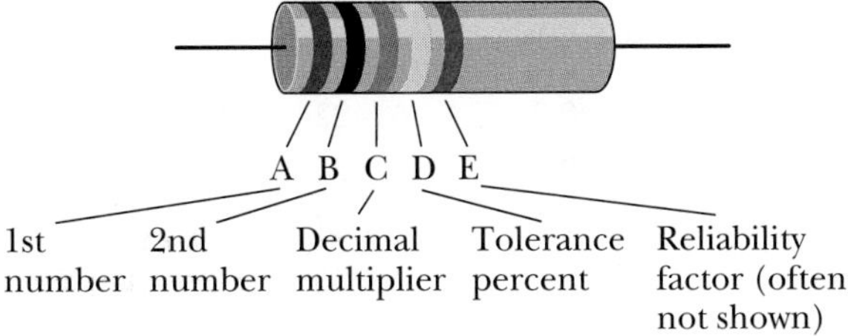

COLOR SIGNIFICANCE CHART

(For resistors and capacitors)

COLOR	NUMBER COLOR REPRESENTS	DECIMAL MULTIPLIER	TOLERANCE PERCENT	VOLTAGE RATING	% CHANGE PER 1,000 HRS OPER.
Black	0	1	—	—	—
Brown	1	10	1*	100*	1%
Red	2	100	2*	200*	0.1%
Orange	3	1,000	3*	300*	0.01%
Yellow	4	10,000	4*	400*	0.001%
Green	5	100,000	5*	500*	—
Blue	6	1,000,000	6*	600*	—
Violet	7	10,000,000	7*	700*	—
Gray	8	100,000,000	8*	800*	—
White	9	1,000,000,000	9*	900*	—
Gold	—	0.1	5	1,000*	—
Silver	—	0.01	10	2,000*	—
No Color	—	—	20	500*	—

(*Applicable to capacitors only)

B. SAMPLE SURFACE-MOUNT TECHNOLOGY (SMT) "CHIP" RESISTOR CODING527

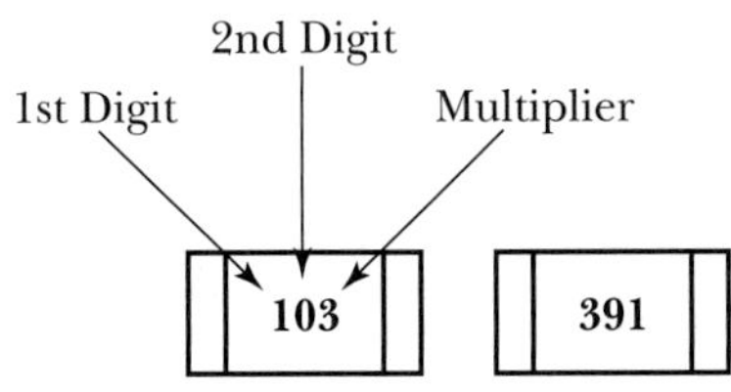

Code: 103 = 1, then 0, then 000 or 10,000 Ω   Code: 391 = 3, then 9, then 0, or 390 Ω

C. TYPICAL 5% AND 10% RESISTOR VALUES

Ω	Ω	KΩ	KΩ	KΩ	MΩ
10	100	1.0	10	100	1.0
11	110	1.1	11	110	1.1
12	120	1.2	12	120	1.2
13	130	1.3	13	130	1.3
15	150	1.5	15	150	1.5
16	160	1.6	16	160	1.6
18	180	1.8	18	180	1.8
20	200	2.0	20	200	2.0
22	220	2.2	22	220	2.2
24	240	2.4	24	240	2.4
27	270	2.7	27	270	2.7
30	300	3.0	30	300	3.0
33	330	3.3	33	330	3.3
36	360	3.6	36	360	3.6
39	390	3.9	39	390	3.9
43	430	4.3	43	430	4.3
47	470	4.7	47	470	4.7
51	510	5.1	51	510	5.1
56	560	5.6	56	560	5.6
62	620	6.2	62	620	6.2
68	680	6.8	68	680	6.8
75	750	7.5	75	750	7.5
82	820	8.2	82	820	8.2
91	910	9.1	91	910	9.1

All values listed are available in ±5% tolerance. Only values in red are available in ±10% tolerances.

D. CAPACITOR COLOR CODES

5th and 6th band = V rating (multiply numbers by 100)

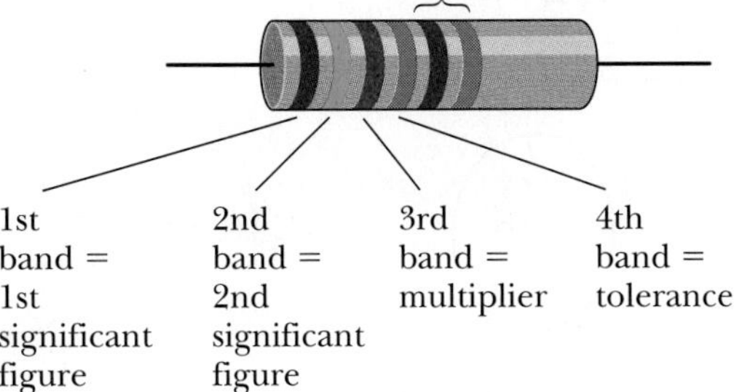

See Color Significance Chart for meaning of colors

Band color coding system for tubular ceramic capacitors

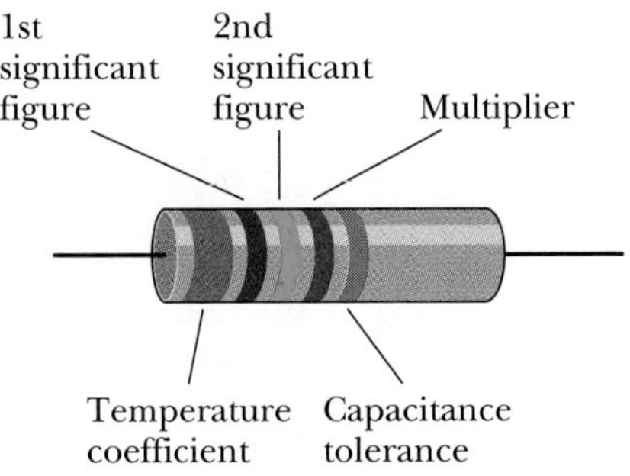

E. SAMPLE CERAMIC DISK CAPACITORS CODED MARKINGS

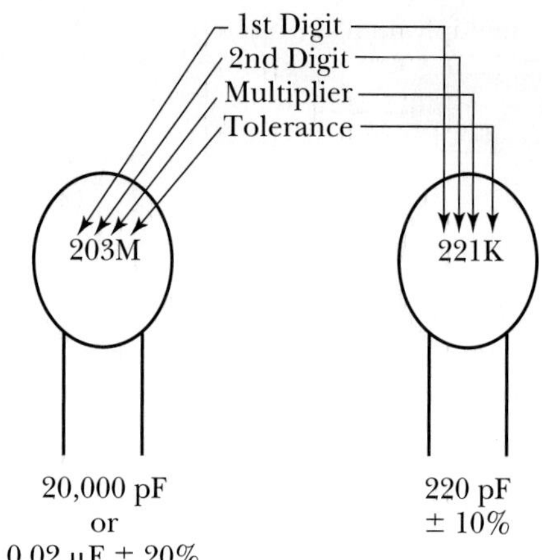

20,000 pF
or
0.02 μF ± 20%

220 pF
± 10%

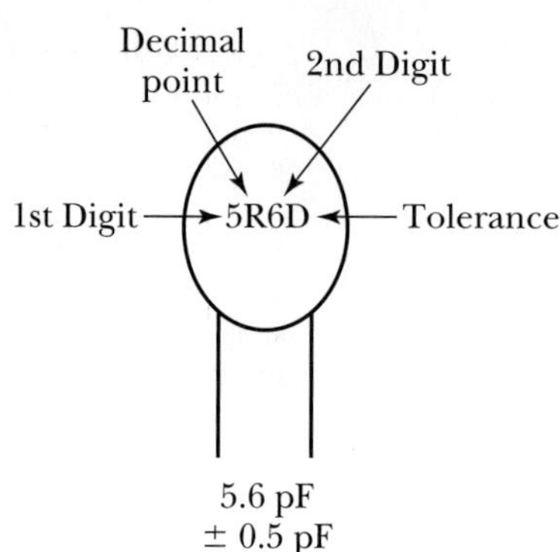

5.6 pF
± 0.5 pF

Third-Digit Multiplier

Number	Multiply by
0	Nothing
1	10
2	100
3	1,000
4	10,000

Letter Tolerance Code

Under 10 pF Values:

Letter	Tolerance
B	± 0.1 pF
C	± 0.25 pF
D	± 0.5 pF
F	± 1.0 pF

Over 10 pF Values:

Letter	Tolerance
E	± 25%
F	± 1%
G	± 2%
H	± 2.5%
J	± 5%
K	± 10%
M	± 20%
P	− 0%, + 100%
S	− 20%, + 50%
W	− 0%, + 200%
X	− 20%, + 40%
Z	− 20%, + 80%

F. SURFACE-MOUNT TECHNOLOGY (SMT) "CHIP" CAPACITOR CODING

SMT Capacitor Significant Figures Letter Code

Character	Significant Figures	Character	Significant Figures
A	1.0	R	4.3
B	1.1	S	4.7
C	1.2	T	5.1
D	1.3	U	5.6
E	1.5	V	6.2
F	1.6	W	6.8
G	1.8	X	7.5
H	2.0	Y	8.2
J	2.2	Z	9.1
K	2.4	a	2.5
L	2.7	b	3.5
M	3.0	d	4.0
N	3.3	e	4.5
P	3.6	f	5.0
Q	3.9	m	6.0
		n	7.0
		t	8.0
		y	9.0

SMT Capacitor Multiplier Code

Number	Decimal Multiplier
0	1
1	10
2	100
3	1,000
4	10,000
5	100,000
6	1,000,000
7	10,000,000
8	100,000,000
9	0.1

Decoding Examples:

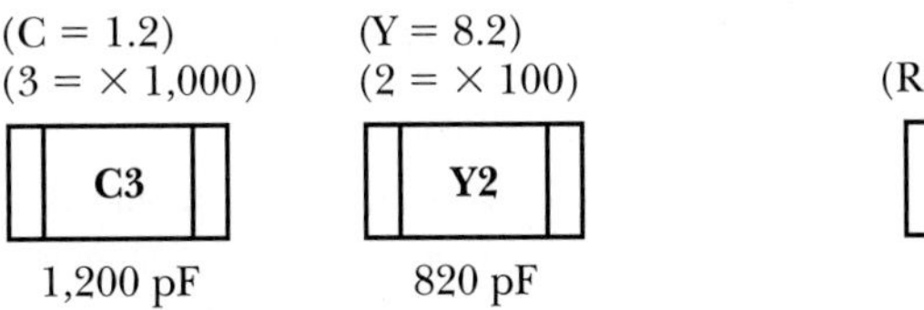

(Letter plus number = above 100 pF) **(Numbers only = below 100 pF)**

G. SEMICONDUCTOR DIODE COLOR CODES(S)

(NOTE: Prefix "1N . . . " is understood.)

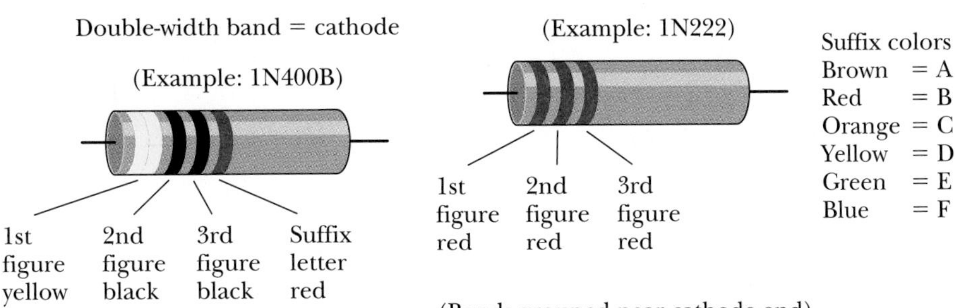

APPENDIX B

Answers

CHAPTER 1

Review Questions

1. c

3. c

5. a

7. a

9. a

11. a

13. T

15. T

17. F

19. T

21. T

23. F

25. F

27. F

29. T

31. F

33. T

35. $V_2 = V_S \times \frac{R_2}{R_1 + R_2}$

37. $P_T = P_1 + P_2 + \ldots$
or $P_T = V_S \times I_S$

39. $I_2 = I_T \times \frac{R_1}{R_1 + R_2}$

41. The current entering a parallel circuit will be divided (split) based on the resistance (impedance) of each branch.

Problems

1. **a.** 11 kΩ (5%)
c. 33 kΩ (20%)
e. 560 kΩ (10%)

3. **a.** $R_1 = 1$ kΩ, $V_1 = 8$ V, $I_1 = 8$ mA, $P_1 = 64$ mW
$R_2 = 1.5$ kΩ, $V_2 = 12$ V, $I_2 = 8$ mA, $P_2 = 96$ mW
c. $R_1 = 400$ Ω, $V_1 = 6$ V, $I_1 = 15$ mA, $P_1 = 90$ mW
$R_2 = 600$ Ω, $V_2 = 9$ V, $I_2 = 15$ mA, $P_2 = 135$ mW

5. **a.** $R_1 = 1$ kΩ, $V_1 = 7.385$ V, $I_1 = 7.385$ mA, $P_1 = 54.54$ mW
$R_2 = 1.5$ kΩ, $V_2 = 7.615$ V, $I_2 = 5.078$ mA, $P_2 = 38.67$ mW
$R_3 = 3.3$ kΩ, $V_3 = 7.615$ V, $I_3 = 2.307$ mA, $P_3 = 17.57$ mW

7. **a.** 0.5 μF
c. 0.06 μF
e. 2 μF
g. 89 pF

CHAPTER 2

In-Process Learning Checks

—1—

1. Before doping (in their pure form) semiconductor materials are sometimes called ***intrinsic*** semiconductors. Once they are doped with tiny amounts of an impurity atom, they are called ***extrinsic*** semiconductors.

2. N-type materials are formed by doping a ***tetra***valent semiconductor material with a ***penta***valent material. P-type materials are formed by doping a ***tetra***valent semiconductor material with a ***tri***valent material.

3. The doping atoms for N-type materials are called ***donor*** atoms because they donate extra electrons to the covalent bonds. The doping atoms for P-type materials are called ***acceptor*** atoms because they accept electrons that will fill the holes in the covalent bonds.

4. The majority carriers in an N-type material are ***electrons,*** and the minority carriers are ***holes.*** The majority carriers in a P-type material are ***holes,*** and the minority carriers are ***electrons.***

Review Questions

1. **a.** A *trivalent* atom has three valence electrons.
b. A *tetravalent* atom has four valence electrons.
c. A *pentavalent* atom has five valence electrons.

3. **a.** Electrons
b. Holes

5. **a.** 0.7 V
b. 0.3 V

7. b

9. P-N junction semiconductor showing an external dc power supply connected for forward biasing.

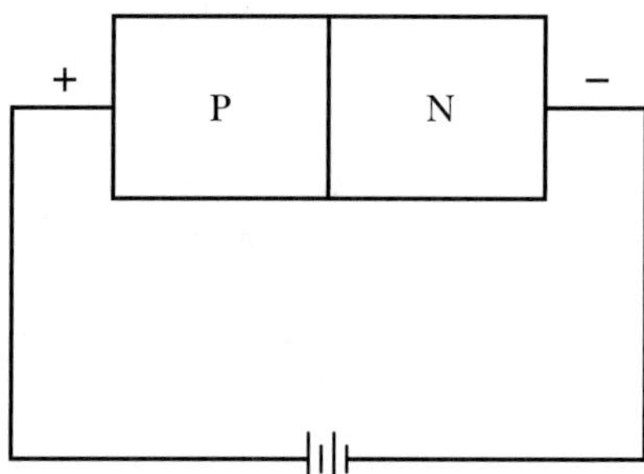

11. Forward-bias portion of the *I–V* curve for a P-N junction.

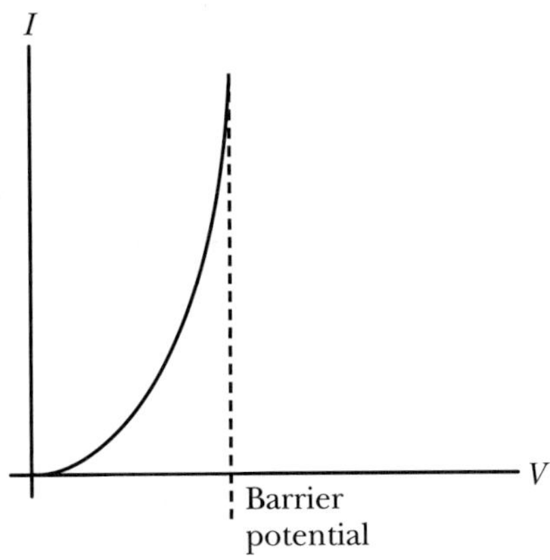

Problems

1. 32
3. 3rd

Analysis Questions

1. Valence-band electrons are bound to their atom and are not free to move from one atom to another in the material. Electrons in the conduction band are free to move. Also, conduction-band electrons possess more energy than their corresponding valence-band electrons.
3. Increasing the amount of forward bias narrows the depletion region. Decreasing the amount of forward bias or applying a reverse bias increases the width of the depletion region.
5. In a semiconductor, holes flow from positive to negative, and electrons flow from negative to positive. Conventional current flow shows current flowing from positive to negative, and the electron-flow convention shows current flowing from negative to positive. So in a manner of speaking, hole flow is similar to conventional current flow, and electron flow is similar to the electron version of current flow.

CHAPTER 3

Practice Problems

—1—

1. –12.7 V
2. 0 V
3. 0.7 V
4. 12 V

—2—

1. 0.7 V
2. 75.3 V
3. 753 μA

—3—

1. 6 V
2. 12 V
3. 5 V, 14 V
4. 5 V, 6 V
5. 5 mA, 6 mA
6. 5 mA, 14 mA
7. 0 mA, 8 mA
8. 0 W, 48 mW

In-Process Learning Checks

—1—

1. To cause conduction in a diode, the diode must be ***forward*** biased.
2. To reverse bias a diode, connect the negative source voltage to the ***P***-type material and the positive source voltage to the *N*-type material.
3. The anode of a diode corresponds to the ***P***-type material, and the cathode corresponds to the *N*-type material.
4. The forward conduction voltage drop across a silicon diode is approximately 0.7 V.
5. When a diode is reverse biased in a circuit, it acts like an ***open*** switch.
6. When the diode is reverse biased, there will be no current flow through the diode.
7. When a diode is connected in series with a resistor, the voltage across the resistor is very nearly equal to the dc source voltage when the diode is ***forward*** biased.
8. The four most general diode ratings are ***forward voltage drop, average forward current, peak reverse voltage, and reverse-breakdown voltage.***

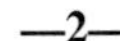

—2—

1. Rectifier diodes are used where it is necessary to change ***ac*** current power into ***dc*** current power.
2. Where additional cooling is necessary, a rectifier diode can be connected to a ***heat sink*** to dissipate heat more efficiently.
3. When an ac waveform is applied to a rectifier diode, the diode's ***reverse breakdown voltage*** rating must be greater than the peak voltage level.
4. The main current specification for rectifier diodes is ***average forward current.***
5. To test the forward conduction of a diode with an ohmmeter, connect the ***negative*** lead of the meter to the cathode and the ***positive*** lead to the anode.
6. The forward resistance of a good diode should be much ***less*** than its reverse resistance.
7. Switching diodes have a ***reverse recovery time*** rating that is hundreds of times less than most rectifier diodes.
8. The type of diode circuit that removes the peaks from an input waveform is called a ***clipper*** circuit.
9. The type of diode circuit that changes the baseline level of an input waveform is called a ***clamp*** circuit.

Datasheet Questions (IN4004)

1. 0.8 V
2. 1.0 A
3. 400 V
4. 0.8 W
5. Axial lead standard recovery rectifiers

Datasheet Questions (IN4148)

1. 1.0 V
2. 10 mA
3. 100 V
4. 10 mW
5. Small signal diode

Datasheet Questions (IN4733A)

1. 5.1 V
2. 49 mA
3. 249.9 mW/1 W
4. Silicon 1-W zener diodes

Review Questions

1.

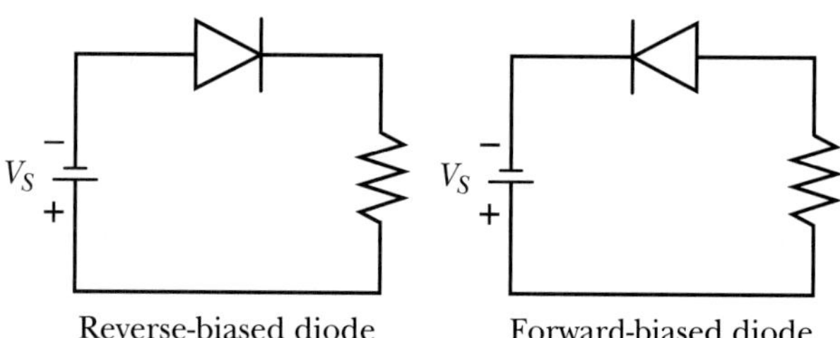

3.

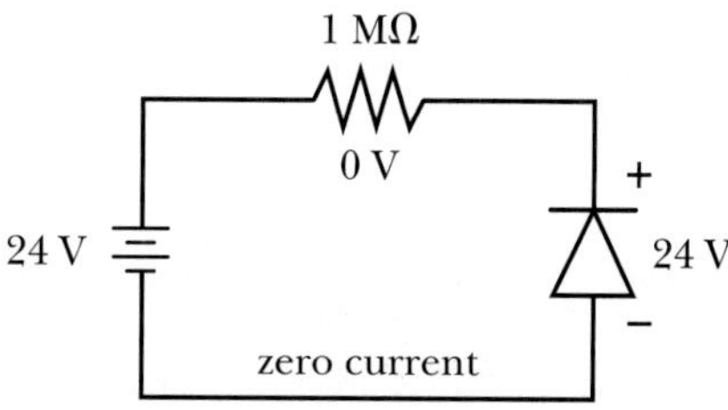

5.

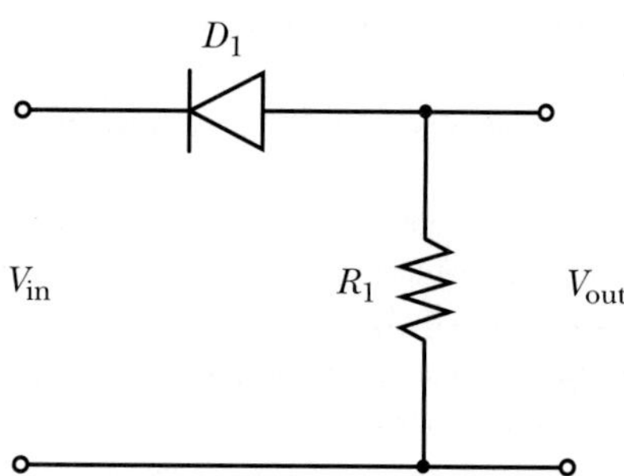

7. A diode clamp circuit shifts the baseline of an input waveform from 0 V to a different level. Whether the baseline is shifted positive or negative depends on the direction the diode is connected into the circuit.
9. **a.** The reverse breakdown voltage is the maximum amount of reverse bias a diode can windstand indefinitely.
   b. The forward conduction voltage is the voltage drop across a diode when it is carrying forward-bias current. This value is fairly constant at 0.3 V for germanium diodes and 0.7 V for silicon diodes.
   c. The average forward current specification is the maximum amount of forward current a diode can carry for an indefinite time.
   d. The reverse recovery time is the time that is required for changing the condition of a diode from reverse bias to forward bias.
11. The voltage across the zener diode will be unchanged. The voltage across the resistor will increase. Increasing the applied voltage will cause the current through the circuit to increase.
13.

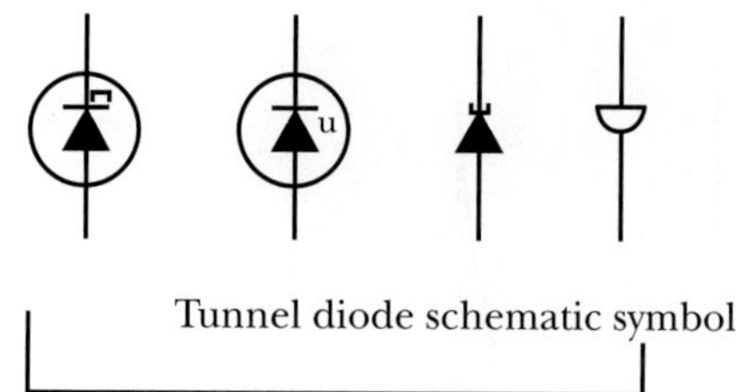

15. As the reverse-bias voltage increases, the depletion region becomes wider. The increased width of the depletion region causes a decrease in the capacitance of the varactor diode. With a decrease of the reverse-bias voltage, the depletion region becomes narrower resulting in an increase of the varactor diode capacitance.

Problems

1. **a.** Forward biased
   b. 5.3 mA
   c. 5.3 V
   d. 5.3 mA
3. **a.** Reverse biased
   b. 0 mA
   c. 0 V
   d. –12 V
5. **a.** 22.6 Ω
   b. 5.65 W
7. 100.7 V
9.

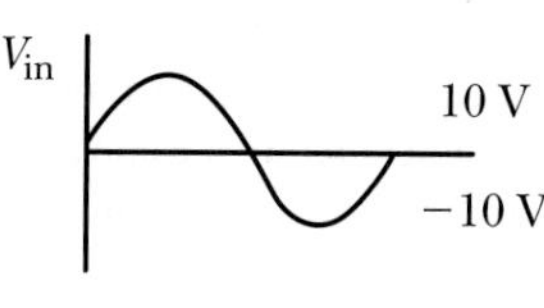

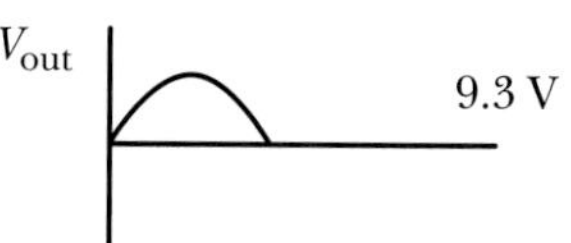

11.

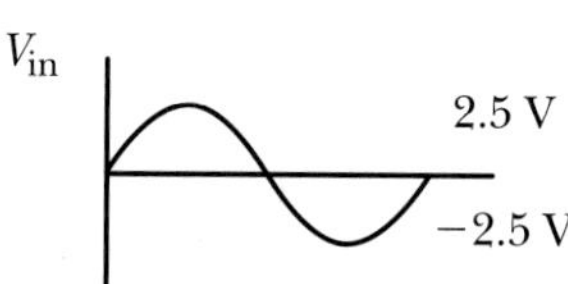

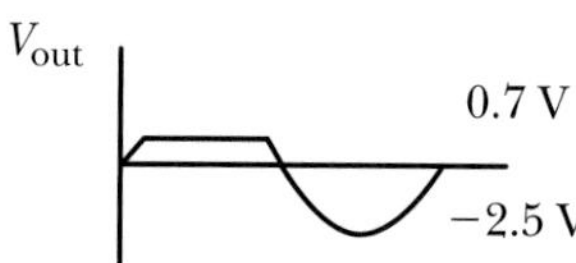

13.

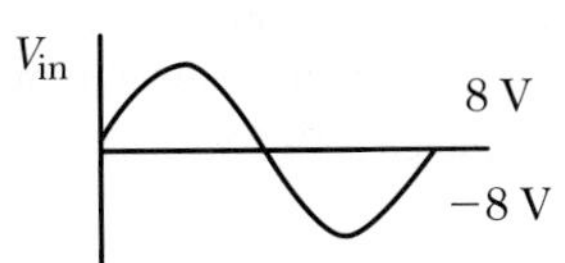

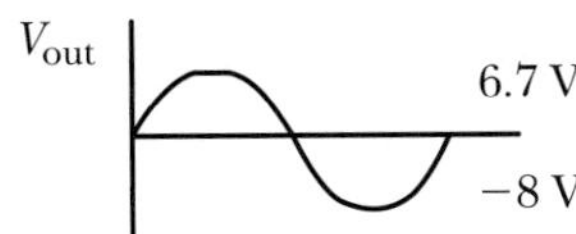

15.

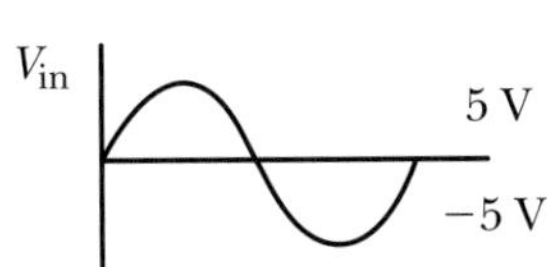

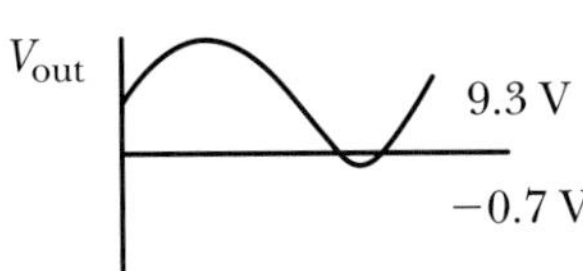

17. **a.** $V_{out} = 8$ V
$I_Z = 0$ mA
b. $V_{out} = 10$ V
$I_Z = 25.6$ mA
$P_Z = 256$ mW

19. **a.** 12 V
b. 8 V
c. 80 mA
d. 960 mW

21. **a.** 3 Ω
b. 3 V
c. 3 W
d. 9 W

23. 2 V
0 V

Analysis Questions

1. Connecting a resistor in series with both diodes ensures that the voltage across each of the parallel branches exceeds the voltage required for forward conduction of both diodes.
3. Answer varies with the results of the research, but should be in the general range of 1.5 to 120 V.
5. Answers will vary.
7. Discrete devices are separate or individual components. Integrated circuit devices are 2 or more circuit elements inseparably associated on or within a substrate.

Datasheet Challenge Problems

1. Answers will vary.
(Fairchild Semiconductor, Vishay Semiconductor, Diodes Inc.)
3. Answers will vary.
(Diodes Inc., Microsemi, Fairchild Semiconductor)
5. Answers will vary—information should include manufacturers, forward voltage drop, average forward current, and reverse voltage as a minimum.
7. Answers will vary—visit *www.diodes.com* and select products, under Diodes—switching diodes (IN4148, IN4150, BAS16).
9. Answers will vary—under *www.diodes.com* find Products, then zener diodes, and then the 6.2-V zener. Click on the part number to obtain a datasheet.

CHAPTER 4

Practice Problems

—1—

$V_{dc} = 0.45 \times V_{rms} = 0.45 \times 200\text{ V} = 90\text{ V}$

—2—

$V_{dc} = 0.45 \times V_{rms} = 0.45 \times 440\text{ V} = 198\text{ V}$
PIV $= 1.414 \times V_{rms} = 1.414 \times 440\text{ V} = 622\text{ V}$
Ripple frequency = same as ac input = 60 pps (pulse per second)
Polarity of output = positive with respect to ground reference

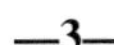

$V_{dc} = 0.9 \times V_{rms}$ of half the secondary $= 0.9 \times 150\text{ V} = 135\text{ V}$
Ripple frequency $= 2 \times$ ac input frequency $= 2 \times 400 = 800$ pps
PIV $= 1.414 \times V_{rms}$ full secondary $= 1.414 \times 300\text{ V} = 424\text{ V}$

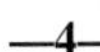

1. $V_{dc} = 0.9 \times V_{rms}$ input $= 0.9 \times 120\text{ V} = 108\text{ V}$
2. $I_{dc} = V_{dc}/R_L = 108\text{ V}/50\ \Omega = 2.16\text{ A}$
3. PIV (Bridge) $= 1.414 \times V_{rms}$ input $= 1.414 \times 120 = 169.7$ V
4. Ripple frequency (Bridge) $= 2 \times$ ac input frequency $= 2 \times 60\text{ Hz} = 120$ pps

In-Process Learning Checks

—1—

1. The simplest rectifier circuit is the ***half-wave*** circuit.
2. The ripple frequency of a bridge rectifier circuit is ***two times*** the ripple frequency of a half-wave rectifier.
3. The circuit having the highest output voltage for a given transformer secondary voltage is the ***bridge*** rectifier.
4. For a given full transformer secondary voltage, the full-wave rectifier circuit unfiltered dc output voltage is ***equal to*** the dc output of a half-wave rectifier that is using the same transformer.
5. To find the average dc output (unfiltered) of a center-tapped, full wave rectifier, multiply the full secondary rms voltage by ***0.45.***

1. In a C-type filter, the output of the rectifier circuit is connected in ***parallel*** with a filter ***capacitor*** and the load resistance.
2. In an L-type filter, the output of the rectifier circuit is connected in ***series*** with a filter ***choke (inductor)*** and the load resistance.
3. In an L-type filter, energy is stored in the form of a ***magnetic*** field.
4. The five main performance characteristics used in the selection and quality of a power supply filter network are ***output ripple, regulation, rectifier peak current limits, load current,*** and ***output voltage.***
5. The two main causes of swings in the output voltage of a power supply are changes in the ***input*** voltage level and changes in the ***output*** current demand.
6. The three terminals on an integrated-circuit voltage regulator are the unregulated ***dc input,*** regulated ***dc output,*** and ***common ground.***

7. The ***lower*** the percent of ripple, the higher the quality of the power supply.
8. Full-load output voltage is taken when the current from the power supply is ***maximum.***
9. The ***higher*** the percent of regulation, the higher the quality of the power supply.

Review Questions

1.

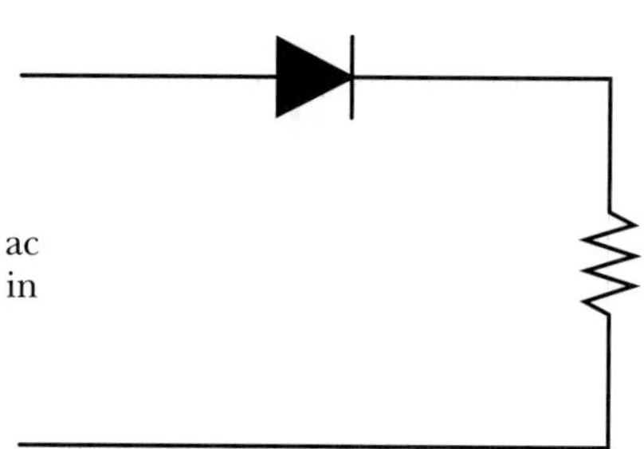

3. When you turn around a rectifier diode, the polarity of the output from a half-wave rectifier is reversed. Switching the connections to the ac source has no effect on the operation of the rectifier, however.

5.

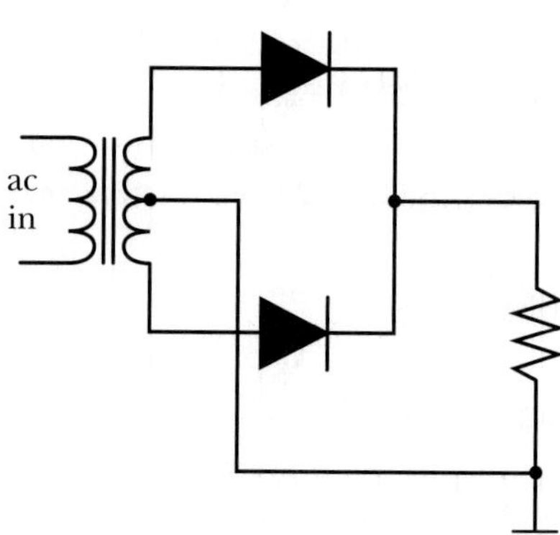

7. 120 pps
9.

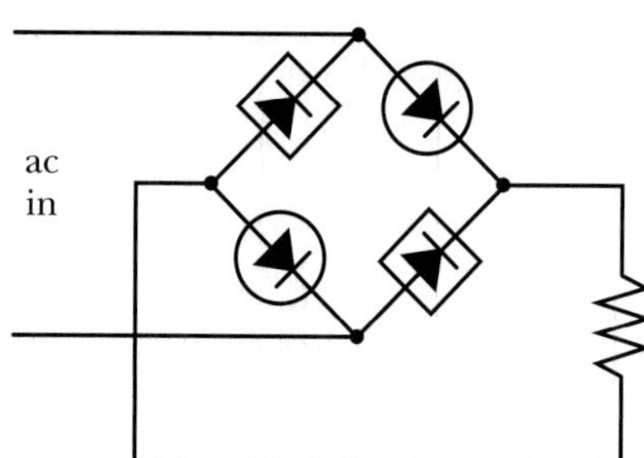

11. 800 pps
13. Excessive ripple is most likely caused by an open capacitor or shorted inductor in the power supply filter network.
15.

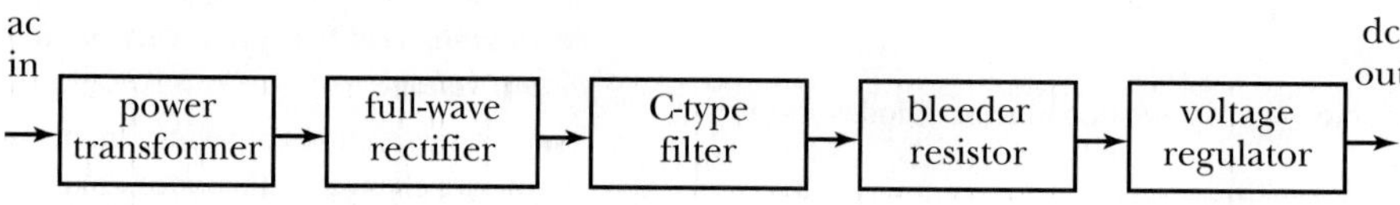

17. A no-load condition occurs when no current is drawn from the power supply. A full-load condition occurs when the maximum allowable amount of current is drawn from the power supply.
19. The RMS input voltage is 212 V.

Problems

1. **a.** 54 Vdc
   b. 170 V (unfiltered)
   c. 60 Hz
3. 1.7 A
5. ≅ 15.45:1 step-down
7. 10%
9. 10.6% regulation
11. **a.** 21.4 mA
    b. 256.8 mW
    c. 40 mA
    d. 320 mW
    e. 18.6 mA
    f. 223.2 mW
    g. 800 mW
    h. 32.1%
13. **a.** 0.214 A
    b. 2.568 W
    c. 0.218 A
    d. 1.798 W
    e. 4.36 W
    f. 58.9%
15. **a.** 0.268 A
    b. 4.02 W
    c. 0.270 A
    d. 1.426 W
    e. 5.4 W
    f. 74.4%
17. **a.** 6.458 V
    b. 64.6 mA

Datasheet Problems

1. Load regulation is 9 mV (typical), 100 mV (maximum) for 5 mA to 1.5 A load current.
   % regulation = 2.04%

Analysis Questions

1.

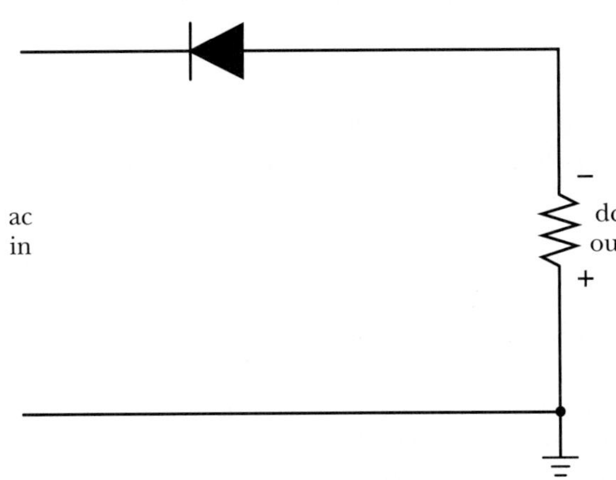

3. Capacitor discharges less between charging pulses.
5. Optimum inductance is that value of the first inductor where peak rectifier current is kept from exceeding the dc load current by more than 10% at maximum load levels.
7. A shorted diode allows ac voltage to reach the filter capacitor. Electrolytic capacitors can handle only dc voltages of a certain polarity. When ac voltage is applied to an electrolytic filter capacitor, when a rectifier diode shorts, for example, the reverse half-cycle of ac voltage shorts the capacitor and quickly destroys it.

CHAPTER 5

Practice Problems

—1—

1. $I_E = I_C + I_B = 10$ mA + 100 μA = 10.1 mA
2. $\beta_{dc} = I_C/I_B = 2$ mA/15 μA = 133 for both NPN and PNP transistors
3. $I_C = \beta_{dc} \times I_B = 100 \times 10$ μA = 1 mA
   $I_E = I_C + I_B = 1$ mA + 10 μA = 1.01 mA
   $\alpha = I_C/I_E = 1$ mA/1.01 mA = 0.99

—2—

Cutoff point is (from Formula 5–31) 5 V.

Saturation current is (from Formula 5–32):

$I_{C_{(sat)}} = V_{CC}/(R_C + R_E) = 5$ V/517 Ω = 9.67 mA

—3—

Divider current: $I = V_{CC}/(R_1 + R_2) = 10$ V/12 kΩ = 0.83 mA

Base voltage: $V_B = I \times R_2 = 0.83$ mA × 2 kΩ = 1.67 V

Emitter voltage: $V_E = V_B - 0.7$ V = 0.97 V

Emitter current: $I_E = V_E/R_E = 0.97$ V/1 kΩ = 0.97 mA

Collector voltage: $V_C = V_{CC} - (I_C \times R_C) = 10$ V – 4.6 V = 5.4 V

Transistor voltage: $V_{CE} = V_{CC} - (V_{RC} + V_{RE}) = 10$ V – 5.57 V = 4.43 V

dynamic emitter resistance: $r'_e = 25.86$ Ω

input impedance: $Z_{in} = 1.64$ kΩ

output impedance: $Z_{out} = 4.7$ kΩ

Voltage gain: $A_V = -4.58$

Current gain: $A_i = 1.58$

Power gain: $A_p = 7.25$

Output voltage: $V_{out} = 0.916\ V_{pp}$

(The negative sign indicates a 180° phase shift)

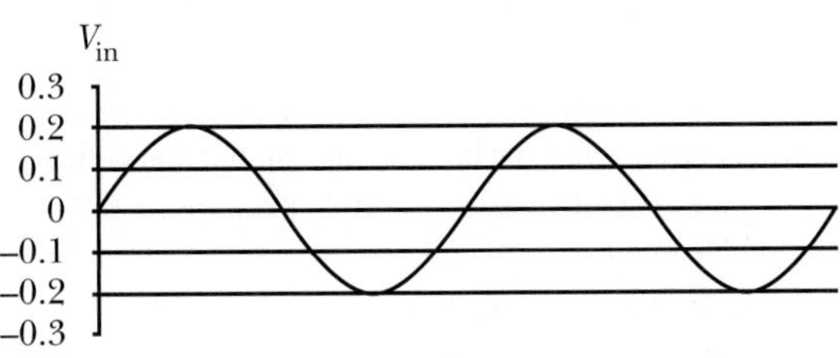

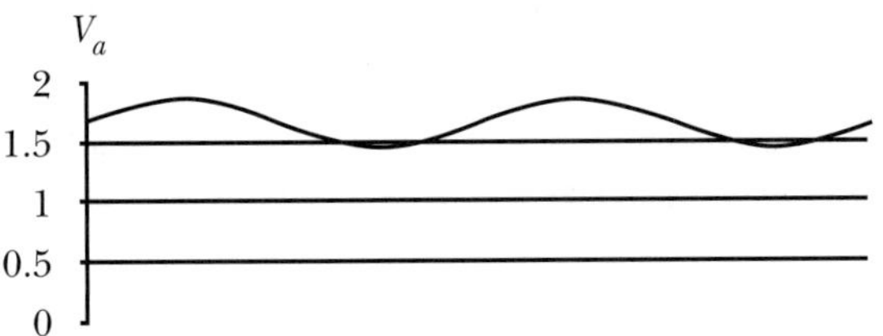

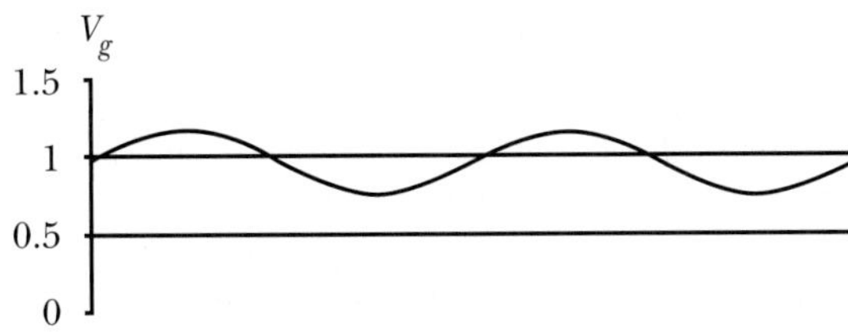

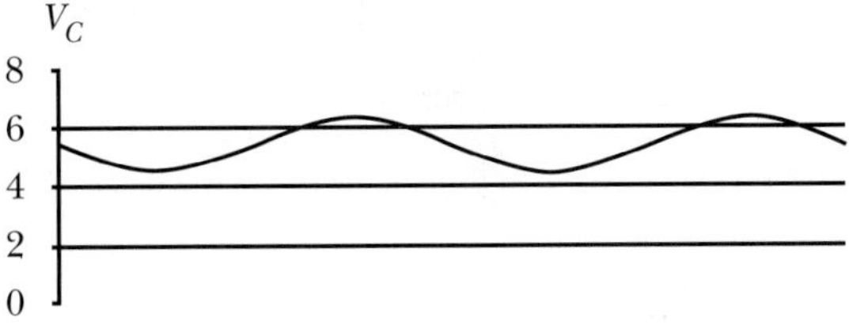

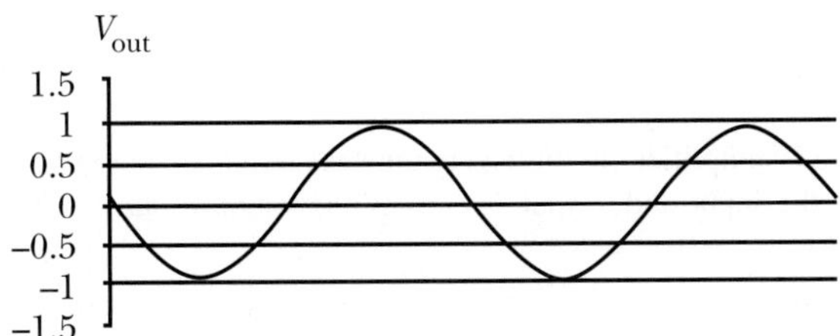

In-Process Learning Checks

—1—

1. The symbol for an NPN transistor shows the emitter element pointing ***away from*** the base element.
2. The symbol for a PNP transistor shows the emitter element pointing ***in toward*** the base element.
3. For BJTs:
   V_{BE} is the ***base-emitter*** voltage
   V_{CB} is the ***collector-base*** voltage
   V_{CE} is the ***collector-emitter*** voltage
4. For a typical BJT circuit:
   I_C stands for the ***collector*** current
   I_B is the ***base*** current
   I_E is the ***emitter*** current

—2—

1. Under normal operating conditions, the base-emitter junction of a BJT must be ***forward*** biased, while the collector-base junction is ***reverse*** biased.
2. Current must be flowing through the ***base-emitter*** junction of a BJT before current can flow between the emitter and collector.
3. Stated in words, the dc beta of a BJT is the ratio of ***collector current*** divided by ***base current.***
4. Stated in words, the alpha of a BJT is the ratio of ***collector current*** divided by ***emitter current.***
5. The α of a BJT is *less* than 1, while the β_{dc} is ***greater*** than 1.

—3—

1. BJTs use a small amount of base ***current*** to control a larger amount of collector ***current.*** It can be said that BJTs are ***current*** controllers.
2. When base current in a BJT is zero, the collector current is *zero.*
3. BJTs are basically used as ***switches*** and ***amplifiers.***
4. When a BJT is being used as a switch (as in Figure 5–17), V_{CE} is maximum when I_B is ***minimum,*** and V_{CE} is minimum when I_B is ***maximum.*** In the same circuit, I_C is maximum when I_B is ***maximum,*** and I_C is minimum when I_B is ***minimum.***
5. When a BJT is being used as voltage amplifier (as in Figure 5–18), an increase in base current causes a ***decrease*** in V_{CE}.
6. On a family of collector characteristic curves, the horizontal axis represents the ***collector emitter voltage*** and the vertical axis represents the ***collector current.*** Each curve in the family represents a different level of ***base current.***

In-Process Learning Checks

—4—

1. The voltage gain of the amplifier = ***250.***
2. The current gain of the amplifier = ***50.***
3. The power gain = 1800.

—5—

1. The BJT amplifier circuit having the base terminal common to both the input and output circuits is the ***common-base (CB)*** amplifier.
2. The BJT amplifier circuit providing both voltage and current gain is the ***common-emitter (CE)*** amplifier.
3. The BJT amplifier circuit having 180° phase difference between input and output signals is the ***common-emitter (CE)*** amplifier.
4. The BJT amplifier circuit having a voltage gain of less than 1 is the ***common-collector (CC)*** amplifier.
5. The BJT amplifier circuit having an input to the base and output from the emitter is called a common-***collector*** amplifier.
6. In an amplifier that uses more than one stage, the overall gain is found by ***multiplying*** the gains of the individual stages.
7. The voltage follower amplifier is another name for a common-***collector*** amplifier.
8. The Darlington amplifier is made up of two common-***collector*** amplifiers.

—6—

1. Class A amplifiers conduct during ***360°*** of the input cycle; Class B amplifiers conduct for ***180°*** of the input cycle; and Class C amplifiers conduct for approximately ***120°*** of the input cycle.
2. A dc load line shows all operating points of a given amplifier from ***cutoff*** to ***saturation.***
3. The amount of saturation current = ***12 mA.*** The value of the cutoff voltage for this circuit = ***12 V.***
4. The Q point for a Class A amplifier is located at the ***half-way point*** on the dc load line (halfway between cutoff and saturation).
5. The Class ***C*** amplifier has the Q point located below the cutoff.
6. Collector current flows in a Class ***A*** amplifier, even when there is no signal applied to the input.

Review Questions

1. Two, two
3. N, P
5. Negative to base; positive (or less negative) to emitter; negative to collector.
7. Approximately 96–99% of a BJT's emitter current appears in the collector circuit.
9. A reverse-biased P-N junction is one where the P-type material is connected to the negative side of the source and the N-type material is connected to the positive side of the source.
11. Reverse
13. False
15. Collector current (I_C) increases as base current (I_B) increases.
17. Increasing the forward-biasing current (I_B) in a typical BJT voltage amplifier causes the collector-emitter voltage drop (V_{CE}) to decrease.
19. Define the following ratings for a BJT:
    - **a.** $I_{C_{(max)}}$ = maximum collector current; the greatest amount of current the transistor can handle for an indefinitely long period.
    - **b.** $P_{D_{(max)}}$ = maximum power dissipation; greatest amount of power the transistor can dissipate without the help of a heat sink.
21. Emitter
23. Typical impedance characteristics of the various amplifier configurations follow:
    - **a.** Common-emitter: input Z = low, output Z = low to medium
    - **b.** Common-base: input Z = low, output Z = fairly high
    - **c.** Common-collector: input Z = high, output Z = low

25. Common-emitter
27. A common-collector amplifier stage followed by a common-emitter stage provides a high input impedance followed by high voltage gain.
29. Small-signal transistor amplifier stages may be used in the first stages of radio and TV receivers; large-signal transistor amplifier stages in audio power amplifier stages for driving loudspeakers.
31. Class A amplifiers have the least distortion.
33. The Q point is usually set at the mid-point of the load line for Class A operation.
35. For Class A operation of a common-emitter circuit, the quiescent value of V_{CE} is about one half the value of V_{CC}, or 10 V.
37. **a.** Divider current: $I = V_{CC}/(R_1 + R_2) = 15\text{ V}/10.1\text{ k}\Omega = 1.49\text{ mA}$
    b. Base voltage: $V_B = I \times R_2 = 1.49\text{ mA} \times 1\text{ k}\Omega = 1.49\text{ V}$
    c. Emitter voltage: $V_E = V_B - 0.7\text{ V} = 1.49\text{ V} - 0.7\text{ V} = 0.79\text{ V}$
    d. Collector voltage: $V_C = V_{CC} - (I_C \times R_C = 15\text{ V} - (0.79\text{ mA} \times 10\text{ k}\Omega) = 15\text{ V} - 7.9\text{ V} = 7.1\text{ V}$
    e. Collector-to-emitter voltage: $V_{CE} = V_{CC} - (V_{RC} + V_{RE}) = 7.1\text{ V} - 0.79\text{ V} = 6.31\text{ V}$
39. A common collector BJT amplifier circuit is needed to match a high impedance to a low impedance.

Problems

1. **a.** $I_C = 40.5$ mA
   b. $\beta_{dc} = 80$
3. $\beta_{dc} = 500$
5. $I_C = 1.6$ mA
7. $\alpha = 0.99$
9. **a.** $I_C = 38.4$ mA
   b. $I_B = 1.6$ mA
   c. $\beta_{dc} = 24$
11. $\beta_{ac} = 100$
13. $P_d = 480$ mW
15. $\beta_{ac} \equiv 100$
17. **a.** 6.59 V
    b. 5.89 mA
19. **a.** 5.46 mA
    b. 6 V
21. **a.** 0 mA
    b. 0 V
    c. 10 V
    d. 0 V
    e. 0 mA
    f. Class C
    g. 10 V
23. 18.3
25. 1.98

Analysis Questions

1. Outline drawings for TO-3, TO-5, TO-92, TO-220, and SOT-89 transistor packages are shown.

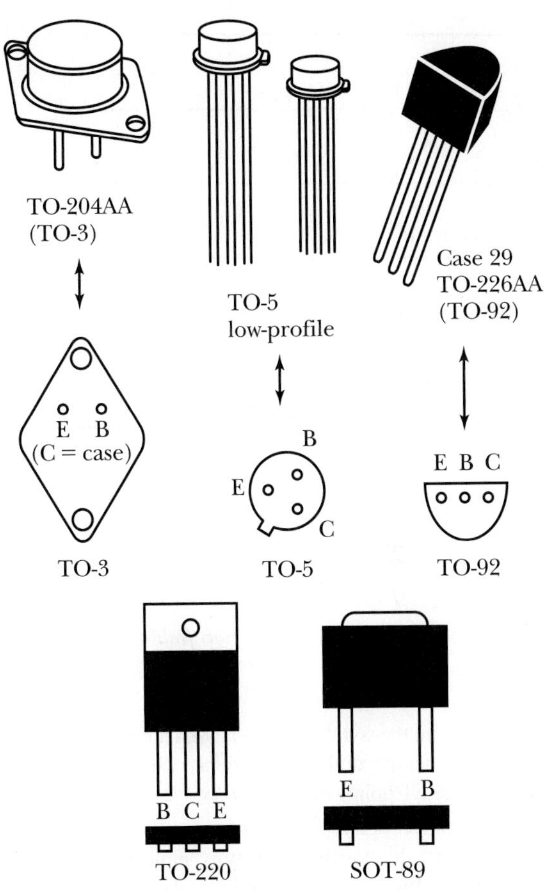

3. The purpose of a heat sink is to draw heat away from a semiconductor device and to dissipate it into the air by means of heat convection. Heat sinks are made from extruded aluminum stock that usually has rows of fins that greatly increase the effective area exposed to surrounding air. The semiconductor device to be called is securely bolted to the heat sink, sometimes with a thin layer of mica serving to isolate electrically the case of the semiconductor from the heat sink. Heat sinks are sometimes anodized with a flat black color to enhance heat dissipation according to the principle of black-body radiation.
5. Formula 25–1 shows that I_E is greater than I_C (unless all terms are zero): $I_E = I_C + I_B$. And since I_E is always greater than I_C, the quotient I_C/I_E has to be less than 1. Therefore, $\alpha = I_C/I_E$ (Formula 25–5) is always less than 1.

Analysis Questions

7. The steeper the slope of the load line, the less change in collector-emitter voltage for a given change in base current. The diagram shows that a steeper curve is created by lowering the collector resistance. You can conclude that the voltage gain of an amplifier changes in proportion to changes in the collector resistance.

9. Cutoff voltage = 15 V; saturation current = 11.7 mA.
11. The ratio of R_1:R_2 should be approximately 6.5:1.

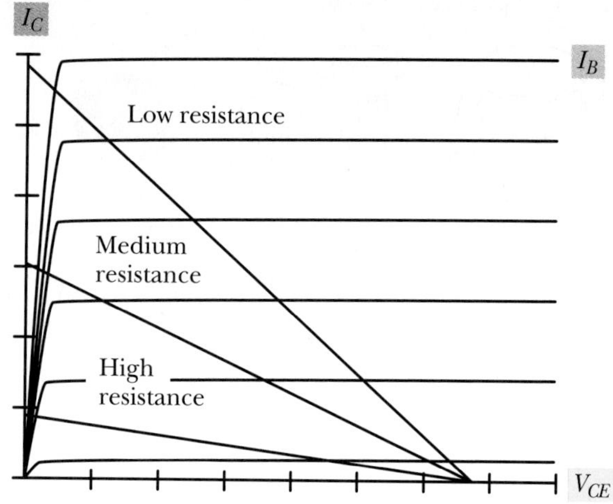

CHAPTER 6

In-Process Learning Checks

—1—

1. *JFET* is the abbreviation for ***junction field-effect transistor.***
2. The three terminals on a JFET are called ***source, gate,*** and ***drain.***
3. The arrow in the symbol for an N-channel JFET points ***in toward*** the source drain connections, while the arrow for a P-channel JFET points ***away from*** source-drain connections.
4. In an N channel JFET, the drain supply voltage must be connected so that the positive polarity is applied to the ***drain*** terminal of the JFET and the negative polarity is applied to the ***source*** terminal.
5. In a P-channel JFET, the drain supply voltage must be connected so that the positive polarity is applied to the ***source*** terminal of the JFET and the negative polarity is applied to the ***drain*** terminal.
6. The charge carriers in the channel material of a JFET always flow from the ***source*** terminal to the ***drain*** terminal.
7. In an N-channel JFET, the gate supply voltage must be connected so that the positive polarity is applied to the ***source*** terminal of the JFET and the negative polarity is applied to the ***gate*** terminal.
8. In a P-channel JFET, the gate supply voltage must be connected so that the positive polarity is applied to the ***gate*** terminal of the JFET and the negative polarity is applied to the ***source*** terminal.
9. V_{DS} stand for ***voltage between the drain and source terminals.***

   V_{GS} stands for ***voltage between the gate and source terminals.***

   I_D stands for ***drain current.***

   g_m stands for ***transconductance.***
10. The greatest amount of current flows through the channel of a JFET when V_{GS} is at its ***zero*** level, while the least amount of drain current flows when V_{GS} is at its ***maximum*** (or $V_{GS_{(off)}}$) level.

—2—

1. When a polarity opposite the drain polarity is applied to the gate of a JFET through a large-value resistor, the bias method is called ***gate*** bias.
2. When the gate and source of JFET are both connected to ground through resistors, the bias method being used is called ***self-bias.***
3. When the gate of a JFET is connected to a voltage divider and the source is connected through a resistor to ground, the bias method being used is called ***voltage-divider*** bias.
4. The common-***gate*** FET amplifier has the input signal applied to the source terminal and the signal taken from the drain.
5. The common-***source*** FET amplifier has the signal applied to the gate terminal and the output signal taken from the drain.
6. The common-***drain*** FET amplifier has the signal applied to the gate terminal and output taken from the source terminal.

—3—

1. The term ***MOS*** stands for ***metal-oxide semiconductor.***
2. The gate terminal in a D-MOSFET is separated from the changed material by a thin layer of ***silicon dioxide*** (or ***metal-oxide insulation***).
3. In a MOSFET, the direction of flow of charge carriers is always from the ***source*** terminal to the ***drain*** terminal.
4. An outward-pointing arrow on a D-MOSFET symbol indicates an *N*-type substrate and a *P*-type channel. An inward-pointing arrow on a D-MOSFET symbol indicates a *P*-type substrate and an *N*-type channel.
5. The proper polarity for V_{DD} of an N-channel D-MOSFET is ***negative*** to the source and ***positive*** to the drain. For a P-channel D-MOSFET, the proper polarity for V_{DD} is ***positive*** to the source and ***negative*** to the drain.
6. When operating an N-channel D-MOSFET in the depletion mode, the polarity of V_{GG} is ***negative*** to the gate terminal. And when V_{GS} is at its 0 V level, I_D is at its ***maximum*** level.
7. The ***E-MOSFET*** is the only FET that is nonconducting when the gate voltage is zero.
8. The proper polarity for V_{DD} of an N-channel E-MOSFET is ***negative*** to the source and ***positive*** to the drain. The proper polarity of V_{GG} is ***positive*** to the gate terminal.

Review Questions

1. V_{DS} = voltage measured between the drain and source terminals.

   V_{GS} = voltage measured between the gate and source terminals.

   I_D = current at the drain terminal (usually the same is found at the source terminal).

   I_{DSS} = drain current that flows when no bias voltage is applied to the gate.

   $V_{GS_{(off)}}$ = the amount of gate-source voltage required to turn off drain current.
3. For N-channel: V_{DD} is positive, and V_{GG} is negative. For P-channel: V_{DD} is negative, and V_{GG} is positive.
5. A JFET uses a voltage (V_{GS}) to control a current (I_D), whereas a BJT uses a current (I_B) to control a current (I_C). Also, a JFET is fully conducting when there is no bias applied to the gate, whereas a BJT is switched off when there is no bias applied to the base.

7. Transconductance is equal to a change in drain current divided by the corresponding change in gate-source voltage.
9. A D-MOSFET makes an ideal Class-A amplifier because the required gate-source bias voltage is zero. This greatly simplifies the required bias circuitry.
11. **a.** For the N-channel depletion mode: V_{DD} is positive and V_{GS} is negative.
    b. For the P-channel depletion mode: V_{DD} is negative and V_{GS} is positive.
    c. For the N-channel enhancement mode: V_{DD} is positive and V_{GS} is positive.
    d. For the P-channel enhancement mode: V_{DD} is negative and V_{GS} is negative.
13. In the depletion mode, increasing the amount of gate-source voltage depletes the channel of charge carriers, thus tending to decrease the amount of drain current. In an enhancement mode, increasing the amount of gate-source voltage enhances the density of charge carriers in the channel, thus tending to increase the amount of drain current.
15. An E-MOSFET makes an ideal electronic switching circuit (Class-B amplifier) because it is off when there is no gate-source voltage applied, and it is turned on when the gate-source voltage is applied.
17. **a.** For an N-channel JFET, $V_{GS_{(off)}}$ is negative. For a P-channel JFET, $V_{GS_{(off)}}$ is positive.
    b. For an N-channel D-MOSFET, $V_{GS_{(off)}}$ is negative. For a P-channel D-MOSFET, $V_{GS_{(off)}}$ is positive.
    c. For an N-channel E-MOSFET, $V_{GS_{(off)}}$ is 0 V. For a P-channel E-MOSFET $V_{GS_{(off)}}$ is 0 V.
19. The zener diodes that are internally connected between the gate and source terminals of some MOSFET devices protect the device from damage that could otherwise be caused by an external build-up of static voltage.

Problems

1. **a.** $V_G = -4$ V
   b. $V_{RD} = 8$ V
   c. $V_{DS} = 4$ V
   d. $I_{RG} = 0$
3. **a.** $g_m = 0.5$ mS (or 500 μS)
   b. $\Delta V_{DS} = 2$ V
5. **a.** $I_D = 1.67$ mA
   b. $V_{DS} = 6.33$ V
   c. $V_G = 0$
   d. $V_{GS} = -2$ V
   e. $V_{RG} = 0$
   f. $I_{RG} = 0$
7. **a.** $I_{R_2} = 90.2$ μA
   b. $V_{R_2} = 2.98$ V
   c. $V_{R_1} = 9.02$ V
   d. $V_{RS} = 4$ V
   e. $V_{GS} = -1.02$ V
   f. $I_D = 8.51$ mA
   g. $V_{DS} = 2.21$ V

Analysis Questions

1. Attempting to operate a JFET in an enhancement mode would forward bias the gate-source junction. This means the device could not operate as a FET, and also there would be a risk of overheating the gate junction due to excessive forward current flow.
3. A D-MOSFET has an uninterrupted channel of N-type or P-type material, which makes it possible to operate it in a depletion mode. It can also be operated in an enhancement mode because of a region of the opposite type of material that can become part of the channel when the gate-source voltage contributes charge carriers to the opposite-type material. An E-MOSFET, on the other hand, does not have an uninterrupted channel of a type of semiconductor material, so it is not possible to operate it in a depletion mode. An E-MOSFET can be operated in the enhancement mode because it has an opposite-type material that can take on charge carriers that form a conductive channel, or inversion layer.
5. Answers will vary.

   When an input is 5V, the FET associated with that input signal will turn ON, providing a ground to that motor winding.

   When an input is 0 V, the FET associated with that input will be OFF. No current flow will occur through that motor winding.

   Sequencing the inputs as provided in Table 6–38b results in a clockwise rotation of the motor. Sequencing the inputs as provided in Table 6–38c results in a counterclockwise rotation of the motor.

CHAPTER 7

Practice Problems

—1—

1. $A_v = -(R_f/R_i) = -(1\ \text{M}\Omega/100\ \text{k}\Omega) = -10$
2. $R_f = A_v \times R_i \times -(-100) = 2\ \text{k}\Omega = 200\ \text{k}\Omega$
3. $R_i = R_f/A_v = -270\ \text{k}\Omega/(-20) = 13.5\ \text{k}\Omega$

—2—

1. A_v (noninverting) $= (R_f/R_i) + 1 = (20{,}000\ \Omega/2{,}000\ \Omega) + 1 = 11$
2. $R_f = R_i(A_v - 1) = (10\ \text{k}\Omega)(100 - 1) = 990\ \text{k}\Omega$
3. $R_i = R_f/(A_v - 1) = 270\ \text{k}\Omega/19 = 14.2\ \text{k}\Omega$

In-Process Learning Check

—1—

1. An op-amp is a high gain ***direct***-coupled amplifier.
2. The input stage of an op-amp uses a ***differential*** amplifier configuration.
3. The name operational amplifier derives from their early use to perform mathematical ***operations.***
4. The open-loop gain of an op-amp is much ***higher*** than the closed-loop gain.
5. To achieve an output that is the inversion of the input, the input signal is fed to the ***negative*** input of the op-amp.
6. Input signals to the op-amp can be fed to the ***inverting*** input, the ***noninverting*** input, or to both inputs.

—2—

1. For an op-amp inverting amplifier the input signal is applied to the ***inverting*** input, and the feedback resistor is connected from the output to the ***inverting*** input.
2. The voltage gain of an inverting amplifier is a negative value in order to indicate ***inversion*** of the signal.
3. In the schematic for an op-amp inverting amplifier, resistor R_f is located between the ***output*** of the circuit and the ***inverting*** input. Resistor R_i is connected to the ***inverting*** terminal.
4. For an op-amp noninverting amplifier the input signal is applied to the ***noninverting*** input, and the feedback resistor is connected from the output to the ***inverting*** input.
5. The special case of a noninverting op-amp circuit that has a voltage gain of 1 is called a ***voltage follower*** circuit.
6. An op-amp circuit that has one signal input and one grounded input is called a ***single***-ended circuit.
7. An op-amp circuit that has different signal sources connected to its inverting and noninverting inputs is called a ***double***-ended circuit.
8. A comparator circuit is an example of an op-amp operating in the ***double***-ended input mode.

Review Questions

1. The amplifier circuit typically used as the input stage for an operational amplifier is a ***differential*** amplifier.
3. The key parameters for the ideal op-amp are infinite voltage gain, infinite bandwidth, infinite input impedance, and zero output impedance.
5. A comparator circuit is an example of an op-amp circuit that is operated in the open-loop gain mode.
7. Decrease
9. 16, 11
11. The output voltage waveform is twice the input voltage and has the same phase.
13. The output signal has a phase that is opposite the input.
15. The feedback component is a capacitor, and the input component is a resistor.
17. Integrator
19. $V_{out} = V_1 + V_2 + V_3$ or $V_{out} = -(V_1 + V_2 + V_3)$

Problems

1. $A_v = -208$
3. $R_i = 8.33\ \text{k}\Omega$
5. $A_v = 2$
7. a. $V_{out} = +5\ \text{V}$ c. $V_{out} = -1\ \text{V}$
   b. $V_{out} = +1\ \text{V}$ d. $V_{out} = -5\ \text{V}$
9. a. $V_{out} = -8\ \text{V}$ c. $V_{out} = 0\ \text{V}$
   b. $V_{out} = 0\ \text{V}$ d. $V_{out} = +8\ \text{V}$
11. a.

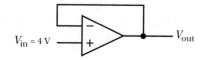

   b. $V_{out} = 4\ \text{V}$
13. 0.227 V
15. 6 V
17. a. 204
   b. 1100 1100
   c. CC
   d. DB7 = H
   DB6 = H
   DB5 = L
   DB4 = L
   DB3 = H
   DB2 = H
   DB1 = L
   DB0 = L

Analysis Questions

1. From Formula 28–2: A_v (noninverting) = $(R_f/R_i) + 1$. For a voltage follower, $R_f = 0$. Substituting this into the formula, we get: A_v (noninverting) = $(0/R_i) + 1 = 0 + 1 = 1$. Thus, the voltage gain of the circuit is 1, or unity.
3. Voltage gain = V_{out}/V_{in} = 2 V/10 mV = 200 $R_f = A_v \times R_i = 200 \times 2.2\ \text{k}\Omega = 440\ \text{k}\Omega$
5. From National Semiconductor ADC0804 datasheet, conversion time is 103 μsec (minimum), 114 μsec (maximum) for a clock of 640 kHz. The conversion time is the amount of time required for the Analog-to-Digital Converter to process the acquired analog input voltage into a digital value.

CHAPTER 8

In-Process Learning Checks

1. As the input signal frequency increases, the op-amp's unity gain bandwidth product causes the open-loop voltage gain of the op-amp to *decrease.*
2. At an input signal frequency of 100 kHz, the op-amp's unity gain bandwidth product causes the actual op-amp circuit gain to be *less than* the ideal op-amp circuit gain.
3. The slew rate is a measure of the output signal *voltage* change with respect to *time.*

Review Questions

1. unity gain bandwidth product
3. True
5. False
7. 0°
9. difference
11. False

Problems

1. 1000
3. a. –5
   c. 1 V_{pp} 100 kHz sine wave with 180° phase shift
5. Inverting gain: actual is –4.717, ideal is –5
   Noninverting gain: actual is 4.717, ideal is 5

Analysis Questions

1. 18
3. The output will be 0 V.
5. Excel graphs

CHAPTER 9

In-Process Learning Check 1

1. For a first order low-pass filter, the *pass*-band is all frequencies less than the cutoff frequency and the *stop*band is all frequencies greater than the cutoff frequency.
2. At the cutoff frequency, the voltage gain for the unity gain first order low-pass filter is *–3 dB.*
3. The order of a filter is determined by the number of *capacitors* in the circuit.
4. For a second order low-pass filter, the gain-frequency slope above the cutoff frequency is *–40 dB/decade.*

In-Process Learning Check 2

1. For a first order high-pass filter, the *stop*band is all frequencies less than the cutoff frequency and the *pass*-band is all frequencies greater than the cutoff frequency.
2. At the cutoff frequency, the voltage gain for the unity gain first order high-pass filter is *–3 dB.*
3. The second order filter is also known as a *Sallen-Key* filter.
4. For a second order high-pass filter, the gain-frequency slope below the cutoff frequency is *–40 dB/decade.*

Review Questions

1. True
3. high pass
5. True
7. True
9. False

Problems

1. –40 dB/decade for all frequencies greater than 10 kHz
   –40 dB
   0.1 V_{rms}
3. 723 Hz
   1.414 V_{rms} 723 Hz sine wave
5. 16.9 kHz
   1.414 V_{rms} 16.9 kHz sine wave
7. 1.78 kHz

Analysis Questions

1. Excel graph

CHAPTER 10

Practice Problems

—1—

1. JFET Q_1 is the active amplifying device.
2. L_{1_A}, L_{1_B}, and C_1 make up the frequency determining part of the circuit.
3. Capacitor C_3 is part of the ***feedback*** portion of this oscillator circuit.
4. From Formula 29–3:
   $L_T = L_{1_A} + L_{1_B} = 0.1 \text{ mH} + 0.1 \text{ mH} = 0.2 \text{ mH}$
5. From Formula 29–2:
   $f_r = 1/2\pi\sqrt{L_TC} = 1/2\pi\sqrt{(0.2 \text{ mH})(1{,}000 \text{ pF})}$
   $f_r = 355.9 \text{ kHz}$
6. Decreasing the value of C increases the oscillating frequency. Increasing the value of L_{1_A} decreases the oscillating frequency.

—2—

1. The active amplifying device is JFET Q_1.
2. C_1, C_2, and L_p the inductance of the primary of transformer T_1 make up the frequency determining part of the circuit.
3. From Formula 29–5:
   $C_T = C_1\,C_2/(C_1 + C_2) = 50 \text{ pF}$
4. From Formula 29–4:
   $f_r = 1/2\pi\sqrt{LC_T}$
   $f_r = 1/2\pi\sqrt{(750\ \mu\text{H})(50 \text{ pF})} = 822 \text{ kHz}$
5. When the total capacitance is decreased, the frequency of oscillation increases. When the value of inductance increases, the frequency of oscillation decreases.

—3—

1. From Formula 29–8:
   $f_r = 1/2\pi RC$
   $f_r = 1/2\pi(4.7\ \text{k}\Omega)(0.002\ \mu\text{F}) = 16.9 \text{ kHz}$
2. Doubling the values of the capacitors reduces the operating frequency by one half. Cutting the resistor values in half causes the operating frequency to double.
3. Rearranging Formula 29–8 to solve for the value of R:
   $R = 1/2\pi f_rC$
   $R = 1/2\pi(1 \text{ kHz})(0.1\ \mu\text{F}) = 1.6\ \text{k}\Omega$

In-Process Learning Check

—1—

1. The *LC* oscillator that has a tapped coil or a set of two coils in the tank circuit is called the ***Hartley*** oscillator.
2. The *LC* oscillator that has a tank circuit of two capacitors and a single inductor is called the ***Colpitts*** oscillator.
3. The *LC* oscillator that has three capacitors in the tank circuit is called the ***Clapp*** oscillator.
4. The output of a ***Colpitts*** *LC* oscillator is often coupled to the next stage by means of a transformer whose primary winding is part of the tank circuit.
5. When $C_1 = 100$ pF:
   $C_T = (100 \text{ pF})(500 \text{ pF})/(100 \text{ pF} + 500 \text{ pF}) = 83.3 \text{ pF}$
   $f_r = 1/2\pi\sqrt{(100\ \mu\text{H})(83.8 \text{ pF})} = 1.74 \text{ MHz}$

   When $C_1 = 850$ pF:
   $C_T = (850 \text{ pF})(500 \text{ pF})/(850 \text{ pF} + 500 \text{ pf}) = 315 \text{ pF}$
   $f_r = 1/2\ \pi\sqrt{(100\ \mu\text{H})(315 \text{ pF})} = 897 \text{ kHz or } 0.897 \text{ MHz}$

—2—

1. Whereas ***LC*** oscillators operate according to resonance, *RC* oscillators operate according to ***phase-shift.***
2. The four basic elements and conditions for starting and sustaining oscillation in *RC* sine-wave oscillators are ***power source, frequency determining elements, amplifier,*** and ***positive feedback.***

3. The sine-wave oscillator that uses a three-stages *RC* network for achieving a 180° phase shift at the frequency of oscillation is the ***phase-shift*** oscillator.
4. The sine-wave oscillator that uses a lead-lag *RC* network to produce 0° phase shift at the frequency of oscillation is the ***Wien-bridge*** oscillator.
5. The amplifier element of a phase-shift oscillator must produce a phase shift of 180°, while the amplifier element of a Wien-bridge oscillator must produce a phase shift of 0°.

Review Questions

1. A power source, a device or components that determine the frequency of oscillation, amplification, and positive feedback.
3. The Hartley oscillator uses a tapped coil. The Colpits oscillator has two capacitors with a ground connection between them. The Clapp oscillator uses three capacitors and a single inductor in the tank circuit.
5. From Formula 29–3:

   $L_T = L_{1_A} + L_{1_B} = 20$ mH

   From Formula 29–2:

   $f_r = 1/2\pi\sqrt{L_T C} = 3.56$ kHz
7. From Formula 29–5:

   $C_T = C_1C_2/(C_1 + C_2) = 0.05\ \mu$F

   From Formula 29–4:

   $f_r = 1/2\pi\sqrt{LC_T} = 7.12$ kHz
9. The phase-shift oscillator uses a network of three identical *RC* elements. The Wien-bridge oscillator uses a lead-lag network.
11. At the frequency of oscillation, the *RC* network in a phase-shift oscillator produces a phase shift of ***180°***, and the amplifier produces a phase shift of ***180°***.
13. From Formula 29–8:

    $f_r = 1/2\pi RC = 159$ Hz
15. In order to decrease the active time of a monostable multivibrator, you must ***decrease*** the value of the timing capacitor and/or ***decrease*** the value of the timing resistors.
17. $T = t_1 + t_2 = 10\ \mu s + 8\ \mu s = 18\ \mu s$

    $f = 1/T = 55.6$ kHz
19. From Formula 29–12:

    $f = 1/0.69C(R_A + 2R_B) = 483$ Hz

Problems

1. 160 kHz
3. 625 Hz
5. 13.5 MHz
7. 7.33 nF
9. 65 Hz

Analysis Questions

1.

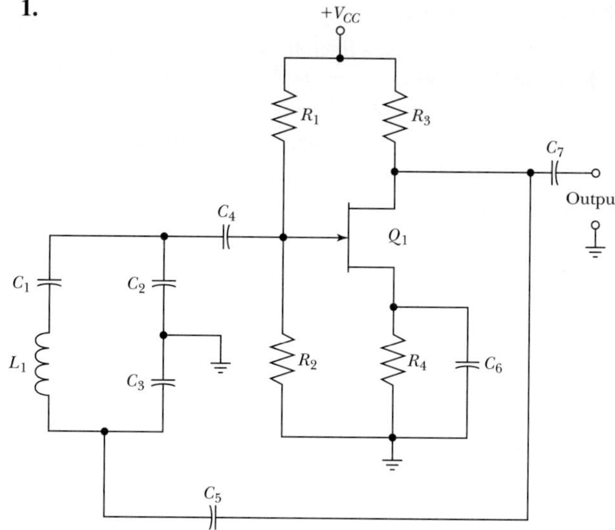

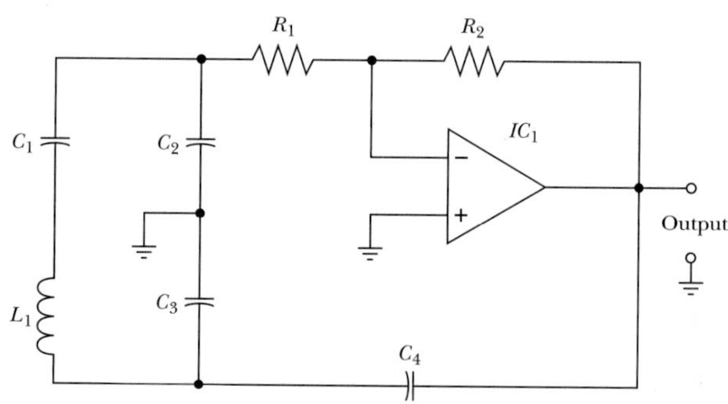

3. The total phase shift from the output to the input of the amplifier element in an oscillator must be 0° (or 360°) at the operating frequency to provide positive feedback. In a phase-shift oscillator, the *RC* network shifts the waveform by 180°, so the amplifier element must also shift the signal by 180° to satisfy the requirement for positive feedback. The *RC* network in a Wien-bridge oscillator, on the other hand, does not shift the phase of the waveform at the oscillator frequency; so the amplifier must not shift the waveform, either, in order to provide a total of 0° for positive feedback.
5. The feedback path through the three-section *RC* network of a phase-shift oscillator reduces the signal level a great deal by the time it reaches the input of the amplifier element. None of the other oscillators described in this chapter reduces the signal so much. The amplifier must have a high gain to overcome the unusual amount of signal loss through the *RC* network.
7. A *bistable multivibrator,* or flip-flop, is a type of multivibrator that is stable in both of its two states—on and off. It changes state each time it is triggered.
9. The operating frequency of a 555 stable multivibrator decreases as the value of the timing capacitor increases. The operating frequency increases as the value of the timing.

CHAPTER 11

In-Process Learning Checks

—1—

1. A thyristor is a ***four***-layered device that has just ***two*** operating states.
2. An SCR is gated on by applying a ***forward***-biasing voltage to the ***gate-cathode*** P-N junction. A ***LASCR*** (or ***light-activated SCR***) can also be gated on by shining a light onto its lens.
3. An SCR will conduct when it is gated and a voltage is applied between the cathode and anode such that the anode polarity is ***positive*** and the cathode is ***negative.***
4. The SCR rating that specifies the maximum forward voltage an SCR can handle is called the ***peak forward blocking voltage.***
5. The SCR rating that specified the maximum reverse voltage an SCR can handle is called the ***peak reverse voltage.***
6. The smallest amount of cathode-anode forward current that can sustain conduction of an SCR is called the ***holding current.***
7. The ***GCS*** (or ***gate-controlled switch***) is a type of SCR that can be gated off as well as gated on.
8. The ***silicon-controlled switch*** is a type of SCR that has two gate terminals.
9. In order to control both cycles of ac power, two SCRs must be connected in reverse with respect to one another in an arrangement that is known as ***inverse parallel*** or ***back-to-back.***
10. When a pair of SCRs are gated halfway through their respective half-cycles of the ac power waveform, the circuit is said to be firing at ***90***°.

—2—

1. Thyristors that can conduct in two directions (such as a diac and triac) are said to be ***bidirectional*** devices.
2. A diac has ***no*** gate terminals, while a triac has ***one*** gate terminal.
3. The only way to get a diac to conduct is by exceeding its ***forward*** breakover and reverse ***breakover*** ratings.
4. A triac operates like a pair of ***SCRs*** connected back-to-back.
5. The most common commercial and industrial applications of ***diacs*** are as triggering devices for triacs.
6. In order to apply full ac power to a load, a triac should be fired at the ***0***° point on each half-cycle.

Review Questions

1. The two possible states of operation of a thyristor are ***fully conducting*** and ***fully nonconducting.***
3. The conditions required for turning on an SCR are (1) forward-biasing voltage between the cathode and anode, and (2) forward-biasing voltage between the cathode and gate. The SCR is turned off only by removing or reversing the cathode-anode voltage.
5. *Holding current* is the least amount of forward-conducting current a thyristor can conduct and remain conducting.
7. GCS—Gate controlled switch

   LASCR—Light-activated silicon-controlled rectifier

   SCR—Silicon-controlled rectifier

   SCS—Silicon-controlled rectifier
9. The SCR requires the anode current to drop below the holding current; this can be accomplished by removing or reversing the voltage applied between the cathode and anode.

   The GCS can also be turned off by applying a reverse-bias voltage across the gate-cathode P-N junction.
11. LASCRs are gated by (1) applying a positive voltage to the gate terminal, or (2) directing light through a lens.
13. Bidirectional is the ability of a device to conduct current in both directions.

    A common resistor is bidirectional.

    A junction diode is not bidirectional.
15. The "peak forward blocking voltage" is the largest amount of forward bias that can be applied to a thyristor that is not on without the thyristor breaking into conduction. The "peak reverse voltage rating" is the largest amount of reverse voltage that can be applied between the cathode and anode without breaking down the junctions.
17. The two conditions required for switching on a triac are:
    1. A voltage is applied between the cathode and anode (polarity is not important).
    2. A forward-bias voltage is applied to the gate-cathode junction (at least momentarily).
19. The main purpose of an ac phase control circuit is to adjust the point at which the triac is triggered within each half-cycle.

Analysis Questions

1. The SCR uses a small amount of gate current to control a larger anode current. This control is for unidirectional operation.
3. Diacs, like zener diodes, will not begin conducting current until they reach their device rating. At this point the diac breaks down to a 0.7-V forward voltage drop, while the zener diode maintains the rated voltage level, increasing its current flow.

CHAPTER 12

Practice Problems

—1—

R = 325Ω

—2—

Closed switches are: S1, S3, S4, S6, and S7

45.5 mA

In-Process Learning Checks

1. An Led emits light energy when an electron in the ***conduction*** band falls to the ***valence*** band.
2. The choice of ***semiconductor material*** determines the color of the light from the LED.
3. The arrow on the LED symbol points ***away from*** the diode.
4. An LED emits light when it is ***forward*** biased.
5. The formula for calculating the value of the resistor to be used in series with an LED and its dc voltage source is R = $(V_S—V_D)/I_F$, where:

   V_S is ***the voltage of the source***

   V_D is ***the voltage across the diode***

   I_F is ***the forward diode current***

6. The reverse-breakdown voltage of an LED is generally ***lower*** than that of a typical rectifier diode, and the forward voltage drop of an LED is ***higher*** than that of a rectifier diode.
7. Light falling onto the depletion region of a reverse biased diode ***increases*** the amount of reverse leakage current.
8. The arrow on a photodiode symbol points ***toward*** the diode.
9. A photodiode is connected into the circuit so that it is ***reverse*** biased.
10. In an opto-isolator, an ***LED*** is the source and a ***photodiode*** is the dector of light energy.

Review Questions

1. The conditions required to get an LED to emit light are:
   a. The LED must be forward biased.
   b. The LED must be conducting correct.
3. The light falling onto the photodiode junction causes a reverse leakage current to flow. An increase in the light intensity results in an increase of the reverse current.
5. For the LED symbol, the arrows point away from the "diode" symbol. For the photodiode symbol, the arrows point into the "diode" symbol.
7. The light energy emitted from the laser diode is more intense than the light energy emitted from LED.

Problems

1. **a.** 25 mA
   b. 25 mA
   c. 113 mW
3. switches S1, S2, S3, and S9 36.4 mA

Analysis Questions

1. Laser diodes provide the source of coherent light for scanning the bar code
3. The DMC-16299 is a 16 × 2 monochrome alphanumeric LCD module. The module does not have any backlighting.

APPENDIX C

Troubleshooting Challenge Test Results

Find the number listed next to the test you chose and record the result.

1. —
2. signal normal
3. 2 kΩ
4. 2 V
5. 1.5 V
6. high
7. —
8. 0 V
9. —
10. —
11. 0 V
12. 7 V
13. —
14. 0 Ω
15. ≅ 0 V
16. 1 kΩ
17. 400 Ω
18. —
19. infinite Ω
20. —
21. (h.w. rect. ⌒– waveform)
22. 10 V
23. no signal
24. circuit operates normally
25. 10 kΩ
26. slightly high
27. —
28. 0 V
29. 0 Ω
30. greatly below max value
31. 5 V
32. 400 Ω
33. infinite Ω
34. 6.4 V
35. noticeably high
36. 4 kΩ
37. 10 kΩ
38. 3.3 V
39. 14 V
40. —
41. ≅ 1.5 mH
42. 3.0 V
43. 7 V
44. —
45. high
46. 0 Ω
47. —
48. 2.3 V
49. 14 V
50. —
51. —
52. 10 V
53. (Inv. sine wave) ∿ (14 VAC)
54. 10 kΩ
55. max value
56. 10 V
57. signal normal
58. infinite Ω
59. —
60. 275 pF
61. 14 V
62. 1 kΩ
63. infinite Ω
64. —
65. ≅ 6 VDC
66. 0 V
67. 7.5 V
68. 10 kΩ
69. low
70. signal normal
71. no change in operation
72. 0 Ω
73. —
74. (sine wave) (120 V) ∿
75. 10 V
76. slightly low
77. —
78. 3.1 V
79. 7.5 V
80. 7 V
81. —
82. 0 V
83. 10 V
84. —
85. slightly below max value
86. 1.5 V
87. 5 Ω
88. 14 V
89. 2.3 V
90. infinite Ω
91. —
92. normal
93. 10 kΩ
94. ≅ 5 V
95. 10 V
96. ≅ 11 V
97. —
98. low
99. —
100. infinite Ω
101. —
102. 0 V
103. 0 V
104. ≅ 5 kΩ
105. —
106. no signal
107. 50 pF
108. —
109. 12 kΩ
110. —
111. no change in operation
112. —
113. signal normal
114. 297 Ω
115. —
116. ≅ 5 V
117. within normal range
118. 150 mA
119. 17 mA
120. 100 Ω
121. distorted
122. clipped sine wave
123. slightly lower
124. slightly high
125. positive saturation
126. normal
127. normal
128. low
129. high
130. high
131. low
132. high
133. high
134. low
135. high
136. high
137. low
138. high
139. high
140. low
141. high
142. low
143. low
144. normal
145. low
146. low
147. normal
148. low
149. low
150. normal
151. low
152. low
153. normal
154. normal
155. low
156. normal
157. normal
158. low
159. normal
160. normal
161. high
162. normal
163. normal

164. normal
165. low
166. normal
167. normal
168. normal
169. normal
170. normal
171. low
172. normal
173. normal
174. normal
175. normal
176. normal
177. low
178. normal
179. normal
180. normal
181. normal
182. high
183. low
184. normal
185. high
186. normal
187. normal
188. normal
189. low
190. normal
191. normal
192. normal
193. high
194. high
195. normal
196. high
197. low
198. normal
199. low
200. low
201. low
202. high
203. low
204. low
205. low
206. high
207. low
208. low
209. high
210. low
211. low
212. high
213. low
214. low
215. high
216. low
217. normal
218. high
219. low
220. low
221. high
222. low
223. normal
224. high
225. low
226. normal
227. high
228. low
229. normal
230. normal
231. normal
232. normal
233. normal
234. high
235. normal
236. high
237. normal
238. low
239. low
240. low
241. normal
242. low
243. normal
244. normal
245. low
246. low
247. normal
248. abnormal
249. low
250. low
251. normal
252. abnormal
253. normal
254. normal
255. normal
256. normal
257. low
258. abnormal
259. low
260. low
261. normal
262. abnormal
263. normal
264. normal
265. normal
266. normal
267. low
268. normal
269. low
270. high
271. abnormal
272. low
273. low
274. normal
275. abnormal
276. high
277. normal
278. high
279. normal
280. normal
281. high
282. high
283. normal
284. normal
285. normal
286. normal
287. normal
288. normal
289. normal
290. none

APPENDIX D

Using a Scientific Calculator

Sample problem: Find R_T, I_T, and V_T for a series circuit containing three 20 kΩ resistors, where V_1 = 10 V. See the diagram below.

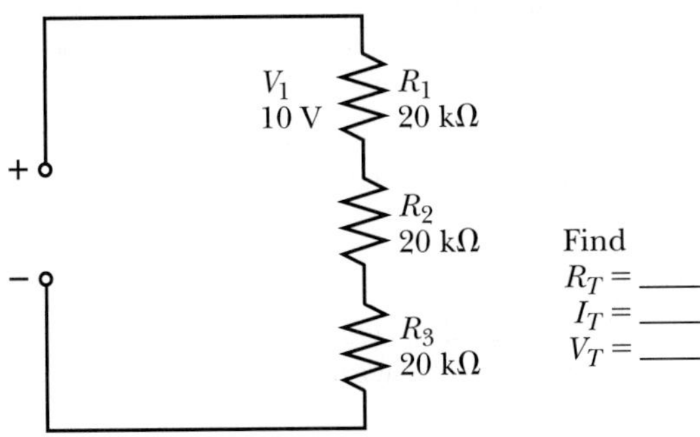

Elemental Method

Step 1: Turn calculator on and clear.

Step 2: Find R_T where $R_T = R_1 + R_2 + R_3$, by entering 20,000 + 20,000 + 20,000, then press the [=] button. The answer should be 60,000. (This represents 60,000 Ω, or 60 kΩ.)

Step 3: Find I by finding the I through R_1, where $I_1 = V_1/R_1$. Clear the calculator by pushing the [CE/C] button. Input as follows: Input 10 (for V_1), then the [÷] symbol button, then 20,000 (for R_1), then the [=] button. The answer should be 0.0005. (This represents 0.0005 A, or 0.5 mA.)

Step 4: Find V_T by using the formula $V_T = I_T \times R_T$. You may clear the calculator or, leave the 0.0005 you had just calculated, since you need it in this calculation). At any rate, be sure that 0.0005 is slowing, then push the [×] symbol button, then input 60,000 (for the R_T value). Push the [=] button. The answer should be 30, representing 30 V.

NOTE: This approach is appropriate for many calculators. Some brands, depending on their notation approach, will not use the exact same sequence of key strokes.

Using the "elemental approach" just described, if you solve for the parameters asked for in the following circuit, your answers should match the ones shown in parenthesis after each of the blanks. Try using your calculator and the "elemental approach" to see if you get the answers, as shown.

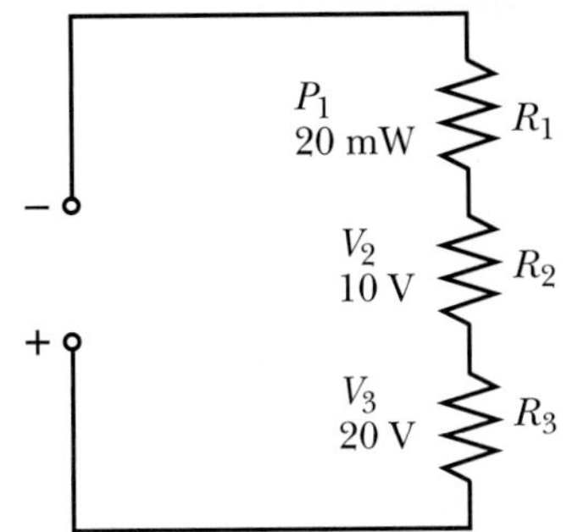

Find:

V_T = ______ (40 V)

P_3 = ______ (40 mW)

R_1 = ______ (5 kΩ)

R_2 = ______ (5 kΩ)

R_3 = ______ (10 kΩ)

Now that you've used the elemental method of entering numeric data, let's look at three other ways you can enter data and use a scientific calculator to advantage. The three ways include:

1. using the [EE] or [EXP] key on a scientific calculator (along with powers of 10);
2. using the [EE] or [EXP] key on a scientific calculator that is in the "engineering mode" (along with powers of 10); and
3. using the [EE] or [EXP] key on a scientific calculator that is in the "scientific notation mode" (along with powers of 10).

All three of these methods will save you from having to enter numerous zeros for very large or very small numbers. For our particular example problem, these approaches save you from having to put in all the zeros for large resistance values or all the zeros after the decimal for the small current value. Any one of these approaches can be helpful for most of your electronic calculations. Once you have learned these techniques, use the one that suits your needs best.

The significant key you will use on the keyboard in conjunction with these modes may be called [EE] on some calculators, or, [EXP] on others. Let's see how this system works.

Special Notations Regarding Scientific Calculators

Numerous brands and types of scientific calculators are on the market. Functions such as squaring a number, finding square roots, finding the arc tangent, and finding antilogarithms may differ slightly from calculator to calculator. For the types of calculations performed throughout the text, the following chart shows some common variations you may see or use. If you are using a different type of scientific calculator than the one used in creating the sample calculator sequences in our text (Calculator #1), the equivalent keystrokes chart, shown on the next page, may be helpful to you.

Using the EE (or EXP) Key with Powers of 10 for Metric Units

Using the same sample problem as before; that is, finding R_T, I_T, and V_T, where there are three series 20-kΩ resistors and $V_1 = 10$ V:

Step 1: Turn calculator on and clear by pressing the [AC/ON] key.

Step 2: Find R_T (where $R_T = R_1 + R_2 + R_3$) by entering the following:

input 20, [EE] (or [EXP]), 3,

(the display reads 20^{03} or 20. 03, meaning 20×10^3)

then input [+],

(the display now reads 20000)

then input 20, [EE] (or [EXP]), 3, a second time

(the display once again reads 20^{03} or 20. 03, showing the number you just entered)

then input [+],

(the display now reads 40000)

then input 20, [EE] (or [EXP]), 3, a third time

(the display will read 20^{03} or 20. 03, again representing the number you just entered)

then enter [=],

(the display will read 60000. This indicate that the value of R_T is 60,000 Ω, or 60 kΩ)

Step 3: Find I by finding the I through R_1, (where $I_1 = V_1/R_1$):

Clear the calculator by pressing the [CE/C] key.

Enter 10 (for the V_1 value),

then enter [÷],

then enter 20, [EE] (or [EXP]), 3 (for the 20-kΩ R_1 value) then press the [=] key.

The display reads 0.0005. (This indicates that current is 0.5 mA.)

Step 4: Find V_T by using the formula $V_T = I_T \times R_T$.

Clear the calculator with the [CE/C] button.

Enter the value of I_T: 0.5, [EE] (or [EXP]), [+/−], 3 (the display should read 0.5^{-03} or 0.5 – 03)

then press the [×] key,

then input 60, [EE] (or [EXP]), 3

then, press the [=] key.

The display reads 30. (This indicates that $V_T = 30$ V.)

Special Notations Chart

TASK DESCRIPTION	CALCULATOR #1 KEYSTROKE(S) TI 36X SOLAR	CALCULATOR #2 KEYSTROKE(S) HP 20S	CALCULATOR #3 KEYSTROKE(S) (Others)
Finding the reciprocal	1/x	1/x	2nd, 1/x
Squaring a number	x^2	↰, x^2	2nd, x^2
Taking a square root	$\sqrt{x}$	$\sqrt{x}$	$\sqrt{x}$
Finding the arctan	2nd, tan	↰, ATAN	INV, tan
Finding the antilog of a natural logarithm (ϵ)	2nd, e^x	e^x	e^x
Converting from polar to rectangular form: Finding the *x*-axis and *y*-axis (*j*) values:	phasor value, X⇕Y, angle value, 2nd (P ▸ R) (Displays *x*-axis value) Follow above sequence with X⇕Y key again (Displays *y*-axis *[j]* value)	phasor value, INPUT, angle value, ↱, → R (Displays *y*-axis *[j]* value) Follow above sequence with ↰, SWAP (Displays *x*-axis value)	phasor value, P-R key, angle value, = key (Displays *x*-axis value) Follow above sequence with X-Y key (Displays *y*-axis *[j]* value)
Converting from rectangular to polar form: Finding the phasor value and the angle	*x*-axis value, X⇕Y, *y*-axis (*j*) value, 3rd, (R ▸ P) Displays the phasor value Press the X⇕Y key again (Displays the angle value)	*x*-axis value, INPUT, *y*-axis (*j*) value, ↰, → P, (Displays angle value) Follow above sequence with ↰, SWAP (Displays the phasor value)	*x*-axis value, R-P key, *y*-axis (*j*) value, = (Displays phasor value) Follow the above sequence with X-Y key, (Displays the angle value)

Now, you try this technique to see if you can match the answers for the circuit parameters called for in the circuit shown below.

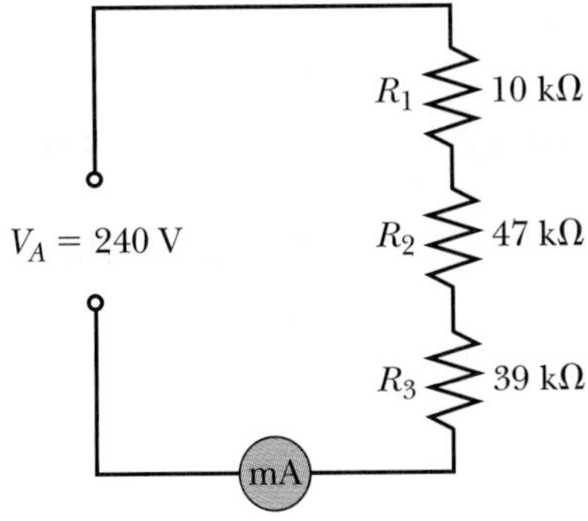

FIND:

R_T = _____ (96 kΩ)

I_T = _____ (2.5 mA)

V_{R_2} = _____ (117.5 V)

Using the Engineering Mode with Powers of 10 for Metric Units

The engineering mode provides a notation that again uses powers of 10. Because most science and engineering calculations use metric measurements that typically have *exponents* that are in multiples or submultiples of three, the data input and answers appear as some number, times a power of 10 with a multiple of three exponent. For example, in electronics, kilohms and kilowatts are expressed in terms of 10^3. Millivolts and milliamperes are expressed in terms of 10^{-3}. Microamperes are stated as 10^{-6}. Nanoamperes are indicated as 10^{-9} and so on.

Using the earlier problem that asked for R_T, I_T and V_T when the circuit was composed of three series 20 kΩ resistors and $V_1 = 10$ V:

Step 1: Turn calculator on and clear by pressing the [AC/ON] key.

Step 2: Press any buttons required to get your calculator into the *engineering mode.* See your calculator's instruction manual for this. NOTE: For the TI 36X solar calculator, the key sequence to get into the engineering mode is [3rd], ENG.

(When you are in this mode, the calculator readout will read $0.^{00}$ or 0.00 before new data are entered.)

Step 3: Find R_T in the *engineering mode* as follows:

Be sure your calculator is in the *engineering mode* and the display reads $0.^{00}$ or 0.00.

Enter 20, [EE] (or [EXP]), 3 (the readout reads $20.^{03}$ or 20.03),

then press [+] (the readout will still read $20.^{03}$ or 20.03),

then enter 20, [EE] (or [EXP]), 3 a second time (the readout will still read $20.^{03}$ or 20.03),

then enter [+] (the readout will read $40.^{03}$ or 40. 03),

then enter 20, [EE] (or [EXP]), 3 a third time (the readout will read $20.^{03}$ or 20.03),

then press the [=] sign (the readout now reads $60.^{03}$ or 60.03).

This indicates 60×10^3, or 60,000 Ω, or 60 kΩ.

Step 4: Find I_T in the *engineering mode* by entering the following (where $I_T = I_1$ and $I_1 = V_1/R_1$):

Be sure your calculator is in the *engineering mode* and the display reads $0.^{00}$ or 0.00.

Enter 10 (for the V_1 value),

then press the [÷] key,

then enter 20, [EE] (or [EXP]), 3,

then press the [=] key

The display should show $500.^{-06}$ or 500.–06.

(This indicates 500×10^{-6}, or 0.0005 A, which is the same as 0.5 mA.

$I_T = 0.5$ mA.)

Step 5: Find V_T in the *engineering mode* by using the formula $V_T = I_T \times R_T$.

Be sure the calculator is in the *engineering mode* and the display reads $0.^{00}$ or 0.00.

Enter the I_T value by entering 0.5, [EE] (or [EXP]), [+/–], 3, then press the [×] key, then enter 60, [EE] (or [EXP]), 3,

then, press the [=] key.

The display reads $30.^{00}$ or 30. 00 (This means 30×10^0, which means that $V_T = 30$ V.)

Get some practice in using the *engineering mode* by seeing if you can match the answers shown for the parameters specified.

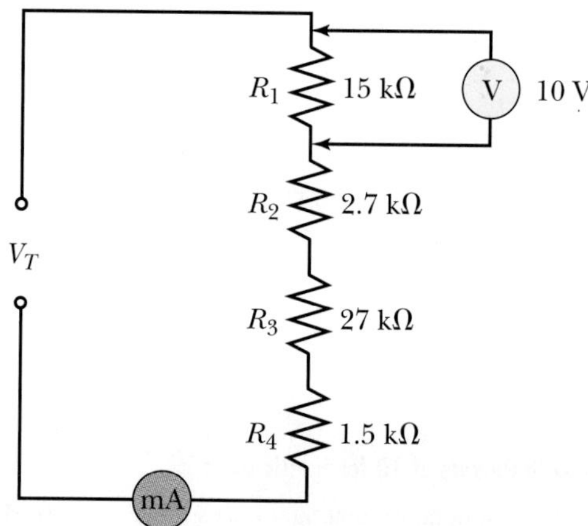

FIND:

$R_T =$ _____ (46.2 kΩ)

$I_T =$ _____ (0.67 mA)

$V_T =$ _____ (30.95 V)

Using the Scientific Notation Mode with Powers of 10 for Metric Units

As you know, scientific notation expresses numerical values as a number between 1 and 10, times the appropriate power of 10. Performing the same calculations as we have been using, the sequence of keys are as follows.

(NOTE: Make sure the calculator is *cleared* and you are in the *scientific notation mode* for each of the following calculations.)

To find R_T for our example circuit:
Enter 20, [EE] (or [EXP]), 3, [+], 20, [EE] (or [EXP]), 3, [+],
20, [EE] (or [EXP]), 3, then press [=].
The display reads $6.^{04}$ or 6. 04, meaning 6×10^4, or 60,000. Again, this is interpreted as 60,000 Ω, or 60 kΩ, for our circuit.

To find I_T (which is equal to I_1):
Enter 10, [÷], 20, [EE] (or [EXP]), 3, then press [=].
The display reads $5.^{-04}$ or 5.–04, interpreted as 5×10^{-4}, or 0.5 mA.

To find V_T:
Enter 5, [EE] (or [EXP]), [+/–], 4, [×], 60, [EE] (or [EXP]), 3, then press [=].
The display reads $3.^{01}$ or 3. 01, meaning 3×10^1, or 30 V.

Get some practice in using the *scientific notation mode* by seeing if you can match the answers shown for the parameters specified.

R_T = _____ (25.9 kΩ)
I_T = _____ (0.5 mA)
V_T = _____ (12.95 V)

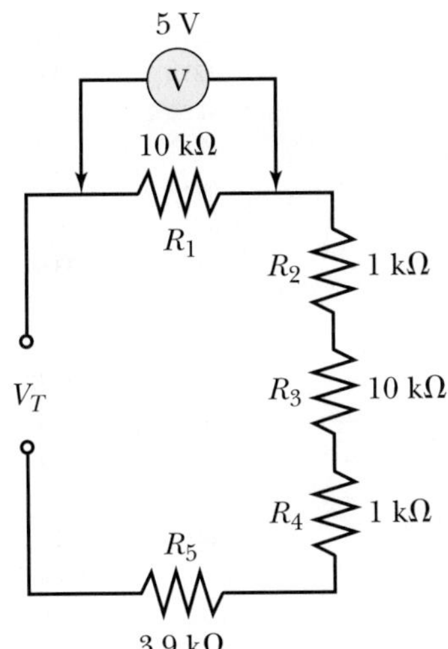

APPENDIX E

Copper-Wire Table

Wire Size A.W.G. (B&S)	Diam. in Mils[1]	Circular Mil Area	Turns per Linear Inch (25.4 mm)[2] Enamel	S.C.E.	D.C.C.	Cont.-duty current[3] single wire in open air	Cont.-duty current[3] wires or cables in conduits or bundles	Feet per Pound (0.45 kg) Bare	Ohms per 1000 ft. 25°C	Current Carrying Capacity[4] at 700 C.M. per Amp.	Diam. in mm.	Nearest British S.W.G. No.
1	289.3	83,690	—	—	—	—	—	3.947	0.1264	119.6	7.348	1
2	257.6	66,370	—	—	—	—	—	4.977	0.1593	94.8	6.544	3
3	229.4	52,640	—	—	—	—	—	6.276	0.2009	75.2	5.827	4
4	204.3	41,740	—	—	—	—	—	7.914	0.2533	59.6	5.189	5
5	181.9	33,100	—	—	—	—	—	9.980	0.3195	47.3	4.621	7
6	162.0	26,250	—	—	—	—	—	12.58	0.4028	37.5	4.115	8
7	144.3	20,820	—	—	—	—	—	15.87	0.5080	29.7	3.665	9
8	128.5	16,510	7.6	—	7.1	73	46	20.01	0.6405	23.6	3.264	10
9	114.4	13,090	8.6	—	7.8	—	—	25.23	0.8077	18.7	2.906	11
10	101.9	10,380	9.6	9.1	8.9	55	33	31.82	1.018	14.8	2.588	12
11	90.7	8,234	10.7	—	9.8	—	—	40.12	1.284	11.8	2.305	13
12	80.8	6,530	12.0	11.3	10.9	41	23	50.59	1.619	9.33	2.053	14
13	72.0	5,178	13.5	—	12.8	—	—	63.80	2.042	7.40	1.828	15
14	64.1	4,107	15.0	14.0	13.8	32	17	80.44	2.575	5.87	1.628	16
15	57.1	3,257	16.8	—	14.7	—	—	101.4	3.247	4.65	1.450	17
16	50.8	2,583	18.9	17.3	16.4	22	13	127.9	4.094	3.69	1.291	18
17	45.3	2,048	21.2	—	18.1	—	—	161.3	5.163	2.93	1.150	18
18	40.3	1,624	23.6	21.2	19.8	16	10	203.4	6.510	2.32	1.024	19
19	35.9	1,288	26.4	—	21.8	—	—	256.5	8.210	1.84	0.912	20
20	32.0	1,022	29.4	25.8	23.8	11	7.5	323.4	10.35	1.46	0.812	21
21	28.5	810	33.1	—	26.0	—	—	407.8	13.05	1.16	0.723	22
22	25.3	642	37.0	31.3	30.0	—	5	514.2	16.46	0.918	0.644	23
23	22.6	510	41.3	—	37.6	—	—	648.4	20.76	0.728	0.573	24
24	20.1	404	46.3	37.6	35.6	—	—	817.7	26.17	0.577	0.511	25
25	17.9	320	51.7	—	38.6	—	—	1,031	33.00	0.458	0.455	26
26	15.9	254	58.0	46.1	41.8	—	—	1,300	41.62	0.363	0.405	27
27	14.2	202	64.9	—	45.0	—	—	1,639	52.48	0.288	0.361	29
28	12.6	160	72.7	54.6	48.5	—	—	2,067	66.17	0.228	0.321	30
29	11.3	127	81.6	—	51.8	—	—	2,607	83.44	0.181	0.286	31
30	10.0	101	90.5	64.1	55.5	—	—	3,287	105.2	0.144	0.255	33
31	8.9	80	101	—	59.2	—	—	4,145	132.7	0.114	0.227	34
32	8.0	63	113	74.1	61.6	—	—	5,227	167.3	0.090	0.202	36
33	7.1	50	127	—	66.3	—	—	6,591	211.0	0.072	0.180	37
34	6.3	40	143	86.2	70.0	—	—	8,310	266.0	0.057	0.160	38
35	5.6	32	158	—	73.5	—	—	10,480	335	0.045	0.143	38–39
36	5.0	25	175	103.1	77.0	—	—	13,210	423	0.036	0.127	39–40
37	4.5	20	198	—	80.3	—	—	16,660	533	0.028	0.113	41
38	4.0	16	224	116.3	83.6	—	—	21,010	673	0.022	0.101	42
39	3.5	12	248	—	86.6	—	—	26,500	848	0.018	0.090	43
40	3.1	10	282	131.6	89.7	—	—	33,410	1,070	0.014	0.080	44

[1]A mil is 0.001 inch. A circular mil is a square mil × $\frac{\pi}{4}$. The circular mil (c.m.) area of a wire is the square of the mil diameter.

[2]Figures given are approximate only; insulation thickness varies with manufacturer.

[3]Max. wire temp. of 212°F (100°C) and max. ambient temp. of 135°F (57°C).

[4]700 circular mils per ampere is a satisfactory design figure for small transformers, but values from 500 to 1,000 c.m. are commonly used.

APPENDIX F

Useful Conversion Factors

To Convert	Into	Multiply By
ampere-turns/cm	amp-turns/inch	2.54
ampere-turns/inch	amp-turns/cm	0.3937
ampere-turns/inch	amp-turns/meter	39.37
ampere-turns/meter	amp-turns/inch	0.0254
Centigrade	Fahrenheit	(°C × 9/5) + 32
centimeters	feet	3.281×10^{-2}
centimeters	inches	0.3937
centimeters/sec	feet/sec	0.03281
circular mils	sq mils	0.7854
circumference	radians	6.283
circular mils	sq inches	7.854×10^{-7}
coulombs	faradays	1.036×10^{5}
degrees	radians	0.01745
dynes	joules/cm	10^{-7}
dynes	joules/meter (newtons)	10^{-5}
ergs	foot-pounds	7.367×10^{-6}
ergs	joules	10^{-7}
farads	microfarads	10^{6}
faradays	ampere-hours	26.8
faradays	coulombs	9.649×10^{4}
feet	centimeters	30.48
foot-pounds	ergs	1.356×10^{7}
gausses	lines/sq inch	6.452
gausses	webers/sq meter	10^{-4}
gilberts	ampere-turns	0.7958
grams	dynes	980.7
grams	pounds	2.205×10^{-3}
henrys	millihenrys	1,000
horsepower	foot-lbs/sec	550
horsepower	kilowatts	0.7457
horsepower	watts	745.7
inches	centimeters	2.54
inches	millimeters	25.4
joules	ergs	10^{7}
joules	watt-hrs	2.778×10^{-4}
kilolines	maxwells	1,000
kilowatts	foot-lbs/sec	737.6
kilowatts	horsepower	1.341
kilowatts	watts	1,000
kilowatt-hrs	joules	3.6×10^{6}
lines/sq inch	webers/sq meter	$1{,}550 \times 10^{-5}$
maxwells	kilolines	0.001
maxwells	webers	10^{-8}
microfarad	farads	10^{-6}
microns	meters	1×10^{-6}
millihenrys	henrys	0.001
mils	inches	0.001
nepers	decibels	8.686
ohms	megohms	10^{-6}
quadrants	degrees	90
quadrants	radians	1.571
radians	degrees	57.3
square inches	sq cms	6.452
temperature (°F)–32	temperature (°C)	5/9
watts	horsepower	1.341×10^{-3}
watts (Abs.)	joules/sec	1
webers	maxwells	10^{8}
webers/sq meter	webers/sq inch	6.452×10^{-4}
yards	meters	0.9144

APPENDX G

Schematic Symbols

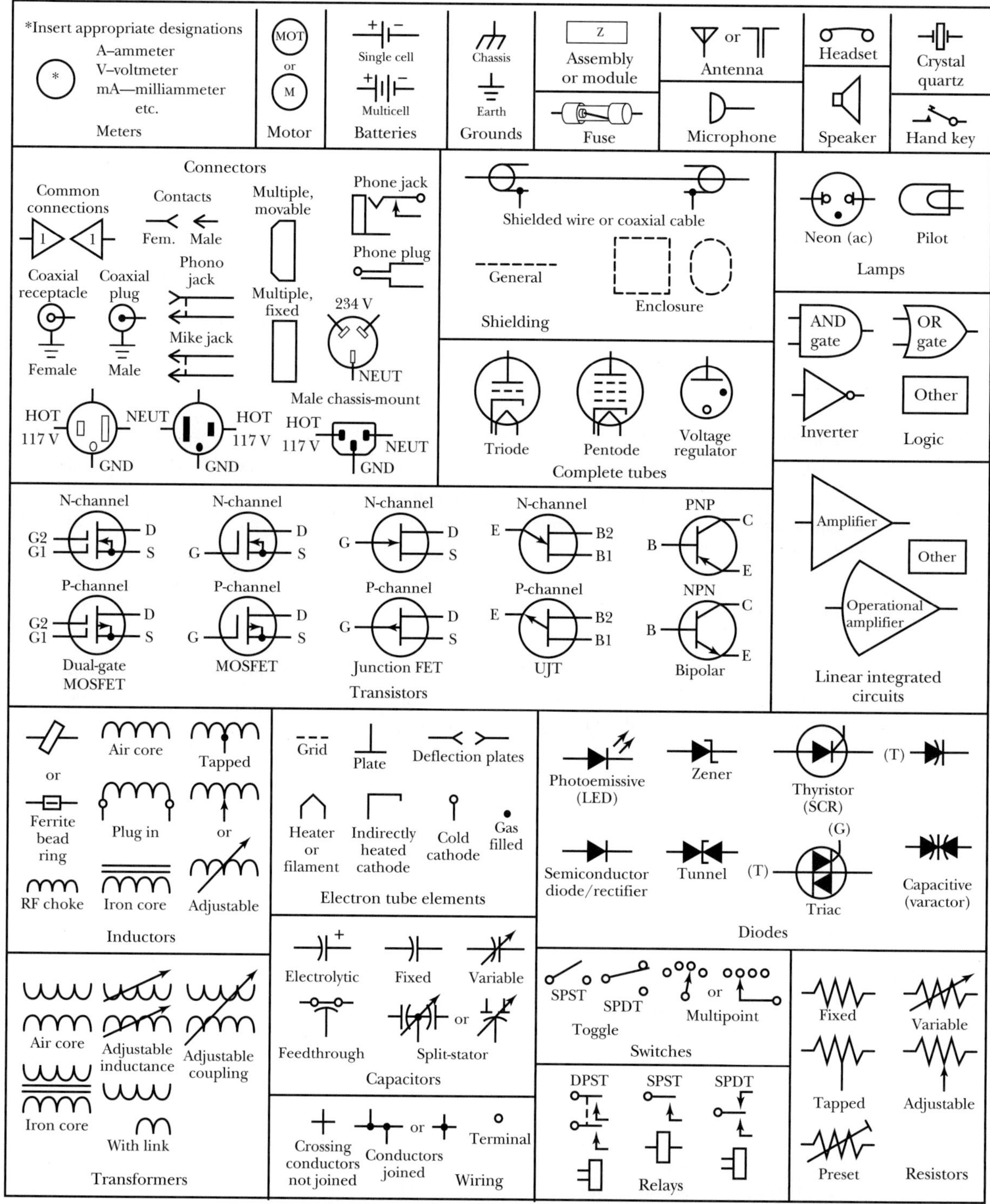

APPENDIX H

Glossary

ac: abbreviation for alternating current. The letters "ac" are also used as a prefix, or modifier to designate voltages, waveforms, and so forth that periodically alternate in polarity.

ac beta (β_{ac}): for a BJT, the amount of change in collector current divided by the corresponding amount of change in base current.

active device: a component or device that either supplies energy to the circuit (such as a battery or power supply) or converts energy from one form to another form (such as a transformer or a transistor).

active limiter: an op amp circuit consisting of nonlinear elements that restricts the voltage levels.

actual closed loop gain: the amplifier circuit's gain that includes the effects of the op-amp's open-loop gain on the amplifier's closed loop gain.

admittance (Y): the ability of a circuit with both resistance and reactance to pass ac current; the reciprocal of impedance.

alpha (α): the ratio of collector current to emitter current.

alternation: one-half of a cycle. Alternations are often identified as the positive alternation or the negative alternation when dealing with ac quantities.

ampere (A): the basic unit of current flow. An ampere of current flow represents electron movement at a rate of one coulomb per second; that amount of current flowing through one ohm of resistance with one volt applied.

ampere-hour rating: a method of rating cells or batteries based on the amount of current they supply over a specified period with specified conditions.

amplifier: a circuit that can increase the peak-to-peak voltage, current, or power of a signal.

analog operation: circuit operation where the outputs vary continuously over a range as a direct result of the changing input levels; also called linear operation.

AND gate: a digital logic gate where both inputs must be high for the output to be high.

anode: the terminal on a diode that is composed of a P-type semiconductor and must be made positive with respect to the cathode in order to cause conduction through a diode. Its symbol is a triangle or arrow.

apparent power (S): the product of *V* and *I* in an ac circuit, without regard to phase difference between *V* and *I*. The unit of measure is volt-amperes. In the power triangle it is the vector resulting from true and reactive power.

astable multivibrator: a type of multivibrator that has no stable state. It is a form of *RC*-controlled oscillator that produces a rectangular waveform. It is also known as a free-running multivibrator.

AT: the abbreviation for ampere-turns or magnetomotive force.

atom: the basic building block of matter composed of different types of particles. Major atom particles are the electron, proton, and neutron.

attenuator: a controllable voltage divider network in oscilloscopes that helps adjust the amplitude of input signals to a useable size.

audio amplifier: an amplifier circuit designed to amplify signals between 20 Hz to 20 kHz (the audio frequency range).

autotransformer: a special single-winding transformer where part of the winding is in both the primary and secondary circuit. These transformers can be made to step up or step down, depending on where on the winding the primary source and the secondary load are connected.

average value (for one-half cycle): in ac sinusoidal values, the average height of the curve above the zero axis, expressed as 0.637 times maximum value. (NOTE: The average height over an *entire* cycle is zero; however, the average height of one alternation is as defined above.)

bandpass: the band of frequencies that passes through a filter with minimal attenuation or degradation; often designated by specific frequencies, where the lower frequency is the frequency at the low-frequency half-power point, and the upper frequency is the frequency at the upper half-power point on the resonance curve; also called pass-band.

bandpass filter: a filter that passes a selected band of frequencies with little loss but greatly attenuates frequencies either above or below the selected band.

band reject filter: also known as bandstop filter and notch filter.

bandstop filter: a filter that stops (greatly attenuates) a selected band of frequencies while allowing all other frequencies to pass with little attenuation; also called band reject filter.

bandwidth: the frequency range between the low cutoff frequency and the high cutoff frequency. The frequency limits where a specified response level is shown by a frequency-sensitive circuit, such as a resonant circuit. The cutoff level is defined at 70.7% of peak response, sometimes called the half-power points on the response curve.

barrier po\tential: a small voltage that is developed across the depletion layer of a P-N junction. For silicon junctions, the barrier potential is about 0.7 V, and for germanium junctions it is about 0.3 V.

base: the center of the NPN or PNP sandwich in a BJT that separates the emitter and the collector transistor regions. The emitter-base junction is forward-biased and shows low impedance to current flow. The collector-base junction is reverse-biased and displays high resistance to current flow.

base current (I_B): the level of current that flows into or out of the base terminal on a BJT. This current is normally due to the forward biasing of the emitter-base P-N junction.

base-emitter voltage (V_{BE}): the level of voltage applied between the base and emitter terminals on a BJT. Forward base current flow is enabled when the polarity of V_{BE} is such that it forward biases the base-emitter junction.

battery: a dc voltage source containing a combination of cells, connected to produce higher voltage or current than a single cell produces alone.

beta (β): the ratio of collector current to emitter current.

beyond pinchoff: the linear region of operation for the FET. Condition of maximum drain current for the applied gate-to-source voltage.

bias circuit: the circuit that establishes the dc circuit conditions for BJT and FET amplifiers.

bidirectional: the ability of a device to conduct current in two directions. Diacs and triacs are bidirectional, whereas SCRs and rectifier diodes are unidirectional (conduction current in only one direction).

bilateral resistance: any resistance having equal resistance in either direction; that is, *R* is the same for current passing either way through the component.

bipolar junction transistor (BJT): a three-terminal, bipolar semiconductor containing both majority and minority current carriers. The device has alternate layers of P- and N-type semiconductor materials (which form two PN junction diodes). This transistor requires the flow of both types of charge carriers (electrons and holes) to complete the path for current flow through it. Examples are the NPN and the PNP transistor.

bleeder: the resistive component assuring that the minimum current drawn from a power supply system is sufficient to achieve reasonable voltage regulation over the range of output currents demanded from the supply; serves to discharge the filter capacitors when the power supply is turned off, which adds safety for people working on electronic equipment fed by the power supply. This resistive component is also called bleeder resistor.

bleeder current: the fixed current through a resistive system that helps keep voltage constant under varying load conditions.

bleeder resistor: the resistor connected in parallel with power supply to draw a bleeder current.

block diagrams: graphic illustrations of systems or subsystems by means of blocks (or boxes) that contain information about the function of each block. These blocks are connected by lines that show direction of flow (signal, fluid, electrical energy, etc.) and how the blocks interact. Block diagrams can illustrate any kind of system: electrical, electronic, hydraulic, mechanical, and so forth. "Flow" generally moves from left to right through the diagram but is not a requirement.

branch: consists of a single component, or two or more components connected in series.

bridge rectifier: a full-wave rectifier circuit that requires only four diodes arranged in series-parallel such that two of them conduct on one ac input half-cycle and the other two conduct on the opposite ac input half-cycle.

bypass capacitors: capacitors used to maintain amplifier stability preventing unwanted high frequency oscillations at the amplifier output.

capacitance: the ability to store electrical energy in the form of an electrostatic field.

capacitive reactance (X_C): the opposition a capacitor gives to ac or pulsating dc current. The symbol used to represent capacitive reactance is X_C.

capacitor: a device consisting of two or more conductors, separated by nonconductor(s). When charged, a capacitor stores electrical energy in the form of an electrostatic field and blocks dc current.

cathode: the part of a diode that is composed of an N-type semiconductor and must be made negative with respect to the anode in order to cause conduction through the diode. Its symbol is a bar with a perpendicular connecting lead. (This bar symbol is also perpendicular to the anode arrow symbol.)

cell: (relates to voltage sources) a single (stand-alone) element or unit within a combination of similar units that convert chemical energy into electrical energy. The chemical action within a cell produces dc voltage at its output terminals or contacts.

cemf (counter-emf or back-emf): the voltage induced in an inductance that opposes a current change and is present due to current changes causing changing flux linkages with the conductor(s) making up the inductor.

charge (electrical): for our purposes, charge can be thought of as the electrical energy present where there is an accumulation of excess electrons, or a deficiency of electrons. For example, a capacitor stores electrical energy in the form of "charge" when one of its plates has more electrons than the other.

charging a capacitor: moving electrons to one plate of a capacitor and moving electrons off the other plate; resultant charge movement is called charging current.

chip: a semiconductor substrate where active devices (transistors) and passive devices (resistors and capacitors) form a complete circuit.

"chip" capacitor: a surface-mount device (SMD), sometimes called a "surface-mount" capacitor. Characterized by its small size, almost zero lead-length connection method, and its use in printed circuit board surface-mount applications.

"chip" resistor: a surface-mount device (SMD), sometimes called "surface-mount" resistor designed to be used in surface-mount technology applications. Typically, very small and rectangular in shape, with metallic end electrodes as the points of connection to the component.

circuit: a combination of elements or components that are connected to provide paths for current flow to perform some useful function.

clamp circuit: a circuit that offsets the zero baseline of an input waveform to a different positive or negative value.

Clapp oscillator: an *LC* resonant-circuit oscillator that requires a single inductor and three capacitors. One capacitor is in the inductor leg of the tank circuit, and two capacitors are connected in series and center-tapped to ground in the other leg of the tank circuit.

class A, B, and C: in reference to amplifier operation, the classifications that denote the operating range of the circuit. Class A operates halfway between cutoff and saturation. Class B operates at cutoff. Class C operates at below cutoff bias.

clipper circuit: a type of circuit that clips off the portion of a waveform that extends beyond a prescribed voltage level. Clipper circuits can be designed to clip positive or negative levels, or both.

coefficient of coupling (k): the amount or degree of coupling between two circuits. In inductively coupled circuits, it is expressed as the fractional amount of the total flux in one circuit that links the other circuit. If all the flux links, the coefficient of coupling is 1. The symbol for coefficient of coupling is k. When k is multiplied by 100, the percentage of total flux linking the circuits is expressed.

collector: the transistor electrode that receives current carriers that have transversed from the emitter through the base region; the electrode that removes carriers from the base-collector junction.

collector-base voltage (V_{CB}): the amount of voltage present between the collector and base of a BJT. For proper operation of the BJT, this voltage must have a polarity that reverse biases the P-N junction between the base and collector regions of the transistor.

collector characteristic curves: a family of curves that shows how BJT collector current increases with an increasing amount of emitter-collector voltage at several different base-current levels. These curves can be used for constructing load lines for BJT amplifier circuits.

collector current (I_C): the value of current measured at the collector terminal of a BJT.

collector-emitter voltage (V_{CE}): the amount of voltage present between the emitter and collector terminals of a BJT. For proper operation of the BJT, this voltage must have a polarity such that the flow of majority charge carriers is from emitter to collector.

Colpitts oscillator: an *LC* resonant-circuit oscillator that uses a single inductor in parallel with two series-connected capacitors that are grounded at their common point.

common-base: relating to transistor amplifiers, the circuit configuration where the base is common to both input and output circuits.

common-base amplifier (CB): a BJT amplifier configuration where the input signal is applied to the emitter and the output is taken from the collector.

common-collector: the transistor amplifier circuit configuration where the collector is common for input and output circuits.

common-collector amplifier (CC): a BJT amplifier configuration where the input signal is applied to the base and the output is taken from the emitter. Also known as an emitter-follower circuit.

common-drain amplifier (CD): an FET amplifier configuration where the input signal is applied to the gate and the output is taken from the source. Also known as a source-follower circuit.

common-emitter: in reference to transistor amplifiers, the circuit configuration where the emitter is common to both input and output circuits.

common-emitter amplifier (CE): a BJT amplifier configuration where the input signal is applied to the base and the output is taken from the collector.

common-gate amplifier (CG): an FET amplifier configuration where the input signal is applied to the source terminal and the output is taken from the drain.

common-mode gain: comparison (ratio) of the output voltage of a differential amplifier to the common-mode input voltage. Ideally, the output should be zero when the same input is fed to both differential amplifier inputs.

common-mode input: an input voltage to a differential amplifier that is common to both its inputs.

common-mode rejection ratio (CMRR): a measure expressing the amount of rejection a differential amplifier shows to common-mode inputs; the ratio between the differential-mode gain and the common-mode gain equals the CMRR.

common-source amplifier (CS): an FET amplifier configuration where the input signal is applied to the gate and the output is taken from the drain.

complex numbers: term that expresses a combination of a *real* and an *imaginary* term which must be added vectorially or with phasors. The real term is either a positive or negative number displayed on the horizontal axis of the rectangular coordinate system. The imaginary term, *j operator,* expresses a value on the vertical or *j*-axis. The value of the real term is expressed first. For example, $5 + j10$ means plus 5 units (to the right) on the horizontal axis and up 10 units on the vertical axis.

compound: a form of matter that can be chemically divided into simpler substances and that has two or more types of atoms.

conductance: the ease with which current flows through a component or circuit; the opposite of resistance.

conduction band: is the upper band of allowed states where electrons when given enough energy are free to move within the crystal.

conductor: a material that has many free electrons due to its atoms' outer rings having less than four electrons, which is less than the eight needed for chemical and/or electrical stability.

constant current source: a current source whose output current is constant even with varying load resistances. Its output voltage varies to enable constant output current, as appropriate.

constant voltage source: a voltage source where output voltage (voltage applied to circuitry connected to source output terminals) remains virtually constant even under varying current demand (load) situations.

coordinate system: in ac circuit analysis, a system to indicate phase angles within four, 90° quadrants. The vertical component is the "Y" axis. The horizontal component is the "X" axis.

coordinates: the vertical and horizontal axes that describe the position and magnitude of a vector.

copper losses: the I^2R losses in the copper wire of the transformer windings. Since current passes through the wire and the wire has resistance to current flow, an I^2R loss is created in the form of heat, rather than this energy being a useful part of energy transfer from primary source to secondary load.

core: soft iron piece to help strengthen magnetic field; helps produce a uniform air gap between the permanent magnet's pole pieces and itself.

core losses: energy dissipated in the form of heat by the transformer core that subtracts from the energy transferred from primary to secondary; basically comprised of eddy-current losses (due to spurious induced currents in the core material creating an I^2R loss) and hysteresis losses (due to energy used in constantly changing the polarity of the magnetic domains in the core material).

cosine: a trigonometric function of an angle of a right triangle; the ratio of the adjacent side to the hypotenuse.

cosine function: a trigonometric function used in ac circuit analysis; relationship of a specific angle of a right triangle to the side adjacent to that angle and the hypotenuse where $\cos\theta$ = adjacent/hypotenuse.

coulomb: the basic unit of charge; the amount of electrical charge represented by $6.25 = 10^{18}$ electrons.

covalent bonding: the pairing of adjacent atom electrons to create a more stable atomic structure.

CRT (cathode-ray tube): the oscilloscope component where electrons strike a luminescent screen material causing light to be emitted. Waveforms are displayed on an oscilloscope's CRT.

crystal oscillator: an oscillator circuit that uses a special element (usually quartz-crystal material and holder) for the primary frequency determining element in the circuit.

current (I): the progressive movement of electrons through a conductor. Current is measured in amperes.

current divider circuit: any parallel circuit divides total current through its branches in inverse relationship to its branch R values.

current leading voltage: sometimes used to describe circuits where the circuit current leads the circuit voltage, or the circuit voltage lags the circuit current.

current ratio: the relationship of the current values of secondary and primary, or vice versa. In an ideal transformer, the current ratio is the *inverse* of the voltage and turns ratios; $I_P/I_S = N_S/N_P = V_S/V_P$.

cutoff: a region of operation for FETs and BJTs where no current is flowing.

cycle: a complete series of values of a periodic quantity that recur over time.

dc beta (β_{dc}): for a BJT, the ratio of collector current (I_C) to the corresponding amount of base current (I_B).

delta (Δ): When used in mathematical equations, a term meaning "a small change in." When used in conjunction with an electrical network configuration, it describes the general triangular appearance of the schematic diagram layout, which depicts the electrical connections between components. Also, sometimes, "delta networks" are laid out in a fashion that they are called pi (π) networks.

depletion mode: a mode of operation for FET devices where the gate bias tends to reduce the availability of charge carriers in the channel.

depletion region: the region near the junction of a P-N junction where the current carriers are depleted due to the combining effect of holes and electrons.

diac: a type of bidirectional thyristor that is switched on by exceeding the peak forward blocking voltage or peak reverse breakdown voltage. Diacs do not require gate terminals.

dielectric: nonconductive material separating capacitor plates.

dielectric constant: the relative permittivity of a given material compared to the permittivity of vacuum or air; the relative ability of a given material to store electrostatic energy (per given volume) to that of vacuum or air.

dielectric strength: the highest voltage a given dielectric material tolerates before electrically breaking down or rupturing; rated in volts per thousandths of an inch, V per mil.

differential amplifier: an amplifier circuit configuration having two inputs and one output; produces an output that is related to the difference (or differential) between the two inputs.

differential-mode gain: the gain of a differential amplifier when operated in the differential mode; the ratio of its output voltage to the differential mode (difference between the two input voltages) input voltage.

differentiator circuit: an op-amp configuration that produces an output voltage proportional to the rate of change of the input signal.

diffusion current: current flowing through the pn junction without an externally applied source of energy.

digital operation: circuit operation where the output varies in discrete steps or in an on-off, high-low, 1-0 category of operation.

discharging a capacitor: balancing (or neutralizing) the charges on capacitor plates by allowing excess electrons on the negative charged plate to move through a circuit to the positive charged plate. When opposite capacitor plates have the same charge, the voltage between them is zero and the capacitor is fully discharged.

doping: introducing specific amounts of P- or N-type elements (called impurities) into a semiconductor crystal (lattice) structure to aid current conduction through the material.

doping material: a material that is added to a pure semiconductor material that enables the existence of charge carriers.

double-ended input: a type of op-amp configuration that uses two inputs.

drain: the portion of an FET that receives charge carriers from the channel.

dropping resistor: a resistor connected in series with a given load that drops the circuit applied voltage to the level required by the load when passing the rated load current through it. (Recall a *voltage drop* is the difference in voltage between two points caused by a loss of pressure (or loss of emf) as current flows through a component offering opposition to current flow.)

effective value (RMS): the value of ac voltage or current that produces the same heating effect in a resistor as that produced by a dc voltage or current of equal value. In ac, this value is 0.707 times maximum value.

efficiency: percentage, or ratio, of useful output (such as power) delivered compared to amount of input required to deliver that output. For example, in electrical circuits, the amount of power input to a device, component stage, or system compared to the amount of power delivered to the load. Generic formula is Efficiency (%) = $(P_{out}/P_{in}) \times 100$.

electrical charge: an excess or deficiency of electrons compared to the normal neutral condition of having an equal number of electrons and protons in each atom of the material.

electrode: generally, one of the electrical terminals that is a conducting path for electrons in or out of a cell, battery, or other electrical or electronic device.

electromagnetic frequency spectrum: the listing of frequencies and/or wavelengths from below (sub) audio frequencies to beyond gamma rays.

electromagnetism: the property of a conductor to produce a magnetic field when current is passing through it.

electron: the negatively charged particle in an atom orbiting the atom's nucleus.

electrostatic (electric) field: the region surrounding electrically charged bodies; area where an electrically charged particle experiences force.

element: a form of matter that cannot be chemically divided into simpler substances and that has only one type of atom.

emitter: the portion of a BJT that emits charge carriers into the device.

emitter current (I_E): the amount of current found at the emitter terminal of a BJT.

energy: the ability to do work, or that which is expended in doing work. The basic unit of energy is the erg, where 980 ergs = 1 gram cm of work. A common unit in electricity/electronics is the joule, which equals 10^7 ergs.

enhancement mode: a mode of operation for FET devices where the gate bias tends to enhance the availability of charge carriers in the channel.

ε: the Greek letter epsilon, used to signify the mathematical value of 2.71828. This number, raised to various exponents that relate to time constants, can be used to determine electrical circuit parameter values in circuits where exponential changes occur.

equation: a mathematical statement of equality between two quantities or sets of quantities. Example, $A = B + C$.

equivalent resistance: the total resistance of two or more resistances in parallel.

feedback: with signal amplifying or signal generating circuits, that small portion of the output signal that is coupled back to the input circuit of the stage.

fiber-optic cable: designed to guide light (photons) from an optical source to an optical detector. The cables consist of a core, cladding and a jacket. The cable core and cladding are constructed with very pure glass or plastic.

field effect transistor (FET): a unipolar transistor that depends on a voltage field to control the flow of current through a channel.

filter: in power supplies, the component or combination of components that reduce the ac component of the pulsating dc output of the rectifier prior to the output fed to a load.

filter: a circuit used to pass a range of frequencies while blocking all other frequencies.

forward bias: the external voltage applied to a semiconductor junction that causes forward conduction, for example, positive to P-type material and negative to N-type material.

free electrons: the electrons in the outer ring of conductor materials that are easily moved from their original atom. Sometimes termed valence electrons.

free-running multivibrator: a type of multivibrator that runs continuously. It is a form of an *RC*-controlled oscillator circuit that produces a rectangular waveform. It is technically known as an astable multivibrator.

frequency *(f)*: the number of cycles occurring in a unit of time. Generally, frequency is the number of cycles per second, called hertz (Hz), which means cycles per second. In formulas, frequency is expressed as *(f)*, where $(f) = 1/T$.

full-wave rectifier: a type of rectifier that produces a dc output for both ac half-cycles of input.

gain: the ratio of the signal output to the signal input expressed as a unit less value or in decibels (dB). Gain is a ratio of the change in a signal's current, voltage, or power, respectively known as current gain, voltage gain, and power gain.

gain controls: oscilloscope controls that adjust the size of the image displayed.

gate: (1) the terminal on a thyristor, such as an SCR or triac, that is used for the voltage that turns on the device. (2) The terminal on an FET where the voltage is applied that will control the amount of current flowing through the channel.

gate-controlled switch (GCS): a type of SCR that can be gated off as well as gated on.

ground reference: a common reference point for electrical/electronic circuits; a common line or conducting surface in electrical circuits used as the point where electrical measurements are made; a common line or conductor where many components make connection to one side of a power source. Typically, ground reference is the chassis, or a common printed circuit "bus" (or line) called chassis ground. In power line circuits, may also be "earth ground."

half-wave rectifier: a type of rectifier that produces a dc output for only one of the half-cycles in each cycle of input.

Hartley oscillator: an *LC* resonant-circuit oscillator that uses a single capacitor in parallel with two series-connected inductors that are grounded at their common point.

heat sink: a mounting surface for semiconductor devices that is intended to carry away excessive heat from the semiconductor, by dissipating it into the surrounding air.

henry: the basic unit of inductance, abbreviated "H." One henry of inductance is that amount of inductance having one volt induced cemf when the rate of current change is one ampere per second.

Hertz (Hz): the unit of frequency expressing a frequency of one cycle per second.

high-pass filter: a filter that passes frequencies *above* a specified cutoff frequency with little attenuation. Frequencies below the cutoff are greatly attenuated.

holding current: the minimum amount of forward current that a thyristor can carry and still remain switched on.

hybrid circuits: generally, circuits composed of components created by more than one technique, such as thin-film, discrete transistors, and so on.

hysteresis: the difference in voltage between the amount of turn-on voltage and turn-off voltage for an amplifier that is used as a switch. A Schmitt trigger amplifier, for example, might switch on when the input exceeds +2 V, but not switch off until the input drops below +1.25 V. The amount of hysteresis in this case is 0.75 V.

imaginary numbers: complex numbers; the imaginary part of a complex number that is a real number multiplied by the square root of minus one.

impedance: the total opposition an ac circuit displays to current at a given frequency; results from both resistance and reactance. Symbol representing impedance is *Z* and unit of measure is the ohm.

impedance ratio: the relationship of impedances for the transformer windings; directly related to the *square* of the turns ratio; $Z_P/Z_S = N_P^2/N_S^2$. Reflected impedance is the impedance reflected back to the primary from the secondary. This value depends on the square of the turns ratio and value of the load impedance connected across the secondary.

inductance (self-inductance): the property of a circuit to oppose a change in current flow.

induction: the ability to induce voltage in a conductor; ability to induce magnetic properties from one object to another by magnetically linking the objects.

inductive reactance: the opposition an inductance gives to ac or pulsating current.

in-line: typically means in series with the source.

instantaneous value: the value of a sinusoidal quantity at a specific moment, expressed with lower case letters, such as *e* or *v* for instantaneous voltage and *i* for instantaneous current.

insulator: a material in which the atoms' outermost ring electrons are tightly bound and not free to move from atom to atom, as in a conductor. Typically, these materials have close to the eight outer shell electrons required for chemical/electrical stability.

integrated circuit (IC): a combination of electronic circuit elements. In relation to semiconductor ICs, this term generally indicates that all of these components are fabricated on a single substrate or piece of semiconductor material.

integrator circuit: an op-amp configuration that produces an output voltage proportional to the "area under the curve" for the input waveform.

inverse parallel: a way of interconnecting two diodes or triacs such that they are in parallel with the anode of one connected to the cathode of the other. Also known as a back-to-back connection.

inverter: a digital logic gate where the output is the opposite of the input. If the input is low, the output is high. If the input is high, the output is low.

inverting amplifiers: amplifier configurations (especially in reference to op-amps) that produce an output waveform having the opposite polarity of the input waveform.

ion: atoms that have gained or lost electrons and are no longer electrically balanced, or neutral atoms. Atoms that lose electrons become positive ions. Atoms that gain electrons become negative ions.

isolation transformer: a 1:1 turns ratio transformer providing electrical isolation between circuits connected to its primary and secondary windings.

joule: the unit of energy. One joule of energy is the amount of energy moving one coulomb of charge between two points with a potential difference of one volt of emf; the work performed by a force of one newton acting through a distance of one meter equals one joule of energy. Also, 3.6×10^6 joules = 1 kilowatt hour.

junction field-effect transistor (JFET): FETs that use the voltage field around a reverse-biased P-N junction to control the flow of charge carriers through the channel region.

Kirchhoff's current law: the value of current entering a point must equal the value of current leaving that same point in the circuit.

Kirchhoff's voltage law: the arithmetic sum of voltage drops around a closed loop equals V applied; the algebraic sum of voltages around the entire closed loop, including the source, equals zero.

L, L_T: L is the abbreviation representing inductance or inductor. L_T is the abbreviation for total inductance. L_1, L_2, and so on represent specific individual inductances.

large-signal amplifiers: amplifiers designed to operate effectively with relatively large ac input signals; e.g., power levels greater than 1 watt.

laser diode: a light-emitting diode that produces light of a single frequency (coherent light) when it is forward conducting.

leading in phase: in ac circuits the electrical quantity that first reaches its maximum positive point is specified as leading the electrical quantity that reaches that point later.

light-activated SCR (LASCR): an SCR that can be gated on by directing light energy onto its gate-cathode junction.

light-emitting diode (LED): a junction diode that emits light energy when it is forward biased and conducting.

limiter circuit: circuit of nonlinear elements that restricts the voltage levels (also known as clipper).

linear amplifier: an amplifier that produces an output waveform that has the same general shape as the input waveform.

linear network: a circuit whose electrical behavior does not change with different voltage or current values.

linear region: the BJT operating region where $I_C = \beta \times I_B$.

load: the amount of current or power drain required from the source by a component, device, or circuit.

loaded voltage divider: a network of resistors designed to create various voltage levels of output from one source voltage; where parallel loads, which demand current, are connected.

loading effect: changing the electrical parameters of an existing circuit when a load component, device, or circuit is connected. With voltmeters, the effect of the meter circuit's resistance when in parallel with the portion of the tested circuit; thus, causing a change in the circuit's operation and some change in the electrical quantity values throughout the circuit.

load line: a straight line drawn between the points of cutoff voltage and saturation current on a family of collector characteristic curves.

loop in electrical networks, a complete electrical circuit or current path.

low-pass filter: a filter that permits all frequencies *below* a specified cutoff frequency to pass with little attenuation. Frequencies above the cutoff frequency are attenuated greatly.

magnitude: the comparative size or amount of one quantity with respect to another quantity of the same type.

majority carriers: the electrons or holes in extrinsic semiconductor materials that enable current flow. In N-type materials, the majority carriers are electrons. In P-type materials, the majority carriers are holes.

maximum power transfer theorem: a theorem that states when the resistance (or impedance) of the load is properly matched to the internal resistance of the power source, maximum power transfer can take place between the source and the load.

mesh: sometimes called loops. In electrical networks, a set of branches that form a complete electrical path. If any branch is omitted, the remainder of the circuit does not form a complete path.

metal-oxide semiconductor FET (MOSFET): a unipolar silicon transistor that uses a very thin layer of metal-oxide insulation between the gate and channel.

microprocessor: typically, a VLSI chip that acts as the central processing unit (CPU) for a computer system.

minority carriers: the electrons in P-type material and the holes in N-type material; called minority since there are fewer electrons in P-type material than holes and fewer holes in N-type material than electrons.

mode: related to semiconductor ICs; the type of operation for which the IC is designed. The linear mode is commonly used for amplifiers. The digital mode is often used for gates.

molecule: the smallest particle of a compound that resembles the compound substance itself.

monolithic: one piece as in a monolithic IC formed on one semiconductor substrate (or one stone).

monostable multivibrator: a type of multivibrator that is stable in only one state. An input trigger pulse sets the circuit to its unstable state where it remains for a period of time determined by an *RC* circuit. This circuit is also known as a one-shot multivibrator.

MOS: abbreviation for metal-oxide semiconductor; relates to the fabrication method.

μ_r: the symbol for relative permeability.

multimeter: an instrument designed to measure several types of electrical quantities, for example, current, voltage, and resistance; commonly found in the forms of the analog VOM and the digital multimeter (DMM).

multiple-source circuit: a circuit containing more than one voltage or current source.

multiplier resistor(s) (voltmeter): the current-limiting series resistance(s) used with analog-type current meter movements, enabling them to be used as voltmeters that measure voltages much greater than the basic movement's voltage drop at full-scale current. The value of the multiplier resistance is such that its resistance (plus the meter's resistance) limits current to the full-scale value at the desired voltage-range level.

multivibrator: a circuit that produces a waveform that represents a condition of being fully switched on or fully switched off. The three basic types of multivibrators are the monostable (one-shot), the bistable, and the astable (free-running) multivibrators.

NAND gate: an AND gate with inverted output. That is, if both inputs are high, the output is low. For all other input conditions, the output is high.

network: a combination of electrically connected components.

neutron: a particle in the nucleus of the atom that displays no charge. Its mass is approximately equal to the proton.

node: a current junction or branching point in a circuit.

noninverting amplifiers: amplifiers that produce an output that has the same polarity or phase as their input.

nonlinear amplifier: an amplifier that produces an output waveform that is different from its input waveform.

NOR gate: an OR gate with inverted output. That is, if either or both inputs are high, the output is low. Only when both inputs are low is output high.

Norton's theorem: a theorem regarding circuit networks stating that the network can be replaced (and/or analyzed) by using an equivalent circuit consisting of a single "constant-current" source (called I_N) and a single shunt resistance (called R_N).

notch filter: also known as bandstop filter and band reject filter.

NPN transistor: a transistor made of P material sandwiched between two outside N material areas. Then, going from end-to-end, the transistor is an NPN type. The symbol for the NPN transistor shows the emitter arrow going away from the base.

N-type material: a semiconductor material that has an excess of electrons as compared to holes.

ohm (Ω): the basic unit of resistance; the amount of electrical resistance limiting the current to one ampere with 1 V applied.

Ohm's law: a mathematical statement describing the relationships among current, voltage, and resistance in electrical circuits. Common equations are $V = I \times R$, $I = V/R$, and $R = V/I$.

ohmic region: an operational region for IFETs, and MOSFETs characterized by the drain-to-source resistance of the component.

one-shot multivibrator: a popular name for a monostable multivibrator. (*See* monostable multivibrator.)

op-amp: abbreviation for operational amplifier; a versatile amplifier circuit with very high gain, high input impedance, and low output impedance.

open circuit: any break in the current path that is undesired, such as a broken wire or component, or designed, such as open switch contacts in a lighting circuit.

operational amplifier (op-amp): a versatile differential amplifier circuit with very high gain, high input impedance, and low output impedance.

optocoupler: also known as n opto-isolator.

optoelectronic device: a group of electronic devices that generate or respond to light.

opto-isolator: a component that consists of a photoemitter (LED) and a photodetector providing electrical isolation between the input and output circuits. The device is also known as a n opto-coupler.

OR gate: a digital logic gate where if either input is high, or if both inputs are high, the output is high.

order: the number of reactive components located within a circuit. The order of a filter establishes the slope of the filter voltage gain for signal frequencies outside the passband frequency range.

oscillator: a circuit that generates and sustains an output signal without an input signal supplied by another circuit or source.

oscilloscope: a device that visually displays signal waveforms so time and amplitude parameters are easily determined.

parallel branch: a single current path within a circuit having two or more current paths where each of the paths are connected to the same voltage points; a single current path within a parallel circuit.

parallel circuit: a circuit with two or more paths for current flow where all components are connected between the same voltage points.

passband: the frequency range where the input signal will appear at (be passed to) the filter output.

passive device: a component or device that only absorbs energy (such as a resistor) or absorbs energy and later returns that energy to the circuit (such as a capacitor or an inductor).

peak detector: a circuit whose output voltage is equal to the input signal's highest voltages.

peak-inverse-voltage (PIV): the maximum voltage across a diode in the reverse (nonconducting) direction. The maximum voltage that a diode tolerates in this direction is the peak-inverse-voltage rating.

peak-to-peak value (p-p): the difference between the positive peak value and the negative peak value of a periodic waveform; computed for the sine wave as 2.828 times effective value.

peak value (V_{pk} or maximum value) (V_{max}): the maximum positive or negative value that a sinusoidal quantity reaches, sometimes referred to as peak value. It is calculated as 1.414 times effective value.

pentavalent atoms: an atom normally having five electrons in the valence shell. Such materials are often used as a donor impurity in the manufacture of N-type semiconductor materials. Examples are phosphorus, arsenic, and antimony.

period (T): the time required for one complete cycle. In formulas, it is expressed as *T,* where $T = 1/f$.

permeability (mu or μ): the ability of a material to pass, conduct, or concentrate magnetic flux.

phase angle (θ): the difference in time or angular degrees between ac electrical quantities (or periodic functions) when the two quantities reach a certain point in their periodic function. This difference is expressed by time difference (the fraction of a period involved), or by angular difference (expressed as degrees or radians). (NOTE: A radian = 57.3°. $2\pi \times$ a radian = 360°.) An example is the difference in phase between circuit voltage and circuit current in ac circuits that contain reactances, such as inductive reactance.

phase-shift oscillator: an *RC* sine-wave oscillator that uses a sequence of three *RC* circuits to achieve a total phase shift of 180° at the frequency of oscillation.

phasor: a quantity that expresses position relative to time.

photodetector: also known as a photo-reactive device

photodiode cell: a device where resistance between terminals is changed by light striking its light-sensitive element; a light-controlled variable resistance. Typically, the more light striking it, the lower its terminal resistance is.

photo-reactive device: a semiconductor device that produces a voltage or current when optical (photons) energy are applied to the device. This device is also known as a photodetector.

piezoelectric effect: an effect enabled by certain kinds of materials whereby applying a voltage to the material causes it to change size and shape, and bending the material causes it to generate a voltage.

pinch-off voltage (V_P): the source-drain voltage level (of an FET) where a further increase in source-drain voltage causes virtually no increase in drain current.

P-N junction: the location in a semiconductor where the P- and N-type materials join; where the transition from one type material to the other type occurs.

P-N junction diode: a semiconductor diode that consists of a single P-N junction.

PNP transistor: a transistor where the middle section is N-type material and the two outside sections are composed of P-type material, thus called a PNP transistor. The symbol for the PNP transistor shows the emitter arrow going toward the base.

polar form notation: expressing circuit parameters with respect to polar coordinates; expressing a point with respect to distance and angle (direction) from a fixed reference point on the polar axis. For example, $25 \angle 45°$.

polarity: in an electrical circuit, a means of designating differences in the electrical charge condition of two points, or a means of designating direction of current flow. In a magnetic system, a means of designating the magnetic differences between two points or locations.

positive clamper: a circuit whose output voltage is totally positive voltages with a peak-to-peak amplitude equal to the input signal's peak-to-peak amplitude. The average voltage level of the circuit output is equal to half of the input signal's peak-to-peak amplitude.

positive feedback: a portion of the output of an amplifier system that is fed back to the input with the same phase or polarity as the original input signal.

positive limiter: a circuit that restricts the positive voltage levels of the signal.

potential: in electrical circuits, the potential, or ability, to move electrons. In electromotive force, the difference in charge levels at two points. This potential difference is measured in volts.

power: the rate of doing work. In mechanical terms, a horsepower equals work done at a rate of 550 foot pounds per second. In terms of electrical power, it is electrical work at a rate of one joule per second.

power factor (p.f.): the ratio of true power to apparent power. It is equal to the cosine of the phase angle (θ).

powers of 10: working with small or large numbers can be simplified by converting the unit to a more manageable number times a power of ten. An example is 0.000006 ampere expressed as 6×10^{-6} amperes. Some of the frequently used powers of ten are 10^{-12} for pico (p) units, 10^{-6} for micro (μ) units, 10^{-3} for milli units, 10^3 for kilo units, and 10^6 for mega units.

power supply: a circuit that converts ac power into dc power required for the operation of most kinds of electronic systems.

precision full-wave rectifier positive output: a circuit in which each half cycle of the input signal is converted to a positive half cycle voltage without a reduction of the signal amplitude.

precision half-wave rectifier negative output: a circuit where only the negative half cycle of the input signal appears at the output without a reduction of the signal amplitude while the input signal's positive half cycle is blocked.

precision half-wave rectifier positive output: a circuit where only the positive half cycle of the input signal appears at the output without a reduction of the signal amplitude while the input signal's negative half cycle is blocked.

product-over-the-sum method: a method of finding total resistance in parallel circuits by using two resistances at a time in the formula $R_T = R_1 \times R_2/(R_1 + R_2)$.

proton: the positively charged particle in an atom located in the nucleus of the atom. Its weight is approximately 1,836 times that of the electron, and approximately the same weight as the neutron.

P-type material: a semiconductor material that has an excess number of holes compared to electrons.

pulsating dc: a varying voltage that is one polarity, as opposed to reversing polarity as ac does.

push-pull amplifier: an amplifier circuit that uses two Class B amplifiers, one to amplify one alternation of the ac input waveform and a second to amplify the opposite alternation. The circuit has the advantage of Class-B efficiency and the low distortion of a Class-A amplifier.

Pythagorean theorem: a theorem expressing the mathematical relationship of the sides of a right triangle, where the square of the hypotenuse is equal to the sum of the squares of the other two sides. That is, $c^2 = a^2 + b^2$. Equation used in electronics is $c = \sqrt{a^2 + b^2}$.

Q (figure of merit): a number indicating the ratio of the amount of stored energy in the inductor magnetic field to the amount of energy dissipated. Found by using the ratio of the inductance's inductive reactance (X_L) to the resistance of the inductor *(R)*. That is, $Q = X_L/R$.

Q factor: the relative quality or figure of merit of a given component or circuit; often related to the ratio of X_L/R of the inductor in the resonant circuit. Other circuit losses can also enter into the effective *Q* of the circuit, such as the loading that a load couples to the resonant circuit causes and the effect of *R* introduced in the circuit. The less the losses are, the higher the *Q* is. The sharpness of the resonant circuit response curve is directly related to the *Q* of the circuit.

Q point: the operating point of an amplifier when there is no input signal.

quadrants: segments of the coordinate system representing 90° angles, starting at a 0° reference plane.

reactance: the opposition a capacitor or inductor gives to alternating current or pulsating direct current.

reactive power (Q): the vertical component of the ac power triangle that equals $VI \sin \theta$; expressed as VAR (voltampere-reactive).

real numbers: in math, any rational or irrational number.

rectangular form notation: expressing a set of parameters with the complex number system. For example, $R \pm jX$ indicates a combination of resistance and reactance. Thus, $5 + j10$ means 5 ohms of resistance in series with 10 ohms of inductive reactance.

rectangular waveforms: a "square" waveform that has only two voltage levels, often one level that is close to the main supply voltage and one that is close to zero volts.

rectifier circuit: a circuit that converts an ac waveform to pulsating dc.

rectifier diode: a junction diode that is designed for the higher current and higher reverse-voltage conditions encountered in power supply rectifier circuits.

rectify (rectification): the process of changing ac to pulsating dc.

regulator: a section of some power supplies that keeps the dc output voltage at a desired level in spite of changes in the input voltage level and/or changes in the output loading.

relative permeability (μ_r): comparing a material's permeability with air or a vacuum; relative permeability is not a constant, since it varies for a given material depending on the degree of magnetic field intensity present; relative permeability

$$(\mu_r) = \frac{\text{flux density with given core}}{\text{flux density with vacuum core.}}$$

resistance (R): in an electrical circuit, the opposition to electron movement or current flow. Its value is affected by many factors, including the dimensions and material of the conductors, the physical makeup and types of the components in the circuit, and temperature.

resistivity (R): a photodetector constant relating the output current to the received optical power.

resistor: an electrical component that provides resistance in ohms and that controls or limits current or distributes or divides voltage.

resistor color code: a system to display the ohmic value and tolerance of a resistor by means of colored stripes or bands around the body of the resistor.

reverse bias: the external voltage applied to a semiconductor junction that causes the semiconductor not to conduct; for example, negative to P-type material and positive to N-type material.

reverse breakdown voltage: a rating for semiconductors that specifies the maximum amount of reverse voltage that a P-N junction can withstand without breaking down and conducting in the reverse direction.

reverse Polish notation (RPN): an operational system, based on a technique derived from the work of a Polish scientist, used in some calculators to simplify and minimize the number of input steps required to solve certain complex mathematical problems. Since the notation method used is somewhat reversed from the scientist's original technique, the system, or mode, is sometimes called reverse Polish notation (RPN).

ripple: the ac component in a pulsating dc signal, such as the output of a power supply.

saturation: the maximum current flow in a BJT circuit.

scalar: a quantity expressing only quantity, such as a "real number."

scale: the face of an analog-type meter containing calibration marks that interpret the pointer's position in appropriate units of measure.

schematic diagrams: graphic illustrations of circuits or subcircuits and/or systems or subsystems. Schematics are more detailed than block diagrams and give specific information regarding how components are connected. Schematics show the types, values, and ratings of component parts, and in many cases show the electrical values of voltage, current, and so forth at various points throughout the circuit.

Schmitt trigger: a nonlinear amplifier circuit that converts an ac input waveform into a rectangular waveform. The output waveform goes to its high level when the input exceeds a certain voltage level, and the output goes to its low level when the input falls below a certain voltage level.

SCR: abbreviation for silicon-controlled rectifier; often used to control the time during an ac cycle the load receives power from the power source. A four-layer, P-N-P-N device where conduction is initiated by a voltage to the gate. Conduction doesn't cease until anode voltage is reversed, removed, or greatly reduced although the initiating signal has been removed.

selectivity: referring to resonant circuits, selectivity indicates the sharpness of the response curve. A narrow and more pointed response curve results in a higher degree of frequency selectiveness or selectivity.

self-bias: a method of biasing an amplifier such that the bias voltage is generated by the main flow of current through the device, itself. The method is usually used where the biasing voltage has a polarity opposite the main supply voltage (as with depletion-mode FETs).

semiconductor: a material with four valence (outermost ring) electrons. It is neither a good conductor, nor a good insulator.

semiconductor diode: a semiconductor P-N junction that passes current easily in one direction while not passing current easily in the other direction.

sensitivity rating: a means of rating analog-type meters, or meter movements to indicate how much current is required to cause full-scale deflection; how much voltage is dropped by the meter movement at full-scale current value. Also, voltmeters are rated by the ohms-per-volt required to limit current to full-scale value. (This is inversely related to the current sensitivity rating. That is, the reciprocal of the full-scale current value yields the ohms-per-volt rating.)

series circuit: a circuit in which components are connected in tandem or in a string providing only one path for current flow.

series-parallel circuit: a circuit comprised of a combination of both series-connected and parallel-connected components and/or subcircuits.

short circuit: generally, an undesired very low resistance path across two points in a circuit. It may be across one component, several components, or across the entire circuit. There can be designed shorts in circuits, such as purposeful "jumpers" or closed switch contacts.

siemens: the unit used to measure or quantify conductance; the reciprocal of the ohm ($G = 1/R$).

silicon-controlled rectifier (SCR): four-layer (P-N-P-N) thyristor device that is gated on by forward-biasing the gate-cathode junction and is turned off only by eliminating or reversing the polarity of the cathode-anode voltage.

silicon-controlled switch (SCS): an SCR that has two gate terminals that can be used for switching on the device.

sine: a trigonometric function of an angle of a right triangle; the ratio of the opposite side to the hypotenuse.

sine function: a trigonometric function used in ac circuit analysis; expresses the relationship of a specified angle of a right triangle to the side opposite that angle and the hypotenuse where sin θ = opposite/hypotenuse.

sine wave: a wave or waveform expressed in terms of the sine function relative to time.

single-ended input: a type of op-amp configuration that uses a single input.

sinusoidal quantity: a quantity that varies as a function of the sine or cosine of an angle.

slew rate: an op-amp parameter indicating the output voltage change over time.

small-signal amplifiers: amplifiers designed to operate effectively with small ac input signals; that is, power levels less than 1 watt, and peak-to-peak ac current values less than 0.1 times the amplifier's input bias current.

source: in electrical circuits, the source is the device supplying the circuit applied voltage, or the electromotive force causing current to flow through the circuit. Sources range from simple flashlight cells to complex electronic circuits.

source: the portion of an FET that emits charge carriers into the channel.

source/gate/drai:n the electrodes in a field effect transistor device; gate controls current flow between the source and drain.

SSI, MSI, LSI, and VLSI: SSI is small-scale integration (less than 10 gates on a single chip). MSI is medium-scale integration (less than 100 gates on a single chip). LSI is large-scale integration (more than 100 gates on a single chip). VLSI is very-large-scale integration (over 1000 gates on a single chip).

steady-state condition: the condition where circuit values and conditions are stable or constant; opposite of transient conditions, such as switch-on or switch-off conditions, when values are changing from one state to another.

stopband: the frequency range where the input signal will be blocked (stopped to) from appearing at the filter output.

subcircuit: a part or portion of a larger circuit.

summing amplifier: an op-amp configuration that mathematically adds (or sums) the voltage levels found at two or more inputs.

superposition theorem: superimposing one set of values on another to create a set of values different from either set of the superimposed values alone; the algebraic sum of the superimposed sets of values.

surface mount technology: the technology in which discrete electrical components or devices (such as resistors and capacitors) are manufactured in a way that allows miniaturization of the component elements and minimizing of lead lengths. These components are designed to be mounted and soldered directly to the "surface" of printed circuit boards.

susceptance (B): the relative ease that ac current passes through a pure reactance; the reciprocal of reactance; the imaginary component of admittance.

switches: devices for making or breaking continuity in a circuit.

switching diode: a junction diode that is designed to allow a very short recovery time when the applied voltage changes polarity.

tangent function: a trigonometric function used in ac circuit analysis; expresses the relationship of a specified angle of a right triangle to the side opposite and side adjacent to that angle where tan θ = opposite/adjacent.

tetravalent atoms: an atom normally having four electrons in the outer, or valence, shell. Such materials are often used as the base material for semiconductors. Examples include silicon and germanium.

Thevenin's theorem: a theorem used to simplify analysis of circuit networks because the network can be replaced by a simplified equivalent circuit consisting of a single voltage source (called V_{TH}) and a single series resistance (called R_{TH}).

thin-film process: a process where thin layers of conductive or nonconductive materials are used to form integrated circuits.

thyristor: a semiconductor device that is composed of four layers, P-N-P-N. The conduction is either all or nothing; it is either switched fully on or turned fully off.

transconductance (g_m): for FETs, the ratio of a change in drain current to the corresponding change in gate voltage. Unit of measurement is siemens.

transformers: devices that transfer energy from one circuit to another through electromagnetic (mutual) induction.

triac: a special semiconductor device, sometimes called a 5-layer N-P-N-P-N device, that controls power over an entire cycle, rather than half the cycle, like the SCR device; equivalent of two SCRs in parallel connected in opposite directions but having a common gate.

trivalent materials: materials that have three valence electrons; used in doping semiconductor material to produce P-type materials where holes are the majority current carriers. Examples include boron, aluminum, gallium, and indium.

true (or real) power (P): the power expended in the ac circuit resistive parts that equals VI cos θ, or apparent power × cos θ. The unit of measure is watts.

tunnel diode: a special PN junction diode that has been doped to exhibit high-speed switching capability and a special negative-resistance over part of its operating range; sometimes used in oscillator and amplifier circuits.

turns ratio: the number of turns on the separate windings of a transformer; expressed as the number of turns on the secondary versus the number of turns on the primary (N_S/N_P); or frequently stated as the ratio of primary-to-secondary turns.

unity gain bandwidth product (GBW): a constant that provides the frequency and open-loop gain limitations of the operational amplifier.

valence electrons: the electrons in the outer shell of an atom.

varactor diode: a silicon, voltage-controlled semiconductor capacitor where capacitance is varied by changing the junction reverse bias; used in various applications where voltage-controlled, variable capacitance can be advantageous.

vector: a quantity that illustrates both magnitude and direction. A vector's direction represents position relative to space.

volt (V): the basic unit of potential difference or electromotive force (emf); the amount of emf causing a current flow of one ampere through a resistance of one ohm.

voltage: the amount of emf available to move a current.

voltage divider action: the dividing of a specific source voltage into various levels of voltages by means of series resistors. The distribution of voltages at various points in the circuit is dependent on what proportion each resistor is of the total resistance in the circuit.

voltage follower: an amplifier configuration where the output voltage is slightly less than the input voltage level, and the output has the same phase or polarity as the input.

voltage rate of change (*dV/dt*): the amount of voltage level change per unit time. For example, a rate of change of one volt per second, 10 V per μ second, and so forth.

voltage ratio: in an ideal transformer with 100% coupling ($k = 1$), the voltage ratio is the same as the turns ratio, since the voltage induced in each turn of the secondary and the primary windings is the same for a given rate of flux change; $V_S/V_P = N_S/N_P$.

voltampere-reactive (VAR): the unit of reactive power in contrast to real power in watts; 1 VAR is equal to one reactive volt-ampere.

watt: the basic unit of electrical power. The basic formula for electrical power is P (in watts) equals V (in volts) times I (in amperes).

weber: the unit of magnetic flux indicating 10^8 maxwells (flux lines).

Wien-bridge oscillator: a type of *RC* sine-wave oscillator that uses a lead-lag network to maintain a zero-degree phase shift of the feedback signal at the frequency of oscillation.

work: the expenditure of energy, where work = force × distance. The basic unit of work is the foot poundal. NOTE: At sea level, 32.16 foot poundals = one foot pound.

zener diode: a junction diode that is designed to allow reverse breakdown at a certain voltage level, thereby acting as a voltage regulator device.

INDEX